T0374239

Random Matrix Methods for Wireless Communications

Blending theoretical results with practical applications, this book provides an introduction to random matrix theory and shows how it can be used to tackle a variety of problems in wireless communications. The Stieltjes transform method, free probability theory, combinatoric approaches, deterministic equivalents, and spectral analysis methods for statistical inference are all covered from a unique engineering perspective. Detailed mathematical derivations are presented throughout, with thorough explanations of the key results and all fundamental lemmas required for the readers to derive similar calculus on their own. These core theoretical concepts are then applied to a wide range of real-world problems in signal processing and wireless communications, including performance analysis of CDMA, MIMO, and multi-cell networks, as well as signal detection and estimation in cognitive radio networks. The rigorous yet intuitive style helps demonstrate to students and researchers alike how to choose the correct approach for obtaining mathematically accurate results.

Romain Couillet is an Assistant Professor at the Chair on System Sciences and the Energy Challenge at Supélec, France. Previously he was an Algorithm Development Engineer for ST-Ericsson, and he received his PhD from Supélec in 2010.

Mérouane Debbah is a Professor at Supélec, where he holds the Alcatel-Lucent Chair on Flexible Radio. He is the recipient of several awards, including the 2007 General Symposium IEEE Globecom best paper award and the Wi-Opt 2009 best paper award.

Random Matrix Methods for Wireless Communications

Romain Couillet and Mérouane Debbah
École Supérieure d'Électricité, Gif sur Yvette, France

CAMBRIDGE
UNIVERSITY PRESS

University Printing House, Cambridge CB2 8BS, United Kingdom

Cambridge University Press is part of the University of Cambridge.

It furthers the University's mission by disseminating knowledge in the pursuit of education, learning and research at the highest international levels of excellence.

www.cambridge.org
Information on this title: www.cambridge.org/9781107011632

First published 2011

A catalogue record for this publication is available from the British Library

Library of Congress Cataloguing in Publication data
Couillet, Romain, 1983–
Random matrix methods for wireless communications / Romain Couillet, Merouane Debbah.
p. cm.
Includes bibliographical references and index.
ISBN 978-1-107-01163-2 (hardback)
1. Wireless communication systems – Mathematics. 2. Matrix analytic methods.
I. Debbah, Merouane, 1975– II. Title.
TK5103.2.C68 2011
621.38401'51–dc23

2011013189

ISBN 978-1-107-01163-2 Hardback

To my family,
– Romain Couillet

To my parents,
– Mérouane Debbah

Contents

Preface

More than sixty years have passed since the 1948 landmark paper of Shannon providing the capacity of a single antenna point-to-point communication channel. The method was based on information theory and led to a revolution in the field, especially on how communication systems were designed. The tools then showed their limits when we wanted to extend the analysis and design to the multi-terminal multiple antenna case, which is the basis of the wireless revolution since the nineties. Indeed, in the design of these networks, engineers frequently stumble on the scalability problem. In other words, as the number of nodes or bandwidth increase, problems become harder to solve and the determination of the precise achievable rate region becomes an intractable problem. Moreover, engineering insight progressively disappears and we can only rely on heavy simulations with all their caveats and limitations. However, when the system is sufficiently large, we may hope that a macroscopic view could provide a more useful abstraction of the network. The properties of the new macroscopic model nonetheless need to account for microscopic considerations, e.g. fading, mobility, etc. We may then sacrifice some structural details of the microscopic view but the macroscopic view will preserve sufficient information to allow for a meaningful network optimization solution and the derivation of insightful results in a wide range of settings.

Recently, a number of research groups around the world have taken this approach and have shown how tools borrowed from physical and mathematical frameworks, e.g. percolation theory, continuum models, game theory, electrostatics, mean field theory, stochastic geometry, just to name a few, can capture most of the complexity of dense random networks in order to unveil some relevant features on network-wide behavior.

The following book falls within this trend and aims to provide a comprehensive understanding on how random matrix theory can model the complexity of the interaction between wireless devices. It has been more than fifteen years since random matrix theory was successfully introduced into the field of wireless communications to analyze CDMA and MIMO systems. One of the useful features, especially of the large dimensional random matrix theory approach, is its ability to predict, under certain conditions, the behavior of the empirical eigenvalue distribution of products and sums of matrices. The results are striking in terms of accuracy compared to simulations with reasonable matrix sizes, and the theory has been shown to be an efficient tool to predict the behavior of wireless systems with only few meaningful parameters. Random matrix theory is also increasingly making its way into the statistical signal processing field with the generalization of detection and inference methods, e.g. array processing, hypothesis tests, parameter estimation, etc., to the multi-variate case. This comes as a small revolution in modern signal processing as legacy estimators, such as the MUSIC method, become increasingly obsolete and unadapted to large sensing arrays with few observations.

The authors are confident and have no doubt on the usefulness of the tool for the engineering community in the upcoming years, especially as networks become denser. They also think that random matrix theory should become sooner or later a major tool for electrical engineers, taught at the graduate level in universities. Indeed, engineering education programs of the twentieth century were mostly focused on the Fourier transform theory due to the omnipresence of frequency spectrum. The twenty-first century engineers know by now that space is the next frontier due to the omnipresence of spatial spectrum modes, which refocuses the programs towards a Stieltjes transform theory.

We sincerely hope that this book will inspire students, teachers, and engineers, and answer their present and future problems.

Romain Couillet and Mérouane Debbah

Acknowledgments

This book is the fruit of many years of the authors' involvement in the field of random matrix theory for wireless communications. This topic, which has gained increasing interest in the last decade, was brought to light in the telecommunication community in particular through the work of Stephen Hanly, Ralf Müller, Shlomo Shamai, Emre Telatar, David Tse, Antonia Tulino, and Sergio Verdú, among others. It then rapidly grew into a joint research framework gathering both telecommunication engineers and mathematicians, among which Zhidong Bai, Vyacheslav L. Girko, Leonid Pastur, and Jack W. Silverstein.

The authors are especially indebted to Prof. Silverstein for the agreeable time spent discussing random matrix matters. Prof. Silverstein has a very insightful approach to random matrices, which it was a delight to share with him. The general point of view taken in this book is mostly influenced by Prof. Silverstein's methodology. The authors are also grateful to the many colleagues working in this field whose knowledge and wisdom about applied random matrix theory contributed significantly to its current popularity and elegance. This book gathers many of their results and intends above all to deliver to the readers this simplified approach to applied random matrix theory. The colleagues involved in long and exciting discussions as well as collaborative works are Florent Benaych-Georges, Pascal Bianchi, Laura Cottatellucci, Maxime Guillaud, Walid Hachem, Philippe Loubaton, Mylène Maïda, Xavier Mestre, Aris Moustakas, Ralf Müller, Jamal Najim, and Øyvind Ryan.

Regarding the book manuscript itself, the authors would also like to sincerely thank the anonymous reviewers for their wise comments which contributed to improve substantially the overall quality of the final book and more importantly the few people who dedicated a long time to thoroughly review the successive drafts and who often came up with inspiring remarks. Among the latter are David Gregoratti, Jakob Hoydis, Xavier Mestre, and Sebastian Wagner.

The success of this book relies in a large part on these people.

Romain Couillet and Mérouane Debbah

Acronyms

AWGN	additive white Gaussian noise
BC	broadcast channel
BPSK	binary pulse shift keying
CDMA	code division multiple access
CI	channel inversion
CSI	channel state information
CSIR	channel state information at receiver
CSIT	channel state information at transmitter
d.f.	distribution function
DPC	dirty paper coding
e.s.d.	empirical spectral distribution
FAR	false alarm rate
GLRT	generalized likelihood ratio test
GOE	Gaussian orthogonal ensemble
GSE	Gaussian symplectic ensemble
GUE	Gaussian unitary ensemble
i.i.d.	independent and identically distributed
l.s.d.	limit spectral distribution
MAC	multiple access channel
MF	matched-filter
MIMO	multiple input multiple output
MISO	multiple input single output
ML	maximum likelihood
LMMSE	linear minimum mean square error
MMSE	minimum mean square error
MMSE-SIC	MMSE and successive interference cancellation
MSE	mean square error
MUSIC	multiple signal classification
NMSE	normalized mean square error
OFDM	orthogonal frequency division multiplexing

OFDMA orthogonal frequency division multiple access
p.d.f. probability density function
QAM quadrature amplitude modulation
QPSK quadrature pulse shift keying
ROC receiver operating characteristic
RZF regularized zero-forcing
SINR signal-to-interference plus noise ratio
SISO single input single output
SNR signal-to-noise ratio
TDMA time division multiple access
ZF zero-forcing

Notation

Linear algebra

$\mathbf{X}$	Matrix
$\mathbf{I}_N$	Identity matrix of size $N \times N$
X_{ij}	Entry (i,j) of matrix $\mathbf{X}$ (unless otherwise stated)
$(X)_{ij}$	Entry (i,j) of matrix $\mathbf{X}$
$[X]_{ij}$	Entry (i,j) of matrix $\mathbf{X}$
$\{f(i,j)\}_{i,j}$	Matrix with (i,j) entry $f(i,j)$
$(X_{ij})_{i,j}$	Matrix with (i,j) entry X_{ij}
$\mathbf{x}$	Vector (column by default)
$\mathbf{x}^*$	Vector of the complex conjugates of the entries of $\mathbf{x}$
x_i	Entry i of vector $\mathbf{x}$
$F^{\mathbf{X}}$	Empirical spectral distribution of the Hermitian $\mathbf{X}$
$\mathbf{X}^{\mathsf{T}}$	Transpose of $\mathbf{X}$
$\mathbf{X}^{\mathsf{H}}$	Hermitian transpose of $\mathbf{X}$
$\operatorname{tr}\mathbf{X}$	Trace of $\mathbf{X}$
$\det\mathbf{X}$	Determinant of $\mathbf{X}$
$\operatorname{rank}(\mathbf{X})$	Rank of $\mathbf{X}$
$\Delta(\mathbf{X})$	Vandermonde determinant of $\mathbf{X}$
$\\|\mathbf{X}\\|$	Spectral norm of the Hermitian matrix $\mathbf{X}$
$\operatorname{diag}(x_1,\ldots,x_n)$	Diagonal matrix with (i,i) entry x_i
$\ker(\mathbf{A})$	Null space of the matrix $\mathbf{A}$, $\ker(\mathbf{A}) = \{\mathbf{x}, \mathbf{Ax} = 0\}$
$\operatorname{span}(\mathbf{A})$	Subspace generated by the columns of the matrix $\mathbf{A}$

Real and complex analysis

$\mathbb{N}$	The space of natural numbers
$\mathbb{R}$	The space of real numbers
$\mathbb{C}$	The space of complex numbers
A^*	The space $A \setminus \{0\}$
x^+	Right-limit of the real x
x^-	Left-limit of the real x

$(x)^+$	For $x \in \mathbb{R}$, $\max(x, 0)$
$\text{sgn}(x)$	Sign of the real x
$\Re[z]$	Real part of z
$\Im[z]$	Imaginary part of z
z^*	Complex conjugate of z
i	Square root of -1 with positive imaginary part
$f'(x)$	First derivative of the function f
$f''(x)$	Second derivative of the function f
$f'''(x)$	Third derivative of the function f
$f^{(p)}(x)$	Derivative of order p of the function f
$\|f\|$	Norm of a function $\|f\| = \sup_x \|f(x)\|$
$1_A(x)$	Indicator function of the set A $1_A(x) = 1$ if $x \in A$, $1_A(x) = 0$ otherwise
$\delta(x)$	Dirac delta function, $\delta(x) = 1_{\{0\}}(x)$
$\Delta(x\|A)$	Convex indicator function $\Delta(x\|A) = 0$ if $x \in A$, $\Delta(x\|A) = \infty$ otherwise
$\text{Supp}(F)$	Support of the distribution function F
$x_1, x_2, \ldots$	Series of general term x_n
$x_n \to \ell$	Simple convergence of the series $x_1, x_2, \ldots$ to ℓ
$x_n = o(y_n)$	Upon existence, $x_n/y_n \to 0$ as $n \to \infty$
$x_n = O(y_n)$	There exists K, such that $x_n \leq K y_n$ for all n
$n/N \to c$	As $n \to \infty$ and $N \to \infty$, $n/N \to c$
$\mathcal{W}(z)$	Lambert-W function satisfying $\mathcal{W}(z)e^{\mathcal{W}(z)} = z$
$\text{Ai}(x)$	Airy function
$\Gamma(x)$	Gamma function, $\Gamma(n) = (n-1)!$ for n integer

Probability theory

$(\Omega, \mathcal{F}, P)$	Probability space Ω with σ-field $\mathcal{F}$ and measure P
$P_X(x)$	Density of the random variable X
$p_X(x)$	Density of the scalar random variable X
$P_{(X_i)}(x)$	Unordered density of the random variable $X_1, \ldots, X_N$
$P^{\geq}_{(X_i)}(x)$	Ordered density of the random variable $X_1 \geq \ldots \geq X_N$
$P^{\leq}_{(X_i)}(x)$	Ordered density of the random variable $X_1 \leq \ldots \leq X_N$
μ_X	Probability measure of X, $\mu_X(A) = P(X(A))$
$\mu_{\mathbf{X}}$	Probability distribution of the eigenvalues of $\mathbf{X}$
$\mu^{\infty}_{\mathbf{X}}$	Probability distribution associated with the l.s.d. of $\mathbf{X}$
$P_X(x)$	Density of the random variable X, $P_X(x)dx = \mu_X(dx)$
$F_X(x)$	Distribution function of X (real), $F_X(x) = \mu_X((-\infty, x])$
$\text{E}[X]$	Expectation of X, $\text{E}[X] = \int_\Omega X(\omega)d\omega$

$\mathrm{E}[f(X)]$	Expectation of $f(X)$, $\mathrm{E}[f(X)] = \int_\Omega f(X(\omega))d\omega$
$\mathrm{var}(X)$	Variance of X, $\mathrm{var}(X) = \mathrm{E}[X^2] - \mathrm{E}[X]^2$
$X \sim \mathcal{L}$	X is a random variable with density $\mathcal{L}$
$\mathcal{N}(\boldsymbol{\mu}, \boldsymbol{\Sigma})$	Real Gaussian distribution of mean $\boldsymbol{\mu}$ and covariance $\boldsymbol{\Sigma}$
$\mathcal{CN}(\boldsymbol{\mu}, \boldsymbol{\Sigma})$	Complex Gaussian distribution of mean $\boldsymbol{\mu}$ and covariance $\boldsymbol{\Sigma}$
$\mathcal{W}_N(n, \mathbf{R})$	Real zero mean Wishart distribution with n degrees of freedom and covariance $\mathbf{R}$
$\mathcal{CW}_N(n, \mathbf{R})$	Complex zero mean Wishart distribution with n degrees of freedom and covariance $\mathbf{R}$
$Q(x)$	Gaussian Q-function, $Q(x) = P(X > x)$, $X \sim \mathcal{N}(0,1)$
F^+	Tracy–Widom distribution function
F^-	Conjugate Tracy–Widom d.f., $F^-(x) = 1 - F^+(-x)$
$x_n \xrightarrow{\text{a.s.}} \ell$	Almost sure convergence of the series $x_1, x_2, \ldots$ to ℓ
$F_n \Rightarrow F$	Weak convergence of the d.f. series $F_1, F_2, \ldots$ to F
$X_n \Rightarrow X$	Weak convergence of the series $X_1, X_2, \ldots$ to the random X

Random Matrix Theory

$m_F(z)$	Stieltjes transform of the function F
$m_{\mathbf{X}}(z)$	Stieltjes transform of the eigenvalue distribution of $\mathbf{X}$
$\mathcal{V}_F(z)$	Shannon transform of the function F
$\mathcal{V}_{\mathbf{X}}(z)$	Shannon transform of the eigenvalue distribution of $\mathbf{X}$
$R_F(z)$	R transform of the function F
$R_{\mathbf{X}}(z)$	R transform of the eigenvalue distribution of $\mathbf{X}$
$S_F(z)$	S transform of the function F
$S_{\mathbf{X}}(z)$	S transform of the eigenvalue distribution of $\mathbf{X}$
$\eta_F(z)$	η-transform of the function F
$\eta_{\mathbf{X}}(z)$	η-transform of the eigenvalue distribution of $\mathbf{X}$
$\psi_F(z)$	ψ-transform of the function F
$\psi_{\mathbf{X}}(z)$	ψ-transform of the eigenvalue distribution of $\mathbf{X}$
$\mu \boxplus \nu$	Additive free convolution of μ and ν
$\mu \boxminus \nu$	Additive free deconvolution of μ and ν
$\mu \boxtimes \nu$	Multiplicative free convolution of μ and ν
$\mu \boxslash \nu$	Multiplicative free deconvolution of μ and ν

Topology

A^c	Complementary of the set A
$\#A$	Cardinality of the discrete set A
$A \oplus B$	Direct sum of the spaces A and B
$\bigoplus_{1 \leq i \leq n} A_i$	Direct sum of the spaces A_i, $1 \leq i \leq n$

$\langle x, A \rangle$	Norm of the orthogonal projection of x on the space A

Miscellaneous

$x \triangleq y$	x is defined as y
$\operatorname{sgn}(\sigma)$	Signature (or parity) of the permutation σ, $\operatorname{sgn}(\sigma) \in \{-1, 1\}$

1 Introduction

1.1 Motivation

We initiate the book with a classical example, which exhibits both the non-obvious behavior of large dimensional random matrices and the motivation behind their study.

Consider a random sequence $x_1, \ldots, x_n$ of n independent and identically distributed (i.i.d.) *observations* of a given random process. The classical law of large numbers states that the sequence $x_1, \frac{x_1+x_2}{2}, \ldots$, with nth term $\frac{1}{n}\sum_{k=1}^{n} x_k$ tends almost surely to the deterministic value $\mathrm{E}[x_1]$, the expectation of this process, as n tends to infinity. Denote $(\Omega, \mathcal{F}, P)$ the probability space that generates the infinite sequence $x_1, x_2, \ldots$. For a given realization $\omega \in \Omega$, we will denote $x_1(\omega), x_2(\omega), \ldots$ the realization of the random sequence $x_1, x_2, \ldots$. We recall that almost sure convergence means that there exists $A \subset \Omega$, with $P(A) = 1$, such that, for $\omega \in A$

$$\frac{1}{n}\sum_{k=1}^{n} x_k(\omega) \to \mathrm{E}[x_1] \triangleq \int_{\Omega} x_1(w)dw.$$

We also remind briefly that the notation $(\Omega, \mathcal{F}, P)$ designates the triplet composed of the space of random realizations Ω, i.e. in our case $\omega \in \Omega$ is the realization of a series $x_1(\omega), x_2(\omega), \ldots$, $\mathcal{F}$ is a σ-field on Ω, which can be seen as the space of the measurable *events* on Ω, e.g. the space $B = \{x_1(\omega) > 0\} \in \mathcal{F}$ is such an event, and P is a probability measure on $\mathcal{F}$, i.e. P is a function that assigns to every event in $\mathcal{F}$ a probability.

This law of large numbers is fundamental in the sense that it provides a deterministic feature for a process ruled by 'chance' (or more precisely, ruled by a deterministic process, the precise nature of which the observer is unaware). This allows the observer to be able to retrieve deterministic information from random variables based on any observed random sequence (within a space of probability one). If, for instance, $x_1, \ldots, x_n$ are successive samples of a stationary zero mean white noise waveform $x(t)$, i.e. $\mathrm{E}[x(t)x(t-\tau)] = \sigma^2\delta(t)$, it is the usual signal processing problem to estimate the power $\sigma^2 = \mathrm{E}[|x_1|^2]$ of the noise process; the empirical variance σ_n^2, i.e.

$$\sigma_n^2 = \frac{1}{n}\sum_{i=1}^{n} |x_i|^2$$

is a classical estimate of σ^2 which, according to the law of large numbers, is such that $\sigma_n^2 \to \sigma^2$ almost surely when $n \to \infty$. It is often said that σ_n^2 is a *consistent* estimator of σ^2 as it is asymptotically and almost surely equal to σ^2. To avoid confusion with the two-dimensional case treated next, we will say instead that σ_n^2 is an *n-consistent* estimator of σ^2, as it is asymptotically accurate as n grows large. Obviously, we are never provided with an infinitely long observation time window, so that n is usually large but finite, and therefore σ_n^2 is merely an approximation of σ^2.

With the emergence of multiple antenna systems, channel spreading codes, sensor networks, etc., signal processing problems have become more and more concerned with vectorial inputs rather than scalar inputs. For a sequence of n i.i.d. random vectors, the law of large numbers still applies. For instance, for $\mathbf{x}_1, \mathbf{x}_2, \ldots \in \mathbb{C}^N$ randomly drawn from a given N-variate zero mean random process

$$\mathbf{R}_n = \frac{1}{n}\sum_{i=1}^{n} \mathbf{x}_i\mathbf{x}_i^{\mathsf{H}} \to \mathbf{R} \triangleq \mathrm{E}[\mathbf{x}_1\mathbf{x}_1^{\mathsf{H}}] \tag{1.1}$$

almost surely as $n \to \infty$, where the convergence is considered for any matrix norm, i.e. $\|\mathbf{R} - \mathbf{R}_n\| \to 0$ on a set of probability one. The matrix $\mathbf{R}_n$ is often referred to as the *empirical covariance matrix* or as the *sample covariance matrix*, as it is computed from observed vector samples. We will use this last phrase throughout the book. Following the same semantic field, the matrix $\mathbf{R}$ will be referred to as the *population covariance matrix*, as it characterizes the innate nature of all stochastic vectors $\mathbf{x}_i$ from the overall population of such vectors. The empirical $\mathbf{R}_n$ is again an n-consistent estimator of $\mathbf{R}$ of $\mathbf{x}_1$ and, as before, as n is taken very large *for N fixed*, $\mathbf{R}_n$ is a good approximation of $\mathbf{R}$ in the sense of the aforementioned matrix norm. However, in practical applications, it might be that the number of available snapshots $\mathbf{x}_k$ is indeed very large but not extremely large compared to the vector size N. This situation arises in diverse application fields, such as biology, finance, and, of course, wireless communications. If this is the case, as will become obvious in the following examples and against intuition, the difference $\|\mathbf{R} - \mathbf{R}_n\|$ can be far from zero even for large n.

Since the DNA of many organisms have now been entirely sequenced, biologists and evolutionary biologists are interested in the correlations between genes, e.g.: How does the presence of a given gene (or gene sequence) in an organism impact the probability of the presence of another given gene? Does the activation of a given gene come along with the activation of several other genes? To be able to study the joint correlation between a large population of the several ten thousands of human genes, call this number N, we need a large sample of genome sequences extracted from human beings, call the number of such samples n. It is therefore typical that the $N \times n$ matrix of the n gene sequence samples does not have many more columns than rows, or, worse, may even have more rows than columns. We see already that, in this case, the sample covariance matrix $\mathbf{R}_n$ is necessarily rank-deficient (of maximum rank $N - n$), while $\mathbf{R}$ has all the chances to be full rank. Therefore, $\mathbf{R}_n$ is obviously no longer a good approximation of $\mathbf{R}$, even if n were very large in the first place, since the eigenvalues of $\mathbf{R}_n$ and $\mathbf{R}$ differ by at least $N - n$ terms.

In the field of finance, the interest of statisticians lies in the interactions between assets in the market and the joint time evolution of their stock market indices. The vectors $\mathbf{x}_1, \ldots, \mathbf{x}_n$ here may be representative of n months of market index evolution of N different brands of a given product, say soda, the ith entry of the column vector $\mathbf{x}_k$ being the evolution of the market index of soda i in month k. Obviously, this case differs from the independent vector case presented up to now since the evolution at month $k+1$ is somewhat correlated to the evolution at previous month k, but let us assume for simplicity that the month evolution is at least an uncorrelated process (which does not imply independence). Similar to the gene case for biologists, it often turns out that the $N \times n$ matrix under study contains few columns compared to the number of rows, although both dimensions are typically large compared to 1. Of specific importance to traders is the largest eigenvalue of the population covariance matrix $\mathbf{R}$ of (a centered and normalized version of) the random process $\mathbf{x}_1$, which is an indicator of the maximal risk against investment returns taken by a trader who constitutes a portfolio from these assets. From the biology example above, it has become clear that the eigenvalues of $\mathbf{R}_n$ may be a very inaccurate estimate of those of $\mathbf{R}$; thus, $\mathbf{R}_n$ cannot be relied on to estimate the largest eigenvalue and hence the trading risk. The case of wireless communications will be thoroughly detailed in Part II, and a first motivation is given in the next paragraph.

Returning to the initial sample covariance matrix model, we have already mentioned that in the scalar case the strong law of large numbers ensures that it suffices for n to be quite large compared to 1 for σ_n^2 to be a good estimator for σ^2. In the case where data samples are vectors, if n is large compared to 1, whatever N, then the (i,j) entry $R_{n,ij}$ of $\mathbf{R}_n$ is a good estimator of the (i,j) entry R_{ij} of $\mathbf{R}$. This might (mis)lead us to assume that as n is much greater than one, $\mathbf{R}_n \simeq \mathbf{R}$ in some sense. However, if both N and n are large compared to 1 but n is not large compared to N, then the peculiar thing happens: the eigenvalue distribution of $\mathbf{R}_n$ (see this as an histogram of the eigenvalues) in general converges, but does not converge to the eigenvalue distribution of $\mathbf{R}$. This has already been pointed out in the degenerated case $N > n$, for which $\mathbf{R}_n$ has $N - n$ null eigenvalues, while $\mathbf{R}$ could be of full rank. This behavior is evidenced in Figure 1.1 in which we consider $\mathbf{x}_1 \sim \mathcal{CN}(0, \mathbf{I}_N)$ and then $\mathbf{R} = \mathbf{I}_N$, for $N = 500$, $n = 2000$. In that case, notice that $\mathbf{R}_n$ converges *point-wise* to $\mathbf{I}_N$

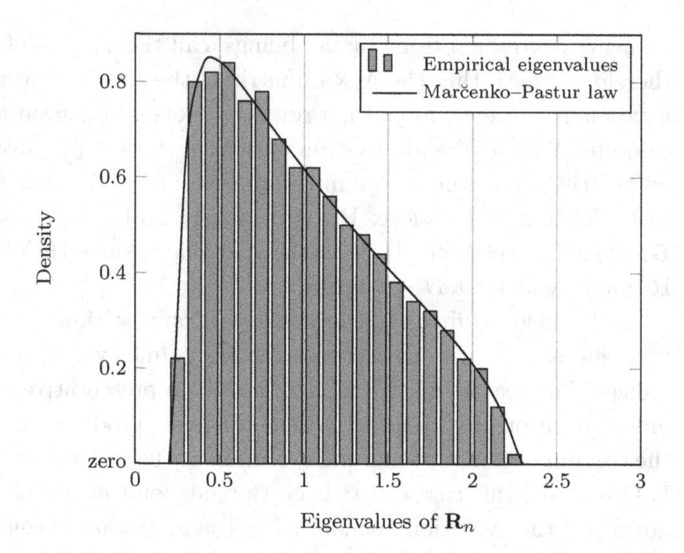

Figure 1.1 Histogram of the eigenvalues of $\mathbf{R}_n = \frac{1}{n}\sum_{k=1}^{n} \mathbf{x}_k\mathbf{x}_k^{\mathsf{H}}$, $\mathbf{x}_k \in \mathbb{C}^N$, for $n = 2000$, $N = 500$.

when n is large, as $R_{n,ij}$, the entry (i,j) of $\mathbf{R}_n$, is given by:

$$R_{n,ij} = \frac{1}{n}\sum_{k=1}^{n} x_{ik}x_{jk}^*$$

which is close to one if $i = j$ and close to zero if $i \neq j$. This is obviously irrespective of N, which is not involved in the calculus here. However, the eigenvalues of $\mathbf{R}_n$ do not converge to a single mass in 1 but are spread around 1. This apparent contradiction is due to the fact that N grows along with n but n/N is never large. We say in that case that, while $\mathbf{R}_n$ is an n-consistent estimator of $\mathbf{R}$, it is not an (n, N)*-consistent* estimator of $\mathbf{R}$. The seemingly paradoxical behavior of the eigenvalues of $\mathbf{R}_n$, while $\mathbf{R}_n$ converges *point-wise* to $\mathbf{I}_N$, lies in fact in the rate convergence of the entries of $\mathbf{R}_n$ towards the entries of $\mathbf{I}_N$. Due to central limit arguments for the sample mean of scalar i.i.d. random variables, $R_{n,ij} - \mathrm{E}[R_{n,ij}]$ is of order $O(1/\sqrt{n})$. When determining the eigenvalues of $\mathbf{R}_n$, the deviations around the means are negligible when n is large and N fixed. However, for N and n both large, these residual deviations of the entries of $\mathbf{R}_n$ (their number is N^2) are no longer negligible and the eigenvalue distribution of $\mathbf{R}_n$ is not a single mass in 1. In some sense, we can see $\mathbf{R}_n$ as a matrix close to the identity but whose entries all contain some small residual "energy," which becomes relevant as much of such small energy is cumulated.

This observation has very important consequences, which motivate the need for singling out the study of large empirical covariance matrices and more generally of large random Hermitian matrices as a unique field of mathematics. Wireless communications may be the one research field in which large matrices have started to play a fundamental role. Indeed, current and more importantly future wireless communication systems are multi-dimensional in several respects (spatial with antennas, temporal with random codes, cellular-wise with large number of users, multiple cooperative network nodes, etc.) and random in other respects (time-varying fading channels, noisy communications, etc.). The study of the behavior of large wireless communication systems therefore calls for advanced mathematical tools that can easily deal with large dimensional random matrices. Consider for instance a multiple input multiple output (MIMO) complex channel matrix $\mathbf{H} \in \mathbb{C}^{N \times n}$ between an n-antenna transmitter and an N-antenna receiver, the entries of which are independent and complex Gaussian with zero mean and variance $1/n$. If uniform power allocation across the antennas is used at the transmit antenna array and the additive channel noise is white and Gaussian, the achievable transmission rates over this channel are all rates less than the channel mutual information

$$\mathcal{I}(\sigma^2) = \mathrm{E}\left[\log_2 \det\left(\mathbf{I}_N + \frac{1}{\sigma^2}\mathbf{H}\mathbf{H}^{\mathsf{H}}\right)\right] \tag{1.2}$$

where σ^{-2} denotes now the signal-to-noise ratio (SNR) at the receiver and the expectation is taken over the realizations of the random channel $\mathbf{H}$, varying according to the Gaussian distribution. Now note that $\mathbf{H}\mathbf{H}^{\mathsf{H}} = \sum_{i=1}^{n} \mathbf{h}_i \mathbf{h}_i^{\mathsf{H}}$ with $\mathbf{h}_i \in \mathbb{C}^N$ the ith column of $\mathbf{H}$, $\mathbf{h}_1, \ldots, \mathbf{h}_n$ being i.i.d. random vectors. The matrix $\mathbf{H}\mathbf{H}^{\mathsf{H}}$ can then be seen as the sample covariance matrix of some hypothetical random N-variate variable $\sqrt{n}\mathbf{h}_1$. From our previous discussion, denoting $\mathbf{H}\mathbf{H}^{\mathsf{H}} = \mathbf{U}\mathbf{\Lambda}\mathbf{U}^{\mathsf{H}}$ the spectral decomposition of $\mathbf{H}\mathbf{H}^{\mathsf{H}}$, we have:

$$\mathcal{I}(\sigma^2) = \mathrm{E}\left[\log_2 \det\left(\mathbf{I}_N + \frac{1}{\sigma^2}\mathbf{\Lambda}\right)\right] = \mathrm{E}\left[\sum_{i=1}^{N} \log_2\left(1 + \frac{\lambda_i}{\sigma^2}\right)\right] \tag{1.3}$$

with $\lambda_1, \ldots, \lambda_N$ the eigenvalues of $\mathbf{H}\mathbf{H}^{\mathsf{H}}$, which again are not all close to one, even for n and N large. The achievable transmission rates are then explicitly dependent on the eigenvalue distribution of $\mathbf{H}\mathbf{H}^{\mathsf{H}}$. More generally, it will be shown in Chapters 12–15 that random matrix theory provides a powerful framework, with multiple methods, to analyze the achievable transmission rates and rate regions of a large range of multi-dimensional setups (MIMO, CDMA, multi-user transmissions, MAC/BC channels, etc.) and to derive the capacity-achieving signal covariance matrices for some of these systems, i.e. determine the non-negative definite matrix $\mathbf{P} \in \mathbb{C}^{N \times N}$, which, under some trace constraint $\operatorname{tr}\mathbf{P} \leq P$, maximizes the expression

$$\mathcal{I}(\sigma^2; \mathbf{P}) = \mathrm{E}\left[\log \det\left(\mathbf{I}_N + \frac{1}{\sigma^2}\mathbf{H}\mathbf{P}\mathbf{H}^{\mathsf{H}}\right)\right]$$

for numerous fading channel models for $\mathbf{H}$.

1.2 History and book outline

The present book is divided into two parts: a first part on the theoretical fundamentals of random matrix theory, and a second part on the applications of random matrix theory to the field of wireless communications. The first part will give a rather broad, although not exhaustive, overview of fundamental and recent results concerning random matrices. However, the main purpose of this part goes beyond a listing of important theorems. Instead, it aims on the one hand at providing the reader with a large, yet incomplete, range of techniques to handle problems dealing with random matrices, and on the other hand at developing sketches of proofs of the most important results in order to provide further intuition to the reader. Part II will be more practical as it will apply most of the results derived in Part I to problems in wireless communications, such as system performance analysis, signal sensing, parameter estimation, receiver design, channel modeling, etc. Every application will be commented on with regard to the theoretical results developed in Part I, for the reader to have a clear understanding of the reasons why the practical results hold, of their main limitations, and of the questions left open. Before moving on to Part I, in the following we introduce in detail the objectives of both parts through a brief historical account of eighty years of random matrix theory.

The origin of the study of random matrices is usually said to date back to 1928 with the pioneering work of the statistician John Wishart [Wishart, 1928]. Wishart was interested in the behavior of sample covariance matrices of i.i.d. random vector processes $\mathbf{x}_1, \ldots, \mathbf{x}_n \in \mathbb{C}^N$, in the form of the matrix $\mathbf{R}_n$ previously introduced

$$\mathbf{R}_n = \frac{1}{n} \sum_{i=1}^{n} \mathbf{x}_i \mathbf{x}_i^{\mathsf{H}}. \tag{1.4}$$

Wishart provided an expression of the joint probability distribution of the entries of such a matrix when its column vector entries are themselves independent and have an identical standard complex Gaussian distribution, i.e. $x_{ij} \sim \mathcal{CN}(0,1)$. These normalized matrices with i.i.d. standard Gaussian entries are now called *Wishart matrices.* Wishart matrices were thereafter generalized and extensively studied. Today there exists in fact a large pool of properties on the joint distribution of the eigenvalues, the distribution of the extremes eigenvalues, the distribution of the ratios between extreme eigenvalues, etc.

The first asymptotic considerations, i.e. the first results on matrices of asymptotically large dimensions, appeared with the work of the physician Eugene Wigner [Wigner, 1955] on nuclear physics, who considered (properly scaled) symmetric matrices with independent entries uniformly distributed in $\{1, -1\}$ and proved the convergence of the marginal probability distribution of its eigenvalues towards the deterministic *semi-circle law,* as the dimension of the matrix grows to infinity. Hermitian $n \times n$ matrices with independent upper-

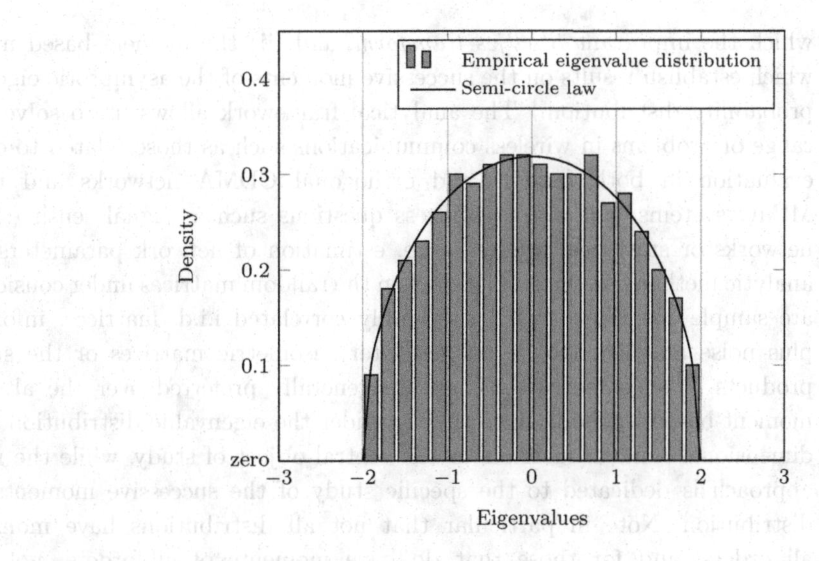

Figure 1.2 Histogram of the eigenvalues of a Wigner matrix and the semi-circle law, for $n = 500$.

triangular entries of zero mean and variance $1/n$ are now referred to as *Wigner matrices*. The empirical eigenvalues of a large Wigner matrix and the semi-circle law are illustrated in Figure 1.2, for a matrix of size $n = 500$. From this time on, infinite size random matrices have drawn increasing attention in many domains of physics [Mehta, 2004] (nuclear physics [Dyson, 1962a], statistical mechanics, etc.), finance [Laloux *et al.*, 2000], evolutionary biology [Arnold *et al.*, 1994], etc. The first accounts of work on large dimensional random matrices for wireless communications are attributed to Tse and Hanly [Tse and Hanly, 1999] on the performance of large multi-user linear receivers, Verdú and Shamai [Verdú and Shamai, 1999] on the capacity of code division multiple access (CDMA) systems, among others. The pioneering work of Telatar [Telatar, 1995] on the transmission rates achievable with multiple antennas, paralleled by Foschini [Foschini and Gans, 1998], is on the contrary a particular example of the use of small dimensional random matrices for capacity considerations. In its final version of 1999, the article also mentions asymptotic laws for capacity [Telatar, 1999]. We will see in Chapter 13 that, while Telatar's original proof of the capacity growth rate for increasing number of antennas in a multiple antenna setup is somewhat painstaking, large random matrix theory provides a straightforward result. In Chapter 2, we will explore some of the aforementioned results on random matrices of small dimension, which will be shown to be difficult to manipulate for simply structured matrices and rather intractable to extend to more structured matrices.

The methods used for random matrix-based calculus are mainly segmented into: (i) the analytical methods, which treat asymptotic eigenvalue distributions of large matrices in a comprehensive framework of analytical tools, among which the important *Stieltjes transform*, and (ii) the moment-based methods, which establish results on the successive moments of the asymptotic eigenvalues probability distribution.[1] The analytical framework allows us to solve a large range of problems in wireless communications such as those related to capacity evaluation in both random and orthogonal CDMA networks and in large MIMO systems, but also to address questions such as signal sensing in large networks or statistical inference, i.e. estimation of network parameters. These analytic methods are mostly used when the random matrices under consideration are sample covariance matrices, doubly correlated i.i.d. matrices, information plus noise matrices (to be defined later), isometric matrices or the sum and products of such matrices. They are generally preferred over the alternative moment-based methods since they consider the eigenvalue distribution of large dimensional random matrices as the central object of study, while the moment approach is dedicated to the specific study of the successive moments of the distribution. Note in particular that not all distributions have moments of all orders, and for those that do have moments of all orders, not all are uniquely defined by the series of their moments. However, in some cases of very structured matrices whose entries are non-trivially correlated, as in the example of Vandermonde matrices [Ryan and Debbah, 2009], the moment-based methods convey a more accessible treatment. Both analytical and moment-based methods are not completely disconnected from one another as they share a common denominator when it comes to dealing with unitarily invariant random matrices, such as standard Gaussian or Haar matrices, i.e. unitarily invariant unitary matrices. This common denominator, namely the field of *free probability theory*, bridges the analytical tools to the moment-based methods via derivatives of the Stieltjes transform, the R-transform, and the S-transform. The latter can be expressed in power series with coefficients intimately linked to moments and cumulants of the underlying random matrix eigenvalue distributions. The free probability tool, due to Voiculescu [Voiculescu *et al.*, 1992], was not initially meant to deal specifically with random matrices but with more abstract non-commutative algebras, large dimensional random matrices being a particular case of such algebras. The extension of classical probability theory to free probability provides interesting and often surprising results, such as a strong equivalence between some classical probability distributions, e.g. Poisson, Gaussian, and the asymptotic probability distribution of the eigenvalues of some random matrix models, e.g. Wishart matrices and Wigner matrices. Some classical probability tools, such as the characteristic function, are also extensible through analytic tools of random matrix theory.

[1] Since the terminology *method of moments* is already dedicated to the specific technique which aims at constructing a distribution function from its moments (under the condition that the moments uniquely determine the distribution), see, e.g. Section 30 of [Billingsley, 1995], we will carefully avoid referring to any random matrix technique based on moments as the method of moments.

The division between analytical and moment-based methods, with free probability theory lying in between, can be seen from another point of view. It will turn out that the analytical methods, and most particularly the Stieltjes transform approach, take full advantage of the independence between the entries of large dimensional random matrices. As for moment-based methods, from a free probability point of view, they take full advantage of the invariance properties of large dimensional matrices, such as the invariance by the left or right product with unitary matrices. The theory of orthogonal polynomials follows the same pattern, as it benefits from the fact that the eigenvalue distribution of unitarily invariant random matrices can be studied regardless of the (uniform) eigenvector distribution. In this book, we will see that the Stieltjes transform approach can solve most problems involving random matrices with invariance properties as well. This makes this distinction between random matrices with independent entries and random matrices with invariance properties not so obvious to us. For this reason, we will keep distinguishing between the analytical approaches that deal with the eigenvalue distribution as the central object of concern and the moment-based approaches that are only concerned with successive moments. We will also briefly introduce the rather old theory of orthogonal polynomials which has received much interest lately regarding the study of limiting laws of largest eigenvalues of random matrices but which requires significant additional mathematical effort for proper usage, while applications to wireless communications are to this day rather limited, although in constant expansion. We will therefore mostly state the important results from this field, particularly in terms of limit theorems of extreme eigenvalues, see Chapter 9, without development of the corresponding proofs.

In Chapter 3, Chapter 4, and Chapter 5, we will introduce the analytical and moment-based methods, as well as notions of free probability theory, which are fundamental to understand the important concept of asymptotic freeness for random matrices. We will also provide in these chapters a sketch of the proof of the convergence of the eigenvalue distribution of the Wishart and Wigner matrices to the Marčenko–Pastur law, depicted in Figure 1.1, and the semi-circle law, depicted in Figure 1.2, using the Stieltjes transform and the method of moments, respectively. Generic methods to determine (almost sure) limiting distributions of the eigenvalues of large dimensional random matrices, as well as other functionals of such large matrices (e.g. log determinant), will be reviewed in detail in these chapters. Chapter 6 will discuss the alternative methods used when the empirical eigenvalue distribution of large random matrices do not necessarily converge when the dimensions increase: in that case, in place of limit distributions, we will introduce the so-called *deterministic equivalents*, which provide deterministic approximations of functionals of random matrices of finite size. These approximations are (almost surely) asymptotically accurate as the matrix dimensions grow to infinity, making them consistent with the methods developed in Chapter 3. In addition to limiting eigenvalue distributions and deterministic equivalents, in Chapter 3 and Chapter 6, central limit theorems that extend the convergence theorems to a higher precision order will be introduced. These central limit theorems constitute a first step into a more thorough analysis of the asymptotic deviations of the spectrum around its almost sure limit or around its deterministic equivalent. Chapter 7 will discuss advanced results on the spectrum of both the sample covariance matrix model and the information plus noise model, which have been extensively studied and for which many results have been provided in the literature, such as the proof of the asymptotic absence of eigenvalues outside the support of the limiting distribution. Beyond the purely mathematical convenience of such a result, being able to characterize where the eigenvalues, and especially the extreme eigenvalues, are expected to lie is of fundamental importance to perform hypothesis testing decisions and in statistical inference. In particular, the characterization of the spectrum of sample covariance matrices will be used to retrieve information on functionals of the population covariance matrix from the observed sample covariance matrix, or functionals of the signal space matrix from the observed information plus noise matrix. Such methods will be referred to as *eigen-inference techniques* and are developed in Chapter 8. The first part will then conclude with Chapter 9, which extends the analysis of Section 7.1 to the expression of the limiting distributions of the extreme eigenvalues. We will also introduce in this chapter the *spiked models*, which have recently received a lot of attention for their many practical implications. These objects are necessary tools for signal sensing in large dimensional networks, which are currently of major interest with regard to the recent incentive for cognitive radios. In Chapter 10, the essential results of Part I will finally be summarized and rediscussed with respect to their applications to the field of wireless communications.

The second part of this book is dedicated to the application of the different methods described in the first chapter to different problems in wireless communications. As already mentioned, the first applications of random matrix theory to wireless communications are exclusively related to asymptotic system performance analysis, and especially channel capacity considerations. The idea of considering asymptotically large matrix approximations was initially linked to studies in CDMA communications, where both the number of users and the length of the spreading codes are potentially very large [Li *et al.*, 2004; Tse and Hanly, 1999; Tse and Verdú, 2000; Tse and Zeitouni, 2000; Zaidel *et al.*, 2001]. It then occurred to researchers that large matrix approximations work rather well when the size of the effective matrix under study is not so large, e.g. for matrices of size 8×8 or even 4×4 (in the case of random unitary matrices, simulations suggest that approximations for matrices of size 2×2 are even acceptable). This motivated further studies in systems where the number of relevant parameters is moderately large. In particular, studies of MIMO communications [Chuah *et al.*, 2002; Hachem *et al.*, 2008b; Mestre *et al.*, 2003; Moustakas and Simon, 2005; Müller, 2002], designs of multi-user receivers [Honig and Xiao, 2001; Müller and Verdú, 2001], multi-cell communications [Abdallah and Debbah, 2004; Couillet *et al.*, 2011a; Peacock *et al.*, 2008], multiple access channels and broadcast channels [Couillet *et al.*, 2011a; Wagner *et al.*, 2011] started to be considered. More recent topics featuring large decentralized systems, such as game theory-based cognitive radios [Meshkati *et al.*, 2005] and *ad-hoc* and mesh networks [Fawaz *et al.*, 2011; Levêque and Telatar, 2005] were also treated using similar random matrix theoretical tools. The initial fundamental reason for the attractiveness of asymptotic results of random matrix theory for the study of system performance lies in the intimate link between the Stieltjes transform and the information-theoretic expression of mutual information. It is only recently, thanks to a new wave of theoretical results, that the wireless communication community has realized that many more questions can be addressed than just mutual information and system performance evaluations.

In Chapter 11, we will start with an introduction to the important results of random matrix theory for wireless communications and their connections to the methods detailed in the first part of this book. In Chapters 12–15, we will present the latest results concerning achievable rates for a wide range of wireless models, mostly taken from the aforementioned references.

From an even more practical, implementation-oriented point of view, we will also introduce some ideas, rather scattered in the literature, which allow engineers to reduce the computational burden of transceivers by anticipating complex calculus and also to reduce the feedback load within communication networks by benefiting from deterministic approximations brought by large dimensional matrix analysis. The former will be introduced in Chapter 13, where we discuss means to reduce the computational difficulty of implementing large minimum mean square error (MMSE) CDMA or MIMO receivers, which require to perform real-time matrix inversions, from basis expansion models. Besides, it will be seen that in a multiple antenna uplink cellular network, the ergodic capacity maximizing precoding matrices of all users, computed at the central base station, can be fed back to the users under the form of partial information contained within a few bits (the number of which do not scale with the number of transmit antennas), which are sufficient for every user to compute their own transmit covariance matrix. These ideas contribute to the cognitive radio incentive for distributed intelligence within large dimensional networks.

Of major interest these days are also the questions of statistical estimation and detection in large decentralized networks or cognitive radio networks, spurred especially by the cognitive radio framework, a trendy example of which is the femto-cell incentive [Calin *et al.*, 2010; Claussen *et al.*, 2008]. Take for example the problem of radar detection using a large antenna array composed of N sensors. Each signal arising at the sensor array originates from a finite number K of sources positioned at specific angles with respect to the sensor support. The population covariance matrix $\mathbf{R} \in \mathbb{C}^{N \times N}$ of the N-dimensional sample vectors is composed of K distinct eigenvalues, with multiplicity linked to the size of every individual detected object, and possibly some null eigenvalues if N is large. Gathering n successive samples of the N-dimensional vectors, we can construct a sample covariance matrix $\mathbf{R}_n$ as in (1.1). We might then be interested in retrieving the individual eigenvalues of $\mathbf{R}$, which translate the distance with respect to each object, or retrieving the multiplicity of every eigenvalue, which is then an indicator of the size of the detected objects, or even detecting the angles of arrival of the incoming waveforms, which is a further indication of the geographical position of the objects. The multiple signal classification (MUSIC) estimator [Schmidt, 1986] has long been considered efficient for such a treatment and was indeed proved n-consistent. However, with N and n of similar order of magnitude, i.e. if we wish to increase the number of sensors, the MUSIC algorithm is largely biased [Mestre, 2008a]. Random matrices considerations, again based on the Stieltjes transform, were recently used to arrive at alternative (N, n)-consistent solutions [Karoui, 2008; Mestre, 2008b; Rao *et al.*, 2008]. These methods of eigenvalue and eigenvector retrieval are referred to as *eigen-inference* methods. In the case of a cognitive radio network, say a femto-cell network, every femto-cell, potentially composed of a large number of sensors, must be capable of discovering its environment, i.e. detecting the presence of surrounding users in communication, in order to exploit spectrum opportunities, i.e. unoccupied transmission bandwidth, unused by the surrounding licensed networks. One of the first objectives of a femto-cell is therefore to evaluate the number of surrounding users and to infer their individual transmit powers. The study of the power estimation capabilities of femto-cells is performed both throughout analytical and moment-based methods in [Couillet *et al.*, 2011c; Rao and Edelman, 2008; Ryan and Debbah, 2007a]. Statistical eigen-inference are also used to address hypothesis testing problems, such as signal sensing. Consider that an array of N sensors captures n samples of the incoming waveform and generates the empirical covariance matrix $\mathbf{R}_N$. The question is whether $\mathbf{R}_N$ indicates the presence of a signal issued by a source or only indicates the presence of noise. It is natural to consider that, if the histogram of the eigenvalues of $\mathbf{R}_N$ is close to that of Figure 1.1, this evidences the presence of noise but the absence of a transmitted signal. A large range of methods have been investigated, in different scenarios such as multi-source detection, multiple antenna signal sensing, known or unknown signal-to-noise ratio (SNR), etc. to come up with (N, n)-consistent detection tests [Bianchi *et al.*, 2011; Cardoso *et al.*, 2008]. These methods are here of particular interest when the system dimensions are extremely large to ensure low rates of false positives (declaring pure noise to be a transmitted signal) and of false negatives (missing the detection of a transmitted signal). When a small number of sensors are used, small dimensional random matrix models, however more involved, may be required [Couillet and Debbah, 2010a]. Note finally that, along with inference methods for signal sensing, small dimensional random matrices are also used in the new field of Bayesian channel modeling [Guillaud *et al.*, 2007], which will be introduced in Chapter 18 and thoroughly detailed.

Chapter 16, Chapter 17 and Chapter 18 discuss the solutions brought by the field of asymptotically large and small random matrices to the problems of estimation, detection, and system modeling, respectively. Chapter 19 will then discuss the perspectives and challenges envisioned for the future of random matrices in topics related to wireless communications. Finally, in Chapter 20, we draw the conclusions of this book.

Part I

Theoretical aspects

2 Random matrices

It is often assumed that random matrices is a field of mathematics which treats matrix models as if matrices were of infinite size and which then approximate functionals of realistic finite size models using asymptotic results. We wish first to insist on the fact that random matrices are necessarily of finite size, so we do not depart from conventional linear algebra. We start this chapter by introducing initial considerations and *exact* results on finite size random matrices. We will see later that, for some matrix models, it is then interesting to study the features of some random matrices with large dimensions. More precisely, we will see that the eigenvalue distribution function $F^{\mathbf{B}_N}$ of some $N \times N$ random Hermitian matrices $\mathbf{B}_N$ converge in distribution (often almost surely so) to some deterministic limit F when N grows to infinity. The results obtained for F can then be turned into approximative results for $F^{\mathbf{B}_N}$, and therefore help to provide approximations of some functionals of $\mathbf{B}_N$. Even if it might seem simpler for some to think of F as the eigenvalue distribution of an infinite size matrix, this does not make much sense in mathematical terms, and we will never deal with such objects as infinite size matrices, but only with sequences of finite dimensional matrices of increasing size.

2.1 Small dimensional random matrices

We start with a formal definition of a *random matrix* and introduce some notations.

2.1.1 Definitions and notations

Definition 2.1. An $N \times n$ matrix $\mathbf{X}$ is said to be a *random matrix* if it is a matrix-valued random variable on some probability space $(\Omega, \mathcal{F}, P)$ with entries in some measurable space $(\mathcal{R}, \mathcal{G})$, where $\mathcal{F}$ is a σ-field on Ω with probability measure P and $\mathcal{G}$ is a σ-field on $\mathcal{R}$. As per conventional notations, we denote $\mathbf{X}(\omega)$ the realization of the variable $\mathbf{X}$ at point $\omega \in \Omega$.

The rigorous introduction of the probability space $(\Omega, \mathcal{F}, P)$ is only necessary here for some details on proofs given in this chapter. When not mentioning either

Ω or the σ-field $\mathcal{F}$ in the rest of this book, it will be clear what implicit probability space is being referred to. In general, though, this formalism is unimportant. Also, unless necessary for clarity, we will in the following equally refer to $\mathbf{X}$ as the random variable and as its random realization $\mathbf{X}(\omega)$ for $\omega \in \Omega$. The space $\mathcal{R}$ will often be taken to be either $\mathbb{R}$ or $\mathbb{C}$, i.e. for all $\omega \in \Omega$, $\mathbf{X}(\omega) \in \mathbb{R}^{N\times n}$ or $\mathbf{X}(\omega) \in \mathbb{C}^{N\times n}$.

We define the *probability distribution* of $\mathbf{X}$ to be $\mu_{\mathbf{X}}$, the joint probability distribution of the entries of $\mathbf{X}$, such that, for $A \in \mathcal{G}^{N\times n}$

$$\mu_{\mathbf{X}}(A) \triangleq P(\{\omega, \mathbf{X}(\omega) \in A\}).$$

In most practical cases, $\mu_{\mathbf{X}}$ will have a *probability density function* (p.d.f.) with respect to whatever measure on $\mathcal{R}^{N\times n}$, which we will denote $P_{\mathbf{X}}$, i.e. for $d\mathbf{Y} \in \mathcal{G}$ an elementary volume around $\mathbf{Y} \in \mathcal{R}^{N\times n}$

$$P_{\mathbf{X}}(\mathbf{Y})d\mathbf{Y} \triangleq \mu_{\mathbf{X}}(d\mathbf{Y}) = P(\{\omega, \mathbf{X}(\omega) \in d\mathbf{Y}\}).$$

In order to differentiate the probability distribution function of real random variables X from that of multivariate entities, we will often denote $p_X \triangleq P_X$ in lowercase characters. For vector-valued real random variables $(X_1, \ldots, X_N)$, we further denote $P_{(X_1,\ldots,X_N)}(x_1, \ldots, x_N)$ or $P_{(X_i)}(x_1, \ldots, x_N)$ the density of the *unordered values* $X_1, \ldots, X_N$ $P^{\leq}_{(X_1,\ldots,X_N)}(x_1, \ldots, x_N)$ or $P^{\leq}_{(X_i)}(x_1, \ldots, x_N)$ the density of the *non-decreasing* values $X_1 \leq \ldots \leq X_N$ and $P^{\geq}_{(X_1,\ldots,X_N)}(x_1, \ldots, x_N)$ or $P^{\geq}_{(X_i)}(x_1, \ldots, x_N)$ the density of the *non-increasing* values $X_1 \geq \ldots \geq X_N$.

The (cumulative) *distribution function* (d.f.) of a real random variable will often be denoted by the letters F, G, or H, e.g. for $x \in \mathbb{R}$

$$F(x) \triangleq \mu_X((-\infty, x])$$

denotes the d.f. of X.

We will in particular often consider the marginal probability distribution function of the eigenvalues of random Hermitian matrices $\mathbf{X}$. Unless otherwise stated, the d.f. of the *real* eigenvalues of $\mathbf{X}$ will be denoted $F^{\mathbf{X}}$.

Remark 2.1. As mentioned earlier, in most applications, the probability space Ω needs not be defined but may have interesting interpretations. In wireless communications, if $\mathbf{H} \in \mathbb{C}^{n_r \times n_t}$ is a random MIMO channel between an n_t-antenna transmitter and an n_r-antenna receiver, Ω can be seen as the space of "possible environments," the elements ω of which being all valid snapshots of the physical world at a given instant. The random value $\mathbf{H}(\omega)$ is therefore the realization of an instantaneous $n_r \times n_t$ multiple antenna propagation channel.

We will say that a sequence $F_1, F_2, \ldots$ of real-supported distribution functions converge *weakly* to the function F, if, for $x \in \mathbb{R}$ a continuity point of F

$$\lim_{n\to\infty} F_n(x) = F(x).$$

This will be denoted as

$$F_n \Rightarrow F.$$

We will also say that a sequence $x_1, x_2, \ldots$ of random variables converges *almost surely* to the constant x if

$$P(\lim_n x_n = x) = P(\{\omega, \lim_n x_n(\omega) = x\}) = 1.$$

The underlying probability space $(\Omega, \mathcal{F}, P)$ here is assumed to be the space that generates the *sequences* $x_1(\omega), x_2(\omega), \ldots$ (and not the individual entries), for $\omega \in \Omega$. This will be denoted

$$x_n \xrightarrow{\text{a.s.}} x.$$

We will in particular be interested in the almost sure weak convergence of distribution functions, i.e. in proving that, for some sequence $F_1, F_2, \ldots$ of d.f. with weak limit F, there exists a space $A \in \mathcal{F}$, such that $P(A) = 1$ and, for all $x \in \mathbb{R}$, $\omega \in A$ implies

$$\lim_n F_n(x; \omega) = F(x)$$

with $F_1(\cdot; \omega), F_2(\cdot; \omega), \ldots$ one realization of the d.f.-valued random sequence $F_1, F_2, \ldots$.

Although this will rarely be used, we also mention the notation for convergence *in probability* of a sequence $x_1, x_2, \ldots$ to x. A sequence $x_1, x_2, \ldots$ is said to converge in probability to x if, for all $\varepsilon > 0$

$$\lim_n P(|x_n - x| > \varepsilon) = 0.$$

2.1.2 Wishart matrices

In the following section, we provide elementary results on the distribution of random matrices with Gaussian entries and of their eigenvalue distributions.

As mentioned in the Introduction, the very first random matrix considerations date back to 1928 [Wishart, 1928], with the expression of the p.d.f. of random matrices $\mathbf{X}\mathbf{X}^\mathsf{H}$, for $\mathbf{X} \in \mathbb{C}^{N \times n}$ with columns $\mathbf{x}_1, \ldots, \mathbf{x}_n \sim \mathcal{CN}(0, \mathbf{R})$. Such a matrix is called a (central) Wishart matrix.

Definition 2.2. The $N \times N$ random matrix $\mathbf{X}\mathbf{X}^\mathsf{H}$ is a (real or complex) *central Wishart matrix* with n degrees of freedom and covariance matrix $\mathbf{R}$ if the columns of the $N \times n$ matrix $\mathbf{X}$ are zero mean independent (real or complex) Gaussian vectors with covariance matrix $\mathbf{R}$. This is denoted

$$\mathbf{X}\mathbf{X}^\mathsf{H} = \mathbf{X}\mathbf{X}^\mathsf{T} \sim \mathcal{W}_N(n, \mathbf{R})$$

for real Wishart matrices and

$$\mathbf{X}\mathbf{X}^\mathsf{H} \sim \mathcal{CW}_N(n, \mathbf{R})$$

for complex Wishart matrices.

Defining the *Gram matrix* associated with any complex matrix $\mathbf{X}$ as being the matrix $\mathbf{X}\mathbf{X}^{\mathsf{H}}$, $\mathbf{X}\mathbf{X}^{\mathsf{H}} \sim \mathcal{CW}_N(n, \mathbf{R})$ is by definition the Gram matrix of a matrix with Gaussian i.i.d. columns of zero mean and covariance $\mathbf{R}$. When $\mathbf{R} = \mathbf{I}_N$, it is usual to refer to $\mathbf{X}$ as a *standard Gaussian matrix.*

The interest of Wishart matrices lies primarily in the following remark.

Remark 2.2. Let $\mathbf{x}_1, \ldots, \mathbf{x}_n \in \mathbb{C}^N$ be n independent samples of the random process $\mathbf{x}_1 \sim \mathcal{CN}(0, \mathbf{R})$. Then, denoting $\mathbf{X} = [\mathbf{x}_1, \ldots, \mathbf{x}_n]$

$$\sum_{i=1}^{n} \mathbf{x}_i \mathbf{x}_i^{\mathsf{H}} = \mathbf{X}\mathbf{X}^{\mathsf{H}}.$$

For this reason, the random matrix $\mathbf{R}_n \triangleq \frac{1}{n}\mathbf{X}\mathbf{X}^{\mathsf{H}}$ is often referred to as an *(empirical) sample covariance matrix* associated with the random process $\mathbf{x}_1$. This is to be contrasted with the *population covariance matrix* $\mathbf{R}$, whose relation to the $\mathbf{R}_n$ was already evidenced as non-trivial, when n and N are of the same order of magnitude. Of particular importance is the case when $\mathbf{R} = \mathbf{I}_N$. In this situation, $\mathbf{X}\mathbf{X}^{\mathsf{H}}$, sometimes referred to as a *zero (or null) Wishart matrix*, is proportional to the sample covariance matrix of a white Gaussian process. The zero (or null) terminology is due to the signal processing problem of hypothesis testing, in which we have to decide whether the observed $\mathbf{X}$ emerges from a white noise process or from an information plus noise process. The noise hypothesis is often referred to as the *null hypothesis.*

Wishart provides us with the p.d.f. of Wishart matrices $\mathbf{W}$ in the space of non-negative definite matrices with elementary volume $d\mathbf{W}$, as follows.

Theorem 2.1 ([Wishart, 1928]). *The p.d.f. of the complex Wishart matrix* $\mathbf{X}\mathbf{X}^{\mathsf{H}} \sim \mathcal{CW}_N(n, \mathbf{R})$, $\mathbf{X} \in \mathbb{C}^{N \times n}$, *in the space of* $N \times N$ *non-negative definite complex matrices, for* $n \geq N$, *is*

$$P_{\mathbf{X}\mathbf{X}^{\mathsf{H}}}(\mathbf{B}) = \frac{\pi^{-\frac{1}{2}N(N-1)}}{\det \mathbf{R}^n \prod_{i=1}^{N}(n-i)!} e^{-\operatorname{tr}(\mathbf{R}^{-1}\mathbf{B})} \det \mathbf{B}^{n-N}. \tag{2.1}$$

Note in particular that, for $N = 1$, this is the distribution of a real random variable X with $2X$ chi-square-distributed with $2n$ degrees of freedom.

Proof. Since $\mathbf{X}$ is a Gaussian matrix of size $N \times n$ with zero mean and covariance $\mathbf{R}$, we know that

$$P_{\mathbf{X}}(\mathbf{A})d\mathbf{X} = \frac{1}{\pi^{Nn} \det \mathbf{R}^n} e^{-\operatorname{tr}(\mathbf{R}^{-1}\mathbf{A}\mathbf{A}^{\mathsf{H}})} d\mathbf{X}$$

which is the probability of an elementary volume $d\mathbf{X}$ around point $\mathbf{A}$ in the space of $N \times n$ complex matrices with measure $d\mathbf{X} \triangleq \prod_{i,j} dX_{ij}$, with X_{ij} the entry (i, j) of $\mathbf{X}$. Now, to derive the probability $P_{\mathbf{X}\mathbf{X}^{\mathsf{H}}}(\mathbf{B})d(\mathbf{X}\mathbf{X}^{\mathsf{H}})$, it suffices to operate a variable change between $\mathbf{X}$ and $\mathbf{X}\mathbf{X}^{\mathsf{H}}$ from the space of $N \times n$ complex matrices to the space of $N \times N$ non-negative definite complex matrices. This is obtained by Wishart in [Wishart, 1928], who shows that

$$d\mathbf{X} = \frac{\pi^{-\frac{1}{2}N(N-1)+Nn}}{\prod_{i=1}^{N}(n-i)!} \det(\mathbf{X}\mathbf{X}^{\mathsf{H}})^{n-N} d(\mathbf{X}\mathbf{X}^{\mathsf{H}})$$

and therefore:

$$P_{\mathbf{X}\mathbf{X}^{\mathsf{H}}}(\mathbf{B}) d(\mathbf{X}\mathbf{X}^{\mathsf{H}}) = \frac{\pi^{-\frac{1}{2}N(N-1)}}{\det \mathbf{R}^{n} \prod_{i=1}^{N}(n-i)!} e^{-\operatorname{tr}(\mathbf{R}^{-1}\mathbf{B})} \det \mathbf{B}^{n-N} d(\mathbf{X}\mathbf{X}^{\mathsf{H}})$$

which is the intended formula. □

In [Ratnarajah and Vaillancourt, 2005, Theorem 3], Ratnarajah and Vaillancourt extend the result from Wishart to the case of singular matrices, i.e. when $n < N$. This is given as follows.

Theorem 2.2 ([Ratnarajah and Vaillancourt, 2005]). *The p.d.f. of the complex Wishart matrix $\mathbf{X}\mathbf{X}^{\mathsf{H}} \sim \mathcal{C}W_N(n, \mathbf{R})$, $\mathbf{X} \in \mathbb{C}^{N \times n}$, in the space of $N \times N$ non-negative definite complex matrices of rank n, for $n < N$, is*

$$P_{\mathbf{X}\mathbf{X}^{\mathsf{H}}}(\mathbf{B}) = \frac{\pi^{-\frac{1}{2}N(N-1)+n(n-N)}}{\det \mathbf{R}^{n} \prod_{i=1}^{n}(n-i)!} e^{-\operatorname{tr}(\mathbf{R}^{-1}\mathbf{B})} \det \mathbf{\Lambda}^{n-N}$$

with $\mathbf{\Lambda} \in \mathbb{C}^{n \times n}$ the diagonal matrix of the positive eigenvalues of $\mathbf{B}$.

As already noticed from Equation (1.3) of the mutual information of MIMO communication channels and from the brief introduction of eigenvalue-based methods for detection and estimation, the center of interest of random matrices often lies in the distribution of their eigenvalues. For null Wishart matrices, notice that $P_{\mathbf{X}\mathbf{X}^{\mathsf{H}}}(\mathbf{B}) = P_{\mathbf{X}\mathbf{X}^{\mathsf{H}}}(\mathbf{U}\mathbf{B}\mathbf{U}^{\mathsf{H}})$, for any unitary $N \times N$ matrix $\mathbf{U}$.[1] Otherwise stated, the eigenvectors of the random variable $\mathbf{R}_N$ are uniformly distributed over the space $\mathcal{U}(N)$ of unitary $N \times N$ matrices. As such, the eigenvectors do not carry relevant information, and $P_{\mathbf{X}\mathbf{X}^{\mathsf{H}}}(\mathbf{B})$ is only a function of the eigenvalues of $\mathbf{B}$.

The joint p.d.f. of the eigenvalues of zero Wishart matrices were studied simultaneously in 1939 by different authors [Fisher, 1939; Girshick, 1939; Hsu, 1939; Roy, 1939]. The main two results are summarized in the following.

Theorem 2.3. *Let the entries of $\mathbf{X} \in \mathbb{C}^{N \times n}$ be i.i.d. Gaussian with zero mean and unit variance. Denote $m = \min(n, N)$ and $M = \max(n, N)$. The joint p.d.f. $P^{\geq}_{(\lambda_i)}$ of the positive ordered eigenvalues $\lambda_1 \geq \ldots \geq \lambda_N$ of the zero Wishart matrix*

[1] We recall that a unitary matrix $\mathbf{U} \in \mathbb{C}^{N \times N}$ is such that $\mathbf{U}\mathbf{U}^{\mathsf{H}} = \mathbf{U}^{\mathsf{H}}\mathbf{U} = \mathbf{I}_N$.

$\mathbf{XX}^{\mathsf{H}}$ *is given by:*

$$P^{\geq}_{(\lambda_i)}(\lambda_1, \ldots, \lambda_N) = e^{-\sum_{i=1}^m \lambda_i} \prod_{i=1}^m \frac{\lambda_i^{M-m}}{(m-i)!(M-i)!} \Delta(\mathbf{\Lambda})^2$$

where, for a Hermitian non-negative $m \times m$ *matrix* $\mathbf{\Lambda}$,[2] $\Delta(\mathbf{\Lambda})$ *denotes the Vandermonde determinant of its eigenvalues* $\lambda_1, \ldots, \lambda_m$

$$\Delta(\mathbf{\Lambda}) \triangleq \prod_{1 \leq i < j \leq m} (\lambda_j - \lambda_i).$$

The marginal p.d.f. p_λ $(\triangleq P_\lambda)$ *of the unordered eigenvalues is*

$$p_\lambda(\lambda) = \frac{1}{m} \sum_{k=0}^{m-1} \frac{k!}{(k+M-n)!} [L_k^{M-m}(\lambda)]^2 \lambda^{M-m} e^{-\lambda}$$

where L_n^k *are the Laguerre polynomials defined as*

$$L_n^k(\lambda) = \frac{e^\lambda}{k! \lambda^n} \frac{d^k}{d\lambda^k}(e^{-\lambda} \lambda^{n+k}).$$

The generalized case of (non-zero) central Wishart matrices is more involved since it requires advanced tools of multivariate analysis, such as the fundamental Harish–Chandra integral [Chandra, 1957]. We will mention the result of Harish–Chandra, which is at the core of major results in signal sensing and channel modeling presented in Chapter 16 and Chapter 18, respectively.

Theorem 2.4 ([Chandra, 1957]). *For non-singular* $N \times N$ *positive definite Hermitian matrices* $\mathbf{A}$ *and* $\mathbf{B}$ *of respective eigenvalues* $a_1, \ldots, a_N$ *and* $b_1, \ldots, b_N$, *such that, for all* $i \neq j$, $a_i \neq a_j$ *and* $b_i \neq b_j$, *we have:*

$$\int_{\mathbf{U} \in \mathcal{U}(N)} e^{\kappa \operatorname{tr}(\mathbf{AUBU}^{\mathsf{H}})} d\mathbf{U} = \left(\prod_{i=1}^{N-1} i! \right) \kappa^{\frac{1}{2}N(N-1)} \frac{\det\left(\{e^{-b_j a_i}\}_{1 \leq i,j \leq N}\right)}{\Delta(\mathbf{A})\Delta(\mathbf{B})}$$

where, for any bivariate function f, $\{f(i,j)\}_{1 \leq i,j \leq N}$ *denotes the* $N \times N$ *matrix of* (i,j) *entry* $f(i,j)$, $\mathcal{U}(N)$ *is the space of* $N \times N$ *unitary matrices, and* $d\mathbf{U}$ *denotes the invariant measure on* $\mathcal{U}(N)$ *normalized to make the total measure unity.*

The remark that $d\mathbf{U}$ is a normalized measure on $\mathcal{U}(N)$ arises from the fact that, contrary to spaces such as $\mathbb{R}^N$, $\mathcal{U}(N)$ is compact. As such, it can be attached a measure $d\mathbf{U}$ such that $\int_{\mathcal{U}(N)} d\mathbf{U} = V$ for V the volume of $\mathcal{U}(N)$ under this measure. We arbitrarily take $V = 1$ here. In other publications, e.g., [Hiai and Petz, 2006], the authors choose other normalizations. As for the invariant measure terminology (called Haar measure in subsequent sections), it refers to

[2] We will respect the convention that x (be it a scalar or a Hermitian matrix) is non-negative if $x \geq 0$, while x is positive if $x > 0$.

the measure such that $d\mathbf{U}$ is a constant (taken to equal 1 here). In the rest of this section, we take this normalization.

Theorem 2.4 enables the calculus of the marginal joint-eigenvalue distribution of (non-zero) central Wishart matrices [Itzykson and Zuber, 2006] given as:

Theorem 2.5 (Section 8.7 of [James, 1964]). *Let the columns of $\mathbf{X} \in \mathbb{C}^{N\times n}$ be i.i.d. zero mean Gaussian with positive definite covariance $\mathbf{R}$, and $n \geq N$. The joint p.d.f. $P^{\geq}_{(\lambda_i)}$ of the ordered positive eigenvalues $\lambda_1 \geq \ldots \geq \lambda_N$ of the central Wishart matrix $\mathbf{X}\mathbf{X}^{\mathsf{H}}$ reads:*

$$P^{\geq}_{(\lambda_i)}(\lambda_1,\ldots,\lambda_N) = (-1)^{\frac{1}{2}N(N-1)} \frac{\det(\{e^{-\frac{\lambda_i}{r_j}}\}_{1\leq i,j\leq N})}{\det \mathbf{R}^n} \frac{\Delta(\mathbf{\Lambda})}{\Delta(\mathbf{R}^{-1})} \prod_{j=1}^{N} \frac{\lambda_j^{n-N}}{(n-j)!}$$

with $r_1 > \ldots > r_N > 0$ the eigenvalues of $\mathbf{R}$ and $\mathbf{\Lambda} = \operatorname{diag}(\lambda_1,\ldots,\lambda_N)$.

Proof. The idea behind the proof is first to move from the space of $\mathbf{X}\mathbf{X}^{\mathsf{H}}$ non-negative definite matrices to the product space of matrices $(\mathbf{U},\mathbf{\Lambda})$, $\mathbf{U}$ being unitary and $\mathbf{\Lambda}$ diagonal with positive entries. For this, it suffices to notice from a Jacobian calculus that

$$d(\mathbf{X}\mathbf{X}^{\mathsf{H}}) = \frac{\pi^{\frac{1}{2}N(N-1)}}{\prod_{i=1}^{N} i!} \prod_{i<j}^{N} (\lambda_j - \lambda_i)^2 d\mathbf{\Lambda} d\mathbf{U}$$

with $\mathbf{X}\mathbf{X}^{\mathsf{H}} = \mathbf{U}\mathbf{\Lambda}\mathbf{U}^{\mathsf{H}}$.[3] It then suffices to integrate out the matrices $\mathbf{U}$ for fixed $\mathbf{\Lambda}$ from the probability distribution $P_{\mathbf{X}\mathbf{X}^{\mathsf{H}}}$. This requires the Harish–Chandra formula, Theorem 2.4. □

Noticing that the eigenvalue labeling is in fact irrelevant, we obtain the joint unordered eigenvalue distribution of central Wishart matrices as follows.

Theorem 2.6 ([James, 1964]). *Let the columns of $\mathbf{X} \in \mathbb{C}^{N\times n}$ be i.i.d. zero mean Gaussian with positive definite covariance $\mathbf{R}$ and $n \geq N$. The joint p.d.f. $P_{(\lambda_i)}$ of the unordered positive eigenvalues $\lambda_1,\ldots,\lambda_N$ of the central Wishart matrix $\mathbf{X}\mathbf{X}^{\mathsf{H}}$, reads:*

$$P_{(\lambda_i)}(\lambda_1,\ldots,\lambda_N) = (-1)^{\frac{1}{2}N(N-1)} \frac{1}{N!} \frac{\det(\{e^{-\frac{\lambda_i}{r_j}}\}_{1\leq i,j\leq N})}{\det \mathbf{R}^n} \frac{\Delta(\mathbf{\Lambda})}{\Delta(\mathbf{R}^{-1})} \prod_{j=1}^{N} \frac{\lambda_j^{n-N}}{(n-j)!}$$

with $r_1 > \ldots > r_N > 0$ the eigenvalues of $\mathbf{R}$ and $\mathbf{\Lambda} = \operatorname{diag}(\lambda_1,\ldots,\lambda_N)$.

Note that it is necessary here that the eigenvalues of $\mathbf{R}$ be distinct. This is because the general expression of Theorem 2.6 involves integration over the

[3] It is often found in the literature that $d(\mathbf{X}\mathbf{X}^{\mathsf{H}}) = \prod_{i<j}^{N} (\lambda_j - \lambda_i)^2 d\mathbf{\Lambda} d\mathbf{U}$. This results from another normalization of the invariant measure on $\mathcal{U}(N)$.

unitary space which expresses in closed-form in this scenario thanks to the Harish–Chandra formula, Theorem 2.4.

A similar expression for the singular case when $n < N$ is provided in [Ratnarajah and Vaillancourt, 2005, Theorem 4], but the final expression is less convenient to express in closed-form as it features hypergeometric functions, which we introduce in the following. Other extensions of the central Wishart p.d.f. to *non-central Wishart matrices*, i.e. originating from a matrix with non-centered Gaussian entries, have also been considered, leading to results on the joint-entry distribution [Anderson, 1946], eigenvalue distribution [Jin *et al.*, 2008], largest and lowest eigenvalue marginal distribution, condition number distribution [Ratnarajah *et al.*, 2005a], etc. The results mentioned so far are of considerable importance when matrices of small dimension have to be considered in concrete applications. We presently introduce the result of [Ratnarajah and Vaillancourt, 2005, Theorem 4] that generalizes Theorem 2.6 and the result that concerns non-central complex Wishart matrices [James, 1964], extended in [Ratnarajah *et al.*, 2005b]. For this, we need to introduce a few definitions.

Definition 2.3. For $\kappa = (k_1, \ldots, k_N)$ with $k_1 + \ldots + k_N = k$ for some (k, N), we denote for $x \in \mathbb{C}$

$$[x]_\kappa = \prod_{i=1}^{N} (x - i + 1)_{k_i}$$

where $(u)_k = u(u+1)\ldots(u+k-1)$. We also denote, for $\mathbf{X} \in \mathbb{C}^{N \times N}$ with complex eigenvalues $\lambda_1, \ldots, \lambda_N$

$$C_\kappa(\mathbf{X}) = \chi_{[\kappa]}(1)\chi_{[\kappa]}(\mathbf{X})$$

the *complex zonal polynomial* of $\mathbf{X}$, where $\chi_{[\kappa]}(1)$ is given by:

$$\chi_{[\kappa]}(1) = k! \frac{\prod_{i<j}^{N}(k_i - k_j - i - j)}{\prod_{i=1}^{N}(k_i + N - i)!}$$

and $\chi_{[\kappa]}(\mathbf{X})$ reads:

$$\chi_{[\kappa]}(\mathbf{X}) = \frac{\det\left(\{\lambda_i^{k_j+N-j}\}_{i,j}\right)}{\det\left(\{\lambda_i^{N-j}\}_{i,j}\right)}.$$

The hypergeometric function of a complex matrix $\mathbf{X} \in \mathbb{C}^{N \times N}$ is then defined by

$${}_pF_q(a_1, \ldots, a_p; b_1, \ldots, b_p; \mathbf{X}) = \sum_{k=0}^{\infty} \sum_{\kappa} \frac{[a_1]_\kappa \ldots [a_p]_\kappa}{[b_1]_\kappa \ldots [b_p]_\kappa} \frac{C_\kappa(\mathbf{X})}{k!}$$

where $a_1, \ldots, a_p, b_1, \ldots, b_p \in \mathbb{C}$, $\sum_\kappa$ is a summation over all partitions κ of k elements into N and $[x]_\kappa = \prod_{i=1}^{N}(x - i + 1)_{k_i}$.

We then define the hypergeometric function of two complex matrices of the same dimension $\mathbf{X}, \mathbf{Y} \in \mathbb{C}^{N \times N}$ to be

$$ {}_pF_q(\mathbf{X}, \mathbf{Y}) = \int_{\mathcal{U}(N)} {}_pF_q(\mathbf{X}\mathbf{U}\mathbf{Y}\mathbf{U}^{\mathsf{H}}) d\mathbf{U}. $$

We have in particular that

$$ {}_0F_0(\mathbf{X}) = e^{\operatorname{tr} \mathbf{X}} $$
$$ {}_1F_0(a; \mathbf{X}) = \det(\mathbf{I}_N - \mathbf{X})^{-a}. $$

From this observation, we remark that the Harish–Chandra formula, Theorem 2.4, actually says that

$$ {}_0F_0(\mathbf{A}, \mathbf{B}) = \int_{\mathbf{U} \in \mathcal{U}(N)} {}_0F_0(\mathbf{A}\mathbf{U}\mathbf{B}\mathbf{U}^{\mathsf{H}}) d\mathbf{U} = \left(\prod_{i=1}^{N-1} i! \right) \frac{\det \left(\{ e^{-b_j a_i} \}_{1 \leq i,j \leq N} \right)}{\Delta(\mathbf{A}) \Delta(\mathbf{B})} $$

as long as the eigenvalues of $\mathbf{A}$ and $\mathbf{B}$ are all distinct. In particular, Theorem 2.6 can be rewritten in its full generality, i.e. without the constraint of distinctness of the eigenvalues of $\mathbf{R}$, when replacing the determinant expressions by the associated hypergeometric function.

Further generalizations of hypergeometric functions of multiple Hermitian matrices, non-necessarily of the same dimensions, exist. In particular, for $\mathbf{A}_1, \ldots, \mathbf{A}_p$ all symmetric with $\mathbf{A}_k \in \mathbb{C}^{N_k \times N_k}$, we mention the definition, see, e.g., [Hanlen and Grant, 2003]

$$ {}_0F_0(\mathbf{A}_1, \ldots, \mathbf{A}_p) = \sum_{k=0}^{\infty} \sum_{\kappa} \left(\frac{C_\kappa(\mathbf{A}_1)}{k!} \prod_{i=2}^{p} \frac{C_\kappa(\mathbf{A}_i)}{C_\kappa(\mathbf{I}_N)} \right) \tag{2.2} $$

with $N = \max_k \{N_k\}$ and the partitions κ have dimension $\min_k \{N_k\}$. For $\mathbf{A} \in \mathbb{C}^{N \times N}$ and $\mathbf{B} \in \mathbb{C}^{n \times n}$, $n \leq N$, we therefore have the general expression of ${}_0F_0(\mathbf{A}, \mathbf{B})$ as

$$ {}_0F_0(\mathbf{A}, \mathbf{B}) = \sum_{k=0}^{\infty} \sum_{\kappa} \frac{C_\kappa(\mathbf{A}) C_\kappa(\mathbf{B})}{k! C_\kappa(\mathbf{I}_N)}. \tag{2.3} $$

Sometimes, the notation ${}_0F_0^{(n)}$ is used where the superscript (n) indicates that the partitions κ are n-dimensional. For a deeper introduction to zonal polynomials and matrix-valued hypergeometric functions, see, e.g., [Muirhead, 1982].

The extension of Theorem 2.6 to the singular case is as follows.

Theorem 2.7 (Theorem 4 of [Ratnarajah and Vaillancourt, 2005]). *Let the columns of $\mathbf{X} \in \mathbb{C}^{N \times n}$ be i.i.d. zero mean Gaussian with non-negative definite covariance $\mathbf{R}$ and $n < N$. The joint p.d.f. $P_{(\lambda_i)}$ of the unordered positive* *eigenvalues* $\lambda_1, \ldots, \lambda_n$ *of the central Wishart matrix* $\mathbf{X}\mathbf{X}^{\mathsf{H}}$, *reads:*

$$P_{(\lambda_i)}(\lambda_1, \ldots, \lambda_N) = \Delta(\mathbf{\Lambda})^2 {}_0F_0(-\mathbf{R}^{-1}, \mathbf{\Lambda}) \prod_{j=1}^{n} \frac{\lambda_j^{N-n}}{(n-j)!(N-j)!}$$

where $\mathbf{\Lambda} = \mathrm{diag}(\lambda_1, \ldots, \lambda_n)$.

Note here that $\mathbf{R}^{-1} \in \mathbb{C}^{N \times N}$ and $\mathbf{\Lambda} \in \mathbb{C}^{n \times n}$, $n < N$, so that ${}_0F_0(-\mathbf{R}^{-1}, \mathbf{\Lambda})$ does *not* take the closed-form conveyed by the Harish–Chandra theorem, Theorem 2.4, but can be evaluated from Equation (2.3).

The result from James on non-central Wishart matrices is as follows.

Theorem 2.8 ([James, 1964]). *If* $\mathbf{X} \in \mathbb{C}^{N \times n}$ *is a complex Gaussian matrix with mean* $\mathbf{M}$ *and invertible covariance* $\mathbf{\Sigma} = \frac{1}{n}\mathrm{E}[(\mathbf{X} - \mathbf{M})(\mathbf{X} - \mathbf{M})^{\mathsf{H}}]$, *then* $\mathbf{X}\mathbf{X}^{\mathsf{H}}$ *is distributed as*

$$P_{\mathbf{X}\mathbf{X}^{\mathsf{H}}}(\mathbf{B}) = \frac{\pi^{-\frac{1}{2}N(N-1)}}{\prod_{i=1}^{N}(n-i)!} \frac{\det(\mathbf{B})^{n-N}}{\det(\mathbf{\Sigma})^n} e^{-\operatorname{tr} \mathbf{\Sigma}^{-1}(\mathbf{M}\mathbf{M}^{\mathsf{H}}+\mathbf{B})} {}_0F_1(n; \mathbf{\Sigma}^{-1}\mathbf{M}\mathbf{M}^{\mathsf{H}}\mathbf{\Sigma}^{-1}\mathbf{B}).$$

Also, the density of the unordered eigenvalues $\lambda_1, \ldots, \lambda_N$ *of* $\mathbf{\Sigma}^{-1}\mathbf{X}\mathbf{X}^{\mathsf{H}}$ *expresses as a function of the unordered eigenvalues* $m_1, \ldots, m_N$ *of* $\mathbf{\Sigma}^{-1}\mathbf{M}\mathbf{M}^{\mathsf{H}}$ *as*

$$P_{(\lambda_i)}(\lambda_1, \ldots, \lambda_N) = \Delta(\mathbf{\Sigma}^{-1}\mathbf{M}\mathbf{M}^{\mathsf{H}})^2 e^{-\sum_{i=1}^{N}(m_i+\lambda_i)} \prod_{i=1}^{N} \frac{\lambda_i^{n-N}}{(n-i)!(N-i)!}$$

if $m_i \neq m_j$ *for all* $i \neq j$.

We complete this section by the introduction of an identity that allows us to extend results such as Theorem 2.6 to the case when an eigenvalue of $\mathbf{R}$ has multiplicity greater than one. As can be observed in the expression of the eigenvalue density in Theorem 2.6, equating population eigenvalues leads in general to bringing numerators and denominators to zero. To overcome this difficulty when some eigenvalues have multiplicities larger than one, we have the following result which naturally extends [Simon *et al.*, 2006, Lemma 6].

Theorem 2.9 ([Couillet and Guillaud, 2011; Simon *et al.*, 2006]). *Let* $f_1, \ldots, f_N$ *be a family of infinitely differentiable functions and let* $x_1, \ldots, x_N \in \mathbb{R}$. *Denote*

$$R(x_1, \ldots, x_N) \triangleq \frac{\det\left(\{f_i(x_j)\}_{i,j}\right)}{\prod_{i<j}(x_j - x_i)}.$$

Then, for $N_1, \ldots, N_p$ *such that* $N_1 + \ldots + N_p = N$ *and for* $y_1, \ldots, y_p \in \mathbb{R}$ *distinct*

$$\lim_{\substack{x_1,\ldots,x_{N_1}\to y_1\\ \cdots \\ x_{N-N_p+1},\ldots,x_N\to y_p}} R(x_1,\ldots,x_N)$$

$$= \frac{\det\left[f_i(y_1), f_i'(y_1), \ldots, f_i^{(N_1-1)}(y_1), \ldots, f_i(y_p), f_i'(y_p), \ldots, f_i^{(N_p-1)}(y_p)\right]}{\prod_{1\leq i<j\leq p}(y_j - y_i)^{N_i N_j} \prod_{l=1}^{p}\prod_{j=1}^{N_l-1} j!}.$$

This observation will be useful when deriving results in Chapter 16 in the context of signal detection, where one of the population eigenvalues, equal to σ^2, the variance of the additive noise, has a multiplicity in general much larger than one. This result is obtained from successive iterations of L'Hospital rule, which we remind below and which will be useful for other purposes in the book.

Theorem 2.10. *Let* $f(x)$ *and* $g(x)$ *be two real-valued continuous functions, differentiable in a neighborhood of* a*, such that*

$$\lim_{x\to a} f(x) = \lim_{x\to a} g(x) = 0$$

or

$$\lim_{x\to a} |f(x)| = \lim_{x\to a} |g(x)| = \infty$$

and such that

$$\lim_{x\to a} \frac{f'(x)}{g'(x)} = l.$$

Then we have:

$$\lim_{x\to a} \frac{f(x)}{g(x)} = l.$$

Theorem 2.9 is in particular obtained by taking for instance

$$R(x_1,\ldots,x_N) = R(y_1+\varepsilon,\ldots,y_1+N_1\varepsilon,\ldots,y_p+\varepsilon,\ldots,y_p+N_p\varepsilon)$$

and using L'Hospital rule in the limit $\varepsilon \to 0$. Indeed, in the simplest scenario of a convergence of x_1, x_2 to y_1, we have

$$\lim_{x_1,x_2\to y_1} R(x_1,\ldots,x_N)$$

$$= \lim_{\varepsilon\to 0} \frac{\det\left[f_i(y_1+\varepsilon), f_i(y_1+2\varepsilon), f_i(x_3), \ldots, f_i(x_N)\right]}{\varepsilon \prod_{i>j>2}(x_i - x_j) \prod_{i=3}^{N}(x_i - y_1 - \varepsilon)(x_i - y_1 - 2\varepsilon)}$$

which, after application of L'Hospital rule, equals (upon existence of the limit), to the limiting ratio between the numerator

$$\sum_{\sigma \in \mathcal{S}_N} \operatorname{sgn}(\sigma) \left[f'_{\sigma_1}(y_1+\varepsilon) f_{\sigma_2}(y_1+2\varepsilon) \prod_{i=3}^{N} f_{\sigma_i}(x_i) \right.$$

$$\left. + 2 f_{\sigma_1}(y_1+\varepsilon) f'_{\sigma_2}(y_1+2\varepsilon) \prod_{i=3}^{N} f_{\sigma_i}(x_i) \right]$$

and the denominator

$$\prod_{i>j>2} (x_i - x_j) \prod_{i=3}^{N} (x_i - y_1 - \varepsilon)(x_i - y_1 - 2\varepsilon)$$

$$+ \varepsilon \left(\prod_{i>j>2} (x_i - x_j) \prod_{i=3}^{N} (x_i - y_1 - \varepsilon)(x_i - y_1 - 2\varepsilon) \right)'$$

as $\varepsilon \to 0$, with $\mathcal{S}_N$ the set of permutations of $\{1, \ldots, N\}$ whose elements $\sigma = (\sigma_1, \ldots, \sigma_N)$ have signature (or parity) $\operatorname{sgn}(\sigma)$. Calling σ' the permutation such that $\sigma'_1 = \sigma_2$, $\sigma'_2 = \sigma_1$ and $\sigma'_i = \sigma_i$ for $i \geq 3$, taking the limit when $\varepsilon \to 0$, we can reorder the sum in the numerator as

$$\sum_{(\sigma,\sigma') \subset \mathcal{S}_N} \operatorname{sgn}(\sigma) \left[f'_{\sigma_1}(y_1) f_{\sigma_2}(y_1) \prod_{i=3}^{N} f_{\sigma_i}(x_i) + 2 f_{\sigma_1}(y_1) f'_{\sigma_2}(y_1) \prod_{i=3}^{N} f_{\sigma_i}(x_i) \right]$$

$$+ \operatorname{sgn}(\sigma') \left[f'_{\sigma'_1}(y_1) f_{\sigma'_2}(y_1) \prod_{i=3}^{N} f_{\sigma'_i}(x_i) + 2 f_{\sigma'_1}(y_1) f'_{\sigma'_2}(y_1) \prod_{i=3}^{N} f_{\sigma'_i}(x_i) \right].$$

But from the definition of σ', we have $\operatorname{sgn}(\sigma') = -\operatorname{sgn}(\sigma)$ and then this becomes

$$\sum_{(\sigma,\sigma') \subset \mathcal{S}_N} \operatorname{sgn}(\sigma) \left[-f'_{\sigma_1}(y_1) f_{\sigma_2}(y_1) \prod_{i=3}^{N} f_{\sigma_i}(x_i) + f_{\sigma_1}(y_1) f'_{\sigma_2}(y_1) \prod_{i=3}^{N} f_{\sigma_i}(x_i) \right]$$

or equivalently

$$\sum_{\sigma \in \mathcal{S}_N} \operatorname{sgn}(\sigma) f_{\sigma_1}(y_1) f'_{\sigma_2}(y_1) \prod_{i=3}^{N} f_{\sigma_i}(x_i)$$

the determinant of the expected numerator. As for the denominator, it clearly converges also to the expected result, which finally proves Theorem 2.9 in this simple case.

As exemplified by the last results given in Theorem 2.7 and Theorem 2.8, the mathematical machinery required to obtain relevant results, already for such a simple case of correlated non-centered Gaussian matrices, is extremely involved. This is the main reason for the increasing attractiveness of *large dimensional random matrices*, which, as is discussed in the following, provide surprisingly simple results in comparison.

2.2 Large dimensional random matrices

2.2.1 Why go to infinity?

When random matrix problems relate to rather large dimensional matrices, it is often convenient to mentally assume that these matrices have in fact extremely large dimensions in order to blindly apply the limiting results for large matrices to the effective finite-case scenario. It turns out that this approximate technique is often stunningly precise and can be applied to approximate scenarios where the effective matrix under consideration is not larger than 8×8, and even sometimes 4×4 and 2×2. Before delving into the core of large dimensional random matrix theory, let us explain further what we mean by "assuming extremely large size and applying results to the finite size case." As was already evidenced, we are largely interested in the eigenvalue structure of random matrices and in particular in the marginal density of their eigenvalues.

Consider an $N \times N$ (non-necessarily random) Hermitian matrix $\mathbf{T}_N$. Define its *empirical spectral distribution* (e.s.d.) $F^{\mathbf{T}_N}$ to be the distribution function of the eigenvalues of $\mathbf{T}_N$, i.e. for $x \in \mathbb{R}$

$$F^{\mathbf{T}_N}(x) = \frac{1}{N} \sum_{j=1}^{N} 1_{\{x, \lambda_j \leq x\}}(x)$$

where $\lambda_1, \ldots, \lambda_N$ are the eigenvalues of $\mathbf{T}_N$.[4] For such problems as signal source detection from a multi-dimensional sensor (e.g. scanning multiple frequencies), $\mathbf{T}_N$ might be a diagonal matrix with K distinct eigenvalues, of multiplicity N/K each, representing the power transmitted by each source on N/K independent frequency bands. Assuming a random signal matrix $\mathbf{S}_N \in \mathbb{C}^{N \times n}$ with i.i.d. entries is transmitted by the joint-source (composed of the concatenation of N-dimensional signal vectors transmitted in n successive time instants), and N, n are both large, then a rather simple model is to assume that the sensor observes a matrix $\mathbf{Y}_N = \mathbf{T}_N^{\frac{1}{2}} \mathbf{S}_N$ from which it needs to infer the K distinct eigenvalues of $\mathbf{T}_N$. This model is in fact unrealistic, as it should consider other factors such as the additive thermal noise, but we will stick here to this easier model for the sake of simplicity. Inferring the K eigenvalues of $\mathbf{T}_N$ is a hard problem in finite size random matrix theory, for which the maximum likelihood method boils down to a K-dimensional space search. If, however, we consider the hypothetical series $\{\mathbf{Y}_{pN} = \mathbf{T}_{pN} \mathbf{S}_{pN}, p \in \mathbb{N}\}$ with $\mathbf{S}_{pN} \in \mathbb{C}^{pN \times pn}$ whose entries follow the same distribution as those of $\mathbf{S}_N$, and $\mathbf{T}_{pN} = \mathbf{T}_N \otimes \mathbf{I}_p$, with $\otimes$ the Kronecker product, then the system has just been made larger without impacting its structural ingredients. Indeed, we have just made the problem larger by growing N into pN and n into pn. The most important fact to observe here is that

[4] The Hermitian property is fundamental to ensure that all eigenvalues of $\mathbf{T}_N$ belong to the real line. However, the extension of the e.s.d. to non-Hermitian matrices is sometimes required; for a definition, see (1.2.2) of [Bai and Silverstein, 2009].

the distribution function of $\mathbf{T}_N$ has not been affected, i.e. $F^{\mathbf{T}_{pN}} = F^{\mathbf{T}_N}$, for each p. It turns out, as will be detailed in Chapter 16, that when $p \to \infty$, $F^{\mathbf{Y}_{pN}\mathbf{Y}_{pN}^{\mathsf{H}}}$ has a deterministic weak limit and there exist computationally efficient techniques to retrieve the exact K distinct eigenvalues of $\mathbf{T}_N$, from the *non-random* limit of the e.s.d. $\lim_{p\to\infty} F^{\mathbf{Y}_{pN}\mathbf{Y}_{pN}^{\mathsf{H}}}$). Going back to the finite (but large enough) N case, we can apply the same deterministic techniques to the random $F^{\mathbf{Y}_N\mathbf{Y}_N^{\mathsf{H}}}$ instead of $\lim_{p\to\infty} F^{\mathbf{Y}_{pN}\mathbf{Y}_{pN}^{\mathsf{H}}}$ to obtain a good approximation of the eigenvalues of $\mathbf{T}_N$. These approximated values are consistent with a proportional growth of n and N, as they are almost surely exact when N and n tend to infinity with positive limiting ratio, and are therefore (n, N)-consistent.

This is basically how most random matrix results work: (i) we artificially let both n, N dimensions grow to infinity with constant ratio, (ii) very often, assuming large dimensions asymptotically leads to deterministic expressions, i.e. independent of the realization ω (at least for ω in a subset of Ω of probability one) which are simpler to manipulate, and (iii) we can then apply the deterministic results obtained in (ii) to the finite dimensional stochastic observation ω at hand and usually have a good approximation of the small dimensional matrix behavior. The fact that small dimensional matrices enjoy similar properties as their large dimensional counterparts makes the above approach extremely attractive, notably to wireless communication engineers, who often deal with not-so-large matrix models.

2.2.2 Limit spectral distributions

Let us now focus on large dimensional random matrices and abandon for now the practical applications, which are discussed in Part II. The relevant aspect of some classes of large $N \times N$ Hermitian random matrices $\mathbf{X}_N$ is that their (random) e.s.d. $F^{\mathbf{X}_N}$ converges, as $N \to \infty$, towards a non-random distribution F. This function F, if it exists, will be called the *limit spectral distribution* (l.s.d.) of $\mathbf{X}_N$. Weak convergence of $F^{\mathbf{X}_N}$ to F, i.e. for all x where F is continuous, $F^{\mathbf{X}_N}(x) - F(x) \to 0$, is often sufficient to obtain relevant results; this is denoted

$$F^{\mathbf{X}_N} \Rightarrow F.$$

In most cases, though, the weak convergence of $F^{\mathbf{X}_N}$ to F will only be true on a set of matrices $\mathbf{X}_N = \mathbf{X}_N(\omega)$ of measure one. This will be mentioned with the phrase $F^{\mathbf{X}_N} \Rightarrow F$ *almost surely*.

We detail in the following the best-known examples of such convergence. The first result on limit spectral distributions is due to Wigner [Wigner, 1955, 1958], who establishes the convergence of the eigenvalue distribution of a particular case of the now-called Wigner matrices. In its generalized form [Arnold, 1967, 1971; Bai and Silverstein, 2009], this is:

Theorem 2.11 (Theorem 2.5 and Theorem 2.9 of [Bai and Silverstein, 2009]). *Consider an* $N \times N$ *Hermitian matrix* $\mathbf{X}_N$, *with independent entries* $\frac{1}{\sqrt{N}} X_{N,ij}$ *such that* $\mathrm{E}[X_{N,ij}] = 0$, $\mathrm{E}[|X_{N,ij}|^2] = 1$ *and there exists* ε *such that the* $X_{N,ij}$ *have a moment of order* $2+\varepsilon$. *Then* $F^{\mathbf{X}_N} \Rightarrow F$ *almost surely, where* F *has density* f *defined as*

$$f(x) = \frac{1}{2\pi}\sqrt{(4-x^2)^+}. \tag{2.4}$$

Moreover, if the $X_{N,ij}$ *are identically distributed, the result holds without the need for existence of a moment of order* $2+\varepsilon$.

The l.s.d. F is the *semi-circle law*,[5] depicted in Figure 1.2. The sketch of a proof of the semi-circle law, based on the method of moments (Section 30 of [Billingsley, 1995]) is presented in Section 5.1.

This result was then followed by limiting results on other types of matrices, such as the *full circle law* for non-symmetric random matrices, which is a largely more involved problem, starting with the fact that the eigenvalues of such matrices are no longer restricted to the real axis [Hwang, 1986; Mehta, 2004]. Although Girko was the first to provide a proof of this result, the most general result is due to Bai in 1997 [Bai, 1997].

Theorem 2.12. *Let* $\mathbf{X}_N \in \mathbb{C}^{N\times N}$ *have i.i.d. entries* $\frac{1}{\sqrt{N}} X_{N,ij}$, $1 \leq i,j \leq N$, *such that* $X_{N,11}$ *has zero mean, unit variance and finite sixth order moment. Additionally, assume that the joint distribution of the real and imaginary parts of* $\frac{1}{\sqrt{N}} X_{N,11}$ *has bounded density. Then, with probability one, the e.s.d. of* $\mathbf{X}_N$ *tends to the uniform distribution on the unit complex disc. This distribution is referred to as the circular law, or full circle law.*

The circular law is depicted in Figure 2.1.

Theorem 2.11 and Theorem 2.12 and the aforementioned results have important consequences in nuclear physics and statistical mechanics, largely documented in [Mehta, 2004]. In wireless communications, though, Theorem 2.11 and Theorem 2.12 are not the most fundamental and widely used results. Instead, in the wireless communications field, we are often interested in sample covariance matrices or even more general matrices such as i.i.d. matrices multiplied both on the left and on the right by deterministic matrices, or i.i.d. matrices with a variance profile, i.e. with independent entries of zero mean but difference variances. Those matrices are treated in problems of detection, estimation and capacity evaluation which we will further discuss in Part II. The best known result with a large range of applications in telecommunications is the convergence of the e.s.d. of the Gram matrix of a random matrix with i.i.d. entries of zero

[5] Note that the semi-circle law sometimes refers, instead of F, to the density f, which is the "semi-circle"-shaped function.

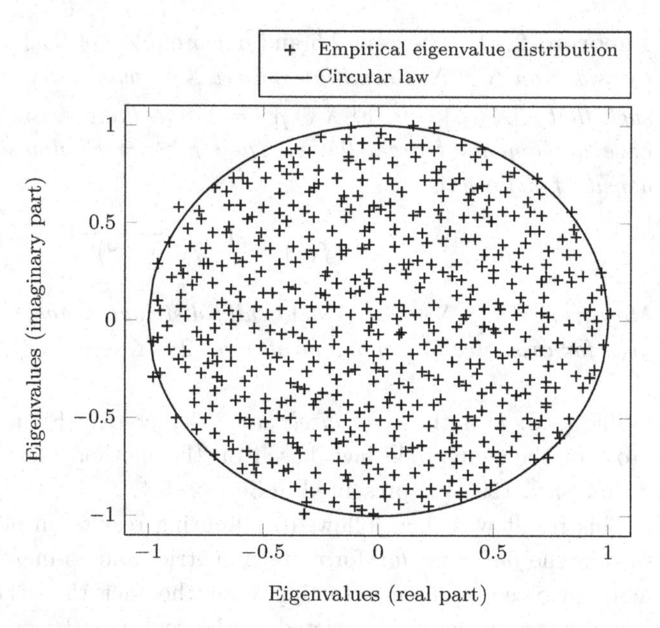

Figure 2.1 Eigenvalues of $\mathbf{X}_N = \left(\frac{1}{\sqrt{N}}X_{ij}^{(N)}\right)_{ij}$ with $X_{ij}^{(N)}$ i.i.d. standard Gaussian, for $N = 500$, against the circular law.

mean and normalized variance. This result is due to Marčenko and Pastur [Marčenko and Pastur, 1967], so that the limiting e.s.d. of the Gram matrix is called the *Marčenko–Pastur law.* The result unfolds as follows.

Theorem 2.13. *Consider a matrix* $\mathbf{X} \in \mathbb{C}^{N \times n}$ *with i.i.d. entries* $\left(\frac{1}{\sqrt{n}}X_{N,ij}\right)$, *such that* $X_{N,11}$ *has zero mean and unit variance. As* $n, N \to \infty$ *with* $\frac{N}{n} \to c \in (0, \infty)$, *the e.s.d. of* $\mathbf{R}_n = \mathbf{X}\mathbf{X}^{\mathsf{H}}$ *converges weakly and almost surely to a non-random distribution function* F_c *with density* f_c *given by:*

$$f_c(x) = (1 - c^{-1})^+\delta(x) + \frac{1}{2\pi c x}\sqrt{(x-a)^+(b-x)^+} \tag{2.5}$$

where $a = (1 - \sqrt{c})^2$, $b = (1 + \sqrt{c})^2$ *and* $\delta(x) = 1_{\{0\}}(x)$.

Note that, similar to the notation introduced previously, $\mathbf{R}_n$ is the sample covariance matrix associated with the random vector $(X_{N,11}, \ldots, X_{N,N1})^{\mathsf{T}}$, with population covariance matrix $\mathbf{I}_N$. The d.f. F_c is named the Marčenko–Pastur law with limiting ratio c.[6] This is depicted in Figure 2.2 for different values of the limiting ratio c. Notice in particular that, as is expected from the discussion in the Preface, when c tends to be small and approaches zero, the Marčenko–Pastur law reduces to a single mass in 1. A proof of Theorem 2.13, which follows a Stieltjes

[6] Similarly as with the semi-circle law, the Marčenko–Pastur law can also refer to the density f_c of F_c.

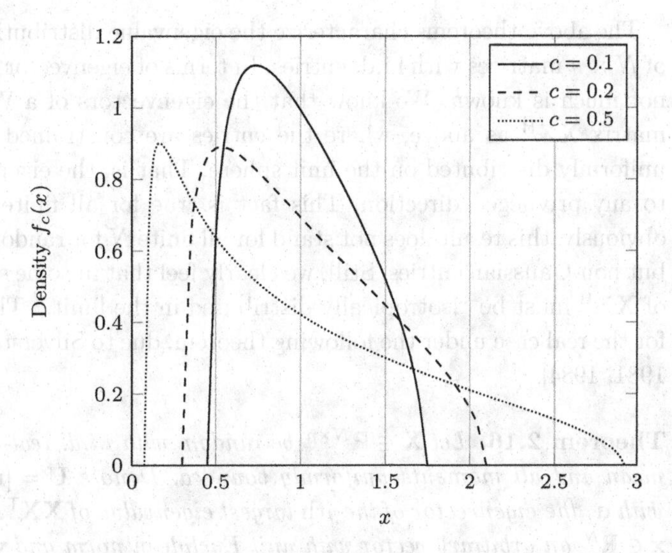

Figure 2.2 Marčenko–Pastur law for different limit ratios $c = \lim N/n$.

transform-based method, is proposed in Section 3.2. Since the Marčenko–Pastur law has bounded support, it has moments of all orders, which are explicitly given in the following result.

Theorem 2.14. *Let F_c be the Marčenko–Pastur law with ratio c and with density f_c given by* (2.5)*. The successive moments $M_1, M_2, \ldots$ of F_c are given, for all integer k, by*

$$M_k = \frac{1}{k}\sum_{i=0}^{k-1}\binom{k}{i}\binom{k}{i+1}c^i.$$

Note that further generalizations of Theorem 2.13 have been provided in the literature, the most general form of which comes as follows.

Theorem 2.15 (Theorem 3.10 of [Bai and Silverstein, 2009])**.** *Consider a matrix $\mathbf{X} \in \mathbb{C}^{N\times n}$ with entries $\left(\frac{1}{\sqrt{n}}X_{N,ij}\right)$, independent for all i, j, n and such that $X_{N,ij}$ has zero mean, unit variance, and finite $2+\varepsilon$ order moment (ε being independent of i, j, n). Then, as $n, N \to \infty$ with $\frac{N}{n} \to c \in (0, \infty)$, the e.s.d. of $\mathbf{R}_n = \mathbf{X}\mathbf{X}^{\mathsf{H}}$ converges almost surely to the Marčenko–Pastur law F_c, with density given by* (2.5)*.*

This last result goes beyond the initial Marčenko–Pastur identity as it does not assume identically distributed entries in $\mathbf{X}$. However, the result would no longer stand if the variances of the entries were different, or were not independent. These scenarios are of relevance in wireless communications to model multi-dimensional channels with correlation or with a variance profile, for instance.

The above theorems characterize the eigenvalue distribution of Gram matrices of $N \times n$ matrices with i.i.d. entries. In terms of eigenvector distribution, though, not much is known. We know that the eigenvectors of a Wishart matrix, i.e. a matrix $\mathbf{X}\mathbf{X}^{\mathsf{H}}$ as above, where the entries are constrained to be Gaussian, are uniformly distributed on the unit sphere. That is, the eigenvectors do not point to any privileged direction. This fact is true for all finite dimension N. Now, obviously, this result does not stand for all finite N for random matrices with i.i.d. but non-Gaussian entries. Still, we clearly feel that in some sense the eigenvectors of $\mathbf{X}\mathbf{X}^{\mathsf{H}}$ must be "isotropically distributed in the limit." This is stated precisely for the real case under the following theorem, due to Silverstein [Silverstein, 1979, 1981, 1984].

Theorem 2.16. *Let $\mathbf{X} \in \mathbb{R}^{N \times n}$ be random with i.i.d. real-valued entries of zero mean and all moments uniformly bounded. Denote $\mathbf{U} = [\mathbf{u}_1, \ldots, \mathbf{u}_N] \in \mathbb{R}^{N \times N}$, with $\mathbf{u}_j$ the eigenvector of the jth largest eigenvalue of $\mathbf{X}\mathbf{X}^{\mathsf{T}}$. Additionally, denote $\mathbf{x} \in \mathbb{R}^N$ an arbitrary vector with unit Euclidean norm and $\mathbf{y} = (y_1, \ldots, y_N)^{\mathsf{T}}$ the random vector defined by*

$$\mathbf{y} = \mathbf{U}\mathbf{x}.$$

Then, as $N, n \to \infty$ with limiting ratio $N/n \to c$, $0 < c \leq 1$, for all $t \in [0, 1]$

$$\sum_{k=1}^{\lfloor tN \rfloor} y_k^2 \xrightarrow{\text{a.s.}} t$$

where $\lfloor x \rfloor$ is the greatest integer smaller than x.

This result indicates some sort of uniformity in the distribution of the eigenvectors of $\mathbf{X}\mathbf{X}^{\mathsf{T}}$. In [Silverstein, 1986], Silverstein extends Theorem 2.16 into a limit theorem of the fluctuations of the random process $\sum_{k=1}^{\lfloor tN \rfloor} y_k^2$, the weak limit being a Brownian bridge. Apart from these results, though, not much more is known about eigenvectors of random matrices. This subject has however recently gained some more interest and recent results will be given later in Chapter 8.

In the following chapters, we introduce the important tools known to this day to characterize the l.s.d. of a large range of random matrix classes. We start with the Stieltjes transform and provide a thorough proof of the Marčenko–Pastur law via the Stieltjes transform method. This proof will allow the reader to have a clear view on the building blocks required for deriving most results of large dimensional random matrix theory for wireless communications.

3 The Stieltjes transform method

This chapter is the first of three chapters dedicated to providing the reader with an overview of the most important tools used in problems related to large random matrices. These tools will form a strong basis for the reader to be able to appreciate the extensions discussed in the more technical Chapters 6–9. We first visit in this chapter the main results proved via the *Stieltjes transform*, to be defined subsequently. The Stieltjes transform tool is at first not very intuitive and not as simple as the moment-based methods developed later. For this reason, we start with a step-by-step proof of the Marčenko–Pastur law, Theorem 2.13, for large dimensional matrices with i.i.d. entries, before we can address more elaborate random matrix models with non-independent or not identically distributed entries. We will then introduce the Stieltjes transform related tools that are the R-transform and the S-transform, which bear interesting properties related to moments of the e.s.d. of some random matrix models. These R-transform and the S-transform, along with the free probability theory from which they originate, are the fundamental link between the Stieltjes transform and the moment-based methods. These are discussed thoroughly in Chapters 4–5.

3.1 Definitions and overview

To be able to handle the powerful tools that are the Stieltjes transform and moment-based methods, we start with some prior definitions of the Stieltjes transform and other related functionals, which will be often used in the subsequent chapters.

We first introduce the Stieltjes transform.

Definition 3.1. Let F be a real-valued bounded measurable function over $\mathbb{R}$. Then the Stieltjes transform $m_F(z)$of F,[1] for $z \in \mathrm{Supp}(F)^c$, the complex space

[1] We borrow here the notation m to a large number of contributions from Bai, Silverstein *et al.* In other works, the notation s or S for the Stieltjes transform is used. However, in this work, the notation S will be reserved to the S-transform to be defined later.

complementary to the support of F,[2] is defined as

$$m_F(z) \triangleq \int_{-\infty}^{\infty} \frac{1}{\lambda - z} dF(\lambda). \tag{3.1}$$

For all F which admits a Stieltjes transform, the inverse transformation exists and formulates as follows.

Theorem 3.1. *If x is a continuity points of F, then:*

$$F(x) = \frac{1}{\pi} \lim_{y \to 0^+} \int_{-\infty}^{x} \Im \left[m_F(x + iy) \right] dx. \tag{3.2}$$

Proof. Since the function $t \mapsto \frac{1}{t-(x+iy)}$ is continuous and tends to zero as $|t| \to \infty$, it has uniformly bounded norm on the support of F and we can then apply Tonelli's theorem, Theorem 3.16, and write

$$\begin{aligned}
&\frac{1}{\pi} \int_a^b \Im \left[m_F(x + iy) \right] dx \\
&= \frac{1}{\pi} \int_a^b \int \frac{y}{(t-x)^2 + y^2} dF(t) dx \\
&= \frac{1}{\pi} \int \int_a^b \frac{y}{(t-x)^2 + y^2} dx dF(t) \\
&= \frac{1}{\pi} \int \left[\tan^{-1} \left(\frac{b-t}{y} \right) - \tan^{-1} \left(\frac{a-t}{y} \right) \right] dF(t).
\end{aligned}$$

As $y \to 0^+$, this tends to $\int 1_{[a,b]}(t) dF(t) = F(b) - F(a)$. □

In all practical applications considered in this book, F will be a distribution function. Therefore, there exists an intimate link between distribution functions and their Stieltjes transforms. More precisely, if F_1 and F_2 are two distribution functions (therefore right-continuous by definition, see, e.g. Section 14 of [Billingsley, 1995]) that have the same Stieltjes transform, then F_1 and F_2 coincide everywhere and the converse is true. As a consequence, m_F uniquely determines F and vice-versa. It will turn out that, while working on the distribution functions of the empirical eigenvalues of large random matrices is often a tedious task, the approach via Stieltjes transforms simplifies greatly the study. The initial intuition behind the Stieltjes transform approach for random

[2] We recall that the support Supp(F) of a d.f. F with density f is the closure of the set $\{x \in \mathbb{R}, f(x) > 0\}$.

matrices lies in the following remark. For a Hermitian matrix $\mathbf{X} \in \mathbb{C}^{N \times N}$

$$\begin{aligned} m_{F^{\mathbf{X}}}(z) &= \int \frac{1}{\lambda - z} dF^{\mathbf{X}}(\lambda) \\ &= \frac{1}{N} \operatorname{tr} (\mathbf{\Lambda} - z\mathbf{I}_N)^{-1} \\ &= \frac{1}{N} \operatorname{tr} (\mathbf{X} - z\mathbf{I}_N)^{-1} \end{aligned}$$

in which we denoted $\mathbf{\Lambda}$ the diagonal matrix of eigenvalues of $\mathbf{X}$. Working with the Stieltjes transform of $F^{\mathbf{X}}$ then boils down to working with the matrix $(\mathbf{X} - z\mathbf{I}_N)^{-1}$, and more specifically the sum of its diagonal entries. From matrix inversion lemmas and several fundamental matrix identities, it is then rather simple to derive limits of traces $\frac{1}{N} \operatorname{tr} (\mathbf{X} - z\mathbf{I}_N)^{-1}$, as N grows large, and therefore to derive a limit of the Stieltjes transform of $F^{\mathbf{X}}$. For instance, in the case of large sample covariance matrices $\mathbf{R}_n$, we will see that it is rather easy to show that $m_{F^{\mathbf{R}_n}}$ tends almost surely to a function m, which is itself the Stieltjes transform of a distribution function F. Thanks to Theorem 3.10, this will prove that $F^{\mathbf{R}_n} \Rightarrow F$ almost surely. For notational simplicity, we may denote $m_{\mathbf{X}} \triangleq m_{F^{\mathbf{X}}}$ the Stieltjes transform of the e.s.d. of the Hermitian matrix $\mathbf{X}$ and call $m_{\mathbf{X}}$ the *Stieltjes transform of* $\mathbf{X}$.

An identity of particular interest is the relation between the Stieltjes transform of $\mathbf{AB}$ and $\mathbf{BA}$ when $\mathbf{AB}$ is Hermitian.

Lemma 3.1. *Let* $\mathbf{A} \in \mathbb{C}^{N \times n}$, $\mathbf{B} \in \mathbb{C}^{n \times N}$, *such that* $\mathbf{AB}$ *is Hermitian. Then, for* $z \in \mathbb{C} \setminus \mathbb{R}$

$$\frac{n}{N} m_{F^{\mathbf{BA}}}(z) = m_{F^{\mathbf{AB}}}(z) + \frac{N-n}{N} \frac{1}{z}.$$

Also, for $\mathbf{X} \in \mathbb{C}^{N \times n}$ *and for* $z \in \mathbb{C} \setminus \mathbb{R}^+$, *we have:*

$$\frac{n}{N} m_{F^{\mathbf{X}^{\mathsf{H}}\mathbf{X}}}(z) = m_{F^{\mathbf{X}\mathbf{X}^{\mathsf{H}}}}(z) + \frac{N-n}{N} \frac{1}{z}.$$

The identity follows directly from the fact that both $\mathbf{AB}$ and $\mathbf{BA}$ have the same eigenvalues except for additional zero eigenvalues for the larger matrix. Hence, say $n \geq N$, the larger matrix has N eigenvalues being the same as the eigenvalues of the smaller matrix, plus additional $(n - N)$ eigenvalues equal to zero. Each one of the latter leads to the addition of a term $1/(0 - z) = -1/z$, hence the identity.

Also, we have the following trivial relation.

Lemma 3.2. *Let* $\mathbf{X} \in \mathbb{C}^{N \times N}$ *be Hermitian and* a *be a non-zero real. Then, for* $z \in \mathbb{C} \setminus \mathbb{R}$

$$m_{a\mathbf{X}}(az) = \frac{1}{a} m_{\mathbf{X}}(z).$$

This unfolds by noticing that

$$m_{a\mathbf{X}}(az) = \int \frac{1}{at - az} dF^{\mathbf{X}}(t) = \frac{1}{a} \int \frac{1}{t - z} dF^{\mathbf{X}}(t).$$

For practical calculus in the derivations of the subsequent chapters, we need to introduce the following important properties for the Stieltjes transform of distribution functions, see, e.g., [Hachem *et al.*, 2007].

Theorem 3.2. *Let m_F be the Stieltjes transform of a distribution function F, then:*

- *m_F is analytic over $\mathbb{C}^+$,*
- *if $z \in \mathbb{C}^+$, then $m_F(z) \in \mathbb{C}^+$,*
- *if $z \in \mathbb{C}^+$, $|m_F(z)| \leq \frac{1}{\Im[z]}$ and $\Im[1/m_F(z)] \leq -\Im[z]$,*
- *if $F(0^-) = 0$,[3] then m_F is analytic over $\mathbb{C} \setminus \mathbb{R}^+$. Moreover, $z \in \mathbb{C}^+$ implies $zm_F(z) \in \mathbb{C}^+$ and we have the inequalities*

$$|m_F(z)| \leq \begin{cases} \frac{1}{|\Im[z]|} & , \; z \in \mathbb{C} \setminus \mathbb{R} \\ \frac{1}{|z|} & , \; z < 0 \\ \frac{1}{\mathrm{dist}(z, \mathbb{R}^+)} & , \; z \in \mathbb{C} \setminus \mathbb{R}^+ \end{cases}$$

with dist *the Euclidean distance.*

Conversely, if m is a function analytical on $\mathbb{C}^+$ such that $m(z) \in \mathbb{C}^+$ if $z \in \mathbb{C}^+$ and

$$\lim_{y \to \infty} -iy \; m(iy) = 1 \tag{3.3}$$

then m is the Stieltjes transform of a distribution function F given by

$$F(b) - F(a) = \lim_{y \to 0} \frac{1}{\pi} \int_a^b \Im[m(x + iy)] dx.$$

If, moreover, $zm(z) \in \mathbb{C}^+$ for $z \in \mathbb{C}^+$, then $F(0^-) = 0$, in which case m has an analytic continuation on $\mathbb{C} \setminus \mathbb{R}^+$.

The first inequalities will often be used when providing limiting results for some large dimensional random matrix models involving the Stieltjes transform. The converse results will be more rarely used, restricted mainly to some technical points in the proof of advanced results. Note that, if the limit in Equation (3.3) is finite but different from 1, then $m(z)$ is said to be the *Stieltjes transform of a finite measure on* $\mathbb{R}^+$.

An interesting corollary of Theorem 3.2, which will often be reused in technical proofs, is the following.

[3] We will denote $F(0^-)$ and $F(0^+)$ the limit of $F(x)$ when x tends to zero from below or from above, respectively.

Corollary 3.1. *Let $t > 0$ and $m_F(z)$ be the Stieltjes transform of a distribution function F. Then, for $z \in \mathbb{C}^+$*

$$\left| \frac{1}{1 + t m_F(z)} \right| \leq \frac{|z|}{\Im[z]}.$$

Proof. It suffices to realize here, from the properties of the Stieltjes transform given in the converse of Theorem 3.2, that

$$\frac{-1}{z(1 + t m_F(z))}$$

is the Stieltjes transform of some distribution function. It therefore unfolds, from the Stieltjes transform inequalities of Theorem 3.2, that

$$\left| \frac{-1}{z(1 + t m_F(z))} \right| \leq \frac{1}{\Im[z]}.$$

□

A further corollary of Corollary 3.1 then reads:

Corollary 3.2. *Let $\mathbf{x} \in \mathbb{C}^N$, $t > 0$ and $\mathbf{A} \in \mathbb{C}^{N \times N}$ be Hermitian non-negative definite. Then, for $z \in \mathbb{C}^+$*

$$\left| \frac{1}{1 + t \mathbf{x}^{\mathsf{H}} (\mathbf{A} - z\mathbf{I}_N)^{-1} \mathbf{x}} \right| \leq \frac{|z|}{\Im[z]}.$$

Proof. This unfolds by writing $\mathbf{A} = \mathbf{U}^{\mathsf{H}} \mathbf{\Lambda} \mathbf{U}$, the spectral decomposition of $\mathbf{A}$, with $\mathbf{\Lambda} = \operatorname{diag}(\lambda_1, \ldots, \lambda_N) \in \mathbb{C}^{N \times N}$ diagonal and $\mathbf{U} \in \mathbb{C}^{N \times N}$. Denoting $\mathbf{y} = \mathbf{U}\mathbf{x}$ and $\mathbf{y} = (y_1, \ldots, y_N)^{\mathsf{T}}$, we have:

$$\mathbf{x}^{\mathsf{H}} (\mathbf{A} - z\mathbf{I}_N)^{-1} \mathbf{x} = \sum_i \frac{|y_i|^2}{\lambda_i - z}.$$

Under this notation, it is clear that the function $f(z) = \mathbf{x}^{\mathsf{H}} (\mathbf{A} - z\mathbf{I}_N)^{-1} \mathbf{x}$ maps $\mathbb{C}^+$ to $\mathbb{C}^+$ and that $\lim_{y \to \infty} -iy f(iy) = \sum_j |y_j|^2 > 0$. Therefore, up to a positive scaling factor, $f(z)$ is the Stieltjes transform of a probability measure. We can therefore use Corollary 3.2, which completes the proof. □

In wireless communications, we are often interested in calculating the data transmission rate achievable on a multi-dimensional $N \times n$ communication channel $\mathbf{H}$. We are therefore often led to evaluate functions in the form of (1.2). It turns out that the Stieltjes transform is directly connected to this expression of mutual information through the so-called *Shannon transform*, initially coined by Tulino and Verdú.

Definition 3.2 (Section 2.3.3 of [Tulino and Verdú, 2004]). Let F be a probability distribution defined on $\mathbb{R}^+$. The Shannon transform $\mathcal{V}_F$ of F is defined, for $x \in \mathbb{R}^+$, as

$$\mathcal{V}_F(x) \triangleq \int_0^\infty \log(1 + x\lambda) dF(\lambda). \tag{3.4}$$

The Shannon transform of F is related to its Stieltjes transform m_F through the expression

$$\mathcal{V}_F(x) = \int_{\frac{1}{x}}^\infty \left(\frac{1}{t} - m_F(-t) \right) dt = \int_0^x \left(\frac{1}{t} - \frac{1}{t^2} m_F\left(-\frac{1}{t} \right) \right) dt. \tag{3.5}$$

The expression in brackets in (1.2) is N times the right-hand side of (3.4) if F is chosen to be the e.s.d. of $\mathbf{H}\mathbf{H}^{\mathsf{H}}$. To evaluate (1.2), it is therefore sufficient to evaluate the Stieltjes transform of F. This is the very starting point of capacity evaluations using the Stieltjes transform.

Another important characteristic of the Stieltjes transform, both from a theoretical and a practical point of view, is its relationship to moments of the underlying distribution. We have in particular the following result.

Theorem 3.3. *If F has compact support included in $[a, b]$, $0 < a < b < \infty$, then, for $z \in \mathbb{C} \setminus \mathbb{R}$ such that $|z| > b$, $m_F(z)$ can be expanded in Laurent series as*

$$m_F(z) = -\frac{1}{z} \sum_{k=0}^\infty \frac{M_k}{z^k} \tag{3.6}$$

where

$$M_k = \int_{-\infty}^\infty \lambda^k dF(\lambda)$$

is the kth *order moment of F.*

By successive differentiations of $zm_F(-1/z)$, we can recover the series of moments $M_1, M_2, \ldots$ of F; the Stieltjes transform is then a moment generating function (at least) for compactly supported probability distributions. If F is the e.s.d. of the Hermitian matrix $\mathbf{X}$, then M_k is also called the kth *order moment of* $\mathbf{X}$. The above result provides therefore a link between the Stieltjes transform of $\mathbf{X}$ and the moments of $\mathbf{X}$. The moments of random Hermitian matrices are in fact of practical interest whenever direct usage of Stieltjes transform-based methods are too difficult. Chapter 5 is dedicated to an account of these moment-based considerations.

Before concluding this section, we introduce a few additional tools, all derived from the Stieltjes transform, which have fundamental properties regarding the moments of Hermitian matrices. From a theoretical point of view, they help bridge classical probability theory to *free probability theory* [Hiai and Petz, 2006; Voiculescu *et al.*, 1992], to be introduced in Chapter 4. For a slightly more exhaustive account of the most commonly used functionals of e.s.d. of large dimensional random matrices and their main properties, refer to [Tulino and Verdú, 2004].

We consider first the *R-transform*, defined as follows.

Definition 3.3. Let F be a distribution function on $\mathbb{R}$ and let m_F be its Stieltjes transform, then the R-transform of F, denoted R_F, is such that

$$m_F(R_F(z) + z^{-1}) = -z \tag{3.7}$$

or equivalently

$$m_F(z) = \frac{1}{R_F(-m_F(z)) - z}. \tag{3.8}$$

If F is associated with a probability measure μ, then R_μ will also denote the R-transform of F. Also, if F is the e.s.d. of the Hermitian matrix $\mathbf{X}$, then R_F will be called the R-transform of $\mathbf{X}$. The importance of the R-transform lies in Theorem 4.6, to be introduced in Chapter 4. Roughly speaking, the R-transform is the random matrix equivalent to the *characteristic function* of scalar random variables in the sense that, under appropriate conditions (independence is not sufficient) on the Hermitian random matrices $\mathbf{A}$ and $\mathbf{B}$, the R-transform of the l.s.d. of $\mathbf{A} + \mathbf{B}$ is the sum of the R-transforms of the l.s.d. of $\mathbf{A}$ and of the l.s.d. of $\mathbf{B}$ (upon existence of these limits).

Similar to the R-transform, which has additive moment properties, we define the S-transform, which has product moment properties.

Definition 3.4. Let F be a distribution function on $\mathbb{R}$ and let m_F be its Stieltjes transform, then the S-transform of F, denoted S_F, satisfies

$$m_F\left(\frac{z+1}{zS_F(z)}\right) = -zS_F(z).$$

If F has probability measure μ, then S_μ denotes also the S-transform of F. Under suitable conditions, the S-transform of the l.s.d. of a matrix product $\mathbf{AB}$ is the product of the S-transforms of the l.s.d. of $\mathbf{A}$ and the l.s.d. of $\mathbf{B}$.

Both R- and S-transforms are particularly useful when dealing with matrix models involving unitary matrices; in particular, they will be used in Chapter 12 to evaluate the capacity of networks using orthogonal CDMA communications.

We also introduce two alternative forms of the Stieltjes transform, namely the η-transform and the ψ-transform, which are sometimes preferred over the Stieltjes transform because they turn out to be more convenient for readability in certain derivations, especially those derivations involving the R- and S-transform. The η-transform is defined as follows.

Definition 3.5. For F a distribution function with support in $\mathbb{R}^+$, we define the η-transform of F, denoted η_F, to be the function defined for $z \in \mathbb{C} \setminus \mathbb{R}^-$ as

$$\eta_F(z) \triangleq \int \frac{1}{1+zt} dF(t).$$

As such, the η-transform can be expressed as a function of the Stieltjes transform as

$$\eta_F(z) = \frac{1}{z} m_F\left(-\frac{1}{z}\right).$$

The ψ-transform is defined similarly in the following.

Definition 3.6. For F a distribution function with support in $\mathbb{R}^+$, we define the ψ-transform of F, denoted ψ_F, to be the function defined for $z \in \mathbb{C} \setminus \mathbb{R}^+$ as

$$\psi_F(z) \triangleq \int \frac{zt}{1-zt} dF(t).$$

Therefore, the ψ-transform can be written as a function of the Stieltjes transform as

$$\psi_F(z) = -1 - \frac{1}{z} m_F\left(\frac{1}{z}\right).$$

These tools are obviously totally equivalent and are used in place of the Stieltjes transform only in order to simplify long derivations.

The next section introduces the proof of one of the pioneering fundamental results in large dimensional random matrix theory. This proof will demonstrate the power of the Stieltjes transform tool.

3.2 The Marčenko–Pastur law

As already mentioned, the Stieltjes transform was used by Marčenko and Pastur in [Marčenko and Pastur, 1967] to derive the Marčenko–Pastur law of large dimensional Gram matrices of random matrices with i.i.d. entries. We start with some reminders before providing the essential steps of the proof of the Marčenko–Pastur law and providing the complete proof.

We recall, from Theorem 3.1, that studying the Stieltjes transform of a Hermitian matrix $\mathbf{X}$ is equivalent to studying the distribution function $F^{\mathbf{X}}$ of the eigenvalues of $\mathbf{X}$. The celebrated result that triggered the now extensive use of the Stieltjes transform is due to Marčenko and Pastur [Marčenko and Pastur, 1967] on the limiting distribution of the e.s.d. of sample covariance matrices with identity population covariance matrix. Although many different proofs exist by now for this result, some using different approaches than the Stieltjes transform method, we will focus here on what we will later refer to as the the Marčenko–Pastur method. This method is both simple and pedagogical for it uses the building blocks of the analytical aspect of large dimensional random matrix theory in an elegant manner. We give in the following first the key steps and then the precise steps of this derivation. This proof will serve as grounds for the derivations of further results, which utilize mostly the same approach, and also the same lemmas and identities. In Chapter 6, we will discuss the drawbacks of the method as it fails to generalize to some more involved matrix models, especially when the e.s.d. of the large matrix under study does not converge. Less intuitive but more powerful methods will then be proposed, among which the *Bai and Silverstein approach* and the *Gaussian methods*.

We recall that the result we want to prove is the following. Let $\mathbf{X} \in \mathbb{C}^{N\times n}$ be a matrix with i.i.d. entries $\left(\frac{1}{\sqrt{n}}X_{N,ij}\right)$, such that $X_{N,11}$ has zero mean and unit variance. As $n, N \to \infty$ with $\frac{N}{n} \to c \in (0,\infty)$, the e.s.d. of $\mathbf{R}_n \triangleq \mathbf{X}\mathbf{X}^{\mathsf{H}}$ converges almost surely to a non-random distribution function F_c with density f_c given by:

$$f_c(x) = (1-c^{-1})^+\delta(x) + \frac{1}{2\pi c x}\sqrt{(x-a)^+(b-x)^+}$$

where $a = (1-\sqrt{c})^2$, $b = (1+\sqrt{c})^2$ and $\delta(x) = 1_{\{0\}}(x)$.

Before providing the extended proof, let us outline the general derivation. The idea behind the proof is to study the Stieltjes transform $\frac{1}{N}\operatorname{tr}(\mathbf{X}\mathbf{X}^{\mathsf{H}} - z\mathbf{I}_N)^{-1}$ of $F^{\mathbf{X}\mathbf{X}^{\mathsf{H}}}$ instead of $F^{\mathbf{X}\mathbf{X}^{\mathsf{H}}}$ itself, and more precisely to study the diagonal entries of the matrix $(\mathbf{X}\mathbf{X}^{\mathsf{H}} - z\mathbf{I}_N)^{-1}$, often called the *resolvent* of $\mathbf{X}$. The main steps consists of the following.

- It will first be observed, through algebraic manipulations, involving matrix inversion lemmas, that the diagonal entry $(1,1)$ of $(\mathbf{X}\mathbf{X}^{\mathsf{H}} - z\mathbf{I}_N)^{-1}$ can be written as a function of a quadratic form $\mathbf{y}^{\mathsf{H}}(\mathbf{Y}^{\mathsf{H}}\mathbf{Y} - z\mathbf{I}_n)^{-1}\mathbf{y}$, with $\mathbf{y}$ the first column of $\mathbf{X}^{\mathsf{H}}$ and $\mathbf{Y}^{\mathsf{H}}$ defined as $\mathbf{X}^{\mathsf{H}}$ with column $\mathbf{y}$ removed. Precisely, we will have

$$\left[(\mathbf{X}\mathbf{X}^{\mathsf{H}} - z\mathbf{I}_N)^{-1}\right]_{11} = \frac{1}{-z - z\mathbf{y}^{\mathsf{H}}(\mathbf{Y}^{\mathsf{H}}\mathbf{Y} - z\mathbf{I}_n)^{-1}\mathbf{y}}.$$

- Due to the important *trace lemma*, which we will introduce and prove below, the quadratic form $\mathbf{y}^{\mathsf{H}}(\mathbf{Y}^{\mathsf{H}}\mathbf{Y} - z\mathbf{I}_n)^{-1}\mathbf{y}$ can then be shown to be asymptotically very close to $\frac{1}{n}\operatorname{tr}(\mathbf{Y}^{\mathsf{H}}\mathbf{Y} - z\mathbf{I}_n)^{-1}$ (asymptotically meaning here for increasingly large N and almost surely). This is:

$$\mathbf{y}^{\mathsf{H}}(\mathbf{Y}^{\mathsf{H}}\mathbf{Y} - z\mathbf{I}_n)^{-1}\mathbf{y} \simeq \frac{1}{n}\operatorname{tr}(\mathbf{Y}^{\mathsf{H}}\mathbf{Y} - z\mathbf{I}_n)^{-1}$$

where we non-rigorously use the symbol "$\simeq$" to mean "almost surely equal in the large N limit."

- Another lemma, the *rank*-1 *perturbation lemma*, will state that a perturbation of rank 1 of the matrix $\mathbf{Y}^{\mathsf{H}}\mathbf{Y}$, e.g. the addition of the rank-1 matrix $\mathbf{y}\mathbf{y}^{\mathsf{H}}$ to $\mathbf{Y}^{\mathsf{H}}\mathbf{Y}$, does not affect the value of $\frac{1}{n}\operatorname{tr}(\mathbf{Y}^{\mathsf{H}}\mathbf{Y} - z\mathbf{I}_n)^{-1}$ in the large dimensional limit. In particular, $\frac{1}{n}\operatorname{tr}(\mathbf{Y}^{\mathsf{H}}\mathbf{Y} - z\mathbf{I}_n)^{-1}$ can be approximated in the large n

limit by $\frac{1}{n}\operatorname{tr}(\mathbf{X}^{\mathsf{H}}\mathbf{X}-z\mathbf{I}_n)^{-1}$. With our non-rigorous formalism, this is:

$$\frac{1}{n}\operatorname{tr}(\mathbf{Y}^{\mathsf{H}}\mathbf{Y}-z\mathbf{I}_n)^{-1} \simeq \frac{1}{n}\operatorname{tr}(\mathbf{X}^{\mathsf{H}}\mathbf{X}-z\mathbf{I}_n)^{-1}$$

an expression which is now independent of $\mathbf{y}$, and in fact independent of the choice of the column of $\mathbf{X}^{\mathsf{H}}$, which is initially taken out. The same derivation therefore holds true for any diagonal entry of $(\mathbf{X}\mathbf{X}^{\mathsf{H}}-z\mathbf{I}_N)^{-1}$.

- But then, we know that $\frac{1}{n}\operatorname{tr}(\mathbf{X}\mathbf{X}^{\mathsf{H}}-z\mathbf{I}_N)^{-1}$ can be written as a function of $\frac{1}{n}\operatorname{tr}(\mathbf{X}^{\mathsf{H}}\mathbf{X}-z\mathbf{I}_n)^{-1}$ from Lemma 3.1. This is:

$$\frac{1}{n}\operatorname{tr}(\mathbf{X}^{\mathsf{H}}\mathbf{X}-z\mathbf{I}_n)^{-1} = \frac{1}{n}\operatorname{tr}(\mathbf{X}\mathbf{X}^{\mathsf{H}}-z\mathbf{I}_N)^{-1} + \frac{N-n}{n}\frac{1}{z}.$$

- It follows that each diagonal entry of $(\mathbf{X}\mathbf{X}^{\mathsf{H}}-z\mathbf{I}_N)^{-1}$ can be written as a function of $\frac{1}{n}\operatorname{tr}(\mathbf{X}\mathbf{X}^{\mathsf{H}}-z\mathbf{I}_N)^{-1}$ itself. By summing up all N diagonal elements and averaging by $1/N$, we end up with an approximated relation between $\frac{1}{N}\operatorname{tr}(\mathbf{X}\mathbf{X}^{\mathsf{H}}-z\mathbf{I}_N)^{-1}$ and itself

$$\frac{1}{N}\operatorname{tr}\left(\mathbf{X}\mathbf{X}^{\mathsf{H}}-z\mathbf{I}_N\right)^{-1} \simeq \frac{1}{1-\frac{N}{n}-z-z\frac{N}{n}\frac{1}{N}\operatorname{tr}\left(\mathbf{X}\mathbf{X}^{\mathsf{H}}-z\mathbf{I}_N\right)^{-1}}$$

which is asymptotically exact almost surely.

- Since this appears to be a second order polynomial in $\frac{1}{N}\operatorname{tr}\left(\mathbf{X}\mathbf{X}^{\mathsf{H}}-z\mathbf{I}_N\right)^{-1}$, this can be solved and we end up with an expression of the limiting Stieltjes transform of $\mathbf{X}\mathbf{X}^{\mathsf{H}}$. From Theorem 3.1, we finally find the explicit form of the l.s.d. of $F^{\mathbf{X}\mathbf{X}^{\mathsf{H}}}$, i.e. the Marčenko–Pastur law.

We now proceed to the thorough proof of Theorem 2.13. For this, we restrict the entries of the random matrix $\mathbf{X}$ to have finite eighth order moment. The extension to the general case can be handled in several ways. We will mention after the proof the ideas behind a powerful technique known as the *truncation, centralization, and rescaling method*, due to Bai and Silverstein [Silverstein and Bai, 1995], which allows us to work with truncated random variables, i.e. random variables on a bounded support (therefore having moments of all orders), in place of the actual random variables. It will be shown why working with the truncated variables is equivalent to working with the variables themselves and why the general version of Theorem 2.13 unfolds.

3.2.1 Proof of the Marčenko–Pastur law

We wish this proof to be complete in the sense that all notions of random matrix theory and probability theory tools are thoroughly detailed. As such, the proof contains many embedded lemmas, with sometimes further embedded results. The reader may skip most of these secondary results for ease of read. Those lemmas are nonetheless essential to the understanding of the basics of random matrix theory using the Stieltjes transform and deserve, as such, a lengthy explanation.

Let $\mathbf{X} \in \mathbb{C}^{N\times n}$ be a random matrix with i.i.d. entries of zero mean, variance $1/n$, and finite eighth order moment, and denote $\mathbf{R}_N = \mathbf{X}\mathbf{X}^{\mathsf{H}}$. We start by singling out the first row $\mathbf{y}^{\mathsf{H}} \in \mathbb{C}^{1\times n}$ of $\mathbf{X}$, and we write

$$\mathbf{X} \triangleq \begin{bmatrix} \mathbf{y}^{\mathsf{H}} \\ \mathbf{Y} \end{bmatrix}.$$

Now, for $z \in \mathbb{C}^+$, we have

$$(\mathbf{R}_N - z\mathbf{I}_N)^{-1} = \begin{bmatrix} \mathbf{y}^{\mathsf{H}}\mathbf{y} - z & \mathbf{y}^{\mathsf{H}}\mathbf{Y}^{\mathsf{H}} \\ \mathbf{Y}\mathbf{y} & \mathbf{Y}\mathbf{Y}^{\mathsf{H}} - z\mathbf{I}_{N-1} \end{bmatrix}^{-1} \tag{3.9}$$

the trace of which is the Stieltjes transform of $F^{\mathbf{R}_N}$. Our interest being to compute this Stieltjes transform, and then the sum of the diagonal elements of the matrix in (3.9), we start by considering the entry $(1,1)$. For this, we need a classical matrix inversion lemma

Lemma 3.3. *Let $\mathbf{A} \in \mathbb{C}^{N\times N}$, $\mathbf{D} \in \mathbb{C}^{n\times n}$ be invertible, and $\mathbf{B} \in \mathbb{C}^{N\times n}$, $\mathbf{C} \in \mathbb{C}^{n\times N}$. Then we have:*

$$\begin{pmatrix} \mathbf{A} & \mathbf{B} \\ \mathbf{C} & \mathbf{D} \end{pmatrix}^{-1} = \begin{pmatrix} (\mathbf{A}-\mathbf{B}\mathbf{D}^{-1}\mathbf{C})^{-1} & -\mathbf{A}^{-1}\mathbf{B}(\mathbf{D}-\mathbf{C}\mathbf{A}^{-1}\mathbf{B})^{-1} \\ -(\mathbf{A}-\mathbf{B}\mathbf{D}^{-1}\mathbf{C})^{-1}\mathbf{C}\mathbf{A}^{-1} & (\mathbf{D}-\mathbf{C}\mathbf{A}^{-1}\mathbf{B})^{-1} \end{pmatrix}. \tag{3.10}$$

We apply Lemma 3.3 to the block matrix (3.9) to obtain the upper left entry

$$\left(-z + \mathbf{y}^{\mathsf{H}}\left(\mathbf{I}_N - \mathbf{Y}^{\mathsf{H}}\left(\mathbf{Y}\mathbf{Y}^{\mathsf{H}} - z\mathbf{I}_{N-1}\right)^{-1}\mathbf{Y}\right)\mathbf{y}\right)^{-1}.$$

From the relation $\mathbf{I}_N - \mathbf{A}_N(\mathbf{I}_N + \mathbf{B}_N\mathbf{A}_N)^{-1}\mathbf{B}_N = (\mathbf{I}_N + \mathbf{A}_N\mathbf{B}_N)^{-1}$, this further expresses as

$$\left(-z - z\mathbf{y}^{\mathsf{H}}(\mathbf{Y}^{\mathsf{H}}\mathbf{Y} - z\mathbf{I}_n)^{-1}\mathbf{y}\right)^{-1}.$$

We then have

$$\left[(\mathbf{R}_N - z\mathbf{I}_N)^{-1}\right]_{11} = \frac{1}{-z - z\mathbf{y}^{\mathsf{H}}(\mathbf{Y}^{\mathsf{H}}\mathbf{Y} - z\mathbf{I}_n)^{-1}\mathbf{y}}. \tag{3.11}$$

To go further, we need an additional result, proved initially in [Bai and Silverstein, 1998], that we formulate in the following theorem.

Theorem 3.4. *Let $\mathbf{A}_1, \mathbf{A}_2, \ldots$, with $\mathbf{A}_N \in \mathbb{C}^{N\times N}$, be a series of matrices with uniformly bounded spectral norm. Let $\mathbf{x}_1, \mathbf{x}_2, \ldots$, with $\mathbf{x}_N \in \mathbb{C}^N$, be random vectors with i.i.d. entries of zero mean, variance $1/N$, and eighth order moment of order $O(1/N^4)$, independent of $\mathbf{A}_N$. Then*

$$\mathbf{x}_N^{\mathsf{H}}\mathbf{A}_N\mathbf{x}_N - \frac{1}{N}\operatorname{tr}\mathbf{A}_N \xrightarrow{\text{a.s.}} 0 \tag{3.12}$$

as $N \to \infty$.

We mention besides that, in the case where the quantities involved are real and not complex, the entries of $\mathbf{x}_N$ have fourth order moment of order $O(1/N^2)$ and $\mathbf{A}_N$ has l.s.d. A, a central limit of the variations of $\mathbf{x}_N^{\mathsf{T}}\mathbf{A}_N\mathbf{x}_N - \frac{1}{N}\operatorname{tr}\mathbf{A}_N$ is proved in [Tse and Zeitouni, 2000], i.e.

$$\sqrt{N}\left[\mathbf{x}_N^{\mathsf{T}}\mathbf{A}_N\mathbf{x}_N - \frac{1}{N}\operatorname{tr}\mathbf{A}_N\right] \Rightarrow Z \tag{3.13}$$

as $N \to \infty$,[4] with $Z \sim \mathcal{N}(0, v)$, for some variance v depending on the l.s.d. F^A of $\mathbf{A}_N$ and on the fourth moment of the entries of $\mathbf{x}_N$ as follows.

$$v = 2\int t^2 dF^A(t) + (\mathrm{E}[x_{11}^4] - 3)\left(\int t dF^A(t)\right)^2.$$

Intuitively, taking for granted that $\mathbf{x}_N^{\mathsf{H}}\mathbf{A}_N\mathbf{x}_N$ does have a limit, this limit must coincide with the limit of $\mathrm{E}[\mathbf{x}_N^{\mathsf{H}}\mathbf{A}_N\mathbf{x}_N]$. But for finite N

$$\begin{aligned}\mathrm{E}[\mathbf{x}_N^{\mathsf{H}}\mathbf{A}_N\mathbf{x}_N] &= \sum_{i=1}^{N}\sum_{j=1}^{N} A_{N,ij}\mathrm{E}[x_{N,i}^* x_{N,j}] \\ &= \frac{1}{N}\sum_{i=1}^{N}\sum_{j=1}^{N} A_{N,ij}\delta_i^j \\ &= \frac{1}{N}\operatorname{tr}\mathbf{A}_N\end{aligned}$$

which is the expected result. We hereafter provide a rigorous proof of the almost sure limit, which has the strong advantage to introduce very classical probability theoretic tools which will be of constant use in the detailed proofs of the important results of this book. This proof may be skipped in order not to disrupt the flow of the proof of the Marčenko–Pastur law.

Proof. We start by introducing the following two fundamental results of probability theory. The first result is known as the *Markov inequality.*

Theorem 3.5 (Markov Inequality, (5.31) of [Billingsley, 1995]). *For X a real random variable, $\alpha > 0$, we have for all integer k*

$$P(\{\omega, |X(\omega)| \geq \alpha\}) \leq \frac{1}{\alpha^k}\mathrm{E}\left[|X|^k\right].$$

The second result is the *first Borel–Cantelli lemma.*

Theorem 3.6 (First Borel–Cantelli Lemma, Theorem 4.3 in [Billingsley, 1995]). *Let $\{A_N\}$ be $\mathcal{F}$-sets of Ω. If $\sum_N P(A_N) < \infty$, then $P(\limsup_N A_N) = 0$. When A_N has a limit, this implies that $P(\lim_N A_N) = 0$.*

The symbol $\limsup_N A_N$ stands for the set $\bigcap_{k\geq 0}\bigcup_{n\geq k} A_n$. An element $\omega \in \Omega$ belongs to $\limsup_N A_N$ if, for all integer k, there exists $N \geq k$ such that $\omega \in A_N$,

[4] The notation $X_N \Rightarrow X$ for X_N and X random variables, with distribution function F_N and F, respectively, is equivalent to $F_N \Rightarrow F$.

i.e. $\omega \in \limsup_N A_N$ if ω belongs to infinitely many sets A_N. Informally, an event A_N such that $P(\limsup_N A_N) = 0$ is an event that, with probability one, does not happen *infinitely often* (denoted i.o.). The set $\limsup_N A_N$ is sometimes written A_N i.o.

The technique to prove Theorem 3.4 consists in finding an integer k such that

$$\mathrm{E}\left[\left|\mathbf{x}_N^{\mathsf{H}}\mathbf{A}_N\mathbf{x}_N - \frac{1}{N}\operatorname{tr}\mathbf{A}_N\right|^k\right] \leq f_N \tag{3.14}$$

where f_N is constant independent of both $\mathbf{A}_N$ and $\mathbf{x}_N$, such that $\sum_N f_N < \infty$. Then, for some $\varepsilon > 0$, from the Markov inequality, we have that

$$P(\{\omega, Y_N(\omega) \geq \varepsilon\}) \leq \frac{1}{\varepsilon^k}\mathrm{E}\left[Y_N^k\right] \tag{3.15}$$

with $Y_N \triangleq \left|\mathbf{x}_N^{\mathsf{H}}\mathbf{A}_N\mathbf{x}_N - (1/N)\operatorname{tr}\mathbf{A}_N\right|$. Since the right-hand side of (3.15) is summable, it follows from the first Borel–Cantelli lemma, Theorem 3.6, that

$$P(\{\omega, Y_N(\omega) \geq \varepsilon \text{ i.o.}\}) = 0.$$

Since $\varepsilon > 0$ was arbitrary, the above is true for all rational $\varepsilon > 0$. Because the countable union of sets of probability zero is still a set of probability zero (see [Billingsley, 1995]), we finally have that

$$P\left(\bigcup_{(p,q)\in(\mathbb{N}^*)^2}\left\{\omega, Y_N(\omega) \geq \frac{p}{q} \text{ i.o.}\right\}\right) = 0.$$

The complementary of the set in parentheses above satisfies: for all (p, q) there exists $N_0(\omega)$ such that, for all $N \geq N_0(\omega)$, $|Y_N(\omega)| \leq \frac{p}{q}$. This set has probability one, and therefore Y_N has limit zero with probability one, i.e. $\mathbf{x}_N^{\mathsf{H}}\mathbf{A}_N\mathbf{x}_N - \frac{1}{N}\operatorname{tr}\mathbf{A}_N \xrightarrow{\text{a.s.}} 0$. It therefore suffices to find an integer k such that (3.14) is satisfied.

For $k = 4$, expanding $\mathbf{x}_N^{\mathsf{H}}\mathbf{A}_N\mathbf{x}_N$ as a sum of terms $x_i^{\mathsf{H}}x_jA_{i,j}$ and distinguishing the cases when $i = j$ or $i \neq j$, we have:

$$\begin{aligned}&\mathrm{E}\left[\left|\mathbf{x}_N^{\mathsf{H}}\mathbf{A}_N\mathbf{x}_N - \frac{1}{N}\operatorname{tr}\mathbf{A}_N\right|^4\right]\\ &\leq \frac{8}{N^4}\left(\mathrm{E}\left[\sum_{i=1}^N A_{N,ii}(|x_{N,i}|^2 - 1)\right]^4 + \mathrm{E}\left[\sum_{i\neq j} A_{N,ij}x_{N,i}^*x_{N,j}\right]^4\right)\end{aligned}$$

where the inequality comes from: $|x+y|^k \leq (|x|+|y|)^k \leq 2^{k-1}(|x|^k + |y|^k)$. The latter arises from the Hölder's inequality, which states that, for $p, q > 0$ such that $1/p + 1/q = 1$, applied to two sets $x_1, \ldots, x_N$ and $y_1, \ldots, y_N$ [Billingsley,

1995]

$$\sum_{n=1}^{N} |x_n y_n| \leq \left(\sum_{n=1}^{N} |x_n|^p \right)^{1/p} \left(\sum_{n=1}^{N} |y_n|^q \right)^{1/q}$$

taking $N = 2$, $x_1 = x$, $x_2 = y$, $y_1 = y_2 = 1$ and $p = 4$, we have immediately the result. Since the x_i have finite eighth order moment (and therefore finite kth order moment of all $k \leq 8$) and that $\mathbf{A}_N$ has uniformly bounded norm, all the terms in the first sum are finite. Now, expanding the sum as a 4-fold sum, the number of terms that are non-identically zeros is of order $O(N^2)$. The second sum is treated identically, with an order of $O(N^2)$ non-identically null terms. Therefore, along with the factor $1/N^4$ in front, there exists $K > 0$ independent of N, such that the sum is less than K/N^2. This is summable and we have proved Theorem 3.4, when the $\mathbf{x}_k$ have finite eighth order moment.

Remark 3.1. Before carrying on the proof of the Marčenko–Pastur law, we take the opportunity of the introduction of the trace lemma to mention the following two additional results. The first result is an extension of the trace lemma to the characterization of $\mathbf{x}^{\mathsf{H}}\mathbf{A}\mathbf{y}$ for independent $\mathbf{x}$, $\mathbf{y}$ vectors.

Theorem 3.7. *For $\mathbf{A}_1, \mathbf{A}_2, \ldots$, with $\mathbf{A}_N \in \mathbb{C}^{N \times N}$ with uniformly bounded spectral norm, $\mathbf{x}_1, \mathbf{x}_2, \ldots$ and $\mathbf{y}_1, \mathbf{y}_2, \ldots$ two series of i.i.d. variables such that $\mathbf{x}_N \in \mathbb{C}^N$ and $\mathbf{y}_N \in \mathbb{C}^N$ have zero mean, variance $1/N$, and fourth order moment of order $O(1/N^2)$, we have:*

$$\mathbf{x}_N^{\mathsf{H}} \mathbf{A}_N \mathbf{y}_N \xrightarrow{\text{a.s.}} 0.$$

Proof. The above unfolds simply by noticing that $\mathrm{E}|\mathbf{x}_N^{\mathsf{H}} \mathbf{A}_N \mathbf{y}_N|^4 \leq c/N^2$ for some constant c. We give below the precise derivation for this rather easy case.

$$\mathrm{E}\left[\left|\mathbf{x}_N^{\mathsf{H}} \mathbf{A}_N \mathbf{y}_N\right|^4\right] = \mathrm{E}\left[\sum_{\substack{i_1,\ldots,i_4 \\ j_1,\ldots,j_4}} x_{i_1}^* x_{i_2} x_{i_3}^* x_{i_4} y_{j_1} y_{j_2}^* y_{j_3} y_{j_4}^* A_{i_1,j_1} A_{i_2,j_2}^* A_{i_3,j_3} A_{i_4,j_4}^* \right].$$

If one of the x_{i_k} or y_{j_k} appears an odd number of times in one of the terms of the sum, then the expectation of this term is zero. We therefore only account for terms x_{i_k} and y_{j_k} that appear two or four times. If $i_1 = i_2 = i_3 = i_4$ and $j_1 = j_2 = j_3 = j_4$, then:

$$\begin{aligned} \mathrm{E}\left[x_{i_1}^* x_{i_2} x_{i_3}^* x_{i_4} y_{j_1} y_{j_2}^* y_{j_3} y_{j_4}^* A_{i_1,j_1} A_{i_2,j_2}^* A_{i_3,j_3} A_{i_4,j_4}^*\right] &= \frac{1}{N^4} |A_{i_1,j_1}|^4 \mathrm{E}[|x_1|^4] \mathrm{E}[|y_1|^4] \\ &= O(1/N^4). \end{aligned}$$

Since there are as many as N^2 such configurations of $i_1, \ldots, i_4$ and $j_1, \ldots, j_4$, these terms contribute to an order $O(1/N^2)$ in the end. If $i_1 = i_2 = i_3 = i_4$ and

$j_1 = j_3 \neq j_2 = j_4$, then:

$$\mathrm{E}\left[x_{i_1}^* x_{i_2} x_{i_3}^* x_{i_4} y_{j_1} y_{j_2}^* y_{j_3} y_{j_4}^* A_{i_1,j_1} A_{i_2,j_2}^* A_{i_3,j_3} A_{i_4,j_4}^*\right] = \frac{1}{N^4} |A_{i_1,j_1}|^2 |A_{i_1,j_3}|^2 \mathrm{E}[|x_1|^4]$$
$$= O(1/N^4).$$

Noticing that $\sum_{i_1,j_1,j_3} |A_{i_1,j_1}|^2 |A_{i_1,j_3}|^2 = \sum_{i_1,j_1} |A_{i_1,j_1}|^2 (\sum_{j_3} |A_{i_1,j_3}|^2)$, and that $\sum_{i,j} |A_{i,j}|^2 = \operatorname{tr} \mathbf{A}_N \mathbf{A}_N^{\mathsf{H}} = O(N)$ from the bounded norm condition on $\mathbf{A}_N$, we finally have that the sum over all possible i_1 and $j_1 \neq j_3$ is of order $O(1/N^2)$. The same is true for the combination $j_1 = j_2 = j_3 = j_4$ and $i_1 = i_3 \neq i_2 = i_4$, which results in a term of order $O(1/N^2)$. It remains the case when $i_1 = i_3 \neq i_2 = i_4$ and $j_1 = j_3 \neq j_2 = j_4$. This leads to the terms

$$\mathrm{E}\left[x_{i_1}^* x_{i_2} x_{i_3}^* x_{i_4} y_{j_1} y_{j_2}^* y_{j_3} y_{j_4}^* A_{i_1,j_1} A_{i_2,j_2}^* A_{i_3,j_3} A_{i_4,j_4}^*\right] = \frac{1}{N^4} |A_{i_1,j_1}|^2 |A_{i_2,j_3}|^2$$
$$= O(1/N^4).$$

Noticing that $\sum_{i_1,i_3,j_1,j_3} |A_{i_1,j_1}|^2 |A_{i_2,j_3}|^2 = \sum_{i_1,j_1} |A_{i_1,j_1}|^2 (\sum_{i_3,j_3} |A_{i_1,j_3}|^2)$ from the same argument as above, we have that the last term is also of order $O(1/N^2)$. Therefore, the total expected sum is of order $O(1/N^2)$. The Markov inequality and the Borel–Cantelli lemma give the final result. □

The second result is a generalized version of the fourth order moment inequality that led to the proof of the trace lemma above, when the entries of $\mathbf{x}_N = (x_{N,1}, \ldots, x_{N,N})^{\mathsf{T}}$ have moments of all orders. This result unfolds from the same combinatorics calculus as presented in the proof of Theorem 3.7.

Theorem 3.8 (Lemma B.26 of [Bai and Silverstein, 2009]). *Under the conditions of Theorem 3.4, if for all N, k, $\mathrm{E}[|\sqrt{N} x_{N,k}|^m] \leq \nu_m$, then, for all $p \geq 1$*

$$\mathrm{E}\left[\left|\mathbf{x}_N^{\mathsf{H}} \mathbf{A}_N \mathbf{x}_N - \frac{1}{N} \operatorname{tr} \mathbf{A}_N\right|^p\right] \leq \frac{C_p}{N^p} \left[(\nu_4 \operatorname{tr}(\mathbf{A}\mathbf{A}^{\mathsf{H}}))^{\frac{p}{2}} + \nu_{2p} \operatorname{tr}(\mathbf{A}\mathbf{A}^{\mathsf{H}})^{\frac{p}{2}}\right]$$

for C_p a constant depending only on p.

□

Returning to the proof of the Marčenko–Pastur law, in (3.11), $\mathbf{y} \in \mathbb{C}^n$ is extracted from $\mathbf{X}$, which has independent columns, so that $\mathbf{y}$ is independent of $\mathbf{Y}$. Besides, $\mathbf{y}$ has i.i.d. entries of variance $1/n$. For large n, we therefore have

$$\left|\left[(\mathbf{R}_N - z\mathbf{I}_N)^{-1}\right]_{11} - \frac{1}{-z - z\frac{1}{n} \operatorname{tr}(\mathbf{Y}^{\mathsf{H}}\mathbf{Y} - z\mathbf{I}_n)^{-1}}\right| \xrightarrow{\text{a.s.}} 0$$

where the convergence is ensured by verifying that the denominator of the difference has imaginary part uniformly away from zero.

We feel at this point that, for large n, the normalized trace $\frac{1}{n} \operatorname{tr}(\mathbf{Y}^{\mathsf{H}}\mathbf{Y} - z\mathbf{I}_n)^{-1}$ should not be much different from $\frac{1}{n} \operatorname{tr}(\mathbf{X}^{\mathsf{H}}\mathbf{X} - z\mathbf{I}_n)^{-1}$, since the difference between both matrices here is merely the rank-1 matrix $\mathbf{y}\mathbf{y}^{\mathsf{H}}$. This is formalized in the following second theorem.

Theorem 3.9 ([Silverstein and Bai, 1995]). *For $z \in \mathbb{C} \setminus \mathbb{R}^+$, we have the following quadratic form identities.*

(i) Let $z \in \mathbb{C} \setminus \mathbb{R}$, $\mathbf{A} \in \mathbb{C}^{N \times N}$, $\mathbf{B} \in \mathbb{C}^{N \times N}$ with $\mathbf{B}$ Hermitian, and $\mathbf{v} \in \mathbb{C}^N$. Then

$$\left|\operatorname{tr}\left((\mathbf{B} - z\mathbf{I}_N)^{-1} - (\mathbf{B} + \mathbf{v}\mathbf{v}^{\mathsf{H}} - z\mathbf{I}_N)^{-1}\right)\mathbf{A}\right| \leq \frac{\|\mathbf{A}\|}{|\Im[z]|}$$

with $\|\mathbf{A}\|$ the spectral norm of $\mathbf{A}$.

(ii) Moreover, if $\mathbf{B}$ is non-negative definite, for $z \in \mathbb{R}^-$

$$\left|\operatorname{tr}\left((\mathbf{B} - z\mathbf{I}_N)^{-1} - (\mathbf{B} + \mathbf{v}\mathbf{v}^{\mathsf{H}} - z\mathbf{I}_N)^{-1}\right)\mathbf{A}\right| \leq \frac{\|\mathbf{A}\|}{|z|}.$$

This theorem can be further refined for $z \in \mathbb{R}^+$. Generally speaking, it is important to take z away from the support of the eigenvalues of $\mathbf{B}$ and $\mathbf{B} + \mathbf{v}\mathbf{v}^{\mathsf{H}}$ to make sure that both matrices remain invertible. With z purely complex, the position of the eigenvalues of $\mathbf{B}$ and $\mathbf{B} + \mathbf{v}\mathbf{v}^{\mathsf{H}}$ on the real line does not matter, and similarly for $\mathbf{B}$ non-negative definite and real $z < 0$, the position of the eigenvalues of $\mathbf{B}$ does not matter.

In the present situation, $\mathbf{A} = \mathbf{I}_n$ and therefore, irrespective of the actual properties of $\mathbf{y}$ (it might even be a degenerated vector with large Euclidean norm), we have

$$\begin{aligned}
&\frac{1}{n}\operatorname{tr}(\mathbf{Y}^{\mathsf{H}}\mathbf{Y} - z\mathbf{I}_n)^{-1} - m_{F^{\mathbf{X}^{\mathsf{H}}\mathbf{X}}}(z) \\
&= \frac{1}{n}\operatorname{tr}(\mathbf{Y}^{\mathsf{H}}\mathbf{Y} - z\mathbf{I}_n)^{-1} - \frac{1}{n}\operatorname{tr}(\mathbf{X}^{\mathsf{H}}\mathbf{X} - z\mathbf{I}_n)^{-1} \\
&= \frac{1}{n}\operatorname{tr}(\mathbf{Y}^{\mathsf{H}}\mathbf{Y} - z\mathbf{I}_n)^{-1} - \frac{1}{n}\operatorname{tr}(\mathbf{Y}^{\mathsf{H}}\mathbf{Y} + \mathbf{y}\mathbf{y}^{\mathsf{H}} - z\mathbf{I}_n)^{-1} \\
&\to 0
\end{aligned}$$

and therefore:

$$\left[(\mathbf{R}_N - z\mathbf{I}_N)^{-1}\right]_{11} - \frac{1}{-z - zm_{F^{\mathbf{X}^{\mathsf{H}}\mathbf{X}}}(z)} \xrightarrow{\text{a.s.}} 0.$$

By the definition of the Stieltjes transform, since the non-zero eigenvalues of $\mathbf{R}_N = \mathbf{X}\mathbf{X}^{\mathsf{H}}$ and $\mathbf{X}^{\mathsf{H}}\mathbf{X}$ are the same, we have from Lemma 3.1

$$m_{F^{\mathbf{X}^{\mathsf{H}}\mathbf{X}}}(z) = \frac{N}{n} m_{F^{\mathbf{R}_N}}(z) + \frac{N-n}{n}\frac{1}{z} \tag{3.16}$$

which leads to

$$\left[(\mathbf{R}_N - z\mathbf{I}_N)^{-1}\right]_{11} - \frac{1}{1 - \frac{N}{n} - z - z\frac{N}{n}m_{F^{\mathbf{R}_N}}(z)} \xrightarrow{\text{a.s.}} 0. \tag{3.17}$$

The second term in the difference is independent on the initial choice of the entry $(1,1)$ in $(\mathbf{R}_N - z\mathbf{I}_N)^{-1}$. Due to the symmetric structure of $\mathbf{X}$, the result is also true for all diagonal entries (i,i), $i = 1,\ldots,N$. Summing them up and averaging, we conclude that[5]

$$m_{F^{\mathbf{R}_N}}(z) - \frac{1}{1 - \frac{N}{n} - z - z\frac{N}{n} m_{F^{\mathbf{R}_N}}(z)} \xrightarrow{\text{a.s.}} 0. \tag{3.18}$$

Take $\mathbf{R}_1, \mathbf{R}_2, \ldots$ a particular sequence for which (3.18) is verified (such sequences lie in a space of probability one). Since $m_{F^{\mathbf{R}_N}}(z) \leq 1/\Im[z]$ from Theorem 3.2, the sequence $m_{F^{\mathbf{R}_1}}(z), m_{F^{\mathbf{R}_2}}(z), \ldots$ is uniformly bounded in a compact set. Consider now any subsequence $m_{F^{\mathbf{R}_{\psi(1)}}}(z), m_{F^{\mathbf{R}_{\psi(2)}}}(z), \ldots$; along this subsequence, (3.18) is still valid. Since $m_{F^{\mathbf{R}_{\psi(n)}}}$ is uniformly bounded from above, we can select a further subsequence $m_{F^{\mathbf{R}_{\phi(\psi(1))}}}, m_{F^{\mathbf{R}_{\phi(\psi(2))}}}, \ldots$ of $m_{F^{\mathbf{R}_{\psi(1)}}}(z), m_{F^{\mathbf{R}_{\psi(2)}}}(z), \ldots$ which converges (this is an immediate consequence of the Bolzano–Weierstrass theorem). Its limit, call it $m(z; \phi, \psi)$ is still a Stieltjes transform, as can be verified from Theorem 3.2, and is one solution of the implicit equation in m

$$m = \frac{1}{1 - c - z - zcm}. \tag{3.19}$$

The form of the implicit Equation (3.19) is often the best we can obtain more involved models than i.i.d. $\mathbf{X}$ matrices. It will indeed often turn out that no explicit equation for the limiting Stieltjes transform m_F (which we have not yet proved exist) will be available. Additional tools will then be required to ensure that (3.19) admits either (i) a unique scalar solution, when seen as an equation in the dummy variable m, or (ii) a unique functional solution, when seen as an equation in the dummy function-variable m of z. The difference is technically important for practical applications. Indeed, if we need to recover the Stieltjes transform of a d.f. F (for instance to evaluate its associated Shannon transform), it is important to know whether a classical fixed-point algorithm is expected to converge to solutions of (3.19) other than the desired solution. This will be discussed further later in this section.

For the problem at hand, though, (3.19) can be rewritten as the second order polynomial $m(1 - c - z - zcm) = 0$ in the variable m, a unique root of which is the Stieltjes transform of a distribution function taken at z. This limit is of course independent of the choice of ϕ and ψ. Therefore, any subsequence of $m_{F^{\mathbf{R}_1}}(z), m_{F^{\mathbf{R}_2}}(z), \ldots$ admits a further subsequence, which converges to some value $m_F(z)$, which is the Stieltjes transform of some distribution function F. Therefore, from the semi-converse of the Bolzano–Weierstrass theorem, $m_{F^{\mathbf{R}_1}}(z), m_{F^{\mathbf{R}_2}}(z), \ldots$ converges to $m_F(z)$. The latter is given explicitly by

$$m_F(z) = \frac{1-c}{2cz} - \frac{1}{2c} - \frac{\sqrt{(1-c-z)^2 - 4cz}}{2cz} \tag{3.20}$$

[5] We use here the fact that the intersection of countably many sets of probability one on which the result holds is itself of probability one.

where the branch of $\sqrt{(1-c-z)^2-4cz}$ is chosen such that $m_F(z) \in \mathbb{C}^+$ for $z \in \mathbb{C}^+$, $m_F(z) \in \mathbb{C}^-$ for $z \in \mathbb{C}^-$ and $m_F(z) > 0$ for $z < 0$.[6]

Using the inverse-Stieltjes transform formula (3.2), we then verify that $m_{F^{\mathbf{R}_N}}$ has a limit m_F, the Stieltjes transform of the Marčenko–Pastur law, with density

$$F'(x) = (1-c^{-1})^+\delta(x) + \frac{1}{2\pi cx}\sqrt{(x-a)^+(b-x)^+}$$

where $a = (1-\sqrt{c})^2$ and $b = (1+\sqrt{c})^2$. The term $\frac{1}{2x}\sqrt{(x-a)^+(b-x)^+}$ is obtained by computing $\lim_{y\to 0} m_F(x+iy)$, and taking its imaginary part. The coefficient c is then retrieved from the fact that we know first how many zero eigenvalues should be added and second that the density should integrate to 1.

To prove that the almost sure convergence of $m_{F^{\mathbf{R}_N}}(z) \xrightarrow{\text{a.s.}} m_F(z)$ induces the weak convergence $F^{\mathbf{R}_N} \Rightarrow F$ with probability one, we finally need the following theorem.

Theorem 3.10 (Theorem B.9 of [Bai and Silverstein, 2009]). *Let $\{F_N\}$ be a set of bounded real functions such that $\lim_{x\to-\infty} F_N(x) = 0$. Then, for all $z \in \mathbb{C}^+$*

$$\lim_{N\to\infty} m_{F_N}(z) = m_F(z)$$

if and only if there exists F such that $\lim_{x\to-\infty} F(x) = 0$ and $|F_N(x) - F(x)| \to 0$ for all $x \in \mathbb{R}$.

Proof. For $z \in \mathbb{C}^+$, the function $f : (x,z) \mapsto \frac{1}{z-x}$ is continuous and tends to zero when $|x| \to \infty$. Therefore, $|F_N(x) - F(x)| \to 0$ for all $x \in \mathbb{R}$ implies that $\int \frac{1}{z-x} d(F_N - F)(x) \to 0$. Conversely, from the inverse Stieltjes transform formula (3.2), for a, b continuity points of F, $|(F_N(b) - F_N(a)) - (F(b) - F(a))| \leq \frac{1}{\pi} \lim_{y\to 0^+} \int_a^b \Im|(m_N - m)(x+iy)|dx$, which tends to zero as N grows large. Therefore, we have:

$$|F_N(x) - F(x)| \leq |(F_N(x) - F(x)) - (F_N(a) - F(a))| + |F_N(a) - F(a)|$$

which tends to zero as we take, e.g. $a = N$ and $N \to \infty$ (since both $F_N(a) \to 1$ and $F(a) \to 1$). □

The sure convergence of the Stieltjes transform m_{F_N} to m_F therefore ensures that $F_N \Rightarrow F$ and conversely. This is the one theorem, along with the inversion formula (3.2), that fully justifies the usage of the Stieltjes transform to study the convergence of probability measures.

Back to the proof of the Marčenko–Pastur law, we have up to now proved that

$$m_{F^{\mathbf{R}_N}}(z) - m_F(z) \xrightarrow{\text{a.s.}} 0$$

[6] We use a minus sign here in front of $\frac{\sqrt{(1-c-z)^2-4cz}}{2cz}$ for coherence with the principal square root branch when $z < 0$.

for some initially fixed $z \in \mathbb{C}^+$. That is, there exists a subspace $C_z \in \mathfrak{F}$, with $(\Omega, \mathfrak{F}, P)$ the probability space generating the series $\mathbf{R}_1, \mathbf{R}_2, \ldots$, such that $P(C_z) = 1$ for which $\omega \in C_z$ implies that $m_{F^{\mathbf{R}_N(\omega)}}(z) - m_F(z) \to 0$. Since we want to prove the almost sure weak convergence of $F^{\mathbf{R}_N}$ to the Marčenko–Pastur law, we need to prove that the convergence of the Stieltjes transform holds *for all* $z \in \mathbb{C} \setminus \mathbb{R}^+$ on a common space of probability one. We then need to show that there exists $C \in \mathfrak{F}$, with $P(C) = 1$ such that, *for all* $z \in \mathbb{C}^+$, $\omega \in C$ implies that $m_{F^{\mathbf{R}_N(\omega)}}(z) - m_F(z) \to 0$. This requires to use Vitali's convergence theorem [Titchmarsh, 1939] on a countable subset of $\mathbb{C}^+$ (or alternatively, Montel's theorem).

Theorem 3.11. *Let $f_1, f_2, \ldots$ be a sequence of functions, analytic on a region $D \subset \mathbb{C}$, such that $|f_n(z)| \leq M$ uniformly on n and $z \in D$. Further, assume that $f_n(z_j)$ converges for a countable set $z_1, z_2, \ldots \in D$ having a limit point inside D. Then $f_n(z)$ converges uniformly in any region bounded by a contour interior to D. This limit is furthermore an analytic function of z.*

The convergence $m_{F^{\mathbf{R}_N}}(z) - m_F(z) \xrightarrow{\text{a.s.}} 0$ is valid for any z inside a bounded region of $\mathbb{C}^+$. Take countably many $z_1, z_2, \ldots$ having a limit point in some compact region of $\mathbb{C}^+$. For each i, we have $m_{F^{\mathbf{R}_N(\omega)}}(z_i) - m_F(z_i) \xrightarrow{\text{a.s.}} 0$ for $\omega \in C_{z_i}$, some set with $P(C_{z_i}) = 1$. The set $C = \bigcup_i C_{z_i} \in \mathfrak{F}$ over which the convergence holds for all z_i has probability one, as the countable union of sets of probability one. As a consequence, for $\omega \in C$, $m_{F^{\mathbf{R}_N(\omega)}(\omega)}(z_i) - m_F(z_i) \to 0$ for any i. From Vitali's convergence theorem, since $m_{F^{\mathbf{R}_N}}(z) - m_F(z)$ is clearly an analytic function of z, this holds true uniformly in all sets interior to regions where $m_{F^{\mathbf{R}_N}}(z)$ and $m_F(z)$ are uniformly bounded, i.e. in all regions that exclude the real positive half-line. From Theorem 3.10, $F^{\mathbf{R}_N(\omega)}(x) - F(x) \to 0$ for all $x \in \mathbb{R}$ and for all $\omega \in C$, so that, for $\omega \in C$, $F^{\mathbf{R}_N(\omega)}(x) - F(x) \to 0$. This ensures that $F^{\mathbf{R}_N} \Rightarrow F$ almost surely. This proves the almost sure weak convergence of the e.s.d. of $\mathbf{R}_N$ to the Marcěnko-Pastur law.

The reason why the proof above constrains the entries of $\mathbf{X}$ to have finite eighth order moment is due to the trace lemma, Theorem 3.4, which is only proved to work with this eighth order moment assumption. We give below a generalization of Theorem 3.4 when the random variables under consideration are uniformly bounded in some sense, no longer requiring the finite eighth order moment assumption. We will then present the *truncation, centralization, and rescaling* steps, which allow us to prove the general version of Theorem 2.13. These steps consist in replacing the variables X_{ij} by *truncated* versions of these variables, i.e. replacing X_{ij} by zero whenever $|X_{ij}|$ exceeds some predefined threshold. The centralization and centering steps are then used to recenter the modified X_{ij} around its mean and to preserve its variance. The main objective here is to replace variables, which may not have bounded moments, by modified versions of these variables that have moments of all orders. The surprising result is that it is asymptotically equivalent to consider X_{ij} or their altered versions;

as a consequence, in many practical situations, it is unnecessary to make any assumptions on the existence of any moments of order higher than 2 for X_{ij}. We will subsequently show how this operates and why it is sufficient to work with supportly compacted random X_{ij} variables.

3.2.2 Truncation, centralization, and rescaling

A convenient trace lemma for truncated variables can be given as follows.

Theorem 3.12. *Let $\{\mathbf{A}_1, \mathbf{A}_2, \dots\}$, $\mathbf{A}_N \in \mathbb{C}^{N \times N}$, be a series of matrices of growing sizes and $\{\mathbf{x}_1, \mathbf{x}_2, \dots\}$, $\mathbf{x}_N \in \mathbb{C}^N$, be random vectors with i.i.d. entries bounded by $N^{-\frac{1}{2}} \log N$, with zero mean and variance $1/N$, independent of $\mathbf{A}_N$. Then*

$$\mathrm{E}\left[\left|\mathbf{x}_N^{\mathsf{H}} \mathbf{A}_N \mathbf{x}_N - \frac{1}{N} \operatorname{tr} \mathbf{A}_N\right|^6\right] \leq K \|\mathbf{A}_N\|^6 \frac{\log^{12} N}{N^3}$$

for some constant K independent of N.

The sixth order moment here is upper bounded by a bound on the values of the entries of the $\mathbf{x}_k$ instead of a bound on the moments. This alleviates the consideration of the existence of any moment on the entries. Note that a similar result for the fourth order moment also exists that is in general sufficient for practical purposes, but, since going to higher order moments is now immaterial, this result is slightly stronger. The proof of Theorem 3.12 unfolds from the same manipulations as for Theorem 3.4.

Obviously, applying the Markov inequality, Theorem 3.5, and the Borel–Cantelli lemma, Theorem 3.6, the result above implies the almost sure convergence of the difference $\mathbf{x}_N^{\mathsf{H}} \mathbf{A}_N \mathbf{x}_N - \frac{1}{N} \operatorname{tr} \mathbf{A}_N$ to zero. A second obvious remark is that, if the elements of $\sqrt{N}\mathbf{x}_N$ are bounded by some constant C instead of $\log(N)$, the result still holds true.

The following explanations follow precisely Section 3.1.3 in [Bai and Silverstein, 2009]. We start by the introduction of two important lemmas.

Lemma 3.4 (Corollary A.41 in [Bai and Silverstein, 2009]). *For $\mathbf{A} \in \mathbb{C}^{N \times n}$ and $\mathbf{B} \in \mathbb{C}^{N \times n}$*

$$L^4\left(F^{\mathbf{A}\mathbf{A}^{\mathsf{H}}}, F^{\mathbf{B}\mathbf{B}^{\mathsf{H}}}\right) \leq \frac{2}{N} \operatorname{tr}(\mathbf{A}\mathbf{A}^{\mathsf{H}} + \mathbf{B}\mathbf{B}^{\mathsf{H}}) \frac{1}{N} \operatorname{tr}([\mathbf{A} - \mathbf{B}][\mathbf{A} - \mathbf{B}]^{\mathsf{H}})$$

where $L(F, G)$ is the Lévy distance between F and G, given by:

$$L(F, G) \triangleq \inf\left\{\varepsilon, \forall x \in \mathbb{R}, F(x - \varepsilon) - \varepsilon \leq G(x) \leq F(x + \varepsilon) + \varepsilon\right\}.$$

The Lévy distance can be thought of as the length of the side of the largest square that can fit between the functions F and G. Of importance to us presently is the property that, for a sequence $F_1, F_2, \ldots$ of d.f., $L(F_N, F) \to 0$ implies the weak convergence of F_N to F.

The second lemma is a rank inequality.

Lemma 3.5 (Theorem A.44 in [Bai and Silverstein, 2009]). *For* $\mathbf{A} \in \mathbb{C}^{N \times n}$ *and* $\mathbf{B} \in \mathbb{C}^{N \times n}$

$$\left\| F^{\mathbf{A}\mathbf{A}^{\mathsf{H}}} - F^{\mathbf{B}\mathbf{B}^{\mathsf{H}}} \right\| \leq \frac{1}{N} \operatorname{rank}(\mathbf{A} - \mathbf{B})$$

with $\|f\| \triangleq \sup_x |f(x)|$.

We take the opportunity of the introduction of this rank inequality to mention also the following useful result.

Lemma 3.6 (Lemma 2.21 and Lemma 2.23 in [Tulino and Verdú, 2004]). *For* $\mathbf{A}, \mathbf{B} \in \mathbb{C}^{N \times N}$ *Hermitian*

$$\left\| F^{\mathbf{A}} - F^{\mathbf{B}} \right\| \leq \frac{1}{N} \operatorname{rank}(\mathbf{A} - \mathbf{B}).$$

Also, denoting $\lambda_1^{\mathbf{X}} \leq \ldots \leq \lambda_N^{\mathbf{X}}$ *the ordered eigenvalues of the Hermitian* $\mathbf{X}$

$$\frac{1}{N} \sum_{i=1}^{N} (\lambda_i^{\mathbf{A}} - \lambda_i^{\mathbf{B}})^2 \leq \frac{1}{N} \operatorname{tr}(\mathbf{A} - \mathbf{B})^2.$$

Returning to our original problem, let C be a fixed positive real number. The truncation step consists first of generating a matrix $\hat{\mathbf{X}}$ with entries $\frac{1}{\sqrt{n}} \hat{X}_{N,ij}$ defined as a function of the entries $\frac{1}{\sqrt{N}} X_{N,ij}$ of $\mathbf{X}$ as follows.

$$\hat{X}_{N,ij} = X_{N,ij} 1_{\{|X_{N,ij}| < C\}}(X_{N,ij}).$$

This is the truncation step that cuts off the tail of the distribution of X_{11}. Now, since the distribution of X_{11} is not necessarily centered around its mean, we recenter it as follows. We create a further matrix $\tilde{\mathbf{X}}$ with entries $\frac{1}{\sqrt{N}} \tilde{X}_{N,ij}$ such that

$$\tilde{X}_{N,ij} = \hat{X}_{N,ij} - \mathrm{E}[\hat{X}_{N,ij}]. \tag{3.21}$$

This is the centralization step.

The remaining problem is that the random variable X_{11} has lost through this process some of its weight in the cut tails. So we need to further rescale the resulting variable. For this, we create the variable $\bar{\mathbf{X}}$, with entries $\bar{X}_{N,ij}$ defined as

$$\bar{X}_{N,ij} = \frac{1}{\sigma(C)} \tilde{X}_{N,ij}$$

with $\sigma(C)$ defined as

$$\sigma(C)^2 = \mathrm{E}[|\tilde{X}_{N,ij}|].$$

The idea now is to show that the limiting distribution of $F^{\mathbf{R}_N}$ is the same whether we use the i.i.d. entries $X_{N,ij}$ or their truncated, centered, and rescaled versions $\bar{X}_{N,ij}$. If this is so, it is equivalent to work with the $\bar{X}_{N,ij}$ or with $X_{N,ij}$ in order to derive the Marčenko–Pastur law, with the strong advantage that in the truncation process above no moment assumption was required. Therefore, if we can prove that $F^{\mathbf{R}_N}$ converges to the Marčenko–Pastur law with $X_{N,ij}$ replaced by $\bar{X}_{N,ij}$, then we prove the convergence of $F^{\mathbf{R}_N}$ for all distributions of $X_{N,ij}$ without any moment constraint of higher order than 2. This last result is straightforward as it simply requires to go through every step of the proof of the Marčenko–Pastur law and replace every call to the trace lemma, Theorem 3.4, by the updated trace lemma, Theorem 3.12. The remainder of this section is dedicated to proving that the limiting spectrum of $\mathbf{R}_N$ remains the same if the $X_{N,ij}$ are replaced by $\bar{X}_{N,ij}$.

We have from Lemma 3.4

$$L^4\left(F^{\mathbf{X}\mathbf{X}^{\mathsf{H}}}, F^{\hat{\mathbf{X}}\hat{\mathbf{X}}^{\mathsf{H}}}\right) \leq 2\left[\frac{1}{Nn}\sum_{i,j}|X_{N,ij}|^2 + |\hat{X}_{N,ij}|^2\right]\left[\frac{1}{Nn}\sum_{i,j}|X_{N,ij} - \hat{X}_{N,ij}|^2\right]$$
$$\leq 4\left[\frac{1}{Nn}\sum_{i,j}|X_{N,ij}|^2\right]\left[\frac{1}{Nn}\sum_{i,j}|X_{N,ij}|^2 1_{\{|X_{N,ij}|>C\}}(X_{N,ij})\right].$$

From the law of large numbers, both terms in the right-hand side tend to their means almost surely, i.e.

$$\left(\frac{1}{Nn}\sum_{i,j}|X_{N,ij}|^2\right)\left(\frac{1}{Nn}\sum_{i,j}|X_{N,ij}|^2 1_{\{|X_{N,ij}|>C\}}(X_{N,ij})\right)$$
$$\xrightarrow{\text{a.s.}} \mathrm{E}\left[|X_{N,11}|^2 1_{\{|X_{N,11}|>C\}}(X_{N,ij})\right].$$

Notice that this goes to zero as C grows large (since the second order moment of $X_{N,11}$ exists). Now, from Lemma 3.5 and Equation (3.21), we also have

$$\left\|F^{\hat{\mathbf{X}}\hat{\mathbf{X}}^{\mathsf{H}}} - F^{\tilde{\mathbf{X}}\tilde{\mathbf{X}}^{\mathsf{H}}}\right\| \leq \frac{1}{N}\mathrm{rank}(\mathrm{E}[\hat{\mathbf{X}}]) = \frac{1}{N}.$$

This is a direct consequence of the fact that the entries are i.i.d. and therefore $\mathrm{E}[\hat{X}_{N,ij}] = \mathrm{E}[\hat{X}_{N,11}]$, entailing that the matrix composed of the $\mathrm{E}[\hat{X}_{N,ij}]$ has unit rank. The right-hand side of the inequality goes to zero as N grows large.

Finally, from Lemma 3.4 again

$$L^4\left(F^{\bar{\mathbf{X}}\bar{\mathbf{X}}^{\mathsf{H}}} - F^{\tilde{\mathbf{X}}\tilde{\mathbf{X}}^{\mathsf{H}}}\right) \leq 2\left(\frac{1+\sigma(C)^2}{Nn}\sum_{i,j}|\tilde{X}_{N,ij}|^2\right)\left(\frac{1-\sigma(C)^2}{nN}\sum_{i,j}|\tilde{X}_{N,ij}|^2\right)$$

the right-hand side of which converges to $2(1-\sigma(C)^4)\frac{N}{n}$ almost surely as N grows large. Notice again that $\sigma(C)$ converges to 1 as C grows large.

At this point, we go over the proof of the Marčenko–Pastur law but with $\bar{\mathbf{X}}$ in place of $\mathbf{X}$. The derivations unfold identically but, thanks to Theorem 3.12, we nowhere need any moment assumption further than the existence of the second order moment of the entries of $\mathbf{X}$. We then have that, almost surely

$$F^{\bar{\mathbf{X}}\bar{\mathbf{X}}^{\mathsf{H}}} \Rightarrow F.$$

But since the constant C, which defines $\bar{\mathbf{X}}$, was arbitrary from the very beginning, it can be set as large as we want. For $x \in \mathbb{R}$ and $\varepsilon > 0$, we can take C large enough to ensure that

$$\limsup_N |F^{\mathbf{X}\mathbf{X}^{\mathsf{H}}}(x) - F^{\bar{\mathbf{X}}\bar{\mathbf{X}}^{\mathsf{H}}}(x)| < \varepsilon$$

for a given realization of $\mathbf{X}$. Therefore

$$\limsup_N |F^{\mathbf{X}\mathbf{X}^{\mathsf{H}}}(x) - F(x)| < \varepsilon.$$

Since ε is arbitrary, we finally have

$$F^{\mathbf{X}\mathbf{X}^{\mathsf{H}}} \Rightarrow F$$

this event having probability one. This is our final result.

The proof of the Marčenko–Pastur law, with or without truncation steps, can be applied to a large range of models involving random matrices with i.i.d. entries. The first known extension of the Marčenko–Pastur law concerns the l.s.d. of a certain class of random matrices, which contains in particular the $N \times N$ sample covariance matrices $\mathbf{R}_n$ of the type (1.4), where the vector samples $\mathbf{x}_i \in \mathbb{C}^N$ have covariance matrix $\mathbf{R}$. This is presented in the subsequent section.

3.3 Stieltjes transform for advanced models

We recall that, if N is fixed and n grows large, then we have the almost sure convergence $F^{\mathbf{R}_n} \Rightarrow F^{\mathbf{R}}$ of the e.s.d. of the sample covariance matrix $\mathbf{R}_n \in \mathbb{C}^{N \times N}$ originating from n observations towards the population covariance matrix $\mathbf{R}$. This is a consequence of the law of large numbers in classical probability theory. This is not so if both n and N grow large with limit ratio $n/N \to c$, such that $0 < c < \infty$. In this situation, we have the following result instead.

Theorem 3.13 ([Silverstein and Bai, 1995]). *Let* $\mathbf{B}_N = \mathbf{A}_N + \mathbf{X}_N^{\mathsf{H}} \mathbf{T}_N \mathbf{X}_N$, *where* $\mathbf{X}_N = \left(\frac{1}{\sqrt{n}} X_{N,ij} \right)_{i,j} \in \mathbb{C}^{N \times n}$ *with the* $X_{N,ij}$ *independent with zero mean, unit variance, and finite moment of order* $2 + \varepsilon$ *for some* $\varepsilon > 0$ *(*ε *is independent of* N, i, j*),* $\mathbf{T}_N \in \mathbb{C}^{N \times N}$ *diagonal with real entries and whose e.s.d.* $F^{\mathbf{T}_N}$ *converges weakly and almost surely to* F^T*, and* $\mathbf{A}_N \in \mathbb{C}^{n \times n}$ *Hermitian whose e.s.d.* $F^{\mathbf{T}_N}$ *converges weakly and almost surely to* F^A*,* N/n *tends to* c*, with* $0 < c < \infty$ *as* n*,* N *grow large. Then, the e.s.d. of* $\mathbf{B}_N$ *converges weakly and almost surely to* F^B *such that, for* $z \in \mathbb{C}^+$*,* $m_{F^B}(z)$ *is the unique solution with* *positive imaginary part of*

$$m_{F^B}(z) = m_{F^A}\left(z - c\int \frac{t}{1 + t m_{F^B}(z)} dF^T(t)\right). \tag{3.22}$$

Moreover, if the $\mathbf{X}_N$ *has identically distributed entries, then the result holds without requiring that a moment of order* $2 + \varepsilon$ *exists.*

Remark 3.2. In [Bai and Silverstein, 2009] it is precisely shown that the non-i.i.d. case holds if the random variables $X_{N,ij}$ meet a Lindeberg-like condition [Billingsley, 1995]. The existence of moments of order $2 + \varepsilon$ implies that the $X_{N,ij}$ meet the condition, hence the result. The Lindeberg-like condition states exactly here that, for any $\varepsilon > 0$

$$\frac{1}{N^2}\sum_{ij} \mathrm{E}\left[|X_{N,ij}|^2 \cdot 1_{\{|X_{N,ij}| \geq \varepsilon\sqrt{N}\}}(X_{N,ij})\right] \to 0.$$

These conditions merely impose that the distribution of $X_{N,ij}$, for all pairs (i, j), has light tails, i.e. large values have sufficiently low probability.

In the case of Gaussian $\mathbf{X}_N$, $\mathbf{T}_N$ can be taken Hermitian non-diagonal. This is because the joint distribution of $\mathbf{X}_N$ is in this particular case invariant by right-unitary product, i.e. $\mathbf{X}_N\mathbf{U}_N$ has the same joint entry distribution as $\mathbf{X}_N$ for any unitary matrix $\mathbf{U}_N \in \mathbb{C}^{N\times N}$. Therefore, $\mathbf{T}_N$ can be replaced by any Hermitian matrix $\mathbf{U}_N\mathbf{T}_N\mathbf{U}_N^{\mathsf{H}}$ for $\mathbf{U}_N$ unitary. As previously anticipated for this simple extension of the Marčenko–Pastur law, m_{F^B} does not have a closed-form expression.

The particular case when $\mathbf{A}_N = 0$ is interesting in many respects. In this case, (3.22) becomes

$$m_{\underline{F}}(z) = -\left(z - c\int \frac{t}{1 + t m_{\underline{F}}(z)} dF^T(t)\right)^{-1} \tag{3.23}$$

where we denoted $\underline{F} \triangleq F^B$. This special notation will often be used in Section 7.1 to differentiate the l.s.d. F of the matrix $\mathbf{T}_N^{\frac{1}{2}}\mathbf{X}_N\mathbf{X}_N^{\mathsf{H}}\mathbf{T}_N^{\frac{1}{2}}$ from the l.s.d. $\underline{F}$ of the reversed Gram matrix $\mathbf{X}_N^{\mathsf{H}}\mathbf{T}_N\mathbf{X}_N$. Note indeed that, similar to (3.16), the Stieltjes transform $m_{\underline{F}}$ of the l.s.d. $\underline{F}$ of $\mathbf{X}_N^{\mathsf{H}}\mathbf{T}_N\mathbf{X}_N$ is linked to the Stieltjes transform m_F of the l.s.d. F of $\mathbf{T}_N^{\frac{1}{2}}\mathbf{X}_N\mathbf{X}_N^{\mathsf{H}}\mathbf{T}_N^{\frac{1}{2}}$ through

$$m_{\underline{F}}(z) = c m_F(z) + (c-1)\frac{1}{z} \tag{3.24}$$

and then we also have access to a characterization of F, which is the asymptotic eigenvalue distribution of the *sample covariance matrix* model introduced earlier in (1.4), when the columns $\mathbf{x}'_1, \ldots, \mathbf{x}'_n$ of $\mathbf{X}'_N = \sqrt{n}\mathbf{T}_N^{\frac{1}{2}}\mathbf{X}_N$ form a sequence of independent vectors with zero mean and covariance matrix $\mathrm{E}[\mathbf{x}'_1\mathbf{x}'^{\mathsf{H}}_1] = \mathbf{T}_N$, with $\mathbf{T}_N^{\frac{1}{2}}$ a Hermitian square root of $\mathbf{T}_N$. Note however that, contrary to the strict definition of the sample covariance matrix model, we do not impose identical distributions of the vectors of $\mathbf{X}'_N$ here, but only identically mean and covariance.

In addition to the uniqueness of the pair $(z, m_{\underline{F}}(z))$ in the set $\{z \in \mathbb{C}^+, m_{\underline{F}}(z) \in \mathbb{C}^+\}$ solution of (3.23), an inverse formula for the Stieltjes transform can be written in closed-form, i.e. we can define a function $z_{\underline{F}}(m)$ on $\{\underline{m} \in \mathbb{C}^+, z_{\underline{F}}(\underline{m}) \in \mathbb{C}^+\}$, such that

$$z_{\underline{F}}(\underline{m}) = -\frac{1}{\underline{m}} + c\int \frac{t}{1+t\underline{m}} dF^T(t). \tag{3.25}$$

This will turn out to be extremely useful to characterize the spectrum of F. More on this topic is discussed in Section 7.1 of Chapter 7. From a wireless communication point of view, even if this is yet far from obvious, (3.25) is the essential ingredient to derive in particular (N,n)-consistent estimates of the diagonal entries with large multiplicities of the diagonal matrix $\mathbf{P}$ for the channel matrix models $\mathbf{Y} = \mathbf{P}^{\frac{1}{2}}\mathbf{X} \in \mathbb{C}^{N\times n}$ and also $\mathbf{Y} = \mathbf{H}\mathbf{P}^{\frac{1}{2}}\mathbf{X} + \mathbf{W} \in \mathbb{C}^{N\times n}$. The latter, in which $\mathbf{H}$, $\mathbf{X}$ and $\mathbf{W}$ have independent entries, can be used to model the n sampled vectorial data $\mathbf{Y} = [\mathbf{y}_1, \ldots, \mathbf{y}_n]$ received at an array of N sensors originating from K sources with respective powers $P_1, \ldots, P_K$ gathered in the diagonal entries of $\mathbf{P}$. In this model, $\mathbf{H} \in \mathbb{C}^{N\times K}$, $\mathbf{X} \in \mathbb{C}^{K\times n}$ and $\mathbf{W} \in \mathbb{C}^{N\times n}$ may denote, respectively, the concatenated K channel vectors (in columns), the concatenated K transmit data (in rows), and the additive noise vectors, respectively. It is too early at this stage to provide any insight on the reason why (3.25) is so fundamental here. More on this subject will be successively discussed in Chapter 7 and Chapter 17.

We do not prove Theorem 3.13 in this section, which is a special case of Theorem 6.1 in Chapter 6, for which we will provide an extended sketch of the proof.

Theorem 3.13 was further extended by different authors for matrix models when either $\mathbf{X}_N$ has i.i.d. non-centered elements [Dozier and Silverstein, 2007a], $\mathbf{X}_N$ has a *variance profile*, i.e. with independent entries of different variances, and centered [Girko, 1990] or non-centered entries [Hachem *et al.*, 2007], $\mathbf{X}_N$ is a sum of Gaussian matrices with separable variance profile [Dupuy and Loubaton, 2009], $\mathbf{B}_N$ is a sum of such matrices [Couillet *et al.*, 2011a; Peacock *et al.*, 2008], etc. We will present first the two best known results of the previous list, which have now been largely generalized in the contributions just mentioned. The first result of importance is due to Girko [Girko, 1990] on $\mathbf{X}_N$ matrices with centered i.i.d. entries with a variance profile.

Theorem 3.14. *Let the complex $N \times n$ random matrix $\mathbf{X}_N$ be composed of independent entries $\left(\frac{1}{\sqrt{n}}X_{N,ij}\right)$, such that $X_{N,ij}$ has zero mean, variance $\sigma^2_{N,ij}$, and the $\sigma_{N,ij}X_{N,ij}$ are identically distributed. Further, assume that the $\sigma^2_{N,ij}$ are uniformly bounded. Assume that the $\sigma^2_{N,ij}$ converge, as N grows large, to a bounded limit density $p_{\sigma^2}(x,y)$, $(x,y) \in [0,1)^2$, as $n, N \to \infty$, $n/N \to c$. That is, defining $p_{N,\sigma^2}(x,y)$ as*

$$p_{N,\sigma^2}(x,y) \triangleq \sigma^2_{N,ij}$$

for $\frac{i-1}{N} \leq x < \frac{i}{N}$ *and* $\frac{y-1}{N} \leq y < \frac{j}{N}$, $p_{N,\sigma^2}(x,y) \to p_{\sigma^2}(x,y)$, *as* $N, n \to \infty$. *Then the e.s.d. of* $\mathbf{B}_N = \mathbf{X}_N \mathbf{X}_N^{\mathsf{H}}$ *converges weakly and almost surely to a distribution function* F *whose Stieltjes transform is given by:*

$$m_F(z) = \int_0^1 u(x,z)dx$$

and $u(x,z)$ *satisfies the fixed-point equation*

$$u(x,z) = \left[-z + \int_0^c \frac{p_{\sigma^2}(x,y)dy}{1 + \int_0^1 u(x',z)p_{\sigma^2}(x',y)dx'} \right]^{-1}$$

the solution of which being unique in the class of functions $u(x,z) \geq 0$, *analytic for* $\Im[z] > 0$ *and continuous on the plan section* $\{(x,y), x \in [0,1]\}$.

Girko's proof is however still not well understood.[7] In Chapter 6, a more generic form of Theorem 3.14 and a sketch of the proof are provided. This result is fundamental for the analysis of the capacity of MIMO Gaussian channels with a variance profile. In a single-user setup, the usual Kronecker channel model is a particular case of such a model, referred to as a channel model with separable variance profile. That is, σ_{ij}^2 can be written in this case as a separable product $r_i t_j$ with $r_1, \ldots, r_N$ the eigenvalues of the receive correlation matrix and $t_1, \ldots, t_n$ the eigenvalues of the transmit correlation matrix. However, the requirement that the variance profile $\{\sigma_{N,ij}^2\}$ converges to a limit density $p_{\sigma^2}(x,y)$ for large N is often an unpractical assumption. More useful and more general results, e.g., [Hachem *et al.*, 2007], that do not require the existence of a limit will be discussed in Chapter 6, when introducing the so-called *deterministic equivalents* for $m_{F^{\mathbf{X}_N}}(z)$.

The second result of importance in wireless communications deals with the *information plus noise* models, as follows.

Theorem 3.15 ([Dozier and Silverstein, 2007a]). *Let* $\mathbf{X}_N$ *be* $N \times n$ *with i.i.d. entries of zero mean and unit variance,* $\mathbf{A}_N$ *be* $N \times n$ *independent of* $\mathbf{X}_N$ *such that* $F^{\frac{1}{n}\mathbf{A}_N\mathbf{A}_N^{\mathsf{H}}} \Rightarrow H$ *almost surely. Let also* σ *be a positive integer and denote*

$$\mathbf{B}_N = \frac{1}{n}(\mathbf{A}_N + \sigma\mathbf{X}_N)(\mathbf{A}_N + \sigma\mathbf{X}_N)^{\mathsf{H}}.$$

Then, for $n, N \to \infty$ *with* $n/N \to c > 0$, $F^{\mathbf{B}_N} \Rightarrow F$ *almost surely, where* F *is a non-random distribution function whose Stieltjes transform* $m_F(z)$, *for* $z \in \mathbb{C}^+$, *is the unique solution of*

$$m_F(z) = \int \frac{dH(t)}{\frac{t}{1+\sigma^2 c m_F(z)} - (1 + \sigma^2 c m_F(z))z + \sigma^2(1-c)}$$

[7] As Bai puts it [Bai and Silverstein, 2009]: "his proofs have puzzled many who attempt to understand, without success, Girko's arguments."

such that $m_F(z) \in \mathbb{C}^+$ *and* $zm_F(z) \in \mathbb{C}^+$.

It is rather clear why the model $\mathbf{B}_N$ is referred to as *information plus noise.* In practical applications, this result is used in various contexts, such as the evaluation of the MIMO capacity with imperfect channel state information [Vallet and Loubaton, 2009], or the capacity of MIMO channels with line-of-sight components [Hachem *et al.*, 2007]. Both results were generalized in [Hachem *et al.*, 2007] for the case of non-centered matrices with i.i.d. entries with a variance profile, of particular appeal in wireless communications since it models completely the so-called *Rician* channels, i.e. non-centered Rayleigh fading channels with a variance profile. The works [Couillet *et al.*, 2011a; Dupuy and Loubaton, 2009; Hachem *et al.*, 2007] will be further discussed in Chapters 13–14 as they provide asymptotic expressions of the capacity of very general wireless models in MIMO point-to-point channels, MIMO frequency selective Rayleigh fading channels, as well as the rate regions of multiple access channels and broadcast channels. The technical tools required to prove the latter no longer rely on the Marčenko–Pastur approach, although the latter can be used to provide an insight on the expected results. The main limitations of the Marčenko–Pastur approach will be evidenced when proving one of the aforementioned results, namely the result of [Couillet *et al.*, 2011a], in Chapter 6.

Before introducing an important central limit theorem, we make a small digression about the Tonelli theorem, also known as Fubini theorem. This result is of interest when we want to extend results that are known to hold for matrix models involving *deterministic matrices* converging weakly to some l.s.d. to the case when those matrices are now random, converging almost surely to some l.s.d.

3.4 Tonelli theorem

The Tonelli theorem for probability spaces can be stated as follows.

Theorem 3.16 (Theorem 18.3 in [Billingsley, 1995]). *If* $(\Omega, \mathcal{F}, P)$ *and* $(\Omega', \mathcal{F}', P')$ *are two probability spaces, then for* f *an integrable function with respect to the product measure* Q *on* $\mathcal{F} \times \mathcal{F}'$

$$\int_{\Omega\times\Omega'} f(x,y)Q(d(x,y)) = \int_{\Omega}\left[\int_{\Omega'} f(x,y)P'(dy)\right]P(dx)$$

and

$$\int_{\Omega\times\Omega'} f(x,y)Q(d(x,y)) = \int_{\Omega'}\left[\int_{\Omega} f(x,y)P(dy)\right]P'(dx).$$

Moreover, the existence of one of the right-hand side values ensures the integrability of f *with respect to the product measure* Q.

As an application, we mention that Theorem 3.13 was originally stated in [Silverstein and Bai, 1995] for deterministic $\mathbf{T}_N$ matrices with l.s.d. F^T (and, as a matter of fact, under the i.i.d. assumption for the entries of $\mathbf{X}_N$). In Theorem 3.13, though, we mentioned that $F^{\mathbf{T}_N}$ is random and converges to F^T only in the almost sure sense. In the following, assuming only the result from [Silverstein and Bai, 1995] is known, i.e. denoting $\mathbf{B}_N = \mathbf{A}_N + \mathbf{X}_N^{\mathsf{H}}\mathbf{T}_N\mathbf{X}_N$, $F^{\mathbf{B}_N}$ converges almost surely to F for $\mathbf{T}_N$ deterministic having l.s.d. F^T, we wish to prove the more general Theorem 3.13 in the i.i.d. case, i.e. $F^{\mathbf{B}_N}$ converges almost surely to F for $\mathbf{T}_N$ *random* having *almost sure* l.s.d. F^T. This result is exactly due to Theorem 3.16.

Indeed, call $(X, \mathcal{X}, P_X)$ the probability space that engenders the series $\mathbf{X}_1, \mathbf{X}_2, \ldots$, of i.i.d. matrices and $(T, \mathcal{T}, P_T)$ the probability space that engenders the random series $\mathbf{T}_1, \mathbf{T}_2, \ldots$. Further denote $(X \times T, \mathcal{X} \times \mathcal{T}, P_{X\times T})$ the product space. Since $\mathbf{B}_N$ is determined by $\mathbf{X}_N$ and $\mathbf{T}_N$ (assume $\mathbf{A}_N$ deterministic), we can write every possible sequence $\mathbf{B}_1, \mathbf{B}_2, \ldots = \mathbf{B}_1(x,t), \mathbf{B}_2(x,t), \ldots$ for some $(x,t) \in X \times T$. Therefore, what we need to prove here is that the space

$$A \triangleq \left\{(x,t) \in X \times T,\ F^{\mathbf{B}_N(x,t)} \text{ converges to } F^B\right\}$$

has probability one in the product space $X \times T$. From the Tonelli theorem, it follows that

$$\begin{aligned} P(A) &= \int_A P_{X\times T}(d(x,t)) \\ &= \int_{X\times T} 1_A(x,t) P_{X\times T}(d(x,t)) \\ &= \int_T \left[\int_X 1_A(x,t) P_X(dx)\right] P_T(dt). \end{aligned}$$

Let $t_0 \in T$ be a realization of $\mathbf{T}_1, \mathbf{T}_2, \ldots = \mathbf{T}_1(t_0), \mathbf{T}_2(t_0), \ldots$ such that $F^{\mathbf{T}_N(t_0)}$ converges to F^T. Since t_0 is such that $F^{\mathbf{T}_N(t_0)}$ converges to F^T, we can apply Theorem 3.13. Namely, the set of x such that $(x,t_0) \in A$ has measure one, and therefore, for this t_0

$$\int_X 1_A(x,t_0) P_X(dx) = 1.$$

Call $B \subset T$ the space of all t such that $F^{\mathbf{T}_N(t)}$ converges weakly to F^T. We then have

$$P(A) = \int_B P_T(dt) + \int_{T\setminus B}\left[\int_X 1_A(x,t) P_X(dx)\right] P_T(dt).$$

Since $F^{\mathbf{T}_N}$ converges to F^T with probability one, $P_T(B) = 1$, and therefore $P(A) \geq 1$, which is what we needed to show.

This trick is interesting as we can quite often reuse existing results stated for deterministic models, and extend them to stochastic models with almost sure statements. It will be further reused in particular to provide a generalization of the trace lemma, Theorem 3.4, and the rank-1 perturbation lemma, Theorem 3.9, in the case when the matrices involved in both results do not have bounded spectral norm but only almost sure bounded spectral norm for all large dimensions. These generalizations are required to study zero-forcing precoders in multiple input single output broadcast channels, see Section 14.1.

This closes this parenthesis on the Tonelli theorem. We return now to further considerations of asymptotic laws of large dimensional matrices, and to the study of (central) limit theorems.

3.5 Central limit theorems

Due to the intimate relation between the Stieltjes and Shannon transforms (3.5), it is now obvious that the capacity of large dimensional communication channels can be approximated using deterministic limits of the Stieltjes transform. For finite dimensional systems, this however only provides a rather rough approximation of quasi-static channel capacities or alternatively a rather accurate approximation of ergodic channel capacities. No information about outage capacities is accessible to this point, since the variations of the deterministic limit F of the Stieltjes transform $F^{\mathbf{X}_N}$ of some matrix $\mathbf{X}_N \in \mathbb{C}^{N\times N}$ under study are unknown. To this end, we need to study more precisely the fluctuations of the random quantity

$$r_N \left[m_{F^{\mathbf{X}_N}}(z) - m_F(z) \right]$$

for some rate r_N, increasing with N. For $\mathbf{X}_N$ a sample covariance matrix, it turns out that under some further assumptions on the moments of the distribution of the random entries of $\mathbf{X}_N$, the random variable $N \left[m_{F^{\mathbf{X}_N}}(z) - m_F(z) \right]$ has a central limit. This central limit generalizes to any well-behaved functional of $\mathbf{X}_N$.

The first central limit result for functionals of large dimensional random matrices is due to Bai and Silverstein for the covariance matrix model, as follows.

Theorem 3.17 ([Bai and Silverstein, 2004]). *Let $\mathbf{X}_N = \left(\frac{1}{\sqrt{n}} X_{N,ij} \right)_{ij} \in \mathbb{C}^{N\times n}$ have i.i.d. entries, such that $X_{N,11}$ has zero mean, unit variance, and finite fourth order moment. Let $\mathbf{T}_N \in \mathbb{C}^{N\times N}$ be non-random Hermitian non-negative definite with uniformly bounded spectral norm (with respect to N) for which we assume that $F^{\mathbf{T}_N} \Rightarrow H$, as $N \to \infty$. We denote $\tau_1 \geq \ldots \geq \tau_N$ the eigenvalues of $\mathbf{T}_N$. Consider the random matrix*

$$\mathbf{B}_N = \mathbf{T}_N^{\frac{1}{2}} \mathbf{X}_N \mathbf{X}_N^{\mathsf{H}} \mathbf{T}_N^{\frac{1}{2}}$$

as well as

$$\underline{\mathbf{B}}_N = \mathbf{X}_N^{\mathsf{H}} \mathbf{T}_N \mathbf{X}_N.$$

We know from Theorem 3.13 that $F^{\mathbf{B}_N} \Rightarrow F$ for some distribution function F, as $N, n \to \infty$ with limit ratio $c = \lim_N N/n$. Denoting F_N this limit distribution if the series $F^{\mathbf{T}_1}, F^{\mathbf{T}_2}, \ldots$ were to converge to $H = F^{\mathbf{T}_N}$, let

$$G_N \triangleq N\left[F^{\mathbf{B}_N} - F_N\right].$$

Consider now k functions $f_1, \ldots, f_k$ defined on $\mathbb{R}$ that are analytic on the segment

$$\left[\liminf_N \tau_N 1_{(0,1)}(c)(1-\sqrt{c})^2, \limsup_n \tau_1 1_{(0,1)}(c)(1+\sqrt{c})^2\right].$$

Then, if (i) $X_{N,11}$ is real, $\mathbf{T}_N$ is real and $\mathrm{E}[(X_{N,11})^4] = 3$, or (ii) if $X_{N,11}$ is complex, $\mathrm{E}[(X_{N,11})^2] = 0$ and $\mathrm{E}[(X_{N,11})^4] = 2$, then the random vector

$$\left(\int f_1(x)dG_N(x), \ldots, \int f_k(x)dG_N(x)\right)$$

converges weakly to a Gaussian vector $(X_{f_1}, \ldots, X_{f_k})$ with means $(\mathrm{E}[X_{f_1}], \ldots, \mathrm{E}[X_{f_k}])$ and covariance matrix $\mathrm{Cov}(X_f, X_g)$, $(f,g) \in \{f_1, \ldots, f_k\}^2$, such that, in case (i)

$$\mathrm{E}[X_f] = -\frac{1}{2\pi i}\oint f(z) \frac{c\int \underline{m}(z)^3 t^2 (1+t\underline{m})^{-3} dH(t)}{(1 - c\int \underline{m}(z)^2 t^2 (1+t\underline{m}(z))^{-2} dH(t))^2} dz$$

and

$$\mathrm{Cov}(X_f, X_g) = -\frac{1}{2\pi i}\oint\oint \frac{f(z_1)g(z_2)}{(\underline{m}(z_1) - \underline{m}(z_2))^2} \underline{m}'(z_1)\underline{m}'(z_2) dz_1 dz_2$$

while in case (ii) $\mathrm{E}[X_f] = 0$ and

$$\mathrm{Cov}(X_f, X_g) = -\frac{1}{4\pi i}\oint\oint \frac{f(z_1)g(z_2)}{(\underline{m}(z_1) - \underline{m}(z_2))^2} \underline{m}'(z_1)\underline{m}'(z_2) dz_1 dz_2 \tag{3.26}$$

for any couple $(f,g) \in \{f_1, \ldots, f_k\}^2$, and for $\underline{m}(z)$ the Stieltjes transform of the l.s.d. of $\underline{\mathbf{B}}_N$. The integration contours are positively defined with winding number one and enclose the support of F.

The mean and covariance expressions of Theorem 3.17 are not easily exploitable except for some specific $f_1, \ldots, f_k$ functions for which the integrals (or an expression of the covariances) can be explicitly computed. For instance, when $f_1, \ldots, f_k$ are taken to be $f_i(x) = x^i$ and $\mathbf{T}_N = \mathbf{I}_N$, Theorem 3.17 gives the central limit of the joint moments of the distribution $N[F^{\mathbf{B}_N} - F]$ with F the Marčenko–Pastur law.

Corollary 3.3 ([Bai and Silverstein, 2004]). *Under the conditions of Theorem 3.17 with $H = 1_{[1,\infty)}$, in case (ii), denote $\mathbf{v}_N$ the vector of jth entry*

$$(v_N)_j = \mathrm{tr}(\mathbf{B}_N^j) - NM_j$$

where M_j is the limiting jth *order moment of the Marčenko–Pastur law, as $N \to \infty$. Then, as $N, n \to \infty$ with limit ratio $N/n \to c$*

$$\mathbf{v}_N \Rightarrow \mathbf{v} \sim \mathcal{CN}(0, \mathbf{Q})$$

where the (i,j)th *entry of* $\mathbf{Q}$ *is*

$$Q_{i,j} = c^{i+j} \sum_{k_1=0}^{i-1} \sum_{k_2=0}^{j} \binom{i}{k_1} \binom{j}{k_2} \left(\frac{1-c}{c}\right)^{k_1+k_2} \times \sum_{l=1}^{i-k_1} l \binom{2i-1-k_1-l}{i-1} \binom{2j-1-k_2+l}{j-1}.$$

We will see in Chapter 5 that, in the Gaussian case, combinatorial moment-based approaches can also be used to derive the above result in a much faster way. In case (ii), for Gaussian $\mathbf{X}_N$, the first coefficients of $\mathbf{Q}$ can be alternatively evaluated as a function of the limiting *free cumulants* $C_1, C_2, \ldots$ of $\mathbf{B}_N$ [Rao *et al.*, 2008], which will be introduced in Section 5.2. Explicitly, we have:

$$\begin{aligned} Q_{11} &= C_2 - C_1^2 \\ Q_{21} &= -4C_1C_2 + 2C_1^3 + 2C_3 \\ Q_{22} &= 16C_1^2C_2 - 6C_2^2 - 6C_1^4 - 8C_1C_3 + 4C_4 \end{aligned} \tag{3.27}$$

with C_k defined as a function of the moments $M_k \triangleq \lim_N \frac{1}{N} \operatorname{tr} \mathbf{B}_N^k$ through Equation (5.3).

Obtaining central limits requires somewhat elaborate tools, involving the theory of martingales in particular, see Section 35 of [Billingsley, 1995]. An important result for wireless communications, for which we will provide a sketch of the proof using martingales, is the central limit for the log determinant of Wishart matrices. In its full form, this is:

Theorem 3.18. *Let $\mathbf{X}_N \in \mathbb{C}^{N\times n}$ have i.i.d. entries $\left(\frac{1}{\sqrt{n}} X_{N,ij}\right)$, such that $X_{N,11}$ has zero mean, unit variance, and finite fourth order moment. Denote $\mathbf{B}_N = \mathbf{X}_N\mathbf{X}_N^{\mathsf{H}} \in \mathbb{C}^{N\times N}$. We know that, as $N, n \to \infty$ with $N/n \to c$, $F^{\mathbf{B}_N} \Rightarrow F$ almost surely with F the Marčenko–Pastur law with ratio c. Then the Shannon transform $\mathcal{V}_{\mathbf{B}_N}(x) \triangleq \int \log(1+xt) dF^{\mathbf{B}_N}(t)$ of $\mathbf{B}_N$ satisfies*

$$N\left(\mathcal{V}_{\mathbf{B}_N}(x) - \mathrm{E}\left[\mathcal{V}_{\mathbf{B}_N}(x)\right]\right) \Rightarrow X \sim \mathcal{N}(0, \Theta^2)$$

with

$$\Theta^2 = -\log\left(1 - \frac{c m_F(-1/x)}{(1 + c m_F(-1/x))^2}\right) + \kappa \frac{c m_F(-1/x)}{(1 + c m_F(-1/x))^2}$$

$\kappa = \mathrm{E}(X_{N,11})^4 - 2$ and $m_F(z)$ the Stieltjes transform of the Marčenko–Pastur law (3.20).

Note that in the case where $X_{N,11}$ is Gaussian, $\kappa = 0$.

We hereafter provide a sketch of the proof of Theorem 3.18, along with a short introduction to martingale theory.

Proof. Martingale theory requires notions of conditional probability.

Definition 3.7 ((33.8) of [Billingsley, 1995])**.** Let $(\Omega, \mathcal{F}, P)$ be a probability space, and X be a random variable on this probability space. Let $\mathcal{G}$ be a σ-field in $\mathcal{F}$. For $A \in \mathcal{F}$, we denote $P[A\|\mathcal{G}]$ the random variable with realizations $P[A\|\mathcal{G}]_\omega$, $\omega \in \Omega$, which is measurable $\mathcal{G}$, integrable, and such that, for all $G \in \mathcal{G}$

$$\int_G P[A\|\mathcal{G}]dP = P(A \cap G).$$

If $\mathcal{G}$ is the finite set of the unions of the $\mathcal{F}$-sets $B_1, \ldots, B_K$, then, for $\omega \in B_i$, $P(B_i)P[A\|\mathcal{G}]_\omega = P(A \cap B_i)$, i.e. $P[A\|\mathcal{G}]_\omega = P(A|B_i)$, consistently with the usual definition of conditional probability. We can therefore see $P[A\|\mathcal{G}]$ as the probability of the event A when the result of the experiment $\mathcal{G}$ is known. As such, the σ-field $\mathcal{G} \subset \mathcal{F}$ can be seen as an information *filter*, in the sense that it brings a rougher vision of the events of Ω. Of interest to us is the extension of this information filtering under the form of a so-called *filtration* $\mathcal{F}_1 \subset \mathcal{F}_2 \subset \ldots$ for a given sequence of σ-field $\mathcal{F}_1, \mathcal{F}_2, \ldots \subset \mathcal{F}$.

Definition 3.8. Consider a probability space $(\Omega, \mathcal{F}, P)$ and a filtration $\mathcal{F}_1 \subset \mathcal{F}_2 \subset \ldots$ with, for each i, $\mathcal{F}_i \subset \mathcal{F}$. Define $X_1, X_2, \ldots$ a sequence of random variables such that X_i is measurable $\mathcal{F}_i$ and integrable. Then $X_1, X_2, \ldots$ is a *martingale* with respect to $\mathcal{F}_1, \mathcal{F}_2, \ldots$ if, with probability one

$$\mathrm{E}\left[X_{n+1}\|\mathcal{F}_n\right] = X_n$$

where $\mathrm{E}[X\|\mathcal{G}]$ is the conditional expectation defined, for $G \in \mathcal{G}$ as

$$\int_G \mathrm{E}[X\|\mathcal{G}]dP = \int_G XdP.$$

If $X_1, X_2, \ldots$ and $X_1', X_2', \ldots$ are both martingales with respect to $\mathcal{F}_1, \mathcal{F}_2, \ldots$, then $X_1 - X_1', X_2 - X_2', \ldots$ is a *martingale difference* relative to $\mathcal{F}_1, \mathcal{F}_2, \ldots$. More generally, we have the following definition.

Definition 3.9. If $Z_1, Z_2, \ldots$ is such that Z_i is measurable $\mathcal{F}_i$ and integrable, for a filtration $\mathcal{F}_1, \mathcal{F}_2, \ldots$, and satisfies

$$\mathrm{E}[Z_{n+1}\|\mathcal{F}_n] = 0$$

then $Z_1, Z_2, \ldots$ is a *martingale difference* with respect to $\mathcal{F}_1, \mathcal{F}_2, \ldots$.

Informally, this implies that, when the experiment $\mathcal{F}_n$ is known before experiment $\mathcal{F}_{n+1}$ has been realized, the expected observation X_{n+1} at time $n+1$ is exactly equal to the observation X_n at time n. Note that, by taking $\mathcal{F} = \mathcal{F}_N = \mathcal{F}_{N+1} = \ldots$, the filtration can be limited to a finite sequence.

The link to random matrix theory is the following: consider a random matrix $\mathbf{X}_N \in \mathbb{C}^{N \times n}$ with independent columns $\mathbf{x}_1, \ldots, \mathbf{x}_n \in \mathbb{C}^N$. The columns

$\mathbf{x}_k(\omega), \ldots, \mathbf{x}_n(\omega)$ of $\mathbf{X}_N(\omega)$, $\omega \in \Omega$, can be thought of as the result of the experiment $\mathcal{F}_{n-k+1}$ that consists in unveiling successive matrix columns from the last to the first: $\mathcal{F}_1$ unveils $\mathbf{x}_n(\omega)$, $\mathcal{F}_2$ unveils $\mathbf{x}_{n-1}(\omega)$, etc. This results in a filtration $\mathcal{F}_0 \subset \mathcal{F}_1 \subset \ldots \subset \mathcal{F}_n$, with $\mathcal{F}_0 = \{\emptyset, \Omega\}$. The reason why considering martingales will be helpful to the current proof takes the form of Theorem 3.19, introduced later, which provides a central limit result for sums of martingale differences. Back to the hypotheses of Theorem 3.18, consider the above filtration built from the column space of $\mathbf{X}_N$ and denote

$$\alpha_{n,j} \triangleq \mathrm{E}_{n-j+1}\left[\log\det\left(\mathbf{I}_N + x\mathbf{X}_N\mathbf{X}_N^{\mathsf{H}}\right)\right] - \mathrm{E}_{n-j}\left[\log\det\left(\mathbf{I}_N + x\mathbf{X}_N\mathbf{X}_N^{\mathsf{H}}\right)\right]$$

where $\mathrm{E}_{n-j+1}[X]$ stands for $\mathrm{E}[X \| \mathcal{F}_{n-j+1}]$ for X measurable $\mathcal{F}_{n-j+1}$. For all j, the $\alpha_{n,j}$ satisfy

$$\mathrm{E}_{n-j+1}\left[\alpha_{n,j}\right] = 0$$

and therefore $\alpha_{n,n}, \ldots, \alpha_{1,n}$ is a martingale difference relative to the filtration $\mathcal{F}_1 \subset \ldots \subset \mathcal{F}_n$. Note now that the variable of interest, namely

$$\beta_n \triangleq \log\det\left(\mathbf{I}_N + x\mathbf{X}_N\mathbf{X}_N^{\mathsf{H}}\right) - \mathrm{E}\left[\log\det\left(\mathbf{I}_N + x\mathbf{X}_N\mathbf{X}_N^{\mathsf{H}}\right)\right]$$

satisfies

$$\beta_n \triangleq \sum_{j=1}^{n} \left(\alpha_{n,j} - \alpha_{n,j+1}\right).$$

Denoting $\mathbf{X}_{(j)} \triangleq [\mathbf{x}_1, \ldots, \mathbf{x}_{j-1}, \mathbf{x}_{j+1}, \ldots, \mathbf{x}_n]$, notice that

$$\mathrm{E}_{n-j+1}\left[\log\det\left(\mathbf{I}_N + x\mathbf{X}_{(j)}\mathbf{X}_{(j)}^{\mathsf{H}}\right)\right] = \mathrm{E}_{n-j}\left[\log\det\left(\mathbf{I}_N + x\mathbf{X}_{(j)}\mathbf{X}_{(j)}^{\mathsf{H}}\right)\right]$$

and therefore, adding and subtracting $\mathrm{E}_{n-j+1}\left[\log\det\left(\mathbf{I}_N + x\mathbf{X}_{(j)}\mathbf{X}_{(j)}^{\mathsf{H}}\right)\right]$

$$\begin{aligned}
\beta_n &= \mathrm{E}_{n-j}\frac{\log\det\left(\mathbf{I}_N + x\mathbf{X}_{(j)}\mathbf{X}_{(j)}^{\mathsf{H}}\right)}{\log\det\left(\mathbf{I}_N + x\mathbf{X}_N\mathbf{X}_N^{\mathsf{H}}\right)} - \sum_{j=1}^{n}\mathrm{E}_{n-j+1}\frac{\log\det\left(\mathbf{I}_N + x\mathbf{X}_{(j)}\mathbf{X}_{(j)}^{\mathsf{H}}\right)}{\log\det\left(\mathbf{I}_N + x\mathbf{X}_N\mathbf{X}_N^{\mathsf{H}}\right)} \\
&= \mathrm{E}_{n-j}\frac{\log\det\left(\mathbf{I}_{n-1} + x\mathbf{X}_{(j)}^{\mathsf{H}}\mathbf{X}_{(j)}\right)}{\log\det\left(\mathbf{I}_n + x\mathbf{X}_N^{\mathsf{H}}\mathbf{X}_N\right)} - \sum_{j=1}^{n}\mathrm{E}_{n-j+1}\frac{\log\det\left(\mathbf{I}_{n-1} + x\mathbf{X}_{(j)}^{\mathsf{H}}\mathbf{X}_{(j)}\right)}{\log\det\left(\mathbf{I}_n + x\mathbf{X}_N^{\mathsf{H}}\mathbf{X}_N\right)} \\
&= \sum_{j=1}^{n}\mathrm{E}_{n-j+1}\log\left(1 + \mathbf{x}_j^{\mathsf{H}}\left(\mathbf{X}_{(j)}\mathbf{X}_{(j)}^{\mathsf{H}} + \frac{1}{x}\mathbf{I}_N\right)^{-1}\mathbf{x}_j\right) \\
&\quad - \mathrm{E}_{n-j}\log\left(1 + \mathbf{x}_j^{\mathsf{H}}\left(\mathbf{X}_{(j)}\mathbf{X}_{(j)}^{\mathsf{H}} + \frac{1}{x}\mathbf{I}_N\right)^{-1}\mathbf{x}_j\right)
\end{aligned}$$

where the last equality comes first from the expression of a matrix inverse as a function of determinant and cofactors

$$\left[\left(\mathbf{X}_N^{\mathsf{H}}\mathbf{X}_N + \frac{1}{x}\mathbf{I}_n\right)^{-1}\right]_{jj} = \frac{\det\left(\mathbf{X}_{(j)}^{\mathsf{H}}\mathbf{X}_{(j)} + \frac{1}{x}\mathbf{I}_{n-1}\right)}{\det\left(\mathbf{X}_N^{\mathsf{H}}\mathbf{X}_N + \frac{1}{x}\mathbf{I}_n\right)}$$

and second by remembering from (3.11) that

$$\left[\left(\mathbf{X}_N^{\mathsf{H}}\mathbf{X}_N + \frac{1}{x}\mathbf{I}_n\right)^{-1}\right]_{jj} = \frac{x}{1+\mathbf{x}_j^{\mathsf{H}}(\mathbf{X}_{(j)}\mathbf{X}_{(j)}^{\mathsf{H}} + \frac{1}{x}\mathbf{I}_N)^{-1}\mathbf{x}_j}.$$

Using now the fact that

$$\begin{aligned}
&\mathrm{E}_{n-j+1}\log\left(1+\frac{1}{n}\operatorname{tr}\left(\mathbf{X}_{(j)}\mathbf{X}_{(j)}^{\mathsf{H}}+\frac{1}{x}\mathbf{I}_N\right)^{-1}\right)\\
&=\mathrm{E}_{n-j}\log\left(1+\frac{1}{n}\operatorname{tr}\left(\mathbf{X}_{(j)}\mathbf{X}_{(j)}^{\mathsf{H}}+\frac{1}{x}\mathbf{I}_N\right)^{-1}\right)
\end{aligned}$$

then adding and subtracting

$$\log(1) = \log\left(\frac{1+\frac{1}{n}\operatorname{tr}\left(\mathbf{X}_{(j)}\mathbf{X}_{(j)}^{\mathsf{H}}+\frac{1}{x}\mathbf{I}_N\right)^{-1}}{1+\frac{1}{n}\operatorname{tr}\left(\mathbf{X}_{(j)}\mathbf{X}_{(j)}^{\mathsf{H}}+\frac{1}{x}\mathbf{I}_N\right)^{-1}}\right)$$

this is further equal to

$$\begin{aligned}
\beta_n &= \sum_{j=1}^n \mathrm{E}_{n-j+1}\log\left(\frac{1+\mathbf{x}_j^{\mathsf{H}}\left(\mathbf{X}_{(j)}\mathbf{X}_{(j)}^{\mathsf{H}}+\frac{1}{x}\mathbf{I}_N\right)^{-1}\mathbf{x}_j}{1+\frac{1}{n}\operatorname{tr}\left(\mathbf{X}_{(j)}\mathbf{X}_{(j)}^{\mathsf{H}}+\frac{1}{x}\mathbf{I}_N\right)^{-1}}\right)\\
&\quad - \mathrm{E}_{n-j}\log\left(\frac{1+\mathbf{x}_j^{\mathsf{H}}\left(\mathbf{X}_{(j)}\mathbf{X}_{(j)}^{\mathsf{H}}+\frac{1}{x}\mathbf{I}_N\right)^{-1}\mathbf{x}_j}{1+\frac{1}{n}\operatorname{tr}\left(\mathbf{X}_{(j)}\mathbf{X}_{(j)}^{\mathsf{H}}+\frac{1}{x}\mathbf{I}_N\right)^{-1}}\right)\\
&= \sum_{j=1}^n \mathrm{E}_{n-j}^{n-j+1}\log\left(\frac{\mathbf{x}_j^{\mathsf{H}}\left(\mathbf{X}_{(j)}\mathbf{X}_{(j)}^{\mathsf{H}}+\frac{1}{x}\mathbf{I}_N\right)^{-1}\mathbf{x}_j-\frac{1}{n}\operatorname{tr}\left(\mathbf{X}_{(j)}\mathbf{X}_{(j)}^{\mathsf{H}}+\frac{1}{x}\mathbf{I}_N\right)^{-1}}{1+\frac{1}{n}\operatorname{tr}\left(\mathbf{X}_{(j)}\mathbf{X}_{(j)}^{\mathsf{H}}+\frac{1}{x}\mathbf{I}_N\right)^{-1}}\right)\\
&= \sum_{j=1}^n \gamma_j
\end{aligned}$$

with $\mathrm{E}_{n-j}^{n-j+1}X = \mathrm{E}_{n-j+1}X - \mathrm{E}_{n-j}X$, and with

$$\begin{aligned}
\gamma_j &= \mathrm{E}_{n-j}^{n-j+1}\log\left(1+A_j\right)\\
&\triangleq \mathrm{E}_{n-j}^{n-j+1}\log\left(\frac{\mathbf{x}_j^{\mathsf{H}}\left(\mathbf{X}_{(j)}\mathbf{X}_{(j)}^{\mathsf{H}}+\frac{1}{x}\mathbf{I}_N\right)^{-1}\mathbf{x}_j-\frac{1}{n}\operatorname{tr}\left(\mathbf{X}_{(j)}\mathbf{X}_{(j)}^{\mathsf{H}}+\frac{1}{x}\mathbf{I}_N\right)^{-1}}{1+\frac{1}{n}\operatorname{tr}\left(\mathbf{X}_{(j)}\mathbf{X}_{(j)}^{\mathsf{H}}+\frac{1}{x}\mathbf{I}_N\right)^{-1}}\right).
\end{aligned}$$

The sequence $\gamma_n,\ldots,\gamma_1$ is still a martingale difference with respect to the filtration $\mathcal{F}_1 \subset \ldots \subset \mathcal{F}_n$.

We now introduce the fundamental theorem, which justifies the previous steps that led to express β_n as a sum of martingale differences.

Theorem 3.19 (Theorem 35.12 of [Billingsley, 1995]). *Let $\delta_1, \ldots, \delta_n$ be a sequence of martingale differences with respect to the filtration $\mathcal{F}_1 \subset \ldots \subset \mathcal{F}_n$. Assume there exists Θ^2 such that*

$$\sum_{j=1}^{n} \mathrm{E}\left[\delta_j^2 \| \mathcal{F}_j\right] \to \Theta^2$$

in probability, as $n \to \infty$. Moreover assume that, for all $\varepsilon > 0$

$$\sum_{j=1}^{n} \mathrm{E}\left[\delta_j^2 1_{\{|\delta_j| \geq \varepsilon\}}\right] \to 0.$$

Then

$$\sum_{j=1}^{n} \delta_j \Rightarrow X \sim \mathcal{N}(0, \Theta^2).$$

It therefore suffices here to prove that the Lindeberg-like condition is satisfied and to determine Θ^2. For simplicity, we only carry out this second step here. Noticing that A_j is close to zero by the trace lemma, Theorem 3.4

$$\begin{aligned}\sum_{j=1}^{n} \mathrm{E}_{n-j}\gamma_j^2 &\simeq \sum_{j=1}^{n} \mathrm{E}_{n-j}\left(\mathrm{E}_{n-j+1}A_j - \mathrm{E}_{n-j}A_j\right)^2 \\ &\simeq \sum_{j=1}^{n} \mathrm{E}_{n-j}\left(\mathrm{E}_{n-j+1}A_j\right)^2\end{aligned}$$

where we use the symbol '$\simeq$' to denote that the difference in the terms on either side tends to zero for large N and where

$$\begin{aligned}&\mathrm{E}_{n-j}\left(\mathrm{E}_{n-j+1}A_j\right)^2 \\ &\simeq \frac{\mathrm{E}_{n-j}\left[\mathbf{x}_j^{\mathsf{H}}\left(\mathbf{X}_N\mathbf{X}_N^{\mathsf{H}} + \frac{1}{x}\mathbf{I}_N\right)^{-1}\mathbf{x}_j - \frac{1}{n}\operatorname{tr}\left(\mathbf{X}_{(j)}\mathbf{X}_{(j)}^{\mathsf{H}} + \frac{1}{x}\mathbf{I}_N\right)^{-1}\right]^2}{\left(1 + \frac{1}{n}\operatorname{tr}\left(\mathbf{X}_N\mathbf{X}_N^{\mathsf{H}} + \frac{1}{x}\mathbf{I}_N\right)^{-1}\right)^2}.\end{aligned}$$

Further calculus shows that the term in the numerator expands as

$$\begin{aligned}&\mathrm{E}_{n-j}\left(\mathbf{x}_j^{\mathsf{H}}\left(\mathbf{X}_N\mathbf{X}_N^{\mathsf{H}} + \frac{1}{x}\mathbf{I}_N\right)^{-1}\mathbf{x}_j - \frac{1}{n}\operatorname{tr}\left(\mathbf{X}_{(j)}\mathbf{X}_{(j)}^{\mathsf{H}} + \frac{1}{x}\mathbf{I}_N\right)^{-1}\right)^2 \\ &\simeq \frac{1}{n^2}\left(\operatorname{tr}\left(\mathrm{E}_{n-j}\mathbf{X}_{(j)}\mathbf{X}_{(j)}^{\mathsf{H}} + \frac{1}{x}\mathbf{I}_N\right)^{-2} + \kappa\sum_{i=1}^{N}\mathrm{E}_{n-j}\left[\left(\mathbf{X}_{(j)}\mathbf{X}_{(j)}^{\mathsf{H}} + \frac{1}{x}\mathbf{I}_N\right)^{-1}\right]_{ii}\right)\end{aligned}$$

where the term κ appears, and where the term $\frac{1}{n^2}\operatorname{tr}\left(\mathrm{E}_{n-j}\mathbf{X}_{(j)}\mathbf{X}_{(j)}^{\mathsf{H}} + \frac{1}{x}\mathbf{I}_N\right)^{-2}$ further develops into

$$\frac{1}{n^2}\operatorname{tr}\left(\mathrm{E}_{n-j}\mathbf{X}_{(j)}\mathbf{X}_{(j)}^{\mathsf{H}} + \frac{1}{x}\mathbf{I}_N\right)^{-2} \simeq \frac{\frac{1}{n}\operatorname{tr}\mathrm{E}\left(\mathbf{X}_N\mathbf{X}_N^{\mathsf{H}} - z\mathbf{I}_N\right)^{-1}}{1 - \frac{n-j-1}{n}\frac{\frac{1}{n}\operatorname{tr}\mathrm{E}\left(\mathbf{X}_N\mathbf{X}_N^{\mathsf{H}} - z\mathbf{I}_N\right)^{-1}}{\left(1 + \frac{1}{n}\operatorname{tr}\mathrm{E}\left(\mathbf{X}_N\mathbf{X}_N^{\mathsf{H}} - z\mathbf{I}_N\right)^{-1}\right)^2}}.$$

Expressing the limits as a function of m_F, we finally obtain

$$\sum_{j=1}^{n} \mathrm{E}_{n-j}(\mathrm{E}_{n-j+1}A_j)^2 \rightarrow \frac{1}{(1+cm_F(-1/x))^2} \int_0^1 \frac{cm_F(-1/x)^2}{1-(1-y)\frac{cm_F(-1/x)}{(1+cm_F(-1/x))^2}} dy + \kappa \frac{cm_F(-1/x)^2}{(1+m_F(-1/x))^2}$$

from which we fall back on the expected result. □

This completes this first chapter on limiting distribution functions of some large dimensional random matrices and the Stieltjes transform method. In the course of this chapter, the Stieltjes transform has been illustrated to be a powerful tool for the study of the limiting spectrum distribution of large dimensional random matrices. A large variety of practical applications in the realm of wireless communications will be further developed in Part II. However, while i.i.d. matrices, with different flavors (with variance profile, non-centered, etc.), and Haar matrices have been extensively studied, no analytic result concerning more structured matrices used in wireless communications, such as Vandermonde and Hankel matrices, has been found so far. In this case, some moment-based approaches are able to fill in the gap by providing results on all successive moments (when they exist). In the subsequent chapter, we will first introduce the basics of *free probability* theory, which encompasses a very different theory of large dimensional random matrices than discussed so far and which provides a rather comprehensive framework for moment-based approaches, which will then be extended in a further chapter. We will come back to the Stieltjes transform methods for determining limiting spectral distributions in Chapter 6 in which we discuss more elaborated techniques than the Marčenko–Pastur method introduced here.

4 Free probability theory

In this chapter, we introduce free probability theory, a different approach to the already introduced tools for random matrix theory. Free probability will be shown to provide a very efficient framework to study limiting distributions of some models of large dimensional random matrices with symmetric features. Although to this day it does not overcome the techniques introduced earlier and can only be applied to very few random matrix models, this approach has the strong advantage of often being faster at determining the l.s.d. for these models. In particular, some results are derived in a few lines of calculus in the following, which are generalized in Chapter 6 using more advanced tools.

It is in general a difficult problem to deduce the eigenvalue distribution of the sum or the product of two generic Hermitian matrices $\mathbf{A}$, $\mathbf{B}$ as a function of the eigenvalue distributions of $\mathbf{A}$ and $\mathbf{B}$. In fact, it is often impossible as the eigenvectors of $\mathbf{A}$ and $\mathbf{B}$ intervene in the expression of the eigenvalues of their sum or product. In Section 3.2, we provided a formula, Theorem 3.13, linking the Stieltjes transform of the l.s.d. of the sum or the product of a deterministic matrix ($\mathbf{A}_N$ or $\mathbf{T}_N$) and a matrix $\mathbf{X}_N\mathbf{X}_N^{\mathsf{H}}$, where $\mathbf{X}_N$ has i.i.d. entries, to the Stieltjes transform of the l.s.d. of $\mathbf{A}_N$ or $\mathbf{T}_N$. In this section, we will see that Theorem 3.13 can be derived in a few lines of calculus under certain symmetry assumptions on $\mathbf{X}_N$. More general results will unfold from this type of calculus under these symmetry constraints on $\mathbf{X}_N$.

The approach originates from the work of Voiculescu [Voiculescu *et al.*, 1992] on a very different subject. Voiculescu was initially interested in the mathematical description of a theory of probability on non-commutative algebras, called *free probability theory*. The random variables here are elements of a non-commutative *probability space* $(\mathcal{A}, \phi)$, with $\mathcal{A}$ a non-commutative algebra and ϕ a given linear functional. The algebra of Hermitian random matrices is a particular case of such a probability space, for which the random variables, i.e. the random matrices, do not commute with respect to the matrix product. We introduce hereafter the basics of free probability theory that are required to understand the link between algebras of non-commutative random variables and random matrices.

4.1 Introduction to free probability theory

We first define non-commutative probability spaces.

Definition 4.1. A *non-commutative probability space* is a couple (A, ϕ) where A is a non-commutative unital algebra, that is an algebra over $\mathbb{C}$ having a unit denoted by 1, and $\phi : A \to \mathbb{C}$ is a linear functional such that $\phi(1) = 1$.

When the functional ϕ satisfies $\phi(ab) = \phi(ba)$, it is also called a *trace*. As will appear below, the role of ϕ can be compared to the role of the expectation in classical probability theory.

Definition 4.2. Let (A, ϕ) be a non-commutative probability space. In the context of free probability, a *random variable* is an element a of A. We call the *distribution* of a the linear functional ρ_a on $\mathbb{C}[X]$, the algebra of complex polynomials in one variable, defined by

$$\begin{aligned} \rho_a : \mathbb{C}[X] &\to \quad \mathbb{C} \\ P &\mapsto \phi\left(P(a)\right). \end{aligned}$$

The distribution of a non-commutative random variable a is characterized by its *moments*, which are defined to be the sequence $\phi(a), \phi(a^2), \ldots$, the successive images by ρ_a of the normalized monomials of $\mathbb{C}[X]$. The distribution of a non-commutative random variable can often be associated with a real probability measure μ_a in the sense that

$$\phi(a^k) = \int_{\mathbb{R}} t^k d\mu_a(t)$$

for each $k \in \mathbb{N}$. In this case, the moments of all orders of μ_a are of course finite. In free probability theory, it is more conventional to use *probability distributions* μ instead of *distribution functions* F (we recall that $F(x) = \mu(-\infty, x]$ if F is the d.f. associated with the measure μ on $\mathbb{R}$), so we will keep these notations in the present section.

Consider the algebra $\mathcal{A}_N$ of $N \times N$ random matrices whose entries are defined on some common probability space (meant in the classical sense) and have all their moments finite. In what follows, we refer to $\mathbf{X} \in \mathcal{A}_N$ as a *random matrix* in the sense of Definition 2.1, and not as a particular realization $\mathbf{X}(\omega)$, $\omega \in \Omega$, of $\mathbf{X}$. A non-commutative probability space of random variables is obtained by associating to $\mathcal{A}_N$ the functional τ_N given by:

$$\tau_N(\mathbf{X}) = \frac{1}{N}\mathrm{E}\left(\mathrm{tr}\,\mathbf{X}\right) = \frac{1}{N}\sum_{i=1}^{N}\mathrm{E}[X_{ii}] \tag{4.1}$$

with X_{ij} the (random variable) entry (i, j) of $\mathbf{X}$. This is obviously a trace in the free probability sense. This space will be denoted by $(\mathcal{A}_N, \tau_N)$.

Suppose $\mathbf{X}$ is a random Hermitian matrix with real random eigenvalues $\lambda_1, \ldots, \lambda_N$. The distribution $\rho_{\mathbf{X}}$ of $\mathbf{X}$ is defined by the fact that its action on each monomial X^k of $\mathbb{C}[X]$ is given by

$$\rho_{\mathbf{X}}(X^k) = \tau_N(\mathbf{X}^k) = \frac{1}{N}\sum_{i=1}^{N} \mathrm{E}[\lambda_i^k].$$

This distribution is of course associated with the probability measure $\mu_{\mathbf{X}}$ defined, for all bounded continuous f, by

$$\int f(t) d\mu_{\mathbf{X}}(t) = \frac{1}{N}\sum_{i=1}^{N} \mathrm{E}[f(\lambda_i^k)].$$

The notion of distribution introduced in Definition 4.2 is subsequently generalized to the case of multiple random variables. Let a_1 and a_2 be two random variables in a non-commutative probability space $(\mathcal{A}, \phi)$. Consider *non-commutative monomials* in two indeterminate variables of the form $X_{i_1}^{k_1} X_{i_2}^{k_2} \ldots X_{i_n}^{k_n}$, where for all j, $i_j \in \{1, 2\}$, $k_j \geq 1$ and $i_j \neq i_{j+1}$. The algebra $\mathbb{C}\langle X_1, X_2\rangle$ of non-commutative polynomials with two indeterminate variables is defined as the linear span of the space containing 1 and the non-commutative monomials. The *joint distribution* of a_1 and a_2 is then defined as the linear functional on $\mathbb{C}\langle X_1, X_2\rangle$ satisfying

$$\begin{aligned} \rho: \quad & \mathbb{C}\langle X_1, X_2\rangle \rightarrow \mathbb{C} \\ & X_{i_1}^{k_1} X_{i_2}^{k_2} \ldots X_{i_n}^{k_n} \mapsto \rho(X_{i_1}^{k_1} X_{i_2}^{k_2} \ldots X_{i_n}^{k_n}) = \phi(a_{i_1}^{k_1} a_{i_2}^{k_2} \ldots a_{i_n}^{k_n}). \end{aligned}$$

More generally, denote by $\mathbb{C}\langle X_i \,|\, i \in \{1, \ldots, I\}\rangle$ the algebra of non-commutative polynomials in I variables, which is the linear span of 1 and the non-commutative monomials of the form $X_{i_1}^{k_1} X_{i_2}^{k_2} \ldots X_{i_n}^{k_n}$, where $k_j \geq 1$ and $i_1 \neq i_2$, $i_2 \neq i_3$, ..., $i_{n-1} \neq i_n$ are smaller than or equal to I. The joint distribution of the random variables $a_1, \ldots, a_I$ in (A, ϕ) is the linear functional

$$\begin{aligned} \rho : \mathbb{C}\langle X_i \,|\, i \in \{1, \ldots, I\}\rangle &\longrightarrow \mathbb{C} \\ X_{i_1}^{k_1} X_{i_2}^{k_2} \ldots X_{i_n}^{k_n} &\longmapsto \rho(X_{i_1}^{k_1} X_{i_2}^{k_2} \ldots X_{i_n}^{k_n}) = \phi(a_{i_1}^{k_1} a_{i_2}^{k_2} \ldots a_{i_n}^{k_n}) \,. \end{aligned}$$

In short, the joint distribution of the non-commutative random variables $a_1, \ldots, a_I$ is completely specified by their joint moments. We now introduce the important notion of freeness, which will be seen as the free probability equivalent to the notion of independence in classical probability theory.

Definition 4.3. Let $(\mathcal{A}, \phi)$ be a non-commutative probability space. A family $\{\mathcal{A}_1, \ldots, \mathcal{A}_I\}$ of unital subalgebras of $\mathcal{A}$ is *free* if $\phi(a_1 a_2 \ldots a_n) = 0$ for all n-uples $(a_1, \ldots, a_n)$ satisfying

1. $a_j \in \mathcal{A}_{i_j}$ for some $i_j \leq I$ and $i_1 \neq i_2$, $i_2 \neq i_3$, ..., $i_{n-1} \neq i_n$.
2. $\phi(a_j) = 0$ for all $j \in \{1, \ldots, n\}$.

A family of subsets of $\mathcal{A}$ is free if the family of unital subalgebras generated by each one of them is free. Random variables $\{a_1, \ldots, a_n\}$ are free if the family of subsets $\{\{a_1\}, \ldots, \{a_n\}\}$ is free.

Note that in the statement of condition 1, only two successive random variables in the argument of $\phi(a_1 a_2 \ldots a_n)$ belong to two different subalgebras. This condition does not forbid the fact that, for instance, $i_1 = i_3$. Note in particular that, if a_1 and a_2 belong to two different free algebras, then $\phi(a_1 a_2 a_1 a_2) = 0$ whenever $\phi(a_1) = \phi(a_2) = 0$. This relation cannot of course hold if a_1 and a_2 are two real-valued independent random variables and if ϕ coincides with the classical mathematical expectation operator. Therefore freeness, often referred to as a free probability equivalent to independence, cannot be considered as a non-commutative *generalization* of independence because algebras generated by independent random variables in the classical sense are not necessarily free.

Let us make a simple computation involving freeness. Let $\mathcal{A}_1$ and $\mathcal{A}_2$ be two free subalgebras in $\mathcal{A}$. Any two elements a_1 and a_2 of $\mathcal{A}_1$ and $\mathcal{A}_2$, respectively, can be written as $a_i = \phi(a_i)1 + a_i'$ (1 is here the unit of $\mathcal{A}$), so $\phi(a_i') = 0$. Now

$$\phi(a_1 a_2) = \phi\left((\phi(a_1)1 + a_1')\,(\phi(a_2)1 + a_2')\right) = \phi(a_1)\phi(a_2).$$

In other words, the expectations of two free random variables factorize. By decomposing a random variable a_i into $\phi(a_i)1 + a_i'$, the principle of this computation can be generalized to the case of more than two random variables and to the case of higher order moments, and we can check that non commutativity plays a central role there.

Theorem 4.1 ([Biane, 2003]). *Let $\mathcal{A}_1, \ldots, \mathcal{A}_I$ be free subalgebras in $(\mathcal{A}, \phi)$ and let $\{a_1, \ldots, a_n\} \subset \mathcal{A}$ be such that, for all $j \in \{1, \ldots, n\}$, $a_j \in \mathcal{A}_{i_j}$, $1 \leq i_j \leq I$. Let Π be the partition of $\{1, \ldots, n\}$ associated with the equivalence relation $j \equiv k \Leftrightarrow i_j = i_k$, i.e. the random variables a_j are gathered together according to the free algebras to which they belong. For each partition π of $\{1, \ldots, n\}$, let*

$$\phi_\pi = \prod_{\substack{\{j_1, \ldots, j_r\} \in \pi \\ j_1 < \ldots < j_r}} \phi(a_{j_1} \ldots a_{j_r}).$$

There exists universal coefficients $c(\pi, \Pi)$ such that

$$\phi(a_1 \ldots a_n) = \sum_{\pi \leq \Pi} c(\pi, \Pi) \phi_\pi$$

where "$\pi \leq \Pi$" stands for "π is finer than Π," i.e. every element of π is a subset of an element of Π.

The main consequence of this result is that, given a family of free algebras $\mathcal{A}_1, \ldots, \mathcal{A}_I$ in $\mathcal{A}$, only restrictions of ϕ to the algebras $\mathcal{A}_i$ are needed to compute $\phi(a_1 \ldots a_n)$ for any $a_1, \ldots, a_n \in \mathcal{A}$ such that, for all $j \in \{1, \ldots, n\}$, we have $a_j \in \mathcal{A}_{i_j}$, $1 \leq i_j \leq I$. The problem of computing explicitly the universal coefficients

$c(\pi, \Pi)$ has been solved using a combinatorial approach and is addressed in Section 5.2.

Let μ and ν be two compactly supported probability measures on $[0, \infty)$. Then, from [Hiai and Petz, 2006], it always exists two free random variables a_1 and a_2 in some non-commutative probability space $(\mathcal{A}, \phi)$ having distributions μ and ν, respectively. We can see that the distributions of the random variables $a_1 + a_2$ and $a_1 a_2$ depend only on μ and on ν. The reason for this is the following: Definition 4.2 states that the distributions of $a_1 + a_2$ and $a_1 a_2$ are fully characterized by the moments $\phi\left((a_1 + a_2)^n\right)$ and $\phi\left((a_1 a_2)^n\right)$, respectively. To compute these moments, we just need the restriction of ϕ to the algebras generated by $\{a_1\}$ and $\{a_2\}$, according to Theorem 4.1. In other words, $\phi\left((a_1 + a_2)^n\right)$ and $\phi\left((a_1 a_2)^n\right)$ depend on the moments of a_1 and a_2 only.

As such, the distributions of $a_1 + a_2$ and $a_1 a_2$ can, respectively, be associated with probability measures called *free additive convolution* and *free multiplicative convolution* of the distributions μ and ν of these variables. The free additive convolution of μ and ν is denoted $\mu \boxplus \nu$, while the free multiplicative convolution of μ and ν is denoted $\mu \boxtimes \nu$. Both $\mu \boxplus \nu$ and $\mu \boxtimes \nu$ are compactly supported on $[0, \infty)$, see, e.g. page 30 of [Voiculescu *et al.*, 1992]. Also, both additive and multiplicative free convolutions are commutative, e.g. $\mu \boxplus \nu = \nu \boxplus \mu$, and the moments of $\mu \boxplus \nu$ and $\mu \boxtimes \nu$ are related in a universal manner to the moments of μ and to those of ν. We similarly denote $\mu \boxminus \nu$ the *free additive deconvolution*, which is such that, if $\eta = \mu \boxplus \nu$, then $\mu = \eta \boxminus \nu$ and $\nu = \eta \boxminus \mu$, and $\mu \boxslash \nu$ the *free multiplicative deconvolution* which is such that, if $\eta = \nu \boxtimes \nu$, then $\mu = \eta \boxslash \nu$ and $\nu = \eta \boxslash \mu$.

In the following, we express the fundamental link between the operations $\mu \boxplus \nu$ and $\mu \boxtimes \nu$ and the R- and S-transforms, respectively.

4.2 R- and S-transforms

The R-transform, introduced in Definition 3.3 from the point of view of its connection to the Stieltjes transform, fully characterizes $\mu \boxplus \nu$ as a function of μ and ν. This is given by the following result.

Theorem 4.2. *Let μ and ν be compactly supported probability measures of $\mathbb{R}$. Define the R-transform R_μ of the probability distribution μ by the formal series*

$$R_\mu(z) \triangleq \sum_{k \geq 1} C_k z^{k-1}$$

where the $C_1, C_2, \ldots$ are iteratively evaluated from

$$M_n = \sum_{\pi \in \mathrm{NC}(n)} \prod_{V \in \pi} C_{|V|} \tag{4.2}$$

with

$$M_n \triangleq \int x^n \mu(dx)$$

the moment of order k of μ *and* NC(n) *the set of non-crossing partitions of* $\{1, \ldots, n\}$. *Then, we have that*

$$R_{\mu \boxplus \nu}(z) = R_\mu(z) + R_\nu(z).$$

If X is a non-commutative random variable with probability measure μ, R_X will also denote the R-transform of μ.

More is said in Section 5.2 about NC(n), which plays an important role in combinatorial approaches for free probability theory, in a similar way as $\mathcal{P}(n)$, the set of partitions of $\{1, \ldots, n\}$, which plays a fundamental role in classical probability theory.

Similarly, the S-transform introduced in Definition 3.4 allows us to turn multiplicative free convolution into a mere multiplication of power series. In the context of free probability, the S-transform is precisely defined through the following result.

Theorem 4.3. *Given a probability measure* μ *on* $\mathbb{R}$ *with compact support, let* $\psi_\mu(z)$ *be the formal power series defined by*

$$\psi_\mu(z) = \sum_{k \geq 1} z^k \int t^k d\mu(t) = \int \frac{zt}{1 - zt} d\mu(t). \tag{4.3}$$

Let χ_μ *be the unique function analytic in a neighborhood of zero, satisfying*

$$\chi_\mu(\psi_\mu(z)) = z$$

for $|z|$ *small enough. Let also*

$$S_\mu(z) = \chi_\mu(z) \frac{1+z}{z}.$$

The function S_μ *is called the* S*-transform of* μ*, introduced in Definition 3.4. Moreover the* S*-transform* $S_{\mu \boxtimes \nu}$ *of* $\mu \boxtimes \nu$ *satisfies*

$$S_{\mu \boxtimes \nu} = S_\mu S_\nu.$$

Similar to the R-transform, if X is a non-commutative random variable with probability measure μ, S_X will also denote the S-transform of μ.

Remark 4.1. We mention additionally that, as in classical probability theory, a limit theorem for free random variables exists, and is given as follows.

Theorem 4.4 ([Bercovici and Pata, 1996]). *Let* $A_1, A_2, \ldots$ *be a sequence of free random variables on the non-commutative probability space* $(\mathcal{A}, \phi)$*, such that, for all* i*,* $\phi(A_i) = 0$*,* $\phi(A_i^2) = 1$ *and for all* k*,* $\sup_i |\phi(A_i^k)| < \infty$*. We then have, as*

$N \to \infty$

$$\frac{1}{\sqrt{N}}(A_1 + \ldots + A_N) \Rightarrow A$$

where A is a random variable whose distribution has R-transform $R_A(z) = z$. This distribution is the semi-circle law of Figure 1.2, with density defined in (2.4).

The semi-circle law is then the free probability equivalent of the Gaussian distribution.

Now that the basis of free probability theory has been laid down, we move to the application of free probability theory to large dimensional random matrix theory, which is the core interest of this chapter.

4.3 Free probability and random matrices

Voiculescu discovered very important relations between free probability theory and random matrix theory. Non-diagonal random Hermitian matrices are clearly non-commutative random variables. In [Voiculescu, 1991], it is shown that certain independent matrix models exhibit asymptotic free relations.

Definition 4.4. Let $\{\mathbf{X}_{N,1}, \ldots, \mathbf{X}_{N,I}\}$ be a family of random $N \times N$ matrices belonging to the non-commutative probability space $(\mathcal{A}_N, \tau_N)$ with τ_N defined in (4.1). The joint distribution has a *limit distribution* ρ on $\mathbb{C}\langle X_i | i \in \{1, \ldots, I\}\rangle$ as $N \to \infty$ if

$$\rho(X_{i_1}^{k_1} \ldots X_{i_n}^{k_n}) = \lim_{N \to \infty} \tau_N(\mathbf{X}_{N,i_1}^{k_1} \ldots \mathbf{X}_{N,i_n}^{k_n})$$

exists for any non-commutative monomial in $\mathbb{C}\langle X_i | i \in \{1, \ldots, I\}\rangle$.

Consider the particular case where $I = 1$ (we replace $\mathbf{X}_{N,1}$ by $\mathbf{X}_N$ to simplify the notations) and assume $\mathbf{X}_N$ has real eigenvalues and that the distribution of $\mathbf{X}_N$ has a limit distribution ρ. Then, for each $k \geq 0$

$$\rho(X^k) = \lim_{N \to \infty} \int t^k d\mu_{\mathbf{X}_N}(t) \tag{4.4}$$

where $\mu_{\mathbf{X}_N}$ is the measure associated with the distribution of $\mathbf{X}_N$ (seen as a random variable in the free probability framework and not as a random variable in the classical random matrix framework).

Remark 4.2. If ρ is associated with a compactly supported probability measure μ, the convergence of the moments of $\mu_{\mathbf{X}_N}$ to the moments of μ expressed by (4.4) implies the weak convergence of the sequence $\mu_{\mathbf{X}_1}, \mu_{\mathbf{X}_2}, \ldots$ to μ, i.e.

$$\int f(t) d\mu(t) = \lim_{N \to \infty} \int f(t) d\mu_{\mathbf{X}_N}(t)$$

for each continuous bounded function $f(t)$, from Theorem 30.2 of [Billingsley, 1995]. This is a particular case of the *method of moments*, see Section 5.1, which allows us to determine the l.s.d. of a random matrix model from the successive limiting moments of the e.s.d.

We now define asymptotic freeness, which is the extension of the notion of freeness for the algebras of Hermitian random matrices.

Definition 4.5. The family $\{\mathbf{X}_{N,1}, \ldots, \mathbf{X}_{N,I}\}$ of random matrices in $(\mathcal{A}_N, \tau_N)$ is said to be *asymptotically free* if the following two conditions are satisfied:

1. For every integer $i \in \{1, \ldots, I\}$, $\mathbf{X}_{N,i}$ has a limit distribution on $\mathbb{C}[X_i]$.
2. For every family $\{i_1, \ldots, i_n\} \subset \{1, \ldots, I\}$ with $i_1 \neq i_2, \ldots, i_{n-1} \neq i_n$, and for every family of polynomials $\{P_1, \ldots, P_n\}$ in one indeterminate variable satisfying

$$\lim_{N\to\infty} \tau_N\left(P_j(\mathbf{X}_{N,i_j})\right) = 0,\ j \in \{1, \ldots, n\} \tag{4.5}$$

we have:

$$\lim_{N\to\infty} \tau_N\left(\prod_{j=1}^{n} P_j(\mathbf{X}_{N,i_j})\right) = 0. \tag{4.6}$$

The conditions 1 and 2 are together equivalent to the following two conditions: the family $\{\mathbf{X}_{N,1}, \ldots, \mathbf{X}_{N,I}\}$ has a joint limit distribution on $\mathbb{C}\langle X_i | i \in \{1, \ldots, I\}\rangle$ that we denote ρ and the family of algebras $\{\mathbb{C}[X_1], \ldots, \mathbb{C}[X_I]\}$ is free in the non-commutative probability space $(\mathbb{C}\langle X_i | i \in \{1, \ldots, I\}\rangle, \rho)$.

The type of asymptotic freeness introduced by Hiai and Petz in [Hiai and Petz, 2006] is also useful because it deals with almost sure convergence under the normalized classical matrix traces instead of convergence under the functionals τ_N. Following [Hiai and Petz, 2006], the family $\{\mathbf{X}_{N,1}, \ldots, \mathbf{X}_{N,I}\}$ in $(\mathcal{A}_N, \tau_N)$ is said to have a limit ρ almost everywhere if

$$\rho(X_{i_1}^{k_1} \ldots X_{i_n}^{k_n}) = \lim_{N\to\infty} \frac{1}{N} \operatorname{tr}(\mathbf{X}_{N,i_1}^{k_1} \ldots \mathbf{X}_{N,i_n}^{k_n})$$

almost surely for any non-commutative monomial in $\mathbb{C}\langle X_i \,|\, i \in \{1, \ldots, I\}\rangle$. In the case where $N = 1$ and $\mathbf{X}_N$ has real eigenvalues, if the almost sure limit distribution of $\mathbf{X}_N$ is associated with a compactly supported probability measure μ, this condition means that

$$\int f(t) d\mu(t) = \lim_{N\to\infty} \frac{1}{N} \sum_{i=1}^{N} f(\lambda_{i,N})$$

almost surely for each continuous bounded function $f(t)$. In other words, the e.s.d. of $\mathbf{X}_N$ converges weakly and almost surely to the d.f. associated with the measure μ.

The family $\{\mathbf{X}_{N,1},\ldots,\mathbf{X}_{N,I}\}$ in $(\mathcal{A}_N,\tau_N)$ is said to be *asymptotically free almost everywhere* if, for every $i\in\{1,\ldots,I\}$, $\mathbf{X}_{N,i}$ has a non-random limit distribution on $\mathbb{C}[X_i]$ almost everywhere and if the condition 2 above is satisfied with the operator $\tau_N(\cdot)$ replaced by $\frac{1}{N}\operatorname{tr}(\cdot)$ in (4.5) and (4.6), where the limits in these equations are understood to hold almost surely. These conditions imply in particular that $\{\mathbf{X}_{N,1},\ldots,\mathbf{X}_{N,I}\}$ has a non-random limit distribution almost everywhere on $\mathbb{C}\langle X_i|i\in\{1,\ldots,I\}\rangle$.

Along with Remark 4.2, we can adapt Theorem 4.3 to the algebra of Hermitian random matrices. In this case, μ and ν are the e.s.d. of two random matrices $\mathbf{X}$ and $\mathbf{Y}$, and the equality $S_{\mu\boxtimes\nu}=S_\mu S_\nu$ is understood to hold in the almost sure sense. The same is true for Theorem 4.2.

It is important however to remember that the (almost everywhere) asymptotic freeness condition only applies to a limited range of random matrices. Remember that, for $\mathbf{A}_N\in\mathbb{C}^{N\times N}$ deterministic with l.s.d. A and $\mathbf{B}_N=\mathbf{X}_N\mathbf{X}_N^{\mathsf{H}}$ with l.s.d. B and $\mathbf{X}_N\in\mathbb{C}^{N\times n}$ with i.i.d. entries of zero mean and variance $1/n$, we saw in Theorem 3.13 that the l.s.d. of $\mathbf{X}_N\mathbf{A}_N\mathbf{X}_N^{\mathsf{H}}$ and therefore the l.s.d. of $\mathbf{A}_N\mathbf{B}_N$ are functions of A and B alone. It is however not obvious, and maybe not true at all, under the mere i.i.d. conditions on the entries of $\mathbf{B}_N$ that $\mathbf{A}_N$ and $\mathbf{B}_N$ are asymptotically free. Free probability theory therefore does not necessarily embed all the results derived in Chapter 3. The exact condition for which the l.s.d. of $\mathbf{A}_N\mathbf{B}_N$ is only a function of A and B is in fact unknown. Free probability theory therefore provides a partial answer to this problem by introducing a *sufficient condition* for this property to occur, which is the (almost everywhere) asymptotic freeness condition. In fact, not so many families of random matrices are known to be asymptotically free. Of importance to applications in Part II is the following asymptotic freeness result.

Theorem 4.5 (Theorem 4.3.11 of [Hiai and Petz, 2006]). *Let $\{\mathbf{X}_{N,1},\ldots,\mathbf{X}_{N,I}\}$ be a family of $N\times N$ complex bi-unitarily invariant random matrices, i.e. whose joint distribution is invariant both by left- and right-unitary products, and let $\{\mathbf{D}_{N,1},\ldots,\mathbf{D}_{N,J}\}$ be a family of non-random diagonal matrices. Suppose that, as N tends to infinity, the e.s.d. of $\mathbf{X}_{N,i}\mathbf{X}_{N,i}^{\mathsf{H}}$ and $\mathbf{D}_{N,j}\mathbf{D}_{N,j}^{\mathsf{H}}$ converge in distribution and almost surely to compactly supported d.f. Then the family*

$$\left\{\{\mathbf{X}_{N,i}\}_{i\in\{1,\ldots,I\}},\{\mathbf{X}_{N,i}^{\mathsf{H}}\}_{i\in\{1,\ldots,I\}},\{\mathbf{D}_{N,j}\}_{j\in\{1,\ldots,J\}},\{\mathbf{D}_{N,j}^{\mathsf{H}}\}_{j\in\{1,\ldots,J\}}\right\}$$

is asymptotically free almost everywhere as $N\to\infty$.

Theorem 4.5 allows us in particular to derive the (almost sure) l.s.d. of the following random matrices:

- sums and products of random (non-necessarily square) Gaussian matrices;
- sums and products of random Gaussian or unitary matrices and deterministic diagonal matrices;

- the sum or product of a Hermitian $\mathbf{A}_N \in \mathbb{C}^{N\times N}$ by $\mathbf{X}_N\mathbf{B}_N\mathbf{X}_N^{\mathsf{H}} \in \mathbb{C}^{N\times N}$, with $\mathbf{B}_N \in \mathbb{C}^{n\times n}$ Hermitian and $\mathbf{X}_N \in \mathbb{C}^{N\times n}$ Gaussian or originating from columns of a unitary matrix.

Theorem 4.5 can be used for sums and products of non-square Gaussian matrices since the $\mathbf{D}_{N,j}$ matrices can be taken such that their first $(N-n)$ diagonal entries are ones and their last n entries are zeros. It can then be used to derive the l.s.d. of $\mathbf{A}_N + \mathbf{X}_N\mathbf{B}_N\mathbf{X}_N^{\mathsf{H}}$ or $\mathbf{A}_N\mathbf{X}_N\mathbf{B}_N\mathbf{X}_N^{\mathsf{H}}$ for non-diagonal Hermitian $\mathbf{A}_N$ and $\mathbf{B}_N$, since the result holds if $\mathbf{X}_N$ is replaced by $\mathbf{U}_N\mathbf{X}_N\mathbf{V}_N$, with $\mathbf{U}_N$ the stacked columns of eigenvectors for $\mathbf{A}_N$ and $\mathbf{V}_N$ the stacked columns of eigenvectors for $\mathbf{B}_N$. Basically, and roughly speaking, for any couple $(\mathbf{X}, \mathbf{Y})$ of matrices, with mutually independent entries and such that at least one of these matrices is unitarily invariant by left and right product, then $\mathbf{X}$ and $\mathbf{Y}$ are asymptotically free almost everywhere. Intuitively, almost everywhere asymptotic freeness of the couple $(\mathbf{X}, \mathbf{Y})$ means that the random variables $\mathbf{X}$ and $\mathbf{Y}$ have independent entries and that their respective eigenspaces are asymptotically "disconnected," as their dimensions grow, in the sense that the eigenvectors of each random matrix are distributed in a maximally uncorrelated way. In the case where one of the matrices has isotropically distributed eigenvectors, necessarily it will be asymptotically free almost everywhere with respect to any other independent matrix, be it random unitarily invariant or deterministic. On the opposite, two deterministic matrices cannot be free as their eigenvectors point in deterministic and therefore "correlated" directions.

Unitarily invariant unitary matrices, often called Haar matrices, are an important class of matrices, along with Gaussian matrices, for which free probability conveys quite a few results regarding their summation or product to Gaussian matrices, deterministic matrices, etc. Haar matrices can be constructed and defined in the following definition-theorem.

Definition 4.6. Let $\mathbf{X} \in \mathbb{C}^{N\times N}$ be a random matrix with independent Gaussian entries of zero mean and unit variance. Then the matrix $\mathbf{W} \in \mathbb{C}^{N\times N}$, defined as

$$\mathbf{W} = \mathbf{X}\left(\mathbf{X}^{\mathsf{H}}\mathbf{X}\right)^{-\frac{1}{2}}$$

is uniformly distributed on the space $\mathcal{U}(N)$ of $N \times N$ complex unitary matrices. This random matrix is called a *Haar matrix*. Moreover, as N grows large, the e.s.d. $F^{\mathbf{W}}$ of $\mathbf{W}$ converges almost surely to the uniform distribution on the complex unit circle.

For the applications to come, we especially need the following two corollaries of Theorem 4.5.

Corollary 4.1. *Let $\{\mathbf{T}_1, \ldots, \mathbf{T}_K\}$ be $N \times N$ Hermitian random matrices, and let $\{\mathbf{W}_1, \ldots, \mathbf{W}_K\}$ be Haar distributed independent from all $\mathbf{T}_k$. Assume that the empirical eigenvalue distributions of all $\mathbf{T}_k$ converge almost surely* *toward compactly supported probability distributions. Then, the random matrices* $\mathbf{W}_1\mathbf{T}_1\mathbf{W}_1^{\mathsf{H}}, \ldots, \mathbf{W}_K\mathbf{T}_K\mathbf{W}_K^{\mathsf{H}}$ *are asymptotically free almost surely as* $N \to \infty$.

Corollary 4.2 (Proposition 4.3.9 of [Hiai and Petz, 2006]). *Let* $\mathbf{A}_N$ *and* $\mathbf{T}_N$ *be* $N \times N$ *Hermitian random matrices, and let* $\mathbf{W}_N$ *be a Haar distributed unitary random matrix independent from* $\mathbf{A}_N$ *and* $\mathbf{T}_N$. *Assume that the empirical eigenvalue distributions of* $\mathbf{A}_N$ *and of* $\mathbf{T}_N$ *converge almost surely toward compactly supported probability distributions. Then, the random matrices* $\mathbf{A}_N$ *and* $\mathbf{W}_N\mathbf{T}_N\mathbf{W}_N^{\mathsf{H}}$ *are asymptotically free almost surely as* $N \to \infty$.

The above propositions, along with the R- and S-transforms of Theorem 4.2 and Theorem 4.3, will be used to derive the l.s.d. of some random matrix models for which the Stieltjes transform approach is more painful to use.

The R-transform and S-transform introduced in the definition-theorem, Theorem 4.2, and the definition-theorem, Theorem 4.3, respectively, can be indeed redefined in terms of transforms of the l.s.d. of random matrices, as they were already quickly introduced in Definition 3.3 and Definition 3.4. In this case, Theorem 4.2 extends to the following result.

Theorem 4.6. *Let* $\mathbf{A}_N \in \mathbb{C}^{N\times N}$ *and* $\mathbf{B}_N \in \mathbb{C}^{N\times N}$ *be two random matrices. If* $\mathbf{A}_N$ *and* $\mathbf{B}_N$ *are asymptotically free almost everywhere and have respective (almost sure) asymptotic eigenvalue probability distribution* μ_A *and* μ_B, *then* $\mathbf{A}_N + \mathbf{B}_N$ *has an asymptotic eigenvalue distribution* μ, *which is such that, if* R_A *and* R_B *are the respective R-transforms of the l.s.d. of* $\mathbf{A}_N$ *and* $\mathbf{B}_N$

$$R_{A+B}(z) = R_A(z) + R_B(z)$$

almost surely, with R_{A+B} *the R-transform of the asymptotic eigenvalue distribution of* $\mathbf{A}_N + \mathbf{B}_N$. *The distribution* μ *is often denoted* μ_{A+B}, *and we write*

$$\mu_{A+B} = \mu_A \boxplus \mu_B.$$

From the definition of the R-transform, we in particular have the following useful property.

Lemma 4.1. *Let* $\mathbf{X} \in \mathbb{C}^{N\times N}$ *be some Hermitian matrix and* $a \in \mathbb{R}$. *Then*

$$R_{a\mathbf{X}}(z) = aR_{\mathbf{X}}(az).$$

Also, if $\mathbf{X}$ *is random with limiting l.s.d.* F, *in the large* N *limit, the R-transform* $R_{F_{(a)}}$ *of the l.s.d.* $F_{(a)}$ *of the random matrix* $a\mathbf{X}$, *for some* $a > 0$, *satisfies*

$$R_{F_{(a)}}(z) = aR_F(az).$$

This is immediate from Lemma 3.2 and Definition 3.3. Indeed, applying the definition of the R-transform at point az, we have:

$$m_{\mathbf{X}}\left(R_{\mathbf{X}}(az) + \frac{1}{az}\right) = -az$$

while $am_{a\mathbf{X}}(ay) = m_{\mathbf{X}}(y)$ from Lemma 3.2. Together, this is therefore:

$$am_{a\mathbf{X}}\left(aR_{\mathbf{X}}(az) + \frac{1}{z}\right) = -az.$$

Removing the a on each side, we have that $aR_{\mathbf{X}}(az)$ is the R-transform associated with the Stieltjes transform of $a\mathbf{X}$, from Definition 3.3 again.

Similarly, Theorem 4.3 for the S-transform extends to the following result.

Theorem 4.7. *Let $\mathbf{A}_N \in \mathbb{C}^{N\times N}$ and $\mathbf{B}_N \in \mathbb{C}^{N\times N}$ be two random matrices. If $\mathbf{A}_N$ and $\mathbf{B}_N$ are asymptotically free almost everywhere and have respective (almost sure) asymptotic eigenvalue distribution μ_A and μ_B, then $\mathbf{A}_N\mathbf{B}_N$ has an asymptotic eigenvalue distribution μ, which is such that, if S_A and S_B are the respective S-transforms of the l.s.d. of $\mathbf{A}_N$ and $\mathbf{B}_N$*

$$S_{AB}(z) = S_A(z)S_B(z)$$

almost surely, with S_{AB} the S-transform of the l.s.d. of $\mathbf{A}_N\mathbf{B}_N$. The distribution μ is often denoted μ_{AB} and we write

$$\mu_{AB} = \mu_A \boxtimes \mu_B.$$

An equivalent scaling result is also valid for the S-transform. Precisely, we have the following lemma.

Lemma 4.2. *Let $\mathbf{X} \in \mathbb{C}^{N\times N}$ be some Hermitian matrix and let a be a non-zero real. Then*

$$S_{a\mathbf{X}}(z) = \frac{1}{a}S_{\mathbf{X}}(z).$$

If $\mathbf{X}$ is random with limiting l.s.d. F, then the S-transform $S_{F_{(a)}}$ of the l.s.d. $F_{(a)}$ of the random matrix $a\mathbf{X}$ satisfies

$$S_{F_{(a)}}(z) = \frac{1}{a}S_F(z).$$

This unfolds here from noticing that, by the (definition-)Theorem 4.3

$$S_{a\mathbf{X}}(z) = \frac{1+z}{z}\psi_{a\mathbf{X}}^{-1}(z)$$

with $\psi_{a\mathbf{X}}$ the ψ-transform of $F^{a\mathbf{X}}$. Now, by definition of the ψ-transform

$$\psi_{a\mathbf{X}}(z) = \int \frac{azt}{1-azt}dF^{\mathbf{X}}(t) = \psi_{\mathbf{X}}(az)$$

and therefore:

$$\psi_{a\mathbf{X}}^{-1}(z) = \frac{1}{a}\psi_{\mathbf{X}}^{-1}(z)$$

which gives the result.

Of interest to practical applications of the S-transform is also the following lemma.

Lemma 4.3. *Let* $\mathbf{A} \in \mathbb{C}^{N\times n}$ *and* $\mathbf{B} \in \mathbb{C}^{n\times N}$ *such that* $\mathbf{AB}$ *is Hermitian non-negative. Then, for* $z \in \mathbb{C} \setminus \mathbb{R}^+$

$$S_{\mathbf{AB}}(z) = \frac{z+1}{z+\frac{n}{N}} S_{\mathbf{BA}}\left(\frac{N}{n}z\right).$$

If $\mathbf{AB}$ *is random with l.s.d. and* $\mathbf{BA}$ *has l.s.d.* $\underline{F}$ *as* $N, n \to \infty$ *with* $N/n \to c$, $0 < c < \infty$, *then we also have*

$$S_F(z) = \frac{z+1}{z+\frac{1}{c}} S_{\underline{F}}(cz).$$

This result can be proved starting from the definition of the S-transform, Definition 3.4, as follows. Notice first from Definition 3.4 that

$$\begin{aligned}
\psi_{\mathbf{AB}}(z) &= -1 - \frac{1}{z} m_{\mathbf{AB}}(z^{-1}) \\
&= \frac{n}{N}\left(-1 - \frac{1}{z} m_{\mathbf{BA}}(z^{-1})\right) \\
&= \frac{n}{N}\psi_{\mathbf{BA}}(z).
\end{aligned}$$

Now, by definition

$$\psi_{\mathbf{AB}}\left(\frac{z}{1+z} S_{\mathbf{AB}}(z)\right) = z$$

from which:

$$\psi_{\mathbf{BA}}\left(\frac{z}{1+z} S_{\mathbf{AB}}(z)\right) = \frac{N}{n}z.$$

Taking $\psi_{\mathbf{BA}}^{-1}$ (with respect to composition) on each side, this leads to

$$\begin{aligned}
S_{\mathbf{AB}}(z) &= \frac{1+z}{z}\psi_{\mathbf{BA}}^{-1}\left(\frac{N}{n}z\right) \\
&= \frac{\frac{N}{n} + \frac{N}{n}z}{1+\frac{N}{n}z} \frac{1+\frac{N}{n}z}{\frac{N}{n}z} \psi_{\mathbf{BA}}^{-1}\left(\frac{N}{n}z\right) \\
&= \frac{\frac{N}{n} + \frac{N}{n}z}{1+\frac{N}{n}z} S_{\mathbf{BA}}\left(\frac{N}{n}z\right)
\end{aligned}$$

which is the final result.

In the next section, we apply the above results on the R- and S-transforms to random matrix models involving Gaussian matrices, before considering models involving Haar matrices in the subsequent section.

4.4 Free probability for Gaussian matrices

In the following, we will demonstrate in a few lines the result of Theorem 3.13 on the limiting distribution of $\mathbf{B}_N = \mathbf{A}_N + \mathbf{X}_N^{\mathsf{H}}\mathbf{T}_N\mathbf{X}_N$ in the case when $\mathbf{T}_N = \mathbf{I}_N$, $\mathbf{A}_N \in \mathbb{C}^{n\times n}$ has uniformly bounded spectral norm and $\mathbf{X}_N \in \mathbb{C}^{N\times n}$ has Gaussian entries of zero mean and variance $1/n$. To begin with, we introduce some classical results on the R- and S-transforms of classical laws in random matrix theory.

Theorem 4.8. *The semi-circle law and Marčenko–Pastur law have the following properties.*

(i) The R-transform $R_{F_c}(z)$ of the Marčenko–Pastur law F_c with ratio c, i.e. the almost sure l.s.d. of $\mathbf{X}_N\mathbf{X}_N^{\mathsf{H}}$, $\mathbf{X}_N \in \mathbb{C}^{N\times n}$, with i.i.d. entries of zero mean and variance $1/n$, as $N/n \to c$, whose density f_c is given by (2.5), reads:

$$R_{F_c}(z) = \frac{1}{1-cz}$$

and the S-transform $S_{F_c}(z)$ of F_c reads:

$$S_{F_c}(z) = \frac{1}{1+cz}.$$

Similarly, the R-transform $R_{\underline{F}_c}(z)$ of the complementary Marčenko–Pastur law $\underline{F}_c$, i.e. the almost sure l.s.d. of $\mathbf{X}_N^{\mathsf{H}}\mathbf{X}_N$, reads:

$$R_{\underline{F}_c}(z) = \frac{c}{1-z}$$

and the S-transform $S_{\underline{F}_c}(z)$ of $\underline{F}_c$ is given by:

$$S_{\underline{F}_c}(z) = \frac{1}{c+z}.$$

(ii) The R-transform $R_F(z)$ of the semi-circle law F, with density f given by (2.4), reads:

$$R_F(z) = z.$$

If the entries of $\mathbf{X}_N$ are Gaussian distributed, from Corollary 4.2, $\mathbf{A}_N$ and $\mathbf{X}_N^{\mathsf{H}}\mathbf{X}_N$ are asymptotically free almost everywhere. Therefore, from Theorem 4.6, almost surely, the R-transform R_B of the l.s.d. of $\mathbf{B}_N$ satisfies

$$R_B(z) = R_A(z) + R_{\underline{F}_c}(z)$$

with A the l.s.d. of $\mathbf{A}_N$. From the earlier Definition 3.3, Equation (3.8), of the R-transform

$$m_B(z) = \frac{1}{R_B(-m_B(z)) - z} = \frac{1}{R_A(-m_B(z)) + \frac{c}{1+m_B(z)} - z}$$

almost surely. This is equivalent to

$$R_A(-m_B(z)) + \frac{1}{m_B(z)} = z - \frac{c}{1+m_B(z)}$$

almost surely.

From Definition 3.3, Equation (3.7), taking the Stieltjes transform of A on both sides

$$m_B(z) = m_A\left(z - \frac{c}{1+m_B(z)}\right)$$

which is consistent with Theorem 3.13. Slightly more work is required to generalize this result to $\mathbf{T}_N$ different from $\mathbf{I}_N$. This study is carried out in Section 4.5 for the case where $\mathbf{X}_N$ is a Haar matrix instead.

We present now one of the important results known so far concerning Gaussian-based models of deep interest for applications in wireless communications. In [Ryan and Debbah, 2007b], Ryan and Debbah provide indeed a free probability expressions of the *free convolution* and *free deconvolution* for the information plus noise model, summarized as follows.

Theorem 4.9 ([Ryan and Debbah, 2007b]). *Let $\mathbf{X}_N \in \mathbb{C}^{N\times n}$ be a random matrix with i.i.d. Gaussian entries of zero mean and unit variance, and $\mathbf{R}_N$ a (non-necessarily random) matrix such that the eigenvalue distribution of $\mathbf{R}_N = \frac{1}{n}\mathbf{R}_N\mathbf{R}_N^{\mathsf{H}}$ converges weakly and almost surely to the compactly supported probability distribution μ_R, as $n, N \to \infty$ with limit ratio $N/n \to c > 0$. Then the eigenvalue distribution of*

$$\mathbf{B}_N = \frac{1}{n}(\mathbf{R}_N + \sigma\mathbf{X}_N)(\mathbf{R}_N + \sigma\mathbf{X}_N)^{\mathsf{H}}$$

converges weakly and almost surely to the compact supported measure μ_B such that

$$\mu_B = ((\mu_R \boxslash \mu_c) \boxplus \delta_{\sigma^2}) \boxtimes \mu_c \tag{4.7}$$

with μ_c the probability distribution with distribution function the Marčenko–Pastur law and δ_{σ^2} the probability distribution of a single mass in σ^2 (with k*th order moment σ^{2k}). Equation (4.7) is the free convolution of the information plus noise model. This can be reverted as*

$$\mu_R = ((\mu_B \boxtimes \mu_c) \boxminus \delta_{\sigma^2}) \boxslash \mu_c$$

which is the free deconvolution of the information plus noise model.

To understand the importance of the result above, remember for instance that we have already shown that matrix models of the type $\mathbf{B}_N = \mathbf{A}_N + \mathbf{X}_N\mathbf{T}_N\mathbf{X}_N^{\mathsf{H}}$ or $\mathbf{B}_N = (\mathbf{A}_N + \mathbf{X}_N)(\mathbf{A}_N + \mathbf{X}_N)^{\mathsf{H}}$, where $\mathbf{X}_N$ has i.i.d. (possibly Gaussian) entries, can be treated by Stieltjes transform methods (Theorem 3.13 and Theorem 3.15), which provide a complete description of the l.s.d. of $\mathbf{B}_N$, through its Stieltjes transform. However, while this powerful method is capable of treating the very loose case of $\mathbf{X}_N$ with i.i.d. entries, there does not yet exist a unifying framework that allows us to derive easily the limit of the Stieltjes transform for more involved models in which successive sums, products and information plus noise models of i.i.d. matrices are taken. In the Stieltjes transform approach, every new model must be dedicated a thorough analysis that consists in (i) deriving an implicit equation for the Stieltjes transform of the l.s.d., (ii) ensuring that the implicit equation has a unique solution, (iii) studying the convergence of fixed-point algorithms that solve the implicit equations. Consider for instance the model

$$\mathbf{B}_N = \left(\mathbf{H}\mathbf{T}^{\frac{1}{2}}\mathbf{X} + \sigma\mathbf{W}\right)\left(\mathbf{H}\mathbf{T}^{\frac{1}{2}}\mathbf{X} + \sigma\mathbf{W}\right)^{\mathsf{H}} \tag{4.8}$$

where $\mathbf{H}$, $\mathbf{X}$, and $\mathbf{W}$ are independent and all have i.i.d. Gaussian entries, $\mathbf{T}$ is deterministic and $\sigma > 0$. This model arises in wireless communications when a Gaussian signal process $\mathbf{T}^{\frac{1}{2}}\mathbf{X} \in \mathbb{C}^{n_t \times L}$ is sent from an n_t-antenna transmitter during L sampling instants through a MIMO $n_r \times n_t$ channel $\mathbf{H} \in \mathbb{C}^{n_r \times n_t}$ with additive white Gaussian noise $\mathbf{W} \in \mathbb{C}^{n_r \times L}$ with entries of variance σ^2. The matrix $\mathbf{X}$ is assumed to have i.i.d. Gaussian entries and therefore the n_t-dimensional signal vector has transmit covariance matrix $\mathbf{T} \in \mathbb{C}^{n_t \times n_t}$. In the framework of free probability, the model (4.8) is not difficult to treat, since it relies on elementary convolution and deconvolution operation. Precisely, it is the succession of the multiplicative free convolution of the l.s.d. of $\mathbf{X}$ and $\mathbf{T}^{\frac{1}{2}}$ and the information plus noise-free convolution of the l.s.d. of $\mathbf{T}^{\frac{1}{2}}\mathbf{X}$ and $\sigma\mathbf{W}$. As we can already guess from the definitions of the free probability framework and as we will see in detail in Chapter 5, this problem can also be treated from elementary combinatorics computations based on the successive moments of the d.f. under study; these computations will be shown in particular to be easily implemented on a modern computer. The analysis of model (4.8) by the Stieltjes transform approach is more difficult and has been treated in [Couillet *et al.*, 2011c], in which it is proved that $m_F(z)$, $z \in \mathbb{C}^+$, the Stieltjes transform of the l.s.d. F of $\mathbf{B}_N$, is the unique solution with positive imaginary part of a given implicit equation, see Theorem 17.5. However, computer trials suggest that the region of convergence of the fixed-point algorithm solving the implicit equation for $m_F(z)$ has a very small radius, which is very unsatisfactory in practice, therefore giving more interest to the combinatorial approach. Moreover, the complete derivation of the limiting spectrum of $\mathbf{B}_N$ requires a similar mathematical machinery as in the proof of the Marčenko–Pastur, this being requested for every new model. These are two strong arguments that motivate further investigations on automated numerical methods based on combinatoric calculus. Nonetheless, the Stieltjes transform approach is able to treat the case when $\mathbf{X}$ has i.i.d. *non-necessarily Gaussian* entries, e.g. M-QAM, M-PSK modulated signals, which are not proved to satisfy the requirement demanded in the free probability approach. Moreover, as will become clear in Part II, the simplicity and the flexibility of the combinatoric methods are rarely any match for the accuracy and efficiency of the Stieltjes transform approach.

In the following section, we derive further examples of applications of Theorem 4.6 and Theorem 4.7 to the free probability framework introduced in this section for Haar random matrices.

4.5 Free probability for Haar matrices

We start with the introduction of a very simple model, from which the application of the R-transform and the S-transform are rather straightforward. Consider the matrix model $\mathbf{B}_N = \mathbf{W}_N\mathbf{T}_N\mathbf{W}_N^{\mathsf{H}} + \mathbf{A}_N$, where $\mathbf{A}_N \in \mathbb{C}^{N\times N}$, $\mathbf{T}_N \in \mathbb{C}^{n\times n}$ and $\mathbf{W}_N \in \mathbb{C}^{N\times n}$ is a Haar random matrix. Note that this model is the Haar equivalent of Theorem 3.13.

From Corollary 4.2, the matrices $\mathbf{A}_N$ and $\mathbf{W}_N\mathbf{T}_N\mathbf{W}_N^{\mathsf{H}}$ are asymptotically free almost everywhere and therefore we can consider retrieving the l.s.d. of $\mathbf{B}_N$ as a function of those of $\mathbf{A}_N$ and $\mathbf{W}_N\mathbf{T}_N\mathbf{W}_N^{\mathsf{H}}$ using the R-transform.

For the problem at hand, denoting F the l.s.d. of $\mathbf{B}_N$, as $N \to \infty$

$$R_F(z) = R_T(z) + R_A(z)$$

almost surely, with T the l.s.d. of $\mathbf{T}_N$ and A the l.s.d. of $\mathbf{A}_N$. Indeed, $\mathbf{W}_N$ being unitary, the successive traces of powers of $\mathbf{W}_N\mathbf{T}_N\mathbf{W}_N^{\mathsf{H}}$ are also the successive traces of powers of $\mathbf{T}_N$, so that the distribution of the free random variable $\mathbf{W}_N\mathbf{T}_N\mathbf{W}_N^{\mathsf{H}}$ is also the distribution of $\mathbf{T}_N$.

Consider the special case when both $\mathbf{T}_N$ and $\mathbf{A}_N$ have l.s.d. T with density $T'(x) = \frac{1}{2}\delta_0(x) + \frac{1}{2}\delta_1(x)$, then:

$$m_T(z) = \frac{1}{2}\frac{2z-1}{z(z-1)}$$

which, from (3.8), leads to

$$R_F(z) = \frac{z-1-\sqrt{z^2+1}}{z}$$

whose Stieltjes transform can be computed into the form

$$m_F(z) = -\frac{1}{\sqrt{z(z-2)}}$$

using (3.7). We finally obtain the l.s.d. using (3.2)

$$F(x) = 1_{[0,2]}(x)\frac{1}{\pi}\arcsin(x-1)$$

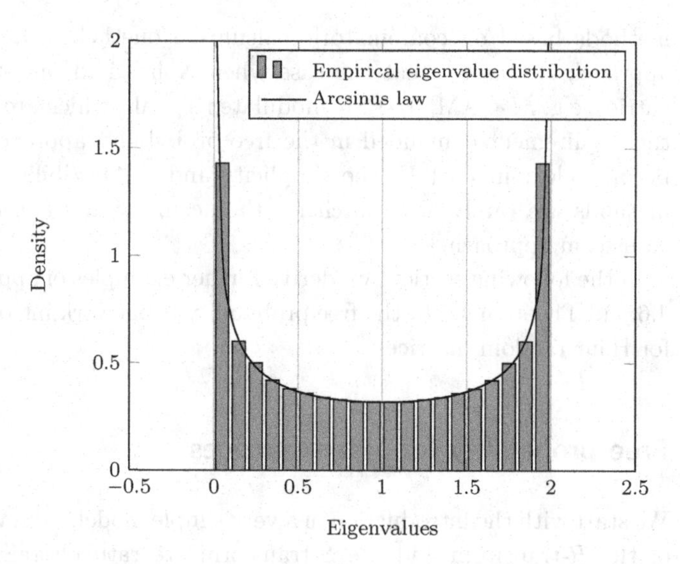

Figure 4.1 Histogram of the eigenvalues of $\mathbf{W}_N\mathbf{T}_N\mathbf{W}_N^{\mathsf{H}} + \mathbf{A}_N$ and the arcsinus law, for $N = 1000$.

which is the arcsinus law, depicted in Figure 4.1.

The same kind of reasoning can be applied to the product of asymptotically free random variables, using the S-transform in place of the R-transform.

For instance, the e.s.d. of $\mathbf{B}_N = \mathbf{W}_N\mathbf{T}_N\mathbf{W}_N^{\mathsf{H}}\mathbf{A}_N$ admits an almost sure l.s.d. F satisfying

$$S_F(z) = S_T(z)S_A(z) \xrightarrow{\text{a.s.}} 0$$

as $N \to \infty$, with T and A the respective l.s.d. of $\mathbf{T}_N$ and $\mathbf{A}_N$.

From the R- and S-transform tools defined and used above, the l.s.d. of more elaborate matrix models involving unitary matrices can be evaluated [Couillet and Debbah, 2009; Hachem, 2008]. From the above Corollary 4.1, for Haar $\mathbf{W}_k$ matrices and deterministic $\mathbf{T}_k$ matrices with limiting compactly supported d.f., the family $\{\mathbf{W}_1\mathbf{T}_1\mathbf{W}_1^{\mathsf{H}}, \ldots, \mathbf{W}_K\mathbf{T}_K\mathbf{W}_K^{\mathsf{H}}\}$ is asymptotically free almost everywhere. Moreover, for $\mathbf{D}$ a Hermitian matrix with compactly supported limited d.f., the family $\{\mathbf{W}_k\mathbf{T}_k\mathbf{W}_k^{\mathsf{H}}, \mathbf{D}\}$ is asymptotically free almost everywhere, due to Corollary 4.2. This allows us to perform seamlessly the operations of S- and R-transforms of Theorem 4.6 and Theorem 4.7, respectively, for all matrix models involving random $\mathbf{W}_k\mathbf{T}_k\mathbf{W}_k^{\mathsf{H}}$ and deterministic $\mathbf{D}$ matrices. We hereafter provide three applications of such large dimensional matrix models.

On the R-transform side, we have the following result.

Theorem 4.10 ([Couillet and Debbah, 2009]). *Let $\mathbf{T}_1, \ldots, \mathbf{T}_K \in \mathbb{C}^{N\times N}$ be diagonal non-negative definite matrices and $\mathbf{W}_1, \ldots, \mathbf{W}_K$, $\mathbf{W}_k \in \mathbb{C}^{N\times N}$, be independent Haar matrices. Denote T_k the l.s.d. of $\mathbf{T}_k$. Finally, let $\mathbf{B}_N \in \mathbb{C}^{N\times N}$*

denote the matrix

$$\mathbf{B}_N = \sum_{k=1}^{K} \mathbf{W}_k \mathbf{T}_k \mathbf{W}_k^{\mathsf{H}}.$$

Then, as N grows large, the e.s.d. of $\mathbf{B}_N$ tends to F whose η-transform η_F is given by:

$$\eta_F(x) = \left(1 + x \sum_{k=1}^{K} \beta_k(x)\right)^{-1}$$

where the functions $\beta_k(x)$, $k \in \{1, \ldots, K\}$, satisfy the K fixed-point equations

$$\beta_k(x) = \int t \left(1 + x \frac{t - \beta_k(x)}{1 + x \sum_{i=1}^{K} \beta_i(x)}\right)^{-1} dT_k(t). \tag{4.9}$$

Also, the Shannon transform $\mathcal{V}_F(x)$ of F is given by:

$$\mathcal{V}_F(x) = \log\left(1 + x \sum_{k=1}^{K} \beta_k(x)\right) + \sum_{k=1}^{K} \int \log(1 + x\eta(x)[t - \beta_k(x)]) dT_k(t). \tag{4.10}$$

This result can be used to characterize the performance of multi-cellular orthogonal CDMA communications in frequency flat channels, see, e.g., [Couillet and Debbah, 2009; Peacock *et al.*, 2008].

Proof. From the R-transform Definition 3.3 and the η-transform Definition 3.5, for a given distribution function G, we have:

$$R_G(-x\eta_G(x)) = -\frac{1}{x}\left(1 - \frac{1}{\eta_G(x)}\right) \tag{4.11}$$

$$\eta_G\left(-\frac{1}{R_G(x) + \frac{1}{x}}\right) = xR_G(x) + 1. \tag{4.12}$$

Denoting R_F the R-transform of F and R_k the R-transform of the l.s.d. of $\mathbf{W}_k \mathbf{T}_k \mathbf{W}_k^{\mathsf{H}}$, we have from Equation (4.12) that

$$xR_k(x) + 1 = \int \frac{1}{1 - \frac{t}{R_k(x) + \frac{1}{x}}} dT_k(t)$$

which is equivalent to

$$R_k(x) = \frac{1}{x} \int \frac{t}{R_k(x) + \frac{1}{x} - t} dT_k(t).$$

Evaluating the above in $-x\eta_F(x)$ and denoting $\beta_k(x) = R_k(-x\eta_F(x))$, we have:

$$\beta_k(x) = \frac{1}{x} \int \frac{t}{1 - x\eta_F(x)\beta_k(x) + x\eta_F(x)t} dT_k(t).$$

Remember now from Corollary 4.1 that the matrices $\mathbf{W}_k\mathbf{T}_k\mathbf{W}_k^{\mathsf{H}}$ are asymptotically free. We can therefore use the fact that

$$R_F(-x\eta_F(x)) = \sum_{k=1}^{K} R_k(-x\eta_F(x)) = \sum_{k=1}^{K} \beta_k(x)$$

from Theorem 4.6. From Equation (4.11), we finally have

$$\sum_{k=1}^{K} \beta_k(x) = -\frac{1}{x}\left(1 - \frac{1}{\eta_F(x)}\right)$$

which is the expected result.

To obtain (4.10), notice from Equation (3.4) that the Shannon transform can be expressed as

$$\mathcal{V}_F(x) = \int_0^x \frac{1}{t}\left(1 - \eta_F(t)\right) dt.$$

We therefore seek an integral form for the η-transform. For this, the strategy is to differentiate logarithm expressions of the $\eta_F(x)$ and $\beta_k(x)$ expressions obtained above and seek for a link between the derivatives. It often turns out that this strategy leads to a compact expression involving only the aforementioned logarithm expressions.

Note first that

$$\frac{1}{x}(1 - \eta_F(x)) = \frac{1}{x}\left(1 - \left(1 + x\sum_{k=1}^{K}\beta_k(x)\right)^{-1}\right) = \sum_{k=1}^{K} \beta_k(x)\eta_F(x).$$

Also note from (4.9) that

$$1 - x\eta_F(x)\beta_k(x) = \int \frac{1 - x\eta_F(x)\beta_k(x)}{1 - x\eta_F(x)\beta_k(x) + x\eta_F(x)t} dT_k(t)$$

and therefore that

$$1 = \int \frac{1}{1 - x\eta_F(x)\beta_k(x) + x\eta_F(x)t} dT_k(t). \tag{4.13}$$

Now, for any k, the derivative along x of

$$C_k(x) \triangleq \int \log(1 - x\eta_F(x)\beta_k(x) + x\eta_F(x)t) dT_k(t)$$

is

$$C_k'(x) = \int \frac{[-\eta_F(x)\beta_k(x) - x\eta_F'(x)\beta_k(x) - x\eta_F(x)\beta_k'(x)] + [\eta_F(x) + x\eta_F'(x)]t}{1 - x\eta_F(x)\beta_k(x) + x\eta_F(x)t} dT_k(t).$$

Recalling now (4.13) and (4.9), this yields

$$\begin{aligned}
&C_k'(x) \\
&= -\eta_F(x)\beta_k(x) - x\eta_F'(x)\beta_k(x) - x\eta_F(x)\beta_k'(x) + (\eta_F(x) + x\eta_F'(x))\beta_k(x) \\
&= -x\eta_F(x)\beta_k'(x).
\end{aligned}$$

We also have

$$\left(\log\left(1 + x\sum_{k=1}^{K}\beta_k(x)\right)\right)' = \eta_F(x)\sum_{k=1}^{K}\beta_k(x) + x\sum_{k=1}^{K}\eta_F(x)\beta_k'(x).$$

Adding this last expression to $\sum_{k=1}^{K} C_k'(x)$, we end up with the desired $\sum_{k=1}^{K}\beta_k(x)\eta_F(x)$. Verifying that $\mathcal{V}_F(0) = 0$, we finally obtain (4.10). □

Note that the strategy employed to obtain the Shannon transform is a very general approach that is very effective for a large range of models. In the case of the sum of Gram matrices of doubly correlated i.i.d. matrices, the Gram matrix of sums of doubly correlated i.i.d. matrices or matrices with a variance profile, this approach was identically used to derive expressions for the Shannon transform, see further Chapter 6.

As for S-transform related derivations, we have the following result.

Theorem 4.11 ([Hachem, 2008]). *Let $\mathbf{D} \in \mathbb{C}^{N\times N}$ and $\mathbf{T} \in \mathbb{C}^{N\times N}$ be diagonal non-negative matrices, and $\mathbf{W} \in \mathbb{C}^{N\times N}$ be a Haar matrix. Denote D and T the respective l.s.d. of $\mathbf{D}$ and $\mathbf{T}$. Denote $\mathbf{B}_N$ the matrix*

$$\mathbf{B}_N = \mathbf{D}^{\frac{1}{2}}\mathbf{W}\mathbf{T}\mathbf{W}^{\mathsf{H}}\mathbf{D}^{\frac{1}{2}}.$$

Then, as N grows large, the e.s.d. of $\mathbf{B}_N$ converges to F whose η-transform $\eta_F(x)$ satisfies

$$\begin{aligned}
\eta_F(x) &= \int (x\gamma(x)t + 1)^{-1}\, dD(t) \\
\gamma(x) &= \int t\,(\eta_F(x) + x\delta(x)t)^{-1}\, dT(t) \\
\delta(x) &= \int t\,(x\gamma(x)t + 1)^{-1}\, dD(t)
\end{aligned}$$

and whose Shannon transform $\mathcal{V}_F(x)$ satisfies

$$\mathcal{V}_F(x) = \int \log(1 + x\gamma(x)t)\, dD(t) + \int \log(\eta_F(x) + x\delta(x)t) dT(t).$$

This last result is proved similarly to Theorem 4.10, using the S-transform identity

$$S_F(x) = S_D(x)S_T(x) \tag{4.14}$$

emerging from the fact that $\mathbf{D}$ and $\mathbf{W}\mathbf{T}\mathbf{W}^{\mathsf{H}}$ are asymptotically free random matrices (from Corollary 4.2) and that the e.s.d. of $\mathbf{W}\mathbf{T}\mathbf{W}^{\mathsf{H}}$ is the e.s.d. of the matrix $\mathbf{T}$.

Proof. The proof requires the introduction of the functional $\zeta(z)$ defined as

$$\zeta(z) \triangleq \frac{\psi(z)}{1+\psi(z)}$$

with $\psi(z)$ introduced in Definition (3.6). It is shown in [Voiculescu, 1987] that ζ is analytical on $\mathbb{C}^+$. We denote its analytical inverse ζ^{-1}. From the definition of the S-transform as a function of ψ, we finally have $S(z) = \frac{1}{z}\zeta^{-1}(z)$.

From the almost sure asymptotic freeness of $\mathbf{D}$ and $\mathbf{W}\mathbf{T}\mathbf{W}^{\mathsf{H}}$, and from (4.14), we have:

$$\zeta_F^{-1}(z) = \zeta_D^{-1}(z) S_T(z).$$

Hence, replacing z by $\zeta_F(-z)$

$$-z = \zeta_D^{-1}(\zeta_F(-z)) S_T(\zeta_F(-z))$$

which, according to the definition of ζ_F, gives

$$\zeta_D^{-1}\left(\frac{\psi_F(-z)}{1+\psi_F(-z)}\right) = \frac{-z}{S_T\left(\frac{\psi_F(-z)}{1+\psi_F(-z)}\right)}.$$

Taking ψ_D on both sides, this is:

$$\psi_F(-z) = \psi_D\left(\frac{-z}{S_T\left(\frac{\psi_F(-z)}{1+\psi_F(-z)}\right)}\right) \tag{4.15}$$

since ζ_D^{-1} maps $\frac{x}{1+x}$ to $\psi_D^{-1}(x)$, and hence $\frac{\psi_F(-z)}{1+\psi_F(-z)}$ to $\psi_D^{-1}(\psi_F(-z))$.

Denoting

$$\gamma(z) = \frac{1}{S_T\left(\frac{\psi_F(-z)}{1+\psi_F(-z)}\right)}$$

this is

$$\psi_F(-z) = \psi_D\left(-z\gamma(z)\right).$$

Computing $\frac{\psi_F}{1+\psi_F}$ and taking ζ_T^{-1} of the resulting expression, we obtain

$$\zeta_T^{-1}\left(\frac{\psi_F(-z)}{1+\psi_F(-z)}\right) = \frac{1}{\gamma(z)}\frac{\psi_F(-z)}{1+\psi_F(-z)}.$$

Taking ψ_T on both sides, this is finally

$$\psi_F(-z) = \psi_T\left(\frac{1}{\gamma(z)}\frac{\psi_F(-z)}{1+\psi_F(-z)}\right). \tag{4.16}$$

To fall back on the system of equations, notice that $\eta_F(z) = 1 - \psi_F(-z)$. Equation (4.15) becomes

$$\eta_F(z) = \eta_D(z\gamma(z))$$

while Equation (4.16) leads to

$$\gamma(z) = \int \frac{t}{\eta_F(z) - \frac{\psi_D(-z\gamma(z))}{\gamma(z)} t} dT(t)$$

which are exactly the desired results with $\delta(z)$ defined as

$$\delta(z) = \frac{\psi_D(-z\gamma(z))}{-z\gamma(z)}.$$

□

Putting together Theorem 4.10 and Theorem 4.11, we can show the following most general result.

Theorem 4.12. *Let $\mathbf{D} \in \mathbb{C}^{N\times N}$, $\mathbf{T}_1, \ldots, \mathbf{T}_K \in \mathbb{C}^{N\times N}$ be diagonal non-negative definite matrices, and $\mathbf{W}_k \in \mathbb{C}^{N\times N}$, $k \in \{1, \ldots, K\}$, be independent Haar matrices. Denote D, T_k the l.s.d. of the matrices $\mathbf{D}$ and $\mathbf{T}_k$, respectively. Finally, denote $\mathbf{B}_N \in \mathbb{C}^{N\times N}$ the matrix*

$$\mathbf{B}_N = \sum_{k=1}^{K} \mathbf{D}^{\frac{1}{2}} \mathbf{W}_k \mathbf{T}_k \mathbf{W}_k^{\mathsf{H}} \mathbf{D}^{\frac{1}{2}}.$$

Then, as N grows large, the e.s.d. of $\mathbf{B}_N$ tends to the F whose η-transform η_F is given by:

$$\eta_F(x) = \left(1 + x\delta(x) \sum_{k=1}^{K} \gamma_k(x)\right)^{-1}$$

where $\gamma_1, \ldots, \gamma_K$ and δ are solutions to the system of equations

$$\delta(x) = \int t \left(1 + xt \sum_{k=1}^{K} \gamma_k(x)\right)^{-1} dD(t)$$

$$\gamma_k(x) = \int t\,(1 - x\delta(x)\gamma_k(x) + x\delta(x)t)^{-1}\, dT_k(t).$$

Also, the Shannon transform $\mathcal{V}_F(x)$ is given by:

$$\mathcal{V}_F(x) = \int \log\left(1 + tx \sum_{k=1}^{K} \gamma_k(x)\right) dD(t)$$
$$+ \sum_{k=1}^{K} \int \log\left(1 - x\delta(x)\gamma_k(x) + x\delta(x)t\right) dT_k(t).$$

To the best of our knowledge, this last result is as far as free probability reasoning can go. In particular, if the products $\mathbf{D}^{\frac{1}{2}}\mathbf{W}_i\mathbf{T}_i^{\frac{1}{2}}$ were replaced by the more general $\mathbf{D}_i^{\frac{1}{2}}\mathbf{W}_i\mathbf{T}_i^{\frac{1}{2}}$, then it is impossible to use free probability results any longer, as the family $\{\mathbf{D}_i^{\frac{1}{2}}\mathbf{W}_i\mathbf{T}_i\mathbf{W}_i^{\mathsf{H}}\mathbf{D}_i^{\frac{1}{2}},\ i \in \{1,\ldots,K\}\}$ is no longer asymptotically free almost everywhere. To deal with this model, which is fundamental to the study of multi-cellular frequency selective orthogonal CDMA systems, we will show in Section 6.2.4 that the Stieltjes transform approach is our only asset. The Stieltjes transform approach is however much less direct and requires more work than the free probability framework, although it is much more flexible.

In this chapter, we introduced free probability theory and its applications to large dimensional random matrices through a definition involving joint moments of spectral distributions with compact supports. Apart from the above analytical results based on the R-transform and S-transform, the common use of free probability theory for random matrices is related to the study of successive moments of distributions, which is in fact more appealing when the models under study are less tractable through analytic approaches. This is the subject of the following chapter, which introduces the combinatoric moment methods and which goes beyond the strict free probability framework to deal with more structured random matrix models enjoying some symmetry properties. To this day, these models have not been addressed through any analytical method but, due to their symmetric structure, combinatorics calculus can be performed.

5 Combinatoric approaches

We start the discussion regarding moment-based approaches by mentioning the well-known *method of moments*, which is a tool aimed at retrieving distribution functions based on moments from a classical probability theory point of view. This method of moments will be opposed to the *moment-based methods* or *moment methods* which is the center of interest of this chapter since it targets specifically combinatoric calculus in non-commutative algebras of random matrices.

5.1 The method of moments

In this section, we will discuss the technique known as the method of moments to derive a probability distribution function from its moments, see, e.g., [Bai and Silverstein, 2009; Billingsley, 1995]. When we are able to infer the limiting spectral distribution of some matrix model, although not able to prove it through classical analytic approaches, it is possible under certain conditions to derive all successive limiting moments of the distribution instead. Under Carleman's condition, to be introduced hereafter, these moments uniquely determine the limiting distribution.

We start by introducing Carleman's condition.

Theorem 5.1. *Let F be a distribution function, and denote $M_1, M_2, \ldots$ its sequence of moments which are assumed to be all finite. If the condition*

$$\sum_{k=1}^{\infty} M_{2k}^{-\frac{1}{2k}} = \infty \tag{5.1}$$

is fulfilled, then F is uniquely determined by the sequence $M_1, M_2, \ldots$.

Therefore, if we only have access to the moments $M_1, M_2, \ldots$ of some distribution F and that Carleman's condition is met, then F is uniquely determined by its moments.

In the specific case where $M_1, M_2, \ldots$ are the moments of the l.s.d. of large Hermitian matrices, we will need the following *moment convergence theorem*.

Theorem 5.2 (Lemma 12.1 and Lemma 12.3 of [Bai and Silverstein, 2009]). *Let $F_1, F_2, \ldots$ be a sequence of distribution functions, such that, for each n, F_n has finite moment $M_{n,k}$ of order k for all k, with $M_{n,k} \to M_k < \infty$ as $n \to \infty$. Assume additionally that the sequence of moments $M_1, M_2, \ldots$ fulfills Carleman's condition* (5.1)*. Then F_n converges to the unique distribution function F with moments M_k.*

To prove that the e.s.d. of a given sequence of Hermitian random matrices tends to a limit distribution function, all that is needed to show is the convergence of the empirical moments to limiting moments that meet Carleman's condition. It then suffices to match these moments to some previously inferred l.s.d., or to try to determine this l.s.d. directly from the moments. In the following, we give the main steps of the proof of the semi-circle law of Theorem 2.11, which we presently recall.

Consider a *Hermitian* matrix $\mathbf{X}_N \in \mathbb{C}^{N\times N}$, with independent entries $\frac{1}{\sqrt{N}}X_{N,ij}$ such that $\mathrm{E}[X_{N,ij}] = 0$, $\mathrm{E}[|X_{N,ij}|^2] = 1$ and there exists ε such that the $X_{N,ij}$ have a moment of order $2+\varepsilon$. Then $F^{\mathbf{X}_N} \Rightarrow F$ almost surely, where F has density f defined as

$$f(x) = \frac{1}{2\pi}\sqrt{(4-x^2)^+}.$$

Moreover, if the $X_{N,ij}$ are identically distributed, the result holds without the need for the existence of a moment of order $2+\varepsilon$.

Proof of Theorem 2.11. We wish to show that, on a space $A \subset \Omega$ of probability one, the empirical moments $M_{N,1}(\omega), M_{N,2}(\omega), \ldots$, $\omega \in A$, of the e.s.d. of a random Wigner matrix with i.i.d. entries converge to the moments $M_1, M_2, \ldots$ of the semi-circle law, and that these moments satisfy Carleman's condition. The moments of the semi-circle law are computed as

$$M_{2k+1} = \frac{1}{2\pi}\int_{-2}^{2} x^{2k+1}\sqrt{4-x^2}dx = 0$$

$$M_{2k} = \frac{1}{2\pi}\int_{-2}^{2} x^{2k}\sqrt{4-x^2}dx = \frac{1}{k+1}\binom{2k}{k}$$

without difficulty, by the change of variable $x = 2\sqrt{y}$. The value of M_{2k} is the kth *Catalan number*. Using Stirling's approximation formula, for k large

$$M_{2k}^{-\frac{1}{2k}} = (k+1)^{\frac{1}{2k}}\frac{(k!)^{\frac{1}{k}}}{((2k)!)^{\frac{1}{2k}}} \sim (k+1)^{\frac{1}{2k}}\frac{1}{2}\frac{(\sqrt{2\pi k})^{\frac{1}{k}}}{(\sqrt{4\pi k})^{\frac{1}{2k}}}$$

which tends to $\frac{1}{2}$. Therefore, there exists $k_0 > 0$, such that $k \geq k_0$ implies $M_{2k}^{-\frac{1}{2k}} > \frac{1}{4}$, which ensures that the series $\sum_k M_{2k}^{-\frac{1}{2k}}$ diverges, and Carleman's condition is satisfied.

The rest of this proof follows from [Anderson *et al.*, 2006, 2010; Bai and Silverstein, 2009], where more details are found. The idea is first to use truncation, centralization, and rescaling steps as described in Section 3.2 so to move from random entries of the Wigner matrix with no moment constraint to random entries with a distribution that admits moments of all orders. By showing that it is equivalent to prove the convergence of $F^{\mathbf{X}_N}$ for either $X_{N,ij}$ or their truncated versions, we can now work with variables with moments of all orders. We need to show that $\mathrm{E}[M_{N,k}] \triangleq \int M_{N,k}(\omega)dP(\omega) \to M_k$ and $\sum_N \mathrm{E}[(M_{N,k} - M_k)^2] < \infty$. This will be sufficient to ensure the almost sure convergence of the empirical moments to the moments of the semi-circle law, by applying the Markov inequality, Theorem 3.5, and the first Borel–Cantelli lemma, Theorem 3.6.

To determine the limit of $\mathrm{E}[M_{N,k}]$, we resort to combinatorial and particularly graph theoretical tools. It is required indeed, when developing the trace of $\mathbf{X}_N^k$

$$\begin{aligned}\mathrm{E}[M_{N,k}] &= \frac{1}{N}\mathrm{E}[\operatorname{tr}(\mathbf{X}_N^k)] \\ &= \frac{1}{N}\sum_{\mathbf{p}\in\{1,\ldots,N\}^k} \mathrm{E}X_{N,p_1p_2}X_{N,p_2p_3}\ldots X_{N,p_kp_1}\end{aligned}$$

with p_i the ith entry of $\mathbf{p}$, to be able to determine which moments of the $X_{N,ij}$ are non-negligible in the limit. These graph theoretical tools are developed in detail in [Bai and Silverstein, 2009] and [Anderson *et al.*, 2006]. What arises in particular is that all odd moments vanish, while it can be shown that, for the even moments

$$\mathrm{E}[M_{N,2k}] = (1 + O(1/N))\frac{1}{N}\sum_{j=0}^{k-1}\mathrm{E}[M_{N,2j}]\mathrm{E}[M_{N,2(k-j-1)}]$$

which establishes an asymptotic recursive equation for $\mathrm{E}[M_{N,2k}]$. For $k = 1$, we have $\mathrm{E}[M_{N,2}] \to 1$. Finally, by comparing the recursive formula to the definition of the *Catalan numbers* C_k, given later in (5.5), we obtain the required result

$$\mathrm{E}M_{N,2k} \to C_k.$$

□

The previous proof provided the successive moments of the Wigner semi-circle law, recalled in the following theorem.

Theorem 5.3. *Let F be the semi-circle law distribution function with density f given by*

$$f(x) = \frac{1}{2\pi}\sqrt{(4 - x^2)^+}.$$

This distribution is compactly supported and therefore has successive moments $M_1, M_2, \ldots$, *given by:*

$$M_{2k+1} = 0,$$
$$M_{2k} = \frac{1}{k+1}\binom{2k}{k}.$$

This general approach can be applied to prove the convergence of the e.s.d. of a random matrix model to a given l.s.d. However, this requires to know a priori the limit distribution sought for. Also, it assumes the existence of moments of all orders, which might already be a stringent assumption, but truncation, centralization, and rescaling steps can be performed prior to using the method of moments to alleviate this condition as was performed above. To the best of our knowledge, for more general random matrix models than Wigner or Wishart matrices, the method of moments leads to rather involved calculus, from which not much can be inferred. We therefore close the method of moments parenthesis here and will never mention it again.

We now move to the moment methods originating from the free probability framework, which are concerned specifically with random matrix models. We start with an introduction of the free moments and free cumulants of random non-commutative variables and random matrices.

5.2 Free moments and cumulants

Remember that we mentioned that Theorem 4.1 is a major result since it allows us to derive the successive moments of the l.s.d. of products of free random matrices from the moments of the l.s.d. of the individual random matrices. For instance, for free random variables A and B, we have:

$$\begin{aligned}
\phi(AB) &= \phi(A)\phi(B) \\
\phi(ABAB) &= \phi(A^2)\phi(B)^2 + \phi(A)^2\phi(B^2) - \phi(A)^2\phi(B)^2 \\
\phi(AB^2A) &= \phi(A^2)\phi(B^2).
\end{aligned}$$

Translated in terms of traces of random matrices, we can compute the so-called *free moments*, i.e. the moments of the (almost sure) l.s.d., of sums and products of random matrices as a function of the free moments of the operands.

It is then possible to derive the limiting spectrum of the sum of asymptotically free random matrices $\mathbf{A}_N$ and $\mathbf{B}_N$ from the sequences of the free moments of the l.s.d. of $\mathbf{A}_N + \mathbf{B}_N$. Theorem 4.2 has already shown that $\mathbf{A}_N + \mathbf{B}_N$ is connected to $\mathbf{A}_N$ and $\mathbf{B}_N$ through their respective R-transforms. Denote A and B two non-commutative random variables, with d.f. the l.s.d. of $\mathbf{A}_N$ and $\mathbf{B}_N$, respectively.

The formal R-transform series of A can be expressed as

$$R_A(z) = \sum_{k=1}^{\infty} C_k z^{k-1} \tag{5.2}$$

where C_k is called the kth *order free cumulant* of A. From Theorem 4.2, we therefore have

$$R_{A+B}(z) = \sum_{k=1}^{\infty} [C_k(A) + C_k(B)] z^{k-1}$$

with $C_k(A)$ and $C_k(B)$ the respective free cumulants of the l.s.d. of $\mathbf{A}_N$ and $\mathbf{B}_N$. In the same way as cumulants of independent random variables add up in classical probability theory, free cumulants of free random variables add up in free probability theory. This summarizes into the following result from Voiculescu [Voiculescu, 1986].

Theorem 5.4. *Let μ_A and μ_B be compactly supported probability distributions, with respective free cumulants $C_1(A), C_2(A), \ldots$ and $C_1(B), C_2(B), \ldots$. Then the free cumulants $C_1(A+B), C_2(A+B), \ldots$ of the distribution $\mu_{A+B} \triangleq \mu_A \boxplus \mu_B$ satisfy*

$$C_k(A+B) = C_k(A) + C_k(B).$$

We recall that, in reference to classical probability theory, the binary additive operation '$\mu_A \boxplus \mu_B$' is called the *free additive convolution* of the distributions μ_A and μ_B. We equivalently defined the binary operation '$\mu_A \boxminus \mu_B$' as the *free additive deconvolution* of μ_B from μ_A.

The cumulants of a given distribution μ_A, with bounded support, can be computed recursively using Theorem 4.2. Equating coefficients of the terms in z^k allows us to derive the kth order free cumulant C_k of A as a function of the first k free moments $M_1, \ldots, M_k$ of A. In particular, the first three cumulants read

$$\begin{aligned} C_1 &= M_1 \\ C_2 &= M_2 - M_1^2 \\ C_3 &= M_3 - 3M_1M_2 + 2M_1^3. \end{aligned}$$

It is therefore possible to evaluate explicitly the successive moments of the probability distribution $\mu_{A+B} = \mu_A \boxplus \mu_B$ of the sum of the random free variables A and B from the successive moments of μ_A and μ_B. The method comes as follows: given μ_A and μ_B, (i) compute the moments and then the cumulants $C_k(A)$ of A and $C_k(B)$ of B, (ii) compute the sum $C_k(A+B) = C_k(A) + C_k(B)$, (iii) from the cumulants $C_k(A+B)$, retrieve the corresponding moments of $A+B$, from which the distribution μ_{A+B}, assumed of compact support, can be found. This approach can be conducted in a combinatorial way, using *non-crossing partitions*. The method is provided by Speicher [Speicher, 1998], which simplifies many free probability calculus introduced by Voiculescu. In particular, the method described above allows us to recover the moments of $A+B$ from the cumulants of A and B in the following result.

Theorem 5.5. *Let A and B be free random variables with respective cumulants $\{C_k(A)\}$ and $\{C_k(B)\}$. The* nth *order free moment $M_n(A+B)$ of the random variable $A+B$ reads:*

$$M_n(A+B) = \sum_{\pi \in \mathrm{NC}(n)} \prod_{V \in \pi} \left(C_{|V|}(A) + C_{|V|}(B) \right)$$

with $\mathrm{NC}(n)$ *the set of non-crossing partitions of* $\{1, \ldots, n\}$ *and* $|V|$ *the cardinality of the subset V in the non-crossing partition partition π.*

Setting $B = 0$, we fall back on the relation between the moments $M_1(A), M_2(A), \ldots$ and the cumulants $C_1(A), C_2(A), \ldots$ of A

$$M_n(A) = \sum_{\pi \in \mathrm{NC}(n)} \prod_{V \in \pi} C_{|V|}(A) \tag{5.3}$$

introduced in Theorem 4.2.

Remark 5.1. Note that, since the limiting distribution functions under study are compactly supported, from Theorem 3.3, the Stieltjes transform m_{F^A} of the d.f. F^A associated with μ_A can be written, for $z \in \mathbb{C}^+$ in the convergence region of the series

$$m_{F^A}(z) = -\sum_{k=0}^{\infty} M_k z^{-k-1}$$

where M_k is the kth order moment of F^A, i.e. the kth order free moment of A. There therefore exists a strong link between the Stieltjes transform of the compactly supported distribution function F^A and the free moments of the non-commutative random variable A.

Contrary to the Stieltjes transform approach, though, the combinatorial method requires to compute all successive cumulants, or a sufficient number of them, to better estimate the underlying distribution. Remember though that, for compactly supported distributions, $M_k/k!$ vanishes fast for large k and then an estimate of only a few first moments might be good enough. This is in particular convenient when the studied distribution μ consists of K masses. If so, assuming the respective weights of each mass are known, the first K cumulants are sufficient to evaluate the full distribution function. Indeed, in the special case of evenly distributed masses, given $M_1, \ldots, M_K$ the first K moments of the distribution, $\mu(x) = \frac{1}{K} \sum_{i=1}^{K} \delta(x - \lambda_i)$ where $\lambda_1, \ldots, \lambda_K$ are the K roots of the polynomial

$$X^K - \Pi_1 X^{K-1} + \Pi_2 X^{K-2} - \ldots + (-1)^K \Pi_K$$

where $\Pi_1, \ldots, \Pi_n$ are the *elementary symmetric polynomials*, recursively computed from the Newton–Girard formula

$$(-1)^K K \Pi_K + \sum_{i=1}^{K} (-1)^{K+i} M_i \Pi_{K-i} = 0. \tag{5.4}$$

See [Séroul, 2000] for more details on the Newton–Girard formula.

Similar relations hold for the product of free random variables. In terms of moments and non-crossing partitions, the result is provided by Nica and Speicher [Nica and Speicher, 1996] in the following theorem.

Theorem 5.6. *Let A and B be free random variables with respective free cumulants $\{C_k(A)\}$ and $\{C_k(B)\}$. Then the nth order free moment $M_n(AB)$ of the random variable AB reads:*

$$M_n(AB) = \sum_{(\pi_1,\pi_2)\in \mathrm{NC}(n)} \prod_{\substack{V_1\in\pi_1 \\ V_2\in\pi_2}} C_{|V_1|}(A) C_{|V_2|}(B).$$

This formula enables the computation of all free moments of the distribution of AB from the free moments of μ_A and μ_B. We recall that the product distribution is denoted $\mu_A \boxtimes \mu_B$ and is called *multiplicative free convolution.* Reverting the polynomial formulas in the free cumulants enables also the computation of the free moments of μ_A from the free moments of μ_{AB} and μ_B. This therefore allows us to recover the distribution of the *multiplicative free deconvolution* $\mu_{AB} \boxslash \mu_B$.

Before debating the important results of free moments and cumulants for wireless communications, we shortly introduce non-crossing partitions and the relations between partitions in classical probability theory and non-crossing partitions in free probability theory.

In (classical) probability theory, we have the following relation between the cumulants c_n and the moments m_n of a given probability distribution

$$m_n = \sum_{\pi\in\mathcal{P}(n)} \prod_{V\in\pi} c_{|V|}$$

where $\mathcal{P}(n)$ is the set of partitions of $\{1, \ldots, n\}$. For instance, $\mathcal{P}(3)$ is composed of the five sets $\{\{1,2,3\}\}$, $\{\{1,2\},\{3\}\}$, $\{\{1,3\},\{2\}\}$, $\{\{2,3\},\{1\}\}$ and $\{\{1\},\{2\},\{3\}\}$. The cardinality of $\mathcal{P}(n)$ is called the *Bell number* B_n, recursively defined by

$$\begin{cases} B_0 & = 1 \\ B_{n+1} & = \sum_{k=0}^{n} \binom{n}{k} B_k. \end{cases}$$

From the example above, $B_3 = 5$.

We recall for instance that the cumulant $c_1 = m_1$ is the distribution mean, $c_2 = m_2 - m_1^2$ is the variance, c_3 is known as the *skewness*, and c_4 is the *kurtosis*.

Free probability theory provides the similar formula (5.3), where the sum is not taken over the partitions of $\{1,\ldots,n\}$ but over the *non-crossing partitions* of $\{1,\ldots,n\}$. Non-crossing partitions are defined as follows.

Definition 5.1. Consider the set $\{1,\ldots,n\}$. The partition $\pi \in \mathcal{P}(n)$ is said to be *non-crossing* if there does not exist $a < b < c < d$ elements of $\{1,\ldots,n\}$ (ordered modulo n), such that both $\{a,c\} \in \pi$ and $\{b,d\} \in \pi$.

Otherwise stated, this means that in a circular graph of the elements of $\{1,\ldots,n\}$ where the elements $V_1,\ldots,V_{|V|}$ of a given $\pi \in \mathcal{P}(n)$ are represented by $|V|$ polygons, with polygon k connecting the elements of V_k and such that the edges of polygon i and polygon j, $i \neq j$, never cross. This is depicted in Figure 5.1 in the case of $\pi = \{\{1,3,4\},\{2\},\{5,6,7\},\{8\}\}$, for $n = 8$. The number of such non-crossing partitions, i.e. the cardinality of $\mathrm{NC}(n)$, is known as the *Catalan number*, denoted C_n, which was seen incidentally to be connected to the moments of the semi-circle law, Theorem 5.3. This is summarized and proved in the following.

Theorem 5.7 ([Anderson *et al.*, 2006]). *The cardinality C_n of $\mathrm{NC}(n)$, for $n \geq 1$, satisfies the recursion equation*

$$C_1 = 1,$$

$$C_n = \sum_{k=1}^{n} C_{n-k} C_{k-1}$$

and is explicitly given by:

$$C_n = \frac{1}{n+1}\binom{2n}{n}. \tag{5.5}$$

We provide below the proof of this result, which is rather short and intuitive.

Proof. Let $\pi \in \mathrm{NC}(n)$ and denote j the smallest element connected to 1 with $j = 1$ if $\{1\} \in \pi$, e.g. $j = 3$ in Figure 5.1. Then necessarily both sets $\{1,\ldots,j-1\}$ and $\{j+1,\ldots,n\}$ are non-crossing and, for fixed link $(1,j)$, the number of non-crossing partitions in $\{1,\ldots,n\}$ is the product between the number of non-crossing partitions in the sets $\{1,\ldots,j-1\}$, i.e. C_{j-1}, and $\{j+1,\ldots,n\}$, i.e. C_{n-j}. We then have the relation

$$C_n = \sum_{j=1}^{n} C_{j-1} C_{n-j}$$

as expected, along with the obvious fact that $C_1 = 1$. By recursion calculus, it is then easy to see that the expression (5.5) satisfies this recursive equality. □

We now return to practical applications of the combinatorial moment framework. All results mentioned so far are indeed of practical use for problems

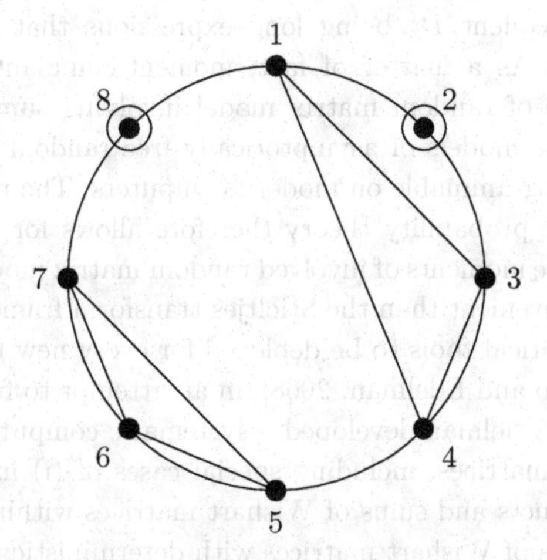

Figure 5.1 Non-crossing partition $\pi = \{\{1,3,4\},\{2\},\{5,6,7\},\{8\}\}$ of NC(8).

related to large dimensional random matrices in wireless communication settings. Roughly speaking, we can now characterize the limiting eigenvalue distributions for sums and products of matrices involving random Gaussian matrices, random unitary matrices, deterministic Hermitian matrices, etc. based on combinatorial calculus of their successive free moments. This is particularly suitable when the full limiting distribution is not required but only a few moments are needed, and when the random matrix model involves a large number of such matrices. The authors believe that all derivations handled by free probability theory can be performed using the Stieltjes transform method, although this requires more work. In particular, random matrix models involving unitary matrices are far more easily handled using free probability approaches than Stieltjes transform tools, as will be demonstrated in the subsequent sections. In contrast, the application range of free probability methods is seriously limited by the need for eigenvalue distributions to be compactly supported probability measures and more importantly by the need for the random matrices under study to be unitarily invariant (or more exactly asymptotically free).

Let us for instance apply the moment-cumulant relations as an application of Theorem 4.9. Denote B_k the kth order moment of μ_B and R_k the kth order moment of μ_R. Equation (4.7) provides a relation between μ_B and μ_R under the form of successive free convolution or deconvolution operations involving in particular the Marčenko–Pastur law. From the moments of the Marčenko–Pastur law, Theorem 2.14, and the above free addition and free product theorems, Theorem 5.5 and Theorem 5.6, we can then obtain polynomial relations between the B_k and the R_k. Following this procedure, Theorem 4.9 entails

$$\begin{aligned}
B_1 &= R_1 + 1, \\
B_2 &= R_2 + (2+2c)R_1 + (1+c), \\
B_3 &= R_3 + (3+3c)R_2 + 3cR_1^2 + (3+9c+3c^2+3)R_1 + (1+3c+c^2)
\end{aligned} \tag{5.6}$$

the subsequent B_k being long expressions that can be derived by computer software. As a matter of fact, moment-cumulant relations can be applied to any type of random matrix model involving sums, products, and information plus noise models of asymptotically free random matrices, the proper calculus being programmable on modern computers. The moment framework that arises from free probability theory therefore allows for very direct derivations of the successive moments of involved random matrix models. In this sense, this is much more convenient than the Stieltjes transform framework which requires involved mathematical tools to be deployed for every new matrix model.

In [Rao and Edelman, 2008], in an attempt to fully exploit the above remark, Rao and Edelman developed a systematic computation framework for a class of random matrices, including special cases of (i) information plus noise models, (ii) products and sums of Wishart matrices within themselves, or (iii) products and sums of Wishart matrices with deterministic matrices. This class is defined by the authors as the class of *algebraic random matrices.* Roughly speaking, this class gathers all random Hermitian matrices $\mathbf{X}$ with l.s.d. F for which there exists a bivariate complex-valued polynomial $L(x, y)$ satisfying

$$L(z, m_F(z)) = 0.$$

The class of algebraic random matrices is large enough to cover many practical applications in wireless communications. However, it does not include even the most basic model $\mathbf{X}_N\mathbf{T}_N\mathbf{X}_N^{\mathsf{H}}$, where $\mathbf{X}_N$ has i.i.d. Gaussian entries and the l.s.d. H of $\mathbf{T}_N$ has a *connected component.* If H is a discrete sum of masses in λ_k, $k = 1, \ldots, K$, then $\mathbf{X}_N\mathbf{T}_N\mathbf{X}_N^{\mathsf{H}}$ is algebraic. Indeed, from Theorem 3.13, the Stieltjes transform $m(z)$ of the l.s.d. of $\mathbf{X}_N\mathbf{T}_N\mathbf{X}_N^{\mathsf{H}}$ satisfies

$$m(z)\left(c\frac{1}{K}\sum_{k=1}^{K}\frac{\lambda_k}{1+\lambda_k m(z)} - z\right) = 1$$

which, after multiplication on both sides by $\prod_k(1 + \lambda_k m(z))$, leads to a bivariate polynomial expression in $(z, m(z))$. This is not true in general when H has a continuous support. Note also that several computer codes for evaluating free moments of algebraic random matrices are provided in [Rao *et al.*, 2008].

As repeatedly mentioned, the free probability framework is very limited in its application scope as it is only applied to large dimensional random matrices with unitarily invariant properties. Nonetheless, the important combinatorial machinery coming along with free probability can be efficiently reused to extend the initial results on Gaussian and Haar matrices to more structured types of matrices that enjoy other symmetry properties. The next chapter introduces the main results obtained for these extended methods in which new partition sets appear.

5.3 Generalization to more structured matrices

In wireless communications, research focuses mainly on Gaussian and Haar random matrices, but not only. Other types of random matrices, more structured, are desirable to study. This is especially the case of random Vandermonde matrices, defined as follows.

Definition 5.2. The *Vandermonde matrix* $\mathbf{V} \in \mathbb{C}^{N\times n}$ generated from the vector $(\alpha_1, \ldots, \alpha_n)^\mathsf{T}$ is the matrix with (i,j)th entry $V_{ij} = \alpha_i^{j-1}$

$$\mathbf{V} = \begin{pmatrix} 1 & 1 & \cdots & 1 \\ \alpha_1 & \alpha_2 & \cdots & \alpha_n \\ \vdots & \vdots & \cdots & \vdots \\ \alpha_1^{N-1} & \alpha_2^{N-1} & \cdots & \alpha_n^{N-1} \end{pmatrix}.$$

A *random Vandermonde matrix* is a normalized (by $\frac{1}{\sqrt{N}}$) Vandermonde matrix whose generating vector $(\alpha_1, \ldots, \alpha_n)^\mathsf{T}$ is a random vector.

In [Ryan and Debbah, 2009], the authors derive the successive moments of the e.s.d. of matrix models involving Vandermonde matrices with generating vector entries drawn uniformly and independently from the complex unit circle. The main result is as follows.

Theorem 5.8 ([Ryan and Debbah, 2009]). *Let $\mathbf{D}_1, \ldots, \mathbf{D}_L$ be L diagonal matrices of size $n \times n$ such that $\mathbf{D}_i$ has an almost sure l.s.d. as $n \to \infty$, for all i. Let $\mathbf{V} \in \mathbb{C}^{N\times n}$ be a random Vandermonde matrix with generators $\alpha_1, \ldots, \alpha_n$ drawn independently and uniformly from the unit complex circle. Call α a random variable on $[0, 2\pi)$ distributed as α_1. For $\rho \in \mathcal{P}(L)$, the set of partitions of $\{1, \ldots, L\}$, define $K_{\rho,\alpha,N}$ as*

$$K_{\rho,\alpha,N} = N^{|\rho|-L-1} \int_{[0,2\pi)^{|\rho|}} \prod_{k=1}^{L} \frac{1 - e^{jN(\alpha_{b(k-1)} - \alpha_{b(k)})}}{1 - e^{j(\alpha_{b(k-1)} - \alpha_{b(k)})}} \prod_{i=1}^{|\rho|} d\alpha_i$$

with $b(k)$ the index of the set of ρ containing k (since the α_i are i.i.d., the set indexing in ρ is arbitrary). If the limit

$$K_{\rho,\alpha} = \lim_{N\to\infty} K_{\rho,\alpha,N}$$

exists, then it is called a Vandermonde mixed moment expansion coefficient. If it exists for all $\rho \in \mathcal{P}(L)$, then we have, as $N, n \to \infty$ with $n/N \to c$, $0 < c < \infty$

$$\frac{1}{n}\operatorname{tr}\left(\prod_{i=1}^{L} \mathbf{D}_i \mathbf{V}^\mathsf{H}\mathbf{V}\right) \to \sum_{\rho\in\mathcal{P}(L)} K_{\rho,\alpha} c^{|\rho|-1} D_\rho$$

almost surely, where, for $\rho = \{\rho_1, \ldots, \rho_K\}$, *we denote*

$$D_{\rho_k} = \lim_{n\to\infty} \frac{1}{n}\operatorname{tr}\left(\prod_{i\in\rho_k} \mathbf{D}_i\right)$$

and

$$D_\rho = \prod_{k=1}^{K} D_{\rho_k}.$$

Contrary to the previous results presented for Gaussian random matrices, the extent of knowledge on the analytical approaches of random matrix theory so far does not enable us to determine the l.s.d. of random Vandermonde matrices in a closed-form. Only the aforementioned free probability approach is known to tackle this problem at this time. Note also that the support of the l.s.d. of such random Vandermonde matrices is not compact. It is therefore a priori uncertain whether the l.s.d. of such matrices can be determined by the limiting moments. For this, we need to verify that Carleman's condition, Theorem 5.1, is met. This has in fact been shown by Tucci and Whiting in [Tucci and Whiting, 2010]. For wireless communication purposes, it is possible to evaluate the capacity of a random Vandermonde channel model under the form of a series of moments. Such a channel arises whenever signals emerging from n sources impinge on an N-fold linear antenna array in line-of-sight. Assuming the sources are sufficiently far from the sensing array, the signals emerging from one particular source are received with equal amplitude by each antenna but with phases rotated proportionally to the difference of the optical path lengths, i.e. proportionally to both the antenna index in the array and the sinus of the incoming angle. Therefore, calling d_i the power of signal source i at the antenna array and $\mathbf{V}$ the Vandermonde matrix with generating vector the n phases of the incoming signals, the matrix $\mathbf{VD}$, $\mathbf{D} = \operatorname{diag}(d_1, \ldots, d_n)$, models the aforementioned communication channel.

Since the moments M_k of the (almost sure) l.s.d. of $\mathbf{VDV}^{\mathsf{H}}$ are only dependent on polynomial expressions of the moments $D_1, \ldots, D_k$ of the l.s.d. of $\mathbf{D}$ (in the notations of Theorem 5.8, $\mathbf{D}_1 = \ldots = \mathbf{D}_L$ so that $D_k \triangleq D_{\{1,\ldots,k\}}$), it is possible to recover the D_k from the M_k and hence obtain an estimate of D_k from the large dimensional observation $\mathbf{VDV}^{\mathsf{H}}$. Assuming a small number of sources, the estimates of $D_1, D_2, \ldots$ provide a further estimate of the respective distance of the signal sources. In [Ryan and Debbah, 2009], the first moments for this setup are provided. Denoting $\bar{M}_k \triangleq cM_k$ and $\bar{D}_k \triangleq cD_k$, we have the relations

$$\begin{aligned}
\bar{M}_1 &= \bar{D}_1 \\
\bar{M}_2 &= \bar{D}_2 + \bar{D}_1^2 \\
\bar{M}_3 &= \bar{D}_3 + 3\bar{D}_2\bar{D}_1 + \bar{D}_1^3 \\
\bar{M}_4 &= \bar{D}_4 + 4\bar{D}_3\bar{D}_1 + \frac{8}{3}\bar{D}_2^2 + 6\bar{D}_2\bar{D}_1^2 + \bar{D}_1^4
\end{aligned}$$

$$\bar{M}_5 = \bar{D}_5 + 5\bar{D}_4\bar{D}_1 + \frac{25}{3}\bar{D}_3\bar{D}_2 + 10\bar{D}_3\bar{D}_1^2 + \frac{40}{3}\bar{D}_2^2\bar{D}_1 + 10\bar{D}_2\bar{D}_1^3 + \bar{D}_1^5$$

from which $\bar{D}_1, \bar{D}_2, \ldots$ can be written as a function of $\bar{M}_1, \bar{M}_2, \ldots$.

The successive moments of other structured matrices can be studied similarly for random Toeplitz or Hankel matrices. The moment calculus can in fact be directly derived from the moment calculus of Vandermonde matrices [Ryan and Debbah, 2011]. We have in particular the following results.

Theorem 5.9. *Let* $\mathbf{X}_N \in \mathbb{R}^{N \times N}$ *be the Toeplitz matrix given by:*

$$\mathbf{X}_N = \frac{1}{\sqrt{N}} \begin{pmatrix} X_0 & X_1 & X_2 & \ldots & X_{N-2} & X_{N-1} \\ X_1 & X_0 & X_1 & & & X_{N-2} \\ X_2 & X_1 & X_0 & & \ddots & \vdots \\ \vdots & & & \ddots & & X_2 \\ X_{N-2} & & & & X_0 & X_1 \\ X_{N-1} & X_{N-2} & \ldots & X_2 & X_1 & X_0 \end{pmatrix}$$

with $X_0, X_1, \ldots$ *real independent Gaussian with zero mean and unit variance. Then the moment* M_k *of order* k *of the l.s.d. of* $\mathbf{X}_N$ *is given by* $M_k = 0$ *for* k *odd and the first even moments are given by:*

$$M_2 = 1$$
$$M_4 = \frac{8}{3}$$
$$M_6 = 11$$
$$M_8 = \frac{1435}{24}.$$

Remember for instance that slow fading frequency selective channels can be modeled by Toeplitz matrices. When the matrices are *Wiener-class* [Gray, 2006], i.e. when the series formed of the elements of the first row is asymptotically summable (this being in particular true when a finite number of elements are non-zero), it is often possible to replace the Toeplitz matrices by circulant matrices without modifying the l.s.d., see Theorem 12.1. Since circulant matrices are diagonalizable in the Fourier basis, their study is simpler, so that this assumption is often considered. However, in many practical cases, the Wiener-class assumption does not hold so results on the l.s.d. of Toeplitz matrices are of major importance.

A similar result holds for Hankel matrices.

Theorem 5.10. *Let $\mathbf{X}_N \in \mathbb{R}^{N\times N}$ be the Hankel matrix defined by*

$$\mathbf{X}_N = \frac{1}{\sqrt{N}} \begin{pmatrix} X_0 & X_1 & X_2 & \dots & X_{N-2} & X_{N-1} \\ X_1 & X_2 & X_3 & & & X_N \\ X_2 & X_3 & X_4 & & \ddots & \vdots \\ \vdots & & & \ddots & & X_{2N-2} \\ X_{N-2} & & & & X_{2N-2} & X_{2N-1} \\ X_{N-1} & X_N & \dots & X_{2N-2} & X_{2N-1} & X_{2N} \end{pmatrix}$$

with $X_0, X_1, \dots$ real independent Gaussian with zero mean and unit variance. Then the free moment M_k of order k of $\mathbf{X}_N$ is given by $M_k = 0$ for k odd and the first even moments are given by:

$$M_2 = 1$$
$$M_4 = \frac{8}{3}$$
$$M_6 = 14$$
$$M_8 = 100.$$

In general, the exact features that random matrices must fulfill for results such as Theorem 5.8 to be easily derived are not yet fully understood. It seems however that any random matrix $\mathbf{X}$ whose joint entry probability distribution is invariant by left product with permutation matrices enters the same scheme as random Vandermonde, Toeplitz and Hankel matrices, i.e. the free moments of matrix products of the type $\prod_{k=1}^{L} \left(\mathbf{D}_k \mathbf{X}\mathbf{X}^{\mathsf{H}}\right)$ can be derived from the moments of $\mathbf{D}_1, \dots, \mathbf{D}_L$. This is a very new, yet immature, field of research.

As already stated in Chapter 4, free probabilistic tools can also be used in place of classical random theoretical tools to derive successive moments of probability distributions of large random matrices. We will mention here an additional usage of moment-based approaches on the results from free probability theory described previously that allows us to obtain exact results on the *expected* eigenvalue distribution of small dimensional random matrices, instead of the almost sure l.s.d. of random matrices.

5.4 Free moments in small dimensional matrices

Thanks to the unitary invariance property of Gaussian matrices, standard combinatorics tools such as non-crossing partitions allow us to further generalize moment results obtained asymptotically, as the matrix dimensions grow large, to exact results on the moments of the expected e.s.d. of matrices for all fixed dimensions. We have in particular the following theorem for small dimensional Wishart and deterministic matrix products.

Theorem 5.11 ([Masucci *et al.*, 2011]). *Let $\mathbf{X}_N \in \mathbb{C}^{N\times n}$ have i.i.d. standard Gaussian entries and $\mathbf{T}_N$ be a (deterministic) $N \times N$ matrix. For any positive integer p, we have:*

$$\mathrm{E}\left[\frac{1}{N}\operatorname{tr}\left(\left(\mathbf{T}_N\frac{1}{n}\mathbf{X}_N\mathbf{X}_N^{\mathsf{H}}\right)^p\right)\right] = \sum_{\pi\in\mathcal{P}(p)} n^{k(\hat{\pi})-p}N^{l(\hat{\pi})-1}T_{\hat{\pi}|odd} \tag{5.7}$$

where $\hat{\pi} \in \mathcal{P}(2p)$ is the permutation such that

$$\begin{cases} \hat{\pi}(2j-1) = 2\pi^{-1}(j), & j\in\{1,2,\ldots,p\} \\ \hat{\pi}(2j) \quad\;\; = 2\pi(j)-1, & j\in\{1,2,\ldots,p\}. \end{cases}$$

Every such $\hat{\pi}$ is attached the equivalence relation $\sim_{\hat{\pi}}$, defined as

$$j \sim_{\hat{\pi}} \hat{\pi}(j)+1.$$

In (5.7), $\hat{\pi}|odd$ is the set consisting in the equivalence classes/blocks of $\hat{\pi}$ which are contained within the odd numbers, $k(\rho)$ is the number of blocks in ρ consisting of only even numbers, $l(\rho)$ is the number of blocks in ρ consisting of only odd numbers, $T_\rho = \prod_{i=1}^k \frac{1}{N}\operatorname{tr}\left(\mathbf{T}_N^{|\rho_i|}\right)$ whenever $\rho = \{\rho_1,\ldots,\rho_k\}$ is a partition with blocks ρ_i, and $|\rho_i|$ is the number of elements in ρ_i.

An information plus noise equivalent to Theorem 4.9 is also provided in [Masucci *et al.*, 2011] which requires further considerations of set partitions. Both results arise from a generic diagrammatic framework to compute successive moments of matrices invariant by row or column permutations.

We complete this introduction on extended combinatorics tools with recent advances in the study of rectangular random matrices from a free probability approach.

5.5 Rectangular free probability

A recent extension of free probability theory for Hermitian random matrices to the most general rectangular matrices has been proposed by Benaych-Georges [Benaych-Georges, 2009]. The quantity of interest is no longer the empirical distribution of the eigenvalues of square Hermitian matrices but the *symmetrized singular law* of rectangular matrices.

Definition 5.3. Let $\mathbf{M} \in \mathbb{C}^{N\times n}$ be a rectangular random matrix on $(\Omega, \mathcal{F}, P)$. The *singular law* μ of $\mathbf{M}$ is the uniform distribution of its singular values $s_1, \ldots, s_{\min(n,N)}$. The kth order moment M_k of μ is defined as

$$M_k = \frac{1}{\min(n,N)}\operatorname{tr}\left(\mathbf{M}\mathbf{M}^{\mathsf{H}}\right)^{\frac{k}{2}}.$$

The *symmetrized singular law* of $\mathbf{M}$ is the probability distribution $\tilde{\mu}$ such that, for any Borel set $A \in \mathcal{F}$, $\tilde{\mu}(A) = \frac{1}{2}(\mu(A) + \mu(-A))$.

We have similar results for rectangular matrices as for Hermitian matrices. In particular, we define a *rectangular additive free convolution* operator '$\boxplus_c$', with $c = \lim_N N/n$, which satisfies the following.

Theorem 5.12. *Let $\mathbf{M}_1$, $\mathbf{M}_2$ be independent bi-unitarily invariant $N \times n$ matrices whose symmetrized singular laws converge, respectively, to the measures μ_1 and μ_2, as n, N grow to infinity with limit ratio $N/n \to c$. Then the symmetrized singular law of $\mathbf{M}_1 + \mathbf{M}_2$ converges to a symmetric probability measure, dependent on μ_1, μ_2, and c only, and which we denote $\mu_1 \boxplus_c \mu_2$.*

The rectangular additive free convolution can be computed explicitly from an equivalent *rectangular R-transform* which has the same property as in the square case of summing cumulants of convolution of symmetrized singular laws.

Theorem 5.13. *For a given symmetric distribution μ, denote*

$$R_{\mu,c}(z) \triangleq \sum_{n=1}^{\infty} C_{2n,c}(\mu) z^n$$

where $C_{2n,c}(\mu)$ are the rectangular free cumulants of μ with ratio c, linked to the free moments $M_n(\mu)$ of μ by

$$M_n(\mu) = \sum_{\pi \in \mathrm{NC}'(2n)} c^{e(\pi)} \prod_{V \in \pi} C_{|V|,c}(\mu)$$

where $\mathrm{NC}'(2n)$ is the subset of non-crossing partitions of $\{1, \ldots, 2n\}$ with all blocks of even cardinality and $e(\pi)$ is the number of blocks of π with even cardinality.

Then, for two distributions μ_1 and μ_2, we have:

$$R_{\mu_1 \boxplus_c \mu_2, c}(z) = R_{\mu_1,c}(z) + R_{\mu_2,c}(z).$$

This is as far as we will go with rectangular random matrices, which is also a very new field of research in mathematics, with still few applications to wireless communications; see [Gregoratti *et al.*, 2010] for an example in the context of relay networks with unitary precoders.

To conclude the last three theoretical chapters, we recollect the different techniques introduced so far in a short conclusion on the methodology to adopt when addressing problems of random matrix theory. The methods to consider heavily depend on the application sought for, on the time we have to invest on the study, and obviously on the feasibility of every individual problem.

5.6 Methodology

It is fundamental to understand why we would address a question regarding some random matrix model from the analytical or the moments approach.

Say our intention is to study the e.s.d. $F^{\mathbf{X}_N}$ of a given random Hermitian matrix $\mathbf{X}_N \in \mathbb{C}^{N\times N}$. Using the analytical methods, $F^{\mathbf{X}_N}$ will be treated as a system parameter and will be shown to satisfy some classical analytic properties, such as: $F^{\mathbf{X}_N}$ has a weak limit F, the Stieltjes transform of F is solution of some implicit equation, etc. The moment-based methods will focus on establishing results on the successive moments $M_1, M_2, \ldots$ of F (or $\mathrm{E}[F^{\mathbf{X}_N}]$) when they exist, such as: M_k is linked to the moments M'_i, $i = 1, \ldots, k$, of another distribution F', M_k vanishes for $k > 2$ for growing dimensions of $\mathbf{X}_N$, etc. Both types of methods will therefore ultimately give mutually consistent results. However, the choice of a particular method over the other is often motivated by the following aspects.

1. **Mathematical attractiveness**. Both methods involve totally different mathematical tools and it often turns out that one method is preferable over the other in this respect. In particular, moment-based methods, while leading to tedious combinatorial computations, are very attractive due to their mechanical and simple way of working. In contrast, the analytical methods are not so flexible in some respects and are not yet able to solve many problems already addressed by different moment-based methods. For instance, in Section 5.3, we introduced results on large dimensional random Vandermonde, Toeplitz, Hankel matrices and bi-unitarily invariant rectangular matrices, which analytical methods are far from being able to provide. However, the converse also holds: some random matrix models can be studied by the Stieltjes transform approach, while moment-based methods are unusable. This is in fact the case for all random matrix models involving matrices with independent entries that are often not unitarily invariant. It might also turn out that the moment-based methods are of no use for the evaluation of certain functionals of F. For instance, the fact that the series expansion of $\log(1 + x)$ has convergence radius 1 implies that the moments of F do not allow us to estimate $\int \log(1 + x\lambda)dF(\lambda)$ for large x. The immediate practical consequence is that most capacity expressions cannot be evaluated from the moments of F.
2. **Application context**. The most important drawback of the moment-based methods lies in their results consisting of a series of properties concerning the individual or joint free moments. If we are interested in studying the limiting distribution F of the e.s.d. of a given random matrix $\mathbf{X}_N \in \mathbb{C}^{N\times N}$, two cases generally occur: (i) F is a step function with K discontinuities, i.e. F has K distinct eigenvalues with large multiplicities, in which case results on the first $M_1, \ldots, M_K$ may be sufficient to obtain (or estimate) F completely (especially if the multiplicities are a priori known), although this estimate is likely to perform poorly if *only* K moments are used, and (ii) F is a non-

trivial function, in which case all moments are in general needed to accurately evaluate it. In case (ii), moment-based methods are not desirable because a large number of moments need to be estimated (this often goes along with high computational complexity) and because the moment estimates are themselves correlated according to some non-trivial joint probability distribution, which is even more computationally complex to evaluate. Typically, a small error in the estimate of the moment of order 1 propagates into a larger error in the estimate of the moment of order 2, which itself propagates forward into higher moment estimates. These errors need to be tracked precisely to optimally exploit the successive estimates. Analytical methods, if numerically solvable, are much more appealing in this case. In general, the result of these methods expresses as an approximation of m_F by another Stieltjes transform, the approximation being asymptotically accurate as the system dimensions grow large. The analytical approach will especially be shown often to rely on computationally inexpensive fixed-point algorithms with proven convergence, see, e.g. Chapters 12–15, while moment-based methods require involved combinatorial calculus when a large number of moments has to be taken into account.

3. **Interpretation purpose.** In general, analytical methods provide very compact expressions, from which the typical behavior of the relevant problem parameters can be understood. The example of the optimality of the water-filling algorithm in the capacity maximization problem is a typical case where this phenomenon appears, see Chapters 13–14. On the moment-based method side, results appear in the form of lengthy combinatorial calculus, from which physical interpretation is not always possible.

This concludes the set of three chapters on the limiting results of large dimensional random matrices, using the Stieltjes transform, free probability theory, and related methods using moments. The next chapter will be dedicated to further extensions of the Stieltjes transform approach, which are of fundamental use in the applicative context of wireless communications. The first of these approaches extends the limiting results of Theorems 3.13, 3.14, 3.15, 4.10, 4.11, etc., to the case where the e.s.d. of the underlying matrices does not necessarily converge and allows us to provide accurate approximations of the empirical Stieltjes transform for all finite matrix dimensions. This will be shown to have crucial consequences from a practical point of view, in particular for the performance study of large dimensional wireless communication systems.

6 Deterministic equivalents

6.1 Introduction to deterministic equivalents

The first applications of random matrix theory to the field of wireless communications, e.g., [Tse and Hanly, 1999; Tse and Verdú, 2000; Verdú and Shamai, 1999], originally dealt with the limiting behavior of some simple random matrix models. In particular, these results are attractive as these limiting behaviors only depend on the limiting eigenvalue distribution of the deterministic matrices of the model. This is in fact the case of all the results we have derived and introduced so far; for instance, Theorem 3.13 unveils the limiting behavior of the e.s.d. of $\mathbf{B}_N = \mathbf{A}_N + \mathbf{X}_N^{\mathsf{H}}\mathbf{T}_N\mathbf{X}_N$ when both e.s.d. of $\mathbf{A}_N$ and $\mathbf{T}_N$ converge toward given deterministic distribution functions and $\mathbf{X}_N$ is random with i.i.d. entries. However, for practical applications, it might turn out that:

(i) the e.s.d. of $\mathbf{A}_N$ or $\mathbf{T}_N$ do not necessarily converge to a limiting distribution;
(ii) even if the e.s.d. of the deterministic matrices in the model do all converge to their respective l.s.d., the e.s.d. of the output matrix $\mathbf{B}_N$ might not converge. This is of course not the case in Theorem 3.13, but we will show that this may happen for more involved models, e.g. the models treated by [Couillet *et al.*, 2011a] and [Hachem *et al.*, 2007].

Let us introduce a simple scenario for which the e.s.d. of the random matrix does not converge. This example is borrowed from [Hachem *et al.*, 2007]. Define $\mathbf{X}_N \in \mathbb{C}^{2N\times 2N}$ as

$$\mathbf{X}_N = \begin{pmatrix} \mathbf{X}'_N & 0 \\ 0 & 0 \end{pmatrix} \tag{6.1}$$

with the entries of $\mathbf{X}'_N$ being i.i.d. with zero mean and variance $\frac{1}{N}$. Consider in addition the matrix $\mathbf{T}_N \in \mathbb{C}^{2N\times 2N}$ defined as

$$\mathbf{T}_N = \begin{cases} \begin{pmatrix} \mathbf{I}_N & 0 \\ 0 & 0 \end{pmatrix}, & N \text{ even} \\ \begin{pmatrix} 0 & 0 \\ 0 & \mathbf{I}_N \end{pmatrix}, & N \text{ odd.} \end{cases} \tag{6.2}$$

Then, taking $\mathbf{B}_N = (\mathbf{T}_N + \mathbf{X}_N)(\mathbf{T}_N + \mathbf{X}_N)^{\mathsf{H}}$, $F^{\mathbf{B}_{2N}}$ and $F^{\mathbf{B}_{2N+1}}$ both converge weakly towards limit distributions, as $N \to \infty$, but those distributions

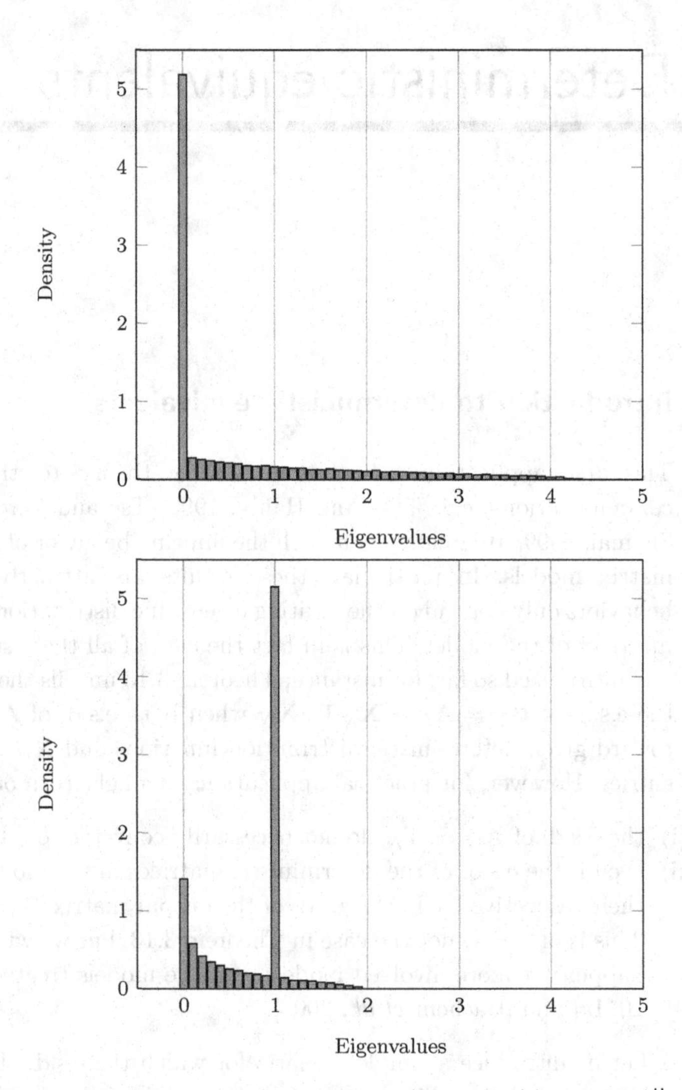

Figure 6.1 Histogram of the eigenvalues of $\mathbf{B}_N = (\mathbf{T}_N + \mathbf{X}_N)(\mathbf{T}_N + \mathbf{X}_N)^{\mathsf{H}}$ modeled in (6.1)–(6.2), for $N = 1000$ (top) and $N = 1001$ (bottom).

differ. Indeed, for N even, half of the spectrum of $\mathbf{B}_N$ is formed of zeros, while for N odd, half of the spectrum of $\mathbf{B}_N$ is formed of ones, the rest of the spectrum being a weighted version of the Marčenko–Pastur law. And therefore there does not exist a limit to $F^{\mathbf{B}_N}$, while $F^{\mathbf{X}_N\mathbf{X}_N^{\mathsf{H}}}$ tends to the uniformly weighted sum of the Marčenko–Pastur law and a mass in zero, and $F^{\mathbf{T}_N\mathbf{T}_N^{\mathsf{H}}}$ tends to the uniformly weighted sum of two masses in zero and one. This is depicted in Figure 6.1.

In such situations, there is therefore no longer any interest in looking at the asymptotic behavior of e.s.d. Instead, we will be interested in finding *deterministic equivalents* for the underlying model.

Definition 6.1. Consider a series of Hermitian random matrices $\mathbf{B}_1, \mathbf{B}_2, \ldots$, with $\mathbf{B}_N \in \mathbb{C}^{N\times N}$ and a series $f_1, f_2, \ldots$ of functionals of $1\times 1, 2\times 2, \ldots$ matrices. A *deterministic equivalent of* $\mathbf{B}_N$ *for the functional* f_N is a series $\mathbf{B}_1^\circ, \mathbf{B}_2^\circ, \ldots$ where $\mathbf{B}_N^\circ \in \mathbb{C}^{N\times N}$, of *deterministic* matrices, such that

$$\lim_{N\to\infty} f_N(\mathbf{B}_N) - f_N(\mathbf{B}_N^\circ) \to 0$$

where the convergence will often be with probability one. Note that $f_N(\mathbf{B}_N^\circ)$ does not need to have a limit as $N\to\infty$. We will similarly call $g_N \triangleq f_N(\mathbf{B}_N^\circ)$ the *deterministic equivalent of* $f_N(\mathbf{B}_N)$, i.e. the deterministic series $g_1, g_2, \ldots$ such that $f_N(\mathbf{B}_N) - g_N \to 0$ in some sense.

We will often take f_N to be the normalized trace of $(\mathbf{B}_N - z\mathbf{I}_N)^{-1}$, i.e. the Stieltjes transform of $F^{\mathbf{B}_N}$. When $f_N(\mathbf{B}_N^\circ)$ does not have a limit, the Marčenko–Pastur method, developed in Section 3.2, will fail. This is because, at some point, all the entries of the underlying matrices will have to be taken into account and not only the diagonal entries, as in the proof we provided in Section 3.2. However, the Marčenko–Pastur method can be tweaked adequately into a technique that can cope with deterministic equivalents. In the following, we first introduce this technique, which we will call the *Bai and Silverstein technique*, and then discuss an alternative technique, known as the *Gaussian method*, which is particularly suited to random matrix models with Gaussian entries. Hereafter, we detail these methods by successively proving two (similar) results of importance in wireless communications, see further Chapters 13–14.

6.2 Techniques for deterministic equivalents

6.2.1 Bai and Silverstein method

We first introduce a deterministic equivalent for the model

$$\mathbf{B}_N = \sum_{k=1}^{K} \mathbf{R}_k^{\frac{1}{2}} \mathbf{X}_k \mathbf{T}_k \mathbf{X}_k^{\mathsf{H}} \mathbf{R}_k^{\frac{1}{2}} + \mathbf{A}$$

where the K matrices $\mathbf{X}_k$ have i.i.d. entries for each k, mutually independent for different k, and the matrices $\mathbf{T}_1, \ldots, \mathbf{T}_K$, $\mathbf{R}_1, \ldots, \mathbf{R}_K$ and $\mathbf{A}$ are 'bounded' in some sense to be defined later. This is more general than the model of Theorem 3.13 in several respects:

(i) left product matrices $\mathbf{R}_k$, $1 \le k \le K$, have been introduced. As an exercise, it can already be verified that a l.s.d. for the model $\mathbf{R}_1^{\frac{1}{2}} \mathbf{X}_1 \mathbf{T}_1 \mathbf{X}_1^{\mathsf{H}} \mathbf{R}_1^{\frac{1}{2}} + \mathbf{A}$ may not exist even if $F^{\mathbf{R}_1}$ and $F^{\mathbf{A}}$ both converge vaguely to deterministic limits, unless some severe additional constraint is put on the eigenvectors of $\mathbf{R}_1$ and $\mathbf{A}$, e.g. $\mathbf{R}_1$ and $\mathbf{A}$ are codiagonalizable. This suggests that the Marčenko–Pastur method will fail to treat this model;

(ii) a sum of K such models is considered (K does not grow along with N here);
(iii) the e.s.d. of the (possibly random) matrices $\mathbf{T}_k$ and $\mathbf{R}_k$ are not required to converge.

While the result to be introduced hereafter is very likely to hold for $\mathbf{X}_1, \ldots, \mathbf{X}_K$ with non-identically distributed entries (as long as they have common mean and variance and some higher order moment condition), we only present here the result where these entries are identically distributed, which is less general than the conditions of Theorem 3.13.

Theorem 6.1 ([Couillet *et al.*, 2011a]). *Let K be some positive integer. For some integer N, let*

$$\mathbf{B}_N = \sum_{k=1}^{K} \mathbf{R}_k^{\frac{1}{2}} \mathbf{X}_k \mathbf{T}_k \mathbf{X}_k^{\mathsf{H}} \mathbf{R}_k^{\frac{1}{2}} + \mathbf{A}$$

be an $N \times N$ matrix with the following hypotheses, for all $k \in \{1, \ldots, K\}$

1. *$\mathbf{X}_k = \left(\frac{1}{\sqrt{n_k}} X_{k,ij}\right) \in \mathbb{C}^{N \times n_k}$ is such that the $X_{k,ij}$ are identically distributed for all N, i, j, independent for each fixed N, and $\mathrm{E}|X_{k,11} - \mathrm{E}X_{k,11}|^2 = 1$;*
2. *$\mathbf{R}_k^{\frac{1}{2}} \in \mathbb{C}^{N \times N}$ is a Hermitian non-negative definite square root of the non-negative definite Hermitian matrix $\mathbf{R}_k$;*
3. *$\mathbf{T}_k = \mathrm{diag}(\tau_{k,1}, \ldots, \tau_{k,n_k}) \in \mathbb{C}^{n_k \times n_k}$, $n_k \in \mathbb{N}^*$, is diagonal with $\tau_{k,i} \geq 0$;*
4. *the sequences $F^{\mathbf{T}_1}, F^{\mathbf{T}_2}, \ldots$ and $F^{\mathbf{R}_1}, F^{\mathbf{R}_2}, \ldots$ are tight, i.e. for all $\varepsilon > 0$, there exists $M > 0$ such that $1 - F^{\mathbf{T}_k}(M) < \varepsilon$ and $1 - F^{\mathbf{R}_k}(M) < \varepsilon$ for all n_k, N;*
5. *$\mathbf{A} \in \mathbb{C}^{N \times N}$ is Hermitian non-negative definite;*
6. *denoting $c_k = N/n_k$, for all k, there exist $0 < a < b < \infty$ for which*

$$a \leq \liminf_N c_k \leq \limsup_N c_k \leq b. \tag{6.3}$$

Then, as all N and n_k grow large, with ratio c_k, for $z \in \mathbb{C} \setminus \mathbb{R}^+$, the Stieltjes transform $m_{\mathbf{B}_N}(z)$ of $\mathbf{B}_N$ satisfies

$$m_{\mathbf{B}_N}(z) - m_N(z) \xrightarrow{\text{a.s.}} 0 \tag{6.4}$$

where

$$m_N(z) = \frac{1}{N} \mathrm{tr} \left(\mathbf{A} + \sum_{k=1}^{K} \int \frac{\tau_k dF^{\mathbf{T}_k}(\tau_k)}{1 + c_k \tau_k e_{N,k}(z)} \mathbf{R}_k - z\mathbf{I}_N \right)^{-1} \tag{6.5}$$

and the set of functions $e_{N,1}(z), \ldots, e_{N,K}(z)$ forms the unique solution to the K equations

$$e_{N,i}(z) = \frac{1}{N} \mathrm{tr}\, \mathbf{R}_i \left(\mathbf{A} + \sum_{k=1}^{K} \int \frac{\tau_k dF^{\mathbf{T}_k}(\tau_k)}{1 + c_k \tau_k e_{N,k}(z)} \mathbf{R}_k - z\mathbf{I}_N \right)^{-1} \tag{6.6}$$

such that $\mathrm{sgn}(\Im[e_{N,i}(z)]) = \mathrm{sgn}(\Im[z])$, if $z \in \mathbb{C} \setminus \mathbb{R}$, and $e_{N,i}(z) > 0$ if z is real negative.

Moreover, for any $\varepsilon > 0$, the convergence of Equation (6.4) is uniform over any region of $\mathbb{C}$ bounded by a contour interior to

$$\mathbb{C} \setminus (\{z : |z| \leq \varepsilon\} \cup \{z = x + iv : x > 0, |v| \leq \varepsilon\}).$$

For all N, the function m_N is the Stieltjes transform of a distribution function F_N, and

$$F^{\mathbf{B}_N} - F_N \Rightarrow 0$$

almost surely as $N \to \infty$.

In [Couillet *et al.*, 2011a], Theorem 6.1 is completed by the following result.

Theorem 6.2. *Under the conditions of Theorem 6.1, the scalars $e_{N,1}(z), \ldots, e_{N,K}(z)$ are also explicitly given by:*

$$e_{N,i}(z) = \lim_{t \to \infty} e_{N,i}^t(z)$$

where, for all i, $e_{N,i}^0(z) = -1/z$ and, for $t \geq 1$

$$e_{N,i}^t(z) = \frac{1}{N} \operatorname{tr} \mathbf{R}_i \left(\mathbf{A} + \sum_{j=1}^K \int \frac{\tau_j dF^{\mathbf{T}_j}(\tau_j)}{1 + c_j \tau_j e_{N,j}^{t-1}(z)} \mathbf{R}_j - z\mathbf{I}_N \right)^{-1}.$$

This result, which ensures the convergence of the classical fixed-point algorithm for an adequate initial condition, is of fundamental importance for practical purposes as it ensures that the $e_{N,1}(z), \ldots, e_{N,K}(z)$ can be determined numerically in a deterministic way. Since the proof of Theorem 6.2 relies heavily on the proof of Theorem 6.1, we will prove Theorem 6.2 later.

Several remarks are in order before we prove Theorem 6.1. We have given much detail on the conditions for Theorem 6.1 to hold. We hereafter discuss the implications of these conditions. Condition 1 requires that the $X_{k,ij}$ be identically distributed across N, i, j, but not necessarily across k. Note that the identical distribution condition could be further released under additional mild conditions (such as all entries must have a moment of order $2 + \varepsilon$, for some $\varepsilon > 0$), see Theorem 3.13. Condition 4 introduces tightness requirements on the e.s.d. of $\mathbf{R}_k$ and $\mathbf{T}_k$. Tightness can be seen as the probabilistic equivalent to boundedness for deterministic variables. Tightness ensures here that no mass of the $F^{\mathbf{R}_k}$ and $F^{\mathbf{T}_k}$ escapes to infinity as n grows large. Condition 6 is more general than the requirement that c_k has a limit as it allows c_k, for all k, to wander between two positive values.

From a practical point of view, $\mathbf{R}_K^{\frac{1}{2}} \mathbf{X}_k \mathbf{T}_k^{\frac{1}{2}}$ will often be used to model a multiple antenna $N \times n_k$ channel with i.i.d. entries with transmit and receive correlations. From the assumptions of Theorem 6.1, the correlation matrices $\mathbf{R}_k$ and $\mathbf{T}_k$ are only required to be 'bounded' in the sense of tightness of their e.s.d. This means that, as the number of antennas grows, the eigenvalues of $\mathbf{R}_k$ and $\mathbf{T}_k$

can only blow up with increasingly low probability. If we increase the number N of antennas on a bounded three-dimensional space, then the rough tendency is for the eigenvalues of $\mathbf{T}_k$ and $\mathbf{R}_k$ to be all small except for a few of them, which grow large but have a probability of order $O(1/N)$, see, e.g., [Pollock *et al.*, 2003]. In that context, Theorem 6.1 holds, i.e. for $N \to \infty$, $F^{\mathbf{B}_N} - F_N \Rightarrow 0$.

It is also important to remark that the matrices $\mathbf{T}_k$ are constrained to be diagonal. This is unimportant when the matrices $\mathbf{X}_k$ are assumed Gaussian in practical applications, as the $\mathbf{X}_k$, being bi-unitarily invariant, can be multiplied on the right by any deterministic unitary matrix without altering the final result. This limitation is linked to the technique used for proving Theorem 6.1. For mathematical completion, though, it would be convenient for the matrices $\mathbf{T}_k$ to be unconstrained. We mention that Zhang and Bai [Zhang, 2006] derive the limiting spectral distribution of the model $\mathbf{B}_N = \mathbf{R}_1^{\frac{1}{2}}\mathbf{X}_1\mathbf{T}_1\mathbf{X}_1^{\mathsf{H}}\mathbf{R}_1^{\frac{1}{2}}$ for unconstrained Hermitian $\mathbf{T}_1$, using a different approach than that presented below.

For practical applications, it will be easier in the following to write (6.6) in a more symmetric way. This is discussed in the following remark.

Remark 6.1. In the particular case where $\mathbf{A} = 0$, the K implicit Equations (6.6) can be developed into the $2K$ linked equations

$$e_{N,i}(z) = \frac{1}{N} \operatorname{tr} \mathbf{R}_i \left(-z \left[\mathbf{I}_N + \sum_{k=1}^{K} \bar{e}_k(z) \mathbf{R}_k \right] \right)^{-1}$$

$$\bar{e}_{N,i}(z) = \frac{1}{n_i} \operatorname{tr} \mathbf{T}_i \left(-z \left[\mathbf{I}_{n_i} + c_i e_{N,i}(z) \mathbf{T}_i \right] \right)^{-1} \tag{6.7}$$

whose symmetric aspect is both more readable and more useful for practical reasons that will be evidenced later in Chapters 13–14. As a consequence, $m_N(z)$ in (6.5) becomes

$$m_N(z) = \frac{1}{N} \operatorname{tr} \left(-z \left[\mathbf{I}_N + \sum_{k=1}^{K} \bar{e}_{N,k}(z) \mathbf{R}_k \right] \right)^{-1}.$$

In the literature and, as a matter of fact, in some deterministic equivalents presented later in this chapter, the variables $e_{N,i}(z)$ may be normalized by $\frac{1}{n_i}$ instead of $\frac{1}{N}$ in order to avoid carrying the factor c_i in front of $e_{N,i}(z)$ in the second fixed-point equation of (6.7). In the application chapters, Chapters 12–15, depending on the situation, either one or the other convention will be taken.

We present hereafter the general techniques, based on the Stieltjes transform, to prove Theorem 6.1 and other similar results introduced in this section. As opposed to the proof of the Marčenko–Pastur law, we cannot prove that that there exists a space of probability one over which $m_{\mathbf{B}_N}(z) \to m(z)$ for all $z \in \mathbb{C} \setminus \mathbb{R}^+$, for a certain limiting function m. Instead, we prove that there exists a space of probability one over which $m_{\mathbf{B}_N}(z) - m_N(z) \to 0$ for all z, for a certain series of Stieltjes transforms $m_1(z), m_2(z), \ldots$. There are in general two main approaches to prove this convergence. The first option is a point-wise approach that consists in proving the convergence for all z in a compact subspace of $\mathbb{C} \setminus \mathbb{R}^+$ having a limit point. Invoking Vitali's convergence theorem, similar to the proof of the Marčenko–Pastur law, we then prove the convergence for all $z \in \mathbb{C} \setminus \mathbb{R}^+$. In the coming proof, we will take $z \in \mathbb{C}^+$. In the proof of Theorem 6.17, we will take z real negative. The second option is a functional approach in which the objects under study are not $m_{\mathbf{B}_N}(z)$ and $m_N(z)$ taken at a precise point $z \in \mathbb{C} \setminus \mathbb{R}^+$ but rather $m_{\mathbf{B}_N}(z)$ and $m_N(z)$ seen as functions lying in the space of Stieltjes transforms of distribution functions with support on $\mathbb{R}^+$. The convergence $m_{\mathbf{B}_N}(z) - m_N(z) \xrightarrow{\text{a.s.}} 0$ is in this case functional and Vitali's convergence theorem is not called for. This is the approach followed in, e.g., [Hachem *et al.*, 2007]. The latter is not detailed in this book.

The first step of the general proof, for either option, consists in determining $m_N(z)$. For this, similar to the Marčenko–Pastur proof, we develop the expression of $m_{\mathbf{B}_N}(z)$, seeking for a limiting result of the kind

$$m_{\mathbf{B}_N}(z) - h_N(m_{\mathbf{B}_N}(z); z) \xrightarrow{\text{a.s.}} 0$$

for some deterministic function h_N, possibly depending on N. Such an expression allows us to infer the nature of a deterministic approximation $m_N(z)$ of $m_{\mathbf{B}_N}(z)$ as a particular solution of the equation in m

$$m - h_N(m; z) = 0. \tag{6.8}$$

This equation rarely has a unique point-wise solution, i.e. for every z, but often has a unique functional solution $z \to m_N(z)$ that is the Stieltjes transform of a distribution function. If the point-wise approach is followed, a unique point-wise solution of (6.8) can often be narrowed down to a certain subspace of $\mathbb{C}$ for z lying in some other subspace of $\mathbb{C}$. In Theorem 6.1, there exists a single solution in $\mathbb{C}^+$ when $z \in \mathbb{C}^+$, a single solution in $\mathbb{C}^-$ when $z \in \mathbb{C}^-$, and a single positive solution when z is real negative. Standard holomorphicity arguments on the function $m_N(z)$ then ensure that $z \to m_N(z)$ is the unique Stieltjes transform satisfying $h_N(m_N(z); z) = m_N(z)$. When using the functional approach, this fact tends to be proved more directly. In the coming proof of Theorem 6.1, we will prove point-wise uniqueness by assuming, as per standard techniques, the alleged existence of two distinct solutions and prove a contradiction. An alternative approach is to prove that the fixed-point algorithm

$$m_0 \in \mathcal{D}$$
$$m_{t+1} = h_N(m_t; z),\ t \geq 0$$

always converges to $m_N(z)$, where $\mathcal{D}$ is taken to be either $\mathbb{R}^-$, $\mathbb{C}^+$ or $\mathbb{C}^-$. This approach, when valid (in some involved cases, convergence may not always arise), is doubly interesting as it allows both (i) to prove point-wise uniqueness for z taken in some subset of $\mathbb{C} \setminus \mathbb{R}^+$, leading to uniqueness of the Stieltjes transform using again holomorphicity arguments, and (ii) to provide an explicit algorithm to compute $m_N(z)$ for $z \in \mathcal{D}$, which is in particular of interest for practical applications when $z = -\sigma^2 < 0$. In the proof of Theorem 6.1, we will introduce both results for completion. In the proof of Theorem 6.17, we will directly proceed to proving the convergence of the fixed-point algorithm for z real negative.

When the uniqueness of the Stieltjes transform $m_N(z)$ has been made clear, the last step is to prove that, in the large N limit

$$m_{\mathbf{B}_N}(z) - m_N(z) \xrightarrow{\text{a.s.}} 0.$$

This step is not so immediate. To this point, we indeed only know that $m_{\mathbf{B}_N}(z) - h_N(m_{\mathbf{B}_N}(z); z) \xrightarrow{\text{a.s.}} 0$ and $m_N(z) - h_N(m_N(z); z) = 0$. This does not imply immediately that $m_{\mathbf{B}_N}(z) - m_N(z) \xrightarrow{\text{a.s.}} 0$. If there are several point-wise solutions to $m - h_N(m; z) = 0$, we need to verify that $m_N(z)$ was chosen to be the one that will eventually satisfy $m_{\mathbf{B}_N}(z) - m_N(z) \xrightarrow{\text{a.s.}} 0$. This will conclude the proof.

We now provide the specific proof of Theorem 6.1. In order to determine the above function h_N, we first develop the Marčenko–Pastur method (for simplicity for $K = 2$ and $\mathbf{A} = 0$). We will realize that this method fails unless all $\mathbf{R}_k$ and $\mathbf{A}$ are constrained to be co-diagonalizable. To cope with this limitation, we will introduce the more powerful Bai and Silverstein method, whose idea is to *guess* along the derivations the suitable form of h_N. In fact, as we will shortly realize, the problem is slightly more difficult here as we will not be able to find such a function h_N (which may actually not exist at all in the first place). We will however be able to find functions $f_{N,i}$ such that, for each i

$$e_{\mathbf{B}_N,i}(z) - f_{N,i}(e_{\mathbf{B}_N,1}(z), \ldots, e_{\mathbf{B}_N,K}(z); z) \xrightarrow{\text{a.s.}} 0$$

where $e_{\mathbf{B}_N,i}(z) \triangleq \frac{1}{N} \operatorname{tr} \mathbf{R}_i(\mathbf{B}_N - z\mathbf{I}_N)^{-1}$. We will then look for a function $e_{N,i}(z)$ that satisfies

$$e_{N,i}(z) = f_{N,i}(e_{N,1}(z), \ldots, e_{N,K}(z); z).$$

From there, it will be easy to determine a further function g_N such that

$$m_{\mathbf{B}_N}(z) - g_N(e_{\mathbf{B}_N,1}(z), \ldots, e_{\mathbf{B}_N,K}(z); z) \xrightarrow{\text{a.s.}} 0$$

and

$$m_N(z) - g_N(e_{N,1}(z), \ldots, e_{N,K}(z); z) = 0.$$

We will therefore have finally

$$m_{\mathbf{B}_N}(z) - m_N(z) \xrightarrow{\text{a.s.}} 0.$$

Proof of Theorem 6.1. In order to have a first insight on what the deterministic equivalent m_N of $m_{\mathbf{B}_N}$ may look like, the Marčenko–Pastur method will be applied with the (strong) additional assumption that $\mathbf{A}$ and all $\mathbf{R}_k$, $1 \leq k \leq K$, are diagonal and that the e.s.d. $F^{\mathbf{T}_k}$, $F^{\mathbf{R}_k}$ converge for all k as N grows large. In this scenario, $m_{\mathbf{B}_N}$ has a limit when $N \to \infty$ and the method, however more tedious than in the proof of the Marčenko–Pastur law, leads naturally to m_N.

Consider the case when $K = 2$, $\mathbf{A} = 0$ for simplicity and denote $\mathbf{H}_k = \mathbf{R}_k^{\frac{1}{2}}\mathbf{X}_k\mathbf{T}_k^{\frac{1}{2}}$. Following similar steps as in the proof of the Marčenko–Pastur law, we start with matrix inversion lemmas

$$\begin{aligned}&\left(\mathbf{H}_1\mathbf{H}_1^{\mathsf{H}} + \mathbf{H}_2\mathbf{H}_2^{\mathsf{H}} - z\mathbf{I}_N\right)^{-1}_{11} \\ &= \left[-z - z[\mathbf{h}_1^{\mathsf{H}}\mathbf{h}_2^{\mathsf{H}}]\left(\begin{bmatrix}\mathbf{U}_1^{\mathsf{H}}\\ \mathbf{U}_2^{\mathsf{H}}\end{bmatrix}[\mathbf{U}_1\mathbf{U}_2] - z\mathbf{I}_{n_1+n_2}\right)^{-1}\begin{bmatrix}\mathbf{h}_1\\ \mathbf{h}_2\end{bmatrix}\right]^{-1}\end{aligned}$$

with the definition $\mathbf{H}_i^{\mathsf{H}} = [\mathbf{h}_i\mathbf{U}_i^{\mathsf{H}}]$. Using the block matrix inversion lemma, the inner inversed matrix in this expression can be decomposed into four submatrices. The upper-left $n_1 \times n_1$ submatrix reads:

$$\left(-z\mathbf{U}_1^{\mathsf{H}}(\mathbf{U}_2\mathbf{U}_2^{\mathsf{H}} - z\mathbf{I}_{N-1})^{-1}\mathbf{U}_1 - z\mathbf{I}_{n_1}\right)^{-1}$$

while, for the second block diagonal entry, it suffices to revert all ones in twos and vice-versa. Taking the limits, using Theorem 3.4 and Theorem 3.9, we observe that the two off-diagonal submatrices will not play a role, and we finally have

$$\begin{aligned}&\left(\mathbf{H}_1\mathbf{H}_1^{\mathsf{H}} + \mathbf{H}_2\mathbf{H}_2^{\mathsf{H}} - z\mathbf{I}_N\right)^{-1}_{11} \\ &\simeq \Bigg[-z - zr_{11}\frac{1}{n_1}\operatorname{tr}\mathbf{T}_1\left(-z\mathbf{H}_1^{\mathsf{H}}(\mathbf{H}_2\mathbf{H}_2^{\mathsf{H}} - z\mathbf{I}_N)^{-1}\mathbf{H}_1 - z\mathbf{I}_{n_1}\right)^{-1} \\ &\quad -zr_{21}\frac{1}{n_2}\operatorname{tr}\mathbf{T}_2\left(-z\mathbf{H}_2^{\mathsf{H}}(\mathbf{H}_1\mathbf{H}_1^{\mathsf{H}} - z\mathbf{I}_N)^{-1}\mathbf{H}_1 - z\mathbf{I}_{n_2}\right)^{-1}\Bigg]^{-1}\end{aligned}$$

where the symbol "$\simeq$" denotes some kind of yet unknown large N convergence and where we denoted r_{ij} the jth diagonal entry of $\mathbf{R}_i$. Observe that we can proceed to a similar derivation for the matrix $\mathbf{T}_1\left(-z\mathbf{H}_1^{\mathsf{H}}(\mathbf{H}_2\mathbf{H}_2^{\mathsf{H}} - z\mathbf{I}_N)^{-1}\mathbf{H}_1 - z\mathbf{I}_{n_1}\right)^{-1}$ that now appears. Denoting now $\mathbf{H}_i = [\tilde{\mathbf{h}}_i\tilde{\mathbf{U}}_i]$, we have indeed

$$\begin{aligned}&\left[\mathbf{T}_1\left(-z\mathbf{H}_1^{\mathsf{H}}(\mathbf{H}_2\mathbf{H}_2^{\mathsf{H}} - z\mathbf{I}_N)^{-1}\mathbf{H}_1 - z\mathbf{I}_{n_1}\right)^{-1}\right]_{11} \\ &= \tau_{11}\left[-z - z\tilde{\mathbf{h}}_1^{\mathsf{H}}\left(\tilde{\mathbf{U}}_1\tilde{\mathbf{U}}_1^{\mathsf{H}} + \mathbf{H}_2\mathbf{H}_2^{\mathsf{H}} - z\mathbf{I}_N\right)^{-1}\tilde{\mathbf{h}}_1\right]^{-1} \\ &\simeq \tau_{11}\left[-z - zc_1\tau_{11}\frac{1}{N}\operatorname{tr}\mathbf{R}_1\left(\mathbf{H}_1\mathbf{H}_1^{\mathsf{H}} + \mathbf{H}_2\mathbf{H}_2^{\mathsf{H}} - z\mathbf{I}_N\right)^{-1}\right]^{-1}\end{aligned}$$

with τ_{ij} the jth diagonal entry of $\mathbf{T}_i$. The limiting result here arises from the trace lemma, Theorem 3.4 along with the rank-1 perturbation lemma, Theorem 3.9. The same result holds when changing ones in twos.

We now denote by e_i and $\bar{e}_i$ the (almost sure) limits of the random quantities

$$e_{\mathbf{B}_N,i} = \frac{1}{N}\operatorname{tr}\mathbf{R}_i\left(\mathbf{H}_1\mathbf{H}_1^{\mathsf{H}} + \mathbf{H}_2\mathbf{H}_2^{\mathsf{H}} - z\mathbf{I}_N\right)^{-1}$$

and

$$\bar{e}_{\mathbf{B}_N,i} = \frac{1}{N}\operatorname{tr}\mathbf{T}_i\left(-z\mathbf{H}_1^{\mathsf{H}}(\mathbf{H}_2\mathbf{H}_2^{\mathsf{H}} - z\mathbf{I}_N)^{-1}\mathbf{H}_1 - z\mathbf{I}_{n_1}\right)^{-1}$$

respectively, as $F^{\mathbf{T}_i}$ and $F^{\mathbf{R}_i}$ converge in the large N limit. These limits exist here since we forced $\mathbf{R}_1$ and $\mathbf{R}_2$ to be co-diagonalizable. We find

$$e_i = \lim_{N\to\infty} \frac{1}{N} \operatorname{tr} \mathbf{R}_i \left(-z\bar{e}_{\mathbf{B}_N,i}\mathbf{R}_1 - z\bar{e}_{\mathbf{B}_N,i}\mathbf{R}_2 - z\mathbf{I}_N\right)^{-1}$$

$$\bar{e}_i = \lim_{N\to\infty} \frac{1}{N} \operatorname{tr} \mathbf{T}_i \left(-zc_i e_{\mathbf{B}_N,i}\mathbf{T}_i - z\mathbf{I}_{n_i}\right)^{-1}$$

where the type of convergence is left to be determined. From this short calculus, we can infer the form of (6.7).

This derivation obviously only provides a hint on the deterministic equivalent for $m_N(z)$. It also provides the aforementioned observation that $m_N(z)$ is not itself solution of a fixed-point equation, although $e_{N,1}(z), \ldots, e_{N,K}(z)$ are. To prove Theorem 6.1, irrespective of the conditions imposed on $\mathbf{R}_1, \ldots, \mathbf{R}_K$, $\mathbf{T}_1, \ldots, \mathbf{T}_K$ and $\mathbf{A}$, we will successively go through four steps, given below. For readability, we consider the case $K = 1$ and discard the useless indexes. The generalization to $K \geq 1$ is rather simple for most of the steps but requires cumbersome additional calculus for some particular aspects. These pieces of calculus are not interesting here, the reader being invited to refer to [Couillet *et al.*, 2011a] for more details. The four-step procedure is detailed below.

- *Step 1.* We first seek a function f_N, such that, for $z \in \mathbb{C}^+$

  $$e_{\mathbf{B}_N}(z) - f_N(e_{\mathbf{B}_N}(z); z) \xrightarrow{\text{a.s.}} 0$$

  as $N \to \infty$, where $e_{\mathbf{B}_N}(z) = \frac{1}{N} \operatorname{tr} \mathbf{R}(\mathbf{B}_N - z\mathbf{I}_N)^{-1}$. This function f_N was already inferred by the Marčenko–Pastur approach. Now, we will make this step rigorous by using the *Bai and Silverstein approach*, as is done in, e.g., [Dozier and Silverstein, 2007a; Silverstein and Bai, 1995]. Basically, the function f_N will be found using an inference procedure. That is, starting from a very general form of f_N, i.e. $f_N = \frac{1}{N} \operatorname{tr} \mathbf{R}\mathbf{D}^{-1}$ for some matrix $\mathbf{D} \in \mathbb{C}^{N\times N}$ (not yet written as a function of z or $e_{\mathbf{B}_N}(z)$), we will evaluate the difference $e_{\mathbf{B}_N}(z) - f_N$ and progressively discover which matrix $\mathbf{D}$ will make this difference increasingly small for large N.
- *Step 2.* For fixed N, we prove the existence of a solution to the implicit equation in the dummy variable e

  $$f_N(e; z) = e. \tag{6.9}$$

  This is often performed by proving the existence of a sequence $e_{N,1}, e_{N,2}, \ldots$, lying in a compact space such that $f_N(e_{N,k}; z) - e_{N,k}$ converges to zero, in which case there exists at least one converging subsequence of $e_{N,1}, e_{N,2}, \ldots$, whose limit e_N satisfies (6.9).
- *Step 3.* Still for fixed N, we prove the uniqueness of the solution of (6.9) lying in some specific space and we call this solution $e_N(z)$. This is classically performed by assuming the existence of a second distinct solution and by exhibiting a contradiction.

- *Step 4.* We finally prove that

$$e_{\mathbf{B}_N}(z) - e_N(z) \xrightarrow{\text{a.s.}} 0$$

and, similarly, that

$$m_{\mathbf{B}_N}(z) - m_N(z) \xrightarrow{\text{a.s.}} 0$$

as $N \to \infty$, with $m_N(z) \triangleq g_N(e_N(z); z)$ for some function g_N.

At first, following the works of Bai and Silverstein, a truncation, centralization, and rescaling step is required to replace the matrices $\mathbf{X}$, $\mathbf{R}$, and $\mathbf{T}$ by truncated versions $\hat{\mathbf{X}}$, $\hat{\mathbf{R}}$, and $\hat{\mathbf{T}}$, respectively, such that the entries of $\hat{\mathbf{X}}$ have zero mean, $\|\hat{\mathbf{X}}\| \leq k \log(N)$, for some constant k, $\|\hat{\mathbf{R}}\| \leq \log(N)$ and $\|\hat{\mathbf{T}}\| \leq \log(N)$. Similar to the truncation steps presented in Section 3.2.2, it is shown in [Couillet *et al.*, 2011a] that these truncations do not restrict the generality of the final result for $\{F^{\mathbf{T}}\}$ and $\{F^{\mathbf{R}}\}$ forming tight sequences, that is:

$$F^{\mathbf{R}^{\frac{1}{2}}\mathbf{X}\mathbf{T}\mathbf{X}^{\mathsf{H}}\mathbf{R}^{\frac{1}{2}}} - F^{\hat{\mathbf{R}}^{\frac{1}{2}}\hat{\mathbf{X}}\hat{\mathbf{T}}\hat{\mathbf{X}}^{\mathsf{H}}\hat{\mathbf{R}}^{\frac{1}{2}}} \Rightarrow 0$$

almost surely, as N grows large. Therefore, we can from now on work with these truncated matrices. We recall that the main interest of this procedure is to be able to derive a deterministic equivalent (or l.s.d.) of the underlying random matrix model without the need for any moment assumption on the entries of $\mathbf{X}$, by replacing the entries of $\mathbf{X}$ by truncated random variables that have moments of all orders. Here, the interest is in fact two-fold, since, in addition to truncating the entries of $\mathbf{X}$, also the entries of $\mathbf{T}$ and $\mathbf{R}$ are truncated in order to be able to prove results for matrices $\mathbf{T}$ and $\mathbf{R}$ that in reality have eigenvalues growing very large but that will be assumed to have entries bounded by $\log(N)$. For readability in the following, we rename $\mathbf{X}$, $\mathbf{T}$, and $\mathbf{R}$ the truncated matrices.

Remark 6.2. Alternatively, expected values can be used to discard the stochastic character. This introduces an additional convergence step, which is the approach followed by Hachem, Najim, and Loubaton in several publications, e.g., [Hachem *et al.*, 2007] and [Dupuy and Loubaton, 2009]. This additional step consists in first proving the almost sure weak convergence of $F^{\mathbf{B}_N} - G_N$ to zero, for G_N some auxiliary deterministic distribution (such as $G_N = \mathrm{E}[F^{\mathbf{B}_N}]$), before proving the convergence $G_N - F_N \Rightarrow 0$.

Step 1. First convergence step

We start with the introduction of two fundamental identities.

Lemma 6.1 (Resolvent identity). *For invertible* $\mathbf{A}$ *and* $\mathbf{B}$ *matrices, we have the identity*

$$\mathbf{A}^{-1} - \mathbf{B}^{-1} = -\mathbf{A}^{-1}(\mathbf{A} - \mathbf{B})\mathbf{B}^{-1}.$$

This can be verified easily by multiplying both sides on the left by $\mathbf{A}$ and on the right by $\mathbf{B}$ (the resulting equality being equivalent to Lemma 6.1 for $\mathbf{A}$ and $\mathbf{B}$ invertible).

Lemma 6.2 (A matrix inversion lemma, (2.2) in [Silverstein and Bai, 1995]). *Let $\mathbf{A} \in \mathbb{C}^{N\times N}$ be Hermitian invertible, then, for any vector $\mathbf{x} \in \mathbb{C}^N$ and any scalar $\tau \in \mathbb{C}$, such that $\mathbf{A} + \tau\mathbf{x}\mathbf{x}^{\mathsf{H}}$ is invertible*

$$\mathbf{x}^{\mathsf{H}}(\mathbf{A} + \tau\mathbf{x}\mathbf{x}^{\mathsf{H}})^{-1} = \frac{\mathbf{x}^{\mathsf{H}}\mathbf{A}^{-1}}{1 + \tau\mathbf{x}^{\mathsf{H}}\mathbf{A}^{-1}\mathbf{x}}.$$

This is verified by multiplying both sides by $\mathbf{A} + \tau\mathbf{x}\mathbf{x}^{\mathsf{H}}$ from the right.

Lemma 6.1 is often referred to as the *resolvent identity*, since it will be mainly used to take the difference between matrices of type $(\mathbf{X} - z\mathbf{I}_N)^{-1}$ and $(\mathbf{Y} - z\mathbf{I}_N)^{-1}$, which we remind are called the *resolvent matrices* of $\mathbf{X}$ and $\mathbf{Y}$, respectively.

The fundamental idea of the approach by Bai and Silverstein is to *guess* the deterministic equivalent of $m_{\mathbf{B}_N}(z)$ by writing it under the form $\frac{1}{N}\operatorname{tr}\mathbf{D}^{-1}$ at first, where $\mathbf{D}$ needs to be determined. This will be performed by taking the difference $m_{\mathbf{B}_N}(z) - \frac{1}{N}\operatorname{tr}\mathbf{D}^{-1}$ and, along the lines of calculus, successively determining the good properties $\mathbf{D}$ must satisfy so that the difference tends to zero almost surely.

We then start by taking $z \in \mathbb{C}^+$ and $\mathbf{D} \in \mathbb{C}^{N\times N}$ some invertible matrix whose normalized trace would ideally be close to $m_{\mathbf{B}_N}(z) = \frac{1}{N}\operatorname{tr}(\mathbf{B}_N - z\mathbf{I}_N)^{-1}$. We then write

$$\mathbf{D}^{-1} - (\mathbf{B}_N - z\mathbf{I}_N)^{-1} = \mathbf{D}^{-1}(\mathbf{A} + \mathbf{R}^{\frac{1}{2}}\mathbf{X}\mathbf{T}\mathbf{X}^{\mathsf{H}}\mathbf{R}^{\frac{1}{2}} - z\mathbf{I}_N - \mathbf{D})(\mathbf{B}_N - z\mathbf{I}_N)^{-1} \tag{6.10}$$

using Lemma 6.1.

Notice here that, since $\mathbf{B}_N$ is Hermitian non-negative definite, and $z \in \mathbb{C}^+$, the term $(\mathbf{B}_N - z\mathbf{I}_N)^{-1}$ has uniformly bounded spectral norm (bounded by $1/\Im[z]$). Since $\mathbf{D}^{-1}$ is desired to be close to $(\mathbf{B}_N - z\mathbf{I}_N)^{-1}$, the same property should also hold for $\mathbf{D}^{-1}$. In order for the normalized trace of (6.10) to be small, we need therefore to focus exclusively on the inner difference on the right-hand side. It seems then interesting at this point to write $\mathbf{D} \triangleq \mathbf{A} - z\mathbf{I}_N + p_N\mathbf{R}$ for p_N left to be defined. This leads to

$$\begin{aligned}
&\mathbf{D}^{-1} - (\mathbf{B}_N - z\mathbf{I}_N)^{-1} \\
&= \mathbf{D}^{-1}\mathbf{R}^{\frac{1}{2}}\left(\mathbf{X}\mathbf{T}\mathbf{X}^{\mathsf{H}}\right)\mathbf{R}^{\frac{1}{2}}(\mathbf{B}_N - z\mathbf{I}_N)^{-1} - p_N\mathbf{D}^{-1}\mathbf{R}(\mathbf{B}_N - z\mathbf{I}_N)^{-1} \\
&= \mathbf{D}^{-1}\sum_{j=1}^{n}\tau_j\mathbf{R}^{\frac{1}{2}}\mathbf{x}_j\mathbf{x}_j^{\mathsf{H}}\mathbf{R}^{\frac{1}{2}}(\mathbf{B}_N - z\mathbf{I}_N)^{-1} - p_N\mathbf{D}^{-1}\mathbf{R}(\mathbf{B}_N - z\mathbf{I}_N)^{-1}
\end{aligned}$$

where in the second equality we used the fact that $\mathbf{X}\mathbf{T}\mathbf{X}^{\mathsf{H}} = \sum_{j=1}^{n}\tau_j\mathbf{x}_j\mathbf{x}_j^{\mathsf{H}}$, with $\mathbf{x}_j \in \mathbb{C}^N$ the jth column of $\mathbf{X}$ and τ_j the jth diagonal element of $\mathbf{T}$. Denoting $\mathbf{B}_{(j)} = \mathbf{B}_N - \tau_j\mathbf{R}^{\frac{1}{2}}\mathbf{x}_j\mathbf{x}_j^{\mathsf{H}}\mathbf{R}^{\frac{1}{2}}$, i.e. $\mathbf{B}_N$ with column j removed, and using Lemma

6.2 for the matrix $\mathbf{B}_{(j)}$, we have:

$$\mathbf{D}^{-1} - (\mathbf{B}_N - z\mathbf{I}_N)^{-1} = \sum_{j=1}^{n} \tau_j \frac{\mathbf{D}^{-1}\mathbf{R}^{\frac{1}{2}}\mathbf{x}_j\mathbf{x}_j^{\mathsf{H}}\mathbf{R}^{\frac{1}{2}}(\mathbf{B}_{(j)} - z\mathbf{I}_N)^{-1}}{1 + \tau_j \mathbf{x}^{\mathsf{H}}\mathbf{R}^{\frac{1}{2}}(\mathbf{B}_{(j)} - z\mathbf{I}_N)^{-1}\mathbf{R}^{\frac{1}{2}}\mathbf{x}_j} - p_N \mathbf{D}^{-1}\mathbf{R}(\mathbf{B}_N - z\mathbf{I}_N)^{-1}.$$

Taking the trace on each side, and recalling that, for a vector $\mathbf{x}$ and a matrix $\mathbf{A}$, $\operatorname{tr}(\mathbf{A}\mathbf{x}\mathbf{x}^{\mathsf{H}}) = \operatorname{tr}(\mathbf{x}^{\mathsf{H}}\mathbf{A}\mathbf{x}) = \mathbf{x}^{\mathsf{H}}\mathbf{A}\mathbf{x}$, this becomes

$$\frac{1}{N}\operatorname{tr}\mathbf{D}^{-1} - \frac{1}{N}\operatorname{tr}(\mathbf{B}_N - z\mathbf{I}_N)^{-1} = \frac{1}{N}\sum_{j=1}^{n} \tau_j \frac{\mathbf{x}_j^{\mathsf{H}}\mathbf{R}^{\frac{1}{2}}(\mathbf{B}_{(j)} - z\mathbf{I}_N)^{-1}\mathbf{D}^{-1}\mathbf{R}^{\frac{1}{2}}\mathbf{x}_j}{1 + \tau_j \mathbf{x}^{\mathsf{H}}\mathbf{R}^{\frac{1}{2}}(\mathbf{B}_{(j)} - z\mathbf{I}_N)^{-1}\mathbf{R}^{\frac{1}{2}}\mathbf{x}_j} - p_N \frac{1}{N}\operatorname{tr}\mathbf{R}(\mathbf{B}_N - z\mathbf{I}_N)^{-1}\mathbf{D}^{-1} \tag{6.11}$$

where quadratic forms of the type $\mathbf{x}^{\mathsf{H}}\mathbf{A}\mathbf{x}$ appear.

Remembering the trace lemma, Theorem 3.4, which can a priori be applied to the terms $\mathbf{x}_j^{\mathsf{H}}\mathbf{R}^{\frac{1}{2}}(\mathbf{B}_{(j)} - z\mathbf{I}_N)^{-1}\mathbf{D}^{-1}\mathbf{R}^{\frac{1}{2}}\mathbf{x}_j$ since $\mathbf{x}_j$ is independent of the matrix $\mathbf{R}^{\frac{1}{2}}(\mathbf{B}_{(j)} - z\mathbf{I}_N)^{-1}\mathbf{D}^{-1}\mathbf{R}^{\frac{1}{2}}$, we notice that by setting

$$p_N = \frac{1}{n}\sum_{j=1}^{n} \frac{\tau_j}{1 + \tau_j c \frac{1}{N}\operatorname{tr}\mathbf{R}(\mathbf{B}_N - z\mathbf{I}_N)^{-1}}.$$

Equation (6.11) becomes

$$\frac{1}{N}\operatorname{tr}\mathbf{D}^{-1} - \frac{1}{N}\operatorname{tr}(\mathbf{B}_N - z\mathbf{I}_N)^{-1} = \frac{1}{N}\sum_{j=1}^{n} \tau_j \left[\frac{\mathbf{x}_j^{\mathsf{H}}\mathbf{R}^{\frac{1}{2}}(\mathbf{B}_{(j)} - z\mathbf{I}_N)^{-1}\mathbf{D}^{-1}\mathbf{R}^{\frac{1}{2}}\mathbf{x}_j}{1 + \tau_j \mathbf{x}^{\mathsf{H}}\mathbf{R}^{\frac{1}{2}}(\mathbf{B}_{(j)} - z\mathbf{I}_N)^{-1}\mathbf{R}^{\frac{1}{2}}\mathbf{x}_j} - \frac{\frac{1}{n}\operatorname{tr}\mathbf{R}(\mathbf{B}_N - z\mathbf{I}_N)^{-1}\mathbf{D}^{-1}}{1 + c\tau_j \frac{1}{N}\operatorname{tr}\mathbf{R}(\mathbf{B}_N - z\mathbf{I}_N)^{-1}} \right] \tag{6.12}$$

which is suspected to converge to zero as N grows large, since both the numerators and the denominators converge to one another. Let us assume for the time being that the difference effectively goes to zero almost surely. Equation (6.12) implies

$$\frac{1}{N}\operatorname{tr}(\mathbf{B}_N - z\mathbf{I}_N)^{-1} - \frac{1}{N}\operatorname{tr}\left(\mathbf{A} + \frac{1}{n}\sum_{j=1}^{n}\frac{\tau_j\mathbf{R}}{1 + \tau_j c \frac{1}{N}\operatorname{tr}\mathbf{R}(\mathbf{B}_N - z\mathbf{I}_N)^{-1}} - z\mathbf{I}_N\right)^{-1} \xrightarrow{\text{a.s.}} 0$$

which determines $m_{\mathbf{B}_N}(z) = \frac{1}{N}\operatorname{tr}(\mathbf{B}_N - z\mathbf{I}_N)^{-1}$ as a function of the trace $\frac{1}{N}\operatorname{tr}\mathbf{R}(\mathbf{B}_N - z\mathbf{I}_N)^{-1}$, and not as a function of itself. This is the observation made earlier when we obtained a first hint on the form of $m_N(z)$ using the Marčenko–Pastur method, according to which we cannot find a function f_N such that $m_{\mathbf{B}_N}(z) - f_N(m_{\mathbf{B}_N}(z), z) \xrightarrow{\text{a.s.}} 0$. Instead, running the same steps as above, it is rather easy now to observe that

$$\frac{1}{N}\operatorname{tr}\mathbf{R}\mathbf{D}^{-1} - \frac{1}{N}\operatorname{tr}\mathbf{R}(\mathbf{B}_N - z\mathbf{I}_N)^{-1}$$
$$= \frac{1}{N}\sum_{j=1}^{n}\tau_j\left[\frac{\mathbf{x}_j^{\mathsf{H}}\mathbf{R}^{\frac{1}{2}}(\mathbf{B}_{(j)} - z\mathbf{I}_N)^{-1}\mathbf{R}\mathbf{D}^{-1}\mathbf{R}^{\frac{1}{2}}\mathbf{x}_j}{1+\tau_j\mathbf{x}_j^{\mathsf{H}}\mathbf{R}^{\frac{1}{2}}(\mathbf{B}_{(j)} - z\mathbf{I}_N)^{-1}\mathbf{R}^{\frac{1}{2}}\mathbf{x}_j} - \frac{\frac{1}{n}\operatorname{tr}\mathbf{R}(\mathbf{B}_N - z\mathbf{I}_N)^{-1}\mathbf{R}\mathbf{D}^{-1}}{1+\tau_j\frac{c}{N}\operatorname{tr}\mathbf{R}(\mathbf{B}_N - z\mathbf{I}_N)^{-1}}\right]$$

where $\|\mathbf{R}\| \leq \log N$. Then, denoting $e_{\mathbf{B}_N}(z) \triangleq \frac{1}{N}\operatorname{tr}\mathbf{R}(\mathbf{B}_N - z\mathbf{I}_N)^{-1}$, we suspect to have also

$$e_{\mathbf{B}_N}(z) - \frac{1}{N}\operatorname{tr}\mathbf{R}\left(\mathbf{A} + \frac{1}{n}\sum_{j=1}^{n}\frac{\tau_j}{1+\tau_j c e_{\mathbf{B}_N}(z)}\mathbf{R} - z\mathbf{I}_N\right)^{-1} \xrightarrow{\text{a.s.}} 0$$

and

$$m_{\mathbf{B}_N}(z) - \frac{1}{N}\operatorname{tr}\left(\mathbf{A} + \frac{1}{n}\sum_{j=1}^{n}\frac{\tau_j}{1+\tau_j c e_{\mathbf{B}_N}(z)}\mathbf{R} - z\mathbf{I}_N\right)^{-1} \xrightarrow{\text{a.s.}} 0$$

which is exactly what was required, i.e. $e_{\mathbf{B}_N}(z) - f_N(e_{\mathbf{B}_N}(z); z) \xrightarrow{\text{a.s.}} 0$ with

$$f_N(e; z) = \frac{1}{N}\operatorname{tr}\mathbf{R}\left(\mathbf{A} + \frac{1}{n}\sum_{j=1}^{n}\frac{\tau_j}{1+\tau_j c e}\mathbf{R} - z\mathbf{I}_N\right)^{-1}$$

and $m_{\mathbf{B}_N}(z) - g_N(e_{\mathbf{B}_N}(z); z) \xrightarrow{\text{a.s.}} 0$ with

$$g_N(e; z) = \frac{1}{N}\operatorname{tr}\left(\mathbf{A} + \frac{1}{n}\sum_{j=1}^{n}\frac{\tau_j}{1+\tau_j c e}\mathbf{R} - z\mathbf{I}_N\right)^{-1}.$$

We now prove that the right-hand side of (6.12) converges to zero almost surely. This rather technical part justifies the use of the truncation steps and is the major difference between the works of Bai and Silverstein [Dozier and Silverstein, 2007a; Silverstein and Bai, 1995] and the works of Hachem *et al.* [Hachem *et al.*, 2007]. We first define

$$w_N \triangleq \sum_{j=1}^{n}\frac{\tau_j}{N}\left[\frac{\mathbf{x}_j^{\mathsf{H}}\mathbf{R}^{\frac{1}{2}}(\mathbf{B}_{(j)} - z\mathbf{I}_N)^{-1}\mathbf{R}\mathbf{D}^{-1}\mathbf{R}^{\frac{1}{2}}\mathbf{x}_j}{1+\tau_j\mathbf{x}_j^{\mathsf{H}}\mathbf{R}^{\frac{1}{2}}(\mathbf{B}_{(j)} - z\mathbf{I}_N)^{-1}\mathbf{R}^{\frac{1}{2}}\mathbf{x}_j} - \frac{\frac{1}{n}\operatorname{tr}\mathbf{R}(\mathbf{B}_N - z\mathbf{I}_N)^{-1}\mathbf{R}\mathbf{D}^{-1}}{1+\tau_j\frac{c}{N}\operatorname{tr}\mathbf{R}(\mathbf{B}_N - z\mathbf{I}_N)^{-1}}\right]$$

which we then divide into four terms, in order to successively prove the convergence of the numerators and the denominators. Write

$$w_N = \frac{1}{N}\sum_{j=1}^{n}\tau_j\left(d_j^1 + d_j^2 + d_j^3 + d_j^4\right)$$

where

$$d_j^1 = \frac{\mathbf{x}_j^{\mathsf{H}}\mathbf{R}^{\frac{1}{2}}(\mathbf{B}_{(j)} - z\mathbf{I}_N)^{-1}\mathbf{R}\mathbf{D}^{-1}\mathbf{R}^{\frac{1}{2}}\mathbf{x}_j}{1+\tau_j\mathbf{x}_j^{\mathsf{H}}\mathbf{R}^{\frac{1}{2}}(\mathbf{B}_{(j)} - z\mathbf{I}_N)^{-1}\mathbf{R}^{\frac{1}{2}}\mathbf{x}_j} - \frac{\mathbf{x}_j^{\mathsf{H}}\mathbf{R}^{\frac{1}{2}}(\mathbf{B}_{(j)} - z\mathbf{I}_N)^{-1}\mathbf{R}\mathbf{D}_{(j)}^{-1}\mathbf{R}^{\frac{1}{2}}\mathbf{x}_j}{1+\tau_j\mathbf{x}_j^{\mathsf{H}}\mathbf{R}^{\frac{1}{2}}(\mathbf{B}_{(j)} - z\mathbf{I}_N)^{-1}\mathbf{R}^{\frac{1}{2}}\mathbf{x}_j}$$

$$d_j^2 = \frac{\mathbf{x}_j^{\mathsf{H}}\mathbf{R}^{\frac{1}{2}}(\mathbf{B}_{(j)} - z\mathbf{I}_N)^{-1}\mathbf{R}\mathbf{D}_{(j)}^{-1}\mathbf{R}^{\frac{1}{2}}\mathbf{x}_j}{1+\tau_j\mathbf{x}_j^{\mathsf{H}}\mathbf{R}^{\frac{1}{2}}(\mathbf{B}_{(j)} - z\mathbf{I}_N)^{-1}\mathbf{R}^{\frac{1}{2}}\mathbf{x}_j} - \frac{\frac{1}{n}\operatorname{tr}\mathbf{R}(\mathbf{B}_{(j)} - z\mathbf{I}_N)^{-1}\mathbf{R}\mathbf{D}_{(j)}^{-1}}{1+\tau_j\mathbf{x}_j^{\mathsf{H}}\mathbf{R}^{\frac{1}{2}}(\mathbf{B}_{(j)} - z\mathbf{I}_N)^{-1}\mathbf{R}^{\frac{1}{2}}\mathbf{x}_j}$$

$$d_j^3 = \frac{\frac{1}{n}\operatorname{tr}\mathbf{R}(\mathbf{B}_{(j)} - z\mathbf{I}_N)^{-1}\mathbf{R}\mathbf{D}_{(j)}^{-1}}{1+\tau_j\mathbf{x}_j^{\mathsf{H}}\mathbf{R}^{\frac{1}{2}}(\mathbf{B}_{(j)} - z\mathbf{I}_N)^{-1}\mathbf{R}^{\frac{1}{2}}\mathbf{x}_j} - \frac{\frac{1}{n}\operatorname{tr}\mathbf{R}(\mathbf{B}_N - z\mathbf{I}_N)^{-1}\mathbf{R}\mathbf{D}^{-1}}{1+\tau_j\mathbf{x}_j^{\mathsf{H}}\mathbf{R}^{\frac{1}{2}}(\mathbf{B}_{(j)} - z\mathbf{I}_N)^{-1}\mathbf{R}^{\frac{1}{2}}\mathbf{x}_j}$$

$$d_j^4 = \frac{\frac{1}{n}\operatorname{tr}\mathbf{R}(\mathbf{B}_N - z\mathbf{I}_N)^{-1}\mathbf{R}\mathbf{D}^{-1}}{1+\tau_j\mathbf{x}_j^{\mathsf{H}}\mathbf{R}^{\frac{1}{2}}(\mathbf{B}_{(j)} - z\mathbf{I}_N)^{-1}\mathbf{R}^{\frac{1}{2}}\mathbf{x}_j} - \frac{\frac{1}{n}\operatorname{tr}\mathbf{R}(\mathbf{B}_N - z\mathbf{I}_N)^{-1}\mathbf{R}\mathbf{D}^{-1}}{1+c\tau_j e_{\mathbf{B}_N}}$$

where we introduced $\mathbf{D}_{(j)} = \mathbf{A} + \frac{1}{n}\sum_{k=1}^{n}\frac{\tau_k}{1+\tau_k c e_{\mathbf{B}_{(j)}}(z)}\mathbf{R} - z\mathbf{I}_N$, i.e. $\mathbf{D}$ with $e_{\mathbf{B}_N}(z)$ replaced by $e_{\mathbf{B}_{(j)}}(z)$. Under these notations, it is simple to show that $w_N \xrightarrow{\text{a.s.}} 0$ since every term d_j^k can be shown to go fast to zero.

One of the difficulties in proving that the d_j^k tends to zero at a sufficiently fast rate lies in providing inequalities for the quadratic terms of the type $\mathbf{y}^{\mathsf{H}}(\mathbf{A} - z\mathbf{I}_N)^{-1}\mathbf{y}$ present in the denominators. For this, we use Corollary 3.2, which states that, for any non-negative definite matrix $\mathbf{A}$, $\mathbf{y} \in \mathbb{C}^N$ and for $z \in \mathbb{C}^+$

$$\left|\frac{1}{1+\tau_j\mathbf{y}^{\mathsf{H}}(\mathbf{A} - z\mathbf{I}_N)^{-1}\mathbf{y}}\right| \leq \frac{|z|}{\Im[z]}. \tag{6.13}$$

Also, we need to ensure that $\mathbf{D}^{-1}$ and $\mathbf{D}_{(j)}^{-1}$ have uniformly bounded spectral norm. This unfolds from the following lemma.

Lemma 6.3 (Lemma 8 of [Couillet *et al.*, 2011a]). *Let* $\mathbf{D} = \mathbf{A} + i\mathbf{B} + iv\mathbf{I}_N$, *with* $\mathbf{A} \in \mathbb{C}^{N\times N}$ *Hermitian,* $\mathbf{B} \in \mathbb{C}^{N\times N}$ *Hermitian non-negative and* $v > 0$. *Then* $\|\mathbf{D}\| \leq v^{-1}$.

Proof. Noticing that $\mathbf{D}\mathbf{D}^{\mathsf{H}} = (\mathbf{A} + i\mathbf{B})(\mathbf{A} - i\mathbf{B}) + v^2\mathbf{I}_N + 2v\mathbf{B}$, the smallest eigenvalue of $\mathbf{D}\mathbf{D}^{\mathsf{H}}$ is greater than or equal to v^2 and therefore $\|\mathbf{D}^{-1}\| \leq v^{-1}$. □

At this step, we need to invoke the generalized trace lemma, Theorem 3.12. From Theorem 3.12, (6.13), Lemma 6.3 and the inequalities due to the truncation steps, we can then show that

$$\tau_j|d_j^1| \leq \|\mathbf{x}_j\|^2\frac{c\log^7 N|z|^3}{N\Im[z]^7}$$

$$\tau_j|d_j^2| \leq \frac{\log N\left|\mathbf{x}_j^{\mathsf{H}}\mathbf{R}^{\frac{1}{2}}(\mathbf{B}_{(j)} - z\mathbf{I}_N)^{-1}\mathbf{R}\mathbf{D}_{(j)}^{-1}\mathbf{R}^{\frac{1}{2}}\mathbf{x}_j - \frac{1}{n}\operatorname{tr}\mathbf{R}(\mathbf{B}_{(j)} - z\mathbf{I}_N)^{-1}\mathbf{R}\mathbf{D}_{(j)}^{-1}\right|}{\Im[z]|z|^{-1}}$$

$$\tau_j|d_j^3| \leq \frac{|z|\log^3 N}{\Im[z]N}\left(\frac{1}{\Im[z]^2} + \frac{c|z|^2\log^3 N}{\Im[z]^6}\right)$$

$$\tau_j|d_j^4| \leq \frac{\log^4 N\left(\left|\mathbf{x}_j^{\mathsf{H}}\mathbf{R}^{\frac{1}{2}}(\mathbf{B}_{(j)} - z\mathbf{I}_N)^{-1}\mathbf{R}^{\frac{1}{2}}\mathbf{x}_j - \frac{1}{n}\operatorname{tr}\mathbf{R}(\mathbf{B}_{(j)} - z\mathbf{I}_N)^{-1}\right| + \frac{\log N}{N\Im[z]}\right)}{\Im[z]^3|z|^{-1}}.$$

Applying the trace lemma for truncated variables, Theorem 3.12, and classical inequalities, there exists $\bar{K} > 0$ such that we have simultaneously

$$\mathrm{E}|\|\mathbf{x}_j\|^2 - 1|^6 \leq \frac{\bar{K}\log^{12} N}{N^3}$$

and

$$\mathrm{E}|\mathbf{x}_j^{\mathsf{H}}\mathbf{R}^{\frac{1}{2}}(\mathbf{B}_{(j)} - z\mathbf{I}_N)^{-1}\mathbf{R}\mathbf{D}_{(j)}^{-1}\mathbf{R}^{\frac{1}{2}}\mathbf{x}_j - \frac{1}{n}\operatorname{tr}\mathbf{R}(\mathbf{B}_{(j)} - z\mathbf{I}_N)^{-1}\mathbf{R}\mathbf{D}_{(j)}^{-1}|^6$$
$$\leq \frac{\bar{K}\log^{24} N}{N^3\Im[z]^{12}}$$

and

$$\mathrm{E}|\mathbf{x}_j^{\mathsf{H}}\mathbf{R}^{\frac{1}{2}}(\mathbf{B}_{(j)} - z\mathbf{I}_N)^{-1}\mathbf{R}^{\frac{1}{2}}\mathbf{x}_j - \frac{1}{n}\operatorname{tr}\mathbf{R}^{\frac{1}{2}}(\mathbf{B}_{(j)} - z\mathbf{I}_N)^{-1}\mathbf{R}^{\frac{1}{2}}|^6$$
$$\leq \frac{\bar{K}\log^{18} N}{N^3\Im[z]^6}.$$

All three moments above, when summed over the n indexes j and multiplied by any power of $\log N$, are summable. Applying the Markov inequality, Theorem 3.5, the Borel–Cantelli lemma, Theorem 3.6, and the line of arguments used in the proof of the Marčenko–Pastur law, we conclude that, for any $k > 0$, $\log^k N \max_{j\leq n} \tau_j d_j \xrightarrow{\text{a.s.}} 0$ as $N \to \infty$, and therefore:

$$e_{\mathbf{B}_N}(z) - f_N(e_{\mathbf{B}_N}(z); z) \xrightarrow{\text{a.s.}} 0$$
$$m_{\mathbf{B}_N}(z) - g_N(e_{\mathbf{B}_N}(z); z) \xrightarrow{\text{a.s.}} 0.$$

This convergence result is similar to that of Theorem (3.22), although in the latter each side of the minus sign converges, when the eigenvalue distributions of the deterministic matrices in the model converge. In the present case, even if the series $\{F^{\mathbf{T}}\}$ and $\{F^{\mathbf{R}}\}$ converge, it is not necessarily true that either $e_{\mathbf{B}_N}(z)$ or $f_N(e_{\mathbf{B}_N}(z), z)$ converges.

We wish to go further here by showing that, for all finite N, $f_N(e; z) = e$ has a solution (Step 2), that this solution is unique in some space (Step 3) and that, denoting $e_N(z)$ this solution, $e_N(z) - e_{\mathbf{B}_N}(z) \xrightarrow{\text{a.s.}} 0$ (Step 4). This will imply naturally that $m_N(z) \triangleq g_N(e_N(z); z)$ satisfies $m_{\mathbf{B}_N}(z) - m_N(z) \xrightarrow{\text{a.s.}} 0$, for all $z \in \mathbb{C}^+$. Vitali's convergence theorem, Theorem 3.11, will conclude the proof by showing that $m_{\mathbf{B}_N}(z) - m_N(z) \xrightarrow{\text{a.s.}} 0$ for all z outside the positive real half-line.

Step 2. Existence of a solution

We now show that the implicit equation $e = f_N(e; z)$ in the dummy variable e has a solution for each finite N. For this, we use a special trick that consists in growing the matrices dimensions asymptotically large while maintaining the deterministic components untouched, i.e. while maintaining $F^{\mathbf{R}}$ and $F^{\mathbf{T}}$ the same. The idea is to fix N and consider for all $j > 0$ the matrices $\mathbf{T}_{[j]} = \mathbf{T} \otimes \mathbf{I}_j \in \mathbb{C}^{jn \times jn}$, $\mathbf{R}_{[j]} =$

$\mathbf{R} \otimes \mathbf{I}_j \in \mathbb{C}^{jN \times jN}$ and $\mathbf{A}_{[j]} = \mathbf{A} \otimes \mathbf{I}_j \in \mathbb{C}^{jN \times jN}$. For a given x

$$f_{[j]}(x;z) \triangleq \frac{1}{jN} \operatorname{tr} \mathbf{R}_{[j]} \left(\mathbf{A}_{[j]} + \int \frac{\tau dF^{\mathbf{T}_{[j]}}(\tau)}{1 + c\tau x} \mathbf{R}_{[j]} - z\mathbf{I}_{Nj} \right)^{-1}$$

which is constant whatever j and equal to $f_N(x;z)$. Defining

$$\mathbf{B}_{[j]} = \mathbf{A}_{[j]} + \mathbf{R}_{[j]}^{\frac{1}{2}} \mathbf{X} \mathbf{T}_{[j]} \mathbf{X}^{\mathsf{H}} \mathbf{R}_{[j]}^{\frac{1}{2}}$$

for $\mathbf{X} \in \mathbb{C}^{Nj \times nj}$ with i.i.d. entries of zero mean and variance $1/(nj)$

$$e_{\mathbf{B}_{[j]}}(z) = \frac{1}{jN} \operatorname{tr} \mathbf{R}_{[j]} (\mathbf{A}_{[j]} + \mathbf{R}_{[j]}^{\frac{1}{2}} \mathbf{X} \mathbf{T}_{[j]} \mathbf{X}^{\mathsf{H}} \mathbf{R}_{[j]}^{\frac{1}{2}} - z\mathbf{I}_{Nj})^{-1}.$$

With the notations of Step 1, $w_{Nj} \to 0$ as $j \to \infty$, for all sequences $\mathbf{B}_{[1]}, \mathbf{B}_{[2]}, \ldots$ in a set of probability one. Take such a sequence. Noticing that both $e_{\mathbf{B}_{[j]}}(z)$ and the integrand $\frac{\tau}{1+c\tau e_{\mathbf{B}_{[j]}}(z)}$ of $f_{[j]}(x,z)$ are uniformly bounded for *fixed N* and growing j, there exists a subsequence of $e_{\mathbf{B}_{[1]}}, e_{\mathbf{B}_{[2]}}, \ldots$ over which they both converge, when $j \to \infty$, to some limits e and $\tau(1 + c\tau e)^{-1}$, respectively. But since $w_{jN} \to 0$ for this realization of $e_{\mathbf{B}_{[1]}}, e_{\mathbf{B}_{[2]}}, \ldots$, for growing j, we have that $e = \lim_j f_{[j]}(e,z)$. But we also have that, for all j, $f_{[j]}(e,z) = f_N(e,z)$. We therefore conclude that $e = f_N(e,z)$ and we have found a solution.

Step 3. Uniqueness of a solution

Uniqueness is shown classically by considering two hypothetical solutions $e \in \mathbb{C}^+$ and $\underline{e} \in \mathbb{C}^+$ to (6.6) and by showing then that $e - \underline{e} = \gamma(e - \underline{e})$, where $|\gamma|$ must be shown to be less than one. Indeed, taking the difference $e - \underline{e}$, we have with the resolvent identity

$$\begin{aligned} e - \underline{e} &= \frac{1}{N} \operatorname{tr} \mathbf{R} \mathbf{D}_e^{-1} - \frac{1}{N} \operatorname{tr} \mathbf{R} \mathbf{D}_{\underline{e}}^{-1} \\ &= \frac{1}{N} \operatorname{tr} \mathbf{R} \mathbf{D}_e^{-1} \left(\int \frac{c\tau^2 (e - \underline{e}) dF^{\mathbf{T}}(\tau)}{(1 + c\tau e)(1 + c\tau \underline{e})} \right) \mathbf{R} \mathbf{D}_{\underline{e}}^{-1} \end{aligned}$$

in which $\mathbf{D}_e$ and $\mathbf{D}_{\underline{e}}$ are the matrix $\mathbf{D}$ with $e_{\mathbf{B}_N}(z)$ replaced by e and $\underline{e}$, respectively. This leads to the expression of γ as follows.

$$\gamma = \int \frac{c\tau^2}{(1 + c\tau e)(1 + c\tau \underline{e})} dF^{\mathbf{T}}(\tau) \frac{1}{N} \operatorname{tr} \mathbf{D}_e^{-1} \mathbf{R} \mathbf{D}_{\underline{e}}^{-1} \mathbf{R}.$$

Applying the Cauchy–Schwarz inequality to the diagonal elements of $\frac{1}{N} \mathbf{D}_e^{-1} \mathbf{R} \int \frac{\sqrt{c\tau}}{1+c\tau e} dF^{\mathbf{T}}(\tau)$ and of $\frac{1}{N} \mathbf{D}_{\underline{e}}^{-1} \mathbf{R} \int \frac{\sqrt{c\tau}}{1+c\tau \underline{e}} dF^{\mathbf{T}}(\tau)$, we then have

$$\begin{aligned} |\gamma| &\leq \sqrt{\int \frac{c\tau^2 dF^{\mathbf{T}}(\tau)}{|1 + c\tau e|^2 N} \operatorname{tr} \mathbf{D}_e^{-1} \mathbf{R} (\mathbf{D}_e^{\mathsf{H}})^{-1} \mathbf{R}} \sqrt{\int \frac{c\tau^2 dF^{\mathbf{T}}(\tau)}{|1 + c\tau \underline{e}|^2 N} \operatorname{tr} \mathbf{D}_{\underline{e}}^{-1} \mathbf{R} (\mathbf{D}_{\underline{e}}{}^{\mathsf{H}})^{-1} \mathbf{R}} \\ &\triangleq \sqrt{\alpha} \sqrt{\underline{\alpha}}. \end{aligned}$$

We now proceed to a parallel computation of $\Im[e]$ and $\Im[\underline{e}]$ in the hope of retrieving both expressions in the right-hand side of the above equation.

Introducing the product $(\mathbf{D}_e^{\mathsf{H}})^{-1}\mathbf{D}_e^{\mathsf{H}}$ in the trace, we first write e under the form

$$e = \frac{1}{N}\operatorname{tr}\left(\mathbf{D}_e^{-1}\mathbf{R}(\mathbf{D}_e^{\mathsf{H}})^{-1}\left(\mathbf{A} + \left[\int \frac{\tau}{1+c\tau e^*}dF^{\mathbf{T}}(\tau)\right]\mathbf{R} - z^*\mathbf{I}_N\right)\right). \tag{6.14}$$

Taking the imaginary part, this is:

$$\begin{aligned}\Im[e] &= \frac{1}{N}\operatorname{tr}\left(\mathbf{D}_e^{-1}\mathbf{R}(\mathbf{D}_e^{\mathsf{H}})^{-1}\left(\left[\int \frac{c\tau^2\Im[e]}{|1+c\tau e|^2}dF^{\mathbf{T}}(\tau)\right]\mathbf{R} + \Im[z]\mathbf{I}_N\right)\right)\\ &= \Im[e]\alpha + \Im[z]\beta\end{aligned}$$

where

$$\beta \triangleq \frac{1}{N}\operatorname{tr}\mathbf{D}_e^{-1}\mathbf{R}(\mathbf{D}_e^{\mathsf{H}})^{-1}$$

is positive whenever $\mathbf{R} \neq 0$, and similarly $\Im[\underline{e}] = \underline{\alpha}\Im[\underline{e}] + \Im[z]\underline{\beta}$, $\underline{\beta} > 0$ with

$$\underline{\beta} \triangleq \frac{1}{N}\operatorname{tr}\mathbf{D}_{\underline{e}}^{-1}\mathbf{R}(\mathbf{D}_{\underline{e}}^{\mathsf{H}})^{-1}.$$

Notice also that

$$\alpha = \frac{\alpha\Im[e]}{\Im[e]} = \frac{\alpha\Im[e]}{\alpha\Im[e] + \beta\Im[z]} < 1$$

and

$$\underline{\alpha} = \frac{\underline{\alpha}\Im[\underline{e}]}{\Im[\underline{e}]} = \frac{\underline{\alpha}\Im[\underline{e}]}{\underline{\alpha}\Im[\underline{e}] + \underline{\beta}\Im[z]} < 1.$$

As a consequence

$$|\gamma| \leq \sqrt{\alpha}\sqrt{\underline{\alpha}} = \sqrt{\frac{\Im[e]\alpha}{\Im[e]\alpha + \Im[z]\beta}}\sqrt{\frac{\Im[\underline{e}]\underline{\alpha}}{\Im[\underline{e}]\underline{\alpha} + \Im[z]\underline{\beta}}} < 1$$

as requested. The case $\mathbf{R} = 0$ is easy to verify.

Remark 6.3. Note that this uniqueness argument is slightly more technical when $K > 1$. In this case, uniqueness of the vector $e_1, \ldots, e_K$ (under the notations of Theorem 6.1) needs be proved. Denoting $\mathbf{e} \triangleq (e_1, \ldots, e_K)^{\mathsf{T}}$, this requires to show that, for two solutions $\mathbf{e}$ and $\underline{\mathbf{e}}$ of the implicit equation, $(\mathbf{e} - \underline{\mathbf{e}}) = \mathbf{\Gamma}(\mathbf{e} - \underline{\mathbf{e}})$, where $\mathbf{\Gamma}$ has spectral radius less than one. To this end, a possible approach is to show that $|\Gamma_{ij}| \leq \alpha_{ij}^{\frac{1}{2}}\underline{\alpha}_{ij}^{\frac{1}{2}}$, for α_{ij} and $\underline{\alpha}_{ij}$ defined similar as in Step 3. Then, applying some classical matrix lemmas (Theorem 8.1.18 of [Horn and Johnson, 1985] and Lemma 5.7.9 of [Horn and Johnson, 1991]), the previous inequality implies that

$$\|\mathbf{\Gamma}\| \leq \|(\alpha_{ij}^{\frac{1}{2}}\underline{\alpha}_{ij}^{\frac{1}{2}})_{ij}\|$$

where $(\alpha_{ij}^{\frac{1}{2}}\underline{\alpha}_{ij}^{\frac{1}{2}})_{ij}$ is the matrix with (i,j) entry $\alpha_{ij}^{\frac{1}{2}}\underline{\alpha}_{ij}^{\frac{1}{2}}$ and the norm is the matrix spectral norm. We further have that

$$\|(\alpha_{ij}^{\frac{1}{2}}\underline{\alpha}_{ij}^{\frac{1}{2}})_{ij}\| \leq \|\mathbf{A}\|^{\frac{1}{2}}\|\underline{\mathbf{A}}\|^{\frac{1}{2}}$$

where $\mathbf{A}$ and $\underline{\mathbf{A}}$ are now matrices with (i,j) entry α_{ij} and $\underline{\alpha}_{ij}$, respectively. The multi-dimensional problem therefore boils down to proving that $\|\mathbf{A}\| < 1$ and $\|\underline{\mathbf{A}}\| < 1$. This unfolds from yet another classical matrix lemma (Theorem 2.1 of [Seneta, 1981]), which states in our current situation that, if we have the vectorial relation

$$\Im[\mathbf{e}] = \mathbf{A}\Im[\mathbf{e}] + \Im[z]\mathbf{b}$$

with $\Im[\mathbf{e}]$ and $\mathbf{b}$ vectors of *positive* entries and $\Im[z] > 0$, then $\|\mathbf{A}\| < 1$. The above relation generalizes, without much difficulty, the relation $\Im[e] = \Im[e]\alpha + \Im[z]\beta$ obtained above.

Step 4. Final convergence step

We finally need to show that $e_N - e_{\mathbf{B}_N}(z) \xrightarrow{\text{a.s.}} 0$. This is performed using a similar argument as for uniqueness, i.e. $e_N - e_{\mathbf{B}_N}(z) = \gamma(e_N - e_{\mathbf{B}_N}(z)) + w_N$, where $w_N \to 0$ as $N \to \infty$ and $|\gamma| < 1$; this is true for any $e_{\mathbf{B}_N}(z)$ taken from a space of probability one such that $w_N \to 0$. The major difficulty compared to the previous proof is to control precisely w_N.

The details are as follows. We will show that, for any $\ell > 0$, almost surely

$$\lim_{N\to\infty} \log^\ell N(e_{\mathbf{B}_N} - e_N) = 0. \tag{6.15}$$

Let α_N, β_N be the values as above for which $\Im[e_N] = \Im[e_N]\alpha_N + \Im[z]\beta_N$. Using truncation inequalities

$$\begin{aligned}
\frac{\Im[e_N]\alpha_N}{\beta_N} &\leq \Im[e_N]c\log N \int \frac{\tau^2}{|1+c\tau e_N|^2} dF^{\mathbf{T}}(\tau) \\
&= -\log N \Im\left[\int \frac{\tau}{1+c\tau e_N} dF^{\mathbf{T}}(\tau)\right] \\
&\leq \log^2 N|z|\Im[z]^{-1}.
\end{aligned}$$

Therefore

$$\begin{aligned}
\alpha_N &= \frac{\Im[e_N]\alpha_N}{\Im[e_N]\alpha_N + \Im[z]\beta_N} \\
&= \frac{\Im[e_N]\frac{\alpha_N}{\beta_N}}{\Im[z] + \Im[e_N]\frac{\alpha_N}{\beta_N}} \\
&\leq \frac{\log^2 N|z|}{\Im[z]^2 + \log^2 N|z|}.
\end{aligned} \tag{6.16}$$

We also have

$$e_{\mathbf{B}_N}(z) = \frac{1}{N} \operatorname{tr} \mathbf{D}^{-1}\mathbf{R} - w_N.$$

We write as in Step 3

$$\begin{aligned}
&\Im[e_{\mathbf{B}_N}]\\
&= \frac{1}{N}\operatorname{tr}\left(\mathbf{D}^{-1}\mathbf{R}(\mathbf{D}^{\mathsf{H}})^{-1}\left(\left[\int \frac{c\tau^2\Im[e_{\mathbf{B}_N}]}{|1+c\tau e_{\mathbf{B}_N}|^2}dF^{\mathbf{T}}(\tau)\right]\mathbf{R}+\Im[z]\mathbf{I}_N\right)\right)-\Im[w_N]\\
&\triangleq \Im[e_{\mathbf{B}_N}]\alpha_{\mathbf{B}_N}+\Im[z]\beta_{\mathbf{B}_N}-\Im[w_N].
\end{aligned}$$

Similarly to Step 3, we have $e_{\mathbf{B}_N} - e_N = \gamma(e_{\mathbf{B}_N} - e_N) + w_N$, where now

$$|\gamma| \leq \sqrt{\alpha_{\mathbf{B}_N}}\sqrt{\alpha_N}.$$

Fix an $\ell > 0$ and consider a realization of $\mathbf{B}_N$ for which $w_N \log^{\ell'} N \to 0$, where $\ell' = \max(\ell+1, 4)$ and N large enough so that

$$|w_N| \leq \frac{\Im[z]^3}{4c|z|^2\log^3 N}. \tag{6.17}$$

As opposed to Step 2, the term $\Im[z]\beta_{\mathbf{B}_N} - \Im[w_N]$ can be negative. The idea is to verify that in both scenarios where $\Im[z]\beta_{\mathbf{B}_N} - \Im[w_N]$ is positive and uniformly away from zero, or is not, the conclusion $|\gamma| < 1$ holds. First suppose $\beta_{\mathbf{B}_N} \leq \frac{\Im[z]^2}{4c|z|^2\log^3 N}$. Then by the truncation inequalities, we get

$$\alpha_{\mathbf{B}_N} \leq c\Im[z]^{-2}|z|^2\log^3 N\beta_{\mathbf{B}_N} \leq \frac{1}{4}$$

which implies $|\gamma| \leq \frac{1}{2}$. Otherwise we get from (6.16) and (6.17)

$$\begin{aligned}
|\gamma| &\leq \sqrt{\alpha_N}\sqrt{\frac{\Im[e_{\mathbf{B}_N}]\alpha_{\mathbf{B}_N}}{\Im[e_{\mathbf{B}_N}]\alpha_{\mathbf{B}_N}+\Im[z]\beta_{\mathbf{B}_N}-\Im[w_N]}}\\
&\leq \sqrt{\frac{\log N|z|}{\Im[z]^2+\log N|z|}}.
\end{aligned}$$

Therefore, for all N large

$$\begin{aligned}
\log^\ell N|e_{\mathbf{B}_N} - e_N| &\leq \frac{(\log^\ell N)w_N}{1-\left(\frac{\log^2 N|z|}{\Im[v]^2+\log^2 N|z|}\right)^{\frac{1}{2}}}\\
&\leq 2\Im[z]^{-2}(\Im[z]^2+\log^2 N|z|)(\log^\ell N)w_N\\
&\to 0
\end{aligned}$$

as $N \to \infty$, and (6.15) follows. Once more, the multi-dimensional case is much more technical; see [Couillet *et al.*, 2011a] for details.

We finally show

$$m_{\mathbf{B}_N} - m_N \xrightarrow{\text{a.s.}} 0 \tag{6.18}$$

as $N \to \infty$. Since $m_{\mathbf{B}_N} = \frac{1}{N}\operatorname{tr}\mathbf{D}_N^{-1} - \tilde{w}_N$ (for some $\tilde{w}_N$ defined similar to w_N), we have

$$m_{\mathbf{B}_N} - m_N = \gamma(e_{\mathbf{B}_N} - e_N) - \tilde{w}_N$$

where now

$$\gamma = \int \frac{c\tau^2}{(1+c\tau e_{\mathbf{B}_N})(1+c\tau e_N)} dF^{\mathbf{T}}(\tau) \frac{1}{N} \operatorname{tr} \mathbf{D}^{-1}\mathbf{R}\mathbf{D}_N^{-1}.$$

From the truncation inequalities, we obtain $|\gamma| \leq c|z|^2 \Im[z]^{-4} \log^3 N$. From (6.15) and the fact that $\log^\ell N \tilde{w}_N \xrightarrow{\text{a.s.}} 0$, we finally have (6.18).

In the proof of Theorem 6.17, we will use another technique for this last convergence part, which, instead of controlling precisely the behavior of w_N, consists in proving the convergence on a subset of $\mathbb{C} \setminus \mathbb{R}^+$ that does not meet strong difficulties. Using Vitali's convergence theorem, we then prove the convergence for all $z \in \mathbb{C} \setminus \mathbb{R}^+$. This approach is usually much simpler and is in general preferred.

Returning to the original non-truncated assumptions on $\mathbf{X}$, $\mathbf{T}$, and $\mathbf{R}$, for each of a countably infinite collection of z with positive imaginary part, possessing a limit point with positive imaginary part, we have (6.18). Therefore, by Vitali's convergence theorem, Theorem 3.11, and similar arguments as for the proof of the Marčenko–Pastur law, for any $\varepsilon > 0$, we have exactly that with probability one $m_{\mathbf{B}_N}(z) - m_N(z) \xrightarrow{\text{a.s.}} 0$ uniformly in any region of $\mathbb{C}$ bounded by a contour interior to

$$\mathbb{C} \setminus (\{z : |z| \leq \varepsilon\} \cup \{z = x + iv : x > 0, |v| \leq \varepsilon\}). \tag{6.19}$$

This completes the proof of Theorem 6.1. □

The previous proof is lengthy and technical, when it comes to precisely working out the inequalities based on the truncation steps. Nonetheless, in spite of these difficulties, the line of reasoning in this example can be generalized to more exotic models, which we will introduce also in this section. Moreover, we will briefly introduce alternative techniques of proof, such as the Gaussian method, which will turn out to be based on similar approaches, most particularly for Step 2 and Step 3.

We now prove Theorem 6.2, which we recall provides a deterministic way to recover the unique solution vector $e_{N,1}(z), \ldots, e_{N,K}(z)$ of the implicit Equation (6.6). The arguments of the proof are again very classical and can be reproduced for different random matrix models.

Proof of Theorem 6.2. The convergence of the fixed-point algorithm follows the same line of proof as the uniqueness (Step 2) of Theorem 6.1. For simplicity, we consider also here that $K = 1$. First assume $\Im[z] > 0$. If we consider the difference $e_N^{t+1} - e_N^t$, instead of $e - \underline{e}$, the same development as in the previous proof leads to

$$e_N^{t+1} - e_N^t = \gamma_t (e_N^t - e_N^{t-1}) \tag{6.20}$$

for $t \geq 1$, with γ_t defined by

$$\gamma_t = \int \frac{c\tau^2}{(1+c\tau e_N^{t-1})(1+c\tau e_N^t)} dF^{\mathbf{T}}(\tau) \frac{1}{N} \operatorname{tr} \mathbf{D}_{t-1}^{-1}\mathbf{R}\mathbf{D}_t^{-1}\mathbf{R} \tag{6.21}$$

where $\mathbf{D}_t$ is defined as $\mathbf{D}$ with $e_{\mathbf{B}_N}(z)$ replaced by $e_N^t(z)$. From the Cauchy–Schwarz inequality and the different truncation bounds on the $\mathbf{D}_t$, $\mathbf{R}$, and $\mathbf{T}$ matrices, we have:

$$\gamma_t \leq \frac{|z|^2 c}{\Im[z]^4} \frac{\log^4 N}{N}. \tag{6.22}$$

This entails

$$\left(e_N^{t+1} - e_N^t\right) < \bar{K} \frac{|z|^2 c}{\Im[z]^4} \frac{\log^4 N}{N} \left(e_N^t - e_N^{t-1}\right) \tag{6.23}$$

for some constant $\bar{K}$.

Let $0 < \varepsilon < 1$, and take now a countable set $z_1, z_2, \ldots$ possessing a limit point, such that

$$\bar{K} \frac{|z_k|^2 c}{\Im[z_k]^4} \frac{\log^4 N}{N} < 1 - \varepsilon$$

for all z_k (this is possible by letting $\Im[z_k] > 0$ be large enough). On this countable set, the sequences $e_N^1, e_N^2, \ldots$ are therefore Cauchy sequences on $\mathbb{C}^K$: they all converge. Since the e_N^t are holomorphic functions of z and bounded on every compact set included in $\mathbb{C} \setminus \mathbb{R}^+$, from Vitali's convergence theorem, Theorem 3.11, e_N^t converges on such compact sets.

From the fact that we forced the initialization step to be $e_N^0 = -1/z$, e_N^0 is the Stieltjes transform of a distribution function at point z. It now suffices to verify that, if $e_N^t = e_N^t(z)$ is the Stieltjes transform of a distribution function at point z, then so is e_N^{t+1}. From Theorem 3.2, this requires to ensure that: (i) $z \in \mathbb{C}^+$ and $e_N^t(z) \in \mathbb{C}^+$ implies $e_N^{t+1}(z) \in \mathbb{C}^+$, (ii) $z \in \mathbb{C}^+$ and $ze_N^t(z) \in \mathbb{C}^+$ implies $ze_N^{t+1}(z) \in \mathbb{C}^+$, and (iii) $\lim_{y \to \infty} -ye_N^t(iy) < \infty$ implies that $\lim_{y \to \infty} -ye_N^t(iy) < \infty$. These properties follow directly from the definition of e_N^t. It is not difficult to show also that the limit of e_N^t is a Stieltjes transform and that it is solution to (6.6) when $K = 1$. From the uniqueness of the Stieltjes transform, solution to (6.6) (this follows from the point-wise uniqueness on $\mathbb{C}^+$ and the fact that the Stieltjes transform is holomorphic on all compact sets of $\mathbb{C} \setminus \mathbb{R}^+$), we then have that e_N^t converges for all j and $z \in \mathbb{C} \setminus \mathbb{R}^+$, if e_N^0 is initialized at a Stieltjes transform. The choice $e_N^0 = -1/z$ follows this rule and the fixed-point algorithm converges to the correct solution.

This concludes the proof of Theorem 6.2. □

From Theorem 6.1, we now wish to provide deterministic equivalents for other functionals of the eigenvalues of $\mathbf{B}_N$ than the Stieltjes transform. In particular, we wish to prove that

$$\int f(x) d(F^{\mathbf{B}_N} - F_N)(x) \xrightarrow{\text{a.s.}} 0$$

for some function f. This is valid for all bounded continuous f from the dominated convergence theorem, which we recall presently.

Theorem 6.3 (Theorem 16.4 in [Billingsley, 1995]). *Let $f_N(x)$ be a sequence of real measurable functions converging point-wise to the measurable function $f(x)$, and such that $|f_N(x)| \leq g(x)$ for some measurable function $g(x)$ with $\int g(x)dx < \infty$. Then, as $N \to \infty$*

$$\int f_N(x)dx \to \int f(x)dx.$$

In particular, if $F_N \Rightarrow F$, the F_N and F being d.f., for any continuous bounded function $h(x)$

$$\int h(x)dF_N(x) \to \int h(x)dF(x).$$

However, for application purposes, such as the calculus of MIMO capacity, see Chapter 13, we would like in particular to take f to be the logarithm function. Proving such convergence results is not at all straightforward since f is here unbounded and because $F^{\mathbf{B}_N}$ may not have bounded support for all large N. This requires additional tools which will be briefly evoked here and which will be introduced in detail in Chapter 7.

We have the following result [Couillet *et al.*, 2011a].

Theorem 6.4. *Let x be some positive real number and f be some continuous function on the positive half-line. Let $\mathbf{B}_N$ be a random Hermitian matrix as defined in Theorem 6.1 with the following additional assumptions.*

1. *There exists $\alpha > 0$ and a sequence r_N, such that, for all N*

$$\max_{1\leq k\leq K} \max(\lambda_{r_N+1}^{\mathbf{T}_k}, \lambda_{r_N+1}^{\mathbf{R}_k}) \leq \alpha$$

   where $\lambda_1^{\mathbf{X}} \geq \ldots \geq \lambda_N^{\mathbf{X}}$ denote the ordered eigenvalues of the $N \times N$ matrix $\mathbf{X}$.
2. *Denoting b_N an upper-bound on the spectral norm of the $\mathbf{T}_k$ and $\mathbf{R}_k$, $k \in \{1, \ldots, K\}$, and β some real, such that $\beta > K(b/a)(1+\sqrt{a})^2$ (with a and b such that $a < \liminf_N c_k \leq \limsup_N c_k < b$ for all k), then $a_N = b_N^2 \beta$ satisfies*

$$r_N f(a_N) = o(N). \tag{6.24}$$

Then, for large N, n_k

$$\int f(x)dF^{\mathbf{B}_N}(x) - \int f(x)dF_N(x) \xrightarrow{\text{a.s.}} 0$$

with F_N defined in Theorem 6.1.

In particular, if $f(x) = \log(x)$, under the assumption that (6.24) is fulfilled, we have the following corollary.

Corollary 6.1. *For* $\mathbf{A} = 0$*, under the conditions of Theorem 6.4 with* $f(t) = \log(1 + xt)$*, the Shannon transform* $\mathcal{V}_{\mathbf{B}_N}$ *of* $\mathbf{B}_N$*, defined for positive* x *as*

$$\mathcal{V}_{\mathbf{B}_N}(x) = \int_0^\infty \log(1 + x\lambda) dF^{\mathbf{B}_N}(\lambda) = \frac{1}{N} \log\det(\mathbf{I}_N + x\mathbf{B}_N) \tag{6.25}$$

satisfies

$$\mathcal{V}_{\mathbf{B}_N}(x) - \mathcal{V}_N(x) \xrightarrow{\text{a.s.}} 0$$

where $\mathcal{V}_N(x)$ *is defined as*

$$\begin{aligned}\mathcal{V}_N(x) =& \frac{1}{N} \log\det\left(\mathbf{I}_N + x\sum_{k=1}^K \mathbf{R}_k \int \frac{\tau_k dF^{\mathbf{T}_k}(\tau_k)}{1 + c_k e_{N,k}(-1/x)\tau_k}\right) \\ &+ \sum_{k=1}^K \frac{1}{c_k} \int \log\left(1 + c_k e_{N,k}(-1/x)\tau_k\right) dF^{\mathbf{T}_k}(\tau_k) \\ &+ \frac{1}{x} m_N(-1/x) - 1\end{aligned}$$

with m_N *and* $e_{N,k}$ *defined by* (6.5) *and* (6.6)*, respectively.*

Again, it is more convenient, for readability and for the sake of practical applications in Chapters 12–15 to remark that

$$\begin{aligned}\mathcal{V}_N(x) = &\frac{1}{N} \log\det\left(\mathbf{I}_N + \sum_{k=1}^K \bar{e}_{N,k}(-1/x)\mathbf{R}_k\right) \\ &+ \sum_{k=1}^K \frac{1}{N} \log\det\left(\mathbf{I}_{n_k} + c_k e_{N,k}(-1/x)\mathbf{T}_k\right) \\ &- \frac{1}{x} \sum_{k=1}^K \bar{e}_{N,k}(-1/x) e_{N,k}(-1/x)\end{aligned} \tag{6.26}$$

with $\bar{e}_{N,k}$ defined in (6.7).

Observe that the constraint

$$\max_{1 \le k \le K} \max(\lambda_{r_N+1}^{\mathbf{T}_k}, \lambda_{r_N+1}^{\mathbf{R}_k}) \le \alpha$$

is in general not strong, as the $F^{\mathbf{T}_k}$ and the $F^{\mathbf{R}_k}$ are already known to form tight sequences as N grows large. Therefore, it is expected that only $o(N)$ largest eigenvalues of the $\mathbf{T}_k$ and $\mathbf{R}_k$ grow large. Here, we impose only a slightly stronger constraint that does not allow for the smallest eigenvalues to exceed a constant α. For practical applications, we will see in Chapter 13 that this constraint is met for all usual channel models, even those exhibiting strong correlation patterns (such as densely packed three-dimensional antenna arrays).

Proof of Theorem 6.4 and Corollary 6.1. The only problem in translating the weak convergence of the distribution function $F^{\mathbf{B}_N} - F_N$ in Theorem 6.1 to the convergence of $\int f d[F^{\mathbf{B}_N} - F_N]$ in Theorem 6.4 is that we must ensure that f behaves nicely. If f were bounded, no restriction in the hypothesis of Theorem 6.1 would be necessary and the weak convergence of $F^{\mathbf{B}_N} - F_N$ to zero gives the result. However, as we are particularly interested in the unbounded, though slowly increasing, logarithm function, this no longer holds. In essence, the proof consists first in taking a realization $\mathbf{B}_1, \mathbf{B}_2, \ldots$ for which the convergence $F^{\mathbf{B}_N} - F_N \Rightarrow 0$ is satisfied. Then we divide the real positive half-line in two sets $[0, d]$ and (d, ∞), with d an upper bound on the $2Kr_N$th largest eigenvalue of $\mathbf{B}_N$ for all large N, which we assume for the moment does exist. For any continuous f, the convergence result is ensured on the compact $[0, d]$; if the largest eigenvalue λ_1 of $\mathbf{B}_N$ is moreover such that $2Kr_N f(\lambda_1) = o(N)$, the integration over (d, ∞) for the measure $dF^{\mathbf{B}_N}$ is of order $o(1)$, which is negligible in the final result for large N. Moreover, since $F_N(d) - F^{\mathbf{B}_N}(d) \to 0$, we also have that, for all large N, $1 - F_N(d) = \int_d^\infty dF_N \leq 2Kr_N/N$, which tends to zero. This finally proves the convergence of $\int f d[F^{\mathbf{B}_N} - F_N]$. The major difficulty here lies in proving that there exists such a bound on the $2Kr_N$th largest eigenvalue of $\mathbf{B}_N$. The essential argument that validates the result is the *asymptotic absence of eigenvalues outside the support of the sample covariance matrix*. This is a result of utmost importance (here, we cannot do without it) which will be presented later in Section 7.1. It can be exactly proved that, almost surely, the largest eigenvalue of $\mathbf{X}_k\mathbf{X}_k^{\mathsf{H}}$ is uniformly bounded by any constant $C > (1 + \sqrt{b})^2$ for all large N, almost surely. In order to use the assumptions of Theorem 6.4, we finally need to introduce the following eigenvalue inequality lemma.

Lemma 6.4 ([Fan, 1951]). *Consider a rectangular matrix* $\mathbf{A}$ *and let* $s_i^{\mathbf{A}}$ *denote the* ith *largest singular value of* $\mathbf{A}$, *with* $s_i^{\mathbf{A}} = 0$ *whenever* $i > \text{rank}(A)$. *Let* m, n *be arbitrary non-negative integers. Then for* $\mathbf{A}, \mathbf{B}$ *rectangular of the same size*

$$s_{m+n+1}^{\mathbf{A}+\mathbf{B}} \leq s_{m+1}^{\mathbf{A}} + s_{n+1}^{\mathbf{B}}$$

and for $\mathbf{A}, \mathbf{B}$ *rectangular for which* $\mathbf{AB}$ *is defined*

$$s_{m+n+1}^{\mathbf{AB}} \leq s_{m+1}^{\mathbf{A}} s_{n+1}^{\mathbf{B}}.$$

As a corollary, for any integer $r \geq 0$ *and rectangular matrices* $\mathbf{A}_1, \ldots, \mathbf{A}_K$, *all of the same size*

$$s_{Kr+1}^{\mathbf{A}_1+\ldots+\mathbf{A}_K} \leq s_{r+1}^{\mathbf{A}_1} + \ldots + s_{r+1}^{\mathbf{A}_K}.$$

Since $\lambda_i^{\mathbf{T}_k}$ and $\lambda_i^{\mathbf{R}_k}$ are bounded by α for $i \geq r_N + 1$ and that $\|\mathbf{X}_k\mathbf{X}_k^{\mathsf{H}}\|$ is bounded by C, we have from Lemma 6.4 that the $2Kr_N$th largest eigenvalue of $\mathbf{B}_N$ is uniformly bounded by $CK\alpha^2$. We can then take d any positive real, such that $d > CK\alpha^2$, which is what we needed to show, up to some fine tuning on the final bound.

As for the explicit form of $\int \log(1 + xt)dF_N(t)$ given in (6.26), it results from a similar calculus as in Theorem 4.10. Precisely, we expect the Shannon transform to be somehow linked to $\frac{1}{N}\log\det\left(\mathbf{I}_N + \sum_{k=1}^K \bar{e}_{N,k}(-z)\mathbf{R}_k\right)$ and $\frac{1}{N}\log\det\left(\mathbf{I}_{n_k} + c_k e_{N,k}(-z)\mathbf{T}_k\right)$. We then need to find a connection between the derivatives of these functions along z and $\frac{1}{z} - m_N(-z)$, i.e. the derivative of the Shannon transform. Notice that

$$\begin{aligned}
\frac{1}{z} - m_N(-z) &= \frac{1}{N}\left((z\mathbf{I}_N)^{-1} - \left(z\left[\mathbf{I}_N + \sum_{k=1}^K \bar{e}_{N,k}\mathbf{R}_k\right]\right)^{-1}\right) \\
&= \sum_{k=1}^K \bar{e}_{N,k}(-z)e_{N,k}(-z).
\end{aligned}$$

Since the Shannon transform $\mathcal{V}_N(x)$ satisfies $\mathcal{V}_N(x) = \int_{1/x}^{\infty}[w^{-1} - m_N(-w)]dw$, we need to find an integral form for $\sum_{k=1}^K \bar{e}_{N,k}(-z)e_{N,k}(-z)$. Notice now that

$$\begin{aligned}
\frac{d}{dz}\frac{1}{N}\log\det\left(\mathbf{I}_N + \sum_{k=1}^K \bar{e}_{N,k}(-z)\mathbf{R}_k\right) &= -z\sum_{k=1}^K e_{N,k}(-z)\bar{e}'_{N,k}(-z) \\
\frac{d}{dz}\frac{1}{N}\log\det\left(\mathbf{I}_{n_k} + c_k e_{N,k}(-z)\mathbf{T}_k\right) &= -ze'_{N,k}(-z)\bar{e}_{N,k}(-z)
\end{aligned}$$

and

$$\begin{aligned}
\frac{d}{dz}\left(z\sum_{k=1}^K \bar{e}_{N,k}(-z)e_{N,k}(-z)\right) &= \sum_{k=1}^K \bar{e}_{N,k}(-z)e_{N,k}(-z) \\
&\quad - z\sum_{k=1}^K \left(\bar{e}'_{N,k}(-z)e_{N,k}(-z) + \bar{e}_{N,k}(-z)e'_{N,k}(-z)\right).
\end{aligned}$$

Combining the last three equations, we have:

$$\begin{aligned}
&\sum_{k=1}^K \bar{e}_{N,k}(-z)e_{N,k}(-z) \\
&= \frac{d}{dz}\left[-\frac{1}{N}\log\det\left(\mathbf{I}_N + \sum_{k=1}^K \bar{e}_{N,k}(-z)\mathbf{R}_k\right)\right. \\
&\quad \left. - \sum_{k=1}^K \frac{1}{N}\log\det\left(\mathbf{I}_{n_k} + c_k e_{N,k}(-z)\mathbf{T}_k\right) + z\sum_{k=1}^K \bar{e}_{N,k}(-z)e_{N,k}(-z)\right]
\end{aligned}$$

which after integration leads to

$$\begin{aligned}
&\int_z^{\infty}\left(\frac{1}{w} - m_N(-w)\right)dw \\
&= \frac{1}{N}\log\det\left(\mathbf{I}_N + \sum_{k=1}^K \bar{e}_{N,k}(-z)\mathbf{R}_k\right)
\end{aligned}$$

$$+\sum_{k=1}^{K}\frac{1}{N}\log\det\left(\mathbf{I}_{n_k}+c_k e_{N,k}(-z)\mathbf{T}_k\right)-z\sum_{k=1}^{K}\bar{e}_{N,k}(-z)e_{N,k}(-z)$$

which is exactly the right-hand side of (6.26) for $z=-1/x$. □

Theorem 6.4 and Corollary 6.1 have obvious direct applications in wireless communications since the Shannon transform $\mathcal{V}_{\mathbf{B}_N}$ defined above is the per-dimension capacity of the multi-dimensional channel, whose model is given by $\sum_{k=1}^{K}\mathbf{R}_k^{\frac{1}{2}}\mathbf{X}_k\mathbf{T}_k^{\frac{1}{2}}$. This is the typical model used for evaluating the rate region of a narrowband multiple antenna multiple access channel. This topic is discussed and extended in Chapter 14, e.g. to the question of finding the transmit covariance matrix that maximizes the deterministic equivalent (hence the asymptotic capacity).

6.2.2 Gaussian method

The second result that we present is very similar in nature to Theorem 6.1 but instead of considering sums of matrices of the type

$$\mathbf{B}_N=\sum_{k=1}^{K}\mathbf{R}_k^{\frac{1}{2}}\mathbf{X}_k\mathbf{T}_k\mathbf{X}_k^{\mathsf{H}}\mathbf{R}_k^{\frac{1}{2}}$$

we treat the question of matrices of the type

$$\mathbf{B}_N=\left(\sum_{k=1}^{K}\mathbf{R}_k^{\frac{1}{2}}\mathbf{X}_k\mathbf{T}_k^{\frac{1}{2}}\right)\left(\sum_{k=1}^{K}\mathbf{R}_k^{\frac{1}{2}}\mathbf{X}_k\mathbf{T}_k^{\frac{1}{2}}\right)^{\mathsf{H}}.$$

To obtain a deterministic equivalent for this model, the same technique as before could be used. Instead, we develop an alternative method, known as the *Gaussian method*, when the $\mathbf{X}_k$ have Gaussian i.i.d. entries, for which fast convergence rates of the functional of the mean e.s.d. can be proved.

Theorem 6.5 ([Dupuy and Loubaton, 2009]). *Let K be some positive integer. For two positive integers N,n, denote*

$$\mathbf{B}_N=\left(\sum_{k=1}\mathbf{R}_k^{\frac{1}{2}}\mathbf{X}_k\mathbf{T}_k^{\frac{1}{2}}\right)\left(\sum_{k=1}\mathbf{R}_k^{\frac{1}{2}}\mathbf{X}_k\mathbf{T}_k^{\frac{1}{2}}\right)^{\mathsf{H}}$$

where the notations are the same as in Theorem 6.1, with the additional assumptions that $n_1=\ldots=n_K=n$, the random matrix $\mathbf{X}_k\in\mathbb{C}^{N\times n_k}$ has independent Gaussian entries (of zero mean and variance $1/n$) and the spectral norms $\|\mathbf{R}_k\|$ and $\|\mathbf{T}_k\|$ are uniformly bounded with N. Note additionally that, from the unitarily invariance of $\mathbf{X}_k$, $\mathbf{T}_k$ is not restricted to be diagonal. Then, denoting as above $m_{\mathbf{B}_N}$ the Stieltjes transform of $\mathbf{B}_N$, we have

$$N\left(\mathrm{E}[m_{\mathbf{B}_N}(z)]-m_N(z)\right)=O\left(1/N\right)$$

with m_N defined, for $z \in \mathbb{C} \setminus \mathbb{R}^+$, as

$$m_N(z) = \frac{1}{N} \operatorname{tr} \left(-z \left[\mathbf{I}_N + \sum_{k=1}^{K} \bar{e}_{N,k}(z) \mathbf{R}_k \right] \right)^{-1}$$

where $(\bar{e}_{N,1}, \ldots, \bar{e}_{N,K})$ is the unique solution of

$$\begin{aligned} e_{N,i}(z) &= \frac{1}{n} \operatorname{tr} \mathbf{R}_i \left(-z \left[\mathbf{I}_N + \sum_{k=1}^{K} \bar{e}_{N,k}(z) \mathbf{R}_k \right] \right)^{-1} \\ \bar{e}_{N,i}(z) &= \frac{1}{n} \operatorname{tr} \mathbf{T}_i \left(-z \left[\mathbf{I}_n + \sum_{k=1}^{K} e_{N,k}(z) \mathbf{T}_k \right] \right)^{-1} \end{aligned} \tag{6.27}$$

all with positive imaginary part if $z \in \mathbb{C}^+$, negative imaginary part if $z \in \mathbb{C}^-$, and positive if $z < 0$.

Remark 6.4. Note that, due to the Gaussian assumption on the entries of $\mathbf{X}_k$, the convergence result $N\left(\mathrm{E}[m_{\mathbf{B}_N}(z)] - m_N(z)\right) \to 0$ is both (i) looser than the convergence result $m_{\mathbf{B}_N}(z) - m_N(z) \xrightarrow{\text{a.s.}} 0$ of Theorem 6.1 in that it is only shown to converge in expectation, and (ii) stronger in the sense that a convergence rate of $O(1/N)$ of the Stieltjes transform is ensured. Obviously, Theorem 6.1 also implies $\mathrm{E}[m_{\mathbf{B}_N}(z)] - m_N(z) \to 0$. In fact, while this was not explicitly mentioned, a convergence rate of $1/(\log(N)^p)$, for all $p > 0$, is ensured in the proof of Theorem 6.1. The main applicative consequence is that, while the conditions of Theorem 6.1 allow us to deal with instantaneous or quasi-static channel models $\mathbf{H}_k = \mathbf{R}_k^{\frac{1}{2}} \mathbf{X}_k \mathbf{T}_k^{\frac{1}{2}}$, the conditions of Theorem 6.5 are only valid from an ergodic point of view. However, while Theorem 6.1 can only deal with the per-antenna capacity of a quasi-static (or ergodic) MIMO channel, Theorem 6.5 can deal with the total ergodic capacity of MIMO channels, see further Theorem 6.8.

Of course, while this has not been explicitly proved in the literature, it is to be expected that Theorem 6.5 holds also under the looser assumptions and conclusions of Theorem 6.1 and conversely.

The proof of Theorem 6.5 needs the introduction of new tools, gathered together into the so-called *Gaussian method.* Basically, the Gaussian method relies on two main ingredients:

- an integration by parts formula, borrowed from mathematical physics [Glimm and Jaffe, 1981]

Theorem 6.6. *Let $\mathbf{x} = [x_1, \ldots, x_N]^\mathsf{T} \sim \mathcal{CN}(0, \mathbf{R})$ be a complex Gaussian random vector and $f(\mathbf{x}) \triangleq f(x_1, \ldots, x_N, x_i^*, \ldots, x_N^*)$ be a continuously differentiable functional, the derivatives of which are all polynomially bounded.*

We then have the integration by parts formula

$$\mathrm{E}[x_k f(\mathbf{x})] = \sum_{i=1}^{N} r_{ki} \mathrm{E}\left[\frac{\partial f(\mathbf{x})}{\partial x_i^*}\right]$$

with r_{ki} the entry (k,i) of $\mathbf{R}$.

This relation will be used to *derive directly* the deterministic equivalent, which substitutes to the 'guess-work' step of the proof of Theorem 6.1. Note in particular that it requires us to use all entries of $\mathbf{R}$ here and not simply its eigenvalues. This generalizes the Marčenko–Pastur method that only handled diagonal entries. However, as already mentioned, the introduction of the expectation in front of $x_k f(\mathbf{x})$ cannot be avoided;

- the Nash–Poincaré inequality

Theorem 6.7 ([Pastur, 1999]). *Let $\mathbf{x}$ and f be as in Theorem 6.6, and let $\nabla_{\mathbf{z}} f = [\partial f/\partial z_1, \ldots, \partial f/\partial z_N]^\mathsf{T}$. Then, we have the following Nash–Poincaré inequality*

$$\mathrm{var}(f(\mathbf{x})) \leq \mathrm{E}\left[\nabla_{\mathbf{x}} f(\mathbf{x})^\mathsf{T} \mathbf{R} (\nabla_{\mathbf{x}} f(\mathbf{x}))^*\right] + \mathrm{E}\left[(\nabla_{\mathbf{x}^*} f(\mathbf{x}))^\mathsf{H} \mathbf{R} \nabla_{\mathbf{x}^*} f(\mathbf{x})\right].$$

This result will be used to bound the deviations of the random matrices under consideration.

For more details on Gaussian methods, see [Hachem *et al.*, 2008a]. We now give the main steps of the proof of Theorem 6.5.

Proof of Theorem 6.5. We first consider $\mathrm{E}(\mathbf{B}_N - z\mathbf{I}_N)^{-1}$. Noting that $-z(\mathbf{B}_N - z\mathbf{I}_N)^{-1} = \mathbf{I}_N - (\mathbf{B}_N - z\mathbf{I}_N)^{-1}\mathbf{B}_N$, we apply the integration by parts, Theorem 6.6, in order to evaluate the matrix

$$\mathrm{E}\left[(\mathbf{B}_N - z\mathbf{I}_N)^{-1}\mathbf{B}_N\right].$$

To this end, we wish to characterize every entry

$$\begin{aligned}
&\mathrm{E}\left[\left((\mathbf{B}_N - z\mathbf{I}_N)^{-1}\mathbf{B}_N\right)_{aa'}\right] \\
&= \sum_{1 \leq k, \bar{k} \leq K} \mathrm{E}\left[\left((\mathbf{B}_N - z\mathbf{I}_N)^{-1}\mathbf{R}_k^{\frac{1}{2}}(\mathbf{X}_k\mathbf{T}_k^{\frac{1}{2}}\mathbf{R}_{\bar{k}}^{\frac{1}{2}})(\mathbf{X}_{\bar{k}}\mathbf{T}_{\bar{k}}^{\frac{1}{2}})^\mathsf{H}\right)_{aa'}\right].
\end{aligned}$$

This is however not so simple and does not lead immediately to a nice form enabling us to use the Gaussian entries of the $\mathbf{X}_k$ as the inputs of Theorem 6.6. Instead, we will consider the multivariate expression

$$\mathrm{E}\left[(\mathbf{B}_N - z\mathbf{I}_N)^{-1}_{ab}(\mathbf{R}_k^{\frac{1}{2}}\mathbf{X}_k\mathbf{T}_k^{\frac{1}{2}})_{cd}(\mathbf{R}_{\bar{k}}^{\frac{1}{2}}\mathbf{X}_{\bar{k}}\mathbf{T}_{\bar{k}}^{\frac{1}{2}})^\mathsf{H}_{ea'}\right]$$

for some $k, \bar{k} \in \{1, \ldots, K\}$ and given a, a', b, c, d, e. This enables us to somehow unfold easily the matrix products before we set $b = c$ and $d = e$, and simplify the management of the Gaussian variables. This being said, we take the vector $\mathbf{x}$ of Theorem 6.6 to be the vector whose entries are denoted

$$x_{k,c,d} \triangleq x_{(k-1)Nn+(c-1)N+d} = (\mathbf{R}_k^{\frac{1}{2}}\mathbf{X}_k\mathbf{T}_k^{\frac{1}{2}})_{cd}$$

for all k, c, d. This is therefore a vector of total dimension KNn that collects the entries of all $\mathbf{X}_k$ and accounts for the (Kronecker-type) correlation profile due to $\mathbf{R}_k$ and $\mathbf{T}_k$. The functional $f(\mathbf{x}) = f_{a,b}(\mathbf{x})$ of Theorem 6.6 is taken to be the KNn-dimensional vector $\mathbf{y}^{(a,b)}$ with entry

$$y^{(a,b)}_{\bar{k},a',e} \triangleq y^{(a,b)}_{(\bar{k}-1)Nn+(a'-1)N+e} = (\mathbf{B}_N - z\mathbf{I}_N)^{-1}_{ab}(\mathbf{R}_{\bar{k}}^{\frac{1}{2}}\mathbf{X}_{\bar{k}}\mathbf{T}_{\bar{k}}^{\frac{1}{2}})^{\mathsf{H}}_{ea'}$$

for all $\bar{k}, e, a'$. This expression depends on $\mathbf{x}$ through $(\mathbf{B}_N - z\mathbf{I}_N)^{-1}_{ab}$ and through $x^*_{\bar{k},a',e} = (\mathbf{R}_{\bar{k}}^{\frac{1}{2}}\mathbf{X}_{\bar{k}}\mathbf{T}_{\bar{k}}^{\frac{1}{2}})^{\mathsf{H}}_{ea'}$.

We therefore no longer take $b = c$ or $d = e$ as matrix products would require. This trick allows us to apply seamlessly the integration by parts formula. Applying Theorem 6.6, we have that the entry $(\bar{k}-1)Nn + (a'-1)N + e$ of $\mathrm{E}[x_{k,c,d}f_{a,b}(\mathbf{x})]$, i.e. $\mathrm{E}[x_{k,c,d}y^{(a,b)}_{\bar{k},a',e}]$, is given by:

$$\begin{aligned}&\mathrm{E}[(\mathbf{B}_N - z\mathbf{I}_N)^{-1}_{ab}(\mathbf{R}_k^{\frac{1}{2}}\mathbf{X}_k\mathbf{T}_k^{\frac{1}{2}})_{cd}(\mathbf{R}_{\bar{k}}^{\frac{1}{2}}\mathbf{X}_{\bar{k}}\mathbf{T}_{\bar{k}}^{\frac{1}{2}})^{\mathsf{H}}_{ea'}] \\ &= \sum_{k',c',d'} \mathrm{E}\left[x_{k,c,d}x^*_{k',c',d'}\right]\mathrm{E}\left[\frac{\partial\left((\mathbf{B}_N - z\mathbf{I}_N)^{-1}_{ab}x^*_{\bar{k},a',e}\right)}{\partial x^*_{k',c',d'}}\right]\end{aligned}$$

for all choices of a, b, c, d, e, a'. At this point, we need to proceed to cumbersome calculus, that eventually leads to a nice form when setting $b = c$ and $d = e$.

This gives an expression of $\mathrm{E}\left[\left((\mathbf{B}_N - z\mathbf{I}_N)^{-1}\mathbf{R}_k^{\frac{1}{2}}\mathbf{X}_k\mathbf{T}_k^{\frac{1}{2}}\mathbf{R}_{\bar{k}}^{\frac{1}{2}}\mathbf{X}_{\bar{k}}\mathbf{T}_{\bar{k}}^{\frac{1}{2}}\right)_{aa'}\right]$, which is then summed over all couples $k, \bar{k}$ to obtain

$$\mathrm{E}\left[\left((\mathbf{B}_N - z\mathbf{I}_N)^{-1}\mathbf{B}_N\right)_{aa'}\right] = -z\sum_{k=1}^{K}\bar{e}_{\mathbf{B}_N,k}(z)\mathrm{E}\left[\left((\mathbf{B}_N - z\mathbf{I}_N)^{-1}\mathbf{R}_k\right)_{aa'}\right] + w_{N,aa'}$$

where we defined

$$\begin{aligned}\bar{e}_{\mathbf{B}_N,k}(z) &\triangleq \frac{1}{n}\operatorname{tr}\mathbf{T}_k\left(-z\left[\mathbf{I}_n + \sum_{k=1}^{K}e_{\mathbf{B}_N,k}\mathbf{T}_k\right]\right)^{-1} \\ e_{\mathbf{B}_N,k}(z) &\triangleq \mathrm{E}\left[\frac{1}{N}\operatorname{tr}\mathbf{R}_k(\mathbf{B}_N - z\mathbf{I}_N)^{-1}\right]\end{aligned}$$

and $w_{N,aa'}$ is a residual term that must be shown to be going to zero at a certain rate for increasing N. Using again the formula $-z(\mathbf{B}_N - z\mathbf{I}_N)^{-1} = \mathbf{I}_N - (\mathbf{B}_N - z\mathbf{I}_N)^{-1}\mathbf{B}_N$, this entails

$$\mathrm{E}\left[(\mathbf{B}_N - z\mathbf{I}_N)^{-1}\right] = -\frac{1}{z}\left(\mathbf{I}_N + \sum_{k=1}^{K}\bar{e}_{\mathbf{B}_N,k}(z)\mathbf{R}_k\right)^{-1}\left[\mathbf{I}_N + \mathbf{W}_N\right]$$

with $\mathbf{W}_N$ the matrix of (a, a') entry $w_{N,aa'}$. Showing that $\mathbf{W}_N$ is negligible with summable entries as $N \to \infty$ is then solved using the Nash–Poincaré inequality, Theorem 6.7, which again leads to cumbersome but doable calculus.

The second main step consists in considering the system (6.27) (the uniqueness of the solution of which is treated as for Theorem 6.1) and showing that, for any uniformly bounded matrix $\mathbf{E}$

$$\mathrm{E}\left[\operatorname{tr}\mathbf{E}(\mathbf{B}_N - z\mathbf{I}_N)^{-1}\right] = \operatorname{tr}\mathbf{E}(-z[\mathbf{I}_N + \sum_{k=1}^{K} \bar{e}_{N,k}(z)\mathbf{R}_k])^{-1} + O\left(\frac{1}{N}\right)$$

from which $N(\mathrm{E}[e_{\mathbf{B}_N,k}(z)] - e_{N,k}(z)) = O(1/N)$ (for $\mathbf{E} = \mathbf{R}_k$) and finally $N(\mathrm{E}[m_{\mathbf{B}_N}(z)] - m_N(z)) = O(1/N)$ (for $\mathbf{E} = \mathbf{I}_N$). This is performed in a similar way as in the proof for Theorem 6.1, with the additional results coming from the Nash–Poincaré inequality. □

The Gaussian method, while requiring more intensive calculus, allows us to unfold naturally the deterministic equivalent under study for all types of matrix combinations involving Gaussian matrices. It might as well be used as a tool to infer the deterministic equivalent of more involved models for which such deterministic equivalents are not obvious to 'guess' or for which the Marčenko–Pastur method for diagonal matrices cannot be used. For the latest results derived from this technique, refer to, e.g., [Hachem *et al.*, 2008a; Khorunzhy *et al.*, 1996; Pastur, 1999]. It is believed that Haar matrices can be treated using the same tools, to the effort of more involved computations but, to the best of our knowledge, there exists no reference of such a work, yet.

In the same way as we derived the expression of the Shannon transform of the model $\mathbf{B}_N$ of Theorem 6.1 in Corollary 6.1, we have the following result for $\mathbf{B}_N$ in Theorem 6.5.

Theorem 6.8 ([Dupuy and Loubaton, 2010]). *Let $\mathbf{B}_N \in \mathbb{C}^{N\times N}$ be defined as in Theorem 6.5. Then the Shannon transform $\mathcal{V}_{\mathbf{B}_N}$ of $\mathbf{B}_N$ satisfies*

$$N(\mathrm{E}[\mathcal{V}_{\mathbf{B}_N}(x)] - \mathcal{V}_N(x)) = O(1/N)$$

where $\mathcal{V}_N(x)$ is defined, for $x > 0$, as

$$\begin{aligned}\mathcal{V}_N(x) &= \frac{1}{N}\log\det\left(\mathbf{I}_N + \sum_{k=1}^{K} \bar{e}_{N,k}(-1/x)\mathbf{R}_k\right)\\ &+ \frac{1}{N}\log\det\left(\mathbf{I}_n + \sum_{k=1}^{K} e_{N,k}(-1/x)\mathbf{T}_k\right)\\ &- \frac{n}{N}\frac{1}{x}\sum_{k=1}^{K} \bar{e}_{N,k}(-1/x)e_{N,k}(-1/x).\end{aligned} \tag{6.28}$$

Note that the expressions of (6.26) and (6.28) are very similar, apart from the position of a summation symbol.

Both Theorem 6.1 and Theorem 6.5 can then be compiled into an even more general result, as follows. This is however not a corollary of Theorem 6.1 and Theorem 6.5, since the complete proof must be derived from the beginning.

Theorem 6.9. *For $k = 1, \ldots, K$, denote $\mathbf{H}_k \in \mathbb{C}^{N \times n_k}$ the random matrix such that, for a given positive L_k*

$$\mathbf{H}_k = \sum_{l=1}^{L_k} \mathbf{R}_{k,l}^{\frac{1}{2}} \mathbf{X}_{k,l} \mathbf{T}_{k,l}^{\frac{1}{2}}$$

for $\mathbf{R}_{k,l}^{\frac{1}{2}}$ a Hermitian non-negative square root of the Hermitian non-negative $\mathbf{R}_{k,l} \in \mathbb{C}^{N \times N}$, $\mathbf{T}_{k,l}^{\frac{1}{2}}$ a Hermitian non-negative square root of the Hermitian non-negative $\mathbf{T}_{k,l} \in \mathbb{C}^{n_k \times n_k}$ and $\mathbf{X}_{k,l} \in \mathbb{C}^{N \times n_k}$ with Gaussian i.i.d. entries of zero mean and variance $1/n_k$. All $\mathbf{R}_{k,l}$ and $\mathbf{T}_{k,l}$ are uniformly bounded with respect to N, n_k. Denote also for all k, $c_k = N/n_k$.

Call $m_{\mathbf{B}_N}(z)$ the Stieltjes transform of $\mathbf{B}_N = \sum_{k=1}^K \mathbf{H}_k \mathbf{H}_k^{\mathsf{H}}$, i.e. for $z \in \mathbb{C} \setminus \mathbb{R}^+$

$$m_{\mathbf{B}_N}(z) = \frac{1}{N} \operatorname{tr} \left(\sum_{k=1}^K \mathbf{H}_k \mathbf{H}_k^{\mathsf{H}} - z \mathbf{I}_N \right)^{-1}.$$

We then have

$$N \left(\mathrm{E}[m_{\mathbf{B}_N}(z)] - m_N(z) \right) \to 0$$

where $m_N(z)$ is defined as

$$m_N(z) = \frac{1}{N} \operatorname{tr} \left(-z \left[\sum_{k=1}^K \sum_{l=1}^{L_k} e_{N;k,l}(z) \mathbf{R}_{k,l} + \mathbf{I}_N \right] \right)^{-1}$$

and $e_{N;k,l}$ solves the fixed-point equations

$$e_{N;k,l}(z) = \frac{1}{n_k} \operatorname{tr} \mathbf{T}_{k,l} \left(-z \left[\sum_{l'=1}^{L_k} \bar{e}_{N;k,l'}(z) \mathbf{T}_{k,l'} + \mathbf{I}_{n_k} \right] \right)^{-1}$$

$$\bar{e}_{N;k,l}(z) = \frac{1}{n_k} \operatorname{tr} \mathbf{R}_{k,l} \left(-z \left[\sum_{k'=1}^K \sum_{l'=1}^{L_{k'}} e_{N;k',l'}(z) \mathbf{R}_{k',l'} + \mathbf{I}_N \right] \right)^{-1}.$$

We also have that the Shannon transform $\mathcal{V}_{\mathbf{B}_N}(x)$ of $\mathbf{B}_N$ satisfies

$$N \left(\mathrm{E}[\mathcal{V}_{\mathbf{B}_N}(x)] - \mathcal{V}_N(x) \right) \to 0$$

where

$$\begin{aligned}
\mathcal{V}_N(x) &= \frac{1}{N} \log\det \left(\sum_{k=1}^K \sum_{l=1}^{L_k} e_{N;k,l}(-1/x) \mathbf{R}_{k,l} + \mathbf{I}_N \right) \\
&\quad + \sum_{k=1}^K \frac{1}{N} \log\det \left(\sum_{l=1}^{L_k} \bar{e}_{N;k,l}(-1/x) \mathbf{T}_{k,l} + \mathbf{I}_{n_k} \right) \\
&\quad - \frac{1}{x} \sum_{k=1}^K \frac{n_k}{N} \sum_{l=1}^{L_k} e_{N;k,l}(-1/x) \bar{e}_{N;k,l}(-1/x).
\end{aligned}$$

For practical applications, this formula provides the whole picture for the *ergodic* rate region of large MIMO multiple access channels, with K multiple antenna users, user k being equipped with n_k antennas, when the different channels into consideration are frequency selective with L_k taps for user k, slow fading in time, and for each tap modeled as Kronecker with receive and transmit correlation $\mathbf{R}_{k,l}$ and $\mathbf{T}_{k,l}$, respectively.

We now move to another type of deterministic equivalents, when the entries of the matrix $\mathbf{X}$ are not necessarily of zero mean and have possibly different variances.

6.2.3 Information plus noise models

In Section 3.2, we introduced an important limiting Stieltjes transform result, Theorem 3.14, for the Gram matrix of a random i.i.d. matrix $\mathbf{X} \in \mathbb{C}^{N \times n}$ with a variance profile $\{\sigma_{ij}^2/n\}$, $1 \leq i \leq N$ and $1 \leq j \leq n$. One hypothesis of Girkos's law is that the profile $\{\sigma_{ij}\}$ converges to a density $\sigma(x, y)$ in the sense that

$$\sigma_{ij} - \int_{\frac{i-1}{N}}^{\frac{i}{N}} \int_{\frac{j-1}{n}}^{\frac{j}{n}} \sigma(x, y) dx dy \to 0.$$

It will turn out in practical applications that such an assumption is in general unusable. Typically, suppose that σ_{ij} is the channel fading between antenna i and antenna j, respectively, at the transmitter and receiver of a multiple antenna channel. As one grows N and n simultaneously, there is no reason for the σ_{ij} to converge in any sense to a density $\sigma(x, y)$. In the following, we therefore rewrite Theorem 3.14 in terms of deterministic equivalents without the need for any assumption of convergence. This result is in fact a corollary of the very general Theorem 6.14, presented later in this section, although the deterministic equivalent is written in a slightly different form. A sketch of the proof using the Bai and Silverstein approach is also provided.

Theorem 6.10. *Let $\mathbf{X}_N \in \mathbb{C}^{N \times n}$ have independent entries x_{ij} with zero mean, variance σ_{ij}^2/n and $4 + \varepsilon$ moment of order $O(1/N^{2+\varepsilon/2})$, for some ε. Assume that the σ_{ij} are deterministic and uniformly bounded, over n, N. Then, as N, n grow large with ratio $c_n \triangleq N/n$ such that $0 < \liminf_n c_n \leq \limsup_n c_n < \infty$, the e.s.d. $F^{\mathbf{B}_N}$ of $\mathbf{B}_N = \mathbf{X}_N \mathbf{X}_N^{\mathsf{H}}$ satisfies*

$$F^{\mathbf{B}_N} - F_N \Rightarrow 0$$

almost surely, where F_N is the distribution function of Stieltjes transform $m_N(z)$, $z \in \mathbb{C} \setminus \mathbb{R}^+$, given by:

$$m_N(z) = \frac{1}{N} \sum_{k=1}^{N} \frac{1}{\frac{1}{n} \sum_{i=1}^{n} \sigma_{ki}^2 \frac{1}{1+e_{N,i}(z)} - z}$$

where $e_{N,1}(z),\ldots,e_{N,n}(z)$ form the unique solution of

$$e_{N,j}(z)=\frac{1}{n}\sum_{k=1}^{N}\frac{\sigma_{kj}^2}{\frac{1}{n}\sum_{i=1}^{n}\sigma_{ki}^2\frac{1}{1+e_{N,i}(z)}-z} \tag{6.29}$$

such that all $e_{N,j}(z)$ are Stieltjes transforms of a distribution function.

The reason why point-wise uniqueness of the $e_{N,j}(z)$ is not provided here is due to the approach of the proof of uniqueness followed by Hachem *et al.* [Hachem *et al.*, 2007] which is a functional proof of uniqueness of the Stieltjes transforms that the applications $z\mapsto e_{N,i}(z)$ define. This does not mean that point-wise uniqueness does not hold but this is as far as this theorem goes.

Theorem 6.10 can then be written is a more compact and symmetric form by rewriting $e_{N,j}(z)$ in (6.29) as

$$\begin{aligned} e_{N,j}(z) &= -\frac{1}{z}\frac{1}{n}\sum_{k=1}^{N}\frac{\sigma_{kj}^2}{1+\bar{e}_{N,k}(z)} \\ \bar{e}_{N,k}(z) &= -\frac{1}{z}\frac{1}{n}\sum_{i=1}^{n}\frac{\sigma_{ki}^2}{1+e_{N,i}(z)}. \end{aligned} \tag{6.30}$$

In this case, $m_N(z)$ is simply

$$m_N(z)=-\frac{1}{z}\frac{1}{N}\sum_{k=1}^{N}\frac{1}{1+\bar{e}_{N,k}(z)}.$$

Note that this version of Girko's law, Theorem 3.14, is both more general in the assumptions made, and more explicit. We readily see in this result that fixed-point algorithms, if they converge at all, allow us to recover the $2n$ coupled Equations (6.30), from which $m_N(z)$ is then explicit.

For the sake of understanding and to further justify the strength of the techniques introduced so far, we provide hereafter the first steps of the proof using the Bai and Silverstein technique. A complete proof can be found as a particular case of [Hachem *et al.*, 2007; Wagner *et al.*, 2011].

Proof. Instead of studying $m_N(z)$, let us consider the more general $e_{\mathbf{A}_N}(z)$, a deterministic equivalent for

$$\frac{1}{N}\operatorname{tr}\mathbf{A}_N\left(\mathbf{X}_N\mathbf{X}_N^{\mathsf{H}}-z\mathbf{I}_N\right)^{-1}.$$

Using Bai and Silverstein approach, we introduce $\mathbf{F}\in\mathbb{C}^{N\times N}$ some matrix yet to be defined, and compute

$$e_{\mathbf{A}_N}(z)=\frac{1}{N}\operatorname{tr}\mathbf{A}_N\left(\mathbf{F}-z\mathbf{I}_N\right)^{-1}.$$

Using the resolvent identity, Lemma 6.1, and writing $\mathbf{X}_N\mathbf{X}_N^{\mathsf{H}} = \sum_{i=1}^n \mathbf{x}_i\mathbf{x}_i^{\mathsf{H}}$, we have:

$$\begin{aligned}
&\frac{1}{N}\operatorname{tr}\mathbf{A}_N\left(\mathbf{X}_N\mathbf{X}_N^{\mathsf{H}} - z\mathbf{I}_N\right)^{-1} - \frac{1}{N}\operatorname{tr}\mathbf{A}_N\left(\mathbf{F} - z\mathbf{I}_N\right)^{-1} \\
&= \frac{1}{N}\operatorname{tr}\mathbf{A}_N\left(\mathbf{X}_N\mathbf{X}_N^{\mathsf{H}} - z\mathbf{I}_N\right)^{-1}\mathbf{F}\left(\mathbf{F} - z\mathbf{I}_N\right)^{-1} \\
&- \frac{1}{N}\sum_{i=1}^n \operatorname{tr}\mathbf{A}_N\left(\mathbf{X}_N\mathbf{X}_N^{\mathsf{H}} - z\mathbf{I}_N\right)^{-1}\mathbf{x}_i\mathbf{x}_i^{\mathsf{H}}\left(\mathbf{F} - z\mathbf{I}_N\right)^{-1}
\end{aligned}$$

from which we then express the second term on the right-hand side under the form of sums for $i \in \{1,\ldots,N\}$ of $\mathbf{x}_i^{\mathsf{H}}\left(\mathbf{F} - z\mathbf{I}_N\right)^{-1}\mathbf{A}_N\left(\mathbf{X}_N\mathbf{X}_N^{\mathsf{H}} - z\mathbf{I}_N\right)^{-1}\mathbf{x}_i$ and we use Lemma 6.2 on the matrix $\left(\mathbf{X}_N\mathbf{X}_N^{\mathsf{H}} - z\mathbf{I}_N\right)^{-1}$ to obtain

$$\begin{aligned}
&\frac{1}{N}\operatorname{tr}\mathbf{A}_N\left(\mathbf{X}_N\mathbf{X}_N^{\mathsf{H}} - z\mathbf{I}_N\right)^{-1} - \frac{1}{N}\operatorname{tr}\mathbf{A}_N\left(\mathbf{F} - z\mathbf{I}_N\right)^{-1} \\
&= \frac{1}{N}\operatorname{tr}\mathbf{A}_N\left(\mathbf{X}_N\mathbf{X}_N^{\mathsf{H}} - z\mathbf{I}_N\right)^{-1}\mathbf{F}\left(\mathbf{F} - z\mathbf{I}_N\right)^{-1} \\
&- \frac{1}{N}\sum_{i=1}^n \frac{\mathbf{x}_i^{\mathsf{H}}\left(\mathbf{F} - z\mathbf{I}_N\right)^{-1}\mathbf{A}_N\left(\mathbf{X}_{(i)}\mathbf{X}_{(i)}^{\mathsf{H}} - z\mathbf{I}_N\right)^{-1}\mathbf{x}_i}{1 + \mathbf{x}_i^{\mathsf{H}}\left(\mathbf{X}_{(i)}\mathbf{X}_{(i)}^{\mathsf{H}} - z\mathbf{I}_N\right)^{-1}\mathbf{x}_i}
\end{aligned} \tag{6.31}$$

with $\mathbf{X}_{(i)} = [\mathbf{x}_1,\ldots,\mathbf{x}_{i-1},\mathbf{x}_{i+1},\ldots,\mathbf{x}_n]$.

Under this form, $\mathbf{x}_i$ and $\left(\mathbf{X}_{(i)}\mathbf{X}_{(i)}^{\mathsf{H}} - z\mathbf{I}_N\right)^{-1}$ have independent entries. However, $\mathbf{x}_i$ does not have identically distributed entries, so that Theorem 3.4 cannot be straightforwardly applied. We therefore define $\mathbf{y}_i \in \mathbb{C}^N$ as

$$\mathbf{x}_i = \boldsymbol{\Sigma}_i\mathbf{y}_i$$

with $\boldsymbol{\Sigma}_i \in \mathbb{C}^{N\times N}$ a diagonal matrix with kth diagonal entry equal to σ_{ki}, and $\mathbf{y}_i$ has identically distributed entries of zero mean and variance $1/n$. Replacing all occurrences of $\mathbf{x}_i$ in (6.31) by $\boldsymbol{\Sigma}_i\mathbf{y}_i$, we have:

$$\begin{aligned}
&\frac{1}{N}\operatorname{tr}\mathbf{A}_N\left(\mathbf{X}_N\mathbf{X}_N^{\mathsf{H}} - z\mathbf{I}_N\right)^{-1} - \frac{1}{N}\operatorname{tr}\mathbf{A}_N\left(\mathbf{F} - z\mathbf{I}_N\right)^{-1} \\
&= \frac{1}{N}\operatorname{tr}\mathbf{A}_N\left(\mathbf{X}_N\mathbf{X}_N^{\mathsf{H}} - z\mathbf{I}_N\right)^{-1}\mathbf{F}\left(\mathbf{F} - z\mathbf{I}_N\right)^{-1} \\
&- \frac{1}{N}\sum_{i=1}^n \frac{\mathbf{y}_i^{\mathsf{H}}\boldsymbol{\Sigma}_i\left(\mathbf{F} - z\mathbf{I}_N\right)^{-1}\mathbf{A}_N\left(\mathbf{X}_{(i)}\mathbf{X}_{(i)}^{\mathsf{H}} - z\mathbf{I}_N\right)^{-1}\boldsymbol{\Sigma}_i\mathbf{y}_i}{1 + \mathbf{y}_i^{\mathsf{H}}\boldsymbol{\Sigma}_i\left(\mathbf{X}_{(i)}\mathbf{X}_{(i)}^{\mathsf{H}} - z\mathbf{I}_N\right)^{-1}\boldsymbol{\Sigma}_i\mathbf{y}_i}.
\end{aligned} \tag{6.32}$$

Applying the trace lemma, Theorem 3.4, the quadratic terms of the form $\mathbf{y}_i^{\mathsf{H}}\mathbf{Y}\mathbf{y}_i$ are close to $\frac{1}{n}\operatorname{tr}\mathbf{Y}$. Therefore, in order for (6.32) to converge to zero, $\mathbf{F}$ *ought to* take the form

$$\mathbf{F} = \frac{1}{n}\sum_{i=1}^n \frac{1}{1 + e_{\mathbf{B}_N,i}(z)}\boldsymbol{\Sigma}_i^2$$

with

$$e_{\mathbf{B}_N,i}(z) = \frac{1}{n} \operatorname{tr} \mathbf{\Sigma}_i^2 \left(\mathbf{X}_N \mathbf{X}_N^{\mathsf{H}} - z\mathbf{I}_N\right)^{-1}.$$

We therefore infer that $e_{N,i}(z)$ takes the form

$$e_{N,i}(z) = \frac{1}{n} \sum_{k=1}^{N} \frac{\sigma_{ki}^2}{\frac{1}{n} \sum_{i=1}^{n} \sigma_{ki}^2 \frac{1}{1+e_{N,i}(z)} - z}$$

by setting $\mathbf{A}_N = \mathbf{\Sigma}_i^2$.

From this point on, the result unfolds by showing the almost sure convergence towards zero of the difference $e_{N,i}(z) - \frac{1}{n} \operatorname{tr} \mathbf{\Sigma}_i^2 \left(\mathbf{X}_N \mathbf{X}_N^{\mathsf{H}} - z\mathbf{I}_N\right)^{-1}$ and the functional uniqueness of the implicit equation for the $e_{N,i}(z)$. □

The symmetric expressions (6.30) make it easy to derive also a deterministic equivalent of the Shannon transform.

Theorem 6.11. *Let $\mathbf{B}_N$ be defined as in Theorem 6.10 and let $x > 0$. Then, as N, n grow large with uniformly bounded ratio $c_n = N/n$, the Shannon transform $\mathcal{V}_{\mathbf{B}_N}(x)$ of $\mathbf{B}_N$, defined as*

$$\mathcal{V}_{\mathbf{B}_N}(x) \triangleq \frac{1}{N} \log\det\left(\mathbf{I}_N + x\mathbf{B}_N\right)$$

satisfies

$$\mathrm{E}[\mathcal{V}_{\mathbf{B}_N}(x)] - \mathcal{V}_N(x) \to 0$$

where $\mathcal{V}_N(x)$ is given by:

$$\mathcal{V}_N(x) = \frac{1}{N} \sum_{k=1}^{N} \log\left(1 + \bar{e}_{N,k}(-\frac{1}{x})\right) + \frac{1}{N} \sum_{i=1}^{n} \log\left(1 + e_{N,i}(-\frac{1}{x})\right)$$
$$- \frac{x}{nN} \sum_{\substack{1 \le k \le N \\ 1 \le i \le n}} \frac{\sigma_{ki}^2}{\left(1 + \bar{e}_{N,k}(-\frac{1}{x})\right)\left(1 + e_{N,i}(-\frac{1}{x})\right)}.$$

It is worth pointing out here that the Shannon transform convergence result is only stated in the mean sense and not, as was the case in Theorem 6.4, in the almost sure sense. Remember indeed that the convergence result of Theorem 6.4 depends strongly on the fact that the empirical matrix $\mathbf{B}_N$ can be proved to have bounded spectral norm for all large N, almost surely. This is a consequence of spectral norm inequalities and of Theorem 7.1. However, it is not known whether Theorem 7.1 holds true for matrices with a variance profile and the derivation of Theorem 6.4 can therefore not be reproduced straightforwardly.

It is in fact not difficult to show the convergence of the Shannon transform in the mean via a simple dominated convergence argument. Indeed, remembering the Shannon transform definition, Definition 3.2, we have:

$$\mathrm{E}[\mathcal{V}_{\mathbf{B}_N}(x)] - \mathcal{V}_N(x) = \int_{\frac{1}{x}}^{\infty} \left(\frac{1}{t} - \mathrm{E}[m_{\mathbf{B}_N}(-t)] \right) dt - \int_{\frac{1}{x}}^{\infty} \left(\frac{1}{t} - m_N(-t) \right) dt \tag{6.33}$$

for which we in particular have

$$\begin{aligned}
&\left| \left(\frac{1}{t} - \mathrm{E}[m_{\mathbf{B}_N}(-t)] \right) - \left(\frac{1}{t} - m_N(-t) \right) \right| \\
&\leq \left| \frac{1}{t} - \mathrm{E}[m_{\mathbf{B}_N}(-t)] \right| + \left| \frac{1}{t} - m_N(-t) \right| \\
&= \left| \int \left(\frac{1}{t} - \frac{1}{\lambda + t} \right) \mathrm{E}[dF^{\mathbf{B}_N}(\lambda)] \right| + \left| \int \left(\frac{1}{t} - \frac{1}{\lambda + t} \right) dF_N(\lambda) \right| \\
&\leq \frac{1}{t^2} \int \lambda \mathrm{E}[dF^{\mathbf{B}_N}(\lambda)] + \frac{1}{t^2} \int \lambda dF_N(\lambda).
\end{aligned}$$

It is now easy to prove from standard expectation calculus that both integrals above are upper-bound by $\limsup_N \sup_i \|\mathbf{R}_i\| < \infty$. Writing Equation (6.33) under the form of a single integral, we have that the integrand tends to zero as $N \to \infty$ and is summable over the integration parameter t. Therefore, from the dominated convergence theorem, Theorem 6.3, $\mathrm{E}[\mathcal{V}_{\mathbf{B}_N}(x)] - \mathcal{V}_N(x) \to 0$.

Note now that, in the proof of Theorem 6.10, there is no actual need for the matrices $\mathbf{\Sigma}_k$ to be diagonal. Also, there is no huge difficulty added by considering the matrix $\mathbf{X}_N\mathbf{X}_N^{\mathsf{H}} + \mathbf{A}_N$, instead of $\mathbf{X}_N\mathbf{X}_N^{\mathsf{H}}$ for any deterministic $\mathbf{A}_N$. As such, Theorem 6.10 can be further generalized as follows.

Theorem 6.12 ([Wagner *et al.*, 2011]). *Let $\mathbf{X}_N \in \mathbb{C}^{N \times n}$ have independent columns $\mathbf{x}_i = \mathbf{H}_i\mathbf{y}_i$, where $\mathbf{y}_i \in \mathbb{C}^{N_i}$ has i.i.d. entries of zero mean, variance $1/n$, and $4+\varepsilon$ moment of order $O(1/n^{2+\varepsilon/2})$, and $\mathbf{H}_i \in \mathbb{C}^{N \times N_i}$ are such that $\mathbf{R}_i \triangleq \mathbf{H}_i\mathbf{H}_i^{\mathsf{H}}$ has uniformly bounded spectral norm over n, N. Let also $\mathbf{A}_N \in \mathbb{C}^{N \times N}$ be Hermitian non-negative and denote $\mathbf{B}_N = \mathbf{X}_N\mathbf{X}_N^{\mathsf{H}} + \mathbf{A}_N$. Then, as N, $N_1, \ldots, N_n$, and n grow large with ratios $c_i \triangleq N_i/n$ and $c_0 \triangleq N/n$ satisfying $0 < \liminf_n c_i \leq \limsup_n c_i < \infty$ for $0 \leq i \leq n$, we have that, for all non-negative Hermitian matrix $\mathbf{C}_N \in \mathbb{C}^{N \times N}$ with uniformly bounded spectral norm*

$$\frac{1}{n} \operatorname{tr} \mathbf{C}_N \left(\mathbf{B}_N - z\mathbf{I}_N \right)^{-1} - \frac{1}{n} \operatorname{tr} \mathbf{C}_N \left(\frac{1}{n} \sum_{i=1}^{n} \frac{1}{1 + e_{N,i}(z)} \mathbf{R}_i + \mathbf{A}_N - z\mathbf{I}_N \right)^{-1} \xrightarrow{\text{a.s.}} 0$$

where $e_{N,1}(z), \ldots, e_{N,n}(z)$ form the unique functional solution of

$$e_{N,j}(z) = \frac{1}{n} \operatorname{tr} \mathbf{R}_j \left(\frac{1}{n} \sum_{i=1}^{n} \frac{1}{1 + e_{N,i}(z)} \mathbf{R}_i + \mathbf{A}_N - z\mathbf{I}_N \right)^{-1} \tag{6.34}$$

such that all $e_{N,j}(z)$ are Stieltjes transforms of a non-negative finite measure on $\mathbb{R}^+$. Moreover, $(e_{N,1}(z), \ldots, e_{N,n}(z))$ is given by $e_{N,i}(z) = lim_{k \to \infty} e_{N,i}^{(k)}(z)$, where

$e_{N,i}^{(0)} = -1/z$ *and, for* $k \geq 0$

$$e_{N,j}^{(k+1)}(z) = \frac{1}{n} \operatorname{tr} \mathbf{R}_j \left(\frac{1}{n} \sum_{i=1}^{n} \frac{1}{1 + e_{N,i}^{(k)}(z)} \mathbf{R}_i + \mathbf{A}_N - z\mathbf{I}_N \right)^{-1}.$$

Also, for $x > 0$, *the Shannon transform* $\mathcal{V}_{\mathbf{B}_N}(x)$ *of* $\mathbf{B}_N$, *defined as*

$$\mathcal{V}_{\mathbf{B}_N}(x) \triangleq \frac{1}{N} \log\det\left(\mathbf{I}_N + x\mathbf{B}_N\right)$$

satisfies

$$\mathrm{E}[\mathcal{V}_{\mathbf{B}_N}(x)] - \mathcal{V}_N(x) \to 0$$

where $\mathcal{V}_N(x)$ *is given by:*

$$\mathcal{V}_N(x) = \frac{1}{N} \log\det\left(\mathbf{I}_N + x\left[\frac{1}{n}\sum_{i=1}^{n} \frac{1}{1 + e_{N,i}(-\frac{1}{x})}\mathbf{R}_i + \mathbf{A}_N\right]\right)$$
$$+ \frac{1}{N}\sum_{i=1}^{n} \log\left(1 + e_{N,i}(-\frac{1}{x})\right) - \frac{1}{N}\sum_{i=1}^{n} \frac{e_{N,i}(-\frac{1}{x})}{1 + e_{N,i}(-\frac{1}{x})}.$$

Remark 6.5. Consider the identically distributed entries $\mathbf{x}_1, \ldots, \mathbf{x}_n$ in Theorem 6.12, and take $n_1, \ldots, n_K$ to be K integers such that $\sum_i n_i = n$. Define $\tilde{\mathbf{R}}_1, \ldots, \tilde{\mathbf{R}}_K \in \mathbb{C}^{N \times N}$ to be K non-negative definite matrices with uniformly bounded spectral norm and $\mathbf{T}_1 \in \mathbb{C}^{n_1 \times n_1}, \ldots, \mathbf{T}_K \in \mathbb{C}^{n_K \times n_K}$ to be K diagonal matrices with positive entries, $\mathbf{T}_k = \operatorname{diag}(t_{k1}, \ldots, t_{kn_k})$. Denote $\mathbf{R}_k = \tilde{\mathbf{R}}_j t_{ji}$, $k \in \{1, \ldots, n\}$, with j the smallest integer such that $k - (n_1 + \ldots + n_{j-1}) > 0$, $n_0 = 0$, and $i = k - (n_1 + \ldots + n_{j-1})$. Under these conditions and notations, up to some hypothesis restrictions, Theorem 6.12 with $\mathbf{H}_i = \tilde{\mathbf{R}}_i^{\frac{1}{2}}$ also generalizes Theorem 6.1 applied to the sum of K Gram matrices with left correlation matrix $\tilde{\mathbf{R}}_1, \ldots, \tilde{\mathbf{R}}_K$ and right correlation matrices $\mathbf{T}_1, \ldots, \mathbf{T}_K$.

From Theorem 6.12, taking $\mathbf{A}_N = 0$, we also immediately have that the distribution function F_N with Stieltjes transform

$$m_N(z) = \frac{1}{N} \operatorname{tr}\left(\frac{1}{n}\sum_{i=1}^{n} \frac{1}{1 + e_{N,i}(z)}\mathbf{R}_i - z\mathbf{I}_N\right)^{-1} \tag{6.35}$$

where

$$e_{N,j}(z) = \frac{1}{n} \operatorname{tr} \mathbf{R}_j \left(\frac{1}{n}\sum_{i=1}^{n} \frac{1}{1 + e_{N,i}(z)}\mathbf{R}_i - z\mathbf{I}_N\right)^{-1} \tag{6.36}$$

is a deterministic equivalent for $F^{\mathbf{X}_N \mathbf{X}_N^{\mathsf{H}}}$. An interesting result with application in low complex filter design, see Section 13.6 of Chapter 13, is the description in closed-form of the successive moments of the distribution function F_N.

Theorem 6.13 ([Hoydis *et al.*, 2011c]). *Let F_N be the d.f. associated with the Stieltjes transform $m_N(z)$ defined by (6.35) with $e_{N,i}(z)$ given by (6.36). Further denote $M_{N,0}, M_{N,1}, \ldots$ the successive moments of F_N*

$$M_{N,p} = \int x^p dF_N(x).$$

Then, these moments are explicitly given by:

$$M_{N,p} = \frac{(-1)^p}{p!} \frac{1}{N} \operatorname{tr} \mathbf{T}_p$$

with $\mathbf{T}_0, \mathbf{T}_1, \ldots$ defined iteratively from the following set of recursive equations initialized with $\mathbf{T}_0 = \mathbf{I}_N$, $f_{k,0} = -1$ and $\delta_{k,0} = \frac{1}{n} \operatorname{tr} \mathbf{R}_k$ for $k \in \{1, \ldots, n\}$

$$\mathbf{T}_{p+1} = \sum_{i=0}^{p} \sum_{j=0}^{i} \binom{p}{i} \binom{i}{j} \mathbf{T}_{p-i} \mathbf{Q}_{i-j+1} \mathbf{T}_j$$

$$\mathbf{Q}_{p+1} = \frac{p+1}{n} \sum_{k=1}^{n} f_{k,p} \mathbf{R}_k$$

$$f_{k,p+1} = \sum_{i=0}^{p} \sum_{j=0}^{i} \binom{p}{i} \binom{i}{j} (p-i+1) f_{k,j} f_{k,i-j} \delta_{k,p-i}$$

$$\delta_{k,p+1} = \frac{1}{n} \operatorname{tr} \mathbf{R}_k \mathbf{T}_{p+1}.$$

Moreover, with $\mathbf{B}_N = \mathbf{X}_N \mathbf{X}_N^{\mathsf{H}}$, $\mathbf{X}_N$ being defined in Theorem 6.12, we have for all integer p

$$\int x^p \mathrm{E}[dF^{\mathbf{B}_N}(x)] - M_{N,p} \to 0$$

as $N, n \to \infty$.

Note that a similar result was established from a combinatorics approach in [Li *et al.*, 2004] which took the form of involved sums over non-crossing partitions, when all $\mathbf{R}_k$ matrices are Toeplitz and of Wiener class [Gray, 2006]. The proof of the almost sure convergence of $\int x^p dF^{\mathbf{B}_N}(x)$ to $M_{N,p}$, claimed in [Li *et al.*, 2004], would require proving that the support $\mathbf{B}_N$ is almost surely uniformly bounded from above for all large N. However, this fact is unknown to this day so that convergence in the mean can be ensured, while almost sure convergence can only be conjectured. It holds true in particular when the family $\{\mathbf{R}_1, \ldots, \mathbf{R}_n\}$ is extracted from a finite set.

Proof. Note that F_N is necessarily compactly supported as the $\|\mathbf{R}_i\|$ are uniformly bounded and that the $e_{N,i}(z)$ are non-negative for $z < 0$. Reminding then that the Stieltjes transform m_N of F_N can be written in that case under the form of a moment generating function by (3.6), the expression of the successive moments unfolds from successive differentiations of $-zm_N(-1/z)$, taken in

$z = 0$. The convergence of the difference of moments is then a direct consequence of the dominated convergence theorem, Theorem 6.3. □

Another generalization of Theorem 6.10 is found in [Hachem *et al.*, 2007], where $\mathbf{X}_N$ still has a variance profile but has non-zero mean. The result in the latter is more involved and expresses as follows.

Theorem 6.14. *Let $\mathbf{X}_N = \mathbf{Y}_N + \mathbf{A}_N \in \mathbb{C}^{N\times n}$ be a random matrix where $\mathbf{Y}_N$ has independent entries y_{ij} with zero mean, variance σ_{ij}^2/N and finite $4+\varepsilon$ moment of order $O(1/N^{2+\varepsilon/2})$, and $\mathbf{A}_N$ is a deterministic matrix. Denote $\mathbf{\Sigma}_j \in \mathbb{C}^{N\times N}$ the diagonal matrix with ith diagonal entry σ_{ij} and $\bar{\mathbf{\Sigma}}_i \in \mathbb{C}^{n\times n}$ the diagonal matrix with jth diagonal entry σ_{ij}. Suppose moreover that the columns of $\mathbf{A}_N$ have uniformly bounded Euclidean norm and that the σ_{ij} are uniformly bounded, with respect to N and n. Then, as N, n grow large with ratio $c_N = N/n$, such that $0 < \liminf_N c_N \leq \limsup_N c_N < \infty$, the e.s.d. $F^{\mathbf{B}_N}$ of $\mathbf{B}_N \triangleq \mathbf{X}_N\mathbf{X}_N^{\mathsf{H}}$ satisfies*

$$F^{\mathbf{B}_N} - F_N \Rightarrow 0$$

almost surely, with F_N the distribution function with Stieltjes transform $m_N(z)$, $z \in \mathbb{C}\setminus\mathbb{R}^+$, given by:

$$m_N(z) = \frac{1}{N}\operatorname{tr}\left(\mathbf{\Psi}^{-1} - z\mathbf{A}_N\bar{\mathbf{\Psi}}\mathbf{A}_N^{\mathsf{T}}\right)^{-1}$$

where $\mathbf{\Psi} \in \mathbb{C}^{N\times N}$ is diagonal with ith entry $\psi_i(z)$, $\bar{\mathbf{\Psi}} \in \mathbb{C}^{n\times n}$ is diagonal with jth entry $\bar{\psi}_j(z)$, with $\psi_i(z)$ and $\bar{\psi}_j(z)$, $1 \leq i \leq N$, $1 \leq j \leq n$, the unique solutions of

$$\psi_i(z) = -\frac{1}{z}\left[1 + \frac{1}{n}\operatorname{tr}\bar{\mathbf{\Sigma}}_i^2\left(\bar{\mathbf{\Psi}}^{-1} - z\mathbf{A}_N^{\mathsf{T}}\mathbf{\Psi}\mathbf{A}_N\right)^{-1}\right]^{-1}$$

$$\bar{\psi}_j(z) = -\frac{1}{z}\left[1 + \frac{1}{n}\operatorname{tr}\mathbf{\Sigma}_j^2\left(\mathbf{\Psi}^{-1} - z\mathbf{A}_N\bar{\mathbf{\Psi}}\mathbf{A}_N^{\mathsf{T}}\right)^{-1}\right]^{-1}$$

which are Stieltjes transforms of distribution functions.

Besides, for $x = -\frac{1}{z} > 0$, let $\mathcal{V}_{\mathbf{B}_N}(x) = \frac{1}{N}\log\det\left(\mathbf{I}_N + x\mathbf{X}_N\mathbf{X}_N^{\mathsf{H}}\right)$ be the Shannon transform of $\mathbf{B}_N$. Then

$$\mathrm{E}[\mathcal{V}_{\mathbf{B}_N}(x)] - \mathcal{V}_N(x) \to 0$$

as N, n grow large, where $\mathcal{V}_N(x)$ is defined by

$$\mathcal{V}_N(x) = \frac{1}{N}\log\det\left[x\mathbf{\Psi}^{-1} + \mathbf{A}\bar{\mathbf{\Psi}}\mathbf{A}^{\mathsf{T}}\right] + \frac{1}{N}\log\det\left(x\bar{\mathbf{\Psi}}^{-1}\right) - \frac{1}{x}\frac{1}{nN}\sum_{i,j}\sigma_{ij}^2 t_i\bar{t}_j$$

with t_i the ith diagonal entry of the diagonal matrix $\left(\mathbf{\Psi}^{-1} + x\mathbf{A}_N\bar{\mathbf{\Psi}}\mathbf{A}_N^{\mathsf{T}}\right)^{-1}$ and $\bar{t}_j$ the jth diagonal entry of the diagonal matrix $\left(\bar{\mathbf{\Psi}}^{-1} + x\mathbf{A}_N^{\mathsf{T}}\mathbf{\Psi}\mathbf{A}_N\right)^{-1}$.

Remark 6.6. In [Hachem *et al.*, 2008b], it is shown in particular that, if the entries of $\mathbf{Y}_N$ are Gaussian distributed, then the difference between the Stieltjes transform of $\mathrm{E}[F^{\mathbf{B}_N}]$ and its deterministic equivalent, as well as the difference between the Shannon transform of $\mathrm{E}[F^{\mathbf{B}_N}]$ and its deterministic equivalent converge to zero at rate $O(1/N^2)$.

6.2.4 Models involving Haar matrices

As evidenced in the previous section, Hermitian random matrices with i.i.d. entries or originating from general sums or products of such matrices are convenient to study using Stieltjes transform-based methods. This is essentially due to the trace lemma, Theorem 3.4, which provides an almost sure limit to $\mathbf{x}^{\mathsf{H}}(\mathbf{X}\mathbf{X}^{\mathsf{H}} - \mathbf{x}\mathbf{x}^{\mathsf{H}} - z\mathbf{I}_N)^{-1}\mathbf{x}$ with $\mathbf{x}$ one of the independent columns of the random matrix $\mathbf{X}$. Such results can actually be found for more structured random matrices, such as the random bi-unitarily invariant unitary $N \times N$ matrices. We recall from Definition 4.6 that these random matrices are often referred to as *Haar matrices* or *isometric matrices*. Among the known properties of interest here of Haar matrices [Petz and Réffy, 2004], we have the following trace lemma [Chaufray *et al.*, 2004; Debbah *et al.*, 2003a], equivalent to Theorem 3.4 for i.i.d. random matrices.

Theorem 6.15. *Let $\mathbf{W}$ be $n < N$ columns of an $N \times N$ Haar matrix and suppose $\mathbf{w}$ is a column of $\mathbf{W}$. Let $\mathbf{B}_N$ be an $N \times N$ random matrix, which is a function of all columns of $\mathbf{W}$ except $\mathbf{w}$. Then, assuming that, for growing N, $c = \sup_n n/N < 1$ and $B = \sup_N \|\mathbf{B}_N\| < \infty$, we have:*

$$\mathrm{E}\left[\left|\mathbf{w}^{\mathsf{H}}\mathbf{B}_N\mathbf{w} - \frac{1}{N-n}\operatorname{tr}(\mathbf{\Pi}\mathbf{B}_N)\right|^4\right] \leq \frac{C}{N^2} \tag{6.37}$$

where $\mathbf{\Pi} = \mathbf{I}_N - \mathbf{W}\mathbf{W}^{\mathsf{H}} + \mathbf{w}\mathbf{w}^{\mathsf{H}}$ and C is a constant which depends only on B and c. If $\sup_N \|\mathbf{B}_N\| < \infty$, by the Markov inequality, Theorem 3.5, and the Borel–Cantelli lemma, Theorem 3.6, this entails

$$\mathbf{w}^{\mathsf{H}}\mathbf{B}_N\mathbf{w} - \frac{1}{N-n}\operatorname{tr}(\mathbf{\Pi}\mathbf{B}_N) \xrightarrow{\text{a.s.}} 0. \tag{6.38}$$

Proof. We provide here an intuitive, yet non-rigorous, sketch of the proof. Let $\mathbf{U} \in \mathbb{C}^{N\times(n-1)}$ be $n-1$ columns of a unitary matrix. We can write all unit-norm vectors $\mathbf{w}$ in the space orthogonal to the space spanned by the columns of $\mathbf{U}$ as $\mathbf{w} = \frac{\mathbf{\Pi}\mathbf{x}}{\|\mathbf{\Pi}\mathbf{x}\|}$, where $\mathbf{\Pi} = \mathbf{I}_N - \mathbf{U}\mathbf{U}^{\mathsf{H}}$ is the projector on the space orthogonal to $\mathbf{U}\mathbf{U}^{\mathsf{H}}$ (and thus $\mathbf{\Pi}\mathbf{\Pi} = \mathbf{\Pi}$) and $\mathbf{x}$ is a Gaussian vector with zero mean and covariance matrix $\mathrm{E}[\mathbf{x}\mathbf{x}^{\mathsf{H}}] = \mathbf{I}_N$ independent of $\mathbf{U}$. This makes $\mathbf{w}$ uniformly distributed in its space. Also, the vector $\mathbf{x}$ is independent of $\mathbf{\Pi}$ by construction. We therefore have from Theorem 3.4 and for N large

$$\mathbf{w}^{\mathsf{H}}\mathbf{B}_N\mathbf{w} = \frac{1}{N}\mathbf{x}^{\mathsf{H}}\mathbf{\Pi}\mathbf{B}_N\mathbf{\Pi}\mathbf{x}\frac{N}{\|\mathbf{\Pi}\mathbf{x}\|^2} \simeq \frac{1}{N}\operatorname{tr}(\mathbf{\Pi}\mathbf{B}_N)\frac{N}{\|\mathbf{\Pi}\mathbf{x}\|^2}.$$

where the symbol "$\simeq$" stands for some approximation in the large N limit. Notice then that $\mathbf{\Pi x}$ is, up to a basis change, a vector composed of $N-n+1$ i.i.d. standard Gaussian entries and $n-1$ zeros. Hence $\frac{\|\mathbf{\Pi x}\|^2}{N-n} \to 1$. Defining now $\mathbf{W}$ such that $\mathbf{W}\mathbf{W}^{\mathsf{H}} - \mathbf{w}\mathbf{w}^{\mathsf{H}} = \mathbf{U}\mathbf{U}^{\mathsf{H}}$, the reasoning remains valid, and this entails (6.38). □

Since $\mathbf{B}_N$ in Theorem 6.15 is assumed of uniformly bounded spectral norm, $\mathbf{w}^{\mathsf{H}}\mathbf{B}_N\mathbf{w}$ is uniformly bounded also. Hence, if N, n grow large with ratio n/N uniformly away from one, the term $\frac{1}{N-n}\mathbf{w}^{\mathsf{H}}\mathbf{B}_N\mathbf{w}$ tends to zero. This therefore entails the following corollary, which can be seen as a rank-1 perturbation of Theorem 6.15.

Corollary 6.2. *Let* $\mathbf{W}$ *and* $\mathbf{B}_N$ *be defined as in Theorem 6.15, with* N *and* n *such that* $\limsup_n \frac{n}{N} < 1$. *Then, as* N, n *grow large, for* $\mathbf{w}$ *any column of* $\mathbf{W}$

$$\mathbf{w}^{\mathsf{H}}\mathbf{B}_N\mathbf{w} - \frac{1}{N-n} \operatorname{tr} \mathbf{B}_N \left(\mathbf{I}_N - \mathbf{W}\mathbf{W}^{\mathsf{H}}\right) \xrightarrow{\text{a.s.}} 0.$$

Corollary 6.2 only differs from Theorem 6.15 by the fact that the projector $\mathbf{\Pi}$ is changed into $\mathbf{I}_N - \mathbf{W}\mathbf{W}^{\mathsf{H}}$.

Also, when $\mathbf{B}_N$ is independent of $\mathbf{W}$, we fall back on the same result as for the i.i.d. case.

Corollary 6.3. *Let* $\mathbf{W}$ *be defined as in Theorem 6.15, and let* $\mathbf{A} \in \mathbb{C}^{N\times N}$ *be independent of* $\mathbf{W}$ *and have uniformly bounded spectral norm. Then, as* N *grows large, for* $\mathbf{w}$ *any column of* $\mathbf{W}$, *we have:*

$$\mathbf{w}^{\mathsf{H}}\mathbf{A}\mathbf{w} - \frac{1}{N} \operatorname{tr} \mathbf{A} \xrightarrow{\text{a.s.}} 0.$$

Theorem 6.15 is the basis for establishing deterministic equivalents involving isometric matrices. In the following, we introduce a result, based on Silverstein and Bai's approach, which generalizes Theorems 4.10, 4.11, and 4.12 to the case when the $\mathbf{W}_i$ matrices are multiplied on the left by different non-necessarily co-diagonalizable matrices. These models are the basis for studying the properties of multi-user or multi-cellular communications both involving unitary precoders and taking into account the frequency selectivity of the channel. From a mathematical point of view, there exists no simple way to study such models using tools extracted solely from free probability theory. In particular, it is interesting to note that in [Peacock *et al.*, 2008], the authors already generalized Theorem 4.12 to the case where the left-product matrices are different but co-diagonalizable. To do so, the authors relied on tools from free probability as the basic instruments and then need some extra matrix manipulation to derive their limiting result, in a sort of hybrid method between free probability and analytical approach. In the results to come, though, no mention will be made to free probability theory, as the result can be derived autonomously from the tools developed in this section.

The following results are taken from [Couillet *et al.*, 2011b], where detailed proofs can be found. We start by introducing the fundamental equations.

Theorem 6.16 ([Couillet *et al.*, 2011b]). *For $i \in \{1,\ldots,K\}$, let $\mathbf{T}_i \in \mathbb{C}^{n_i \times n_i}$ be Hermitian diagonal and let $\mathbf{H}_i \in \mathbb{C}^{N\times N_i}$. Define $\mathbf{R}_i \triangleq \mathbf{H}_i\mathbf{H}_i^{\mathsf{H}} \in \mathbb{C}^{N\times N}$, $c_i = \frac{n_i}{N_i}$ and $\bar{c}_i = \frac{N_i}{N}$. Then the following system of equations in $(\bar{e}_1(z),\ldots,\bar{e}_K(z))$:*

$$\begin{aligned}\bar{e}_i(z) &= \frac{1}{N}\operatorname{tr}\mathbf{T}_i\left(e_i(z)\mathbf{T}_i + [\bar{c}_i - e_i(z)\bar{e}_i(z)]\mathbf{I}_{n_i}\right)^{-1}\\ e_i(z) &= \frac{1}{N}\operatorname{tr}\mathbf{R}_i\left(\sum_{j=1}^K \bar{e}_j(z)\mathbf{R}_j - z\mathbf{I}_N\right)^{-1}\end{aligned} \tag{6.39}$$

has a unique solution $(\bar{e}_1(z),\ldots,\bar{e}_K(z)) \in \mathcal{C}(\mathbb{C},\mathbb{C})$ satisfying $(e_1(z),\ldots,e_K(z)) \in \mathcal{S}(\mathbb{R}^+)^K$ and, for z real negative, $0 \leq e_i(z) < c_i\bar{c}_i/\bar{e}_i(z)$ for all i. Moreover, for each real negative z

$$\bar{e}_i(z) = \lim_{t\to\infty} \bar{e}_i^{(t)}(z)$$

where $\bar{e}_i^{(t)}(z)$ is the unique solution of

$$\bar{e}_i^{(t)}(z) = \frac{1}{N}\operatorname{tr}\mathbf{T}_i\left(e_i^{(t)}(z)\mathbf{T}_i + [\bar{c}_i - e_i^{(t)}(z)\bar{e}_i^{(t)}(z)]\mathbf{I}_{n_i}\right)^{-1}$$

within the interval $[0, c_i\bar{c}_i/e_i^{(t)}(z))$, $e_i^{(0)}(z)$ can take any positive value and $e_i^{(t)}(z)$ is recursively defined by

$$e_i^{(t)}(z) = \frac{1}{N}\operatorname{tr}\mathbf{R}_i\left(\sum_{j=1}^K \bar{e}_j^{(t-1)}(z)\mathbf{R}_j - z\mathbf{I}_N\right)^{-1}.$$

We then have the following theorem on a deterministic equivalent for the e.s.d. of the model $\mathbf{B}_N = \sum_{k=1}^K \mathbf{H}_i\mathbf{W}_i\mathbf{T}_i\mathbf{W}_i^{\mathsf{H}}\mathbf{H}_i^{\mathsf{H}}$.

Theorem 6.17 ([Couillet *et al.*, 2011b]). *For $i \in \{1,\ldots,K\}$, let $\mathbf{T}_i \in \mathbb{C}^{n_i\times n_i}$ be a Hermitian non-negative matrix with spectral norm bounded uniformly along n_i and $\mathbf{W}_i \in \mathbb{C}^{N_i\times n_i}$ be $n_i \leq N_i$ columns of a unitary Haar distributed random matrix. Consider $\mathbf{H}_i \in \mathbb{C}^{N\times N_i}$ a random matrix such that $\mathbf{R}_i \triangleq \mathbf{H}_i\mathbf{H}_i^{\mathsf{H}} \in \mathbb{C}^{N\times N}$ has uniformly bounded spectral norm along N, almost surely. Define $c_i = \frac{n_i}{N_i}$ and $\bar{c}_i = \frac{N_i}{N}$ and denote*

$$\mathbf{B}_N = \sum_{i=1}^K \mathbf{H}_i\mathbf{W}_i\mathbf{T}_i\mathbf{W}_i^{\mathsf{H}}\mathbf{H}_i^{\mathsf{H}}.$$

Then, as $N, N_1,\ldots,N_K, n_1,\ldots,n_K$ grow to infinity with ratios $\bar{c}_i$ satisfying $0 < \liminf \bar{c}_i \leq \limsup \bar{c}_i < \infty$ and $0 \leq c_i \leq 1$ for all i, the following limit holds *true almost surely*

$$F^{\mathbf{B}_N} - F_N \Rightarrow 0$$

where F_N is the distribution function with Stieltjes transform $m_N(z)$ defined by

$$m_N(z) = \frac{1}{N} \operatorname{tr} \left(\sum_{i=1}^{K} \bar{e}_i(z) \mathbf{R}_i - z \mathbf{I}_N \right)^{-1}$$

where $(\bar{e}_1(z), \ldots, \bar{e}_K(z))$ are given by Theorem 6.16.

Consider the case when, for each i, $\bar{c}_i = 1$ and $\mathbf{H}_i = \mathbf{R}_i^{\frac{1}{2}}$ for some square Hermitian non-negative square root $\mathbf{R}_i^{\frac{1}{2}}$ of $\mathbf{R}_i$. We observe that the system of Equations (6.39) is very similar to the system of Equations (6.7) established for the case of i.i.d. random matrices. The noticeable difference here is the addition of the extra term $-e_i \bar{e}_i$ in the expression of $\bar{e}_i$. Without this term, we fall back on the i.i.d. case. Notice also that the case $K = 1$ corresponds exactly to Theorem 4.11, which was treated for $c_1 = 1$.

Another point worth commenting on here is that, when $z < 0$, the fixed-point algorithm to determine e_i can be initialized at any positive value, while the fixed-point algorithm to determine $\bar{e}_i$ *must* be initialized properly. If not, it is possible that $\bar{e}_i^{(t)}$ diverges. Also, if we naively run the fixed-point algorithm jointly over e_i and $\bar{e}_i$, we may end up not converging to the correct solution at all. Based on experience, this case arises sometimes if no particular care is taken.

We hereafter provide both a sketch of the proof and a rather extensive derivation, which explains how (6.39) is derived and how uniqueness is proved. We will only treat the case where, for all i, $\limsup c_i < 1$, $\bar{c}_i = 1$, $\mathbf{H}_i = \mathbf{R}_i^{\frac{1}{2}}$ and the $\mathbf{R}_i$ are deterministic with uniformly bounded spectral norm in order both to simplify notations and for the derivations to be close in nature to those proposed in the proof of Theorem 6.1. The case where in particular $\limsup c_i = 1$ for a certain i only demands some additional technicalities, which are not necessary here. Nonetheless, note that, for practical applications, all these hypotheses are essential, as unitary precoding systems such as code division or space division multiple access systems, e.g. CDMA and SDMA, may require square unitary precoding matrices (hence $c_i = 1$) and may involve rectangular multiple antenna channel matrices $\mathbf{H}_i$; these channels being modeled as Gaussian i.i.d.-based matrices with almost surely bounded spectral norm. The proof follows the derivation in [Couillet *et al.*, 2011b], where a detailed derivation can be found.

The main steps of the proof are similar to those developed for the proof of Theorem 6.1. In order to propose different approaches than in previous derivations, we will work almost exclusively with real negative z, instead of z with positive imaginary part. We will also provide a shorter proof of the final convergence step $m_{\mathbf{B}_N}(z) - m_N(z) \xrightarrow{\text{a.s.}} 0$, relying on restrictions of the domain of z along with arguments from Vitali's convergence theorem. These approaches are valid here because upper bounds on the spectral norms of $\mathbf{R}_i$ and

$\mathbf{T}_i$ are considered, which was not the case for Theorem 6.1. Apart from these technical considerations, the main noticeable difference between the deterministic equivalent approaches proposed for matrices with independent entries and for Haar matrices lies in the first convergence step, which is much more intricate.

Proof. We first provide a sketch of the proof for better understanding, which will enhance the aforementioned main novelty. As usual, we wish to prove that there exists a matrix $\mathbf{F} = \sum_{i=1}^{K} \bar{f}_i \mathbf{R}_i$, such that, for all non-negative $\mathbf{A}$ with $\|\mathbf{A}\| < \infty$

$$\frac{1}{N} \operatorname{tr} \mathbf{A} \left(\mathbf{B}_N - z\mathbf{I}_N\right)^{-1} - \frac{1}{N} \operatorname{tr} \mathbf{A} \left(\mathbf{F} - z\mathbf{I}_N\right)^{-1} \xrightarrow{\text{a.s.}} 0.$$

Contrary to classical deterministic equivalent approaches for random matrices with i.i.d. entries, finding a deterministic equivalent for $\frac{1}{N} \operatorname{tr} \mathbf{A} \left(\mathbf{B}_N - z\mathbf{I}_N\right)^{-1}$ is not straightforward. The reason is that during the derivation, terms such as $\frac{1}{N-n_i} \operatorname{tr} \left(\mathbf{I}_N - \mathbf{W}_i \mathbf{W}_i^{\mathsf{H}}\right) \mathbf{A}^{\frac{1}{2}} \left(\mathbf{B}_N - z\mathbf{I}_N\right)^{-1} \mathbf{A}^{\frac{1}{2}}$, with the $\left(\mathbf{I}_N - \mathbf{W}_i \mathbf{W}_i^{\mathsf{H}}\right)$ prefix will naturally appear, as a result of applying the trace lemma, Theorem 6.15, that will be required to be controlled. We proceed as follows.

- We first denote for all i, $\delta_i \triangleq \frac{1}{N-n_i} \operatorname{tr} \left(\mathbf{I}_N - \mathbf{W}_i \mathbf{W}_i^{\mathsf{H}}\right) \mathbf{R}_i^{\frac{1}{2}} \left(\mathbf{B}_N - z\mathbf{I}_N\right)^{-1} \mathbf{R}_i^{\frac{1}{2}}$ some auxiliary variable. Then, using the same techniques as in the proof of Theorem 6.1, denoting further $f_i \triangleq \frac{1}{N} \operatorname{tr} \mathbf{R}_i \left(\mathbf{B}_N - z\mathbf{I}_N\right)^{-1}$, we prove

$$f_i - \frac{1}{N} \operatorname{tr} \mathbf{R}_i \left(\mathbf{G} - z\mathbf{I}_N\right)^{-1} \xrightarrow{\text{a.s.}} 0$$

with $\mathbf{G} = \sum_{j=1}^{K} \bar{g}_j \mathbf{R}_j$ and

$$\bar{g}_i = \frac{1}{1 - c_i + \frac{1}{N} \sum_{l=1}^{n_i} \frac{1}{1+t_{il}\delta_i}} \frac{1}{N} \sum_{l=1}^{n_i} \frac{t_{il}}{1 + t_{il}\delta_i}$$

where $t_{i1}, \ldots, t_{in_i}$ are the eigenvalues of $\mathbf{T}_i$. Noticing additionally that

$$(1 - c_i)\delta_i - f_i + \frac{1}{N} \sum_{l=1}^{n_i} \frac{\delta_i}{1 + t_{il}\delta_i} \xrightarrow{\text{a.s.}} 0$$

we have a first hint on a first deterministic equivalent for f_i. Precisely, we expect to obtain the set of fundamental equations

$$\Delta_i = \frac{1}{1 - c_i} \left[e_i - \frac{1}{N} \sum_{l=1}^{n_i} \frac{\Delta_i}{1 + t_{il}\Delta_i} \right]$$

$$e_i = \frac{1}{N} \operatorname{tr} \mathbf{R}_i \left(\sum_{j=1}^{K} \frac{1}{1 - c_j + \frac{1}{N} \sum_{l=1}^{n_j} \frac{1}{1+t_{jl}\Delta_j}} \frac{1}{N} \sum_{l=1}^{n_j} \frac{t_{jl}}{1 + t_{jl}\Delta_j} \mathbf{R}_j - z\mathbf{I}_N \right)^{-1}.$$

- The expressions of $\bar{g}_i$ and their deterministic equivalents are however not very convenient under this form. It is then shown that

$$\bar{g}_i - \frac{1}{N} \sum_{l=1}^{n_i} \frac{t_{il}}{1 + t_{il} f_i - f_i \bar{g}_i} = \bar{g}_i - \frac{1}{N} \operatorname{tr} \mathbf{T}_i \left(f_i \mathbf{T}_i + [1 - f_i \bar{g}_i] \mathbf{I}_{n_i}\right)^{-1} \xrightarrow{\text{a.s.}} 0$$

which induces the $2K$-equation system

$$f_i - \frac{1}{N}\operatorname{tr} \mathbf{R}_i \left(\sum_{j=1}^K \bar{g}_j \mathbf{R}_j - z\mathbf{I}_N\right)^{-1} \xrightarrow{\text{a.s.}} 0$$

$$\bar{g}_i - \frac{1}{N}\operatorname{tr} \mathbf{T}_i \left(\bar{g}_i \mathbf{T}_i + [1 - f_i \bar{g}_i]\right)^{-1} \xrightarrow{\text{a.s.}} 0.$$

- These relations are sufficient to infer the deterministic equivalent but will be made more attractive for further considerations by introducing $\mathbf{F} = \sum_{i=1}^K \bar{f}_i \mathbf{R}_i$ and proving that

$$f_i - \frac{1}{N}\operatorname{tr} \mathbf{R}_i \left(\sum_{j=1}^K \bar{f}_j \mathbf{R}_j - z\mathbf{I}_N\right)^{-1} \xrightarrow{\text{a.s.}} 0$$

$$\bar{f}_i - \frac{1}{N}\operatorname{tr} \mathbf{T}_i \left(\bar{f}_i \mathbf{T}_i + [1 - f_i \bar{f}_i]\right)^{-1} = 0$$

where, for $z < 0$, $\bar{f}_i$ lies in $[0, c_i/f_i)$ and is now uniquely determined by f_i. In particular, this step provides an explicit expression $\bar{f}_i$ as a function of f_i, which will be translated into an explicit expression of $\bar{e}_i$ as a function of e_i.

This is the very technical part of the proof. We then prove the existence and uniqueness of a solution to the fixed-point equation

$$e_i - \frac{1}{N}\operatorname{tr} \mathbf{R}_i \left(\sum_{j=1}^K \bar{e}_j \mathbf{R}_j - z\mathbf{I}_N\right)^{-1} = 0$$

$$\bar{e}_i - \frac{1}{N}\operatorname{tr} \mathbf{T}_i \left(\bar{e}_i \mathbf{T}_i + [1 - e_i \bar{e}_i]\right)^{-1} = 0$$

for all finite N, z real negative, and for $\bar{e}_i \in [0, c_i/f_i]$. Here, instead of following the approach of the proof of uniqueness for the fundamental equations of Theorem 6.1, we use a property of so-called *standard functions*. We will show precisely that the vector application $\mathbf{h} = (h_1, \ldots, h_K)$ with

$$h_i : (x_1, \ldots, x_K) \mapsto \frac{1}{N}\operatorname{tr} \mathbf{R}_i \left(\sum_{j=1}^K \bar{x}_j \mathbf{R}_j - z\mathbf{I}_N\right)^{-1}$$

where $\bar{x}_i$ is the unique solution to

$$\bar{x}_i = \frac{1}{N}\operatorname{tr} \mathbf{T}_i \left(\bar{x}_i \mathbf{T}_i + [1 - x_i \bar{x}_i]\right)^{-1}$$

lying in $[0, c_i/x_i)$, is a standard function. It will unfold that the fixed-point equation in $(e_1, \ldots, e_K)$ has a unique solution with positive entries and that this solution can be determined as the limiting iteration of a classical fixed-point algorithm.

The last step proves that the unique solution $(e_1, \ldots, e_N)$ is such that

$$e_i - f_i \xrightarrow{\text{a.s.}} 0$$

which is solved by arguments borrowed from the work of Hachem *et al.* [Hachem *et al.*, 2007], using a restriction on the definition domain of z, which simplifies greatly the calculus.

We now turn to the precise proof. We use again the Bai and Silverstein steps: the convergence $f_i - \frac{1}{N}\operatorname{tr}\mathbf{R}_i\left(\sum_{j=1}^K \bar{f}_j\mathbf{R}_j - z\mathbf{I}_N\right)^{-1} \xrightarrow{\text{a.s.}} 0$ in a first step, the existence and uniqueness of a solution to $e_i = \frac{1}{N}\operatorname{tr}\mathbf{R}_i\left(\sum_{j=1}^K \bar{e}_j\mathbf{R}_j - z\mathbf{I}_N\right)^{-1}$ in a second, and the convergence $e_i - f_i \xrightarrow{\text{a.s.}} 0$ in a third. Although precise control of the random variables involved needs be carried out, as is detailed in [Couillet *et al.*, 2011b], we hereafter elude most technical parts for simplicity and understanding.

Step 1: First convergence step

In this section, we take $z < 0$, until further notice. Let us first introduce the following parameters. We will denote $T = \max_i\{\limsup \|\mathbf{T}_i\|\}$, $R = \max_i\{\limsup \|\mathbf{R}_i\|\}$ and $c = \max_i\{\limsup c_i\}$.

We start with classical deterministic equivalent techniques. Let $\mathbf{A} \in \mathbb{C}^{N\times N}$ be a Hermitian non-negative definite matrix with spectral norm uniformly bounded by A. Taking $\mathbf{G} = \sum_{j=1}^K \bar{g}_j\mathbf{R}_j$, with $\bar{g}_1,\ldots,\bar{g}_K$ left undefined for the moment, we have:

$$\begin{aligned}
&\frac{1}{N}\operatorname{tr}\mathbf{A}(\mathbf{B}_N - z\mathbf{I}_N)^{-1} - \frac{1}{N}\operatorname{tr}\mathbf{A}(\mathbf{G} - z\mathbf{I}_N)^{-1} \\
&= \frac{1}{N}\operatorname{tr}\left[\mathbf{A}(\mathbf{B}_N - z\mathbf{I}_N)^{-1}\sum_{i=1}^K \mathbf{R}_i^{\frac{1}{2}}\left(-\mathbf{W}_i\mathbf{T}_i\mathbf{W}_i^{\mathsf{H}} + \bar{g}_i\mathbf{I}_N\right)\mathbf{R}_i^{\frac{1}{2}}(\mathbf{G} - z\mathbf{I}_N)^{-1}\right] \\
&= \sum_{i=1}^K \bar{g}_i\frac{1}{N}\operatorname{tr}\mathbf{A}(\mathbf{B}_N - z\mathbf{I}_N)^{-1}\mathbf{R}_i(\mathbf{G} - z\mathbf{I}_N)^{-1} \\
&\quad - \frac{1}{N}\sum_{i=1}^K\sum_{l=1}^{n_i} t_{il}\mathbf{w}_{il}^{\mathsf{H}}\mathbf{R}_i^{\frac{1}{2}}(\mathbf{G} - z\mathbf{I}_N)^{-1}\mathbf{A}(\mathbf{B}_N - z\mathbf{I}_N)^{-1}\mathbf{R}_i^{\frac{1}{2}}\mathbf{w}_{il} \\
&= \sum_{i=1}^K \bar{g}_i\frac{1}{N}\operatorname{tr}\mathbf{A}(\mathbf{B}_N - z\mathbf{I}_N)^{-1}\mathbf{R}_i(\mathbf{G} - z\mathbf{I}_N)^{-1} \\
&\quad - \frac{1}{N}\sum_{i=1}^K\sum_{l=1}^{n_i} \frac{t_{il}\mathbf{w}_{il}^{\mathsf{H}}\mathbf{R}_i^{\frac{1}{2}}(\mathbf{G} - z\mathbf{I}_N)^{-1}\mathbf{A}(\mathbf{B}_{(i,l)} - z\mathbf{I}_N)^{-1}\mathbf{R}_i^{\frac{1}{2}}\mathbf{w}_{il}}{1 + t_{il}\mathbf{w}_{il}^{\mathsf{H}}\mathbf{R}_i^{\frac{1}{2}}(\mathbf{B}_{(i,l)} - z\mathbf{I}_N)^{-1}\mathbf{R}_i^{\frac{1}{2}}\mathbf{w}_{il}},
\end{aligned} \tag{6.40}$$

with $t_{i1},\ldots,t_{in_i}$ the eigenvalues of $\mathbf{T}_i$.

The quadratic forms $\mathbf{w}_{il}^{\mathsf{H}}\mathbf{R}_i^{\frac{1}{2}}(\mathbf{G} - z\mathbf{I}_N)^{-1}\mathbf{A}(\mathbf{B}_{(i,l)} - z\mathbf{I}_N)^{-1}\mathbf{R}_i^{\frac{1}{2}}\mathbf{w}_{il}$ and $\mathbf{w}_{il}^{\mathsf{H}}\mathbf{R}_i^{\frac{1}{2}}(\mathbf{B}_{(i,l)} - z\mathbf{I}_N)^{-1}\mathbf{R}_i^{\frac{1}{2}}\mathbf{w}_{il}$ are not asymptotically close to the trace of the inner matrix, as in the i.i.d. case, but to the trace of the inner matrix multiplied by $(\mathbf{I}_N - \mathbf{W}_i\mathbf{W}_i^{\mathsf{H}})$. This complicates the calculus. In the following, we will therefore study the following stochastic quantities, namely the random variables δ_i, β_i and f_i, introduced below.

For every $i \in \{1, \ldots, K\}$, denote

$$\delta_i \triangleq \frac{1}{N - n_i} \operatorname{tr} \left(\mathbf{I}_N - \mathbf{W}_i \mathbf{W}_i^{\mathsf{H}} \right) \mathbf{R}_i^{\frac{1}{2}} \left(\mathbf{B}_N - z \mathbf{I}_N \right)^{-1} \mathbf{R}_i^{\frac{1}{2}}$$
$$f_i \triangleq \frac{1}{N} \operatorname{tr} \mathbf{R}_i \left(\mathbf{B}_N - z \mathbf{I}_N \right)^{-1}$$

both being clearly non-negative. We may already recognize that f_i is a key quantity for the subsequent derivations, as it will be shown to be asymptotically close to e_i, the central parameter of our deterministic equivalent.

Writing $\mathbf{W}_i = [\mathbf{w}_{i,1}, \ldots, \mathbf{w}_{i,n_i}]$ and $\mathbf{W}_i \mathbf{W}_i^{\mathsf{H}} = \sum_{l=1}^{n_i} \mathbf{w}_{il} \mathbf{w}_{il}^{\mathsf{H}}$, we have from standard calculus and the matrix inversion lemma, Lemma 6.2, that

$$\begin{aligned} (1 - c_i)\delta_i &= f_i - \frac{1}{N} \sum_{l=1}^{n_i} \mathbf{w}_{il}^{\mathsf{H}} \mathbf{R}_i^{\frac{1}{2}} \left(\mathbf{B}_N - z \mathbf{I}_N \right)^{-1} \mathbf{R}_i^{\frac{1}{2}} \mathbf{w}_{il} \\ &= f_i - \frac{1}{N} \sum_{l=1}^{n_i} \frac{\mathbf{w}_{il}^{\mathsf{H}} \mathbf{R}_i^{\frac{1}{2}} \left(\mathbf{B}_{(i,l)} - z \mathbf{I}_N \right)^{-1} \mathbf{R}_i^{\frac{1}{2}} \mathbf{w}_{il}}{1 + t_{il} \mathbf{w}_{il}^{\mathsf{H}} \mathbf{R}_i^{\frac{1}{2}} \left(\mathbf{B}_{(i,l)} - z \mathbf{I}_N \right)^{-1} \mathbf{R}_i^{\frac{1}{2}} \mathbf{w}_{il}} \end{aligned} \tag{6.41}$$

with $\mathbf{B}_{(i,l)} = \mathbf{B}_N - t_{il} \mathbf{R}_i^{\frac{1}{2}} \mathbf{w}_{il} \mathbf{w}_{il}^{\mathsf{H}} \mathbf{R}_i^{\frac{1}{2}}$.

Since $z < 0$, $\delta_i \geq 0$, so that $\frac{1}{1 + t_{il}\delta_i}$ is well defined. We recognize already from Theorem 6.15 that each quadratic term $\mathbf{w}_{il}^{\mathsf{H}} \mathbf{R}_i^{\frac{1}{2}} \left(\mathbf{B}_{(i,l)} - z \mathbf{I}_N \right)^{-1} \mathbf{R}_i^{\frac{1}{2}} \mathbf{w}_{il}$ is asymptotically close to δ_i. By adding the term $\frac{1}{N} \sum_{l=1}^{n_i} \frac{\delta_i}{1 + t_{il}\delta_i}$ on both sides, (6.41) can further be rewritten

$$\begin{aligned} &(1 - c_i)\delta_i - f_i + \frac{1}{N} \sum_{l=1}^{n_i} \frac{\delta_i}{1 + t_{il}\delta_i} \\ &= \frac{1}{N} \sum_{l=1}^{n_i} \left[\frac{\delta_i}{1 + t_{il}\delta_i} - \frac{\mathbf{w}_{il}^{\mathsf{H}} \mathbf{R}_i^{\frac{1}{2}} \left(\mathbf{B}_{(i,l)} - z \mathbf{I}_N \right)^{-1} \mathbf{R}_i^{\frac{1}{2}} \mathbf{w}_{il}}{1 + t_{il} \mathbf{w}_{il}^{\mathsf{H}} \mathbf{R}_i^{\frac{1}{2}} \left(\mathbf{B}_{(i,l)} - z \mathbf{I}_N \right)^{-1} \mathbf{R}_i^{\frac{1}{2}} \mathbf{w}_{il}} \right]. \end{aligned}$$

We now apply the trace lemma, Theorem 6.15, which ensures that

$$\mathrm{E} \left[\left| (1 - c_i)\delta_i - f_i + \frac{1}{N} \sum_{l=1}^{n_i} \frac{\delta_i}{1 + t_{il}\delta_i} \right|^4 \right] = O\left(\frac{1}{N^2} \right). \tag{6.42}$$

We do not provide the precise derivations of the fourth order moment inequalities here and in all the equations that follow, our main purpose being concentrated on the fundamental steps of the proof. Precise calculus and upper bounds can be found in [Couillet *et al.*, 2011b]. This is our first relation that links δ_i to $f_i = \frac{1}{N} \operatorname{tr} \mathbf{R}_i \left(\mathbf{B}_N - z \mathbf{I}_N \right)^{-1}$.

Introducing now an additional $\mathbf{A}(\mathbf{G} - z\mathbf{I}_N)^{-1}$ matrix in the argument of the trace of δ_i, with $\mathbf{G}, \mathbf{A} \in \mathbb{C}^{N \times N}$ any non-negative definite matrices, $\|\mathbf{A}\| \leq A$, we denote

$$\beta_i \triangleq \frac{1}{N - n_i} \operatorname{tr} \left(\mathbf{I}_N - \mathbf{W}_i \mathbf{W}_i^{\mathsf{H}} \right) \mathbf{R}_i^{\frac{1}{2}} \left(\mathbf{G} - z \mathbf{I}_N \right)^{-1} \mathbf{A} \left(\mathbf{B}_N - z \mathbf{I}_N \right)^{-1} \mathbf{R}_i^{\frac{1}{2}}.$$

We then proceed similarly as for δ_i by showing

$$\beta_i = \frac{1}{N-n_i} \operatorname{tr} \mathbf{R}_i^{\frac{1}{2}} (\mathbf{G} - z\mathbf{I}_N)^{-1} \mathbf{A} (\mathbf{B}_N - z\mathbf{I}_N)^{-1} \mathbf{R}_i^{\frac{1}{2}}$$
$$= \frac{1}{N-n_i} \sum_{l=1}^{n_i} \frac{\mathbf{w}_{il}^{\mathsf{H}} \mathbf{R}_i^{\frac{1}{2}} (\mathbf{G} - z\mathbf{I}_N)^{-1} \mathbf{A} \left(\mathbf{B}_{(i,l)} - z\mathbf{I}_N\right)^{-1} \mathbf{R}_i^{\frac{1}{2}} \mathbf{w}_{il}}{1 + t_{il} \mathbf{w}_{il}^{\mathsf{H}} \mathbf{R}_i^{\frac{1}{2}} \left(\mathbf{B}_{(i,l)} - z\mathbf{I}_N\right)^{-1} \mathbf{R}_i^{\frac{1}{2}} \mathbf{w}_{il}}$$

from which we have:

$$\frac{1}{N-n_i} \operatorname{tr} \mathbf{R}_i^{\frac{1}{2}} (\mathbf{G} - z\mathbf{I}_N)^{-1} \mathbf{A} (\mathbf{B}_N - z\mathbf{I}_N)^{-1} \mathbf{R}_i^{\frac{1}{2}} - \frac{1}{N-n_i} \sum_{l=1}^{n_i} \frac{\beta_i}{1 + t_{il}\delta_i} - \beta_i$$
$$= \frac{1}{N-n_i} \sum_{l=1}^{n_i} \left[\frac{\mathbf{w}_{il}^{\mathsf{H}} \mathbf{R}_i^{\frac{1}{2}} (\mathbf{G} - z\mathbf{I}_N)^{-1} \mathbf{A} \left(\mathbf{B}_{(i,l)} - z\mathbf{I}_N\right)^{-1} \mathbf{R}_i^{\frac{1}{2}} \mathbf{w}_{il}}{1 + t_{il} \mathbf{w}_{il}^{\mathsf{H}} \mathbf{R}_i^{\frac{1}{2}} \left(\mathbf{B}_{(i,l)} - z\mathbf{I}_N\right)^{-1} \mathbf{R}_i^{\frac{1}{2}} \mathbf{w}_{il}} - \frac{\beta_i}{1 + t_{il}\delta_i} \right].$$

Since numerators and denominators converge again to one another, we can show from Theorem 6.15 again that

$$\mathrm{E}\left[\left| \frac{\mathbf{w}_{il}^{\mathsf{H}} \mathbf{R}_i^{\frac{1}{2}} (\mathbf{G} - z\mathbf{I}_N)^{-1} \mathbf{A} \left(\mathbf{B}_{(i,l)} - z\mathbf{I}_N\right)^{-1} \mathbf{R}_i^{\frac{1}{2}} \mathbf{w}_{il}}{1 + t_{il} \mathbf{w}_{il}^{\mathsf{H}} \mathbf{R}_i^{\frac{1}{2}} \left(\mathbf{B}_{(i,l)} - z\mathbf{I}_N\right)^{-1} \mathbf{R}_i^{\frac{1}{2}} \mathbf{w}_{il}} - \frac{\beta_i}{1 + t_{il}\delta_i} \right|^4 \right] = O\left(\frac{1}{N^2}\right). \tag{6.43}$$

Hence

$$\mathrm{E}\left[\left| \frac{1}{N} \operatorname{tr} \mathbf{R}_i (\mathbf{G} - z\mathbf{I}_N)^{-1} \mathbf{A} (\mathbf{B}_N - z\mathbf{I}_N)^{-1} - \beta_i \left(1 - c_i + \frac{1}{N} \sum_{l=1}^{n_i} \frac{1}{1 + t_{il}\delta_i} \right) \right|^4 \right]$$
$$= O\left(\frac{1}{N^2}\right). \tag{6.44}$$

This provides us with the second relation that links β_i to $\frac{1}{N} \operatorname{tr} \mathbf{R}_i^{\frac{1}{2}} (\mathbf{G} - z\mathbf{I}_N)^{-1} \mathbf{A} (\mathbf{B}_N - z\mathbf{I}_N)^{-1} \mathbf{R}_i^{\frac{1}{2}}$. That is, we have expressed both δ_i and β_i as a function of the traces $\frac{1}{N} \operatorname{tr} \mathbf{R}_i^{\frac{1}{2}} (\mathbf{B}_N - z\mathbf{I}_N)^{-1} \mathbf{R}_i^{\frac{1}{2}}$ and $\frac{1}{N} \operatorname{tr} \mathbf{R}_i^{\frac{1}{2}} (\mathbf{G} - z\mathbf{I}_N)^{-1} \mathbf{A} (\mathbf{B}_N - z\mathbf{I}_N)^{-1} \mathbf{R}_i^{\frac{1}{2}}$, which are more conventional to work with.

We are now in position to determine adequate expressions for $\bar{g}_1, \ldots, \bar{g}_K$. From the fact that $\mathbf{w}_{il}^{\mathsf{H}} \mathbf{R}_i^{\frac{1}{2}} (\mathbf{B}_{(i,l)} - z\mathbf{I}_N)^{-1} \mathbf{R}_i^{\frac{1}{2}} \mathbf{w}_{il}$ is asymptotically close to δ_i and that $\mathbf{w}_{il}^{\mathsf{H}} \mathbf{R}_i^{\frac{1}{2}} (\mathbf{G} - z\mathbf{I}_N)^{-1} \mathbf{A} (\mathbf{B}_{(i,l)} - z\mathbf{I}_N)^{-1} \mathbf{R}_i^{\frac{1}{2}} \mathbf{w}_{il}$ is asymptotically close to β_i, we choose, based on (6.44) especially

$$\bar{g}_i = \frac{1}{1 - c_i + \frac{1}{N} \sum_{l=1}^{n_i} \frac{1}{1 + t_{il}\delta_i}} \frac{1}{N} \sum_{l=1}^{n_i} \frac{t_{il}}{1 + t_{il}\delta_i}.$$

We then have

$$\begin{aligned}
&\frac{1}{N}\operatorname{tr}\mathbf{A}(\mathbf{B}_N - z\mathbf{I}_N)^{-1} - \frac{1}{N}\operatorname{tr}\mathbf{A}(\mathbf{G} - z\mathbf{I}_N)^{-1} \\
&= \sum_{i=1}^{K} \frac{\frac{1}{N}\sum_{l=1}^{n_i}\frac{t_{il}}{1+t_{il}\delta_i}\frac{1}{N}\operatorname{tr}\mathbf{R}_i\left(\mathbf{G} - z\mathbf{I}_N\right)^{-1}\mathbf{A}\left(\mathbf{B}_N - z\mathbf{I}_N\right)^{-1}}{1 - c_i + \frac{1}{N}\sum_{l=1}^{n_i}\frac{1}{1+t_{il}\delta_i}} \\
&- \frac{1}{N}\sum_{i=1}^{K}\sum_{l=1}^{n_i}\frac{t_{il}\mathbf{w}_{il}^{\mathsf{H}}\mathbf{R}_i^{\frac{1}{2}}(\mathbf{G} - z\mathbf{I}_N)^{-1}\mathbf{A}(\mathbf{B}_{(i,l)} - z\mathbf{I}_N)^{-1}\mathbf{R}_i^{\frac{1}{2}}\mathbf{w}_{il}}{1 + t_{il}\mathbf{w}_{il}^{\mathsf{H}}\mathbf{R}_i^{\frac{1}{2}}(\mathbf{B}_{(i,l)} - z\mathbf{I}_N)^{-1}\mathbf{R}_i^{\frac{1}{2}}\mathbf{w}_{il}} \\
&= \sum_{i=1}^{K}\frac{1}{N}\sum_{l=1}^{n_i} t_{il}\left[\frac{\frac{1}{N}\operatorname{tr}\mathbf{R}_i\left(\mathbf{G} - z\mathbf{I}_N\right)^{-1}\mathbf{A}\left(\mathbf{B}_N - z\mathbf{I}_N\right)^{-1}}{(1 - c_i + \frac{1}{N}\sum_{l'=1}^{n_i}\frac{1}{1+t_{i,l'}\delta_i})(1 + t_{il}\delta_i)}\right. \\
&\left. - \frac{\mathbf{w}_{il}^{\mathsf{H}}\mathbf{R}_i^{\frac{1}{2}}(\mathbf{G} - z\mathbf{I}_N)^{-1}\mathbf{A}(\mathbf{B}_{(i,l)} - z\mathbf{I}_N)^{-1}\mathbf{R}_i^{\frac{1}{2}}\mathbf{w}_{il}}{1 + t_{il}\mathbf{w}_{il}^{\mathsf{H}}\mathbf{R}_i^{\frac{1}{2}}(\mathbf{B}_{(i,l)} - z\mathbf{I}_N)^{-1}\mathbf{R}_i^{\frac{1}{2}}\mathbf{w}_{il}}\right].
\end{aligned}$$

To show that this last difference tends to zero, notice that $1 + t_{il}\delta_i \geq 1$ and

$$1 - c_i \leq 1 - c_i + \frac{1}{N}\sum_{l=1}^{n_i}\frac{1}{1 + t_{il}\delta_i} \leq 1$$

which ensure that we can divide the term in the expectation in the left-hand side of (6.44) by $1 + t_{il}\delta_i$ and $1 - c_i + \frac{1}{N}\sum_{l=1}^{n_i}\frac{1}{1+t_{il}\delta_i}$ without risking altering the order of convergence. This results in

$$\mathrm{E}\left[\left|\frac{\beta_i}{1 + t_{il}\delta_i} - \frac{\frac{1}{N}\operatorname{tr}\mathbf{R}_i^{\frac{1}{2}}\left(\mathbf{G} - z\mathbf{I}_N\right)^{-1}\mathbf{A}\left(\mathbf{B}_N - z\mathbf{I}_N\right)^{-1}\mathbf{R}_i^{\frac{1}{2}}}{\left(1 - c_i + \frac{1}{N}\sum_{l=1}^{n_i}\frac{1}{1+t_{il}\delta_i}\right)(1 + t_{il}\delta_i)}\right|^4\right] = \mathcal{O}\left(\frac{1}{N^2}\right). \tag{6.45}$$

From (6.43) and (6.45), we finally have that

$$\begin{aligned}
\mathrm{E}\Bigg[\Bigg|&\frac{\frac{1}{N}\operatorname{tr}\mathbf{R}_i\left(\mathbf{G} - z\mathbf{I}_N\right)^{-1}\mathbf{A}\left(\mathbf{B}_N - z\mathbf{I}_N\right)^{-1}}{\left(1 - c_i + \frac{1}{N}\sum_{l=1}^{n_i}\frac{1}{1+t_{il}\delta_i}\right)(1 + t_{il}\delta_i)} \\
&- \frac{\mathbf{w}_{il}^{\mathsf{H}}\mathbf{R}_i^{\frac{1}{2}}\left(\mathbf{G} - z\mathbf{I}_N\right)^{-1}\mathbf{A}\left(\mathbf{B}_{(i,l)} - z\mathbf{I}_N\right)^{-1}\mathbf{R}_i^{\frac{1}{2}}\mathbf{w}_{il}}{1 + t_{il}\mathbf{w}_{il}^{\mathsf{H}}\mathbf{R}_i^{\frac{1}{2}}\left(\mathbf{B}_{(i,l)} - z\mathbf{I}_N\right)^{-1}\mathbf{R}_i^{\frac{1}{2}}\mathbf{w}_{il}}\Bigg|^4\Bigg] = \mathcal{O}\left(\frac{1}{N^2}\right)
\end{aligned} \tag{6.46}$$

from which we obtain

$$\mathrm{E}\left[\left|\frac{1}{N}\operatorname{tr}\mathbf{A}(\mathbf{B}_N - z\mathbf{I}_N)^{-1} - \frac{1}{N}\operatorname{tr}\mathbf{A}(\mathbf{G} - z\mathbf{I}_N)^{-1}\right|^4\right] = \mathcal{O}\left(\frac{1}{N^2}\right). \tag{6.47}$$

This provides us with a first interesting result, from which we could infer a deterministic equivalent of $m_{\mathbf{B}_N}(z)$, which would be written as a function of deterministic equivalents of δ_i and deterministic equivalents of f_i, for $i = \{1, \ldots, K\}$. However this form is impractical to work with and we need to go further in the study of $\bar{g}_i$.

Observe that $\bar{g}_i$ can be written under the form

$$\bar{g}_i = \frac{1}{N}\sum_{l=1}^{n_i} \frac{t_{il}}{(1 - c_i + \frac{1}{N}\sum_{l=1}^{n_i}\frac{1}{1+t_{il}\delta_i}) + t_{il}\delta_i(1 - c_i + \frac{1}{N}\sum_{l=1}^{n_i}\frac{1}{1+t_{il}\delta_i})}.$$

We will study the denominator of the above expression and show that it can be synthesized into a much more attractive form.

From (6.42), we first have

$$\mathrm{E}\left[\left|f_i - \delta_i\left(1 - c_i + \frac{1}{N}\sum_{l=1}^{n_i}\frac{1}{1+t_{il}\delta_i}\right)\right|^4\right] = O\left(\frac{1}{N^2}\right).$$

Noticing that

$$1 - \bar{g}_i\delta_i\left(1 - c_i + \frac{1}{N}\sum_{l=1}^{n_i}\frac{1}{1+t_{il}\delta_i}\right) = 1 - c_i + \frac{1}{N}\sum_{l=1}^{n_i}\frac{1}{1+t_{il}\delta_i}$$

we therefore also have

$$\mathrm{E}\left[\left|(1 - \bar{g}_i f_i) - \left(1 - c_i + \frac{1}{N}\sum_{l=1}^{n_i}\frac{1}{1+t_{il}\delta_i}\right)\right|^4\right] = O\left(\frac{1}{N^2}\right).$$

The two relations above lead to

$$\begin{aligned}
&\mathrm{E}\left[\left|\bar{g}_i - \frac{1}{N}\sum_{l=1}^{n_i}\frac{t_{il}}{t_{il}f_i + 1 - f_i\bar{g}_i}\right|^4\right] \\
&= \mathrm{E}\left[\left|\frac{1}{N}\sum_{l=1}^{n_i} t_{il}\frac{t_{il}\left[f_i - \delta_i\kappa_i\right] + \left[1 - f_i\bar{g}_i - \kappa_i\right]}{\left[\kappa_i + t_{il}\delta_i\kappa_i\right]\left[t_{il}f_i + 1 - f_i\bar{g}_i\right]}\right|^4\right]
\end{aligned} \tag{6.48}$$

where we denoted $\kappa_i \triangleq 1 - c_i + \frac{1}{N}\sum_{l=1}^{n_i}\frac{1}{1+t_{il}\delta_i}$.

Again, all differences in the numerator converge to zero at a rate $O(1/N^2)$. However, the denominator presents now the term $t_{il}f_i + 1 - f_i\bar{g}_i$, which must be controlled and ensured to be away from zero. For this, we can notice that $\bar{g}_i \leq T/(1-c)$ by definition, while $f_i \leq R/|z|$, also by definition. It is therefore possible, by taking $z < 0$ sufficiently small, to ensure that $1 - f_i\bar{g}_i > 0$. We therefore from now on assume that such z are considered.

Equation (6.48) becomes in this case

$$\mathrm{E}\left[\left|\bar{g}_i - \frac{1}{N}\sum_{l=1}^{n_i}\frac{t_{il}}{t_{il}f_i + 1 - f_i\bar{g}_i}\right|^4\right] = O\left(\frac{1}{N^2}\right).$$

We are now ready to introduce the matrix $\mathbf{F}$. Consider

$$\mathbf{F} = \sum_{i=1}^{K}\bar{f}_i\mathbf{R}_i,$$

with $\bar{f}_i$ defined as the unique solution to the equation in x

$$x = \frac{1}{N}\sum_{l=1}^{n_i} \frac{t_{il}}{1 - f_i x + f_i t_{il}} \tag{6.49}$$

within the interval $0 \leq x < c_i/f_i$. To prove the uniqueness of the solution within this interval, note simply that

$$\frac{c_i}{f_i} \geq \frac{1}{N}\sum_{l=1}^{n_i} \frac{t_{il}}{1 - f_i(c_i/f_i) + f_i t_{il}}$$
$$0 \leq \frac{1}{N}\sum_{l=1}^{n_i} \frac{t_{il}}{1 - f_i \cdot 0 + f_i t_{il}}$$

and that the function $x \mapsto \frac{1}{N}\sum_{l=1}^{n_i} \frac{t_{il}}{1-f_i x+f_i t_{il}}$ is convex. Hence the uniqueness of the solution in $[0, c_i/f_i]$. We also show that this solution is an attractor of the fixed-point algorithm, when correctly initialized. Indeed, let $x_0, x_1, \ldots$ be defined by

$$x_{n+1} = \frac{1}{N}\sum_{l=1}^{n_i} \frac{t_{il}}{1 - f_i x_n + f_i t_{il}}$$

with $x_0 \in [0, c_i/f_i]$. Then, $x_n \in [0, c_i/f_i]$ implies $1 - f_i x_n + f_i t_{il} \geq 1 - c_i + f_i t_{il} > f_i t_{il}$ and therefore $f_i x_{n+1} \leq c_i$, so $x_0, x_1, \ldots$ are all contained in $[0, c_i/f_i]$. Now observe that

$$x_{n+1} - x_n = \frac{1}{N}\sum_{l=1}^{n_i} \frac{f_i(x_n - x_{n-1})}{(1 + t_{il} f_i - f_i x_n)(1 + t_{il} f_i - f_i x_{n-1})}$$

so that the differences $x_{n+1} - x_n$ and $x_n - x_{n-1}$ have the same sign. The sequence $x_0, x_1, \ldots$ is therefore monotonic and bounded: it converges. Calling x_∞ this limit, we have:

$$x_\infty = \frac{1}{N}\sum_{l=1}^{n_i} \frac{t_{il}}{1 + t_{il} f_i - f_i x_\infty}$$

as required.

To finally prove that $\frac{1}{N}\operatorname{tr}\mathbf{A}(\mathbf{B}_N - z\mathbf{I}_N)^{-1} - \frac{1}{N}\operatorname{tr}\mathbf{A}(\mathbf{F} - z\mathbf{I}_N)^{-1} \xrightarrow{\text{a.s.}} 0$, we want now to show that $\bar{g}_i - \bar{f}_i$ tends to zero at a sufficiently fast rate. For this, we write

$$\mathrm{E}\left[\left|\bar{g}_i - \bar{f}_i\right|^4\right] \leq 8\mathrm{E}\left[\left|\bar{g}_i - \frac{1}{N}\sum_{l=1}^{n_i} \frac{t_{il}}{t_{il} f_i + 1 - f_i \bar{g}_i}\right|^4\right]$$
$$+ 8\mathrm{E}\left[\left|\frac{1}{N}\sum_{l=1}^{n_i} \frac{t_{il}}{t_{il} f_i + 1 - f_i \bar{g}_i} - \frac{1}{N}\sum_{l=1}^{n_i} \frac{t_{il}}{t_{il} f_i + 1 - f_i \bar{f}_i}\right|^4\right]$$

$$= 8\mathrm{E}\left[\left|\bar{g}_i - \frac{1}{N}\sum_{l=1}^{n_i}\frac{t_{il}}{t_{il}f_i + 1 - f_i\bar{g}_i}\right|^4\right]$$

$$+\mathrm{E}\left[|\bar{g}_i - \bar{f}_i|^4\left|\frac{1}{N}\sum_{l=1}^{n_i}\frac{t_{il}f_i}{(t_{il}f_i + 1 - f_i\bar{f}_i)(t_{il}f_i + 1 - f_i\bar{g}_i)}\right|^4\right]. \tag{6.50}$$

We only need to ensure now that the coefficient multiplying $|\bar{g}_i - \bar{f}_i|$ in the right-hand side term is uniformly smaller than one. This unfolds again from noticing that the numerator can be made very small, with the denominator kept away from zero, for sufficiently small $z < 0$. For these z, we can therefore prove that

$$\mathrm{E}\left[|\bar{g}_i - \bar{f}_i|^4\right] = O\left(\frac{1}{N^2}\right).$$

It is important to notice that this holds essentially because we took $\bar{f}_i$ to be the unique solution of (6.49) lying in the interval $[0, c_i/f_i)$. The other solution (that happens to equal $1/f_i$ for $c_i = 1$) does not satisfy this fourth moment inequality.

Finally, we can proceed to proving the deterministic equivalent relations.

$$\frac{1}{N}\operatorname{tr}\mathbf{A}(\mathbf{G} - z\mathbf{I}_N)^{-1} - \frac{1}{N}\operatorname{tr}\mathbf{A}(\mathbf{F} - z\mathbf{I}_N)^{-1}$$

$$= \sum_{i=1}^{K}\frac{1}{N}\sum_{l=1}^{n_i} t_{il}\left[\frac{\frac{1}{N}\operatorname{tr}\mathbf{R}_i\mathbf{A}(\mathbf{G} - z\mathbf{I}_N)^{-1}(\mathbf{F} - z\mathbf{I}_N)^{-1}}{(1 - c_i + \frac{1}{N}\sum_{l'=1}^{n_i}\frac{1}{1 + t_{i,l'}\delta_i})(1 + t_{il}\delta_i)}\right.$$
$$\left. - \frac{\frac{1}{N}\operatorname{tr}\mathbf{R}_i\mathbf{A}(\mathbf{G} - z\mathbf{I}_N)^{-1}(\mathbf{F} - z\mathbf{I}_N)^{-1}}{1 - f_i\bar{f}_i + t_{il}f_i}\right]$$

$$= \sum_{i=1}^{K}\frac{1}{N}\sum_{l=1}^{n_i} t_{il}\left[\left(\frac{1}{(1 - c_i + \frac{1}{N}\sum_{l'=1}^{n_i}\frac{1}{1 + t_{i,l'}\delta_i})(1 + t_{il}\delta_i)} - \frac{1}{1 - f_i\bar{g}_i + t_{il}f_i}\right)\right.$$
$$\left. + \left(\frac{1}{1 - f_i\bar{g}_i + t_{il}f_i} - \frac{1}{1 - f_i\bar{f}_i + t_{il}f_i}\right)\right]\frac{1}{N}\operatorname{tr}\mathbf{R}_i\mathbf{A}(\mathbf{G} - z\mathbf{I}_N)^{-1}(\mathbf{F} - z\mathbf{I}_N)^{-1}.$$

The first difference in brackets is already known to be small from previous considerations on the relations between $\bar{g}_i$ and δ_i. As for the second difference, it also goes to zero fast as $\mathrm{E}[|\bar{g}_i - \bar{f}_i|^4]$ is summable. We therefore have

$$\mathrm{E}\left[\left|\frac{1}{N}\operatorname{tr}\mathbf{A}(\mathbf{G} - z\mathbf{I}_N)^{-1} - \frac{1}{N}\operatorname{tr}\mathbf{A}(\mathbf{F} - z\mathbf{I}_N)^{-1}\right|^4\right] = O\left(\frac{1}{N^2}\right).$$

Together with (6.47), we finally have

$$\mathrm{E}\left[\left|\frac{1}{N}\operatorname{tr}\mathbf{A}(\mathbf{B}_N - z\mathbf{I}_N)^{-1} - \frac{1}{N}\operatorname{tr}\mathbf{A}(\mathbf{F} - z\mathbf{I}_N)^{-1}\right|^4\right] = O\left(\frac{1}{N^2}\right).$$

Applying the Markov inequality, Theorem 3.5, and the Borel–Cantelli lemma, Theorem 3.6, this entails

$$\frac{1}{N}\operatorname{tr}\mathbf{A}\left(\mathbf{B}_N - z\mathbf{I}_N\right)^{-1} - \frac{1}{N}\operatorname{tr}\mathbf{A}\left(\mathbf{F} - z\mathbf{I}_N\right)^{-1} \xrightarrow{\text{a.s.}} 0 \tag{6.51}$$

as N grows large. This holds however to this point for a restricted set of negative z. But now, from the Vitali convergence theorem, Theorem 3.11, and the fact that $\frac{1}{N}\operatorname{tr}\mathbf{A}\left(\mathbf{B}_N - z\mathbf{I}_N\right)^{-1}$ and $\frac{1}{N}\operatorname{tr}\mathbf{A}\left(\mathbf{F} - z\mathbf{I}_N\right)^{-1}$ are uniformly bounded on all closed subset of $\mathbb{C}$ not containing the positive real half-line, we have that the convergence (6.51) holds true for all $z \in \mathbb{C}\setminus\mathbb{R}^+$, and that this convergence is uniform on all closed subsets of $\mathbb{C}\setminus\mathbb{R}^+$.

Applying the result for $\mathbf{A} = \mathbf{R}_j$, this is in particular

$$f_j - \frac{1}{N}\operatorname{tr}\mathbf{R}_j\left(\sum_{i=1}^{K}\bar{f}_i\mathbf{R}_i - z\mathbf{I}_N\right)^{-1} \xrightarrow{\text{a.s.}} 0$$

where we recall that $\bar{f}_i$ is the unique solution to

$$x = \frac{1}{N}\sum_{i=1}^{n_i}\frac{t_{il}}{1 - f_i x + t_{il} f_i}$$

within the set $[0, c_i/f_i)$.

For $\mathbf{A} = \mathbf{I}_N$, this says that

$$m_{\mathbf{B}_N}(z) - \frac{1}{N}\operatorname{tr}\mathbf{R}_j\left(\sum_{i=1}^{K}\bar{f}_i\mathbf{R}_i - z\mathbf{I}_N\right)^{-1} \xrightarrow{\text{a.s.}} 0$$

which proves the sought convergence of the Stieltjes transform. We now move to proving the existence and uniqueness of the set $(e_1,\ldots,e_K) = (e_1(z),\ldots,e_K(z))$.

Step 2: Existence and uniqueness

The existence step unfolds similarly as in the proof of Theorem 6.1. It suffices to consider the matrices $\mathbf{T}_{[p],i} \in \mathbb{C}^{n_i p}$ and $\mathbf{R}_{[p],i} \in \mathbb{C}^{Np}$ for all i defined as the Kronecker products $\mathbf{T}_{[p],i} \triangleq \mathbf{T}_i \otimes \mathbf{I}_p$, $\mathbf{R}_{[p],i} \triangleq \mathbf{R}_i \otimes \mathbf{I}_p$, which have, respectively, the d.f. $F^{\mathbf{T}_i}$ and $F^{\mathbf{R}_i}$ for all p. Similar to the i.i.d. case, it is easy to see that e_i is unchanged by substituting the $\mathbf{T}_{[p],i}$ and $\mathbf{R}_{[p],i}$ to the $\mathbf{T}_i$ and $\mathbf{R}_i$, respectively. Denoting in the same way $f_{[p],i}$ the equivalent of f_i for $\mathbf{T}_{[p],i}$ and $\mathbf{R}_{[p],i}$, from the convergence result of Step 1, we can choose $f_{[1],i}, f_{[2],i}, \ldots$ a sequence of the set of probability one where convergence is ensured as p grows large (N and the n_i are kept fixed). This sequence is uniformly bounded (by $R/|z|$) in $\mathbb{C}\setminus\mathbb{R}^+$, and therefore we can extract a converging subsequence out of it. The limit over this subsequence satisfies the fixed-point equation, which therefore proves existence. It is easy to see that the limit is also the Stieltjes transform of a finite measure on $\mathbb{R}^+$ by verifying the conditions of Theorem 3.2.

We will prove uniqueness of positive solutions $e_1,\ldots,e_K > 0$ for $z < 0$ and the convergence of the classical fixed-point algorithm to these values. We first introduce some notations and useful identities. Notice that, similar to Step 1 with the δ_i terms, we can define, for any pair of variables x_i and $\bar{x}_i$, with $\bar{x}_i$ defined as the solution y to $y = \frac{1}{N}\sum_{l=1}^{n_i} \frac{t_{il}}{1+x_j t_{il} - x_j y}$ such that $0 \leq y < c_j/x_j$, the auxiliary variables $\Delta_1, \ldots, \Delta_K$, with the properties

$$x_i = \Delta_i \left(1 - c_i + \frac{1}{N}\sum_{l=1}^{n_i} \frac{1}{1+t_{il}\Delta_i}\right) = \Delta_i \left(1 - \frac{1}{N}\sum_{l=1}^{n_i} \frac{t_{il}\Delta_i}{1+t_{il}\Delta_i}\right)$$

and

$$1 - x_i\bar{x}_i = 1 - c_i + \frac{1}{N}\sum_{l=1}^{n_i} \frac{1}{1+t_{il}\Delta_i} = 1 - \frac{1}{N}\sum_{l=1}^{n_i} \frac{t_{il}\Delta_i}{1+t_{il}\Delta_i}.$$

The uniqueness of the mapping between the x_i and Δ_i can be proved. In fact, it turns out that Δ_i is a monotonically increasing function of x_i with $\Delta_i = 0$ for $x_i = 0$.

We take the opportunity of the above definitions to notice that, for $x_i > x_i'$ and $\bar{x}_i'$, Δ_i' defined similarly as $\bar{x}_i$ and Δ_i

$$x_i\bar{x}_i - x_i'\bar{x}_i' = \frac{1}{N}\sum_{l=1}^{n_i} \frac{t_{il}(\Delta_i - \Delta_i')}{(1+t_{il}\Delta_i)(1+t_{il}\Delta_i')} > 0 \tag{6.52}$$

whenever $\mathbf{T}_i \neq 0$. Therefore $x_i\bar{x}_i$ is a growing function of x_i (or equivalently of Δ_i). This will turn out a useful remark later.

We are now in position to prove the step of uniqueness. Define, for $i \in \{1, \ldots, K\}$, the functions

$$h_i : (x_1, \ldots, x_K) \mapsto \frac{1}{N} \operatorname{tr} \mathbf{R}_i \left(\sum_{j=1}^{K} \bar{x}_j \mathbf{R}_j - z\mathbf{I}_N\right)^{-1}$$

with $\bar{x}_j$ the unique solution of the equation in y

$$y = \frac{1}{N}\sum_{l=1}^{n_j} \frac{t_{jl}}{1 + x_j t_{jl} - x_j y} \tag{6.53}$$

such that $0 \leq y \leq c_j/x_j$.

We will prove in the following that the multivariate function $\mathbf{h} = (h_1, \ldots, h_K)$ is a *standard function*, defined in [Yates, 1995], as follows.

Definition 6.2. A function $\mathbf{h}(x_1, \ldots, x_K) \in \mathbb{R}^K$, $\mathbf{h} = (h_1, \ldots, h_K)$, is said to be a standard function or a standard interference function if it fulfills the following conditions

1. *Positivity:* for all j, if $x_1, \ldots, x_K > 0$ then $h_j(x_1, \ldots, x_K) > 0$,
2. *Monotonicity:* if $x_1 > x_1', \ldots, x_K > x_K'$, then $h_j(x_1, \ldots, x_K) > h_j(x_1', \ldots, x_K')$, for all j,
3. *Scalability:* for all $\alpha > 1$ and j, $\alpha h_j(x_1, \ldots, x_K) > h_j(\alpha x_1, \ldots, \alpha x_K)$.

The important result regarding standard functions [Yates, 1995] is given as follows.

Theorem 6.18. *If a K-variate function $\mathbf{h}(x_1,\ldots,x_K)$ is standard and there exists $(x_1,\ldots,x_K)$ such that, for all j, $x_j \geq h_j(x_1,\ldots,x_K)$, then the fixed-point algorithm that consists in setting*

$$x_j^{(t+1)} = h_j(x_1^{(t)},\ldots,x_K^{(t)})$$

for $t \geq 1$ and for any initial values $x_1^{(0)},\ldots,x_K^{(0)} > 0$ converges to the unique jointly positive solution of the system of K equations

$$x_j = h_j(x_1,\ldots,x_K)$$

with $j \in \{1,\ldots,K\}$.

Proof. The proof of the uniqueness unfolds easily from the standard function assumptions. Take $(x_1,\ldots,x_K)$ and $(x'_1,\ldots,x'_K)$ two sets of supposedly distinct all positive solutions. Then there exists j such that $x_j < x'_j$, $\alpha x_j = x'_j$, and $\alpha x_i \geq x'_i$ for $i \neq j$. From monotonicity and scalability, it follows that

$$x'_j = h_j(x'_1,\ldots,x'_K) \leq h_j(\alpha x_1,\ldots,\alpha x_K) < \alpha h_j(x_1,\ldots,x_K) = \alpha x_j$$

a contradiction. The convergence of the fixed-point algorithm from any point $(x_1,\ldots,x_K)$ unfolds from similar arguments, see [Yates, 1995] for more details. □

Therefore, by showing that $\mathbf{h} \triangleq (h_1,\ldots,h_K)$ is standard, we will prove that the classical fixed-point algorithm converges to the unique set of positive solutions $e_1,\ldots,e_K$, when $z < 0$.

The positivity condition is straightforward as $\bar{x}_i$ is positive for x_i positive and therefore $h_j(x_1,\ldots,x_K)$ is always positive whenever $x_1,\ldots,x_K$ are.

The scalability is also rather direct. Let $\alpha > 1$, then:

$$\begin{aligned}
&\alpha h_j(x_1,\ldots,x_K) - h_j(\alpha x_1,\ldots,\alpha x_K)\\
&= \frac{1}{N}\operatorname{tr}\mathbf{R}_j\left(\sum_{k=1}^K \frac{\bar{x}_k}{\alpha}\mathbf{R}_k - \frac{z}{\alpha}\mathbf{I}_N\right)^{-1} - \frac{1}{N}\operatorname{tr}\mathbf{R}_j\left(\sum_{k=1}^K \bar{x}_k^{(a)}\mathbf{R}_k - z\mathbf{I}_N\right)^{-1}
\end{aligned}$$

where we denoted $\bar{x}_j^{(a)}$ the unique solution to (6.53) with x_j replaced by αx_j, within the set $[0, c_j/(\alpha x_j))$. Since $\alpha x_i > x_i$, from the property (6.52), we have $\alpha x_k \bar{x}_k^{(\alpha)} > x_k \bar{x}_k$ or equivalently $\bar{x}_k^{(a)} - \frac{\bar{x}_k}{\alpha} > 0$. We now define the two matrices $\mathbf{A} \triangleq \sum_{k=1}^k \frac{\bar{x}_k}{\alpha}\mathbf{R}_k - \frac{z}{\alpha}\mathbf{I}_N$ and $\mathbf{A}^{(\alpha)} \triangleq \sum_{k=1}^k \bar{x}_k^{(\alpha)}\mathbf{R}_k - z\mathbf{I}_N$. For any vector $\mathbf{a} \in \mathbb{C}^N$

$$\mathbf{a}^{\mathsf{H}}\left(\mathbf{A} - \mathbf{A}^{(\alpha)}\right)\mathbf{a} = \sum_{k=1}^K \left(\frac{\bar{x}_k}{\alpha} - \bar{x}_k^{(\alpha)}\right)\mathbf{a}^{\mathsf{H}}\mathbf{R}_k\mathbf{a} + z\left(1 - \frac{1}{\alpha}\right)\mathbf{a}^{\mathsf{H}}\mathbf{a} \leq 0$$

since $z < 0$, $1 - \frac{1}{\alpha} > 0$ and $\frac{\bar{x}_k}{\alpha} - \bar{x}_k^{(\alpha)} < 0$. Therefore $\mathbf{A} - \mathbf{A}^{(\alpha)}$ is non-positive definite. Now, from [Horn and Johnson, 1985, Corollary 7.7.4], this implies that $\mathbf{A}^{-1} - (\mathbf{A}^{(\alpha)})^{-1}$ is non-negative definite. Writing

$$\frac{1}{N}\operatorname{tr}\mathbf{R}_j\left(\mathbf{A}^{-1} - (\mathbf{A}^{(\alpha)})^{-1}\right) = \frac{1}{N}\sum_{i=1}^{N}\mathbf{r}_{j,i}^{\mathsf{H}}\left(\mathbf{A}^{-1} - (\mathbf{A}^{(\alpha)})^{-1}\right)\mathbf{r}_{j,i}$$

with $\mathbf{r}_{j,i}$ the ith column of $\mathbf{R}_j$, this ensures $\alpha h_j(x_1,\ldots,x_K) > h_j(\alpha x_1,\ldots,\alpha x_K)$.

The monotonicity requires some more lines of calculus. This unfolds from considering $\bar{x}_i$ as a function of Δ_i, by verifying that $\frac{d}{d\Delta_i}\bar{x}_i$ is negative.

$$\begin{aligned}
\frac{d}{d\Delta_i}\bar{x}_i &= \frac{1}{\Delta_i^2}\left(1 - \frac{1}{1 - \frac{1}{N}\sum_{l=1}^{n_i}\frac{t_{il}\Delta_i}{1+t_{il}\Delta_i}}\right) + \frac{1}{\Delta_i^2}\left(\frac{\frac{1}{N}\sum_{l=1}^{n_i}\frac{t_{il}\Delta_i}{(1+t_{il}\Delta_i)^2}}{\left(1 - \frac{1}{N}\sum_{l=1}^{n_i}\frac{t_{il}\Delta_i}{1+t_{il}\Delta_i}\right)^2}\right) \\
&= \frac{-\frac{1}{N}\sum_{l=1}^{n_i}\frac{t_{il}\Delta_i}{1+t_{il}\Delta_i}\left(1 - \frac{1}{N}\sum_{l=1}^{n_i}\frac{t_{il}\Delta_i}{1+t_{il}\Delta_i}\right) + \frac{1}{N}\sum_{l=1}^{n_i}\frac{t_{il}\Delta_i}{(1+t_{il}\Delta_i)^2}}{\Delta_i^2\left(1 - \frac{1}{N}\sum_{l=1}^{n_i}\frac{t_{il}\Delta_i}{1+t_{il}\Delta_i}\right)^2} \\
&= \frac{\left(\frac{1}{N}\sum_{l=1}^{n_i}\frac{t_{il}\Delta_i}{1+t_{il}\Delta_i}\right)^2 - \frac{1}{N}\sum_{l=1}^{n_i}\frac{t_{il}\Delta_i}{1+t_{il}\Delta_i} + \frac{1}{N}\sum_{l=1}^{n_i}\frac{t_{il}\Delta_i}{(1+t_{il}\Delta_i)^2}}{\Delta_i^2\left(1 - \frac{1}{N}\sum_{l=1}^{n_i}\frac{t_{il}\Delta_i}{1+t_{il}\Delta_i}\right)^2} \\
&= \frac{\left(\frac{1}{N}\sum_{l=1}^{n_i}\frac{t_{il}\Delta_i}{1+t_{il}\Delta_i}\right)^2 - \frac{1}{N}\sum_{l=1}^{n_i}\frac{(t_{il}\Delta_i)^2}{(1+t_{il}\Delta_i)^2}}{\Delta_i^2\left(1 - \frac{1}{N}\sum_{l=1}^{n_i}\frac{t_{il}\Delta_i}{1+t_{il}\Delta_i}\right)^2}.
\end{aligned}$$

From the Cauchy–Schwarz inequality, we have:

$$\begin{aligned}
\left(\sum_{l=1}^{n_i}\frac{1}{N}\frac{t_{il}\Delta_i}{1+t_{il}\Delta_i}\right)^2 &\leq \sum_{l=1}^{n_i}\frac{1}{N^2}\sum_{l=1}^{n_i}\frac{(t_{il}\Delta_i)^2}{(1+t_{il}\Delta_i)^2} \\
&= c_i\frac{1}{N}\sum_{l=1}^{n_i}\frac{(t_{il}\Delta_i)^2}{(1+t_{il}\Delta_i)^2} \\
&< \frac{1}{N}\sum_{l=1}^{n_i}\frac{(t_{il}\Delta_i)^2}{(1+t_{il}\Delta_i)^2}
\end{aligned}$$

which is sufficient to conclude that $\frac{d}{d\Delta_i}\bar{x}_i < 0$. Since Δ_i is an increasing function of x_i, we have that $\bar{x}_i$ is a decreasing function of x_i, i.e. $\frac{d}{dx_i}\bar{x}_i < 0$. This being said, using the same line of reasoning as for scalability, we finally have that, for two sets $x_1,\ldots,x_K$ and $x'_1,\ldots,x'_K$ of positive values such that $x_j > x'_j$

$$\begin{aligned}
&h_j(x_1,\ldots,x_K) - h(x'_1,\ldots,x'_K) \\
&= \frac{1}{N}\operatorname{tr}\mathbf{R}_j\left[\left(\sum_{k=1}^{K}\bar{x}_k\mathbf{R}_k - z\mathbf{I}_N\right)^{-1} - \left(\sum_{k=1}^{K}\bar{x}'_k\mathbf{R}_k - z\mathbf{I}_N\right)^{-1}\right] > 0
\end{aligned}$$

with $\bar{x}'_j$ defined equivalently as $\bar{x}_j$, and where the terms $(\bar{x}'_k - \bar{x}_k)$ are all positive due to negativity of $\frac{d}{dx_i}\bar{x}_i$. This proves the monotonicity condition.

We finally have from Theorem 6.18 that $(e_1, \ldots, e_K)$ is uniquely defined and that the classical fixed-point algorithm converges to this solution from any initialization point (remember that, at each step of the algorithm, the set $\bar{e}_1, \ldots, \bar{e}_K$ must be evaluated, possibly thanks to a further fixed-point algorithm).

Consider now two sets of *Stieltjes transforms* $(e_1(z), \ldots, e_K(z))$ and $(e'_1(z), \ldots, e'_K(z))$, $z \in \mathbb{C} \setminus \mathbb{R}^+$, functional solutions of the fixed-point Equation (6.39). Since $e_i(z) - e'_i(z) = 0$ for all i and for all $z < 0$, and $e_i(z) - e'_i(z)$ is holomorphic on $\mathbb{C} \setminus \mathbb{R}^+$ as the difference of Stieltjes transforms, by analytic continuation (see, e.g., [Rudin, 1986]), $e_i(z) - e'_i(z) = 0$ over $\mathbb{C} \setminus \mathbb{R}^+$. This therefore proves, in addition to *point-wise* uniqueness on the negative half-line, the uniqueness of the *Stieltjes transform* solution of the functional implicit equation and defined over $\mathbb{C} \setminus \mathbb{R}^+$.

We finally complete the proof by showing that the stochastic f_i and the deterministic e_i are asymptotically close to one another as N grows large.

Step 3: Convergence of $e_i - f_i$

For this step, we follow the approach in [Hachem *et al.*, 2007]. Denote

$$\varepsilon_N^i \triangleq f_i - \frac{1}{N} \operatorname{tr} \mathbf{R}_i \left(\sum_{k=1}^{K} \bar{f}_k \mathbf{R}_k - z\mathbf{I}_N \right)^{-1}$$

and recall the definitions of f_i, e_i, $\bar{f}_i$ and $\bar{e}_i$:

$$\begin{aligned}
f_i &= \frac{1}{N} \operatorname{tr} \mathbf{R}_i \left(\mathbf{B}_N - z\mathbf{I}_N\right)^{-1} \\
e_i &= \frac{1}{N} \operatorname{tr} \mathbf{R}_i \left(\sum_{j-1}^{K} \bar{e}_j \mathbf{R}_j - z\mathbf{I}_N \right)^{-1} \\
\bar{f}_i &= \frac{1}{N} \sum_{l=1}^{n_i} \frac{t_{i,l}}{1 - f_i \bar{f}_i + t_{i,l} f_i}, \qquad \bar{f}_i \in [0, c_i/f_i] \\
\bar{e}_i &= \frac{1}{N} \sum_{l=1}^{n_i} \frac{t_{i,l}}{1 - e_i \bar{e}_i + t_{i,l} e_i}, \qquad \bar{e}_i \in [0, c_i/e_i].
\end{aligned}$$

From the definitions above, we have the following set of inequalities

$$f_i \leq \frac{R}{|z|}, \quad e_i \leq \frac{R}{|z|}, \quad \bar{f}_i \leq \frac{T}{1 - c_i}, \quad \bar{e}_i \leq \frac{T}{1 - c_i}. \tag{6.54}$$

We will show in the sequel that

$$e_i - f_i \xrightarrow{\text{a.s.}} 0 \tag{6.55}$$

for all $i \in \{1,\ldots,N\}$. Write the following differences

$$f_i - e_i = \sum_{j=1}^{K} (\bar{e}_j - \bar{f}_j) \frac{1}{N} \operatorname{tr} \mathbf{R}_i \left[\sum_{k=1}^{K} \bar{e}_k \mathbf{R}_k - z\mathbf{I}_N \right]^{-1} \mathbf{R}_j \left[\sum_{k=1}^{K} \bar{f}_k \mathbf{R}_k - z\mathbf{I}_N \right]^{-1} + \varepsilon_N^i$$

$$\bar{e}_i - \bar{f}_i = \frac{1}{N} \sum_{l=1}^{n_i} \frac{t_{i,l}^2 (f_i - e_i) - t_{i,l} \left[f_i \bar{f}_i - e_i \bar{e}_i \right]}{(1 + t_{i,l} e_i - \bar{e}_i e_i)(1 + t_{i,l} f_i - \bar{f}_i f_i)}$$

and

$$f_i \bar{f}_i - e_i \bar{e}_i = \bar{f}_i (f_i - e_i) + e_i (\bar{f}_i - \bar{e}_i).$$

For notational convenience, we define the following values

$$\alpha \triangleq \sup_i \mathrm{E}\left[|f_i - e_i|^4 \right]$$

$$\bar{\alpha} \triangleq \sup_i \mathrm{E}\left[|\bar{f}_i - \bar{e}_i|^4 \right].$$

It is thus sufficient to show that α is summable to prove (6.55). By applying (6.54) to the absolute of the first difference, we obtain

$$|f_i - e_i| \leq \frac{KR^2}{|z|^2} \sup_i |\bar{f}_i - \bar{e}_i| + \sup_i |\varepsilon_N^i|$$

and hence

$$\alpha \leq \frac{8K^4 R^8}{|z|^8} \bar{\alpha} + \frac{8C}{N^2} \tag{6.56}$$

for some constant $C > 0$ such that $\mathrm{E}[|\sup_i \varepsilon_N^i|^4] \leq C/N^2$. This is possible since $\mathrm{E}[|\sup_i \varepsilon_N^i|^4] \leq 8K \sup_i \mathrm{E}[|\varepsilon_N^i|^4]$ and $\mathrm{E}[|\varepsilon_N^i|^4]$ has been proved to be of order $O(1/N^2)$. Similarly, we have for the third difference

$$\begin{aligned} |f_i \bar{f}_i - e_i \bar{e}_i| &\leq |\bar{f}_i| |f_i - e_i| + |e_i| |\bar{f}_i - \bar{e}_i| \\ &\leq \frac{T}{1-c} \sup_i |f_i - e_i| + \frac{R}{|z|} \sup_i |\bar{f}_i - \bar{e}_i| \end{aligned}$$

with c an upper bound on $\max_i \limsup_n c_i$, known to be inferior to one. This result can be used to upper bound the second difference term, which writes

$$\begin{aligned} |\bar{f}_i - \bar{e}_i| &\leq \frac{1}{(1-c)^2} \left(T^2 \sup_i |f_i - e_i| + T |f_i \bar{f}_i - e_i \bar{e}_i| \right) \\ &\leq \frac{1}{(1-c)^2} \left(T^2 \sup_i |f_i - e_i| + T \left[\frac{T}{1-c} \sup_i |f_i - e_i| + \frac{R}{|z|} \sup_i |\bar{f}_i - \bar{e}_i| \right] \right) \\ &= \frac{T^2 (2-c)}{(1-c)^3} \sup_i |f_i - e_i| + \frac{RT}{|z|(1-c)^2} \sup_i |\bar{f}_i - \bar{e}_i|. \end{aligned}$$

Hence

$$\bar{\alpha} \leq \frac{8T^8 (2-c)^4}{(1-c)^{12}} \alpha + \frac{8R^4 T^4}{|z|^4 (1-c)^8} \bar{\alpha}. \tag{6.57}$$

For a suitable z, satisfying $|z| > \frac{2RT}{(1-c)^2}$, we have $\frac{8R^4T^4}{|z|^4(1-c)^8} < 1/2$ and, thus, moving all terms proportional to α on the left

$$\bar{\alpha} < \frac{16T^8(2-c)^4}{(1-c)^{12}}\alpha.$$

Plugging this result into (6.56) yields

$$\alpha \leq \frac{128K^4R^8T^8(2-c)^4}{|z|^8(1-c)^{12}}\alpha + \frac{8C}{N^2}.$$

Take $0 < \varepsilon < 1$. It is easy to check that, for $|z| > \frac{128^{1/8}RT\sqrt{K(2-c)}}{(1-c)^{3/2}(1-\varepsilon)^{1/8}}$, $\frac{128K^4R^8T^8(2-c)^4}{|z|^8(1-c)^{12}} < 1-\varepsilon$ and thus

$$\alpha < \frac{8C}{\varepsilon N^2}. \tag{6.58}$$

Since C does not depend on N, α is clearly summable which, along with the Markov inequality and the Borel–Cantelli lemma, concludes the proof.

Finally, taking the same steps as above, we also have

$$\mathrm{E}\left[|m_{\mathbf{B}_N}(z) - m_N(z)|^4\right] \leq \frac{8C}{\varepsilon N^2}$$

for some $|z|$ large enough. The same conclusion therefore holds: for these z, $m_{\mathbf{B}_N}(z) - m_N(z) \xrightarrow{\text{a.s.}} 0$. From Vitali convergence theorem, since f_i and e_i are uniformly bounded on all closed sets of $\mathbb{C} \setminus \mathbb{R}^+$, we finally have that the convergence is true for all $z \in \mathbb{C} \setminus \mathbb{R}^+$. The almost sure convergence of the Stieltjes transform implies the almost sure weak convergence of $F^{\mathbf{B}_N} - F_N$ to zero, which is our final result. □

As a (not immediate) corollary of the proof above, we have the following result, important for application purposes, see Section 12.2.

Theorem 6.19. *Under the assumptions of Theorem 6.17 with $\mathbf{T}_i$ diagonal for all i, denoting $\mathbf{w}_{ij}$ the jth column of $\mathbf{W}_i$, t_{ij} the jth diagonal entry of $\mathbf{T}_i$, and $z \in \mathbb{C} \setminus \mathbb{R}^+$*

$$\mathbf{w}_{ij}^{\mathsf{H}}\mathbf{H}_i^{\mathsf{H}}\left(\mathbf{B}_N - t_{ij}\mathbf{H}_i\mathbf{w}_{ij}\mathbf{w}_{ij}^{\mathsf{H}}\mathbf{H}_i^{\mathsf{H}} - z\mathbf{I}_N\right)^{-1}\mathbf{H}_i\mathbf{w}_{ij} - \frac{e_i(z)}{\bar{c}_i - e_i(z)\bar{e}_i(z)} \xrightarrow{\text{a.s.}} 0. \tag{6.59}$$

where $e_i(z)$ and $\bar{e}_i(z)$ are defined in Theorem 6.17.

Similar to the i.i.d. case, a deterministic equivalent for the Shannon transform can be derived. This is given by the following proposition.

Theorem 6.20. *Under the assumptions of Theorem 6.17 with $z = -1/x$, for $x > 0$, denoting*

$$\mathcal{V}_{\mathbf{B}_N}(x) = \frac{1}{N}\log\det\left(x\mathbf{B}_N + \mathbf{I}_N\right)$$

the Shannon transform of $\mathbf{B}_N$, *we have:*

$$\mathcal{V}_{\mathbf{B}_N}(x) - \mathcal{V}_N(x) \xrightarrow{\text{a.s.}} 0$$

where

$$\mathcal{V}_N(x) = \frac{1}{N} \log\det\left(\mathbf{I}_N + x\sum_{i=1}^{K} \bar{e}_i \mathbf{R}_i\right) + \sum_{i=1}^{K} \frac{1}{N} \log\det\left([\bar{c}_i - e_i\bar{e}_i]\mathbf{I}_{n_i} + e_i\mathbf{T}_i\right)$$
$$+ \sum_{i=1}^{K} \left[(1-c_i)\log(\bar{c}_i - e_i\bar{e}_i) - \bar{c}_i \log(\bar{c}_i)\right]. \tag{6.60}$$

The proof for the deterministic equivalent of the Shannon transform follows from similar considerations as for the i.i.d. case, see Theorem 6.4 and Corollary 6.1, and is detailed below.

Proof. For the proof of Theorem 6.20, we again take $\bar{c}_i = 1$, $\mathbf{R}_i$ deterministic of bounded spectral norm for simplicity and we assume $c_i \leq 1$ from the beginning, the trace lemma, Theorem 6.15, being unused here. First note that the system of Equations (6.39) is unchanged if we extend the $\mathbf{T}_i$ matrices into $N \times N$ diagonal matrices filled with $N - n_i$ zero eigenvalues. Therefore, we can assume that all $\mathbf{T}_i$ have size $N \times N$, although we restrict the $F^{\mathbf{T}_i}$ to have a mass $1 - c_i$ in zero. Since this does not alter the Equations (6.39), we have in particular $\bar{e}_i < 1/e_i$. This being said, (6.60) now needs to be rewritten

$$\mathcal{V}_N(x) = \frac{1}{N} \log\det\left(\mathbf{I}_N + x\sum_{i=1}^{K} \bar{e}_i \mathbf{R}_i\right) + \sum_{i=1}^{K} \frac{1}{N} \log\det\left([1 - e_i\bar{e}_i]\mathbf{I}_N + e_i\mathbf{T}_i\right).$$

Calling V the function

$$V : (x_1, \ldots, x_K, \bar{x}_1, \ldots, \bar{x}_K, x) \mapsto \frac{1}{N} \log\det\left(\mathbf{I}_N + x\sum_{i=1}^{K} \bar{x}_i \mathbf{R}_i\right)$$
$$+ \sum_{i=1}^{K} \frac{1}{N} \log\det\left([1 - x_i\bar{x}_i]\mathbf{I}_N + x_i\mathbf{T}_i\right)$$

we have:

$$\frac{\partial V}{\partial x_i}(e_1, \ldots, e_K, \bar{e}_1, \ldots, \bar{e}_K, x) = \bar{e}_i - \bar{e}_i \frac{1}{N}\sum_{l=1}^{N} \frac{1}{1 - e_i\bar{e}_i + e_i t_{il}}$$

$$\frac{\partial V}{\partial \bar{x}_i}(e_1, \ldots, e_K, \bar{e}_1, \ldots, \bar{e}_K, x) = e_i - e_i \frac{1}{N}\sum_{l=1}^{N} \frac{1}{1 - e_i\bar{e}_i + e_i t_{il}}.$$

Noticing now that

$$1 = \frac{1}{N}\sum_{l=1}^{N} \frac{1 - e_i\bar{e}_i + e_i t_{il}}{1 - e_i\bar{e}_i + e_i t_{il}} = (1 - e_i\bar{e}_i)\frac{1}{N}\sum_{l=1}^{N} \frac{1}{1 - e_i\bar{e}_i + e_i t_{il}} + e_i\bar{e}_i$$

we have:

$$(1 - e_i\bar{e}_i)\left(1 - \frac{1}{N}\sum_{l=1}^{N}\frac{1}{1 - e_i\bar{e}_i + e_i t_{il}}\right) = 0.$$

But we also know that $0 \leq \bar{e}_i < 1/e_i$ and therefore $1 - e_i\bar{e}_i > 0$. This entails

$$\frac{1}{N}\sum_{l=1}^{N}\frac{1}{1 - e_i\bar{e}_i + e_i t_{il}} = 1. \tag{6.61}$$

From (6.61), we conclude that

$$\frac{\partial V}{\partial x_i}(e_1, \ldots, e_K, \bar{e}_1, \ldots, \bar{e}_K, x) = 0$$
$$\frac{\partial V}{\partial \bar{x}_i}(e_1, \ldots, e_K, \bar{e}_1, \ldots, \bar{e}_K, x) = 0.$$

We therefore have that

$$\begin{aligned}\frac{d}{dx}\mathcal{V}_N(x) &= \sum_{i=1}^{K}\left[\frac{\partial V}{\partial e_i}\frac{\partial e_i}{\partial x} + \frac{\partial V}{\partial \bar{e}_i}\frac{\partial \bar{e}_i}{\partial x}\right] + \frac{\partial V}{\partial x}\\ &= \frac{\partial V}{\partial x}\\ &= \sum_{i=1}^{K}\bar{e}_i\frac{1}{N}\operatorname{tr}\mathbf{R}_i\left(\mathbf{I}_N + x\sum_{j=1}^{K}\bar{e}_j\mathbf{R}_j\right)^{-1}\\ &= \frac{1}{x} - \frac{1}{x^2}\frac{1}{N}\operatorname{tr}\left(\frac{1}{x}\mathbf{I}_N + \sum_{j=1}^{K}\bar{e}_j\mathbf{R}_j\right)^{-1}.\end{aligned}$$

Therefore, along with the fact that $\mathcal{V}_N(0) = 0$, we have:

$$\mathcal{V}_N(x) = \int_0^x\left(\frac{1}{t} - \frac{1}{t^2}m_N\left(-\frac{1}{t}\right)\right)dt$$

and therefore $\mathcal{V}_N(x)$ is the Shannon transform of F_N, according to Definition 3.2.

In order to prove the almost sure convergence $\mathcal{V}_{\mathbf{B}_N}(x) - \mathcal{V}_N(x) \xrightarrow{\text{a.s.}} 0$, we need simply to notice that the support of the eigenvalues of $\mathbf{B}_N$ is bounded. Indeed, the non-zero eigenvalues of $\mathbf{W}_i\mathbf{W}_i^{\mathsf{H}}$ have unit modulus and therefore $\|\mathbf{B}_N\| \leq KTR$. Similarly, thc support of F_N is the support of the eigenvalues of $\sum_{i=1}^{K}\bar{e}_i\mathbf{R}_i$, which are bounded by KTR as well.

As a consequence, for $\mathbf{B}_1, \mathbf{B}_2, \ldots$ a realization for which $F^{\mathbf{B}_N} - F_N \Rightarrow 0$, we have, from the dominated convergence theorem, Theorem 6.3

$$\int_0^{\infty}\log(1 + xt)\, d[F^{\mathbf{B}_N} - F_N](t) \to 0.$$

Hence the almost sure convergence. □

Applications of the above results are found in various telecommunication systems employing random isometric precoders, such as random CDMA, SDMA [Couillet *et al.*, 2011b]. A specific application to assess the optimal number of stream transmissions in multi-antenna interference channels is in particular provided in [Hoydis *et al.*, 2011a], where an extension of Theorem 6.17 to correlated i.i.d. channel matrices $\mathbf{H}_i$ is provided. It is worth mentioning that the approach followed in [Hoydis *et al.*, 2011a] to prove this extension relies on an "inclusion" of the deterministic equivalent of Theorem 6.12 into the deterministic equivalent of Theorem 6.17. The final result takes a surprisingly simple expression and the proof of existence, uniqueness, and convergence of the implicit equations obtained do not require much effort. This "deterministic equivalent of a deterministic equivalent" approach is very natural and is expected to lead to very simple results even for intricate communication models; recall e.g. Theorem 6.9.

We conclude this chapter on deterministic equivalents by a central limit theorem for the Shannon transform of the non-centered random matrix with variance profile of Theorem 6.14.

6.3 A central limit theorem

Central limit theorems are also demanded for more general models than the sample covariance matrix of Theorem 3.17. In wireless communications, it is particularly interesting to study the limiting distribution of the Shannon transform of doubly correlated random matrices, e.g. to mimic Kronecker models, or even more generally matrices of i.i.d. entries with a variance profile. Indeed, the later allows us to study, in addition to the large dimensional ergodic capacity of Rician MIMO channels, as provided by Theorem 6.14, the large dimensional *outage* mutual information of such channels. In [Hachem *et al.*, 2008b], Hachem *et al.* provide the central limit theorem for the Shannon transform of this model.

Theorem 6.21 ([Hachem *et al.*, 2008b]). *Let $\mathbf{Y}_N$ be $N \times n$ whose (i,j)th entry is given by:*

$$Y_{N,ij} = \frac{\sigma_{ij}(n)}{\sqrt{n}} X_{N,ij}$$

with $\{\sigma_{ij}(n)\}_{ij}$ uniformly bounded with respect to n, and $X_{N,ij}$ is the (i,j)th entry of an $N \times n$ matrix $\mathbf{X}_N$ with i.i.d. entries of zero mean, unit variance, and finite eighth order moment. Denote $\mathbf{B}_N = \mathbf{Y}_N \mathbf{Y}_N^{\mathsf{H}}$. We then have, as $N, n \to \infty$ with limit ratio $c = \lim_N N/n$, that the Shannon transform

$$\mathcal{V}_{\mathbf{B}_N}(x) \triangleq \frac{1}{N} \log\det(\mathbf{I}_N + x\mathbf{B}_N)$$

of $\mathbf{B}_N$ *satisfies*

$$\frac{N}{\theta_n}(\mathcal{V}_{\mathbf{B}_N}(x) - \mathrm{E}[\mathcal{V}_{\mathbf{B}_N}(x)]) \Rightarrow X \sim \mathcal{N}(0,1)$$

with

$$\theta_n^2 = -\log\det(\mathbf{I}_n - \mathbf{J}_n) + \kappa\,\mathrm{tr}(\mathbf{J}_n)$$

$\kappa = \mathrm{E}[(X_{N,11})^4] - 3\mathrm{E}[(X_{N,11})^3]$ *for real* $X_{N,11}$, $\kappa = \mathrm{E}[|X_{N,11}|^4] - 2\mathrm{E}[|X_{N,11}|^2]$ *for complex* $X_{N,11}$, *and* $\mathbf{J}_n$ *the matrix with* (i,j)th *entry*

$$J_{n,ij} = \frac{1}{n}\frac{\frac{1}{n}\sum_{k=1}^N \sigma_{ki}^2(n)\sigma_{kj}^2(n)t_k(-1/x)^2}{\left(1+\frac{1}{n}\sum_{k=1}^N \sigma_{ki}^2(n)t_k(-1/x)\right)^2}$$

with $t_i(z)$ *such that* $(t_1(z),\ldots,t_N(z))$ *is the unique Stieltjes transform vector solution of*

$$t_i(z) = \left(-z + \frac{1}{n}\sum_{j=1}^n \frac{\sigma_{ij}^2(n)}{1+\frac{1}{n}\sum_{l=1}^N \sigma_{lj}^2(n)t_l(z)}\right)^{-1}.$$

Observe that the matrix $\mathbf{J}_n$ is in fact the Jacobian matrix associated with the fundamental equations in the $e_{N,i}(z)$, defined in the implicit relations (6.29) of Theorem 6.10 as

$$e_{N,i}(z) = \frac{1}{n}\sum_{k=1}^N \sigma_{ki}^2(n)t_k(z) = \frac{1}{n}\sum_{k=1}^N \sigma_{ki}^2(n)\frac{1}{-z+\frac{1}{n}\sum_{l=1}^n \frac{\sigma_{kl}^2(n)}{1+e_{N,l}(z)}}.$$

Indeed, for all $e_{N,k}(-1/x)$ fixed but $e_{N,j}(-1/x)$, we have:

$$\begin{aligned}
&\frac{\partial}{\partial e_{N,j}(-1/x)}\left[\frac{1}{n}\sum_{k=1}^N \sigma_{ki}^2(n)\frac{1}{\frac{1}{x}+\frac{1}{n}\sum_{l=1}^n \frac{\sigma_{kl}^2(n)}{1+e_{N,l}(-1/x)}}\right] \\
&= \frac{1}{n}\sum_{k=1}^N \sigma_{ki}^2(n)\frac{\frac{1}{n}\sigma_{kj}^2(n)}{(1+e_{N,j}(-1/x))^2}\frac{1}{\left(\frac{1}{x}+\frac{1}{n}\sum_{l=1}^n \frac{\sigma_{kl}^2(n)}{1+e_{N,l}(-1/x)}\right)^2} \\
&= \frac{1}{n}\sum_{k=1}^N \frac{\frac{1}{n}\sigma_{ki}^2(n)\sigma_{kj}^2(n)t_k(-1/x)^2}{(1+e_{N,j}(-1/x))^2} \\
&= J_{n,ji}.
\end{aligned}$$

So far, this observation seems to generalize to all central limits derived for random matrix models with independent entries. This is however only an intriguing but yet unproven fact.

Similar to Theorem 3.17, [Hachem *et al.*, 2008b] provides more than an asymptotic central limit theorem for the Shannon transform of the information plus noise model $\mathcal{V}_{\mathbf{B}_N} - \mathrm{E}[\mathcal{V}_{\mathbf{B}_N}]$, but also the fluctuations for N large of the difference between $\mathcal{V}_{\mathbf{B}_N}$ and its deterministic equivalent $\mathcal{V}_N$, provided in [Hachem *et al.*, 2007]. In the case where $\mathbf{X}_N$ has Gaussian entries, this takes a very compact expression.

Theorem 6.22. *Under the conditions of Theorem 6.21 with the additional assumption that the entries of $\mathbf{X}_N$ are complex Gaussian, we have:*

$$\frac{N}{\sqrt{-\log\det\left(\mathbf{I}_n - \mathbf{J}_n\right)}}\left(\mathcal{V}_{\mathbf{B}_N}(x) - \mathcal{V}_N(x)\right) \Rightarrow X \sim \mathcal{N}(0,1)$$

where $\mathcal{V}_N$ is defined as

$$\mathcal{V}_N(x) = \frac{1}{N}\sum_{i=1}^{N}\log\left(\frac{x}{t_i(-1/x)}\right) + \frac{1}{N}\sum_{j=1}^{n}\log\left(1 + \frac{1}{n}\sum_{l=1}^{N}\sigma_{lj}^2(n)t_l(-1/x)\right)$$
$$- \frac{1}{Nn}\sum_{\substack{1\leq i\leq N\\ 1\leq j\leq n}}\frac{\sigma_{ij}^2(n)t_i(-1/x)}{1+\frac{1}{n}\sum_{l=1}^{N}\sigma_{lj}^2(n)t_l(-1/x)}$$

with $t_1,\ldots,t_N$ and $\mathbf{J}_n$ defined as in Theorem 6.21.

The generalization to distributions of the entries of $\mathbf{X}_N$ with a non-zero kurtosis κ introduces an additional bias term corresponding to the limiting variations of $N(\mathrm{E}[\mathcal{V}_{\mathbf{B}_N}(x)] - \mathcal{V}_N(x))$. This converges instead to zero in the Gaussian case or, as a matter of fact, in the case of any distribution with null kurtosis.

This concludes this short section on central limit theorems for deterministic equivalents.

This also closes this chapter on the classical techniques used for deterministic equivalents, when there exists no limit to the e.s.d. of the random matrix under study. Those deterministic equivalents are seen today as one of the most powerful tools to evaluate the performance of large wireless communication systems encompassing multiple antennas, multiple users, multiple cells, random codes, fast fading channels, etc. which are studied with scrutiny in Part II. In order to study complicated system models involving e.g. doubly-scattering channels, multi-hop channels, random precoders in random channels, etc., the current trend is to study nested deterministic equivalents; that is, deterministic equivalents that account for the stochasticity of multiple independent random matrices, see e.g. Hoydis *et al.* [2011a,b].

In the following, we turn to a rather different subject and study more deeply the limiting spectra of the sample covariance matrix model and of the information plus noise model. For these, much more than limiting spectral densities is known. It has especially been proved that, under some mild conditions, the extreme eigenvalues for both models do not escape the support of the l.s.d. and that a precise characterization of the position of some eigenvalues can be determined. Some additional study will characterize precisely the links between the population covariance matrix (or the information matrix) and the sample covariance matrix (or the information plus noise matrix), which are fundamental to address the questions of inverse problems and more precisely statistical eigen-inference for large dimensional random matrix models. These questions are at the core of the very recent signal processing tools, which enable novel signal sensing techniques and (N, n)-consistent estimation procedures adapted to large dimensional networks.

7 Spectrum analysis

In this chapter, we further study the spectra of the important random matrix models for wireless communications that are the sample covariance matrix and the information plus noise models. It has already been shown in Chapter 3 that, as the e.s.d. of the population covariance matrix (or of the information matrix) converges, the e.s.d. of the sample covariance matrix (or the information plus noise matrix) converges almost surely. The limiting d.f. can then be fully characterized as a function of the l.s.d. of the population covariance matrix (or of the information matrix). It is however not convenient to invert the problem and to describe the l.s.d. of the population covariance matrix (or of the information matrix) as a function of the l.s.d. of the observed matrices. The answer to this inverse problem is provided in Chapter 8, which however requires some effort to be fully accessible. The development of the tools necessary for the statistical eigen-inference methods of Chapter 8 is one of the motivations of the current chapter.

The starting motivation, initiated by the work of Silverstein and Choi [Silverstein and Choi, 1995], which resulted in the important Theorem 7.4 (accompanied later by an important corollary, due to Mestre [Mestre, 2008a], Theorem 7.5), was to characterize the l.s.d. of the sample covariance matrix in closed-form. Remember that, up to this point, we can only characterize the l.s.d. F of a sample covariance matrix through the expression of its Stieltjes transform, as the unique solution $m_F(z)$ of some fixed-point equation for all $z \in \mathbb{C} \setminus \mathbb{R}^+$. To obtain an explicit expression of F, it therefore suffices to use the inverse Stieltjes transform formula (3.2). However, this suggests having a closer look at the limiting behavior of $m_F(z)$ as z approaches the positive real half-line, about which we do not know much yet. Therefore, up to this point in our analysis, it is impossible to describe the support of the l.s.d., apart from rough estimations based on the expression of $\Im[m_F(z)]$, for $z = x + iy$, y being small. It is also not convenient to depict $F'(x)$: the solution is to take $z = x + iy$, with y small and x spanning from zero to infinity, and to draw the curve $z \mapsto \frac{1}{\pi}\Im[m_F(z)]$ for such z. In the following, we will show that, as z tends to $x > 0$, $m_F(z)$ has a limit which can be characterized in two different ways, depending on whether x belongs to the support of F or not. In any case, this limit is also characterized as the solution to an implicit equation, although particular care must be taken as to which of the multiple solutions of this implicit equation needs to be considered.

Before we detail this advanced spectrum characterization, we provide a different set of results, fundamental to the validation of the eigen-inference methods proposed in Chapter 8. These results, namely the asymptotic absence of eigenvalues outside the support of F, Theorem 7.1, and the exact separation of the support into disjoints clusters, Theorem 7.2, are once more due to Bai and Silverstein [Bai and Silverstein, 1998, 1999]. Their object is the analysis, on top of the characterization of F, of the behavior of the particular eigenvalues of the e.s.d. of the sample covariance matrix as the dimensions grow large. It is fundamental to understand here, and this will be reminded again in the next section, that the convergence of the e.s.d. toward F, as the matrix size N grows large, does *not* imply the convergence of the largest eigenvalue of the sample covariance matrix towards the right edge of the support. Indeed, the largest eigenvalues, having weight $1/N$ in the spectrum, do not contribute asymptotically to the support of F. As such, it may well be found outside the support of F for all finite N, without invalidating Theorem 3.13. This particular case in the Marčenko–Pastur model where eigenvalues are found outside the support almost surely when the entries of the random i.i.d. matrix $\mathbf{X}_N$ in Theorem 3.13, $\mathbf{T}_N = \mathbf{I}_N$, have infinite fourth order moment. In this scenario, it is even proved in [Silverstein *et al.*, 1988] that the largest eigenvalue grows without bound as the system dimensions grow to infinity, while all the mass of the l.s.d. is asymptotically kept in the support; if the fourth order moment is finite. Under finite fourth moment assumption though [Bai and Silverstein, 1998; Yin *et al.*, 1988], the important result to be detailed below is that no eigenvalue is to be found outside the limiting support and that the eigenvalues are found where they ought to be. This last statement is in fact slightly erroneous and will be adequately corrected when discussing the *spiked* models that lead some eigenvalues to leave the limiting support. To be more precise, when the moment of order four of the entries of $\mathbf{X}_N$ exists, we can characterize exactly the subsets of $\mathbb{R}^+$ where no eigenvalue is asymptotically found, almost surely. Further discussions on the extreme eigenvalues of sample covariance matrices are provided in Chapter 9, where (non-central) limiting theorems for the distribution of these eigenvalues are provided.

7.1 Sample covariance matrix

7.1.1 No eigenvalues outside the support

As observed in the previous sections, most early results of random matrix theory dealt with the limiting behavior of e.s.d. For instance, the Marčenko–Pastur law ensures that the e.s.d. of the sample covariance matrix $\mathbf{R}_N$ of vectors with i.i.d. entries of zero mean and unit variance converges almost surely towards a limit distribution function F. However, the Marčenko–Pastur law does not say anything about the behavior of any specific eigenvalue, say for instance the extreme lowest and largest eigenvalues $\lambda_{\min}$ and $\lambda_{\max}$ of $\mathbf{R}_N$. It is relevant in particular to wonder whether $\lambda_{\min}$ and $\lambda_{\max}$ can be asymptotically found *outside* the support of F. Indeed, if all eigenvalues but the extreme two are in the support of F, then the l.s.d. of $\mathbf{R}_N$ is still F, which is still consistent with the Marčenko–Pastur law. It turns out that this is not the case in general. Under some mild assumption on the entries of the sample covariance matrix, no eigenvalue is found outside the support. We specifically have the following theorem.

Theorem 7.1 ([Bai and Silverstein, 1998; Yin *et al.*, 1988]). *Let the matrix* $\mathbf{X}_N = \left(\frac{1}{\sqrt{n}} X_{N,ij}\right) \in \mathbb{C}^{N\times n}$ *have i.i.d. entries, such that* $X_{N,11}$ *has zero mean, unit variance, and finite fourth order moment. Let* $\mathbf{T}_N \in \mathbb{C}^{N\times N}$ *be non-random, with uniformly bounded spectral norm* $\|\mathbf{T}_N\|$, *whose e.s.d.* $F^{\mathbf{T}_N}$ *converge weakly to* H. *From Theorem 3.13, the e.s.d. of* $\mathbf{B}_N = \mathbf{T}_N^{\frac{1}{2}} \mathbf{X}_N \mathbf{X}_N^{\mathsf{H}} \mathbf{T}_N^{\frac{1}{2}} \in \mathbb{C}^{N\times N}$ *converges weakly and almost surely towards some distribution function* F, *as* N, n *go to infinity with ratio* $c_N = N/n \to c$, $0 < c < \infty$. *Similarly, the e.s.d. of* $\underline{\mathbf{B}}_N = \mathbf{X}_N^{\mathsf{H}} \mathbf{T}_N \mathbf{X}_N \in \mathbb{C}^{n\times n}$ *converges towards* $\underline{F}$ *given by:*

$$\underline{F}(x) = cF(x) + (1-c)1_{[0,\infty)}(x).$$

Denote $\underline{F}_N$ *the distribution with Stieltjes transform* $m_{\underline{F}_N}(z)$, *which is solution, for* $z \in \mathbb{C}^+$, *of the following equation in* m

$$m = -\left(z - \frac{N}{n}\int \frac{\tau}{1+\tau m} dF^{\mathbf{T}_N}(\tau)\right)^{-1} \tag{7.1}$$

and define F_N *the d.f. such that*

$$\underline{F}_N(x) = \frac{N}{n} F_N(x) + \left(1 - \frac{N}{n}\right) 1_{[0,\infty)}(x).$$

Let $N_0 \in \mathbb{N}$, *and choose an interval* $[a,b]$, $a, b \in (0,\infty]$, *lying in an open interval outside the union of the supports of* F *and* F_N *for all* $N \geq N_0$. *For* $\omega \in \Omega$, *the random space generating the series* $\mathbf{X}_1, \mathbf{X}_2, \ldots$, *denote* $\mathcal{L}_N(\omega)$ *the set of eigenvalues of* $\mathbf{B}_N(\omega)$. *Then*

$$P(\{\omega, \mathcal{L}_N(\omega) \cap [a,b] \neq \emptyset \text{ i.o.}\}) = 0.$$

This means concretely that, given a segment $[a,b]$ outside the union of the supports of F and $F_{N_0}, F_{N_0+1}, \ldots$, for all series $\mathbf{B}_1(\omega), \mathbf{B}_2(\omega), \ldots$, with ω in some set of probability one, there exists $M(\omega)$ such that, for all $N \geq M(\omega)$, there will be no eigenvalue of $\mathbf{B}_N(\omega)$ in $[a,b]$. By definition, F_K is the l.s.d. of an hypothetical $\mathbf{B}_N$ with $H = F^{\mathbf{T}_K}$. The necessity to consider the supports of $F_{N_0}, F_{N_0+1}, \ldots$ is essential when a few eigenvalues of $\mathbf{T}_N$ are isolated and eventually contribute with probability zero to the l.s.d. H. Indeed, it is rather intuitive that, if the largest eigenvalue of $\mathbf{T}_N$ is large compared to the rest, at least one eigenvalue of $\mathbf{B}_N$ will also be large compared to the rest (take $n \gg N$ to be convinced). Theorem 7.1 states exactly here that there will be neither any eigenvalue outside the support of the main mass of $F^{\mathbf{B}_N}$, nor any eigenvalue around the largest one. Those models in which some eigenvalues of $\mathbf{T}_N$ are isolated are referred to as *spiked models*. These are thoroughly discussed in Chapter 9. In wireless communications and modern signal processing, Theorem 7.1 is of key importance for signal sensing and hypothesis testing methods since it allows us to verify whether the eigenvalues empirically found in sample covariance matrix spectra originate either from noise contributions or from signal sources. In the simple case where signals sensed at an antenna array originate either from white noise or from a coherent signal source impaired by white noise, this can be performed by simply verifying if the extreme eigenvalue of the sample covariance matrix is inside or outside the support of the Marčenko–Pastur law (Figure 1.1); see further Chapter 16.

We give hereafter a sketch of the proof, which again only involves the Stieltjes transform.

Proof. Surprisingly, the proof unfolds from a mere (though non-trivial) refinement of the Stieltjes transform relation proved in Theorem 3.13. Let F_N be defined as above and let m_N be its Stieltjes transform. It is possible to show that, for $z = x + iv_N$, with $v_N = N^{-1/68}$

$$\sup_{x\in[a,b]} |m_{\mathbf{B}_N}(z) - m_N(z)| = o\left(\frac{1}{N}v_N\right)$$

almost surely. This result is in fact also true when $\Im[z]$ equals $\sqrt{2}v_N, \sqrt{3}v_N, \ldots$ or $\sqrt{34}v_N$. Note that this refines the known statement that the difference is of order $o(1)$. We take this property, which requires more than ten pages of calculus, for granted. We now have that

$$\max_{1\leq k\leq 34} \sup_{x\in[a,b]} \left| m_{\mathbf{B}_N}(x + ik^{\frac{1}{2}}v_N) - m_N(x + ik^{\frac{1}{2}}v_N)\right| = o(v_N^{67})$$

almost surely. Expanding the Stieltjes transforms and considering only the imaginary parts, we obtain

$$\max_{1\leq k\leq 34} \sup_{x\in[a,b]} \left| \int \frac{d(F^{\mathbf{B}_N}(\lambda) - F_N(\lambda))}{(x-\lambda)^2 + kv_N^2} \right| = o(v_N^{66})$$

almost surely. Taking successive differences over the 34 values of k, we end up with

$$\sup_{x\in[a,b]} \left| \int \frac{(v_N^2)^{33} d(F^{\mathbf{B}_N}(\lambda) - F_N(\lambda))}{\prod_{k=1}^{34}((x-\lambda)^2 + kv_N^2)} \right| = o(v_N^{66}) \tag{7.2}$$

almost surely, from which the term v_N^{66} simplifies on both sides. Consider now $a' < a$ and $b' > b$ such that $[a', b']$ is outside the support of F. We then divide

(7.2) into two terms, as (remember that $1/N = v_N^{68}$)

$$\sup_{x\in[a,b]} \left| \int \frac{1_{\mathbb{R}^+\setminus[a',b']}(\lambda)d(F^{\mathbf{B}_N}(\lambda) - F_N(\lambda))}{\prod_{k=1}^{34}((x-\lambda)^2 + kv_N^2)} + \sum_{\lambda_j\in[a',b']} \frac{v_N^{68}}{\prod_{k=1}^{34}((x-\lambda_j)^2 + kv_N^2)} \right|$$
$$= o(1)$$

almost surely. Assume now that, for a subsequence $\phi(1), \phi(2), \ldots$ of $1, 2, \ldots$, there always exists at least one eigenvalue of $\mathbf{B}_{\phi(N)}$ in $[a,b]$. Then, for x taken equal to this eigenvalue, one term of the discrete sum above (whose summands are all non-negative) is exactly $1/34!$, which is uniformly bounded away from zero. This implies that the integral must also be bounded away from zero. However the integrand of the integral is clearly uniformly bounded on $[a',b']$ and, from Theorem 3.13, $F^{\mathbf{B}_N} - F \Rightarrow 0$. Therefore the integral tends to zero as $N \to \infty$. This is a contradiction. Therefore, the probability that there is an eigenvalue of $\mathbf{B}_N$ in $[a,b]$ infinitely often is null. Now, from [Yin *et al.*, 1988], the largest eigenvalue of $\frac{1}{n}\mathbf{X}_N\mathbf{X}_N^{\mathsf{H}}$ is almost surely asymptotically bounded. Therefore, since $\|\mathbf{T}_N\|$ is also bounded by hypothesis, the theorem applies also to $b = \infty$. □

Note that the finiteness of the fourth order moment of the entries $X_{N,ij}$ is fundamental for the validity of Theorem 7.1. It is indeed proved in [Yin *et al.*, 1988] and [Silverstein *et al.*, 1988] that:

- if the entries $X_{N,ij}$ have finite fourth order moment, with probability one, the largest eigenvalue of $\mathbf{X}_N\mathbf{X}_N^{\mathsf{H}}$ tends to the edge $(1+\sqrt{c})^2$, $c = \lim_N N/n$ of the support of the Marčenko–Pastur law, which is an immediate corollary of Theorem 7.1 with $\mathbf{T}_N = \mathbf{I}_N$;
- if the entries $X_{N,ij}$ do not have a finite fourth order moment then, with probability one, the limit superior of the largest eigenvalue of $\mathbf{X}_N\mathbf{X}_N^{\mathsf{H}}$ is infinite, i.e. with probability one, for all $A > 0$, there exists N such that the largest eigenvalue of $\mathbf{X}_N\mathbf{X}_N^{\mathsf{H}}$ is larger than A. It is therefore important never to forget the underlying assumption made on the tails of the distribution of the entries in $\mathbf{X}_N$.

We now move to an extension of Theorem 7.1.

7.1.2 Exact spectrum separation

Now assume that the e.s.d. of $\mathbf{T}_N$ converges to the distribution function of, say, three evenly weighted masses in λ_1, λ_2, and λ_3. For not-too-large ratios $c_N = N/n$, it is observed that the support of F is divided into up to three clusters of eigenvalues. In particular, when n becomes large while N is kept fixed, the clusters consist of three punctual masses in λ_1, λ_2, and λ_3, as required by classical probability theory. This is illustrated in Figure 7.1 in the case of a three-fold clustered and a two-fold clustered support of F. The reason why we observe sometimes three and sometimes less clusters is linked to the spreading of each cluster due to the limiting ratio c; the smaller c, the thinner the clusters, as already observed in the simple case of the Marčenko–Pastur law, Figure 2.2. Considering Theorem 7.1, it is tempting to assume that, in addition to each cluster of F being composed of one third of the total spectrum mass, each cluster of $\mathbf{B}_N$ contains exactly one third of the eigenvalues of $\mathbf{B}_N$. However, Theorem 7.1 only ensures that no eigenvalue is found outside the support of F for all N larger than a given M, and does not say how the eigenvalues of $\mathbf{B}_N$ are distributed in the various clusters. The answer to this question is provided in [Bai and Silverstein, 1999] in which the exact separation properties of the l.s.d. of such matrices $\mathbf{B}_N$ is discussed.

Theorem 7.2 ([Bai and Silverstein, 1999]). *Assume the hypothesis of Theorem 7.1 with $\mathbf{T}_N$ non-negative definite. Consider similarly $0 < a < b < \infty$ such that $[a, b]$ lies in an open interval outside the support of F and F_N for all large N. Denote additionally λ_k and τ_k the* kth *eigenvalues of $\mathbf{B}_N$ and $\mathbf{T}_N$ in decreasing order, respectively. Then we have:*

1. *If $c(1 - H(0)) > 1$, then the smallest value x_0 in the support of F is positive and $\lambda_N \to x_0$ almost surely, as $N \to \infty$.*
2. *If $c(1 - H(0)) \leq 1$, or $c(1 - H(0)) > 1$ but $[a, b]$ is not contained in $[0, x_0]$, then*[1]

$$P(\lambda_{i_N} > b, \lambda_{i_N+1} < a \text{ for all large } N) = 1$$

where i_N is the unique integer such that

$$\tau_{i_N} > -1/m_F(b),$$
$$\tau_{i_N+1} < -1/m_F(a).$$

Theorem 7.2 ensures in particular the exact separation of the spectrum when $\tau_1, \ldots, \tau_N$ take values in a finite set. Consider for instance the first plot in Figure 7.1 and an interval $[a, b]$ comprised between the second and third clusters. What Theorem 7.2 claims is that, if i_N and $i_N + 1$ are the indexes of the right and left eigenvalues when $F^{\mathbf{B}_N}$ jumps from one cluster to the next, and N is large enough, then there is an associated jump from the corresponding i_Nth and $(i_N + 1)$th eigenvalues of $\mathbf{T}_N$ (for instance, at the position of the discontinuity from eigenvalue 7 to eigenvalue 3).

This bears some importance for signal detection. Indeed, consider the problem of the transmission of information plus noise. Given the dimension p of the signal

[1] The expression "$P(A_N$ for all large $N) = 1$" is used in place of "there exists $B \subset \Omega$, with $P(B) = 1$, such that, for $\omega \in B$, there exists $N_0(\omega)$ for which $N > N_0(\omega)$ implies $\omega \in A_N$." It is particularly important to note that "for all large N" is somewhat misleading as it does *not* indicate the existence of a universal N_0 such that $N > N_0$ implies $\omega \in A_N$ for all $\omega \in B$, but rather the existence of an $N_0(\omega)$ for each such ω. Here, $A_N = \{\omega, \lambda_{i_N}(\omega) > b, \lambda_{i_{N+1}}(\omega) < a\}$ and the space Ω is the generator of the series $\mathbf{B}_1(\omega), \mathbf{B}_2(\omega), \ldots$.

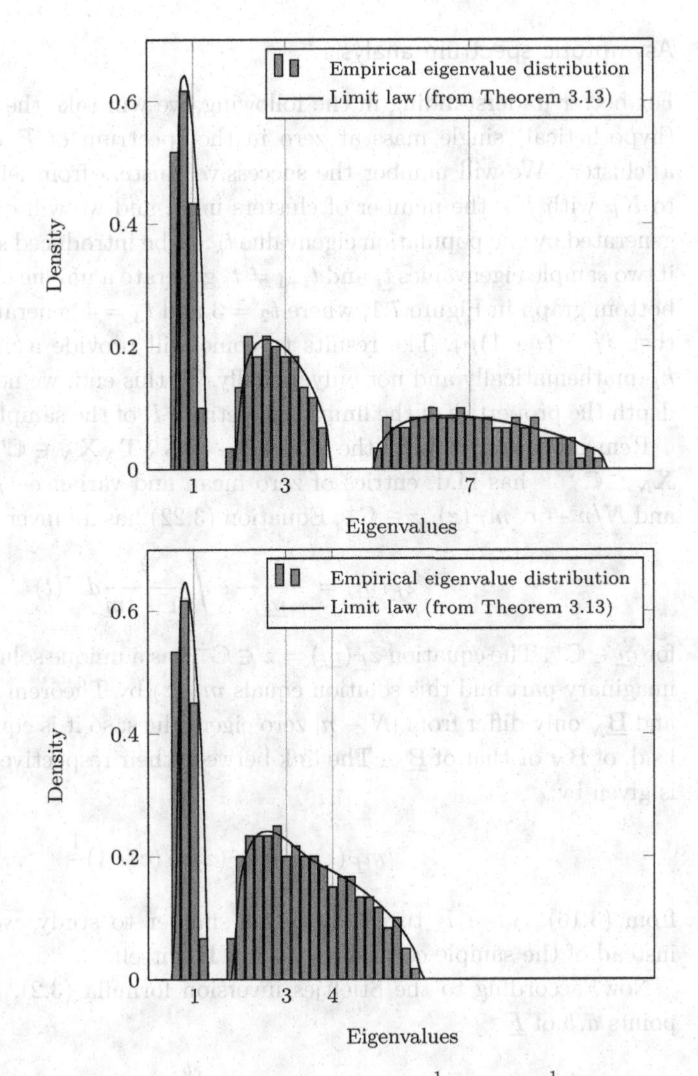

Figure 7.1 Histogram of the eigenvalues of $\mathbf{B}_N = \mathbf{T}_N^{\frac{1}{2}}\mathbf{X}_N\mathbf{X}_N^{\mathsf{H}}\mathbf{T}_N^{\frac{1}{2}}$, $N = 300$, $n = 3000$, with $\mathbf{T}_N$ diagonal composed of three evenly weighted masses in (i) 1, 3, and 7 on top, (ii) 1, 3, and 4 on the bottom.

space and $n - p$ of the noise space, for large c, Theorem 7.2 allows us to isolate the eigenvalues corresponding to the signal space from those corresponding to the noise space. If both eigenvalue spaces are isolated in two distinct clusters, then we can exactly determine the dimension of each space and infer, e.g. the number of transmitting entities. The next question that then naturally arises is to determine for which values of $c = \lim_N n/N$ the support of F separates into 1, 2, or more clusters.

7.1.3 Asymptotic spectrum analysis

For better understanding in the following, we will take the convention that the (hypothetical) single mass at zero in the spectrum of F is *not* considered as a 'cluster'. We will number the successive clusters from left to right, from one to K_F with K_F the number of clusters in F, and we will denote k_F the cluster generated by the population eigenvalue t_k, to be introduced shortly. For instance, if two sample eigenvalues t_i and $t_{i+1} \neq t_i$ generate a unique cluster in F (as in the bottom graph in Figure 7.1, where $t_2 = 3$ and $t_3 = 4$ generate the same cluster), then $i_F = (i+1)_F$). The results to come will provide a unique way to define k_F mathematically and not only visually. To this end, we need to study in more depth the properties of the limiting spectrum F of the sample covariance matrix.

Remember first that, for the model $\underline{\mathbf{B}}_N = \mathbf{X}_N^{\mathsf{H}} \mathbf{T}_N \mathbf{X}_N \in \mathbb{C}^{n \times n}$ of l.s.d. $\underline{F}$, where $\mathbf{X}_N \in \mathbb{C}^{N \times n}$ has i.i.d. entries of zero mean and variance $1/n$, $\mathbf{T}_N$ has l.s.d. H and $N/n \to c$, $m_{\underline{F}}(z)$, $z \in \mathbb{C}^+$, Equation (3.22) has an inverse formula, given by:

$$z_{\underline{F}}(\underline{m}) = -\frac{1}{\underline{m}} + c \int \frac{t}{1 + t\underline{m}} dH(t) \tag{7.3}$$

for $\underline{m} \in \mathbb{C}^+$. The equation $z_{\underline{F}}(\underline{m}) = z \in \mathbb{C}^+$ has a unique solution $\underline{m}$ with positive imaginary part and this solution equals $m_{\underline{F}}(z)$ by Theorem 3.13. Of course, $\mathbf{B}_N$ and $\underline{\mathbf{B}}_N$ only differ from $|N - n|$ zero eigenvalues, so it is equivalent to study the l.s.d. of $\mathbf{B}_N$ or that of $\underline{\mathbf{B}}_N$. The link between their respective Stieltjes transforms is given by:

$$m_{\underline{F}}(z) = c m_F(z) + (c - 1)\frac{1}{z}$$

from (3.16). Since $\underline{F}$ turns out to be simpler to study, we will focus on $\underline{\mathbf{B}}_N$ instead of the sample covariance matrix $\mathbf{B}_N$ itself.

Now, according to the Stieltjes inversion formula (3.2), for every continuity points a, b of $\underline{F}$

$$\underline{F}(b) - \underline{F}(a) = \lim_{y \to 0^+} \frac{1}{\pi} \int_a^b \Im[m_{\underline{F}}(x + iy)] dx.$$

To determine the distribution $\underline{F}$, and therefore the distribution F, we must determine the limit of $m_{\underline{F}}(z)$ as $z \in \mathbb{C}^+$ tends to $x \in \mathbb{R}^*$. It can in fact be shown that this limit exists.

Theorem 7.3 ([Silverstein and Choi, 1995]). *Let $\underline{\mathbf{B}}_N \in \mathbb{C}^{n \times n}$ be defined as previously, with almost sure l.s.d. $\underline{F}$. Then, for $x \in \mathbb{R}^*$*

$$\lim_{\substack{z \to x \\ z \in \mathbb{C}^+}} m_{\underline{F}}(z) \triangleq m^\circ(x) \tag{7.4}$$

exists and the function m° is continuous on $\mathbb{R}^*$. For x in the support of $\underline{F}$, the density $\underline{f}(x) \triangleq \underline{F}'(x)$ equals $\frac{1}{\pi}\Im[m^\circ(x)]$. Moreover, $\underline{f}$ is analytic for all $x \in \mathbb{R}^*$ such that $\underline{f}(x) > 0$.

The study of m° makes it therefore possible to describe the complete support $S_{\underline{F}}$ of $\underline{F}$ as well as the limiting density $\underline{f}$. Since $S_{\underline{F}}$ equals S_F but for an additional mass in zero, this is equivalent to determining the support of S_F. Choi and Silverstein provided an accurate description of the function m°, as follows.

Theorem 7.4 ([Silverstein and Choi, 1995]). *Let $B = \{\underline{m} \mid \underline{m} \neq 0, -1/\underline{m} \in S_H^c\}$, with S_H^c the complementary of S_H, and $x_{\underline{F}}$ be the function defined on B by*

$$x_{\underline{F}}(\underline{m}) = -\frac{1}{\underline{m}} + c\int \frac{t}{1+t\underline{m}} dH(t). \tag{7.5}$$

For $x \in \mathbb{R}^*$, we can determine the limit $m^\circ(x)$ of $m_{\underline{F}}(z)$ as $z \to x$, $z \in \mathbb{C}^+$, along the following rules:

1. *If $x \in S_{\underline{F}}$, then $m^\circ(x)$ is the unique solution in B with positive imaginary part of the equation $x = x_{\underline{F}}(\underline{m})$ in the dummy variable $\underline{m}$.*
2. *If $x \in S_{\underline{F}}^c$, then $m^\circ(x)$ is the unique real solution in B of the equation $x = x_{\underline{F}}(\underline{m})$ in the dummy variable $\underline{m}$ such that $x'_{\underline{F}}(m_0) > 0$. Conversely, for $\underline{m} \in B$, if $x'_{\underline{F}}(\underline{m}) > 0$, then $x_{\underline{F}}(\underline{m}) \in S_{\underline{F}}^c$.*

From rule 1, along with Theorem 7.3, we can evaluate for every $x > 0$ the limiting density $\underline{f}(x)$, hence $F(x)$, by finding the complex solution with positive imaginary part of $x = x_{\underline{F}}(\underline{m})$.

Rule 2 makes it simple to determine analytically the exact support of F. It indeed suffices to draw $x_{\underline{F}}(\underline{m})$ for $-1/\underline{m} \in S_H^c$. Whenever $x_{\underline{F}}$ is increasing on an interval I, $x_{\underline{F}}(I)$ is outside $S_{\underline{F}}$. The support $S_{\underline{F}}$ of $\underline{F}$, and therefore of F (modulo the mass in zero), is then defined exactly by the complementary set

$$S_{\underline{F}} = \mathbb{R} \setminus \bigcup_{\substack{a,b\in\mathbb{R} \\ a<b}} \left\{ x_{\underline{F}}((a,b)) \mid \forall \underline{m} \in (a,b), x'_{\underline{F}}(\underline{m}) > 0 \right\}.$$

This is depicted in Figure 7.2 in the case when H is composed of three evenly weighted masses t_1, t_2, t_3 in $\{1, 3, 5\}$ or $\{1, 3, 10\}$ and $c = 1/10$. Notice that, in the case where $t_3 = 10$, F is divided into three clusters, while, when $t_3 = 5$, F is divided into only two clusters, which is due to the fact that $x_{\underline{F}}$ is non-increasing in the interval $(-1/3, -1/5)$. For applicative purposes, we will see in Chapter 17 that it might be essential that the consecutive clusters be disjoint. This is one reason why Theorem 7.6 is so important.

We do not provide a rigorous proof of Theorem 7.4. In fact, while thoroughly proved in 1995, this result was already intuited by Marčenko and Pastur in 1967 [Marčenko and Pastur, 1967]. The fact that $x_{\underline{F}}(\underline{m})$ increases outside the spectrum of $\underline{F}$ and is not increasing elsewhere is indeed very intuitive, and is not actually limited to the sample covariance matrix case. Observe indeed that, for any F, and any $x_0 \in \mathbb{R}^*$ outside the support of F, $m_F(x_0)$ is clearly well defined

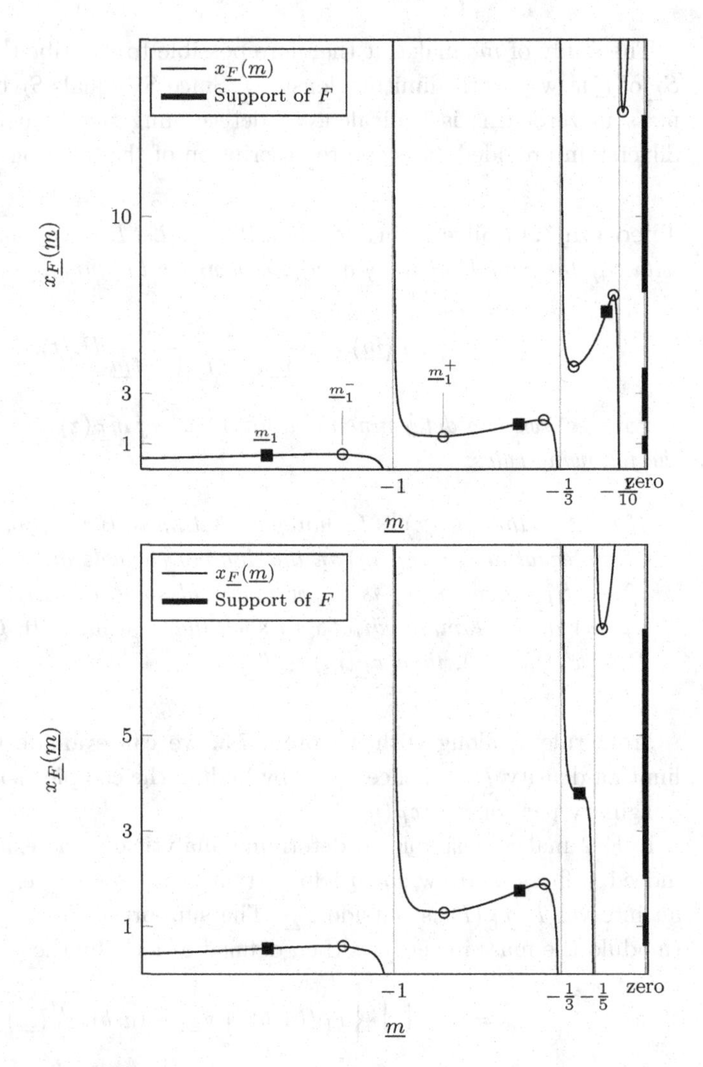

Figure 7.2 $x_F(\underline{m})$ for $\underline{m}$ real, $\mathbf{T}_N$ diagonal composed of three evenly weighted masses in 1, 3, and 10 (top) and 1, 3, and 5 (bottom), $c = 1/10$ in both cases. Local extrema are marked in circles, inflexion points are marked in squares. The support of F can be read on the right vertical axises.

and

$$m_F'(x_0) = \int \frac{1}{(\lambda - x_0)^2} dF(\lambda) > 0.$$

Therefore $m_F(x)$ is continuous and increasing on an open neighborhood of x_0. This implies that it is locally a one-to-one mapping on this neighborhood and therefore admits an inverse $x_F(m)$, which is also continuous and increasing. This explains why $x_F(m)$ increases when its image is outside the spectrum of F. Now, if for some real m_0, $x_F(m_0)$ is continuous and increasing, then it is locally invertible and its inverse ought to be $m_F(x)$, continuous and increasing, in which case x is outside the spectrum of F. Obviously, this reasoning is far from being a proof (at least the converse requires much more work).

From Figure 7.2 and Theorem 7.4, we now observe that, when the e.s.d. of population matrix is composed of a few masses, $x'_{\underline{F}}(\underline{m}) = 0$ has exactly $2K_F$ solutions with K_F the number of clusters in F. Denote these roots in increasing order $\underline{m}_1^- < \underline{m}_1^+ \leq \underline{m}_2^- < \underline{m}_2^+ < \ldots \leq \underline{m}_{K_F}^- < \underline{m}_{K_F}^+$. Each pair $(\underline{m}_j^-, \underline{m}_j^+)$ is such that $x_{\underline{F}}([\underline{m}_j^-, \underline{m}_j^+])$ is the jth cluster in F. We therefore have a way to determine the support of the asymptotic spectrum through the function $x'_{\underline{F}}$. This is presented in the following result.

Theorem 7.5 ([Couillet *et al.*, 2011c; Mestre, 2008a]). *Let $\mathbf{B}_N \in \mathbb{C}^{N \times N}$ be defined as in Theorem 7.1. Then the support S_F of the l.s.d. F of $\mathbf{B}_N$ is*

$$S_F = \bigcup_{j=1}^{K_F} [x_j^-, x_j^+]$$

where $x_1^-, x_1^+, \ldots, x_{K_F}^-, x_{K_F}^+$ are defined by

$$x_j^- = -\frac{1}{\underline{m}_j^-} + \sum_{r=1}^{K} c_r \frac{t_r}{1 + t_r \underline{m}_j^-}$$

$$x_j^+ = -\frac{1}{\underline{m}_j^-} + \sum_{r=1}^{K} c_r \frac{t_r}{1 + t_r \underline{m}_j^+}$$

with $\underline{m}_1^- < \underline{m}_1^+ \leq \underline{m}_2^- < \underline{m}_2^+ \leq \ldots \leq \underline{m}_{K_F}^- < \underline{m}_{K_F}^+$ the $2K_F$ (possibly counted with multiplicity) real roots of the equation in $\underline{m}$

$$\sum_{r=1}^{K} c_r \frac{t_r^2 \underline{m}^2}{(1 + t_r \underline{m}^2)^2} = 1.$$

Note further from Figure 7.2 that, while $x'_{\underline{F}}(\underline{m})$ might not have roots on some intervals $(-1/t_{k-1}, -1/t_k)$, it always has a unique inflexion point there. This is proved in [Couillet *et al.*, 2011c] by observing that $x''_{\underline{F}}(\underline{m}) = 0$ is equivalent to

$$\sum_{r=1}^{K} c_r \frac{t_r^3 \underline{m}^3}{(1 + t_r \underline{m})^3} - 1 = 0$$

the left-hand side of which has always positive derivative and shows asymptotes in the neighborhood of t_r; hence the existence of a unique inflexion point on every interval $(-1/t_{k-1}, -1/t_k)$, for $1 \leq k \leq K$, with convention $t_0 = 0^+$. When $x_{\underline{F}}$ increases on an interval $(-1/t_{k-1}, -1/t_k)$, it must have its inflexion point in a point of positive derivative (from the concavity change induced by the asymptotes). Therefore, to verify that cluster k_F is disjoint from clusters $(k-1)_F$ and $(k+1)_F$ (when they exist), it suffices to verify that the $(k-1)$th and kth roots $\underline{m}_{k-1}$ and $\underline{m}_k$ of $x''_{\underline{F}}(\underline{m})$ are such that $x'_{\underline{F}}(\underline{m}_{k-1}) > 0$ and $x'_{\underline{F}}(\underline{m}_k) > 0$. From this observation, we therefore have the following result.

Theorem 7.6 ([Couillet *et al.*, 2011c; Mestre, 2008b]). *Let $\mathbf{B}_N$ be defined as in Theorem 7.1, with $\mathbf{T}_N = \mathrm{diag}(\tau_1, \ldots, \tau_N) \in \mathbb{R}^{N \times N}$, diagonal containing K distinct eigenvalues $0 < t_1 < \ldots < t_K$, for some fixed K. Denote N_k the multiplicity of the* kth *largest distinct eigenvalue (assuming ordering of the τ_i, we may then have $\tau_1 = \ldots = \tau_{N_1} = t_1, \ldots, \tau_{N-N_K+1} = \ldots = \tau_N = t_K$). Assume also that, for all $1 \leq r \leq K$, $N_r/n \to c_r > 0$, and $N/n \to c$, with $0 < c < \infty$. Then the cluster k_F associated with the eigenvalue t_k in the l.s.d. F of $\mathbf{B}_N$ is distinct from the clusters $(k-1)_F$ and $(k+1)_F$ (when they exist), associated with t_{k-1} and t_{k+1} in F, respectively, if and only if*

$$\sum_{r=1}^{K} c_r \frac{t_r^2 \underline{m}_k^2}{(1 + t_r \underline{m}_k^2)^2} < 1$$
$$\sum_{r=1}^{K} c_r \frac{t_r^2 \underline{m}_{k+1}^2}{(1 + t_r \underline{m}_{k+1}^2)^2} < 1 \tag{7.6}$$

where $\underline{m}_1, \ldots, \underline{m}_K$ are such that $\underline{m}_{K+1} = 0$ and $\underline{m}_1 < \underline{m}_2 < \ldots < \underline{m}_K$ are the K solutions of the equation in $\underline{m}$

$$\sum_{r=1}^{K} c_r \frac{t_r^3 \underline{m}^3}{(1 + t_r \underline{m})^3} = 1.$$

For $k = 1$, this condition ensures $1_F = 2_F - 1$. For $k = K$, this ensures $K_F = (K-1)_F + 1$. For $1 < k < K$, this ensures $(k-1)_F + 1 = k_F = (k+1)_F - 1$.

Remark now that the conditions of Equation (7.6) are left unchanged if all $t_1, \ldots, t_K$ are scaled by a common constant. Indeed, if t_j becomes αt_j for all j, then $\underline{m}_1, \ldots, \underline{m}_K$ become $\underline{m}_1/\alpha, \ldots, \underline{m}_K/\alpha$ and the scaling effects cancel out in Equation (7.6). Therefore, in the case $K = 2$, the separability condition only depends on the ratios c_1, c_2 and on t_1/t_2. If $c_1 = c_2 = c/2$, then we can depict the plot of the critical ratio $1/c$ as a function of t_1/t_2 for which cluster separability happens. This is depicted in Figure 7.3. Since $1/c$ is the limit of the ratio n/N, Figure 7.3 determines, for a fixed observation size N, the limiting number of samples per observation size required to achieve cluster separability. Observe how steeply the plot of $1/c$ increases when t_1 gets close to t_2; this suggests that the tools to be presented later that require this cluster separability will be very inefficient when it comes to separate close sources (the definition of 'closeness' depending on each specific study, e.g. close directions of signal arrivals in radar applications, close transmit powers in signal sensing, etc.). Figure 7.4 depicts the regions of separability of all clusters in the case $K = 3$, for fixed $c = 0.1$, $c_1 = c_2 = c_3$, as a function of the ratios t_3/t_1 and t_2/t_1. Observe that the triplets $(1, 3, 7)$ and $(1, 3, 10)$ are well inside the separability region as suggested, respectively, by Figure 7.1 (top) and Figure 7.2 (top); on the contrary, notice that the triplets $(1, 3, 4)$ and $(1, 3, 5)$ are outside the separability region, confirming then the observations of Figure 7.1 (bottom) and Figure 7.2 (bottom).

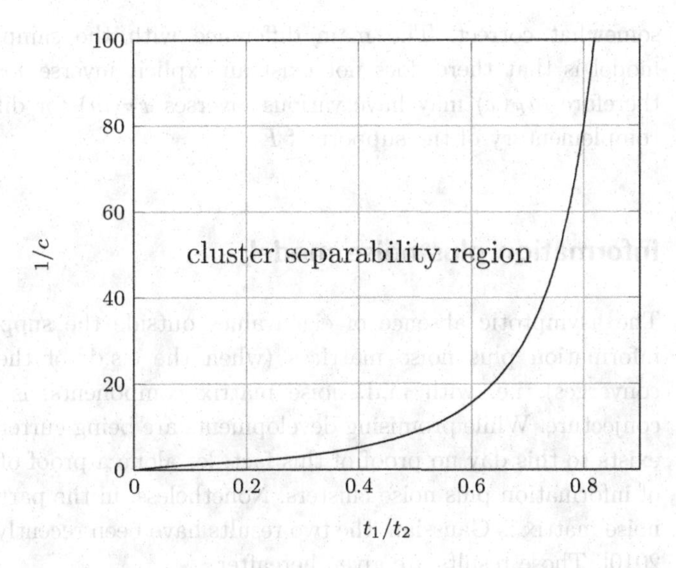

Figure 7.3 Limiting ratio c to ensure separability of (t_1, t_2), $t_1 \leq t_2$, $K = 2$, $c_1 = c_2$.

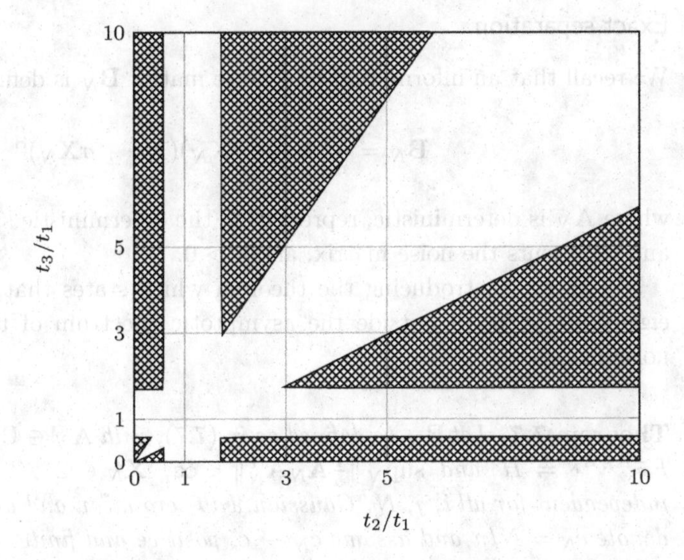

Figure 7.4 Subset of (t_1, t_2, t_3) that satisfy cluster separability condition, $c_1 = c_2 = c_3$, $c = 0.1$, in crosshatched pattern.

After establishing these primary results for the sample covariance matrix models, we now move to the information plus noise model. According to the previous remark borrowed from Marčenko and Pastur in [Marčenko and Pastur, 1967], we infer that it will still be the case that the Stieltjes transform $m_F(x)$, extended to the real axis, has a local inverse $x_F(m)$, which is continuous and increasing, and that the range where $x_F(m)$ increases is exactly the complementary to the support of F. This statement will be shown to be somewhat correct. The main difference with the sample covariance matrix model is that there does not exist an explicit inverse $x_F(m)$, as in (7.5) and therefore $m_F(x)$ may have various inverses $x_F(m)$ for different subsets in the complementary of the support of F.

7.2 Information plus noise model

The asymptotic absence of eigenvalues outside the support of unconstrained information plus noise matrices (when the e.s.d. of the information matrix converges), i.e. with i.i.d. noise matrix components, is still at the stage of conjecture. While promising developments are being currently carried out, there exists to this day no proof of this fact, let alone a proof of the exact separation of information plus noise clusters. Nonetheless, in the particular case where the noise matrix is Gaussian, the two results have been recently proved [Vallet *et al.*, 2010]. Those results are given hereafter.

7.2.1 Exact separation

We recall that an information plus noise matrix $\mathbf{B}_N$ is defined by

$$\mathbf{B}_N = \frac{1}{n}(\mathbf{A}_N + \sigma\mathbf{X}_N)(\mathbf{A}_N + \sigma\mathbf{X}_N)^{\mathsf{H}} \tag{7.7}$$

where $\mathbf{A}_N$ is deterministic, representing the deterministic signal, $\mathbf{X}_N$ is random and represents the noise matrix, and $\sigma > 0$.

We start by introducing the theorem which states that, for all large N, no eigenvalue is found outside the asymptotic spectrum of the information plus noise model.

Theorem 7.7. *Let $\mathbf{B}_N$ be defined as in (7.7), with $\mathbf{A}_N \in \mathbb{C}^{N\times n}$ such that $H_N \triangleq F^{\frac{1}{n}\mathbf{A}_N\mathbf{A}_N^{\mathsf{H}}} \Rightarrow H$ and $\sup_N \|\frac{1}{n}\mathbf{A}_N\mathbf{A}_N^{\mathsf{H}}\| < \infty$, $\mathbf{X}_N \in \mathbb{C}^{N\times n}$ with entries $X_{N,ij}$ independent for all i, j, N, Gaussian with zero mean and unit variance. Further denote $c_N = N/n$ and assume $c_N \to c$, positive and finite. From Theorem 3.15, we know that $F^{\mathbf{B}_N}$ converges almost surely to a limit distribution F with Stieltjes transform $m_F(z)$ solution of the equation in m*

$$\frac{m}{1+\sigma^2 c_N m} = m_H\left(z(1+\sigma^2 c_N m)^2 - \sigma^2(1-c_N)(1+\sigma^2 c_N m)\right) \tag{7.8}$$

this solution being unique for $z \in \mathbb{C}^+$, $m \in \mathbb{C}^+$ and $\Im[zm] \geq 0$. Denote now $m_N(z)$ this solution when m_H is replaced by m_{H_N} and F_N the distribution function with Stieltjes transform $m_N(z)$.

Let $N_0 \in \mathbb{N}$, and choose an interval $[a, b]$ outside the union of the supports of F and F_N for all $N \geq N_0$. For $\omega \in \Omega$, the probability space generating the noise *sequences* $\mathbf{X}_1, \mathbf{X}_2, \ldots$, *denote* $\mathfrak{L}_N(\omega)$ *the set of eigenvalues of* $\mathbf{B}_N(\omega)$. *Then*

$$P(\omega, \mathfrak{L}_N(\omega) \cap [a,b] \neq \emptyset \text{ i.o.}) = 0.$$

The next theorem ensures that the repartition of the eigenvalues in the consecutive clusters is exactly as expected.

Theorem 7.8 ([Vallet *et al.*, 2010]). *Let* $\mathbf{B}_N$ *be as in Theorem 7.7. Let* $a < b$ *be such that* $[a,b]$ *lies outside the support of* F. *Denote* λ_k *and* a_k *the* kth *eigenvalues smallest of* $\mathbf{B}_N$ *and* $\frac{1}{n}\mathbf{A}_N\mathbf{A}_N^{\mathsf{H}}$, *respectively. Then we have:*

1. *If* $c(1-H(0)) > 1$, *then the smallest eigenvalue* x_0 *of the support of* F *is positive and* $\lambda_N \to x_0$ *almost surely, as* $N \to \infty$.
2. *If* $c(1-H(0)) \leq 1$, *or* $c(1-H(0)) > 1$ *but* $[a,b]$ *is not contained in* $[0, x_0]$, *then:*

$$P(\lambda_{i_N} > b, \lambda_{i_N+1} < a \text{ for all large } N) = 1$$

where i_N *is the unique integer such that*

$$\tau_{i_N} > -1/m_F(b)$$

$$\tau_{i_N+1} < -1/m_F(a).$$

We provide hereafter a sketch of the proofs of both Theorem 7.7 and Theorem 7.8 where considerations of complex integration play a fundamental role. In the following chapter, Chapter 8, we introduce in detail the methods of complex integration for random matrix theory and particularly for statistical inference.

Proof of Theorem 7.7 and Theorem 7.8. As already mentioned, these results are only known to hold for the Gaussian case for the time being. The way these results are achieved is similar to the way Theorem 7.1 and Theorem 7.2 were obtained, although the techniques are radically different. Indeed, somewhat similarly to Theorem 7.1, the first objective is to show that the difference $m_N(z) - \mathrm{E}[m_{\mathbf{B}_N}(z)]$ between the deterministic equivalent $m_N(z)$ of the empirical Stieltjes transform $m_{\mathbf{B}_N}(z)$ and $\mathrm{E}[m_{\mathbf{B}_N}(z)]$ goes to zero at a sufficiently fast rate. In the Gaussian case, this rate is of order $O(1/N^2)$. Remember from Theorem 6.5 that such a convergence rate was already observed for doubly correlated Gaussian models and allowed us to ensure that $N(m_N(z) - \mathrm{E}[m_{\mathbf{B}_N}(z)]) \to 0$. Using the fact, established precisely in Chapter 8, that, for holomorphic functions f and a distribution function G

$$\int f(x)dG(x) = -\frac{1}{2\pi i}\oint f(z)m_G(z)dz$$

on a positively oriented contour encircling the support of F, we can infer the recent result from [Haagerup *et al.*, 2006]

$$\mathrm{E}\left[\int f(x)[F^{\mathbf{B}_N} - F_N](dx)\right] = O\left(\frac{1}{N^2}\right).$$

Take f any infinitely differentiable function that is identically one on $[a, b] \subset \mathbb{R}$ and identically zero outside $(a - \varepsilon, b + \varepsilon)$ for some small positive ε, such that $(a - \varepsilon, b + \varepsilon)$ is outside the support of F. From the convergence rate above, we first have.

$$\mathrm{E}\left[\sum_{k=1}^{N} \lambda_k 1_{(a-\varepsilon,b+\varepsilon)}(\lambda_k)\right] = N(F_N(b) - F_N(a)) + O\left(\frac{1}{N}\right)$$

and therefore, for large N, we have in expectation the correct mass of eigenvalues in $(a - \varepsilon, b + \varepsilon)$. But we obviously want more than that: i.e., we want to determine the asymptotic exact number of these eigenvalues. Using the Nash–Poincaré inequality, Theorem 6.7, we can in fact show that, for this choice of f

$$\mathrm{E}\left[\left(\int f(x)[F^{\mathbf{B}_N} - F_N](dx)\right)^2\right] = O\left(\frac{1}{N^4}\right).$$

This is enough to prove, thanks to the Markov inequality, Theorem 3.5, that

$$P\left(\left|\int f(x)[F^{\mathbf{B}_N} - F_N](dx)\right| > \frac{1}{N^{\frac{4}{3}}}\right) < \frac{K}{N^{\frac{4}{3}}}$$

for some constant K. From there, the Borel–Cantelli lemma, Theorem 3.6, ensures that the above event is infinitely often true with probability zero; i.e. the event

$$\left|\sum_{k=1}^{N} \lambda_k 1_{(a-\varepsilon,b+\varepsilon)}(\lambda_k) - N(F_N(b) - F_N(a))\right| > \frac{K}{N^{\frac{1}{3}}}$$

is infinitely often true with probability zero. Therefore, with probability one, there exists N_0 such that, for $N > N_0$ there is no eigenvalue in $(a - \varepsilon, b + \varepsilon)$. This proves the first result.

Take now $[a, b]$ not necessarily outside the support of F and ε such that $(a - \varepsilon, a) \cup (b, b + \varepsilon)$ is outside the support of F. Then, repeating the same procedure as above but to characterize now

$$\left|\sum_{k=1}^{N} \lambda_k 1_{[a,b]}(\lambda_k) - N(F_N(b) - F_N(a))\right|$$

we find that this term equals

$$\left|\sum_{k=1}^{N} \lambda_k 1_{(a-\varepsilon,b+\varepsilon)}(\lambda_k) - N(F_N(b) - F_N(a))\right|$$

almost surely in the large N limit since there is asymptotically no eigenvalue in $(a - \varepsilon, a) \cup (b, b + \varepsilon)$. This now says that the asymptotic number of eigenvalues in $[a, b]$ is $N(F_N(b) - F_N(a))$ almost surely. The fact that the indexes of these eigenvalues are those expected is obvious. If it were not the case, then we can always find an interval on the left or on the right of $[a, b]$ which does not contain

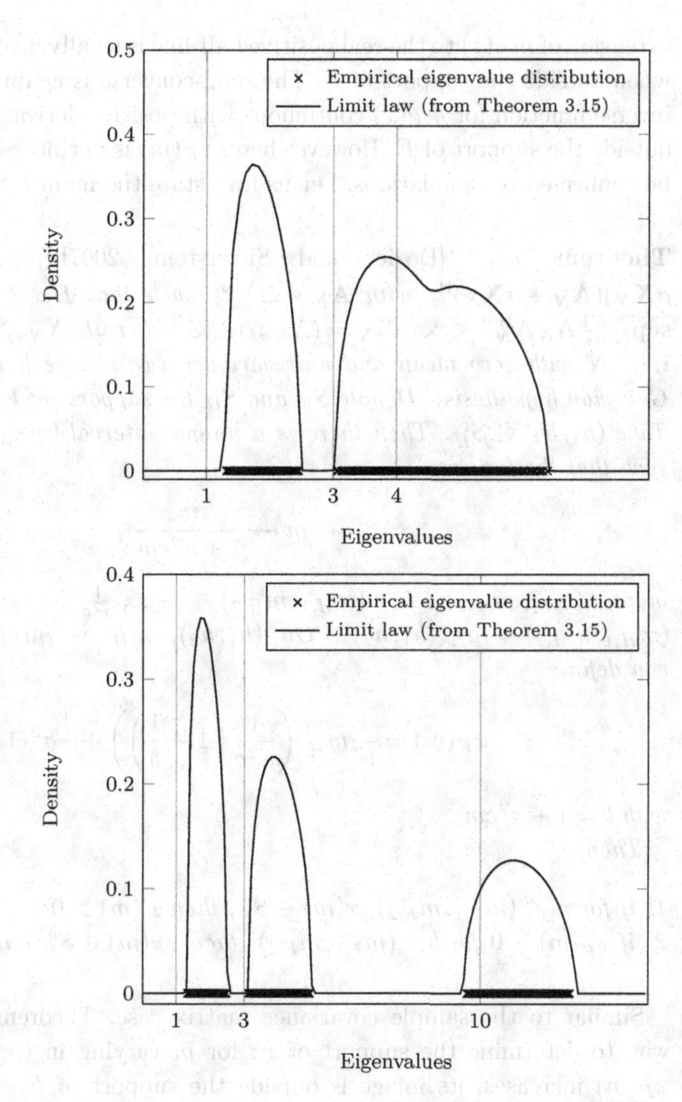

Figure 7.5 Empirical and limit eigenvalue distribution of the information plus noise model $\mathbf{B}_N = \frac{1}{n}(\mathbf{A}_N + \sigma\mathbf{X}_N)(\mathbf{A}_N + \sigma\mathbf{X}_N)^{\mathsf{H}}$, $N = 300$, $n = 3000$ ($c = 1/10$), $F^{\frac{1}{N}\mathbf{A}_N\mathbf{A}_N^{\mathsf{H}}}$ has three evenly weighted masses at $1, 3, 4$ (top) and $1, 3, 10$ (bottom).

the right amount of eigenvalues, which is contradictory from this proof. This completes the proof of both results. □

7.2.2 Asymptotic spectrum analysis

A similar spectrum analysis as in the case of sample covariance matrices when the population covariance matrix has a finite number of distinct eigenvalues can be performed for the information plus noise model. As discussed previously, the extension of $m_F(z)$ to the real positive half-line is locally invertible and increasing when outside the support of F. The semi-converse is again true: if $x_F(m)$ is an inverse function for $m_F(x)$ continuous with positive derivative, then its image is outside the support of F. However here, $x_F(m)$ is not necessarily unique, as will be confirmed by simulations. Let us first state the main result.

Theorem 7.9 ([Dozier and Silverstein, 2007b]). *Let $\mathbf{B}_N = \frac{1}{n}(\mathbf{A}_N + \sigma\mathbf{X}_N)(\mathbf{A}_N + \sigma\mathbf{X}_N)^{\mathsf{H}}$, with $\mathbf{A}_N \in \mathbb{C}^{N\times n}$ such that $H_N \triangleq F^{\frac{1}{n}\mathbf{A}_N\mathbf{A}_N^{\mathsf{H}}} \Rightarrow H$ and $\sup_N \|\frac{1}{n}\mathbf{A}_N\mathbf{A}_N^{\mathsf{H}}\| < \infty$, $\mathbf{X}_N = (X_{N,ij}) \in \mathbb{C}^{N\times n}$ with $X_{N,ij}$ independent for all i, j, N with zero mean and unit variance (we release here the non-necessary Gaussian hypothesis). Denote S_F and S_H the supports of F and H, respectively. Take $(h_1, h_2) \subset S_H^{\mathsf{c}}$. Then there is a unique interval $(m_{F,1}, m_{F,2}) \subset (-\frac{1}{\sigma^2 c}, \infty)$ such that the function*

$$m \mapsto \frac{m}{1 + \sigma^2 cm}$$

maps $(m_{F,1}, m_{F,2})$ to $(m_{H,1}, m_{H,2}) \subset (-\infty, \frac{1}{\sigma^2 c})$, where we introduced $(m_{H,1}, m_{H,2}) = m_H((h_1, h_2))$. On (h_1, h_2), m_H is invertible, and then we can define

$$x_F(m) = \frac{1}{b^2} m_H^{-1}\left(\frac{1}{\sigma^2 c}\left(1 - \frac{1}{b}\right)\right) + \frac{1}{b}\sigma^2(1 - c)$$

with $b = 1 + \sigma^2 cm$.

Then:

1. *if for $m \in (m_{F,1}, m_{F,2})$, $x(m) \in S_F^{\mathsf{c}}$, then $x'(m) > 0$;*
2. *if $x_F'(m) > 0$ for $b \in (m_{F,1}, m_{F,2})$, then $x_F(m) \in S_F^{\mathsf{c}}$ and $m = m_F(x_F(m))$.*

Similar to the sample covariance matrix case, Theorem 7.9 gives readily a way to determine the support of F: for m varying in $(m_{F,1}, m_{F,2})$, whenever $x_F(m)$ increases, its image is outside the support of F. The support of F is therefore the complementary set to the union of all such intervals. We must nonetheless be aware that the definition of $x_F(m)$ is actually linked to the choice of the interval $(h_1, h_2) \subset S_H^{\mathsf{c}}$. In Theorem 7.4, we had a unique explicit inverse for $x_F(m)$ as a function of m, whatever the choice of the pre-image of m_H (the Stieltjes transform of the l.s.d. of the population covariance matrix); this statement no longer holds here.

In fact, if S_H is subdivided into K_H clusters, we can expect at most $K_H + 1$ different local inverses for $x_F(m)$ as m varies along $\mathbb{R}$. This is in fact exactly what is observed. Figure 7.6 depicts the situation when H is composed of three evenly weighted masses in $(1, 3, 4)$, then $(1, 3, 10)$. Observe that $K_H + 1$ different inverses exist that have the aforementioned behavior.

Now, also similar to the sample covariance matrix model, a lot more can be said in the case where H is composed of a finite number of masses. The exact

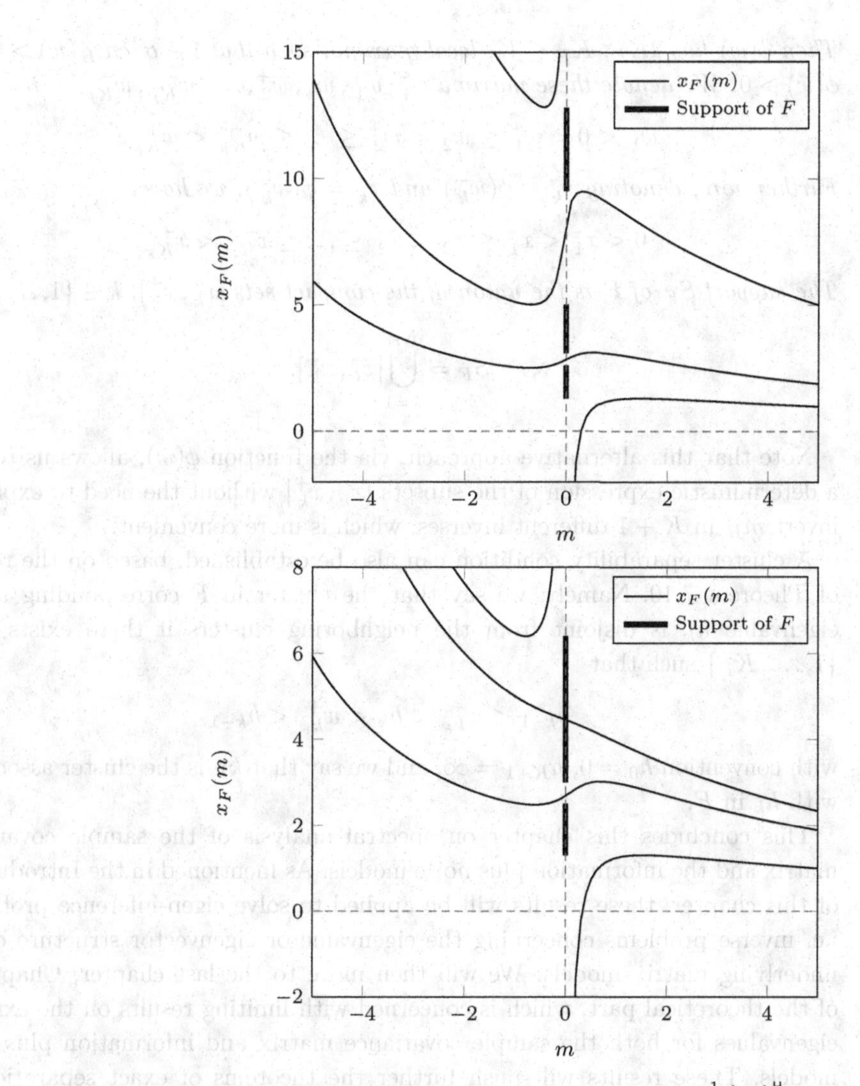

Figure 7.6 Information plus noise model, $x_F(m)$ for m real, $F^{\frac{1}{N}\mathbf{A}_N\mathbf{A}_N^{\mathsf{H}}} \Rightarrow H$, where H has three evenly weighted masses in 1, 3, and 10 (top) and 1, 3, and 4 (bottom), $c = 1/10$, $\sigma = 0.1$ in both cases. The support of F can be read on the central vertical axises.

determination of the boundary of F can be determined. The result is summarized as follows.

Theorem 7.10 ([Vallet *et al.*, 2010]). *Let $\mathbf{B}_N$ be defined as in Theorem 7.9, where $F^{\frac{1}{n}\mathbf{A}_N\mathbf{A}_N^{\mathsf{H}}} = H$ is composed of K eigenvalues $h_1, \ldots, h_K$ (we implicitly assume N takes only values consistent with $F^{\frac{1}{n}\mathbf{A}_N\mathbf{A}_N^{\mathsf{H}}} = H$). Let ϕ be the function on $\mathbb{R} \setminus \{h_1, \ldots, h_K\}$ defined by*

$$\phi(w) = w(1 - \sigma^2 c m_H(w))^2 + (1 - c)\sigma^2(1 - \sigma^2 c m_H(w)).$$

Then $\phi(w)$ has $2K_F$, $K_F \leq K$, local maxima, such that $1 - \sigma^2 c m_H(w) > 0$ and $\phi(w) > 0$. We denote these maxima $w_1^-, w_1^+, w_2^-, w_2^+, \ldots, w_{K_F}^-, w_{K_F}^+$ in the order

$$w_1^- < 0 < w_1^+ \leq w_2^- < w_2^+ \leq \ldots \leq w_{K_F}^- < w_{K_F}^+.$$

Furthermore, denoting $x_k^- = \phi(w_k^-)$ and $x_k^+ = \phi(w_k^+)$, we have:

$$0 < x_1^- < x_1^+ \leq x_2^- < x_2^+ \leq \ldots \leq x_{K_F}^- < x_{K_F}^+.$$

The support S_F of F is the union of the compact sets $[x_k^-, x_k^+]$, $k \in \{1, \ldots, K_F\}$

$$S_F = \bigcup_{k=1}^{K_F} [x_k^-, x_k^+].$$

Note that this alternative approach, via the function $\phi(w)$, allows us to give a deterministic expression of the subsets $[x_k^-, x_k^+]$ without the need to explicitly invert m_H in $K + 1$ different inverses, which is more convenient.

A cluster separability condition can also be established, based on the results of Theorem 7.10. Namely, we say that the cluster in F corresponding to the eigenvalue h_k is disjoint from the neighboring clusters if there exists $k_F \in \{1, \ldots, K_F\}$ such that

$$h_{k-1} < w_{k_F}^- < h_k < w_{k_F}^+ < h_{k+1}$$

with convention $h_0 = 0$, $h_{K+1} = \infty$, and we say that k_F is the cluster associated with h_k in F.

This concludes this chapter on spectral analysis of the sample covariance matrix and the information plus noise models. As mentioned in the Introduction of this chapter, these results will be applied to solve eigen-inference problems, i.e. inverse problems concerning the eigenvalue or eigenvector structure of the underlying matrix models. We will then move to the last chapter, Chapter 9, of the theoretical part, which is concerned with limiting results on the extreme eigenvalues for both the sample covariance matrix and information plus noise models. These results will push further the theorems of exact separation by establishing the limiting distributions of the extreme eigenvalues (although solely in the Gaussian case) and also some properties on the corresponding eigenvectors.

8 Eigen-inference

In the introductory chapter of this book, we mentioned that the sample covariance matrix $\mathbf{R}_n \in \mathbb{C}^{N \times N}$ obtained from n independent samples of a random process $\mathbf{x} \in \mathbb{C}^N$ is a consistent estimate of $\mathbf{R} \triangleq \mathrm{E}[\mathbf{x}\mathbf{x}^{\mathsf{H}}]$ as $n \to \infty$ for N fixed, in the sense that, for any given matrix norm $\|\mathbf{R}_n - \mathbf{R}\| \to 0$ as $n \to \infty$ (the convergence being almost sure under mild assumptions). As such, $\mathbf{R}_n$ was referred to as an n-consistent estimator of $\mathbf{R}$. However, it was then shown by means of the Marčenko–Pastur law that $\mathbf{R}_n$ is *not* an (n, N)-consistent estimator for $\mathbf{R}$ in the sense that, as both (n, N) grow large with ratio bounded away from zero and ∞, the spectral norm of the matrix difference stays often away from zero. We then provided an explicit expression for the asymptotic l.s.d. of $\mathbf{R}_n$ in this case. However, in most estimation problems, we are actually interested in knowing $\mathbf{R}$ itself, and not $\mathbf{R}_n$ (or its limit). That is, we are more interested in the *inverse* problem of finding $\mathbf{R}$ given $\mathbf{R}_n$, rather than in the *direct* problem of finding the l.s.d. of $\mathbf{R}_n$ given $\mathbf{R}$.

8.1 G-estimation

8.1.1 Girko G-estimators

The first well-known examples of (n, N)-consistent estimators were provided by Girko, see, e.g., [Girko], who derived more than fifty (n, N)-consistent estimators for various functionals of random matrices. Those estimators are called G-estimators after Girko's name, and are numbered in sequence as G_1, G_2, etc.

The G_1, G_3, and G_4 estimators may be rather useful in the context of wireless communications and are given hereafter. The first estimator G_1 is a consistent estimator of the log determinant of the population covariance matrix $\mathbf{R}$, also referred to as the *generalized variance*.

Theorem 8.1. *Let $\mathbf{x}_1, \ldots, \mathbf{x}_n \in \mathbb{R}^N$ be n i.i.d. realizations of a given random process with covariance $\mathrm{E}[(\mathbf{x}_1 - \mathrm{E}[\mathbf{x}_1])(\mathbf{x}_1 - \mathrm{E}[\mathbf{x}_1])^{\mathsf{H}}] = \mathbf{R}$ and $n > N$. Denote $\mathbf{R}_n$ the sample covariance matrix defined as*

$$\mathbf{R}_n \triangleq \frac{1}{n} \sum_{i=1}^{n} (\mathbf{x}_i - \hat{\mathbf{x}})(\mathbf{x}_i - \hat{\mathbf{x}})^{\mathsf{H}}$$

where $\hat{\mathbf{x}} \triangleq \frac{1}{n}\sum_{i=1}^N \mathbf{x}_i$. *Define* G_1 *the functional*

$$G_1(\mathbf{R}_n) = \alpha_n^{-1}\left[\log\det(\mathbf{R}_n) + \log\frac{n(n-1)^N}{(n-N)\prod_{k=1}^N(n-k)}\right]$$

with α_n *any sequence such that* $\alpha_n^{-2}\log(n/(n-N)) \to 0$. *We then have*

$$G_1(\mathbf{R}_n) - \alpha_n^{-1}\log\det(\mathbf{R}) \to 0$$

in probability.

The G_3 estimator deals with the inverse covariance matrix. The result here is surprisingly simple.

Theorem 8.2. *Let* $\mathbf{R} \in \mathbb{R}^{N\times N}$ *invertible and* $\mathbf{R}_n \in \mathbb{R}^{N\times N}$ *be defined as in Theorem 8.1. Define* G_3 *as the function*

$$G_3(\mathbf{R}_n) = \left(1 - \frac{N}{n}\right)\mathbf{R}_n^{-1}.$$

Then, for $\mathbf{a} \in \mathbb{R}^N$, $\mathbf{b} \in \mathbb{R}^N$ *of uniformly bounded norm, we have:*

$$\mathbf{a}^\mathsf{T} G_3(\mathbf{R}_n)\mathbf{b} - \mathbf{a}^\mathsf{T}\mathbf{R}^{-1}\mathbf{b} \to 0$$

in probability.

The G_4 estimator is a consistent estimator of the second order moment of the population covariance matrix, in a sample covariance matrix model. This unfolds from Theorem 3.13 and is given in the following.

Theorem 8.3. *Let* $\mathbf{R} \in \mathbb{R}^{N\times N}$ *and* $\mathbf{R}_n \in \mathbb{R}^{N\times N}$ *be defined as in Theorem 8.1. Define* G_4 *the function*

$$G_4(\mathbf{R}_n) = \frac{1}{N}\operatorname{tr}\mathbf{R}_n^2 - \frac{1}{nN}\left(\operatorname{tr}\mathbf{R}_n\right)^2.$$

Then

$$G_4(\mathbf{R}_n) - \frac{1}{N}\operatorname{tr}\mathbf{R}^2 \to 0$$

in probability.

This last result is compliant with the free probability estimator, for less stringent hypotheses on $\mathbf{R}_n$.

It is then possible to derive some functionals of $\mathbf{R}$ based on the observation $\mathbf{R}_n$. Note in particular that the multiplicative free deconvolution operation presented in Chapters 4 and 5 allows us to obtain (n, N)-consistent estimates of the successive moments of the eigenvalue distribution of $\mathbf{R}$ as a function of the moments of the e.s.d. of $\mathbf{R}_n$. Those can therefore be seen as G-estimators of the moments of the l.s.d. of $\mathbf{R}$. Now, we may be interested in an even more difficult inverse problem: provide an (n, N)-consistent estimator of every eigenvalue in $\mathbf{R}$. In the case where $\mathbf{R}$ has eigenvalues with large multiplicities, this problem has been recently solved by Mestre in [Mestre, 2008b]. The following section presents this recent G-estimation result and details the mathematical approach used by Mestre to determine this estimator.

8.1.2 G-estimation of population eigenvalues and eigenvectors

For ease of read, we come back to the notations of Section 7.1. In this case, we have the following result

Theorem 8.4 ([Mestre, 2008b]). *Let* $\mathbf{B}_N = \mathbf{T}_N^{\frac{1}{2}} \mathbf{X}_N \mathbf{X}_N^{\mathsf{H}} \mathbf{T}_N^{\frac{1}{2}} \in \mathbb{C}^{N \times N}$ *be defined as in Theorem 7.6, i.e.* $\mathbf{T}_N$ *has* K *distinct eigenvalues* $t_1 < \ldots < t_K$ *with multiplicities* $N_1, \ldots, N_K$*, respectively, for all* r*,* $N_r/n \to c_r$*,* $0 < c_r < \infty$*. Further denote* $\lambda_1 \leq \ldots \leq \lambda_N$ *the eigenvalues of* $\mathbf{B}_N$ *and* $\boldsymbol{\lambda} = (\lambda_1, \ldots, \lambda_N)^{\mathsf{T}}$*. Let* $k \in \{1, \ldots, K\}$ *and define*

$$\hat{t}_k = \frac{n}{N_k} \sum_{m \in \mathcal{N}_k} (\lambda_m - \mu_m) \tag{8.1}$$

with $\mathcal{N}_k = \{\sum_{j=1}^{k-1} N_j + 1, \ldots, \sum_{j=1}^{k} N_j\}$ *and* $\mu_1 \leq \ldots \leq \mu_N$ *are the ordered eigenvalues of the matrix* $\operatorname{diag}(\boldsymbol{\lambda}) - \frac{1}{n}\sqrt{\boldsymbol{\lambda}}\sqrt{\boldsymbol{\lambda}}^{\mathsf{T}}$.

Then, if condition (7.6) of Theorem 7.6 is fulfilled for k*, i.e. cluster* k_F *in* F *is mapped to* t_k *only, we have:*

$$\hat{t}_k - t_k \to 0$$

almost surely as $N, n \to \infty$*,* $N/n \to c$*,* $0 < c < \infty$*.*

The performance of the estimator of Theorem 8.4 is demonstrated in Figure 8.1 for $K = 3$, $t_1 = 1$, $t_2 = 3$, $t_3 = 10$ in the cases when $N = 6$, $n = 18$, and $N = 30$, $n = 90$. Remember from Figure 7.4 that the set $(1, 3, 10)$ fulfills condition (7.6), so that Theorem 8.4 is valid. Observe how accurate the G-estimates of the t_k are already for very small dimensions. We will see both in the current chapter and in Chapter 17 that, under the assumption that a cluster separability condition is met (here, condition (7.6)), this method largely outperforms the moment-based approach that consists in deriving consistent estimates for the first order moments and inferring t_k from these moments. Note already that the naive approach that would consist in taking the mean of the eigenvalues inside each cluster (bottom of Figure 8.1) shows a potentially large bias in the estimated eigenvalue, although the estimator variance seems to be smaller than with the consistent G-estimator.

The reason why condition (7.6) must be fulfilled is far from obvious but will become evident once we understand the proof of Theorem 8.4. Before proceeding,

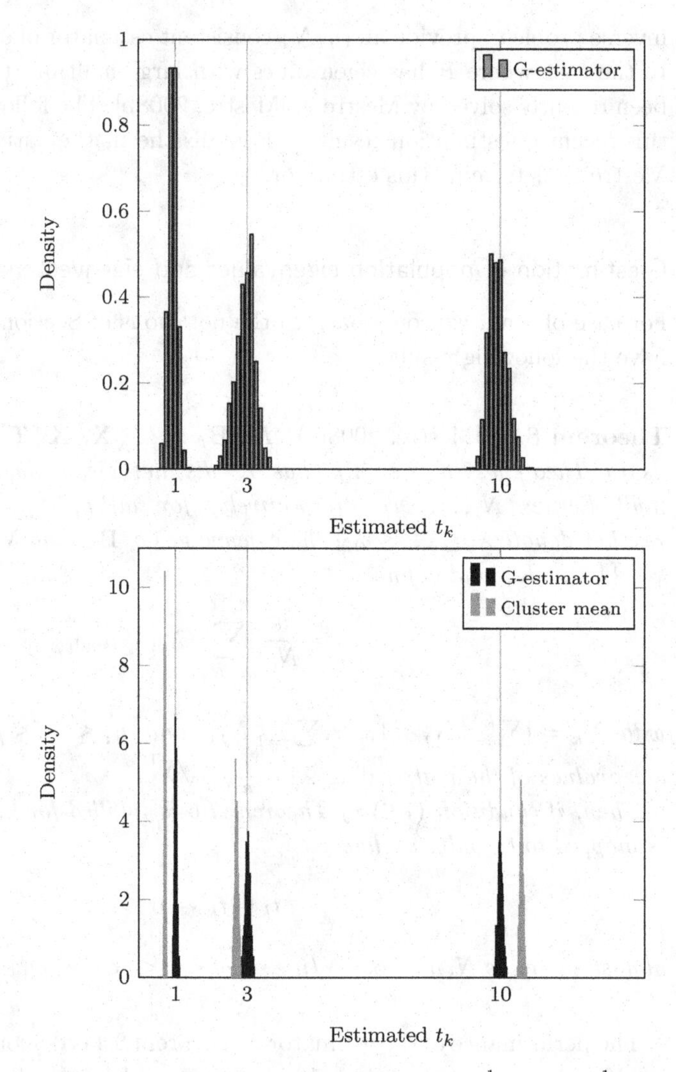

Figure 8.1 G-estimation of t_1, t_2, t_3 in the model $\mathbf{B}_N = \mathbf{T}_N^{\frac{1}{2}}\mathbf{X}_N\mathbf{X}_N^{\mathsf{H}}\mathbf{T}_N^{\frac{1}{2}}$, for $N_1/N = N_2/N = N_3/N = 1/3$,$N/n = 1/10$, for 100 000 simulation runs; Top $N = 6$, $n = 18$, bottom $N = 30$, $n = 90$.

we need to introduce some notions of complex analysis, and its link to the Stieltjes transform.

We first consider the fundamental theorem of complex analysis [Rudin, 1986].

Theorem 8.5 (Cauchy's integral formula.). *Suppose $\mathcal{C}$ is a closed positively oriented path (i.e. counter-clockwise oriented) in an open subset $U \subset \mathbb{C}$ with winding number one (i.e. describing a $360°$ rotation).*

(i) If $z \in U$ is contained in the surface described by $\mathcal{C}$, then for any f holomorphic on U

$$\frac{1}{2\pi i}\oint_{\mathcal{C}} \frac{f(\omega)}{\omega - z} d\omega = f(z). \tag{8.2}$$

If the contour $\mathcal{C}$ is negatively oriented, then the right-hand side becomes $-f(z)$.

(ii) If $z \in U$ is outside the surface described by $\mathcal{C}$, then:

$$\frac{1}{2\pi i}\oint_{\mathcal{C}} \frac{f(\omega)}{\omega - z} d\omega = 0. \tag{8.3}$$

Note that this second result is compliant with the fact that, for f continuous, defined on the real axis, the integral of f along a closed contour $\mathcal{C} \subset \mathbb{R}$ (i.e. a contour that would go from a to b and backwards from b to a) is null.

Consider f some complex holomorphic function on $U \subset \mathbb{C}$, H a distribution function, and denote G the functional

$$G(f) = \int f(z) dH(z).$$

From Theorem 8.5, we then have, for a *negatively* oriented closed path $\mathcal{C}$ enclosing the support of H and with winding number one

$$\begin{aligned} G(f) &= \frac{1}{2\pi i}\int\oint_{\mathcal{C}} \frac{f(\omega)}{z-\omega} d\omega dH(z) \\ &= \frac{1}{2\pi i}\oint_{\mathcal{C}}\int \frac{f(\omega)}{z-\omega} dH(z) d\omega \\ &= \frac{1}{2\pi i}\oint_{\mathcal{C}} f(\omega) m_H(\omega) d\omega \end{aligned} \tag{8.4}$$

the integral inversion being valid since $f(\omega)/(z-\omega)$ is bounded for $\omega \in \mathcal{C}$. Note that the sign inversion due to the negative contour orientation is compensated by the sign reversal of $(\omega - z)$ in the denominator.

If dH is a sum of finite or countable masses and we are interested in evaluating $f(t_k)$, with t_k the value of the kth mass with weight l_k, then on a negatively oriented contour $\mathcal{C}_k$ enclosing t_k and excluding t_j, $j \neq k$

$$l_k f(t_k) = \frac{1}{2\pi i}\oint_{\mathcal{C}_k} f(\omega) m_H(\omega) d\omega. \tag{8.5}$$

This last expression is particularly convenient when we have access to t_k only through an expression of the Stieltjes transform of H.

Now, in terms of random matrices, for the sample covariance matrix model $\mathbf{B}_N = \mathbf{T}_N^{\frac{1}{2}}\mathbf{X}_N\mathbf{X}_N^{\mathsf{H}}\mathbf{T}_N^{\frac{1}{2}}$, we have already noticed that the l.s.d. F of $\mathbf{B}_N$ (or equivalently the l.s.d. $\underline{F}$ of $\underline{\mathbf{B}}_N = \mathbf{X}_N^{\mathsf{H}}\mathbf{T}_N\mathbf{X}_N$) can be rewritten under the form (3.23), which can be further rewritten

$$\frac{c}{m_{\underline{F}}(z)} m_H\left(-\frac{1}{m_{\underline{F}}(z)}\right) = -z m_{\underline{F}}(z) + (c-1) \tag{8.6}$$

where H (previously denoted F^T) is the l.s.d. of $\mathbf{T}_N$. Note that it is allowed to evaluate m_H in $-1/m_{\underline{F}}(z)$ for $z \in \mathbb{C}^+$ since $-1/m_{\underline{F}}(z) \in \mathbb{C}^+$.

As a consequence, if we only have access to $F^{\mathbf{B}_N}$ (from the observation of $\mathbf{B}_N$), then the only link between the observation of $\mathbf{B}_N$ and H is obtained by (i) the fact that $F^{\underline{\mathbf{B}}_N} \Rightarrow \underline{F}$ almost surely and (ii) the fact that $\underline{F}$ and H are related through (8.6). Evaluating a functional f of the eigenvalue t_k of $\mathbf{T}_N$ is then made possible by (8.5). The relations (8.5) and (8.6) are the essential ingredients behind the proof of Theorem 8.4, which we detail below.

Proof of Theorem 8.4. We have from Equation (8.5) that, for any continuous f and for any *negatively oriented* contour $\mathcal{C}_k$ that circles around t_k but none of the t_j for $j \neq k$, $f(t_k)$ can be written under the form

$$\begin{aligned}\frac{N_k}{N} f(t_k) &= \frac{1}{2\pi i} \oint_{\mathcal{C}_k} f(\omega) m_H(\omega) d\omega \\ &= \frac{1}{2\pi i} \oint_{\mathcal{C}_k} \frac{1}{N} \sum_{r=1}^{K} N_r \frac{f(\omega)}{t_r - \omega} d\omega\end{aligned}$$

with H the limit $F^{\mathbf{T}_N} \Rightarrow H$. This provides a link between $f(t_k)$ for all continuous f and the Stieltjes transform $m_H(z)$.

Letting $f(x) = x$ and taking the limit $N \to \infty$, $N_k/N \to c_k/c$, with $c \triangleq c_1 + \ldots + c_K$ the limit of N/n, we have:

$$\frac{c_k}{c} t_k = \frac{1}{2\pi i} \oint_{\mathcal{C}_k} \omega m_H(\omega) d\omega. \tag{8.7}$$

We now want to express m_H as a function of m_F, the Stieltjes transform of the l.s.d. F of $\mathbf{B}_N$. For this, we have the two relations (3.24), i.e.

$$m_{\underline{F}}(z) = c m_F(z) + (c-1)\frac{1}{z}$$

and (8.6) with $F^T = H$, i.e.

$$\frac{c}{m_{\underline{F}}(z)} m_H\left(-\frac{1}{m_{\underline{F}}(z)}\right) = -z m_{\underline{F}}(z) + (c-1).$$

Together, those two equations give the simpler expression

$$m_H\left(-\frac{1}{m_{\underline{F}}(z)}\right) = -z m_{\underline{F}}(z) m_F(z). \tag{8.8}$$

Applying the variable change $\omega = -1/m_{\underline{F}}(z)$ in (8.7), we end up with

$$\begin{aligned}\frac{c_k}{c} t_k &= \frac{1}{2\pi i} \oint_{\mathcal{C}_{\underline{F},k}} z \frac{m'_{\underline{F}}(z)}{m_{\underline{F}}(z) c} + \frac{1-c}{c} \frac{m_{\underline{F}}(z)'}{m^2_{\underline{F}}(z)} dz \\ &= \frac{1}{c} \frac{1}{2\pi i} \oint_{\mathcal{C}_{\underline{F},k}} z \frac{m'_{\underline{F}}(z)}{m_{\underline{F}}(z)} dz,\end{aligned} \tag{8.9}$$

where $\mathcal{C}_{\underline{F},k}$ is the (well-defined) preimage of $\mathcal{C}_k$ by $-1/m_{\underline{F}}$. The second equality (8.9) comes from the fact that the second term in the previous relation is the derivative of $(c-1)/(cm_{\underline{F}}(z))$, which therefore integrates to zero on a closed path, as per classical real or complex integration rules [Rudin, 1986]. Obviously, since $z \in \mathbb{C}^+$ is equivalent to $-1/m_{\underline{F}}(z) \in \mathbb{C}^+$ (the same being true if $\mathbb{C}^+$ is replaced by $\mathbb{C}^-$), $\mathcal{C}_{\underline{F},k}$ is clearly continuous and of non-zero imaginary part whenever $\Im[z] \neq 0$. Now, we must be careful about the exact choice of $\mathcal{C}_{\underline{F},k}$.

Since k is assumed to satisfy the separability conditions of Theorem 7.6, the cluster k_F associated with k in F is distinct from the clusters $(k-1)_F$ and $(k+1)_F$ (whenever they exist). Let us then pick $x_F^{(l)}$ and $x_F^{(r)}$ two real values such that

$$x^+_{(k-1)_F} < x_F^{(l)} < x^-_{k_F} < x^+_{k_F} < x_F^{(r)} < x^-_{(k+1)_F}$$

with $\{x_1^-, x_1^+, \ldots, x_{K_F}^-, x_{K_F}^+\}$ the support boundary of F, as defined in Theorem 7.5. That is, we take a point $x_F^{(l)}$ right on the left side of cluster k_F and a point $x_F^{(r)}$ right on the right side of cluster k_F. Now remember Theorem 7.4 and Figure 7.2; for $x_F^{(l)}$ as defined previously, $m_{\underline{F}}(z)$ has a limit $m^{(l)} \in \mathbb{R}$ as $z \to x_F^{(l)}$, $z \in \mathbb{C}^+$, and a limit $m^{(r)} \in \mathbb{R}$ as $z \to x_F^{(r)}$, $z \in \mathbb{C}^+$, those two limits verifying

$$t_{k-1} < x^{(l)} < t_k < x^{(r)} < t_{k+1} \tag{8.10}$$

with $x^{(l)} \triangleq -1/m^{(l)}$ and $x^{(r)} \triangleq -1/m^{(r)}$.

This is the most important outcome of our integration process. Let us choose $\mathcal{C}_{\underline{F},k}$ to be *any* continuously differentiable contour surrounding cluster k_F such that $\mathcal{C}_{\underline{F},k}$ crosses the real axis in only two points, namely $x_F^{(l)}$ and $x_F^{(r)}$. Since $-1/m_{\underline{F}}(\mathbb{C}^+) \subset \mathbb{C}^+$ and $-1/m_{\underline{F}}(\mathbb{C}^-) \subset \mathbb{C}^-$, $\mathcal{C}_k$ does not cross the real axis whenever $\mathcal{C}_{\underline{F},k}$ is purely complex and is obviously continuously differentiable there; now $\mathcal{C}_k$ crosses the real axis in $x^{(l)}$ and $x^{(r)}$, and is in fact continuous there. Because of (8.10), we then have that $\mathcal{C}_k$ is (at least) continuous and piecewise continuously differentiable and encloses *only* t_k. This is what is required to ensure the validity of (8.9). In Figure 8.2, we consider the case where $\mathbf{T}_N$ is formed of three evenly weighted eigenvalues $t_1 = 1$, $t_2 = 3$ and $t_3 = 10$, and we depict the contours $\mathcal{C}_k$, preimages of $\mathcal{C}_{\underline{F},k}$, $k \in \{1, 2, 3\}$, circular contours around the clusters k_F such that they cross the real line in the positions $x_F^{(l)}$ and $x_F^{(r)}$, corresponding to the inflexion points of $x_F(\underline{m})$ (and an arbitrary large value for the extreme right point).

The difficult part of the proof is completed. The rest will unfold more naturally. We start by considering the following expression

$$\begin{aligned}\hat{t}_k &\triangleq \frac{1}{2\pi i}\frac{n}{N_k}\oint_{\mathcal{C}_{\underline{F},k}} z\frac{m'_{\underline{\mathbf{B}}_N}(z)}{m_{\underline{\mathbf{B}}_N}(z)}dz \\ &= \frac{1}{2\pi i}\frac{n}{N_k}\oint_{\mathcal{C}_{\underline{F},k}} z\frac{\frac{1}{n}\sum_{i=1}^n \frac{1}{(\lambda_i - z)^2}}{\frac{1}{n}\sum_{i=1}^n \frac{1}{\lambda_i - z}}dz, \end{aligned} \tag{8.11}$$

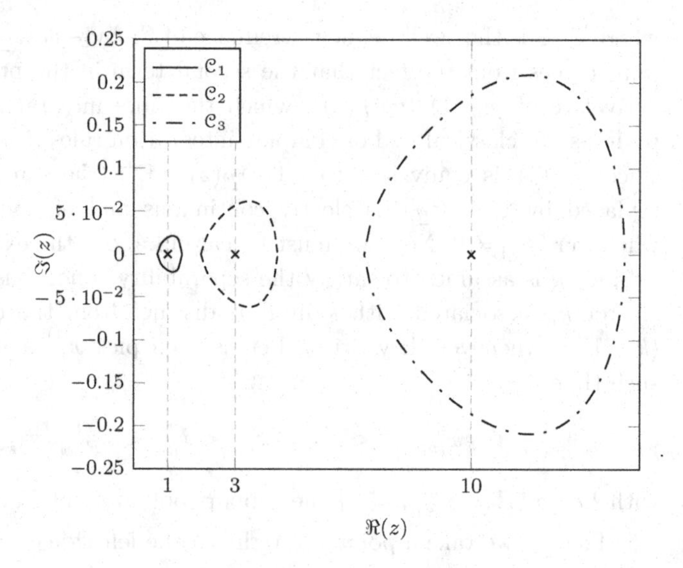

Figure 8.2 Integration contours $\mathcal{C}_k$, $k \in \{1, 2, 3\}$, preimage of $\mathcal{C}_{\underline{F},k}$ by $-1/m_{\underline{F}}$, for $\mathcal{C}_{\underline{F},k}$ a circular contour around cluster k_F, when $\mathbf{T}_N$ composed of three distinct entries, $t_1 = 1$, $t_2 = 3$, $t_3 = 10$, $N_1 = N_2 = N_3$, $N/n = 1/10$.

where we recall that $\underline{\mathbf{B}}_N \triangleq \mathbf{X}_N^{\mathsf{H}} \mathbf{T}_N \mathbf{X}_N$ and, if $n \geq N$, $\lambda_{N+1} = \ldots = \lambda_n = 0$.

The value $\hat{t}_k$ can be viewed as the empirical counterpart of t_k. Now, we know from Theorem 3.13 that $m_{\mathbf{B}_N}(z) \xrightarrow{\text{a.s.}} m_F(z)$ and $m_{\underline{\mathbf{B}}_N}(z) \xrightarrow{\text{a.s.}} m_{\underline{F}}(z)$. It is not difficult to verify, from the fact that $m_{\underline{F}}$ is holomorphic, that the same convergence holds for the successive derivatives.

At this point, we need the two fundamental results that are Theorem 7.1 and Theorem 7.2. We know that, for all matrices $\mathbf{B}_N$ in a set of probability one, all the eigenvalues of $\mathbf{B}_N$ are contained in the support of F for all large N, and that the eigenvalues of $\mathbf{B}_N$ contained in cluster k_F are exactly $\{\lambda_i, i \in \mathcal{N}_k\}$ for these large N. Take such a $\mathbf{B}_N$. For all large N, $m_{\mathbf{B}_N}(z)$ is uniformly bounded over N and $z \in \mathcal{C}_{\underline{F},k}$, since $\mathcal{C}_{\underline{F},k}$ is away from the support of F. The integrand in the right-hand side of (8.11) is then uniformly bounded for all large N and for all $z \in \mathcal{C}_{\underline{F},k}$. By the dominated convergence theorem, Theorem 6.3, we then have that $\hat{t}_k - t_k \xrightarrow{\text{a.s.}} 0$.

It then remains to prove that $\hat{t}_k$ takes the form (8.1). This is performed by residue calculus [Rudin, 1986], i.e. by determining the poles in the expanded expression of $\hat{t}_k$ (when developing $m_{\mathbf{B}_N}(z)$ in its full expression).

For this, we open a short parenthesis to introduce the basic rules of complex integration, required here. First, we need to define poles and residues.

Definition 8.1. Let γ be a continuous, piecewise continuously differentiable contour on $\mathbb{C}$. If f is holomorphic inside γ but on a, i.e.

$$\lim_{z \to a} |f(z)| = \infty$$

then a is a *pole* of f. We then define the *order* of a as being the smallest integer k such that

$$\lim_{z\to a} |(z-a)^k f(z)| < \infty.$$

The *residue* $\text{Res}(f,a)$ of f in a, is then defined as

$$\text{Res}(f,a) \triangleq \lim_{z\to a} \frac{d^{k-1}}{dz^{k-1}} \left[(z-a)^k f(z)\right].$$

This being defined, we have the following fundamental result of complex analysis.

Theorem 8.6. *Let γ be a continuous, piecewise continuously differentiable positively oriented contour on $\mathbb{C}$. For f holomorphic inside γ but on a discrete number of points, we have that*

$$\frac{1}{2\pi i}\oint_\gamma f(z)dz = \sum_{a \text{ pole of } f} \text{Res}(f,a).$$

If γ is negatively oriented, then the right-hand side term must be multiplied by -1.

The calculus procedure of residues is then as follows:

1. we first determine the poles of f lying inside the surface formed by γ,
2. we then determine the order of each pole,
3. we finally compute the residues of f at the poles and evaluate the integral as the sum of all residues.

In our problem, the poles of the integrand of (8.11) are found to be $\lambda_1,\ldots,\lambda_N$ (indeed, the integrand of (8.11) behaves like $O(1/(\lambda_i - z))$ for $z \simeq \lambda_i$) and $\mu_1,\ldots,\mu_N$, the N real roots of the equation in μ, $m_{\underline{\mathbf{B}}_N}(\mu) = 0$ (indeed, the denominator of the integrand cancels for $z = \mu_i$, while the numerator is non-zero). Since $\mathcal{C}_{\underline{F},k}$ encloses only those values λ_i such that $i \in \mathcal{N}_k$, the other poles are discarded. Noticing now that $m_{\mathbf{B}_N}(\mu) \to \pm\infty$ as $\mu \to \lambda_i$, we deduce that $\mu_1 < \lambda_1 < \mu_2 < \ldots < \mu_N < \lambda_N$, and therefore we have that μ_i, $i \in \mathcal{N}_k$ are all in $\mathcal{C}_{\underline{F},k}$ but maybe for μ_j, $j = \min \mathcal{N}_k$. It can in fact be shown that μ_j is also in $\mathcal{C}_{\underline{F},k}$. To notice this last remaining fact, observe simply that

$$\frac{1}{2\pi i}\oint_{\mathcal{C}_k} \frac{1}{\omega} d\omega = 0$$

since zero is not contained in the contour $\mathcal{C}_k$. Applying the variable change $\omega = -1/m_{\underline{F}}(z)$ as previously, this gives

$$\oint_{\mathcal{C}_{\underline{F},k}} \frac{m'_{\underline{F}}(z)}{m^2_{\underline{F}}(z)} dz = 0. \tag{8.12}$$

From the same reasoning as above, with the dominated convergence theorem argument, Theorem 6.3, we have that, for sufficiently large N and almost surely

$$\left| \oint_{\mathcal{C}_{\underline{F},k}} \frac{m'_{\underline{\mathbf{B}}_N}(z)}{m^2_{\underline{\mathbf{B}}_N}(z)} dz \right| < \frac{1}{2}. \tag{8.13}$$

We now proceed to residue calculus in order to compute the integral in the left-hand side of (8.13). Following the above procedure, notice that the poles of (8.12) are the λ_i and the μ_i that lie inside the integration contour $\mathcal{C}_{\underline{F},k}$, all of order one with residues equal to -1 and 1, respectively. These residues are obtained using in particular L'Hospital rule, Theorem 2.10, as detailed below for the final calculus. Therefore, (8.12) equals the number of such λ_i minus the number of such μ_i (remember that the integration contour is negatively oriented, so we need to reverse the signs). We however already know that this difference, for large N, equals either zero or one, since only the position of the leftmost μ_i is unknown yet. But since the integral is asymptotically less than 1/2, this implies that it is identically zero, and therefore the leftmost μ_i (indexed by $\min \mathcal{N}_k$) also lies inside the integration contour.

We have therefore precisely characterized $\mathcal{N}_k$. We can now evaluate (8.11). This calls again for residue calculus, the steps of which are detailed below. Denoting

$$f(z) = z \frac{m'_{\underline{\mathbf{B}}_N}(z)}{m_{\underline{\mathbf{B}}_N}(z)},$$

we find that λ_i (inside $\mathcal{C}_{\underline{F},k}$) is a pole of order 1 with residue

$$\lim_{z \to \lambda_i} (z - \lambda_i) f(z) = -\lambda_i$$

which is straightforwardly obtained from the fact that $f(z) \sim \frac{1}{\lambda_i - z}$ as $z \sim \lambda_i$. Also μ_i (inside $\mathcal{C}_{\underline{F},k}$) is a pole of order 1 with residue

$$\lim_{z \to \mu_i} (z - \mu_i) f(z) = \mu_i$$

which is obtained using L'Hospital rule: upon existence of a limit, we indeed have

$$\lim_{z \to \mu_i} (z - \mu_i) f(z) = \lim_{z \to \mu_i} \frac{\frac{d}{dz}\left[(z - \mu_i) z m'_{\underline{\mathbf{B}}_N}(z)\right]}{\frac{d}{dz}\left[m_{\underline{\mathbf{B}}_N}(z)\right]}$$

which expands as

$$\lim_{z \to \mu_i} (z - \mu_i) f(z) = \lim_{z \to \mu_i} \frac{z m'_{\underline{\mathbf{B}}_N}(z) + z(z - \mu_i) m''_{\underline{\mathbf{B}}_N}(z) + (z - \mu_i) m'_{\underline{\mathbf{B}}_N}(z)}{m'_{\underline{\mathbf{B}}_N}(z)}.$$

Notice now that $|m'_{\underline{\mathbf{B}}_N}(z)|$ is positive and uniformly bounded by $1/\varepsilon^2$ for $\min_i\{|\lambda_i - z|\} > \varepsilon$. Therefore, the ratio is always well defined and, for $z \to \mu_i$ with μ_i poven away from all λ_i, we finally have

$$\lim_{z \to \mu_i} (z - \mu_i) f(z) = \lim_{z \to \mu_i} z = \mu_i.$$

Since the integration contour is chosen to be *negatively oriented*, it must be kept in mind that the signs of the residues need be inverted in the final relation.

It now remains to verify that $\mu_1, \ldots, \mu_N$ are also the eigenvalues of $\text{diag}(\boldsymbol{\lambda}) - \frac{1}{n}\sqrt{\boldsymbol{\lambda}}\sqrt{\boldsymbol{\lambda}}^{\mathsf{T}}$. This is immediate from the following lemma.

Lemma 8.1 ([Couillet *et al.*, 2011c],[Gregoratti and Mestre, 2009]). *Let $\mathbf{A} \in \mathbb{C}^{N\times N}$ be diagonal with entries $\lambda_1, \ldots, \lambda_N$ and $\mathbf{y} \in \mathbb{C}^N$. Then the eigenvalues of $(\mathbf{A} - \mathbf{y}\mathbf{y}^*)$ are the N real solutions in x of*

$$\sum_{i=1}^{N} \frac{y_i^2}{\lambda_i - x} = 1.$$

Proof. Let $\mathbf{A} \in \mathbb{C}^{N\times N}$ be a Hermitian matrix and $\mathbf{y} \in \mathbb{C}^N$. If μ is an eigenvalue of $(\mathbf{A} - \mathbf{y}\mathbf{y}^*)$ with eigenvector $\mathbf{x}$, we have the equivalent relations

$$\begin{aligned}
(\mathbf{A} - \mathbf{y}\mathbf{y}^*)\mathbf{x} &= \mu\mathbf{x}, \\
(\mathbf{A} - \mu\mathbf{I}_N)\mathbf{x} &= \mathbf{y}^*\mathbf{x}\mathbf{y}, \\
\mathbf{x} &= \mathbf{y}^*\mathbf{x}(\mathbf{A} - \mu\mathbf{I}_N)^{-1}\mathbf{y}, \\
\mathbf{y}^*\mathbf{x} &= \mathbf{y}^*\mathbf{x}\mathbf{y}^*(\mathbf{A} - \mu\mathbf{I}_N)^{-1}\mathbf{y}, \\
1 &= \mathbf{y}^*(\mathbf{A} - \mu\mathbf{I}_N)^{-1}\mathbf{y}.
\end{aligned}$$

Take $\mathbf{A}$ diagonal with entries $\lambda_1, \ldots, \lambda_N$, we then have

$$\sum_{i=1}^{N} \frac{y_i^2}{\lambda_i - \mu} = 1.$$

□

Taking $\mathbf{A} = \text{diag}(\boldsymbol{\lambda})$ and $y_i = \sqrt{\lambda_i}/\sqrt{n}$, we have the expected result. This completes the proof of Theorem 8.4. □

Other G-estimators can be derived from this technique. In particular, note that, for $\mathbf{x}, \mathbf{y} \in \mathbb{C}^N$ given vectors, and $\mathbf{T}_N = \sum_{k=1}^{K} t_k \mathbf{U}_k \mathbf{U}_k^{\mathsf{H}} \in \mathbb{C}^{N\times N}$ the spectral distribution of $\mathbf{T}_N$ in Theorem 8.4, we have from residue calculus

$$\mathbf{x}^{\mathsf{H}}\mathbf{U}_k\mathbf{U}_k^{\mathsf{H}}\mathbf{y} = \frac{1}{2\pi i}\oint_{\mathcal{C}_k} \mathbf{x}^{\mathsf{H}}(\mathbf{T}_N - z\mathbf{I}_N)^{-1}\mathbf{y}dz$$

with $\mathcal{C}_k$ a negatively oriented contour enclosing t_k, but none of the t_i, $i \neq k$. From this remark, using similar derivations as above for the quadratic form $\mathbf{x}^{\mathsf{H}}(\mathbf{T}_N - z\mathbf{I}_N)^{-1}\mathbf{y}$ instead of the Stieltjes transform $\frac{1}{N}\operatorname{tr}(\mathbf{T}_N - z\mathbf{I}_N)^{-1}$, we then have the following result.

Theorem 8.7 ([Mestre, 2008b]). *Let $\mathbf{B}_N$ be defined as in Theorem 8.4, and denote $\mathbf{B}_N = \sum_{k=1}^{N} \lambda_k \mathbf{b}_k \mathbf{b}_k^{\mathsf{H}}$, $\mathbf{b}_k^{\mathsf{H}}\mathbf{b}_i = \delta_k^i$, the spectral decomposition of $\mathbf{B}_N$. Similarly, denote $\mathbf{T}_N = \sum_{k=1}^{K} t_k \mathbf{U}_k \mathbf{U}_k^{\mathsf{H}}$, $\mathbf{U}_k^{\mathsf{H}}\mathbf{U}_k = \mathbf{I}_{n_k}$, with $\mathbf{U}_k \in \mathbb{C}^{N\times N_k}$ the eigenspace associated with t_k. For given vectors $\mathbf{x}, \mathbf{y} \in \mathbb{C}^N$, denote*

$$u(k; \mathbf{x}, \mathbf{y}) \triangleq \mathbf{x}^{\mathsf{H}}\mathbf{U}_k\mathbf{U}_k^{\mathsf{H}}\mathbf{y}.$$

Then we have:

$$\hat{u}(k;\mathbf{x},\mathbf{y}) - u(k;\mathbf{x},\mathbf{y}) \xrightarrow{\text{a.s.}} 0$$

as $N, n \to \infty$ *with ratio* $c_N = N/n \to c$, *where*

$$\hat{u}(k;\mathbf{x},\mathbf{y}) \triangleq \sum_{i=1}^{N} \theta_k(i)\mathbf{x}^{\mathsf{H}}\mathbf{b}_k\mathbf{b}_k^{\mathsf{H}}\mathbf{y}$$

and $\theta_k(i)$ *is defined by*

$$\theta_i(k) = \begin{cases} -\phi_k(i) & , \ i \notin \mathcal{N}_k \\ 1 + \psi_k(i) & , \ i \in \mathcal{N}_k \end{cases}$$

with

$$\phi_k(i) = \sum_{r \in \mathcal{N}_k} \left(\frac{\lambda_r}{\lambda_i - \lambda_r} - \frac{\mu_r}{\lambda_i - \mu_r} \right)$$

$$\psi_k(i) = \sum_{r \notin \mathcal{N}_k} \left(\frac{\lambda_r}{\lambda_i - \lambda_r} - \frac{\mu_r}{\lambda_i - \mu_r} \right)$$

and $\mathcal{N}_k$, $\mu_1, \ldots, \mu_N$ *defined as in Theorem 8.4.*

This result will be shown to be appealing in problems of direction of arrival (DoA) detection, see Chapter 17.

We complete this section with modified versions of Girko's G-1 estimator, Theorem 8.1, which are obtained from similar sample covariance arguments as above. The first result is merely a generalization of the convergence in probability of Theorem 8.1 to almost sure convergence.

Theorem 8.8 (Theorem 1 in [Kammoun *et al.*, 2011]). *Define the matrix* $\mathbf{Y}_N = \mathbf{T}_N\mathbf{X}_N + \frac{1}{\sqrt{x}}\mathbf{W}_N \in \mathbb{C}^{N\times M}$ *for* $x > 0$, *with* $\mathbf{X}_N \in \mathbb{C}^{n\times M}$ *and* $\mathbf{W}_N \in \mathbb{C}^{N\times M}$ *random matrices with independent entries of zero mean, unit variance and finite* $2+\varepsilon$ *order moment for some* $\varepsilon > 0$ *and* $\mathbf{T}_N \in \mathbb{C}^{N\times n}$ *deterministic such that* $\mathbf{T}_N\mathbf{T}_N^{\mathsf{H}}$ *has uniformly bounded spectral norm along growing* N. *Assume that the e.s.d. of* $\mathbf{T}_N\mathbf{T}_N^{\mathsf{H}}$ *converges weakly to* H *as* $N \to \infty$. *Denote* $\mathbf{B}_N = \frac{1}{M}\mathbf{Y}_N\mathbf{Y}_N^{\mathsf{H}}$. *Then, as* $N, n, M \to \infty$, *with* $\frac{M}{N} \to c > 1$ *and* $\frac{N}{n} \to c_0$

$$\frac{1}{N}\log\det\left(\mathbf{I}_N + x\mathbf{T}_N\mathbf{T}_N^{\mathsf{H}}\right) - \left[\frac{1}{N}\log\det\left(x\mathbf{B}_N\right) + \frac{M-N}{N}\log\left(\frac{M-N}{M}\right) + 1\right] \xrightarrow{\text{a.s.}} 0.$$

Under this setting, the G-estimator is exactly the estimator of the Shannon transform of $\mathbf{T}_N\mathbf{T}_N^{\mathsf{H}}$ at point x or equivalently of the capacity of a deterministic multiple antenna link $\mathbf{T}_N$ under additive noise variance $1/x$ from the observations of the data vectors $\mathbf{y}_1, \ldots, \mathbf{y}_M$ such that $\mathbf{Y} = [\mathbf{y}_1, \ldots, \mathbf{y}_M]$. A simple way to derive this result is to use the Shannon transform relation of Equation (3.5)

$$\frac{1}{N}\log\det(\mathbf{I}_N + x\mathbf{T}_N\mathbf{T}_N^{\mathsf{H}}) = \int_0^x \left(\frac{1}{t} - \frac{1}{t^2}m_H\left(-\frac{1}{t}\right)\right)dt \tag{8.14}$$

along with the fact, similar to Equation (8.8), that, for $z \in \mathbb{C}\setminus\mathbb{R}^+$

$$m_H\left(-\frac{1}{m_{\underline{F}}(z)} - \frac{1}{x}\right) = -zm_F(z)m_{\underline{F}}(z)$$

with F the l.s.d. of $\mathbf{B}_N = \frac{1}{M}\mathbf{Y}_N\mathbf{Y}_N^{\mathsf{H}}$ and $\underline{F}$ the l.s.d. of $\frac{1}{M}\mathbf{Y}_N^{\mathsf{H}}\mathbf{Y}_N$. The change of variable $t = (1/m_{\underline{F}}(u) + 1/x)^{-1}$ in Equation (8.14) allows us to write $\mathcal{V}_{\mathbf{T}_N\mathbf{T}_N^{\mathsf{H}}}(x)$ as a function of $m_{\underline{F}}$ from which we obtain directly the above estimator.

The second result introduces an additional deterministic matrix $\mathbf{R}_N$, which in an applicative sensing context can be used to infer the achievable communication rate over a channel under unknown interference pattern. We precisely have the following.

Theorem 8.9 (Theorem 2 in [Kammoun *et al.*, 2011]). *Define the matrix* $\mathbf{Y}_N = \mathbf{T}_N\mathbf{X}_N + \frac{1}{\sqrt{x}}\mathbf{W}_N \in \mathbb{C}^{N\times M}$ *for* $x > 0$ *where* $\mathbf{X}_N \in \mathbb{C}^{n\times M}$*, and* $\mathbf{W}_N \in \mathbb{C}^{N\times M}$ *are random matrices with Gaussian independent entries of zero mean and unit variance,* $\mathbf{T}_N \in \mathbb{C}^{n\times n}$ *is deterministic with uniformly bounded spectral norm for which the e.s.d. of* $\mathbf{T}_N\mathbf{T}_N^{\mathsf{H}}$ *converges weakly, and let* $\mathbf{R}_N \in \mathbb{C}^{N\times N}$ *be a deterministic non-negative Hermitian matrix. Then, as* $N, n, M \to \infty$ *with* $1 < \liminf M/N \leq \limsup M/N < \infty$ *and* $0 < \liminf N/n \leq \limsup N/n < \infty$*, we have:*

$$\frac{1}{N}\log\det\left(\mathbf{I}_N + x\left[\mathbf{R}_N + \mathbf{T}_N\mathbf{T}_N^{\mathsf{H}}\right]\right) - \left[\frac{1}{N}\log\det\left(x[\mathbf{B}_N + y_N\mathbf{R}_N]\right) + \frac{M-N}{N}\log(y_N) + \frac{M}{N}(1-y_N)\right] \xrightarrow{\text{a.s.}} 0.$$

with y_N *the unique positive solution of the equation in* y

$$y = \frac{1}{M}\operatorname{tr} y\mathbf{R}_N\left(y\mathbf{R}_N + \mathbf{B}_N\right)^{-1} + \frac{M-N}{M}. \tag{8.15}$$

This result is particularly useful in a rate inference scenario when $\mathbf{R}_N = \mathbf{H}\mathbf{H}^{\mathsf{H}}$ for some multiple antenna channel matrix $\mathbf{H}$ but unknown colored interference $\mathbf{T}_N\mathbf{x}_k + \frac{1}{\sqrt{x}}\mathbf{w}_k$. Theorem 8.9 along with Theorem 8.8 allow for a consistent estimation of the capacity of the MIMO channel $\mathbf{H}$ based on M successive observations of noise-only signals (or the residual terms after data decoding).

The proof of Theorem 8.9 arises first from the fact that, for given y and $\mathbf{R}_N$, a deterministic equivalent for

$$\frac{1}{N}\log\det\left(y\mathbf{R}_N + \mathbf{B}_N\right)$$

was derived in [Vallet and Loubaton, 2009] and expresses as

$$\frac{1}{N}\log\det\left(y\mathbf{R}_N+\mathbf{B}_N\right)$$
$$-\left[\frac{1}{N}\log\det\left(y\mathbf{R}_N+\frac{\mathbf{T}_N\mathbf{T}_N^{\mathsf{H}}+x\mathbf{I}_N}{1+\kappa(y)}\right)+\frac{M}{N}\log(1+\kappa(y))-\frac{\frac{M}{N}\kappa(y)}{1+\kappa(y)}\right]\xrightarrow{\text{a.s.}}0$$

with $\kappa(y)$ the unique positive solution for $y>0$ of

$$\kappa(y)=\frac{1}{M}\operatorname{tr}[\mathbf{T}_N\mathbf{T}_N^{\mathsf{H}}+x\mathbf{I}_N]\left(y\mathbf{R}_N+\frac{1}{1+\kappa(y)}[\mathbf{T}_N\mathbf{T}_N^{\mathsf{H}}+x\mathbf{I}_N]\right)^{-1}.$$

This last result is obtained rather directly using deterministic equivalent methods detailed in Chapter 6. Now, we observe that, if $y=\frac{1}{1+\kappa(y)}$, the term $\frac{1}{N}\log\det(y\mathbf{R}_N+y[\mathbf{T}_N\mathbf{T}_N^{\mathsf{H}}+x\mathbf{I}_N])$ appears, which is very close to what we need. Observing that this has a unique solution, asymptotically close to the unique solution of Equation (8.15), we easily infer the final result. More details are given in [Kammoun *et al.*, 2011].

Similar results are also available beyond the restricted case of sample covariance matrices, in particular for the information plus noise models. We mention especially the information plus noise equivalent to Theorem 8.7, provided in [Vallet *et al.*, 2010], whose study is based on Theorem 7.10.

Theorem 8.10. *Let $\mathbf{B}_N$ be defined as in Theorem 7.8, where we assume that $F^{\frac{1}{n}\mathbf{A}_N\mathbf{A}_N^{\mathsf{H}}}=H$ for all N of practical interest, i.e. we assume $F^{\frac{1}{n}\mathbf{A}_N\mathbf{A}_N^{\mathsf{H}}}$ is composed of K masses in $h_1<\ldots<h_K$ with respective multiplicities $N_1,\ldots,N_K$. Further suppose that $h_1=0$ and let $\mathbf{\Pi}$ be the associated eigenspace of h_1 (the kernel of $\frac{1}{n}\mathbf{A}_N\mathbf{A}_N^{\mathsf{H}}$). Denote $\mathbf{B}_N=\sum_{k=1}^{N}\lambda_k\mathbf{u}_k\mathbf{u}_k^{\mathsf{H}}$ the spectral decomposition of $\mathbf{B}_N$, with $\mathbf{u}_k^{\mathsf{H}}\mathbf{u}_j=\delta_k^j$, and denote*

$$\pi(\mathbf{x})\triangleq\mathbf{x}^{\mathsf{H}}\mathbf{\Pi}\mathbf{x}.$$

Then, we have that

$$\pi(\mathbf{x})-\hat{\pi}(\mathbf{x})\xrightarrow{\text{a.s.}}0$$

where $\hat{\pi}(\mathbf{x})$ is defined as

$$\hat{\pi}(\mathbf{x})\triangleq\sum_{k=1}^{N}\beta_k\mathbf{x}^{\mathsf{H}}\mathbf{u}_k\mathbf{u}_k^{\mathsf{H}}\mathbf{x}$$

with β_k defined as

$$\beta_k=1+\frac{\sigma^2}{N}\sum_{l=N-N_1+1}^{N}\frac{1}{\lambda_l-\lambda_k}+\frac{2\sigma^2}{N}\sum_{l=N-N_1+1}^{N}\frac{\lambda_k}{(\lambda_k-\lambda_l)^2}$$
$$-\sigma^2(1-c)\left[\sum_{l=N-N_1+1}^{N}\frac{1}{\lambda_l-\lambda_k}-\sum_{l=N-N_1+1}^{N}\frac{1}{\mu_l-\lambda_k}\right],\ 1\leq k\leq N-N_1$$

$$\beta_k = 1 + \frac{\sigma^2}{N} \sum_{l=1}^{N-N_1} \frac{1}{\lambda_l - \lambda_k} + \frac{2\sigma^2}{N} \sum_{l=1}^{N-N_1} \frac{\lambda_k}{(\lambda_k - \lambda_l)^2}$$
$$- \sigma^2(1-c) \left[\sum_{l=1}^{N-N_1} \frac{1}{\lambda_k - \mu_l} - \sum_{l=1}^{N-N_1} \frac{1}{\lambda_k - \lambda_l} \right], \; N - N_1 + 1 \leq k \leq N$$

with $\mu_1, \ldots, \mu_N$ *the* N *real roots of* $m_{\mathbf{B}_N}(x) = -1/\sigma^2$.

We do not further develop information plus noise model considerations in this section and move now to second order statistics for G-estimators.

8.1.3 Central limit for G-estimators

The G-estimators derived above are consistent with increasingly large system dimensions but are applied to systems of finite, sometimes small, dimensions. This implies some inevitable inaccuracy in the successive estimates, as observed for instance in Figure 8.1. For application purposes, it is fundamental to be able to assess the quality of these estimates. In mathematical terms, this implies computing statistics of the estimates. Various central limit theorems for estimators of functionals of sample covariance matrix can be found in the literature, notably in Girko's work, see, e.g., [Girko], where central limits for G-estimators are provided.

This section introduces instead a recent result on the limiting distribution of $n(\hat{t}_k - t_k)$ where t_k and $\hat{t}_k$ are defined in Theorem 8.4 as the entries of $\mathbf{T}_N$ for the sample covariance matrix $\mathbf{B}_N = \mathbf{T}_N^{\frac{1}{2}} \mathbf{X}_N \mathbf{X}_N^{\mathsf{H}} \mathbf{T}_N^{\frac{1}{2}} \in \mathbb{C}^{N \times N}$, $\mathbf{X}_N \in \mathbb{C}^{N \times n}$ with entries $\frac{1}{\sqrt{n}} X_{ij}$, i.i.d., such that $\mathbf{X}_{11}$ has zero mean, unit variance, and fourth order moment $\mathrm{E}[|X_{11}|^4] = 2$ and $\mathbf{T}_N \in \mathbb{C}^{N \times N}$ with distinct eigenvalues $t_1 < \ldots < t_K$ of multiplicity $N_1, \ldots, N_K$, respectively. Specifically, we will show that the vector $(n(\hat{t}_k - t_k))_{1 \leq k \leq K}$ is asymptotically Gaussian with zero mean and a covariance which we will evaluate, as $N \to \infty$.

The final result, due to Yao, unfolds as follows.

Theorem 8.11 ([Yao *et al.*, 2011]). *Let* $\mathbf{B}_N$ *be defined as in Theorem 8.4 with* $\mathrm{E}[|X_{N,ij}|^4] = 2$ *and* $N_i/N = c_i + o(1/N)$ *for all* i, $0 < c_i < \infty$. *Denote* $\mathcal{I} \subset \{1, \ldots, K\}$ *the set of indexes* k *such that* k *satisfies the separability condition of Theorem 7.6. Then, for every set* $\mathcal{J} = \{j_1, \ldots, j_p\} \subset \mathcal{I}$, *as* N, n *grow large*

$$\left(n(\hat{t}_k - t_k)\right)_{k \in \mathcal{J}} \Rightarrow X$$

with X *a Gaussian* p*-dimensional vector with zero mean and covariance* $\mathbf{\Theta}^{\mathcal{J}}$ *with* (k, k') *entry* $\Theta_{k,k'}^{\mathcal{J}}$ *defined as*

$$\Theta_{k,k'}^{\mathcal{J}} \triangleq -\frac{1}{4\pi^2 c^2 c_i c_j} \oint_{\mathcal{C}_{j_k}} \oint_{\mathcal{C}_{j_{k'}}} \left[\frac{\underline{m}'(z_1)\underline{m}'(z_2)}{(\underline{m}(z_1) - \underline{m}(z_2))^2} - \frac{1}{(z_1 - z_2)^2} \right] \frac{dz_1 dz_2}{\underline{m}(z_1)\underline{m}(z_2)} \tag{8.16}$$

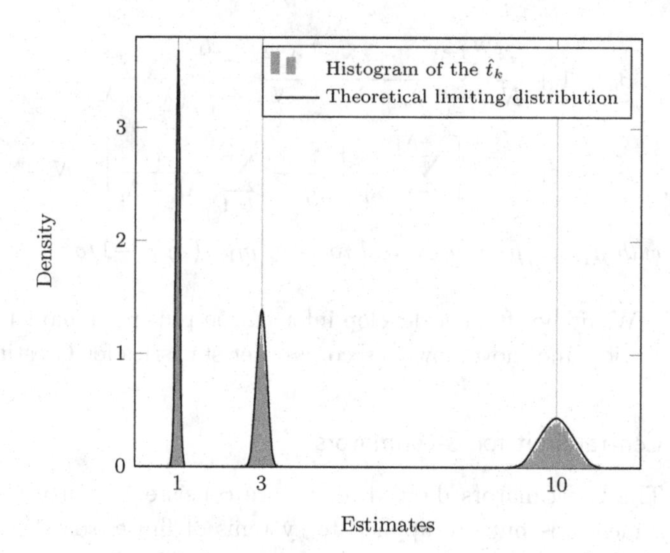

Figure 8.3 Comparison of empirical against theoretical variances for the estimator of Theorem 8.4, based on Theorem 8.11, $K = 3$, $t_1 = 1$, $t_2 = 3$, $t_3 = 10$, $N_1 = N_2 = N_3 = 20$, $n = 600$.

where the contour $\mathcal{C}_k$ encloses the limiting support of the eigenvalues of $\mathbf{B}_N$ indexed by $\mathcal{N}_k = \{\sum_{j=1}^{k-1} N_j + 1, \ldots, \sum_{j=1}^{k} N_j\}$, only, i.e. the cluster k_F as defined in Theorem 7.6. Moreover

$$\hat{\Theta}^{\mathfrak{d}}_{k,k'} - \Theta^{\mathfrak{d}}_{k,k'} \xrightarrow{\text{a.s.}} 0$$

as $N, n \to \infty$, where $\hat{\Theta}^{\mathfrak{d}}_{j_k, j_{k'}}$ is defined by

$$\hat{\Theta}^{\mathfrak{d}}_{k,k'} \triangleq \frac{n^2}{N_k N_{k'}} \left[\sum_{(i,j) \in \mathcal{N}_{j_k} \times \mathcal{N}_{j_{k'}}} \frac{-1}{(\mu_i - \mu_j)^2 m'_{\mathbf{B}_N}(\mu_i) m'_{\mathbf{B}_N}(\mu_j)} + \delta_{kk'} \sum_{i \in \mathcal{N}_k} \left(\frac{m'''_{\mathbf{B}_N}(\mu_i)}{6 m'_{\mathbf{B}_N}(\mu_i)^3} - \frac{m''_{\mathbf{B}_N}(\mu_i)^2}{4 m'_{\mathbf{B}_N}(\mu_i)^4} \right) \right] \tag{8.17}$$

with the quantities $\mu_1, \ldots, \mu_N$ defined as in Theorem 8.4.

In Figure 8.3, the performance of Theorem 8.11 is evaluated against 10 000 Monte Carlo simulations of a scenario of three users, with $t_1 = 1, t_2 = 3, t_3 = 10$, $N_1 = N_2 = N_3 = 20$, $N = 60$, and $n = 600$. It appears that the limiting distribution is very accurate for these values of N, n. Further simulations to obtain empirical estimates $\hat{\Theta}^{\mathfrak{d}}_{k,k}$ of $\Theta^{\mathfrak{d}}_{k,k}$ suggest that $\hat{\Theta}_{k,k}$ is an accurate estimator as well.

We provide hereafter a sketch of the proof of Theorem 8.11.

Proof. The idea of the proof relies on the following remarks:

- from Theorem 3.17, well-behaved functionals of $\mathbf{B}_N$ have a Gaussian limit;

- then, from the proof of Theorem 8.4 and in particular Equation (8.9), the estimator $\hat{t}_k$ of t_k expresses as an integral function of the Stieltjes transform of $F^{\mathbf{B}_N}$ and of its derivative. Applying Theorem 3.17, the variations of $N(m_{\mathbf{B}_N}(z) - m_F(z))$ and of $N(m'_{\mathbf{B}_N}(z) - m'_F(z))$, with F the l.s.d. of $\mathbf{B}_N$ can be proved to be asymptotically Gaussian;
- from there, the Gaussian limits of both $N(m_{\mathbf{B}_N}(z) - m_F(z))$ and $N(m'_{\mathbf{B}_N}(z) - m'_F(z))$ can be further extended to the fluctuations of the integrand in the expression of $\hat{t}_k$ in Equation (8.11), using the so-called *delta method*, to be introduced subsequently;
- final tightness arguments then ensure that the limiting Gaussian fluctuations of the integrand propagate to the fluctuations of the integral, i.e. to $n(\hat{t}_k - t_k)$.

This is the general framework of the proof, for which we provide a sketch hereafter.

Following Theorem 3.17, denote $N(F^{\mathbf{B}_N} - F_N)$ the difference between the e.s.d. of $\mathbf{B}_N$ and the l.s.d. of $\mathbf{B}_N$ modified in such a way that $F^{\mathbf{T}_N}$ replaces the limiting law of $\mathbf{T}_1, \mathbf{T}_2, \ldots$ and N_i/n replaces c_i. Then, for a family $f_1, \ldots, f_p$ of functions holomorphic on $\mathbb{R}^+$, the vector

$$\left(n \int f_i(x) d(F^{\mathbf{B}_N} - F_N)(x)\right)_{1 \leq i \leq p} \tag{8.18}$$

converges to a Gaussian random variable with zero mean and covariance $\mathbf{V}$ with (i,j) entry V_{ij} given by:

$$V_{ij} = -\frac{1}{4\pi^2 c^2} \oint \oint f_i(z_1) f_j(z_2) v_{ij}(z_1, z_2) dz_1 dz_2$$

with

$$v_{ij}(z_1, z_2) = \frac{\underline{m}'(z_1)\underline{m}'(z_2)}{(\underline{m}(z_1) - \underline{m}(z_2))^2} - \frac{1}{(z_1 - z_2)^2}$$

where the integration is over positively oriented contours that circle around the intersection of the supports of F_N for all large N. If we ensure a sufficiently fast convergence of the spectral law of $\mathbf{T}_N$ and of N_i/n, then we can replace F_N by F, the almost sure l.s.d. of $\mathbf{B}_N$, in (8.18). This explains the assumption $N_i/n = c_i + o(1/N)$.

Now, consider Equation (8.9), where t_k is expressed under the form of a complex integral of the Stieltjes transform $m_{\underline{F}}(z)$ of $\underline{F}$ and of its derivative $m'_{\underline{F}}(z)$ (we remind that $\underline{F}$ is the l.s.d. of $\underline{\mathbf{B}}_N = \mathbf{X}_N^{\mathsf{H}} \mathbf{T}_N \mathbf{X}_N$) over a contour $\mathcal{C}_{\underline{F},k}$ that encloses k_F only. Since the functions $(x-z)^{-1}$ and $(x-z)^{-2}$ at any point z of the integration contour $\mathcal{C}_{\underline{F},k}$ are holomorphic on $\mathbb{R}^+$, we can apply straightforwardly Theorem 3.17 to ensure that any vector with entries $n(m_{\mathbf{B}_N}(z_i) - m_{\underline{F}}(z_i))$ and $n(m'_{\mathbf{B}_N}(z_i) - m'_{\underline{F}}(z_i))$, for any finite set of z_i away from the support of F, is asymptotically Gaussian with zero mean and a certain covariance. Then, notice that

$$n\left[z\frac{m'_{\underline{\mathbf{B}}_N}(z_i)}{m_{\underline{\mathbf{B}}_N}(z_i)} - z\frac{m'_{\underline{F}}(z_i)}{m_{\underline{F}}(z_i)}\right] = n\left[\frac{z_i m'_{\underline{\mathbf{B}}_N}(z_i)m_{\underline{F}}(z_i) - z_i m'_{\underline{F}}(z_i)m_{\underline{\mathbf{B}}_N}(z_i)}{m_{\underline{\mathbf{B}}_N}(z_i)m_{\underline{F}}(z_i)}\right]$$

which we would like to express in terms of the differences $n(m_{\underline{\mathbf{B}}_N}(z_i) - m_{\underline{F}}(z_i))$ and $n(m'_{\underline{\mathbf{B}}_N}(z_i) - m'_{\underline{F}}(z_i))$.

To this end, we apply *Slutsky's lemma*, given as follows.

Theorem 8.12 ([Van der Vaart, 2000]). *Let $X_1, X_2, \ldots$ be a sequence of random variables converging weakly to a random variable X and $Y_1, Y_2, \ldots$ converging in probability to a constant c. Then, as $n \to \infty$*

$$Y_n X_n \Rightarrow cX.$$

Applying Theorem 8.12 first to the variables $Y_n = m_{\underline{\mathbf{B}}_N}(z) \xrightarrow{\text{a.s.}} m_{\underline{F}}(z)$ and $X_n = n(m_{\underline{\mathbf{B}}_N}(z) - m_{\underline{F}}(z))$, and then to the variables $Y_n = m'_{\underline{\mathbf{B}}_N}(z) \xrightarrow{\text{a.s.}} m'_{\underline{F}}(z)$ and $X_n = n(m'_{\underline{\mathbf{B}}_N}(z) - m'_{\underline{F}}(z))$, we have rather immediately that

$$n\left[\frac{z_i m'_{\underline{\mathbf{B}}_N}(z_i)m_{\underline{F}}(z_i) - z_i m'_{\underline{F}}(z_i)m_{\underline{\mathbf{B}}_N}(z_i)}{m_{\underline{\mathbf{B}}_N}(z_i)m_{\underline{F}}(z_i)}\right] \Rightarrow z_i\frac{m'_{\underline{F}}(z_i)X - m_{\underline{F}}(z_i)Y}{m_{\underline{F}}(z_i)^2}$$

with X and Y two random variables such that $n[m_{\underline{\mathbf{B}}_N}(z_i) - m_{\underline{F}}(z_i)] \Rightarrow X$ and $n[m'_{\underline{\mathbf{B}}_N}(z_i) - m'_{\underline{F}}(z_i)] \Rightarrow Y$. This last form can be rewritten

$$f(X,Y) = f(X-0, Y-0) = z_i\frac{m'_{\underline{F}}(z_i)}{m_{\underline{F}}(z_i)^2}X - z_i\frac{m_{\underline{F}}(z_i)}{m_{\underline{F}}(z_i)^2}Y$$

where f is therefore a linear function in (X,Y), differentiable at $(0,0)$. In order to pursue, we then introduce the fundamental tool required in this proof, the *delta method*. The delta method allows us to transfer Gaussian behavior from a random variable to a functional of it, according to the following theorem.

Theorem 8.13. *Let $X_1, X_2, \ldots \in \mathbb{R}^n$ be a random sequence such that*

$$a_n(X_n - \mu) \Rightarrow X \sim \mathcal{N}(0, \mathbf{V})$$

for some sequence $a_1, a_2, \ldots \uparrow \infty$. Then for $f : \mathbb{R}^n \to \mathbb{R}^N$, a function differentiable at μ

$$a_n(f(X_n) - f(\mu)) \Rightarrow \mathbf{J}(f)X$$

with $\mathbf{J}(f)$ the Jacobian matrix of f.

Using the delta method on the variables X and Y for different z_i, and applied to the function f, we have that the vector

$$\left(n\left[z_i\frac{m'_{\underline{\mathbf{B}}_N}(z_i)}{m_{\underline{\mathbf{B}}_N}(z_i)} - z_i\frac{m'_{\underline{F}}(z_i)}{m_{\underline{F}}(z_i)}\right]\right)_{1\leq i\leq p}$$

i.e. the deviation of p points of the integrands in (8.11), is asymptotically Gaussian.

In order to propagate the Gaussian limit of the deviations in the integrands of (8.11) to the deviations in $\hat{t}_k$ itself, it suffices to study the behavior of the sum of Gaussian variables over the integration contour $\mathcal{C}_{\underline{F},k}$. Since the integral can be written as the limit of a finite Riemann sum and that a finite Riemann sum of Gaussian random variable is still Gaussian, it suffices to ensure that the finite Riemann sum is still Gaussian in the limit. This requires an additional ingredient: the tightness of the sequences

$$n\left(z\frac{m'_{\underline{\mathbf{B}}_N}(z)}{m_{\underline{\mathbf{B}}_N}(z)} - z\frac{m'_{\underline{F}}(z)}{m_{\underline{F}}(z)}\right)$$

for growing n and for all z in the contour, see [Billingsley, 1968, Theorem 13.1]. This naturally unfolds from a direct application of [Billingsley, 1968, Theorem 13.2], following a similar proof as in [Bai and Silverstein, 2004], and we have proven the Gaussian limit of vectors $(n(\hat{t}_k - t_k))_{k\in\mathcal{J}}$.

The last step of the proof is the calculus of the covariance of the Gaussian limit. This requires to evaluate for all k, k'

$$n^2\mathrm{E}\oint_{\mathcal{C}_{j_k}}\oint_{\mathcal{C}_{j_{k'}}}\left[\frac{z_k m'_{\underline{\mathbf{B}}_N}(z_k)}{m_{\underline{\mathbf{B}}_N}(z_k)} - \frac{z_k m'_{\underline{F}}(z_k)}{m_{\underline{F}}(z_k)}\right]\left[\frac{z_{k'} m'_{\underline{\mathbf{B}}_N}(z_{k'})}{m_{\underline{\mathbf{B}}_N}(z_{k'})} - \frac{z_{k'} m'_{\underline{F}}(z_{k'})}{m_{\underline{F}}(z_{k'})}\right] dz_k dz_{k'}.$$

Integrations by parts simplify the result and lead to (8.16). In order to obtain (8.17), residue calculus is finally performed similar to the proof of Theorem 8.4. □

Note that the proof relies primarily on the central limit theorem of Bai and Silverstein, Theorem 3.17. In particular, for other models more involved than the sample covariance matrix model, the Gaussian limit of the deviations of functionals of the e.s.d. of $\mathbf{B}_N$ must be proven in order both to prove asymptotic central limit of the estimator and even to derive the asymptotic variance of the estimator. This calls for a generalization of Bai and Silverstein central limit theorem to advanced random matrix models, see examples of such models in the context of statistical inference for cognitive radios in Chapter 17.

This recent incentive for eigen-inference based on the Stieltjes transform is therefore strongly constrained by the limited amount of central limit theorems available today. As an alternative to the Stieltjes transform method for statistical inference, we have already mentioned that free probability and methods derived from moments can perform similar inference, based on consistent estimation of the moments only. From the information on the estimated moments, assuming that these moments alone describe the l.s.d., an estimate of the functional under study can be determined. These moment approaches can well substitute Stieltjes transform methods when (i) the model under study is too involved to proceed to complex integration or (ii) when the analytic approach fails, as in the case when clusters mapped to population eigenvalues are not disjoint. Note in particular that the Stieltjes transform method requires exact separation properties, which to this day is known only for very few models.

8.2 Moment deconvolution approach

Remember that the free probability framework allows us to evaluate the successive moments of compactly supported l.s.d. of products, sums, and information plus noise models of asymptotically free random matrices based on the moments of the individual random matrix l.s.d. The operation that evaluates the moments of the output random matrices from the moments of the input deterministic matrices was called free convolution. It was also shown that the moments of an input matrix can be retrieved from those of the resulting output matrix and the other operands (this assumes large matrix dimensions): the associated operation was called free deconvolution. From combinatorics on non-crossing partitions, we stated that it is rather easy to automatize the calculus of free convolved and deconvolved moments. We therefore have already a straightforward way to perform eigen-inference on the successive moments of the l.s.d. of the population covariance matrix from the observed sample covariance matrix, or on the moments of the l.s.d. of the information matrix from the observed information plus noise matrix and so on. Since the l.s.d. are compactly supported, the moments determine the distribution and therefore can be used to perform eigen-inference on various functionals of the l.s.d. However, since moments are unbounded functionals of the eigenvalue spectrum, the moment estimates are usually very inaccurate, more particularly so for high order moments. When estimating the eigenvalues of population covariance matrices $\mathbf{T}_N$, as in Theorem 8.4, moment approaches can be derived although rather impractical, as we will presently see. This is the main drawback of this approach, which, although much more simple and systematic than the previously introduced methods, is fundamentally inaccurate in practice.

Consider again the inference on the individual eigenvalues of $\mathbf{T}_N$ in the model $\mathbf{B}_N = \mathbf{T}_N^{\frac{1}{2}}\mathbf{X}_N\mathbf{X}_N^{\mathsf{H}}\mathbf{T}_N^{\frac{1}{2}} \in \mathbb{C}^{N\times N}$ of Theorem 8.4. For simplicity, we assume that the K distinct eigenvalues of $\mathbf{T}_N$ have the same mass; this fact being known to the experimenter. Since $\mathbf{T}_N$ has K distinct positive eigenvalues $t_1 < t_2 < \ldots < t_K$, we have already mentioned that we can recover these eigenvalues from the first K moments of the l.s.d. H of $\mathbf{T}_N$, recovered from the first K (free deconlvolved) moments of the l.s.d. F of $\mathbf{B}_N$. Those moments are the K roots of the Newton–Girard polynomial (5.2), computed from the moments of H. A naive approach might therefore consist in estimating the moments of $\mathbf{T}_N$ by free deconvolution of the e.s.d. of $\mathbf{B}_N = \mathbf{B}_N(\omega)$, for N finite, and then solving the Newton–Girard polynomial for the estimated moments. We provide hereafter an example for the case $K = 3$.

From the method described in Section 5.2, we obtain that the moments B_1, B_2, B_3 of F are given as a function of the moments T_1, T_2, T_3 of H, as

$$B_1 = T_1,$$
$$B_2 = T_2 + cT_1^2,$$

$$B_3 = T_3 + 3cT_2T_1 + c^2T_1^3$$

with $c = \lim_{n\to\infty} N/n$. Note that, from the extension to finite size random matrices presented in Section 5.4, we could consider the expected e.s.d. of $\mathbf{B}_N$ in place of the l.s.d. of $\mathbf{B}_N$, in which case B_1 and B_2 remain the same, and B_3 becomes

$$B_3 = (1 + n^{-2})T_3 + 3cT_2T_1 + c^2T_1^3.$$

We can then obtain an expression of the T_k by reverting the above equations, as

$$T_1 = B_1,$$
$$T_2 = B_2 - cB_1^2,$$
$$T_3 = (1 + n^{-2})^{-1}\left(B_3 - 3cB_2B_1 + 2c^2B_1^3\right). \tag{8.19}$$

By deconvolving the empirical moments $\hat{B}_k \triangleq \frac{1}{N}\operatorname{tr}\mathbf{B}_N^k$, $1 \leq k \leq 3$, of $\mathbf{B}_N$ with the method above, we obtain estimates $\hat{T}_1, \hat{T}_2, \hat{T}_3$ of the moments T_1, T_2, T_3, in place of T_1, T_2, T_3 themselves. We then obtain estimates $\hat{t}_1, \hat{t}_2, \hat{t}_3$ of t_1, t_2, t_3 by solving the system of equations

$$\hat{T}_1 = \frac{1}{3}\left(\hat{t}_1 + \hat{t}_2 + \hat{t}_3\right),$$
$$\hat{T}_2 = \frac{1}{3}\left(\hat{t}_1^2 + \hat{t}_2^2 + \hat{t}_3^2\right),$$
$$\hat{T}_3 = \frac{1}{3}\left(\hat{t}_1^3 + \hat{t}_2^3 + \hat{t}_3^3\right).$$

We recover the Newton–Girard polynomial by computing the successive elementary symmetric polynomials Π_1, Π_2, Π_3, using (5.4)

$$\Pi_1 = 3\hat{T}_1,$$
$$\Pi_2 = -\frac{1}{2}\hat{T}_2 + \frac{9}{2}\hat{T}_1^2,$$
$$\Pi_3 = \hat{T}_3 - \frac{15}{2}\hat{T}_2\hat{T}_1 + \frac{27}{2}\hat{T}_1^3.$$

The three roots of the equation

$$X^3 - \Pi_1X^2 + \Pi_2X - \Pi_3 = 0$$

are the estimates $\hat{t}_1, \hat{t}_2, \hat{t}_3$ of t_1, t_2, t_3.

However, this method has several major practical drawbacks.

- Inverting the Newton–Girard equation does not ensure that the solutions are all real, since the estimator is not constrained to be real. When running the previous algorithm for not too large N, a large portion of the estimated eigenvalues are indeed returned as purely complex. When this happens, it is difficult to decide what to do with the algorithm output. The G-estimator of Theorem 8.4 on the opposite, being an averaged sum of the eigenvalues of non-negative definite, necessarily provides real positive estimates;

- from the system of Equations (8.19), it turns out that the kth order moment T_k is determined by the moments $B_1, \ldots, B_k$. As a consequence, when substituting $\hat{B}_i$ to B_i, $1 \leq i \leq k$, in (8.19), a small difference between the limiting B_i and the empirical $\hat{B}_i$ entails possibly large estimation errors in all T_k, $k \geq i$. This engenders a snowball effect on the resulting estimates $\hat{T}_k$ for larger k, this effect being increased by the intrinsic growing error between B_i and $\hat{B}_i$ for growing i.

On the positive side, while the G-estimators based on complex analysis are to this day not capable of coping with situations when successive clusters overlap, the moment approach is immune against such situations. The performance of the moment technique against the G-estimator proposed in Section 8.1.2 is provided in Figure 8.4 for the same scenario as in Figure 8.1. We can observe that, although asymptotically unbiased (as H is uniquely determined by its first free moments), the moment-based estimator performs very inaccurately compared to the G-estimator of Theorem 8.4. Note that in the case $N = 30$, $n = 90$, some of the estimates were purely complex and were discarded; running simulations for the scenario $N = 6$, $n = 18$, as in Figure 8.1 leads to even worse results, most of which being purely complex. Now, the limitations of this approach can be corrected by paying more attention on the aforementioned snowball effect for the moment estimates. In particular, thanks to Theorem 3.17, we know that, for the sample covariance matrix model under study, the k-multivariate random variable

$$\left[N \int x d[F - F^{\mathbf{B}_N}](x) \ , \ N \int x^2 d[F - F^{\mathbf{B}_N}](x) \ , \ldots, \ N \int x^k d[F - F^{\mathbf{B}_N}](x)\right]^{\mathsf{T}}$$

has a central limit with covariance matrix given in Corollary 3.3. As a consequence, an alternative estimate $\hat{\mathbf{t}}_{\mathrm{ML}}^{(k)} \triangleq (\hat{t}_{\mathrm{ML},1}^{(k)}, \ldots, \hat{t}_{\mathrm{ML},K}^{(k)})^{\mathsf{T}}$ of the K masses in H is the maximum likelihood (ML) estimate for $(t_1, \ldots, t_K)^{\mathsf{T}}$ based on the observation of k successive moments of $\mathbf{B}_N$. This is given by:

$$\hat{\mathbf{t}}_{\mathrm{ML}}^{(k)} = \arg\min_{\mathbf{t}} (\hat{\mathbf{b}} - \mathbf{b}(\mathbf{t}))^{\mathsf{T}} \mathbf{Q}(\mathbf{t})^{-1} (\hat{\mathbf{b}} - \mathbf{b}(\mathbf{t})) + \log\det \mathbf{Q}(\mathbf{t})$$

where $\mathbf{t} = (T_1, \ldots, T_k)^{\mathsf{T}}$, $\hat{\mathbf{b}} = (\hat{B}_1, \ldots, \hat{B}_k)^{\mathsf{T}}$, $\mathbf{b}(\mathbf{t}) = (B_1, \ldots, B_k)$ assuming $T_k = \frac{1}{K}\sum_{i=1}^K t_i^k$, and $\mathbf{Q}(\mathbf{t})$ is obtained as in (3.27); see [Masucci *et al.*, 2011; Rao *et al.*, 2008] for more details. This method is however computationally expensive in this form, since all vectors $\mathbf{t}$ must be tested, for which every time $\mathbf{Q}(\mathbf{t})$ has to be evaluated. Suboptimal methods are usually envisioned to reduce the computational complexity of the ML estimate down to a reasonable level. In Chapter 17, such methods will be discussed for more elaborate models.

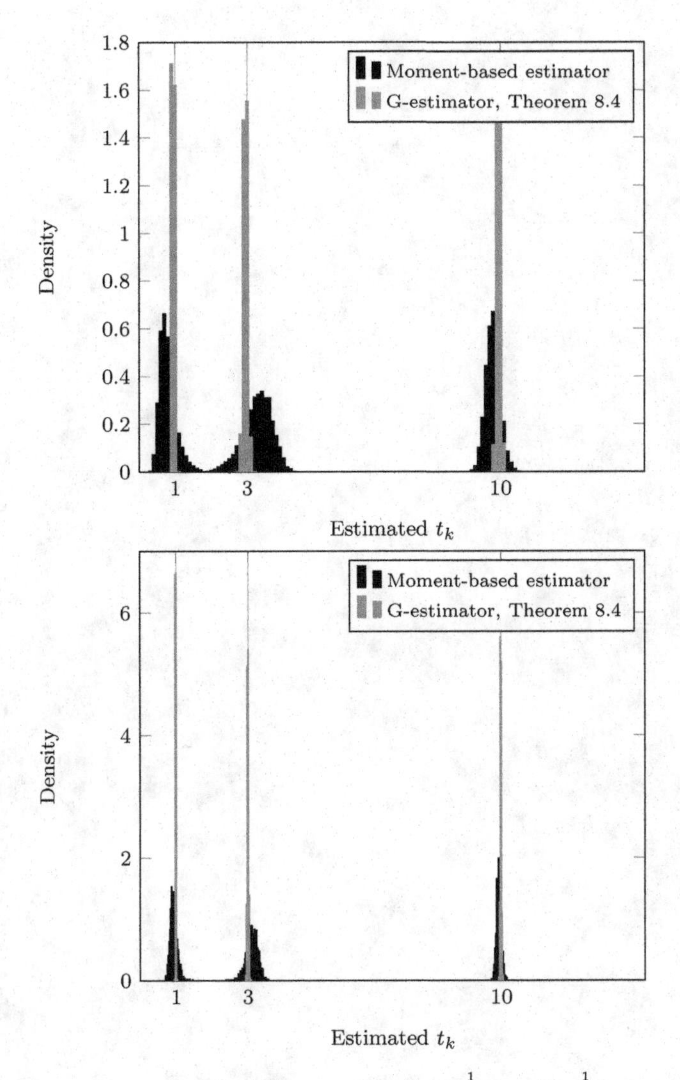

Figure 8.4 Estimation of t_1, t_2, t_3 in the model $\mathbf{B}_N = \mathbf{T}_N^{\frac{1}{2}}\mathbf{X}_N\mathbf{X}_N^{\mathsf{H}}\mathbf{T}_N^{\frac{1}{2}}$ based on first three empirical moments of $\mathbf{B}_N$ and Newton–Girard inversion, for $N_1/N = N_2/N = N_3/N = 1/3$,$N/n = 1/10$, for 100 000 simulation runs; Top $N = 30$, $n = 90$, bottom $N = 90$, $n = 270$. Comparison is made against the G-estimator of Theorem 8.4.

9 Extreme eigenvalues

This last chapter of Part I introduces very recent mathematical advances of deep interest to the field of wireless communications, related to the limiting behavior of the extreme eigenvalues and of their corresponding eigenvectors. Again, the main objects which have been extensively studied in this respect are derivatives of the sample covariance matrix and of the information plus noise matrix.

This chapter will be divided into two sections, whose results emerge from two very different random matrix approaches. The first results, about the limiting extreme eigenvalues of the spiked models, unfold from the previous exact separation results described in Chapter 7. It will in particular be proved that in a sample covariance matrix model, when all population eigenvalues are equal but for the few largest ones, the l.s.d. of the sample covariance matrix is still the Marčenko–Pastur law, but a few eigenvalues may now be found outside the support of the l.s.d. The second set of results concerns mostly random matrix models with Gaussian entries, for which limiting results on the behavior of extreme eigenvalues are available. These results use very different approaches than those proposed so far, namely the theory of orthogonal polynomials and determinantal representations. This subject, which requires many additional tools, is briefly introduced in this chapter. For more information about these tools, see, e.g. the tutorial [Johnstone, 2006] or the book [Mehta, 2004].

We start this section with the spiked models.

9.1 Spiked models

We first discuss the sample covariance matrix model, which can be seen under the spiked model assumption as a *perturbed* sample covariance matrix with identity population covariance matrix. We will then move to a different set of models, using free probability arguments in random matrix models with rotational invariance properties.

9.1.1 Perturbed sample covariance matrix

Let us consider the so-called *spiked models* for the sample covariance matrix model $\mathbf{B}_N = \mathbf{T}_N^{\frac{1}{2}} \mathbf{X}_N \mathbf{X}_N^{\mathsf{H}} \mathbf{T}_N$, with $\mathbf{X}_N \in \mathbb{C}^{N \times n}$ random with i.i.d. entries of zero mean and variance $1/n$, which arise whenever the e.s.d. of the population covariance matrix $\mathbf{T}_N \in \mathbb{C}^{N \times N}$ contains a few outlying eigenvalues. What we mean by "a few outlying eigenvalues" is described in the following. Assume the e.s.d. of the series of matrices $\mathbf{T}_1, \mathbf{T}_2, \ldots$ converges weakly to some d.f. H and denote $\tau_1, \ldots, \tau_N$ the eigenvalues of $\mathbf{T}_N$. Consider now M integers $k_1, \ldots, k_M$. Consider also a set $\alpha_1, \ldots, \alpha_M$ of non-negative reals taken outside the union of the sets $\{\tau_1, \ldots, \tau_N\}$ for all N. Then the e.s.d. of the series $\bar{\mathbf{T}}_1, \bar{\mathbf{T}}_2, \ldots$ of diagonal matrices given by:

$$\bar{\mathbf{T}}_N = \operatorname{diag}(\underbrace{\alpha_1, \ldots, \alpha_1}_{k_1}, \ldots, \underbrace{\alpha_M, \ldots, \alpha_M}_{k_M}, \tau_1, \ldots, \tau_{N - \sum_{i=1}^M k_i})$$

also converges to H as $N \to \infty$, with M and the α_k kept fixed. Indeed, the finitely many eigenvalues $\alpha_1, \ldots, \alpha_M$ of finite multiplicities will have null measure in the asymptotic set of eigenvalues of $\bar{\mathbf{T}}_N$ when $N \to \infty$. These eigenvalues will however lie outside the support of H.

The question that now arises is whether those $\alpha_1, \ldots, \alpha_K$ will induce the presence of some eigenvalues of $\bar{\mathbf{B}}_N \triangleq \bar{\mathbf{T}}_N^{\frac{1}{2}} \mathbf{X}_N \mathbf{X}_N^{\mathsf{H}} \bar{\mathbf{T}}_N^{\frac{1}{2}}$ outside the limiting support of the l.s.d. $\bar{F}$ of $\bar{\mathbf{B}}_N$. First, it is clear that $\bar{F} = F$. Indeed, from Theorem 3.13, F is uniquely determined by H, and therefore the limiting distribution $\bar{F}$ of $\bar{\mathbf{B}}_N$ is nothing but F itself. If there are eigenvalues found outside the support of F, they asymptotically contribute with no mass. These isolated eigenvalues will then be referred to as *spikes* in the following, as they will be outlying asymptotically zero weight eigenvalues. It is important at this point to remind that the existence of eigenvalues outside the support of F is not at all in contradiction with Theorem 7.1. Indeed, Theorem 7.1 precisely states that (with probability one), for all large N, there is no eigenvalue of $\bar{\mathbf{B}}_N$ in a segment $[a, b]$ contained both in the complementary of the support of F *and* in the complementary of the supports of $\bar{F}_N$, determined by the solutions of (7.1), for all large N. In the case where $\tau_j = 1$ for all $j \geq \sum_{i=1}^M k_i$, $\bar{F}(x) = \lim_N \bar{F}_N(x)$ is the Marčenko–Pastur law, which does not exclude $\bar{F}_N$ from containing large eigenvalues of mass $k_i/N \to 0$; therefore, it is possible for $\bar{\mathbf{B}}_N$ to asymptotically have eigenvalues outside the support of F.

The importance of spiked models in wireless communications arises when performing signal sensing, where the (hypothetical) signal space has small dimension compared to the noise space. In this case, we might wish to be able to decide on the presence of a signal based on the spectrum of the sample covariance matrix $\bar{\mathbf{B}}_N$. Typically, if an eigenvalue is found outside the predicted noise spectrum of $\bar{\mathbf{B}}_N$, then this must indicate the presence of a signal bearing informative data, while if all the eigenvalues are inside the limiting support, then this should indicate the absence of such a signal. More on this is discussed in detail in Chapter 16.

However, as we will show in the sequel, it might not always be true that a *spike* in $\bar{\mathbf{T}}_N$ results in a spike in $\bar{\mathbf{B}}_N$ found outside the support of F, in the sense that the support of $\bar{F}_N$ may "hide" the spike in some sense. This is especially true when the size of the main clusters of eigenvalues (linked to the ratio N/n) is large enough to "absorb" the spike of $\bar{\mathbf{B}}_N$ that would have resulted from the population spike of $\bar{\mathbf{T}}_N$. In this case, for signal detection purposes, whether a signal bearing informative data is present or not, there is no way to decide on the presence of this signal by simply looking at the asymptotic spectrum. The condition for decidability when $\mathbf{T}_N = \mathbf{I}_N$ is given in the following result.

Theorem 9.1 ([Baik and Silverstein, 2006]). *Let $\bar{\mathbf{B}}_N = \bar{\mathbf{T}}_N^{\frac{1}{2}}\mathbf{X}_N\mathbf{X}_N^{\mathsf{H}}\bar{\mathbf{T}}_N^{\frac{1}{2}}$, where $\mathbf{X}_N \in \mathbb{C}^{N\times n}$ has i.i.d. entries of zero mean, variance $1/n$, and fourth order moment of order $O(1/n^2)$, and $\bar{\mathbf{T}}_N \in \mathbb{R}^{N\times N}$ is diagonal given by:*

$$\bar{\mathbf{T}}_N = \mathrm{diag}(\underbrace{\alpha_1,\ldots,\alpha_1}_{k_1},\ldots,\underbrace{\alpha_M,\ldots,\alpha_M}_{k_M},\underbrace{1,\ldots,1}_{N-\sum_{i=1}^M k_i})$$

with $\alpha_1 > \ldots > \alpha_M > 0$ for some M. We denote here $c = \lim_N N/n$. Call $M_0 = \#\{j, \alpha_j > 1+\sqrt{c}\}$. For $c<1$, take also M_1 to be such that $M - M_1 = \#\{j, \alpha_j < 1-\sqrt{c}\}$. Denote additionally $\lambda_1,\ldots,\lambda_N$ the eigenvalues of $\bar{\mathbf{B}}_N$, ordered as $\lambda_1 \geq \ldots \geq \lambda_N$. We then have

- *for $1 \leq j \leq M_0$, $1 \leq i \leq k_j$*

$$\lambda_{k_1+\ldots+k_{j-1}+i} \xrightarrow{\text{a.s.}} \alpha_j + \frac{c\alpha_j}{\alpha_j - 1}$$

- *for the other eigenvalues, we must discriminate upon c*
  - *if $c<1$*
    - *for $M_1+1 \leq j \leq M$, $1 \leq i \leq k_j$*

$$\lambda_{N-k_j-\ldots-k_M+i} \xrightarrow{\text{a.s.}} \alpha_j + \frac{c\alpha_j}{\alpha_j - 1}$$

    - *for the indexes of eigenvalues of $\bar{\mathbf{T}}_N$ inside $[1-\sqrt{c}, 1+\sqrt{c}]$*

$$\lambda_{k_1+\ldots+k_{M_0}+1} \xrightarrow{\text{a.s.}} (1+\sqrt{c})^2$$

$$\lambda_{N-k_{M_1+1}-\ldots-k_M} \xrightarrow{\text{a.s.}} (1-\sqrt{c})^2$$

  - *if $c>1$*

$$\lambda_n \xrightarrow{\text{a.s.}} (1-\sqrt{c})^2$$

$$\lambda_{n+1} = \ldots = \lambda_N = 0$$

  - *if $c = 1$*

$$\lambda_{\min(n,N)} \xrightarrow{\text{a.s.}} 0.$$

Therefore, when c is large enough, the segment $[\max(0, 1-\sqrt{c}), 1+\sqrt{c}]$ will contain some of the largest eigenvalues of $\bar{\mathbf{T}}_N$ (those closest to one). If this occurs for a given α_k, the corresponding eigenvalues of $\bar{\mathbf{B}}_N$ will be "attracted" by the left or right end of the support of the l.s.d. of $\bar{\mathbf{B}}_N$. If $c < 1$, small population spikes α_k in $\bar{\mathbf{T}}_N$ may generate spikes of $\bar{\mathbf{B}}_N$ in the interval $(0, (1-\sqrt{c})^2)$; when $c > 1$, though, the α_k smaller than one will result in null eigenvalues of $\bar{\mathbf{B}}_N$.

Remember from the exact separation theorem, Theorem 7.2, that there is a correspondence between the eigenvalues of $\bar{\mathbf{T}}_N$ and those of $\bar{\mathbf{B}}_N$ inside each cluster. Since exactly $\{\alpha_1, \ldots, \alpha_{M_0}\}$ are above $1+\sqrt{c}$ then, asymptotically, exactly $k_1 + \ldots + k_{M_0}$ eigenvalues will lie on the right-end side of the support of the Marčenko–Pastur law. This is depicted in Figure 9.1 where we consider $M = 2$ spikes $\alpha_1 = 2$ and $\alpha_2 = 3$, both of multiplicity $k_1 = k_2 = 2$. We illustrate the decidability condition depending on c by considering first $c = 1/3$, in which case $1 + \sqrt{c} \simeq 1.57 < \alpha_1 < \alpha_2$ and then we expect two spikes of $\bar{\mathbf{B}}_N$ at position $\alpha_1 + c\alpha_1(\alpha_1 - 1)^{-1} \simeq 2.67$ and two spikes of $\bar{\mathbf{B}}_N$ at position $\alpha_2 + c\alpha_2(\alpha_2 - 2)^{-1} = 3.5$. We then move c to $c = 5/4$ for which $\alpha_1 < 1 + \sqrt{c} \simeq 2.12 < \alpha_2$; we therefore expect only the two eigenvalues associated with α_2 at position $\alpha_2 + c\alpha_2(\alpha_2 - 2)^{-1} \simeq 4.88$ to lie outside the spectrum of F. This is approximately what is observed.

The fact that spikes are non-discernible for large c leads to a seemingly paradoxical situation. Consider indeed that the sample space if fixed to n samples while the population space of dimension N increases, so that we increase the collection of input data to improve the quality of the experiment. In the context of signal sensing, if we rely only on a global analysis of the empirical eigenvalues of the input covariance matrix to declare that "if eigenvalues are found outside the support, a signal is detected," then we are better off limiting N to a minimal value and therefore we are better off with a mediocre quality of the experiment; otherwise the decidability threshold is severely impacted. This point is critical and it is essential to understand that the problem here lies in the non-suitability of the decision criterion (that consists just in looking at the eigenvalues outside or inside the support) rather than in the intrinsic non-decidable nature of the problem, which for finite N is not true. If N is large and such that there is no spike outside the support of F while $\bar{\mathbf{T}}_N$ does have spikes, then we will need to look more closely into the tail of the Marčenko–Pastur law, which, for fixed N, contains more than the usual amount of eigenvalues; however, we will see that even this strategy is bound to fail, for very large N. In this case, we may have to resort to studying the joint eigenvalue distribution of $\bar{\mathbf{B}}_N$, which contains the full information. In Chapter 16, we will present a scheme for signal sensing, which aims at providing an optimal sensing decision, based on the complete joint eigenvalue distribution of the input signals, instead of assuming large dimensional assumptions. In these scenarios, the rule of thumb that suggests that small dimensional systems are well approximated by large dimensional analysis now fails, and a significant advantage is provided by the small dimensional analysis.

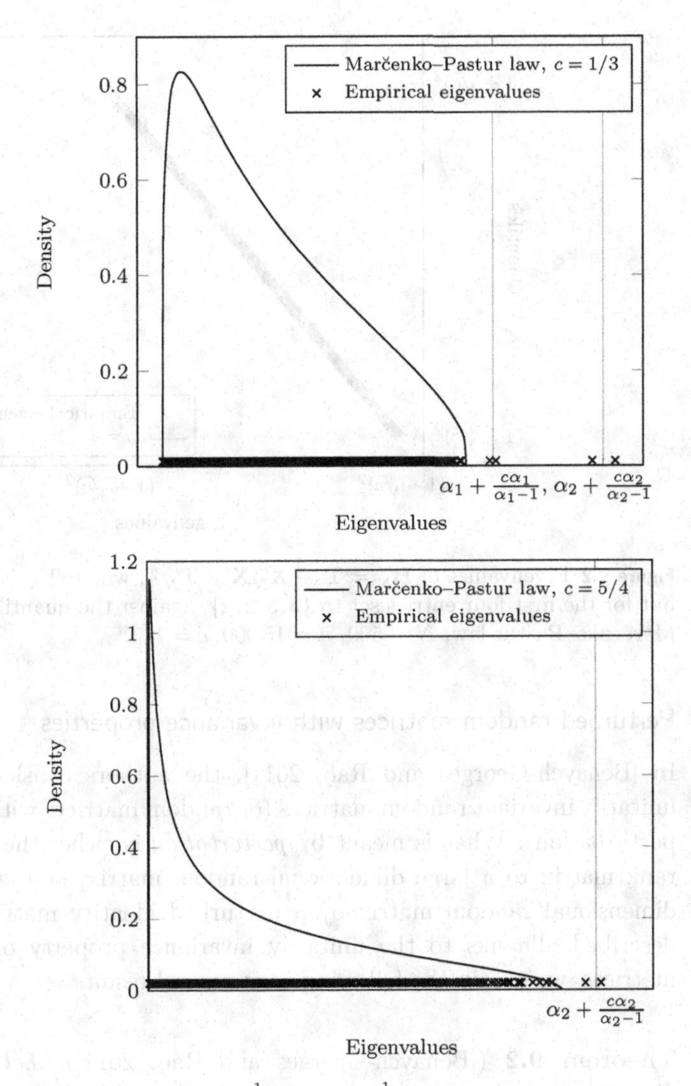

Figure 9.1 Eigenvalues of $\bar{\mathbf{B}}_N = \bar{\mathbf{T}}_N^{\frac{1}{2}} \mathbf{X}_N \mathbf{X}_N{}^{\mathsf{H}} \bar{\mathbf{T}}_N^{\frac{1}{2}}$, where $\bar{\mathbf{T}}_N$ is a diagonal of ones but for the first four entries set to $\{3, 3, 2, 2\}$. On top, $N = 500$, $n = 1500$. One the bottom, $N = 500$, $n = 400$. Theoretical limit eigenvalues of $\bar{\mathbf{B}}_N$ are stressed.

Another way of observing practically when the e.s.d. at hand is close to the Marčenko–Pastur law F is to plot the empirical eigenvalues against the quantiles $F^{-1}(\frac{k-1/2}{N})$ for $k = 1, \ldots, N$. This is depicted in Figure 9.2, for the case $c = 1/3$ with the same set of population spikes $\{2, 2, 3, 3\}$ in $\bar{\mathbf{T}}_N$ as before. We observe again the presence of four outlying eigenvalues in the e.s.d. of $\bar{\mathbf{B}}_N$.

We subsequently move to a different type of results, dealing with the characterization of the extreme eigenvalues and eigenvectors of some perturbed unitarily invariant matrices. These recent results are due to the work of Benaych-Georges and Rao [Benaych-Georges and Rao, 2011].

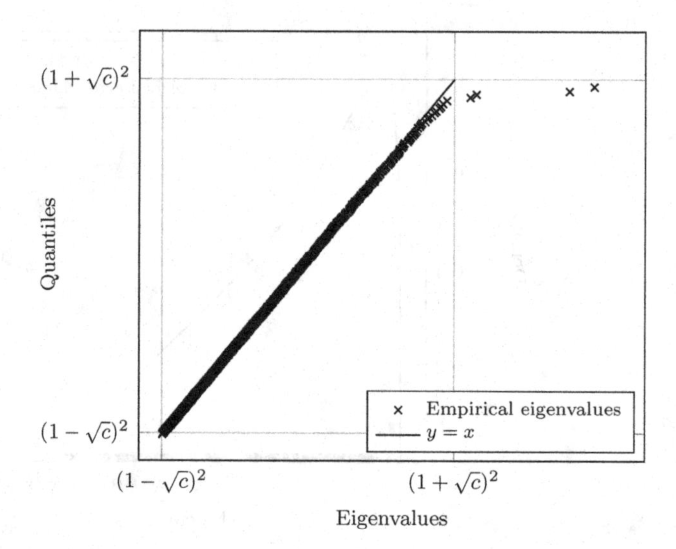

Figure 9.2 Eigenvalues of $\mathbf{B}'_N = \mathbf{T}'^{\frac{1}{2}}_N \mathbf{X}'_N \mathbf{X}'^{\mathsf{H}}_N \mathbf{T}'^{\frac{1}{2}}_N$, where $\mathbf{T}'_N$ is a diagonal of ones but for the first four entries set to $\{3, 3, 2, 2\}$, against the quantiles of the Marčenko–Pastur law, $N = 500$, $n = 15\,000$, $c = 1/3$.

9.1.2 Perturbed random matrices with invariance properties

In [Benaych-Georges and Rao, 2011], the authors consider perturbations of unitarily invariant random matrices (or random matrices with unitarily invariant perturbation). What is meant by *perturbation* is either the addition of a small rank matrix to a large dimensional random matrix, or the product of a large dimensional random matrix by a perturbed identity matrix in the sense just described. Thanks to the unitarily invariance property of either of the two matrices, we obtain the following very general results.

Theorem 9.2 ([Benaych-Georges and Rao, 2011]). *Let $\mathbf{X}_N \in \mathbb{C}^{N\times N}$ be a Hermitian random matrix with ordered eigenvalues $\lambda_1^N \geq \ldots \geq \lambda_N^N$ for which we assume that the e.s.d. $F^{\mathbf{X}_N}$ converges almost surely toward F with compact support with infimum a and supremum b, such that $\lambda_1^N \xrightarrow{\text{a.s.}} b$ and $\lambda_N^N \xrightarrow{\text{a.s.}} a$. Consider also a perturbation matrix $\mathbf{A}_N$ of rank r, with ordered non-zero eigenvalues $a_1^N \geq \ldots \geq a_r^N$. Denote s the integer such that $a_s > 0 > a_{s+1}$. We further assume that either $\mathbf{X}_N$ or $\mathbf{A}_N$ (or both) are bi-unitarily invariant. Denote $\mathbf{Y}_N$ the matrix defined as*

$$\mathbf{Y}_N = \mathbf{X}_N + \mathbf{A}_N$$

with ordered eigenvalues $\nu_1^N \geq \ldots \geq \nu_N^N$. Then, as N grows large, for $i \geq 1$

$$\nu_i^N \xrightarrow{\text{a.s.}} \begin{cases} -m_F^{-1}(1/a_i) & \text{, if } 1 \leq i \leq r \text{ and } 1/a_i < -m_F(b^+) \\ b & \text{, otherwise} \end{cases}$$

where the inverse is with respect to composition. Also, for $i \geq 0$

$$\nu_{n-i}^N \xrightarrow{\text{a.s.}} \begin{cases} -m_F^{-1}(1/a_{r-i}), & \text{if } i < r - s \text{ and } 1/a_i > -m_F(a^-) \\ a & \text{, otherwise.} \end{cases}$$

We also have the same result for multiplicative matrix perturbations, as follows.

Theorem 9.3. *Let $\mathbf{X}_N$ and $\mathbf{A}_N$ be defined as in Theorem 9.2. Denote $\mathbf{Z}_N$ the matrix*

$$\mathbf{Z}_N = \mathbf{X}_N (\mathbf{I}_N + \mathbf{A}_N)$$

with ordered eigenvalues $\mu_1^N \geq \ldots \geq \mu_N^N$. Then, as N grows large, for $i \geq 1$

$$\mu_i^N \xrightarrow{\text{a.s.}} \begin{cases} \psi_F^{-1}(a_i), & \text{if } 1 \leq i \leq s \text{ and } 1/a_i < \psi_F(b^+) \\ b & \text{, otherwise} \end{cases}$$

and, for $i \geq 0$

$$\mu_{n-r+i}^N \xrightarrow{\text{a.s.}} \begin{cases} \psi_F^{-1}(a_i), & \text{if } i < r - s \text{ and } 1/a_i > \psi_F(a^-) \\ a & \text{, otherwise} \end{cases}$$

where ψ_F is the ψ-transform of F, defined in (4.3) as

$$\psi_F(z) = \int \frac{t}{z-t} dF(t) = -1 - \frac{1}{z} m_F\left(\frac{1}{z}\right).$$

This result in particular encompasses the case when $\mathbf{X}_N = \mathbf{W}_N \mathbf{W}_N^{\mathsf{H}}$, with $\mathbf{W}_N$ filled with i.i.d. Gaussian entries, perturbed in the sense of Theorem 9.1. In this sense, this result generalizes Theorem 9.1 for unitarily invariant matrices, although it does not encompass the general i.i.d. case of Theorem 9.1. Recent extensions of the above results on the second order fluctuations of the extreme eigenvalues can be found in [Benaych-Georges *et al.*, 2010].

As noticed above, the study of extreme eigenvalues carries some importance in problems of detection of signals embedded in white noise, but not only. Fields such as speech recognition, statistical learning, or finance also have interests in extreme eigenvalues of covariance matrices. For the particular case of finance, see, e.g., [Laloux *et al.*, 2000; Plerous *et al.*, 2002], consider $\mathbf{X}_N$ is the $N \times n$ matrix in which each row stands for a market product, while every column stands for a time period, say a month, as already presented in Chapter 1. The (i, j)th entry of $\mathbf{X}_N$ contains the evolution of the market index for product i in time period j. If all time-product evolutions are independent in the sense that the evolution of the value of product A for a given month does not impact the evolution of the value of product B, then it is expected that the rows of $\mathbf{X}_N$ are statistically independent. Also, if the time scale is chosen such that the evolution of the price of product A over a given time period is roughly uncorrelated with its evolution on the subsequent time period, then the columns will also be statistically independent.

Therefore, after proper centralization and normalization of the entries (to ensure constant variance), it is expected that, if N, n are large, the empirical eigenvalue distribution of $\mathbf{X}_N\mathbf{X}_N^{\mathsf{H}}$ follows the Marčenko–Pastur law. If not, i.e. if some eigenvalues are found outside the support, then there exist some non-trivial correlation patterns in $\mathbf{X}_N$. The largest eigenvalue here allows the trader to anticipate the largest possible gain to be made if he aligns his portfolio on the corresponding eigenvector. Also, as is the basic rule in finance, the largest eigenvalue gives a rough idea of the largest possible risk of investment in the market (the more gain we expect, the more risky).

Studies about the limiting distribution of the largest eigenvalue are carried out in the subsequent section, in the specific case where the random matrix under study has Gaussian entries.

9.2 Distribution of extreme eigenvalues

In practical applications, since the stochastic matrix observations have sometimes small dimensions, it is fundamental to study the statistical variations of the extreme eigenvalues. Indeed, let us consider the case of the observation of a sample covariance matrix, upon which the experimenter would like to decide whether the population eigenvalue matrix is either an identity matrix or a perturbed identity matrix. A first idea is then to decide whether the observed largest eigenvalue is inside or outside the support of the Marčenko–Pastur law. However, due to the finite dimensionality of the matrix, there is a non-zero probability for the largest observed eigenvalue to lie outside the support. Further information on the statistical distribution of the largest eigenvalue of sample covariance matrices is then required, to be able to design adequate hypothesis tests. We will come back to these practical considerations in Chapter 16.

The study of the second order statistics, i.e. limit theorems on the largest eigenvalues, took off in the mid-nineties initiated by the work of Tracy and Widom [Tracy and Widom, 1996] on $N \times N$ (Wigner) Hermitian matrices with i.i.d. Gaussian entries above the diagonal. Before introducing the results from Tracy and Widom, though, we hereafter provide some notions of orthogonal polynomials in order to understand how these results are derived. This introduction is based on the tutorial [Fyodorov, 2005].

9.2.1 Introduction to the method of orthogonal polynomials

To study the behavior of some particular eigenvalue of a random matrix, it is required to study its marginal distribution. Calling $P^{\leq}_{(\lambda_1^N,\ldots,\lambda_N^N)}(\lambda_1^N,\ldots,\lambda_N^N)$ the joint density of the ordered eigenvalues $\lambda_1^N \leq \ldots \leq \lambda_N^N$ of some Hermitian random matrix $\mathbf{X}_N \in \mathbb{C}^{N\times N}$, the largest eigenvalue λ_N^N has density

$$P_{\lambda_N^N}(\lambda_N^N) = \int_{\lambda_1^N} \cdots \int_{\lambda_{N-1}^N} P^{\leq}_{(\lambda_1^N,\ldots,\lambda_N^N)}(\lambda_1^N,\ldots,\lambda_N^N) d\lambda_1^N \ldots d\lambda_N^N. \quad (9.1)$$

In the case where the order of the eigenvalues is irrelevant, we have that

$$P_{(\lambda_1^N,\ldots,\lambda_N^N)}(\lambda_1^N,\ldots,\lambda_N^N) = \frac{1}{N!} P^{\leq}_{(\lambda_1^N,\ldots,\lambda_N^N)}(\lambda_1^N,\ldots,\lambda_N^N)$$

with $P_{(\lambda_1^N,\ldots,\lambda_N^N)}$ the density of the unordered eigenvalues.

From now on, the eigenvalue indexes $1,\ldots,N$ are considered to be just labels instead of ordering indexes. From the above equality, it is equivalent, and as will turn out actually simpler, to study the unordered eigenvalue distribution rather than the ordered eigenvalue distribution. In the particular case of a zero Wishart matrix with $n \geq N$ degrees of freedom, this property holds and we have from Theorem 2.3 that

$$P_{(\lambda_1^N,\ldots,\lambda_N^N)}(\lambda_1^N,\ldots,\lambda_N^N) = e^{-\sum_{i=1}^N \lambda_i^N} \prod_{i=1}^N \frac{(\lambda_i^N)^{n-N}}{(n-i)!i!} \prod_{i<j} (\lambda_i^N - \lambda_j^N)^2.$$

Similarly, we have for Gaussian Wigner matrices [Tulino and Verdú, 2004], i.e. Wigner matrices with upper-diagonal entries complex standard Gaussian and diagonal entries real standard Gaussian

$$P_{(\lambda_1^N,\ldots,\lambda_N^N)}(\lambda_1^N,\ldots,\lambda_N^N) = \frac{1}{(2\pi)^{\frac{N}{2}}} e^{-\sum_{i=1}^N (\lambda_i^N)^2} \prod_{i=1}^N \frac{1}{i!} \prod_{i<j} (\lambda_i^N - \lambda_j^N)^2. \quad (9.2)$$

The problem now is to be able to compute the multi-dimensional marginalization for either of the above distributions, or for more involved distributions. We concentrate on the simpler Gaussian Wigner case in what follows.

To be able to handle the marginalization procedure, we will use the *reproducing kernel* property, given below which can be found in [Deift, 2000].

Theorem 9.4. *Let $\mathbf{K}_n \in \mathbb{C}^{n\times n}$ with (i,j) entry $K_{ij} = f(x_i, x_j)$ for some complex-valued function f of two real variables and a real vector $\mathbf{x} = (x_1,\ldots,x_n)$. The function f is said to satisfy the reproducing kernel property with respect to a real measure μ if*

$$\int f(x,y)f(y,z)d\mu(y) = f(x,z).$$

Under this condition, we have that

$$\int \det \mathbf{K}_n d\mu(x_n) = (q - (n-1)) \det \mathbf{K}_{n-1} \quad (9.3)$$

with

$$q = \int f(x,x)d\mu(x).$$

The above property is interesting in the sense that, if such a reproducing kernel property can be exhibited, then we can successively iterate (9.3) in order to perform marginalization calculus such as in (9.1).

By working on the expression of the eigenvalue distribution of Gaussian Wigner matrices (9.2), it is possible to write $P_{(\lambda_1^N,\ldots,\lambda_N^N)}$ under the form

$$P_{(\lambda_1^N,\ldots,\lambda_N^N)}(\lambda_1^N,\ldots,\lambda_N^N) = C\det\left(\left\{e^{-\frac{1}{2}(\lambda_j^N)^2}\pi_{i-1}(\lambda_j^N)\right\}_{1\leq i,j\leq N}\right)^2$$

for *any* set of polynomials $(\pi_0,\ldots,\pi_{N-1})$ with π_k of degree k and leading coefficient 1, and for some normalizing constant C. A proof of this fact stems from similar arguments as for the proof of Lemma 16.1, namely that the matrix above can be written under the form of the product of a diagonal matrix with entries $e^{\frac{1}{2}(\lambda_j^N)^2}$ and a matrix with polynomial entries $\pi_i(x_j)$, the determinant of which is proportional to the product of $e^{\sum_j(\lambda_j^N)^2}$ times the Vandermonde determinant $\prod_{i<j}(\lambda_i^N-\lambda_j^N)$.

Now, since we have the freedom to take any set of polynomials $(\pi_0,\ldots,\pi_{N-1})$ with leading coefficient 1, we choose a set of *orthogonal polynomials* with respect to the weighting coefficient e^{-x^2}, i.e. we define $(\pi_0,\ldots,\pi_{N-1})$ to be such that

$$\int e^{-x^2}\pi_i(x)\pi_j(x)dx = \delta_i^j.$$

Denoting now $\mathbf{K}_N \in \mathbb{C}^{N\times N}$ the matrix with (i,j) entry

$$K_{ij} = k_N(\lambda_i^N,\lambda_j^N) \triangleq \sum_{k=0}^{N-1}\left[e^{-\frac{1}{2}(\lambda_i^N)^2}\pi_k(\lambda_i^N)\right]\left[e^{-\frac{1}{2}(\lambda_j^N)^2}\pi_k(\lambda_j^N)\right]$$

we observe easily, from the fact that $\det(\mathbf{A}^2) = \det(\mathbf{A}^\mathsf{T}\mathbf{A})$, that

$$P_{(\lambda_1^N,\ldots,\lambda_N^N)}(\lambda_1^N,\ldots,\lambda_N^N) = C\det\mathbf{K}_N.$$

From the construction of $\mathbf{K}_N$, through the orthogonality of the polynomials $\pi_0(x),\ldots,\pi_{N-1}(x)$, we have that

$$\int k_N(x,y)k_N(y,z)dy = k_N(x,y)$$

and the function k_N has the reproducing kernel property.

This ensures that

$$\int\ldots\int \det\mathbf{K}_N dx_{k+1}\ldots dx_N = (N-k)!\det\mathbf{K}_k$$

where the term $(N-k)!$ follows from the computation of $\int k_n(x,x)dx$ for $n\in\{k+1,\ldots,N\}$.

To finally compute the probability distribution of the largest eigenvalue, note that the probability that it is greater than ξ is complementary to the probability that there is no eigenvalue in $B=(\xi,\infty)$. The latter, called the *hole probability*, expresses as

$$Q_N(B) = \int \cdots \int P_{(\lambda_1^N,\ldots,\lambda_N^N)}(\lambda_1^N,\ldots,\lambda_N^N) \prod_{k=1}^N (1 - 1_B(\lambda_k^N)) d\lambda_1^N \ldots d\lambda_N^N.$$

From the above discussion, expanding the product term, this can be shown to express as

$$Q_N(B) = \sum_{i=0}^N (-1)^i \frac{1}{i!} \int_B \cdots \int_B \det \mathbf{K}_i \, d\lambda_1^N \ldots d\lambda_i^N.$$

This last expression is in fact a *Fredholm determinant*, denoted $\det(\mathcal{I} - \mathcal{K}_N)$, where $\mathcal{K}_N$ is called an integral operator with kernel k_N acting on square integrable functions on B. These Fredholm determinants are well-studied objects, and it is in particular possible to derive the limiting behavior of $Q_N(B)$ as N grows large, which leads presently to the complementary of the Tracy–Widom distribution, and to results such as Theorem 9.5.

Before completing this short introduction, we also mention that alternative contributions such as the recent work of Tucci [Tucci, 2010] establish expressions for functionals of eigenvalue distributions, without resorting to the orthogonal polynomial machinery. In [Tucci, 2010], Tucci provides in particular a closed-form formula for the quantity

$$\int f(t) dF^{\mathbf{X}_N \mathbf{T}_N \mathbf{X}_N^{\mathsf{H}}}(t)$$

for $\mathbf{X}_N \in \mathbb{C}^{N \times n}$ a random matrix with Gaussian entries of zero mean and unit variance and $\mathbf{T}_N$ a deterministic matrix, given under the form of the determinant of a matrix with entries given in an integral form of f and the eigenvalues of $\mathbf{T}_N$. This form is not convenient in its full expression, although it constitutes a first step towards the generalization of the average spectral analysis of Gaussian random matrices. Incidentally, Tucci provides a novel integral expression of the ergodic capacity of a point-to-point 2×2 Rayleigh fading MIMO channel.

Further information on the tools above can be found in the early book from Mehta [Mehta, 2004], the very clear tutorial from Fyodorov [Fyodorov, 2005], and the course notes from Guionnet [Guionnet, 2006], among others. In the following, we introduce the main results concerning limit laws of extreme eigenvalues known to this day.

9.2.2 Limiting laws of the extreme eigenvalues

The major result on the limiting density of extreme eigenvalues is due to Tracy and Widom. It comes as follows.

Theorem 9.5 ([Tracy and Widom, 1996]). *Let $\mathbf{X}_N \in \mathbb{C}^{N \times N}$ be Hermitian with independent Gaussian off-diagonal entries of zero mean and variance $1/N$.*

Denote λ_N^- and λ_N^+ the smallest and largest eigenvalues of $\mathbf{X}_N$, respectively. Then, as $N \to \infty$

$$N^{\frac{2}{3}}\left(\lambda_N^+ - 2\right) \Rightarrow X^+ \sim F^+$$
$$N^{\frac{2}{3}}\left(\lambda_N^- + 2\right) \Rightarrow X^- \sim F^-$$

where F^+ is the Tracy–Widom law given by:

$$F^+(t) = \exp\left(-\int_t^\infty (x-t)^2 q^2(x)dx\right) \tag{9.4}$$

with q the Painlevé II function that solves the differential equation

$$q''(x) = xq(x) + 2q^3(x)$$
$$q(x) \sim_{x\to\infty} \mathrm{Ai}(x)$$

in which $\mathrm{Ai}(x)$ is the Airy function, and F^- is defined as

$$F^-(x) \triangleq 1 - F^+(-x).$$

This theorem is in fact extended in [Tracy and Widom, 1996] to a more general class of matrix spaces, including the space of real symmetric matrices and that of quaternion-valued symmetric matrices, with Gaussian i.i.d. entries. Those are therefore all special cases of Wigner matrices. The space of Gaussian real symmetric matrices is referred to as the *Gaussian orthogonal ensemble* (denoted GOE), that of complex Gaussian Hermitian matrices is referred to as the *Gaussian unitary ensemble* (GUE), and that of quaternion-valued symmetric Gaussian matrices is referred to as the *Gaussian symplectic ensemble* (GSE). The seemingly strange "orthogonal" and "unitary" denominations arise from deeper considerations on these ensembles, involving orthogonal polynomials, see, e.g., [Faraut, 2006] for details.

It was later shown [Bianchi *et al.*, 2010] that the random variables λ_N^+ and λ_N^- are asymptotically independent, giving therefore a simple description of their ratio, the *condition number* of $\mathbf{X}_N$.

Theorem 9.6 ([Bianchi *et al.*, 2010]). *Under the assumptions of Theorem 9.5*

$$\left(N^{\frac{2}{3}}\left(\lambda_N^+ - 2\right), N^{\frac{3}{2}}\left(\lambda_N^- + 2\right)\right) \Rightarrow (X^+, X^-)$$

where X^+ and X^- are independent random variables with respective distributions F^+ and F^-. The random variable λ_N^+/λ_N^- satisfies

$$N^{\frac{2}{3}}\left(\frac{\lambda_N^+}{\lambda_N^-} + 1\right) \Rightarrow -\frac{1}{2}\left(X^+ + X^-\right).$$

The result of interest to our study of extreme eigenvalues of Wishart and perturbed Wishart matrices was proposed later on by Johansson for the largest eigenvalue in the complex case [Johansson, 2000], followed by Johnstone

[Johnstone, 2001] for the largest eigenvalue in the real case, while it took ten years before Feldheim and Sodin provided a proof of the result on the smallest eigenvalue in both real and complex cases [Feldheim and Sodin, 2010]. We only mention here the complex case.

Theorem 9.7 ([Feldheim and Sodin, 2010; Johansson, 2000]). *Let $\mathbf{X}_N \in \mathbb{C}^{N\times n}$ be a random matrix with i.i.d. Gaussian entries of zero mean and variance $1/n$. Denoting λ_N^+ and λ_N^- the largest and smallest eigenvalues of $\mathbf{X}_N\mathbf{X}_N^{\mathsf{H}}$, respectively, we have:*

$$N^{\frac{2}{3}}\frac{\lambda_N^+ - (1+\sqrt{c})^2}{(1+\sqrt{c})^{\frac{4}{3}}\sqrt{c}} \Rightarrow X \sim F^+$$
$$N^{\frac{2}{3}}\frac{\lambda_N^- - (1-\sqrt{c})^2}{-(1-\sqrt{c})^{\frac{4}{3}}\sqrt{c}} \Rightarrow X \sim F^+$$

as $N,n\to\infty$ with $c=\lim_N N/n<1$ and F^+ the Tracy–Widom distribution defined in Theorem 9.5. Moreover, the convergence result for λ_N^+ holds also for $c\geq 1$.

The empirical against theoretical distributions of the largest eigenvalues of $\mathbf{X}_N\mathbf{X}_N^{\mathsf{H}}$ are depicted in Figure 9.3, for $N=500$, $c=1/3$.

Observe that the Tracy–Widom law is largely weighted on the negative half line. This means that the largest eigenvalue of $\mathbf{X}_N\mathbf{X}_N^{\mathsf{H}}$ has a strong tendency to lie much inside the support of the l.s.d. rather than outside. For the same scenario $N=500$, $c=1/3$, we now depict in Figure 9.4 the Tracy–Widom law against the empirical distribution of the largest eigenvalue of $\mathbf{T}_N^{\frac{1}{2}}\mathbf{X}_N\mathbf{X}_N^{\mathsf{H}}\mathbf{T}_N^{\frac{1}{2}}$ in the case where $\mathbf{T}_N\in\mathbb{R}^{N\times N}$ is diagonal composed of all ones but for $T_{11}=1.5$. From Theorem 7.2, no eigenvalue is found outside the asymptotic spectrum of the Marčenko–Pastur law. Figure 9.4 suggests that the largest eigenvalue of $\mathbf{T}_N^{\frac{1}{2}}\mathbf{X}_N\mathbf{X}_N^{\mathsf{H}}\mathbf{T}_N^{\frac{1}{2}}$ does not converge to the Tracy–Widom law since it shows a much heavier tail in the positive side; this is however not true asymptotically. The asymptotic limiting distribution of the largest eigenvalue of $\mathbf{T}_N^{\frac{1}{2}}\mathbf{X}_N\mathbf{X}_N^{\mathsf{H}}\mathbf{T}_N^{\frac{1}{2}}$ is still the Tracy–Widom law, but the convergence towards the second order limit arises at a seemingly much slower rate. This is proved in the following theorem. To appreciate the convergence towards the Tracy–Widom law, N must then be taken much larger.

Theorem 9.8 ([Baik *et al.*, 2005]). *Let $\mathbf{X}_N\in\mathbb{C}^{N\times n}$ have i.i.d. Gaussian entries of zero mean and variance $1/n$ and $\mathbf{T}_N=\operatorname{diag}(\tau_1,\ldots,\tau_N)\in\mathbb{R}^{N\times N}$. Assume, for some fixed r and k, $\tau_{r+1}=\ldots=\tau_N=1$ and $\tau_1=\ldots=\tau_k$ while $\tau_{k+1},\ldots,\tau_r$ lie in a compact subset of $(0,\tau_1)$. Assume further that the ratio N/n is constant, equal to $c<1$ as N,n grow. Denoting λ_N^+ the largest eigenvalue of $\mathbf{T}_N^{\frac{1}{2}}\mathbf{X}_N\mathbf{X}_N^{\mathsf{H}}\mathbf{T}_N^{\frac{1}{2}}$, we have:*

- *If* $\tau_1 < 1 + \sqrt{c}$

$$N^{\frac{2}{3}} \frac{\lambda_N^+ - (1+\sqrt{c})^2}{(1+\sqrt{c})^{\frac{4}{3}}\sqrt{c}} \Rightarrow X^+ \sim F^+$$

with F^+ *the Tracy–Widom distribution.*

- *If* $\tau_1 > 1 + \sqrt{c}$

$$\left(\tau_1^2 - \frac{\tau_1^2 c}{(\tau_1 - 1)^2}\right)^{\frac{1}{2}} n^{\frac{1}{2}} \left[\lambda_N^+ - \left(\tau_1 + \frac{\tau_1 c}{\tau_1 - 1}\right)\right] \Rightarrow X_k \sim G_k$$

with G_k *the distribution function of the largest eigenvalue of the* $k \times k$ *GUE, given by:*

$$G_k(x) = \frac{1}{Z_k} \int_{-\infty}^{x} \cdots \int_{-\infty}^{x} \prod_{1 \leq i < j \leq k} |\xi_i - \xi_j|^2 \prod_{i=1}^{k} e^{-\frac{1}{2}\xi_i^2} d\xi_1 \ldots d\xi_k.$$

with Z_k *a normalization constant. In particular,* $G_1(x)$ *is the Gaussian distribution function.*

The result on spiked eigenvalues was recently extended by Bai and Yao [Bai and Yao, 2008a] to the analysis of the fluctuations of the smallest eigenvalues, when $c < 1$. Precisely, taking now $\tau_{k+1}, \ldots, \tau_r$ in a compact subset of (τ_1, ∞), if $\tau_1 < 1 - \sqrt{c}$, the smallest eigenvalue λ_N^- of $\mathbf{T}_N^{\frac{1}{2}} \mathbf{X}_N \mathbf{X}_N^{\mathsf{H}} \mathbf{T}_N^{\frac{1}{2}}$ follows the distribution G_k after identical centering and scaling as in Theorem 9.8. A further generalization is provided in [Bai and Yao, 2008b] for the case where the weak limit of the e.s.d. is not restricted to the Marčenko–Pastur law.

The corollary of Theorem 9.8 is that, if the largest population eigenvalue (i.e. the largest population spike) is not large enough for any eigenvalue of the sample covariance matrix to escape the support of the Marčenko–Pastur law, whatever its multiplicity k, then the asymptotic distribution of the largest eigenvalue is the Tracy–Widom law. This confirms that the behavior observed in Figure 9.4 has not reached its asymptotic limit. Theorem 9.8 goes further by stating that, if on the contrary τ_1 is larger than the transition limit $1 + \sqrt{c}$, where an eigenvalue will be found outside the support of the Marčenko–Pastur law, then, if $k = 1$, the largest eigenvalue in $\mathbf{T}_N^{\frac{1}{2}} \mathbf{X}_N \mathbf{X}_N^{\mathsf{H}} \mathbf{T}_N^{\frac{1}{2}}$ has a central limit with convergence rate $O(n^{\frac{1}{2}})$ instead of the Tracy–Widom rate $O(n^{\frac{2}{3}})$ when $\tau_1 < 1 + \sqrt{c}$. This sudden convergence rate change is referred to by the author in [Baik *et al.*, 2005] as a *phase transition.* The case $\tau_1 = 1 + \sqrt{c}$ is also treated in [Baik *et al.*, 2005] that shows that λ_N^+ converges in distribution to yet another law F_k, depending on the multiplicity k of τ_1, with rate $O(n^{\frac{2}{3}})$; the law F_k does not have a simple expression though. Since the case $\tau_1 = 1 + \sqrt{c}$ is highly improbable for practical applications, this case was deliberately discarded from Theorem 9.8.

On the other hand, this nice result is yet another disappointment for signal detection applications. Remember that one of our initial motivations to investigate further the asymptotic distribution of the largest eigenvalue of

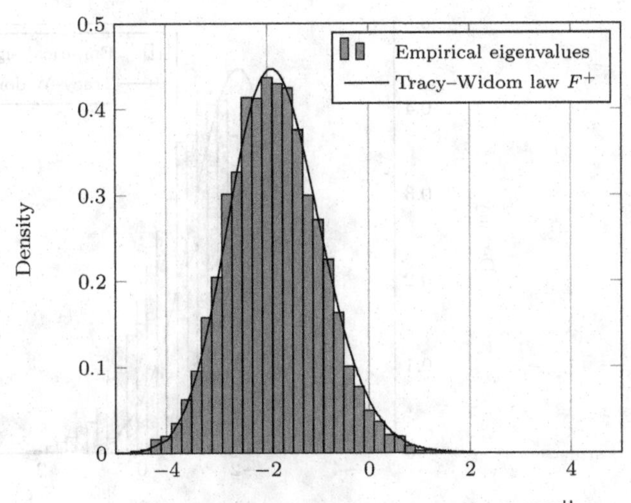

Figure 9.3 Density of $N^{\frac{2}{3}}c^{-\frac{1}{2}}(1+\sqrt{c})^{-\frac{4}{3}}\left[\lambda_N^+ - (1+\sqrt{c})^2\right]$ against the Tracy–Widom law for $N = 500$, $n = 1500$, $c = 1/3$, for the covariance matrix model $\mathbf{XX}^\mathsf{H}$ of Theorem 9.6. Empirical distribution taken over 10 000 Monte-Carlo simulations.

$\mathbf{T}_N^{\frac{1}{2}}\mathbf{X}_N\mathbf{X}_N^\mathsf{H}\mathbf{T}_N^{\frac{1}{2}}$ was the inability to visually determine the presence of a spike $\tau_1 < 1+\sqrt{c}$ from the asymptotic spectrum of $\mathbf{T}_N^{\frac{1}{2}}\mathbf{X}_N\mathbf{X}_N^\mathsf{H}\mathbf{T}_N^{\frac{1}{2}}$. Now, it turns out that even the distribution of the largest eigenvalue in that case is asymptotically the same as that when $\mathbf{T}_N = \mathbf{I}_N$. There is therefore not much left to be done in the asymptotic regime to perform signal detection under the detection threshold $1+\sqrt{c}$. In that case, we may resort to further limit orders, or derive exact expressions of the largest eigenvalue distribution. Similar considerations are addressed in Chapter 16.

We also mention that, in the real case $\mathbf{X}_N \in \mathbb{R}^{N\times n}$, if $\tau_1 > 1+\sqrt{c}$ has multiplicity one, Paul proves that the limiting distribution of $\lambda_N^+ - \left(\tau_1 + \frac{\tau_1 c}{\tau_1 - 1}\right)$ is still Gaussian but with variance double that of the complex case, i.e. $\frac{2}{n}\left(\tau_1^2 - \frac{\tau_1^2 c}{(\tau_1-1)^2}\right)$ [Paul, 2007]. Theorem 9.8 is also extended in [Karoui, 2007] to more general Gaussian sample covariance matrix models, where it is proved that under some conditions on the population covariance matrix, for any integer k fixed, the largest k eigenvalues of the sample covariance matrix have a Tracy–Widom distribution with the same scaling factor but different centering and scaling coefficients.

9.3 Random matrix theory and eigenvectors

Fewer results are known relative to the limiting distribution of the largest eigenvectors. We mention in the following the limiting distribution of the largest

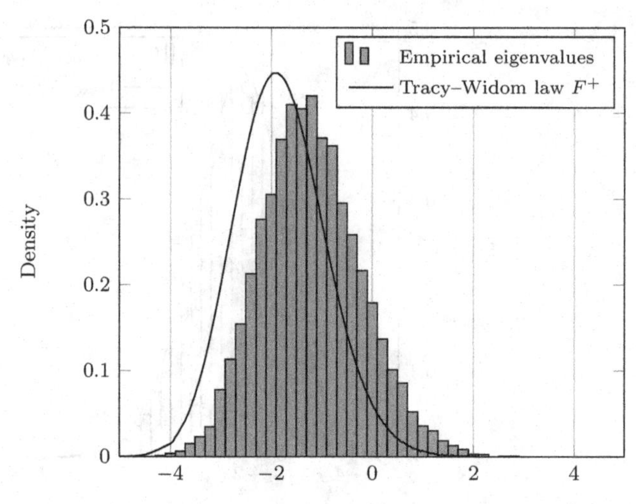

Figure 9.4 Distribution of $N^{\frac{2}{3}}c^{-\frac{1}{2}}(1+\sqrt{c})^{-\frac{4}{3}}\left[\lambda_N^+ - (1+\sqrt{c})^2\right]$ against the Tracy–Widom law for $N = 500$, $n = 1500$, $c = 1/3$, for the covariance matrix model $\mathbf{T}^{\frac{1}{2}}\mathbf{XX}^\mathsf{H}\mathbf{T}^{\frac{1}{2}}$ with $\mathbf{T}$ diagonal with all entries 1 but for $T_{11} = 1.5$. Empirical distribution taken over 10 000 Monte-Carlo simulations.

eigenvector in the spiked model for Gaussian sample covariance matrices, i.e. normalized Wishart matrices with population covariance matrix composed of eigenvalues that are all ones but for a few larger eigenvalues. This is given in the following.

Theorem 9.9 ([Paul, 2007]). *Let $\mathbf{X}_N \in \mathbb{R}^{N\times n}$ have i.i.d. real Gaussian entries of zero mean and variance $1/n$ and $\bar{\mathbf{T}}_N \in \mathbb{R}^{N\times N}$ be defined as*

$$\bar{\mathbf{T}}_N = \mathrm{diag}(\underbrace{\alpha_1,\ldots,\alpha_1}_{k_1},\ldots,\underbrace{\alpha_M,\ldots,\alpha_M}_{k_M},\ \underbrace{1,\ldots,1}_{N-\sum_{i=1}^M k_i}\ )$$

with $\alpha_1 > \ldots > \alpha_M > 0$ for some positive integer M. Then, as $n, N \to \infty$ with limit ratio $N/n \to c$, $0 < c < 1$, for all $i \in \{k_1 + \ldots + k_{j-1} + 1, \ldots, k_1 + \ldots + k_{j-1} + k_j\}$, the eigenvector $\mathbf{p}_i$ associated with the ith largest eigenvalue λ_i of $\bar{\mathbf{T}}_N^{\frac{1}{2}}\mathbf{X}_N\mathbf{X}_N^\mathsf{H}\bar{\mathbf{T}}_N^{\frac{1}{2}}$ satisfies

$$\left|\mathbf{p}_i^\mathsf{T}\mathbf{e}_{N,i}\right|^2 \xrightarrow{\text{a.s.}} \begin{cases} \frac{1-\frac{c}{(\alpha_j-1)^2}}{1+\frac{c}{\alpha_j-1}} & \text{, if } \alpha_j > 1+\sqrt{c} \\ 0 & \text{, otherwise} \end{cases}$$

where $\mathbf{e}_{N,i} \in \mathbb{R}^N$ denotes the vector with all zeros but a one in position i.

Also, if $\alpha_j > 1+\sqrt{c}$ has multiplicity one, denoting $k \triangleq \sum_{l=1}^M k_l$, we write $\mathbf{p}_i = (\mathbf{p}_{A,i}^\mathsf{T}, \mathbf{p}_{B,i}^\mathsf{T})^\mathsf{T}$, with $\mathbf{p}_{A,i} \in \mathbb{R}^k$ the vector of the first k coordinates and $\mathbf{p}_{B,i} \in \mathbb{R}^k$ the vector of the last $N-k$ coordinates. We further take the convention that the *coordinate i of $\mathbf{p}_i$ is non-negative. Then we have, for $i = k_1 + \ldots + k_{j-1} + 1$, as $N, n \to \infty$ with $N/n - c = o(1/\sqrt{n})$*

(i) the vector $\mathbf{p}_{A,i}$ satisfies

$$\sqrt{n}\left(\frac{\mathbf{p}_{A,i}}{\|\mathbf{p}_{A,i}\|} - \mathbf{e}_{M,i}\right) \Rightarrow X$$

where X is an M-variate Gaussian vector with zero mean and covariance $\mathbf{\Sigma}_j$ given by:

$$\mathbf{\Sigma}_j = \left(1 - \frac{c}{(\alpha_j - 1)^2}\right)^{-1} \sum_{\substack{1 \leq l \leq M \\ l \neq j}} \frac{\alpha_l \alpha_j}{(\alpha_l - \alpha_j)^2} \mathbf{e}_{M,l} \mathbf{e}_{M,l}^{\mathsf{T}}$$

(ii) the vector $\mathbf{p}_{B,i}/\|\mathbf{p}_{B,i}\|$ is uniformly distributed on the unit sphere of dimension $N - k - 1$ and is independent of $\mathbf{p}_{A,i}$.

Note that (ii) is valid for all finite dimensions. As a matter of fact, (ii) is valid for any $i \in \{1, \ldots, \min(n, N)\}$. Theorem 9.9 is important as it states in essence that only some of the eigenvectors corresponding to the largest eigenvalues of a perturbed Wishart matrix carry information. Obviously, as α_j tends to $1 + \sqrt{c}$, the almost sure limit of $|\mathbf{p}_i^{\mathsf{T}} \mathbf{e}_{N,i}|$ tends to zero, while the variance of the second order statistics tends to infinity, meaning that increasingly less information can be retrieved as $\alpha_j \to 1 + \sqrt{c}$.

In Figure 9.5, the situation of a single population spike $\alpha_1 = \alpha$ with multiplicity one is considered. The matrix dimensions N and n are taken to be such that $N/n = 1/3$, and $N \in \{100, 200, 400\}$. We compare the averaged empirical projections $|\mathbf{p}_i^{\mathsf{T}} \mathbf{e}_{N,i}|$ against Theorem 9.9. That is, we evaluate the averaged absolute value of the first entry in the eigenvector matrix $\mathbf{U}_N$ in the spectral decomposition of $\bar{\mathbf{T}}_N^{\frac{1}{2}} \mathbf{X}_N \mathbf{X}_N^{\mathsf{H}} \bar{\mathbf{T}}_N^{\frac{1}{2}} = \mathbf{U}_N \operatorname{diag}(\lambda_1, \ldots, \lambda_N) \mathbf{U}_N^{\mathsf{H}}$. We observe that the convergence rate of the limiting projection is very slow. Therefore, although nothing can be said asymptotically on the eigenvectors of a spiked model, when $\alpha < 1 + \sqrt{c}$, there exists a large range of values of N and n for which this is not so.

We also mention the recent result from Benaych-Georges and Rao [Benaych-Georges and Rao, 2011] which, in addition to providing limiting positions for the eigenvalues of some unitarily invariant perturbed random matrix models, Theorem 9.2, provides projection results à la Paul. The main result is as follows.

Theorem 9.10. *Let $\mathbf{X}_N \in \mathbb{C}^{N \times N}$ be a Hermitian random matrix with ordered eigenvalues $\lambda_1 \geq \ldots \geq \lambda_N$. We assume that the e.s.d. $F^{\mathbf{X}_N}$ converges weakly and almost surely toward F with compact support with infimum a and supremum b, such that $\lambda_1 \xrightarrow{\text{a.s.}} b$ and $\lambda_N \xrightarrow{\text{a.s.}} a$. Consider also a perturbation Hermitian matrix $\mathbf{A}_N$ of rank r, with ordered non-zero eigenvalues $a_1 \geq \ldots \geq a_r$. Finally, assume that either $\mathbf{X}_N$ or $\mathbf{A}_N$ (or both) are bi-unitarily invariant. Denote $\mathbf{Y}_N$ the matrix*

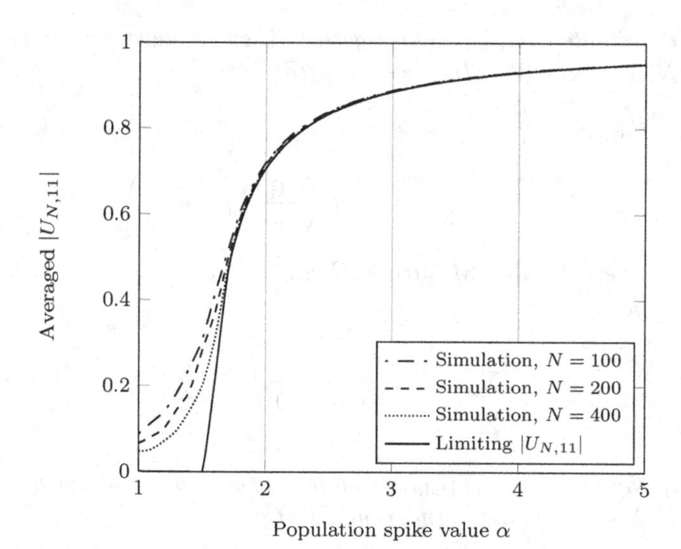

Figure 9.5 Averaged absolute first entry $|\mathbf{U}_{N,11}|$ of the eigenvector corresponding to the largest eigenvalue in $\bar{\mathbf{T}}_N^{\frac{1}{2}}\mathbf{X}_N\mathbf{X}_N^{\mathsf{H}}\bar{\mathbf{T}}_N^{\frac{1}{2}} = \mathbf{U}\,\mathrm{diag}(\lambda_1,\ldots,\lambda_N)\mathbf{U}^{\mathsf{H}}$, with $\mathbf{X}_N$ filled with i.i.d. Gaussian entries $\mathcal{CN}(0,1/n)$ and $\bar{\mathbf{T}}_N \in \mathbb{R}^{N\times N}$ diagonal with all entries one but for the first entry equal to α, $N/n = 1/3$, for varying N.

defined as

$$\mathbf{Y}_N = \mathbf{X}_N + \mathbf{A}_N$$

with order eigenvalues $\nu_1 \geq \ldots \geq \nu_N$. *For* $i \in \{1,\ldots,r\}$ *such that* $1/a_i \in (-m_F(a^-), -m_F(b^+))$, *call* $z_i = \nu_i$ *if* $a_i > 0$ *or* $z_i = \nu_{N-r+i}$ *if* $a_i < 0$, *and* $\mathbf{v}_i$ *an eigenvector associated with* z_i *in the spectral decomposition of* $\mathbf{Y}_N$. *As* N *grows large, we have:*

(i)

$$\langle \mathbf{v}_i, \ker(a_i\mathbf{I}_N - \mathbf{A}_N)\rangle^2 \xrightarrow{\text{a.s.}} \frac{1}{a_i^2 m_F'\left(-\frac{1}{m_F^{-1}(1/a_i)}\right)}$$

(ii)

$$\langle \mathbf{v}_i, \oplus_{j\neq i}\ker(a_j\mathbf{I}_N - \mathbf{A}_N)\rangle^2 \xrightarrow{\text{a.s.}} 0$$

where the notation $\ker(\mathbf{X})$ *denotes the kernel or nullspace of* $\mathbf{X}$, *i.e. the space of vectors* $\mathbf{y}$ *such that* $\mathbf{X}\mathbf{y} = 0$ *and* $\langle x, A\rangle$ *is the norm of the orthogonal projection of* x *on* A.

A similar result for multiplicative matrix perturbations is also available.

Theorem 9.11. *Let* $\mathbf{X}_N$ *and* $\mathbf{A}_N$ *be defined as in Theorem 9.10. Denote* $\mathbf{Z}_N$ *the matrix*

$$\mathbf{Z}_N = \mathbf{X}_N (\mathbf{I}_N + \mathbf{A}_N)$$

with ordered eigenvalues $\mu_1 \geq \ldots \geq \mu_N$. *For* $i \in \{1, \ldots, r\}$ *such that* $1/a_i \in (\psi_F^{-1}(a^-), \psi_F^{-1}(b^+))$ *with* ψ_F *the* ψ*-transform of* F *(see Definition 3.6), call* $z_i = \nu_i$ *if* $a_i > 0$ *or* $z_i = \nu_{N-r+i}$ *if* $a_i < 0$, *and* $\mathbf{v}_i$ *an eigenvector associated with* z_i *in the spectral decomposition of* $\mathbf{Z}_N$. *Then, as* N *grows large, for* $i \geq 1$:

(i)

$$\langle \mathbf{v}_i, \ker(a_i \mathbf{I}_N - \mathbf{A}_N) \rangle^2 \xrightarrow{\text{a.s.}} -\frac{1}{a_i^2 \psi_F^{-1}(1/a_i) \psi_F' \left(\psi_F^{-1}(1/a_i) \right) + a_i}$$

(ii)

$$\langle \mathbf{v}_i, \oplus_{j \neq i} \ker(a_j \mathbf{I}_N - \mathbf{A}_N) \rangle^2 \xrightarrow{\text{a.s.}} 0.$$

The results above have recently been extended by Couillet and Hachem [Couillet and Hachem, 2011], who provide a central limit theorem for the joint fluctuations of the spiky sample eigenvalues and eigenvector projections, for the product perturbation model of Theorem 9.11. These results are particularly interesting in the applicative context of local failure localization in large dimensional systems, e.g. sensor failure or sudden parameter change in large sensor networks, or link failure in a large interconnected graphs. The underlying idea is that a local failure may change the network topology, modeled though the covariance matrix of successive nodal observations, by a small rank perturbation. The perturbation matrix is a *signature of the failure* which is often easier to identify from its eigenvector properties than from its eigenvalues, particularly so in homogeneous networks where each failure leads to similar amplitudes of the extreme eigenvalues.

This completes this short section on extreme eigenvectors. Many more results are expected to be available on this subject in the near future. Before moving to the application part, we summarize the first part of this book and the important results introduced so far.

10 Summary and partial conclusions

In this first part, we started by introducing random matrices as nothing more than a multi-dimensional random variable characterized by the joint probability distribution of its entries. We then observed that the marginal distribution of the eigenvalues of such matrices often carries a lot of information and may even determine the complete stochastic behavior of the random matrix. For instance, we pointed out that the marginal eigenvalue distribution of the Gram matrix $\mathbf{H}\mathbf{H}^{\mathsf{H}}$ associated with the multiple antenna channel $\mathbf{H} \in \mathbb{C}^{N \times n}$ is a sufficient statistic for the characterization of the maximum achievable rate. However, we then realized that, for channels $\mathbf{H}$ more structured than standard Gaussian i.i.d., it is very difficult to provide a general characterization of the marginal empirical eigenvalue distribution of $\mathbf{H}\mathbf{H}^{\mathsf{H}}$.

It was then shown that the eigenvalue distribution of certain classes of random matrices converges weakly to a deterministic limiting distribution function, as the matrix dimensions grow to infinity, the convergence being proved in the almost sure sense. These classes of matrices encompass in particular Wigner matrices, sample covariance matrices, models of the type $\mathbf{A}_N + \mathbf{X}_N\mathbf{T}_N\mathbf{X}_N^{\mathsf{H}}$ when the e.s.d. of $\mathbf{T}_N$ and $\mathbf{A}_N$ have an almost sure limit and Gram matrices $\mathbf{X}_N\mathbf{X}_N^{\mathsf{H}}$ where $\mathbf{X}_N$ has independent entries with a variance profile that has a joint limiting distribution function. The assumptions made on the distribution of the entries of such matrices were in general very mild, while for small dimensional characterization exact results can usually only be derived for matrices with Gaussian independent or loosely correlated entries. While there exist now several different proofs for the major results introduced earlier (e.g. the Marčenko–Pastur law has now been proved from several different techniques using moments [Bai and Silverstein, 2009], the Stieltjes transform [Marčenko and Pastur, 1967], the characteristic function [Faraut, 2006], etc.), we have shown that the Stieltjes transform approach is a very handy tool to determine the limit distribution (and limiting second order statistics) of such matrix models. This is especially suited to the spectrum characterization of matrices with independent entries.

When the matrices involved in the aforementioned models are no longer matrices with i.i.d. entries, but are more structured, the Stieltjes transform approach is more difficult to implement (although still possible according to some recent results introduced earlier). The Stieltjes transform is sometimes better substituted for by simpler tools, namely the R- and S-transforms, when the matrices have some nice symmetric structure. It is at this point that we bridged random matrix theory and free probability theory. The latter provides different tools to study sums and products of a certain class of large dimensional random matrices. This class of random matrices is characterized by the asymptotic freeness property, which is linked to rotational invariance properties. From a practical point of view, asymptotic freeness arises only for random matrix models based on Haar matrices and on Gaussian random matrices. We also introduced the free probability theory from a moment-cumulant viewpoint, which was illustrated to be a very convenient tool to study the l.s.d. of involved random matrix models through their successive free moments. In particular, we showed that, through combinatorics calculus that can be automated on a modern computer, the successive moments of the l.s.d. of (potentially involved) sums, products, and information plus noise models of asymptotically free random matrices can be easily derived. This bears some advantages compared to the Stieltjes transform approach for which case-by-case treatment of every random matrix model must be performed.

From the basic combinatorial grounds of free probability theory, it was then shown that we can go one step further into the study of more structured random matrices. Specifically, it was shown that, by exploiting softer invariance structures of some random matrices, such as the left permutation invariance of some types of random matrices, it is possible to derive expressions of the successive moments of more involved random matrices, such as random Vandermonde matrices. Moreover, the rotational invariance of these matrix models allows us to extend expressions of the moments of the l.s.d. to expressions of the moments of the expected e.s.d. for all finite dimensions. This allows us to refine the moment estimates when dealing with small dimensional matrices. This extrapolation of free probability however is still in its infancy, and is expected to produce a larger number of results in the coming years.

However, the moment-based methods, despite their inherent simplicity, suffer from several shortcomings. First, apart from the very recent results on Vandermonde random matrices, which can be seen as a noticeable exception, moment-based approaches are only useful in practice for dealing with Gaussian and Haar matrices. This is extremely restrictive compared to the models treated with the analytical Stieltjes transform approach, which encompass to this day matrix models based on random matrices with independent entries (sample covariance matrices, matrices with a variance profile, doubly correlated sums of such matrices, etc.) as well as models based on Haar matrices. Secondly, providing an expression of successive moments to approximate a distribution function assumes that the distribution function under study is uniquely characterized by its moments, and more importantly that these moments do exist. If the former assumption fails to be satisfied, the computed moments are mostly unusable; if the latter assumption fails, then these moments cannot even be computed. Due to these important limitations, we dedicated most of Part I to a deep study of the Stieltjes transform tool and of the Stieltjes transform-based methods used to determine the l.s.d. of involved random matrix models, rather than moments methods.

To be complete, we must mention that recent considerations, mostly spurred by Pastur, suggest that, for most classical matrix models discussed so far, it is possible to prove that the l.s.d. of matrix models with independent entries or with Gaussian independent entries are asymptotically the same. This can be proved by using the Gaussian method, introduced in Chapter 6, along with a generalized integration by parts formula and Nash–Poincaré inequality for *generic* matrices with independent entries. Thanks to this method, Pastur also shows that central limit theorems for matrices with independent entries can be recovered from central limit theorems for Gaussian matrices (again accessible through the Gaussian method). Since the latter is much more convenient and much more powerful, as it relies on appreciable properties of the Gaussian distribution, this last approach may adequately replace in the future the Stieltjes transform method, which is sometimes rather difficult to handle. The tool that allows for an extension of the results obtained for matrices with Gaussian entries to unconstrained random matrices with independent entries is referred to as the *interpolation trick*, see, e.g., [Lytova and Pastur, 2009]. Incidentally, for simple random matrix models, such as $\mathbf{X}_N\mathbf{X}_N^{\mathsf{H}}$, where $\sqrt{n}\mathbf{X}_N \in \mathbb{C}^{N\times n}$ has i.i.d. entries of zero mean, unit variance, and some order four cumulant κ, but not only, it can be shown that the variance of the central limit for linear statistics of the eigenvalues can be written under the form $\sigma^2_{\text{Gauss}} + \kappa\sigma^2$, where σ^2_{Gauss} is the variance of the central limit for the Gaussian case and σ^2 is some additional parameter (remember that $\kappa = 0$ in the Gaussian case).

Regarding random matrix models, for which not only the eigenvalue but also the eigenvector distribution plays a role, we further refined the concept of l.s.d. to the concept of deterministic equivalents of the e.s.d. These deterministic equivalents are first motivated by the fact that there might not exist a l.s.d. in the first place and that the deterministic matrices involved in the models (such as the side correlation matrices $\mathbf{R}_k$ and $\mathbf{T}_k$ in Theorem 6.1) may not have a l.s.d. as they grow large. Also, even if there is a limiting d.f. F, it is rather inconvenient that two different series of e.s.d. $F^{\mathbf{B}_1}, F^{\mathbf{B}_2}, \ldots$ and $F^{\tilde{\mathbf{B}}_1}, F^{\tilde{\mathbf{B}}_2}, \ldots$, both converging to the same l.s.d. F, are attributed the same deterministic approximation. Instead, we introduced the concept of deterministic equivalents which provide a specific approximate d.f. F_N for $F^{\mathbf{B}_N}$ such that, as N grows large, $F_N - F^{\mathbf{B}_N} \Rightarrow 0$ almost surely; this way, F_N can approximate $F^{\mathbf{B}_N}$ more closely than would F and the limiting result would still be valid even if $F^{\mathbf{B}_N}$ does not have a limit d.f.. We then introduced different methodologies (i) to determine, through its Stieltjes transform m_N, the deterministic equivalent F_N, (ii) to show that the fixed-point equation to which $m_N(z)$ is a solution admits a unique solution on some restriction of the complex plane, and (iii) to show that $m_N(z) - m_{F^{\mathbf{B}_N}}(z) \xrightarrow{\text{a.s.}} 0$ and therefore $F_N - F^{\mathbf{B}_N} \Rightarrow 0$ almost surely. This leads to a much more involved work than just showing that there exists a l.s.d. to $\mathbf{B}_N$, but it is necessary to ensure the stability of the applications derived from these results. Deterministic equivalents were presented for various models, such as the sum of doubly correlated Gram matrices or general Rician models. For further application purposes, we also derived deterministic equivalents for the Shannon transform of these models.

We then used the Stieltjes transform tool to dig deeper into the study of the empirical eigenvalue distribution of some matrix models, and especially the empirical eigenvalue distribution of sample covariance matrices and information plus noise matrices. While it is already known that the e.s.d. of a sample covariance matrix has a limit d.f. whenever the e.s.d. of the population covariance matrix has a limit, we showed that more can be said about the e.s.d. We observed first that, as the matrix dimensions grow large, no eigenvalue is found outside the support of the limiting d.f. with probability one (under mild assumptions). Then we observed that, when the l.s.d. is formed of a finite union of compact sets, as the matrix dimensions grow large, the number of eigenvalues found in every set (called a cluster) is exactly what we would expect. In particular, if the population covariance matrix in the sample covariance matrix model (or the information matrix in the information plus noise matrix model) is formed of a finite number K of distinct eigenvalues, with respective multiplicities $N_1, \ldots, N_K$ (each multiplicity growing with the system dimensions), then the number of eigenvalues found in every cluster of the e.s.d. exactly matches each N_k or an exact sum of consecutive N_k with high probability. Many properties for sample covariance matrix models were then presented: determination of the exact limiting support, condition for cluster separability, etc. The same types of results were presented for the information plus noise matrix models. However, in this particular case, only exact separation for the Gaussian case has been established so far.

For both sample covariance matrix and information plus noise matrix models, eigen-inference, i.e. inverse problems based on the matrix eigenstructure, was performed to provide consistent estimates for some functionals of the population eigenvalues. Those estimators, that are consistent in the sense of being asymptotically unbiased as both matrix dimensions grow large with comparable sizes, were named G-estimators after Girko who calculated a large number of such consistent estimators.

We then introduced models of sample covariance matrices whose population covariance matrix has a finite number of distinct eigenvalues, but for which some of the eigenvalues have finite multiplicity, not growing with the system dimensions. These models were referred to as spiked models, as these eigenvalues with small multiplicity have an outlying behavior. For these types of matrices, the previous analysis no longer holds and specific study was made. It was shown that these models exhibit a so-called 'phase transition' effect in the sense that, if a population eigenvalue with small multiplicity is greater than a given threshold, then eigenvalues of the sample covariance matrix will be found with high probability *outside* the support of the l.s.d., the number of which being equal to the multiplicity of the population eigenvalue. On the contrary, if the population eigenvalue is smaller than this threshold, then no sample eigenvalue is found outside the support of the l.s.d., again with high probability, and it was shown that the outlying population eigenvalue was then mapped to a sample eigenvalue that converges to the right edge of the spectrum. It was therefore concluded that, from the observation of the l.s.d. (or in practice, from the observation of the e.s.d. for very large matrices), it is impossible to infer on the presence of an outlying eigenvalue in the population covariance matrix. We then dug deeper again to study the behavior of the right edge of the Marčenko–Pastur law and more specifically the behavior of the largest eigenvalue in the e.s.d. of normalized Wishart matrices.

The study of the largest eigenvalue requires different tools than the Stieltjes transform method, among which large deviation analysis, orthogonal polynomials, etc. We observed a peculiar behavior of the largest eigenvalue of the e.s.d. of uncorrelated normalized Wishart matrices, which does not have a classical central limit with convergence rate $O(N^{\frac{1}{2}})$ but a Tracy–Widom limit with convergence rate $O(N^{\frac{2}{3}})$. It was then shown that spiked models in which the isolated population eigenvalues are below the aforementioned critical threshold do not depart from this asymptotic behavior, i.e. the largest eigenvalue converges in distribution to the Tracy–Widom law, although the matrix size must be noticeably larger than in the non-spiked model for the asymptotic behavior to match the empirical distribution. When the largest isolated population eigenvalue is larger than the threshold and of multiplicity one, then it has a central limit with rate $O(N^{\frac{1}{2}})$.

If the results given in Part I have not all been introduced in view of practical wireless communications applications, at least most of them have. Some of them have in fact been designed especially for telecommunication purposes, e.g. Theorems 5.8, 6.4, 6.5, etc. We show in Part II that the techniques presented in the present part can be used for a large variety of problems in wireless communications, aside from the obvious applications to MIMO capacity, MAC and BC rate regions, signal sensing already mentioned. Of particular interest will be the adaption of deterministic equivalent methods to the performance characterization of linearly precoded broadcast channels with transmit correlation, general user path-loss pattern, imperfect channel state information, channel quantization errors, etc. Very compact deterministic equivalent expressions will be obtained, where exact results are mathematically intractable. This is to say the level of details and the flexibility that random matrix theory has reached to cope with more and more realistic system models. Interesting blind direction of arrival and distance estimators will also be presented that use and extend the analysis of the l.s.d. of sample covariance matrix models developed in Section 7.1 and Chapter 8. These examples translate the aptitude of random matrix theory to deal today with more general questions than merely capacity evaluation.

Part II

Applications to wireless communications

11 Introduction to applications in telecommunications

In the preface of [Bai and Silverstein, 2009], Silverstein and Bai provide a table of the number of scientific publications in the domain of random matrix theory for ten-year periods. The table reveals that the number of publications roughly doubled from one period to the next, with an impressive total of more than twelve hundred publications for the 1995-2004 period. This trend is partly due to the mathematical tools developed over the years that allow for more and more possibilities for matrix model analysis. The major reason though is related to the increasing complexity of the system models employed in many fields of physics which demand low complexity analysis. We have already mentioned in the introductory chapter that nuclear physics, biology, finance, and telecommunications are among the fields in which the system complexity involved in the daily work of engineers is growing at a rapid pace. The second part of this book is entirely devoted to wireless communications and to some related signal processing topics. The reader must nonetheless be aware that many models developed here can be adapted to other fields of research, the typical example of such models being the sample covariance matrix model.

In the following section, we provide a brief historical account of the publications in wireless communications dealing with random matrices (from both small and large dimensional viewpoints), from the earlier results in ideal transmission channels down to recent refined examples reflecting more realistic communication environments. It will appear in particular to the reader that, in the latest works, the hypotheses made on channel conditions are precise enough to take into account (sometimes simultaneously) multi-user transmissions, very general channel models with both transmit and receive correlations, Rician models, imperfect channel state information at the sources and the receivers, integration of linear precoders and decoders, inter-cell interference, etc.

11.1 Historical account of major results

It is often mentioned that Tse and Hanly [Tse and Hanly, 1999] initiated the first contribution of random matrix theory to information theory. We would like to insist, following our point that random matrix theory deals both with large dimensional random matrices and small dimensional random matrices, that the important work from Telatar on the multiplexing gain of multiple antenna communications [Telatar, 1995, 1999] was also part of this multi-dimensional system analysis trend of the late nineties. From this time on, the interest in random matrix theory grew vividly, to such an extent that more and more research laboratories dedicated their time to various applications of random matrix theory in large systems. The major driver for this dramatic increase of work in applied random matrix theory is the recent growth of all system dimensions in recent telecommunication systems. The now ten-year-old story of random matrices for wireless communications started with the performance study of code division multiple access.

11.1.1 Rate performance of multi-dimensional systems

The first large dimensional system which was approached by asymptotic analysis is the code division multiple access (CDMA) technology that came along with the third generation of mobile phone communications. We remind that CDMA succeeded the time division multiple access (TDMA) technology used for the second generation of mobile phone communications. In a network with TDMA resource sharing policy, users are successively allocated an exclusive amount of time to exchange data with the access points. Due to the established fixed pattern of time division, one of the major issues of the standard was then that each user could only be allocated a unique time slot, while at the same time a very strict maximal number of users could be accepted by a given access point, regardless of the users' requests in terms of quality of service. In an effort to increase the number of users for a given access point, while dynamically balancing the quality of service offered to each terminal, the CDMA system was selected for the subsequent mobile phone generation. In a CDMA system, each user is allocated a (usually long) *spreading code* that is made roughly orthogonal to the other users' codes, in such a way that all users can simultaneously receive data while experiencing a limited amount of interference from concurrent communications, due to code orthogonality. Equivalently, in the uplink, the users can simultaneously transmit orthogonal streams that can be decoded free of interference at the receiver. Since the spreading codes are rarely fully orthogonal (unless orthogonal codes such as Hadamard codes are used), the more users served by an access point, the more the interference and then the less the quality of service; but at no time is a user rejected for lack of available resource (unless an excessive number of users wishes to access the network). While the achievable transmission data rate for TDMA systems is rather easy to evaluate, the capacity for CDMA networks depends on the precoding strategy applied. One strategy is to build purely orthogonal codes so that all users do not interfere with each other; we refer to this precoding policy as *orthogonal CDMA*. This has the strong advantage of making decoding easy at the receiver and discards the so-called *near-far effect* that leads non-orthogonal users that transmit much power to interfere with more than other users that transmit less power. However, this is not better than TDMA in terms of achievable rates, is more demanding in terms of time synchronization (i.e. the codes of two users not properly synchronous are no longer orthogonal), and suffers significantly from the frequency selectivity of the transmission medium (which induces convolutions of the orthogonal codes, breaking then the orthogonality to some extent). For all these reasons, it is often more sensible to use random i.i.d. codes. This second precoding policy is called *random CDMA*. In practice, codes may originate from random vector generators tailored so to mitigate inter-user interference; these are called pseudo-random CDMA codes. Now the problem is to evaluate the communication rates achieved by such precoders. Indeed, in the orthogonal CDMA approach, assuming frequency flat channel conditions for all users and channel stability over a large number of successive symbol periods, the rates achieved in the uplink (from user terminals to access points) are maximal when the orthogonal codes are as long as the number of users N, and we have the system capacity $C_{\mathrm{orth}}(\sigma^2)$ for a noise power σ^2 given by:

$$C_{\mathrm{orth}}(\sigma^2) = \frac{1}{N} \log\det\left(\mathbf{I}_N + \frac{1}{\sigma^2}\mathbf{W}\mathbf{H}\mathbf{H}^{\mathsf{H}}\mathbf{W}^{\mathsf{H}}\right)$$

where $\mathbf{W} \in \mathbb{C}^{N\times N}$ is the unitary matrix whose columns are the CDMA codes and $\mathbf{H} = \mathrm{diag}(h_1,\ldots,h_N)$ is the diagonal matrix of the channel gains of the users $1,\ldots,N$. By the property $\det(\mathbf{I}+\mathbf{AB}) = \det(\mathbf{I}+\mathbf{BA})$ for matrices $\mathbf{A}$, $\mathbf{B}$ such that both $\mathbf{AB}$ and $\mathbf{BA}$ are square matrices and the fact that $\mathbf{W}$ is unitary, this reduces to

$$C_{\mathrm{orth}}(\sigma^2) = \frac{1}{N} \log\det\left(\mathbf{I}_N + \frac{1}{\sigma^2}\mathbf{H}\mathbf{H}^{\mathsf{H}}\right) = \frac{1}{N}\sum_{i=1}^{N} \log\left(1 + \frac{|h_i|^2}{\sigma^2}\right).$$

This justifies our previous statement on the equivalence between TDMA and CDMA rate performance. When it comes to evaluate the capacity $C_{\mathrm{rand}}(\sigma^2)$ of random CDMA systems, under the same conditions, we have:

$$C_{\mathrm{rand}}(\sigma^2) = \frac{1}{N} \log\det\left(\mathbf{I}_N + \frac{1}{\sigma^2}\mathbf{X}\mathbf{H}\mathbf{H}^{\mathsf{H}}\mathbf{X}^{\mathsf{H}}\right)$$

with $\mathbf{X} \in \mathbb{C}^{N\times N}$ the matrix whose columns are the users' random codes. The result here is no longer trivial and appeals to random matrix analysis. Since the number of users attached to a given access point is usually assumed large, we can use results from large dimensional random matrix theory. The analysis of such systems in the large dimensional regime was performed by Shamai, Tse, and Verdú in [Tse and Verdú, 2000; Verdú and Shamai, 1999].

These capacity expressions may however not be realistic achievable rates in practice, in the sense that they imply non-linear processing at the receiving access points (e.g. decoding based on successive interference cancellation). For complexity reasons, such non-linear processing might not be feasible, so that linear precoders or decoders are often preferred. For random CDMA codes, the capacity achieved by linear decoders in CDMA systems such as matched-filters, linear minimum mean square error (LMMSE) decoders, etc. have been extensively studied from the earlier work of Tse and Hanly [Tse and Hanly, 1999] in frequency flat channels, Evans and Tse [Evans and Tse, 2000] in frequency selective channels, Li and Tulino for reduced-rank LMMSE decoders [Li *et al.*, 2004], Verdú and Shamai [Verdú and Shamai, 1999] for several receivers in frequency flat channels, etc.

While the single-cell *orthogonal* CDMA capacity does not require the random matrix machinery, the study of random CDMA systems is more difficult. Paradoxically, when it comes to linear decoders, the tendency is opposite, as the performance study is in general more involved for the orthogonal case than for the random i.i.d. case. The first substantial work on the performance of linear orthogonal precoded systems arises in 2003 with Debbah and Hachem [Debbah *et al.*, 2003a] on the performance of MMSE decoders in orthogonal CDMA frequency flat fading channels. The subsequent work of Chaufray and Hachem [Chaufray *et al.*, 2004] deals with more realistic sub-optimum MMSE receivers when only partial channel state information is available at the receiver. This comes as a first attempt to consider more realistic transmission conditions. Both aforementioned works have the particularity of providing genuine mathematical results that were at the time not part of the available random matrix literature. As it will often turn out in the intricate models introduced hereafter, the rapid involvement of the wireless communication community in random matrices came along with a fast exhaustion of all the "plug-and-play" results available in the mathematical literature. As a consequence, most results discussed hereafter involving non-trivial channel models often require a deep mathematical introduction, which fully justifies Part I of the present book.

In the same vein as linear systems, linear precoders and decoders for multi-user detection have been given a lot of attention in the past ten years, with the increasing demand for the study of practical scenarios involving a large number of users. The motivation for using random matrix theory here is the increase in the number of users, as well as the increase in the number of antennas used for multi-user transmissions and decoding. Among the notable works in this domain, we mention the analysis of Tse and Zeitouni [Tse and Zeitouni, 2000] on linear multi-user receivers for random CDMA systems. The recent work by Wagner and Couillet [Wagner *et al.*, 2011] derives deterministic equivalents for the sum rate in multi-user broadcast channels with linear precoding and imperfect channel state information at the transmitter. The sum rate expressions obtained in [Wagner *et al.*, 2011] provide different system characterizations. In particular, the optimal training time for channel estimation in quasi-static channels can be evaluated, the optimal cell coverage or number of users to be served can be assessed, etc. We come back to the performance of such linear systems in Chapter 12 and Chapter 14.

One of the major contributions to the analysis of multi-dimensional system performance concerns the capacity derivation of multiple antenna technologies when several transmit and receive antennas are assumed. The derivation from Telatar [Telatar, 1999] of the capacity of frequency flat point-to-point multiple antenna transmissions with Gaussian i.i.d. channel matrix involves complicated small dimensional random matrix calculus. The generalization of Telatar's calculus to non-Gaussian or correlated Gaussian models is even more involved and cannot be treated in general. Large dimensional random matrices help deriving large dimensional capacity expressions for these more exotic channel models. The first results for correlated channels are due to Chuah and Tse [Chuah *et al.*, 2002], Mestre and Fonollosa [Mestre *et al.*, 2003], and Tulino and Verdú [Tulino and Verdú, 2005]. In the last mentioned article, an implicit expression for the asymptotic ergodic capacity of Kronecker channel models is derived, which does not require integral calculus. We recall that a Kronecker-modeled channel is defined as a matrix $\mathbf{H} = \mathbf{R}^{\frac{1}{2}}\mathbf{X}\mathbf{T}^{\frac{1}{2}} \in \mathbb{C}^{n_r \times n_t}$, where $\mathbf{R}^{\frac{1}{2}} \in \mathbb{C}^{n_r \times n_r}$ and $\mathbf{T}^{\frac{1}{2}} \in \mathbb{C}^{n_t \times n_t}$ are deterministic non-negative Hermitian matrices, and $\mathbf{X} \in \mathbb{C}^{n_r \times n_t}$ has Gaussian independent entries of zero mean and normalized variance. This is somewhat less general than similar models where $\mathbf{X}$ has non-necessarily Gaussian entries. We must therefore make the distinction between the Kronecker model and the doubly correlated i.i.d. matrix model, when necessary. Along with the Kronecker model of [Tulino and Verdú, 2005], the capacity of the very general Rician channel model $\mathbf{H} = \mathbf{A} + \mathbf{X}$ is derived in [Hachem *et al.*, 2007], where $\mathbf{A} \in \mathbb{C}^{n_r \times n_t}$ is deterministic and $\mathbf{X} \in n_r \times n_t$ has entries $X_{ij} = \sigma_{ij} Y_{ij}$, where the elements Y_{ij} are i.i.d. (non-necessarily Gaussian) and the factors σ_{ij}^2 form the deterministic *variance profile*.

These results on point-to-point communications in frequency flat channels are further generalized to multi-user communications, to frequency selective communications, and to the most general multi-user frequency selective communications, in doubly correlated channels. The frequency selective channel results are successively due to Moustakas and Simon [Moustakas and Simon, 2007], who conjecture the capacity of Kronecker multi-path channels using the replica method briefly discussed in Section 19.2, followed by Dupuy and Loubaton, who prove those earlier results and additionally determine the capacity achieving signal precoding matrix in [Dupuy and Loubaton, 2010]. On the multi-user side, using tools from free probability theory, Peacock and Honig [Peacock *et al.*, 2008] derive the limit capacity of multi-user communications. Their analysis is based on the assumptions that the number of antennas per user grows large and that all user channels are modeled as Kronecker, such that the left correlations matrices are *co-diagonalizable*. In a parallel work Couillet *et al.* [Couillet *et al.*, 2011a] relax the constraints on the channels of [Peacock *et al.*, 2008] and provide a deterministic equivalent for the points in the rate region of multi-user multiple access and broadcast channels corresponding to *deterministic precoders*, with general doubly correlated channels. Moreover, [Couillet *et al.*, 2011a] provides an expression of the *ergodic* capacity maximizing precoding matrix for each user in the multiple access uplink channel and therefore provides an expression for the boundary of the ergodic multiple access rate region. Both results can then be combined to derive a deterministic equivalent of the rate region for multi-user communications on frequency selective Kronecker channels. To establish results on outage capacity for multiple antenna transmissions, we need to go beyond the expression of deterministic equivalents of the ergodic capacity and study limiting results on the capacity around the mean. Most results are central limits. We mention in particular the important result from Hachem and Najim [Hachem *et al.*, 2008b] on a central limit theorem for the capacity of Rician multiple antenna channels, which follows from their previous work in [Hachem *et al.*, 2007].

The previous multi-user multiple antenna considerations may then be extended to multi-cellular systems. In [Zaidel *et al.*, 2001], the achievable rates in multi-cellular CDMA networks are studied in Wyner's infinite linear cell array model [Wyner, 1994]. Abdallah and Debbah [Abdallah and Debbah, 2004] provide conditions on network planning to improve system capacity of a CDMA network with matched-filter decoding, assuming an infinite number of cells in the network. Circular cell array models are studied by Hoydis *et al.* [Hoydis *et al.*, 2011d], where the optimal number of serving base stations per user is derived in a multi-cell multiple antenna channel model with finite channel coherence time. The finite channel coherence duration parameter involves a limitation of the gain of multi-cell cooperation due to the time required for synchronization (linked to the fundamental diversity-multiplexing trade-off [Tse and Zheng, 2003]), hence a non-trivial rate optimum. Limiting results on the sum capacity in large cellular multiple antenna networks are also studied in [Aktas *et al.*, 2006] and [Huh *et al.*, 2010].

Networks with relay and *ad-hoc* networks are also studied using tools from random matrix theory. We mention in particular the work of Levêque and Telatar [Levêque and Telatar, 2005] who derive scaling laws for large *ad-hoc* networks. On the relay network side Fawaz *et al.* [Fawaz *et al.*, 2011] analyze the asymptotic capacity of relay networks in which relays perform decode and forward. Game theoretical aspects of large dimensional systems were also investigated in light of the results from random matrix theory, among which is the work of Bonneau *et al.* [Bonneau *et al.*, 2007] on power allocation in large CDMA networks.

Chapters 13–15 develop the subject of large multi-user and multiple antenna channels in detail.

11.1.2 Detection and estimation in large dimensional systems

One of the recent hot topics in applied random matrix theory for wireless communications deals with signal detection and source separation capabilities in large dimensional systems. The sudden interest in the late nineties for signal detection using large matrices was spurred simultaneously by the recent mathematical developments and by the recent need for detection capabilities in large dimensional networks. The mathematical milestone might well be the early work of Geman in 1980 [Geman, 1980] who proved that the largest eigenvalue of the e.s.d. of $\mathbf{X}\mathbf{X}^{\mathsf{H}}$, where $\mathbf{X}$ has properly normalized central i.i.d. entries with some assumption on the moments, converges almost surely to the right edge of the Marčenko–Pastur law. This work was followed by generalizations with less constrained assumptions and then by the work of Tracy and Widom [Tracy and Widom, 1996] on the limit distribution of the largest eigenvalue for this model when $\mathbf{X}$ is Gaussian.

On the application side, when source detection is to be performed with the help of a single sensing device, i.e. a single antenna, the optimal decision criterion is given by the Neyman–Pearson test, which was originally derived by Urkowitz for Gaussian channels [Urkowitz, 1967]. This method was then refined to more realistic channel models in, e.g., [Kostylev, 2002; Simon *et al.*, 2003]. The Neyman–Pearson test assumes the occurrence of two possible events $\mathcal{H}_0$ and $\mathcal{H}_1$ with respective probability p_0 and $p_1 = 1 - p_0$. The event $\mathcal{H}_0$ is called the *null hypothesis*, which in our context corresponds to the case where only noise is received at the sensing device, while $\mathcal{H}_1$ is the complementary event, which corresponds to the case where a source is emitting. Upon reception of a sequence of n symbols gathered in the vector $\mathbf{y} = (y_1, \ldots, y_n)^{\mathsf{T}}$ at the sensing device, the Neyman–Pearson criterion states that $\mathcal{H}_1$ must be decided if

$$\frac{P(\mathcal{H}_1|\mathbf{y})}{P(\mathcal{H}_0|\mathbf{y})} = \frac{P(\mathbf{y}|\mathcal{H}_1)p_1}{P(\mathbf{y}|\mathcal{H}_0)p_0} > \gamma$$

for a given threshold γ. If this condition is not met, $\mathcal{H}_0$ must be decided. In the Gaussian channel context, the approach is simple as it merely consists in computing the empirical received power, i.e. the averaged square amplitude $\frac{1}{n}\sum_i |y_i|^2$, and deciding the presence or the absence of a signal source based on whether the empirical received power is greater or less than the threshold γ. Due to the Gaussian assumption, it is then easy to derive the probability of false negatives, i.e. the probability of missing the presence of a source, or of false positives, i.e. the probability of declaring the presence of a source when there is none, which are classical criteria to evaluate the performance of a source detector. Observe already that estimating these performances requires to know the statistics of the y_i or, in the large n hypothesis, to know limiting results on $\sum_i |y_i|^2$.

The generalization to multi-dimensional data $\mathbf{y}_1, \ldots, \mathbf{y}_n \in \mathbb{C}^N$ could consist in summing the total empirical power received across the N sensors, i.e. $\sum_{i=1}^n \|\mathbf{y}_i\|^2$, and then comparing this value to some threshold. This is what is usually done for arrays of N sensors, as it achieves an N-fold increase of performance. Calling $\mathbf{Y} = [\mathbf{y}_1, \ldots, \mathbf{y}_n] \in \mathbb{C}^{N \times n}$, the empirical power reduces to the normalized trace of $\mathbf{Y}\mathbf{Y}^{\mathsf{H}}$. However simple and widely used, see, e.g., [Meshkati *et al.*, 2005], this solution may not always be Neyman–Pearson optimal. Couillet and Debbah [Couillet and Debbah, 2010a] derive the Neyman–Pearson optimal detector in finitely large multiple antenna channels. Their work assumes that both transmit symbols and channel fading links are Gaussian, which are shown to be optimal assumptions in the maximum-entropy principle sense [Jaynes,

1957a,b] when limited prior information is known about the communication channel. The derived Neyman–Pearson criterion for this model turns out to be a rather complex formula involving all eigenvalues of $\mathbf{YY}^{\mathsf{H}}$ and not only their sum. Moreover, under the realistic assumption that the signal-to-noise ratio is not known a priori, the decision criterion expresses under a non-convenient integral form. This is presented in detail in Section 16.3. Simpler solutions are however sought for which may benefit from asymptotic considerations on the eigenvalue distribution of $\mathbf{YY}^{\mathsf{H}}$. A first approach, initially formulated by Zeng and Liang [Zeng and Liang, 2009] and accurately studied by Cardoso and Debbah [Cardoso *et al.*, 2008], consists in considering the ratio of the largest to the lowest eigenvalue in $\mathbf{YY}^{\mathsf{H}}$ and deciding the presence of a signal when this ratio is more than some threshold. This approach is simple and does not assume any prior knowledge on the signal-to-noise ratio. A further approach, actually more powerful, consists in replacing the difficult Neyman–Pearson test by the suboptimal though simpler generalized likelihood ratio test. This is performed by Bianchi *et al.* [Bianchi *et al.*, 2011]. These methods are studied in detail in Section 16.4.

Another non-trivial problem of large matrices with both dimensions growing at a similar rate is the consistent estimation of functionals of eigenvalues and eigenvectors. As already largely introduced in Chapter 8, in many disciplines, we are interested in estimating, e.g. population covariance matrices, eigenvalues of such matrices, etc., from the observation of the empirical sample covariance matrix. With both number of samples n and sample size N growing large at the same rate, it is rather obvious that no good estimate for the whole population covariance matrix can be relied on. However, consistent estimates of functionals of the eigenvalues and eigenvectors have been shown to be possible to obtain. In the context of wireless communications, sample covariance matrices can be used to model the downlink of multiple antenna Gaussian channels with directions of arrivals. Using subspace methods, the directions of arrival can be evaluated as a functional of the eigenvectors of the population null space, i.e. the space orthogonal to the space spanned by the transmit vectors. Being able to evaluate these directions of arrival makes it possible to characterize the position of a source. In [Mestre, 2008a], Mestre proves that the classical n-consistent approach MUSIC method from Schmidt [Schmidt, 1986] to estimate the directions of arrival is largely biased when (n, N) grow large at a similar rate. Mestre then offers an alternative estimator in [Mestre, 2008b] which unfolds directly from Theorem 8.4, under separability assumption of the clusters in the l.s.d. of a sample covariance matrix. This method is shown to be (n, N)-consistent, while showing slightly larger estimate variances. A similar direction of arrival estimation is performed by Vallet *et al.* [Vallet *et al.*, 2010] in the more realistic scenario where the received data vectors are still deterministic but unknown to the sensors. This approach uses an information plus noise model in place of the sample covariance matrix model. The problem of detecting multiple multiple antenna signal sources and estimating the respective distances or powers of each source based on the data collected from an array of sensors enters the same category of problems. Couillet and Silverstein [Couillet *et al.*, 2011c] derive an (n, N)-consistent estimator to this purpose. This estimator assumes simultaneous transmissions over i.i.d. fading channels and is sufficiently general to allow the transmitted signals to originate from different symbol constellations. The blind estimation by an array of sensors of the distances to sources is particularly suited to the recently introduced self-configurable femto-cells. In a few words, femto-cells are required to co-exist with neighboring networks, by reusing spectrum opportunities left available by these networks. The blind detection of adjacent licensed users therefore allows for a dynamical update of the femto-cell coverage area that ensures a minimum interference to the licensed network. See, e.g., [Calin *et al.*, 2010; Claussen *et al.*, 2008] for an introduction to femto-cells and [Chandrasekhar *et al.*, 2009] for an understanding of the need for femto-cells to evaluate the distance to neighboring users.

The aforementioned methods, although very efficient, are however often constrained by strong requirements that restrict their usage. Suboptimal methods that do not fall into these shortcomings are then sought for. Among them, we mention inversion methods based on convex optimization, an example of which being the inversion algorithm for the sample covariance matrix derived by El Karoui [Karoui, 2008], and moment-based approaches, as derived by Couillet and Debbah [Couillet and Debbah, 2008], Rao and Edelman [Rao *et al.*, 2008], Ryan and Debbah [Ryan and Debbah, 2007a], etc. These methods are usually largely more complex and are no match both in complexity and performance for the analytical approaches mentioned before. However, they are much more reliable as they are not constrained by strong prior assumptions on asymptotic spectrum separability. These different problems are discussed at large in Chapter 17.

In the following section, we open a short parenthesis to introduce the currently active field of cognitive radios, which provides numerous open problems related to signal sensing, estimation, and optimization in possibly large dimensional networks.

11.1.3 Random matrices and flexible radio

The field of cognitive radios, also referred to as flexible radios, has known an increasing interest in the past ten years, spurred by the concept of software defined radios, coined by Mitola [Mitola III and Maguire Jr, 1999]. Software defined radios are reconfigurable telecommunication service providers that are meant to dynamically adapt to the client demand. That is, in order to increase the total throughput of multi-protocol communications in a given geographical area [Akyildiz *et al.*, 2006; Tian and Giannakis, 2006], software defined radios will provide various protocol services to satisfy all users in a cell. For instance, cellular phone users in a cognitive radio network will be able to browse the Internet seamlessly through either WiFi, WiMAX, LTE, or 3G protocols, depending on the available spectral resources. This idea, although not yet fully accepted in the industrial world, attempts to efficiently reuse the largely unused (we might say "wasted") bandwidth, which is of fundamental importance in days when no more room is left in the electromagnetic frequency spectrum for future high-speed technologies, see, e.g., [Tandra *et al.*, 2009].

The initial concept of software defined radios has then grown into the more general idea of intelligent, flexible, and self-reconfigurable radios [Hur *et al.*, 2006]. Such radios are composed of smart systems at all places of the communication networks, from the macroscopic infrastructure that spans across thousands of kilometers and that must be globally optimized in some sense, down to the microscopic local in-house networks that must ensure a high quality of service to local users with minimum harm to neighboring communications. From the macroscopic viewpoint, the network must harmonize a large number of users at the network and MAC layers. These aspects may be studied in large game-theoretical frameworks, which demand for large dimensional analysis. Random matrices, but also mean field theory (see, e.g., [Bordenave *et al.*, 2005; Buchegger and Le Boudec, 2005; Sharma *et al.*, 2006]), are common tools to analyze such large decentralized networks. The random matrix approach allows us to characterize deterministic behaviors in large decentralized games. The study of games with a large number of users invokes concepts such as that of Wardrop equilibrium, [Haurie and Marcotte, 1985; Wardrop, 1952]. The first applications to large wireless communication networks mixing both large random matrix analysis and games are due to Bonneau *et al.* [Bonneau *et al.*, 2007, 2008] who study the equilibrium of distributed games for power allocation in the uplink of large CDMA networks. The addition of random matrix tools to the game theoretic settings of such large CDMA networks allows the players in the game, i.e. the CDMA users, to derive deterministic approximations of their functions of merit or cost functions.

On the microscopic side, individual terminals are now required to be smart in the sense that they ought to be able to concur for available spectral resources in decentralized networks. The interest for the decentralized approach typically appears in large networks of extremely mobile users where centralized resource allocation is computationally prohibitive. In such a scenario, mobile users must be able to (i) understand their environment and its evolution by developing sensing abilities (this phase of the communication process is often called *exploration*) and (ii) take optimal decisions concerning resource sharing (this phase is referred to as *exploitation*). These rather novel constraints on the mobile terminals demand for an original technologically disruptive framework for future mobile communications. Part of the exploration requirements for smart terminals is the ability to detect on-going communications in the available frequency resources. This point has already been discussed and is thoroughly dealt with in Chapter 16. Now, another requirement for smart terminals, actually prior to signal sensing, is that of channel modeling. Given some prior information on the environment, such as the geographical location or the number of usually surrounding base stations, the terminal must be able to derive a rough model for the expected communication channel; in this way, sensing procedures will be made faster and more accurate. This requirement again calls for a new framework for channel modeling. This is introduced in Chapter 18, where, upon statistical prior information known to the user terminal, various a priori channel probability distributions are derived. This chapter uses extensively the works from Guillaud, Müller, and Debbah [Debbah and Müller, 2005; Guillaud *et al.*, 2007] and relies primarily on small dimensional random matrix tools.

The chapter ordering somewhat parallels the ordering of Part I, by chronologically introducing models that only require limit distribution theorems, then models that require the introduction of deterministic equivalents and finally detection criteria that rely either on extreme eigenvalue distributions or on asymptotic spectrum considerations and eigen-inference methods. The next chapter deals with system performance evaluation in CDMA communication systems, that is the most documented subject to this day, gathering more than ten years of progress in wireless communications.

12 System performance of CDMA technologies

12.1 Introduction

The following four chapters are dedicated to the analysis of the rate at which information can be reliably transmitted over a physical channel characterized by its space, time, and frequency dimensions. We will often consider situations where one or several of these dimensions can be considered large (and complex) in some sense. Notably, the space dimension will be said to be large when multiple transmit sources or receive sensors are used, which will be the case of multiple antenna, multi-user, or multi-cell communications; the complexity of these channels arises here from the joint statistical behavior of all point-to-point channel links. The time dimension can be said to be large and complex when the inputs to the physical channel exhibit a correlated behavior, such as when signal precoders are used. It may be argued here that the communication channel *itself* is not complex, but for simplified information-theoretic treatment, the transmission channel in such a case is assumed to be the virtual medium formed by the ensemble *precoder and physical channel*. Finally, the frequency dimension exhibits complexity when the channel shows fluctuations in the frequency domain, i.e. frequency selectivity, which typically arises as multi-path reflections come into play in the context of wireless communications or when multi-modal transmissions are used in fiber optics.

To respect historical progress in the use of random matrix theory in wireless communications, we should start with the work of Telatar [Telatar, 1995] on multiple antenna communication channels. However, asymptotic considerations were not immediately applied to multiple antenna systems and, as a matter of fact, not directly applied to evaluate mutual information through Shannon transform formulas. Instead, the first applications of large dimensional random matrix theory [Biglieri *et al.*, 2000; Shamai and Verdú, 2001; Tse and Hanly, 1999; Verdú and Shamai, 1999] were motivated by the similarity between expressions of the signal-to-interference plus noise ratio (SINR) in CDMA precoded transmissions and the Stieltjes transform of some distribution functions. We therefore start by considering the performance of such CDMA systems either with random or unitary codes.

12.2 Performance of random CDMA technologies

We briefly recall that code division multiple access consists in the allocation of (potentially large) orthogonal or quasi-orthogonal codes $\mathbf{w}_1, \ldots, \mathbf{w}_K \in \mathbb{C}^N$ to a series of K users competing for spectral resource access. Every transmitted symbol, either in the downlink (from the access point to the users) or in the uplink (from the users to the access point), is then modulated by the code of the intended user. By making the codes quasi-orthogonal in the time domain, i.e. in the downlink, by ensuring $\mathbf{w}_i^{\mathsf{H}} \mathbf{w}_j \simeq 0$ for $i \neq j$, user k receives its dedicated message though a classical matched-filter (i.e. by multiplying the input signal by $\mathbf{w}_k^{\mathsf{H}}$), while being minimally interfered with by the messages intended for other users. In the uplink, the access point can easily recover the message from all users by matched-filtering successively the input signal by $\mathbf{w}_1^{\mathsf{H}}, \ldots, \mathbf{w}_K^{\mathsf{H}}$. Matched-filtering is however only optimal (from an output SINR viewpoint) when the codes are perfectly orthogonal and when the communication channel does not break orthogonality, i.e. frequency flat channels. A refined filter in that case is the minimum mean square error (MMSE) decoder, which is optimal in terms of SINR experienced at the receiver. Ideally, though, regardless of the code being used, the sum rate optimal solution consists either in the uplink or in the downlink in proceeding to joint decoding at the receiver. Since this requires a lot of information about all user channel conditions, these optimal filters are never considered in the downlink. The performance of all aforementioned filters are presented in the following under more and more constraining communication scenarios.

We first start with uplink CDMA communications before proceeding to the downlink scenario, and consider random i.i.d. CDMA codes before studying orthogonal CDMA codes.

12.2.1 Random CDMA in uplink frequency flat channels

We consider the uplink of a random code division multiple access transmission, with K users sending simultaneously their signals to a unique access point or base station, as in, e.g., [Biglieri *et al.*, 2001; Grant and Alexander, 1998; Madhow and Honig, 1994; Müller, 2001; Rapajic and Popescu, 2000; Schramm and Müller, 1999]. Denote N the length of each CDMA spreading code. User k, $k \in \{1, \ldots, K\}$, has code $\mathbf{w}_k \in \mathbb{C}^N$, which has i.i.d. entries of zero mean and variance $1/N$. At time l, user k transmits the Gaussian symbol $s_k^{(l)}$. The channel from user k to the base station is assumed to be non-selective in the frequency domain, constant over the spreading code length, and is written as the product $h_k \sqrt{P_k}$ of a fast varying parameter h_k, and a long-term parameter $\sqrt{P_k}$ accounting for the power transmitted by user k and the shadowing effect. We refer for simplicity to P_k as *the power of user* k. At time l, we therefore have the transmission model

$$\mathbf{y}^{(l)} = \sum_{k=1}^{K} h_k \mathbf{w}_k \sqrt{P_k} s_k^{(l)} + \mathbf{n}^{(l)} \tag{12.1}$$

where $\mathbf{n}^{(l)} \in \mathbb{C}^N$ denotes the additive Gaussian noise vector with entries of zero mean and variance σ^2, and $\mathbf{y}^{(l)} \in \mathbb{C}^N$ is the signal vector received at the base station.

The expression (12.1) can be written in the more compact form

$$\mathbf{y}^{(l)} = \mathbf{W}\mathbf{H}\mathbf{P}^{\frac{1}{2}}\mathbf{s}^{(l)} + \mathbf{n}^{(l)}$$

with $\mathbf{s}^{(l)} = [s_1^{(l)}, \ldots, s_K^{(l)}]^\mathsf{T} \in \mathbb{C}^K$ a Gaussian vector of zero mean and covariance $\mathrm{E}[\mathbf{s}^{(l)}\mathbf{s}^{(l)\mathsf{H}}] = \mathbf{I}_K$, $\mathbf{W} = [\mathbf{w}_1, \ldots, \mathbf{w}_K] \in \mathbb{C}^{N \times K}$, $\mathbf{P} \in \mathbb{C}^{K \times K}$ is diagonal with kth entry P_k, and $\mathbf{H} \in \mathbb{C}^{K \times K}$ is diagonal with kth entry h_k.

In the following, we will successively study the general expressions of the deterministic equivalents for the capacity achieved by the matched-filter, the minimum-mean square error filter and the optimal decoder. Then, we apply our results to the particular cases of the additive white Gaussian noise (AWGN) channel and the Rayleigh fading channels. The AWGN channel will correspond to the channel for which $h_k = 1$ for all k. The Rayleigh channel is the channel for which $|h_k|$ has Rayleigh distribution with real and imaginary parts of variance $1/2$, and therefore $|h_k|^2$ is χ_2^2-distributed with density $p(|h_k|^2) = e^{-|h_k|^2}$.

12.2.1.1 Matched-filter

We first assume that, at the base station, a matched-filter is applied to the received signal. That is, the base station takes the product $\mathbf{w}_k^\mathsf{H}\mathbf{y}^{(l)}$ to retrieve the data transmitted by user k. In this case, the signal power after matched-filtering reads:

$$P_k |h_k|^2 |\mathbf{w}_k^\mathsf{H}\mathbf{w}_k|^2$$

while the interference plus noise power reads:

$$\mathbf{w}_k^\mathsf{H}\left(\mathbf{W}\mathbf{H}\mathbf{P}\mathbf{H}^\mathsf{H}\mathbf{W}^\mathsf{H} - P_k|h_k|^2\mathbf{w}_k\mathbf{w}_k^\mathsf{H} + \sigma^2\mathbf{I}_N\right)\mathbf{w}_k$$

from which the signal-to-interference ratio $\gamma_k^{(\mathrm{MF})}$ relative to the data of user k is

$$\gamma_k^{(\mathrm{MF})} = \frac{P_k|h_k|^2|\mathbf{w}_k^\mathsf{H}\mathbf{w}_k|^2}{\mathbf{w}_k^\mathsf{H}\left(\mathbf{W}\mathbf{H}\mathbf{P}\mathbf{H}^\mathsf{H}\mathbf{W}^\mathsf{H} - P_k|h_k|^2\mathbf{w}_k\mathbf{w}_k^\mathsf{H} + \sigma^2\mathbf{I}_N\right)\mathbf{w}_k}.$$

Clearly, from the law of large numbers, as $N \to \infty$, $\mathbf{w}_k^\mathsf{H}\mathbf{w}_k \xrightarrow{\text{a.s.}} 1$. Also, from Theorem 3.4, for large N, K such that $0 < \liminf_N K/N \leq \limsup_N K/N < \infty$, we expect to have

$$\mathbf{w}_k^\mathsf{H}\left(\mathbf{W}\mathbf{H}\mathbf{P}\mathbf{H}^\mathsf{H}\mathbf{W}^\mathsf{H} - P_k|h_k|^2\mathbf{w}_k\mathbf{w}_k^\mathsf{H} + \sigma^2\mathbf{I}_N\right)\mathbf{w}_k$$
$$- \frac{1}{N}\operatorname{tr}\left(\mathbf{W}\mathbf{H}\mathbf{P}\mathbf{H}^\mathsf{H}\mathbf{W}^\mathsf{H} - P_k|h_k|^2\mathbf{w}_k\mathbf{w}_k^\mathsf{H} + \sigma^2\mathbf{I}_N\right) \to 0$$

almost surely. However, the conditions of Theorem 3.4 impose that the inner matrix term $\left(\mathbf{WHPH}^{\mathsf{H}}\mathbf{W}^{\mathsf{H}} - P_k|h_k|^2\mathbf{w}_k\mathbf{w}_k^{\mathsf{H}} + \sigma^2\mathbf{I}_N\right)$ be uniformly bounded in spectral norm. This is not the case here since there is a non-zero probability for the largest eigenvalue of $\mathbf{WHPH}^{\mathsf{H}}\mathbf{W}^{\mathsf{H}}$ to grow large. Nonetheless, Theorem 3.4 can be generalized to the case where the inner matrix $\left(\mathbf{WHPH}^{\mathsf{H}}\mathbf{W}^{\mathsf{H}} - P_k|h_k|^2\mathbf{w}_k\mathbf{w}_k^{\mathsf{H}} + \sigma^2\mathbf{I}_N\right)$ only has almost surely bounded spectral norm from a simple application of Tonelli's theorem. Lemma 14.2 provides the exact statement and the proof of this result, which is even more critical for applications in multi-user MIMO communications, see Section 14.1. From Theorem 7.1, the almost sure uniform bounded norm of $\left(\mathbf{WHPH}^{\mathsf{H}}\mathbf{W}^{\mathsf{H}} - P_k|h_k|^2\mathbf{w}_k\mathbf{w}_k^{\mathsf{H}} + \sigma^2\mathbf{I}_N\right)$ is valid here for uniformly bounded $\mathbf{HPH}^{\mathsf{H}}$. Precautions must however be taken for unbounded $\mathbf{HH}^{\mathsf{H}}$, such as for Rayleigh channels.

Now

$$\begin{aligned}
&\frac{1}{N}\operatorname{tr}\left(\mathbf{WHPH}^{\mathsf{H}}\mathbf{W}^{\mathsf{H}} - P_k|h_k|^2\mathbf{w}_k\mathbf{w}_k^{\mathsf{H}} + \sigma^2\mathbf{I}_N\right) - \frac{1}{N}\operatorname{tr}\left(\mathbf{WHPH}^{\mathsf{H}}\mathbf{W}^{\mathsf{H}} + \sigma^2\mathbf{I}_N\right)\\
&= -\frac{1}{N}P_k|h_k|^2\mathbf{w}_k\mathbf{w}_k^{\mathsf{H}}\\
&\to 0
\end{aligned}$$

almost surely. Together, this leads to

$$\begin{aligned}
&\mathbf{w}_k^{\mathsf{H}}\left(\mathbf{WHPH}^{\mathsf{H}}\mathbf{W}^{\mathsf{H}} - P_k|h_k|^2\mathbf{w}_k\mathbf{w}_k^{\mathsf{H}} + \sigma^2\mathbf{I}_N\right)\mathbf{w}_k - \frac{1}{N}\operatorname{tr}\left(\mathbf{WHPH}^{\mathsf{H}}\mathbf{W}^{\mathsf{H}} + \sigma^2\mathbf{I}_N\right)\\
&\to 0
\end{aligned}$$

almost surely. The second term on the left-hand side can be divided into σ^2 and the normalized trace of $\mathbf{WHPH}^{\mathsf{H}}\mathbf{W}^{\mathsf{H}}$. The trace can be further rewritten

$$\frac{1}{N}\operatorname{tr}\left(\mathbf{WHPH}^{\mathsf{H}}\mathbf{W}^{\mathsf{H}}\right) = \frac{1}{N}\sum_{i=1}^{K} P_i|h_i|^2\mathbf{w}_i^{\mathsf{H}}\mathbf{w}_i.$$

Assume additionally that the entries of $\mathbf{w}_i$ have finite eighth order moment. Then, from the trace lemma, Theorem 3.4

$$\frac{1}{N}\operatorname{tr}\left(\mathbf{WHPH}^{\mathsf{H}}\mathbf{W}^{\mathsf{H}}\right) - \frac{1}{N}\sum_{i=1}^{K} P_i|h_i|^2 \xrightarrow{\text{a.s.}} 0.$$

We finally have that the signal-to-interference plus noise ratio $\gamma_k^{(\mathrm{MF})}$ for the signal of user k after the matched-filtering process satisfies

$$\gamma_k^{(\mathrm{MF})} - \frac{P_k|h_k|^2}{\frac{1}{N}\sum_{i=1}^{K} P_i|h_i|^2 + \sigma^2} \xrightarrow{\text{a.s.}} 0. \tag{12.2}$$

If $P_1 = \ldots = P_K \triangleq P$, $K/N \to c$, and the $|h_i|$ are now Rayleigh distributed, the denominator of the deterministic equivalent in (12.2) converges to

$$\sigma^2 + c\int Pte^{-t}dt = \sigma^2 + Pc.$$

If the simpler case when $P_1 = \ldots = P_K \triangleq P$, $K/N \to c$, and $h_i = 1$ for all i, we have simply

$$\gamma_k^{(\mathrm{MF})} \xrightarrow{\text{a.s.}} \frac{P}{Pc + \sigma^2}.$$

The spectral efficiency associated with the matched-filter, denoted $C_{\mathrm{MF}}(\sigma^2)$, is the maximum number of bits that can be reliably transmitted per second and per Hertz of the transmission bandwidth. In the case of large dimensions, the interference being Gaussian, the spectral efficiency in bits/s/Hz is well approximated by

$$C_{\mathrm{MF}}(\sigma^2) = \frac{1}{N} \sum_{k=1}^{K} \log_2\left(1 + \gamma_k^{(\mathrm{MF})}\right).$$

From the discussion above, we therefore have, for N, K large, that

$$C_{\mathrm{MF}}(\sigma^2) - \frac{1}{N} \sum_{k=1}^{K} \log_2\left(1 + \frac{P_k |h_k|^2}{\frac{1}{N}\sum_{i=1}^{K} P_i |h_i|^2 + \sigma^2}\right) \xrightarrow{\text{a.s.}} 0$$

and therefore we have exhibited a deterministic equivalent of the spectral efficiency of the matched-filter for all deterministic channels.

When the $|h_i|$ arise from a Rayleigh distribution, all the P_i equal P and $K/N \to c$, we therefore infer that

$$\begin{aligned} C_{\mathrm{MF}}(\sigma^2) &\xrightarrow{\text{a.s.}} c \int \log_2\left(1 + \frac{Pt}{Pc + \sigma^2}\right) e^{-t} dt \\ &= -c \log_2(e) e^{\frac{Pc+\sigma^2}{P}} \mathrm{Ei}\left(-\frac{Pc + \sigma^2}{P}\right) \end{aligned}$$

with $\mathrm{Ei}(x)$ the exponential integral function

$$\mathrm{Ei}(x) = -\int_{-x}^{\infty} \frac{1}{t} e^{-t} dt.$$

To obtain this expression, particular care must be taken, since, as mentioned previously, it is hazardous to take $\mathbf{H}\mathbf{H}^{\mathsf{H}}$ non-uniformly bounded in spectral norm. However, if the queue of the Rayleigh distribution is truncated at $C > 0$, a mere application of the dominated convergence theorem, Theorem 6.3, ensures the convergence. Growing C large leads to the result.

In the AWGN case, where $h_i = 1$, $P_i = P$ for all i and $K/N \to c$, this is instead

$$C_{\mathrm{MF}}(\sigma^2) \xrightarrow{\text{a.s.}} c \log_2\left(1 + \frac{P}{Pc + \sigma^2}\right).$$

12.2.1.2 MMSE receiver

We assume now that the base station performs the more elaborate minimum mean square error (MMSE) decoding. The signal $\mathbf{y}^{(l)}$ received at the base station at time l is now filtered by multiplying it by the MMSE decoder, as

$$\sqrt{P_k} h_k^* \mathbf{w}_k^{\mathsf{H}} \left(\mathbf{W}\mathbf{H}\mathbf{P}\mathbf{H}^{\mathsf{H}}\mathbf{W}^{\mathsf{H}} + \sigma^2 \mathbf{I}_N \right)^{-1} \mathbf{y}^{(l)}.$$

In this scenario, the signal-to-interference plus noise ratio $\gamma_k^{(\mathrm{MMSE})}$ relative to the signal of user k is slightly more involved to obtain. The signal power is

$$\left[P_k |h_k|^2 \mathbf{w}_k^{\mathsf{H}} \left(\sum_{1 \leq i \leq K} P_i |h_i|^2 \mathbf{w}_i \mathbf{w}_i^{\mathsf{H}} + \sigma^2 \mathbf{I}_N \right)^{-1} \mathbf{w}_k \right]^2$$

while the interference power is

$$P_k |h_k|^2 \sum_{j \neq k} P_j |h_j|^2 \left| \mathbf{w}_k^{\mathsf{H}} \left(\sum_i P_i |h_i|^2 \mathbf{w}_i \mathbf{w}_i^{\mathsf{H}} + \sigma^2 \mathbf{I}_N \right)^{-1} \mathbf{w}_j \right|^2$$
$$+ \sigma^2 P_k |h_k|^2 \mathbf{w}_k^{\mathsf{H}} \left(\sum_i P_i |h_i|^2 \mathbf{w}_i \mathbf{w}_i^{\mathsf{H}} + \sigma^2 \mathbf{I}_N \right)^{-2} \mathbf{w}_k.$$

Working on the interference power, by writing

$$\sum_{j \neq k} P_j |h_j|^2 \mathbf{w}_j \mathbf{w}_j^{\mathsf{H}} = \left(\sum_{1 \leq j \leq K} P_j |h_j|^2 \mathbf{w}_j \mathbf{w}_j^{\mathsf{H}} + \sigma^2 \mathbf{I}_N \right) - \sigma^2 \mathbf{I}_N - P_k |h_k|^2 \mathbf{w}_k \mathbf{w}_k^{\mathsf{H}}$$

we obtain for the interference power

$$P_k |h_k|^2 \mathbf{w}_k^{\mathsf{H}} \left(\sum_i P_i |h_i|^2 \mathbf{w}_i \mathbf{w}_i^{\mathsf{H}} + \sigma^2 \mathbf{I}_N \right)^{-1} \mathbf{w}_k$$
$$- \left[P_k |h_k|^2 \mathbf{w}_k^{\mathsf{H}} \left(\sum_i P_i |h_i|^2 \mathbf{w}_i \mathbf{w}_i^{\mathsf{H}} + \sigma^2 \mathbf{I}_N \right)^{-1} \mathbf{w}_k \right]^2.$$

This simplifies the expression of the ratio of signal power against interference power as

$$\gamma_k^{(\mathrm{MMSE})} = \frac{P_k |h_k|^2 \mathbf{w}_k^{\mathsf{H}} \left(\sum_{1 \leq i \leq K} P_i |h_i|^2 \mathbf{w}_i \mathbf{w}_i^{\mathsf{H}} + \sigma^2 \mathbf{I}_N \right)^{-1} \mathbf{w}_k}{1 - P_k |h_k|^2 \mathbf{w}_k^{\mathsf{H}} \left(\sum_{1 \leq i \leq K} P_i |h_i|^2 \mathbf{w}_i \mathbf{w}_i^{\mathsf{H}} + \sigma^2 \mathbf{I}_N \right)^{-1} \mathbf{w}_k}.$$

Applying the matrix inversion lemma, Lemma 6.2, on $\sqrt{P_k} h_k \mathbf{w}_k$, we finally have the compact form of the MMSE SINR

$$\gamma_k^{(\mathrm{MMSE})} = P_k |h_k|^2 \mathbf{w}_k^{\mathsf{H}} \left(\sum_{\substack{1 \leq i \leq K \\ i \neq k}} P_i |h_i|^2 \mathbf{w}_i \mathbf{w}_i^{\mathsf{H}} + \sigma^2 \mathbf{I}_N \right)^{-1} \mathbf{w}_k.$$

Now, from Theorem 3.4, as N and K grow large with ratio K/N such that $0 < \liminf_N K/N \leq \limsup_N K/N < \infty$

$$\mathbf{w}_k^{\mathsf{H}} \left(\sum_{\substack{1 \leq i \leq K \\ i \neq k}} P_i |h_i|^2 \mathbf{w}_i \mathbf{w}_i^{\mathsf{H}} + \sigma^2 \mathbf{I}_N \right)^{-1} \mathbf{w}_k$$
$$- \frac{1}{N} \operatorname{tr} \left(\mathbf{W}\mathbf{H}\mathbf{P}\mathbf{H}^{\mathsf{H}}\mathbf{W}^{\mathsf{H}} - P_i |h_i|^2 \mathbf{w}_i \mathbf{w}_i^{\mathsf{H}} + \sigma^2 \mathbf{I}_N \right)^{-1} \xrightarrow{\text{a.s.}} 0.$$

From Theorem 3.9, we also have that

$$\frac{1}{N} \operatorname{tr} \left(\mathbf{W}\mathbf{H}\mathbf{P}\mathbf{H}^{\mathsf{H}}\mathbf{W}^{\mathsf{H}} - P_i |h_i|^2 \mathbf{w}_i \mathbf{w}_i^{\mathsf{H}} + \sigma^2 \mathbf{I}_N \right)^{-1}$$
$$- \frac{1}{N} \operatorname{tr} \left(\mathbf{W}\mathbf{H}\mathbf{P}\mathbf{H}^{\mathsf{H}}\mathbf{W}^{\mathsf{H}} + \sigma^2 \mathbf{I}_N \right)^{-1} \to 0$$

where the convergence is sure. Together, the last two equations entail

$$\mathbf{w}_k^{\mathsf{H}} \left[\sum_{i \neq k} P_i |h_i|^2 \mathbf{w}_i \mathbf{w}_i^{\mathsf{H}} + \sigma^2 \mathbf{I}_N \right]^{-1} \mathbf{w}_k - \frac{1}{N} \operatorname{tr} \left(\mathbf{W}\mathbf{H}\mathbf{P}\mathbf{H}^{\mathsf{H}}\mathbf{W}^{\mathsf{H}} + \sigma^2 \mathbf{I}_N \right)^{-1} \xrightarrow{\text{a.s.}} 0.$$

This last expression is the Stieltjes transform $m_{\mathbf{W}\mathbf{H}\mathbf{P}\mathbf{H}^{\mathsf{H}}\mathbf{W}^{\mathsf{H}}}(-\sigma^2)$ of the e.s.d. of $\mathbf{W}\mathbf{H}\mathbf{P}\mathbf{H}^{\mathsf{H}}\mathbf{W}^{\mathsf{H}}$ evaluated at $-\sigma^2 < 0$. Notice that the Stieltjes transform here is independent of the choice of k. Now, if the e.s.d. of $\mathbf{H}\mathbf{P}\mathbf{H}^{\mathsf{H}}$ for all N form a tight sequence, since $\mathbf{W}$ has i.i.d. entries of zero mean and variance $1/N$, we are in the conditions of Theorem 6.1. The Stieltjes transform $m_{\mathbf{W}\mathbf{H}\mathbf{P}\mathbf{H}^{\mathsf{H}}\mathbf{W}^{\mathsf{H}}}(-\sigma^2)$ therefore satisfies

$$m_{\mathbf{W}\mathbf{H}\mathbf{P}\mathbf{H}^{\mathsf{H}}\mathbf{W}^{\mathsf{H}}}(-\sigma^2) - m_N(-\sigma^2) \xrightarrow{\text{a.s.}} 0$$

where $m_N(-\sigma^2)$ is the unique positive solution of the equation in m (see Theorem 3.13, under different hypotheses)

$$m = \left[\frac{1}{N} \operatorname{tr} \mathbf{H}\mathbf{P}\mathbf{H}^{\mathsf{H}} \left(m\mathbf{H}\mathbf{P}\mathbf{H}^{\mathsf{H}} + \mathbf{I}_K \right)^{-1} + \sigma^2 \right]^{-1}. \tag{12.3}$$

We then have that the signal-to-interference plus noise ratio for the signal originating from user k is close to

$$P_k |h_k|^2 m_N(-\sigma^2)$$

for all large N, K, where $m_N(-\sigma^2)$ is the unique positive solution to

$$m = \left[\sigma^2 + \frac{1}{N} \sum_{1 \leq i \leq K} \frac{P_i |h_i|^2}{1 + m P_i |h_i|^2} \right]^{-1}. \tag{12.4}$$

In the Rayleigh case, the e.s.d. of $\mathbf{H}\mathbf{H}^{\mathsf{H}}$ is also Rayleigh distributed, so that $\{F^{\mathbf{H}\mathbf{H}^{\mathsf{H}}}\}$ (indexed by N) forms a tight sequence, which implies the tightness of $\{F^{\mathbf{H}\mathbf{P}\mathbf{H}^{\mathsf{H}}}\}$. As a consequence, $m_N(-\sigma^2)$ has a deterministic almost sure limit

$m(-\sigma^2)$ as $N, K \to \infty$, $K/N \to c$, which is the unique positive solution to the equation in m

$$m = \left[\sigma^2 + c \int \frac{Pt}{1 + Ptm} e^{-t} dt\right]^{-1}. \tag{12.5}$$

The corresponding spectral efficiency $C_{\text{MMSE}}(\sigma^2)$ of the MMSE receiver, for noise variance equal to σ^2, takes the form

$$C_{\text{MMSE}}(\sigma^2) = \frac{1}{N} \sum_{k=1}^{K} \log_2 \left(1 + P_k |h_k|^2 \mathbf{w}_k^{\mathsf{H}} \left[\sum_{i \neq k} P_i |h_i|^2 \mathbf{w}_i \mathbf{w}_i^{\mathsf{H}} + \sigma^2 \mathbf{I}_N\right]^{-1} \mathbf{w}_k\right).$$

For increasing N, K, $K/N \to c$, $P_1 = \ldots = P_K = P$, and $|h_i|$ Rayleigh distributed for every i, we have:

$$C_{\text{MMSE}}(\sigma^2) \xrightarrow{\text{a.s.}} c \int \log_2 \left(1 + Ptm(-\sigma^2)\right) e^{-t} dt$$

which can be verified once more by an adequate truncation of the queue of the Rayleigh distributed at $C > 0$ and then taking $C \to \infty$.

In the AWGN scenario, (12.4) becomes

$$m = \frac{1 + mP}{\sigma^2 + cP + mP\sigma^2}$$

which can be expressed as a second order polynomial in m, whose unique positive solution $m(-\sigma^2)$ reads:

$$m(-\sigma^2) = \frac{-(\sigma^2 + (c-1)P) + \sqrt{(\sigma^2 + (c-1)P)^2 + 4P\sigma^2}}{2P\sigma^2} \tag{12.6}$$

which is therefore the almost sure limit of $m_N(-\sigma^2)$, for $N, K \to \infty$. The spectral efficiency therefore has the deterministic limit

$$C_{\text{MMSE}}(\sigma^2) \xrightarrow{\text{a.s.}} c \log_2 \left(1 + \frac{-(\sigma^2 + (c-1)P) + \sqrt{(\sigma^2 + (c-1)P)^2 + 4P\sigma^2}}{2\sigma^2}\right).$$

12.2.1.3 Optimal receiver

The optimal receiver jointly decodes the data streams from all users. Its spectral efficiency $C_{\text{opt}}(\sigma^2)$ is simply

$$C_{\text{opt}}(\sigma^2) = \frac{1}{N} \log\det \left(\mathbf{I}_N + \frac{1}{\sigma^2} \mathbf{WHPH}^{\mathsf{H}} \mathbf{W}^{\mathsf{H}}\right).$$

A straightforward application of Corollary 6.1 leads to the limiting result

$$C_{\rm opt}(\sigma^2) - \log_2\left(1 + \frac{1}{\sigma^2 N}\sum_{k=1}^{K}\frac{P_k|h_k|^2}{1+cP_k|h_k|^2 m_N(-\sigma^2)}\right)$$
$$-\frac{1}{N}\sum_{k=1}^{K}\log_2\left(1+cP_k|h_k|^2 m_N(-\sigma^2)\right)$$
$$-\log_2(e)\left(\sigma^2 m_N(-\sigma^2)-1\right) \xrightarrow{\text{a.s.}} 0.$$

for $P_1|h_1|^2,\ldots,P_K|h_K|^2$ uniformly bounded across N, where $m_N(-\sigma^2)$ is the unique positive solution of (12.4).

In the case of growing system dimensions and Rayleigh fading on all links, with similar arguments as previously, we have the almost sure convergence

$$C_{\rm opt}(\sigma^2) \xrightarrow{\text{a.s.}} \log_2\left(1+\frac{c}{\sigma^2}\int\frac{Pt}{1+Ptm(-\sigma^2)}e^{-t}dt\right)$$
$$+c\int\log_2\left(1+Ptm(-\sigma^2)\right)e^{-t}dt$$
$$+\log_2(e)\left(\sigma^2 m(-\sigma^2)-1\right).$$

with $m(-\sigma^2)$ defined here as the unique positive solution to (12.5).

In the AWGN scenario, this is instead

$$C_{\rm opt}(\sigma^2) \xrightarrow{\text{a.s.}} \log_2\left(1+\frac{c}{\sigma^2(1+Pm(-\sigma^2))}\right)$$
$$+c\log_2\left(1+Pm(-\sigma^2)\right)$$
$$+\log_2(e)\left(\sigma^2 m(-\sigma^2)-1\right).$$

with $m(-\sigma^2)$ defined here by (12.6).

In Figure 12.1 and Figure 12.3, a comparison is made between the matched-filter, the MMSE filter, and the optimum decoder, for $K = 16$ users and $N = 32$ chips per CDMA code, for different SNR values, for the scenarios of AWGN channels and Rayleigh fading channels, respectively. In these graphs, theoretical expressions are compared against Monte Carlo simulations, the average and standard deviation of which are displayed. We observe, already for these small values of K and N, that the deterministic equivalents for the matched-filter, the minimum mean square error filter, and the optimal decoder fall within (plus or minus) one standard deviation of the empirical spectral efficiency.

In Figure 12.2 and Figure 12.4, the analysis of the spectral efficiency for different limiting ratios $c = \lim K/N$ is performed for these very decoders and for AWGN and Rayleigh fading channels, respectively. The SNR is set to 10 dB. Note that the rate achieved by the optimal decoder grows unbounded as c grows large. This emerges naturally from the fact that the longer the CDMA codes, the more important the data redundancy. The matched-filter also benefits from short code length. On the opposite, the MMSE decoder benefits only from moderate ratios K/N and especially suffers from large K/N. This can be interpreted from

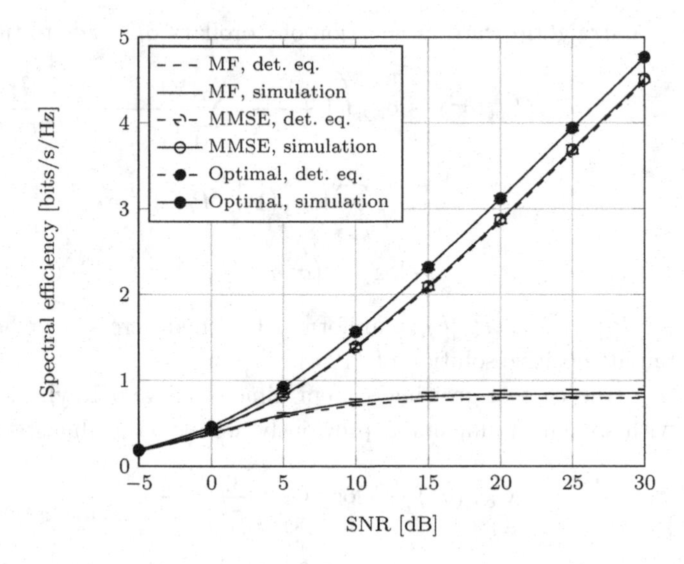

Figure 12.1 Spectral efficiency of random CDMA decoders, AWGN channels. Comparison between simulations and deterministic equivalents (det. eq.), for the matched-filter, the MMSE decoder, and the optimal decoder, $K = 16$ users, $N = 32$ chips per code. Rayleigh channels. Error bars indicate two standard deviations.

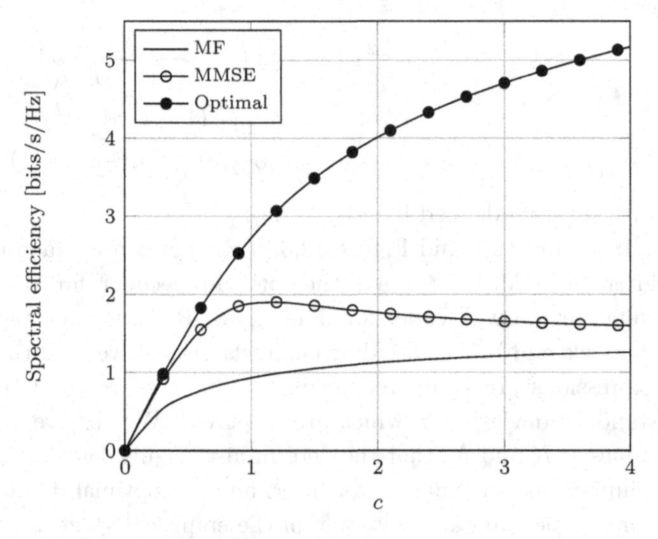

Figure 12.2 Spectral efficiency of random CDMA decoders, for different asymptotic ratios $c = \lim K/N$, SNR=10 dB, AWGN channels. Deterministic equivalents for the matched-filter, the MMSE decoder, and the optimal decoder. Rayleigh channels.

the fact that, for N small compared to K, every user data stream at the output of the MMSE filter is strongly interfered with by inter-code interference from other users. The SINR for every stream is therefore very impacted, to the extent that the spectral efficiency of the MMSE decoder is significantly reduced.

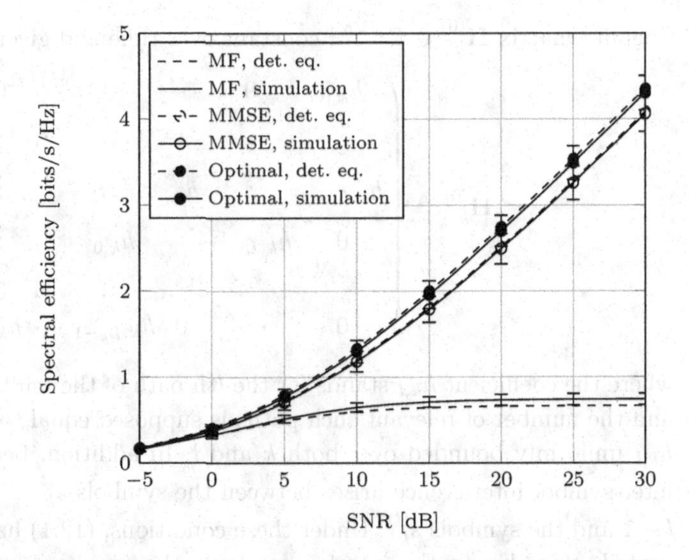

Figure 12.3 Spectral efficiency of random CDMA decoders, Rayleigh fading channels. Comparison between simulations and deterministic equivalents (det. eq.), for the matched-filter, the MMSE decoder, and the optimal decoder, $K = 16$ users, $N = 32$ chips per code. Rayleigh channels. Error bars indicate two standard deviations.

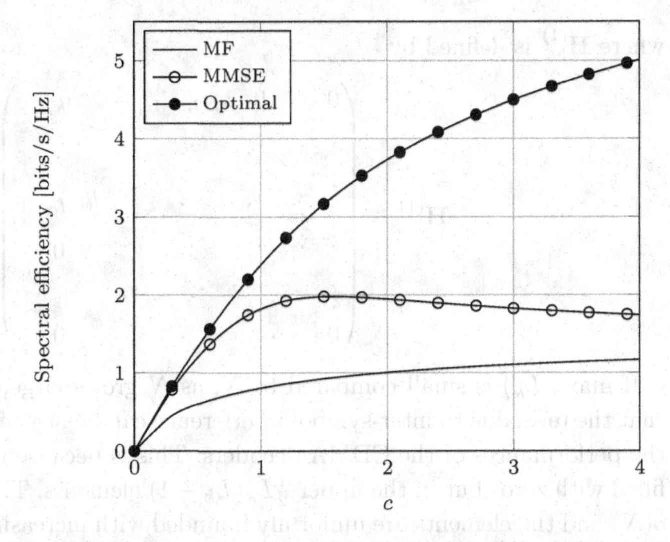

Figure 12.4 Spectral efficiency of random CDMA decoders, for different asymptotic ratios $c = \lim K/N$, SNR=10 dB, Rayleigh fading channels. Deterministic equivalents for the matched-filter, the MMSE decoder, and the optimal decoder.

12.2.2 Random CDMA in uplink frequency selective channels

We consider the same transmission model as in Section 12.2.1, but the transmission channels are now frequency selective, i.e. contain multiple paths. We then replace the flat channel fading h_k coefficients in (12.1) by the convolution

Toeplitz matrix $\mathbf{H}_k^{(0)} \in \mathbb{C}^{N \times N}$, constant over time and given by:

$$\mathbf{H}_k^{(0)} \triangleq \begin{pmatrix} h_{k,0} & 0 & \cdots & \cdots & \cdots & 0 \\ \vdots & \ddots & \ddots & \cdots & \ddots & \vdots \\ h_{k,L_k-1} & \ddots & h_{k_0} & \ddots & \vdots & \vdots \\ 0 & h_{k,L_k-1} & \ddots & h_{k,0} & \ddots & \vdots \\ \vdots & \ddots & \ddots & \ddots & \ddots & 0 \\ 0 & \cdots & 0 & h_{k,L_k-1} & \cdots & h_{k,0} \end{pmatrix}$$

where the coefficient $h_{k,l}$ stands for the lth path of the multi-path channel $\mathbf{H}_k^{(0)}$, and the number of relevant such paths is supposed equal to L_k. We assume the $h_{k,l}$ uniformly bounded over both k and l. In addition, because of multi-path, inter-symbol interference arises between the symbols $s_k^{(l-1)}$ transmitted at time $l-1$ and the symbols $s_k^{(l)}$. Under these conditions, (12.1) has now an additional contribution due to the interference from the previously sent symbol and we finally have

$$\mathbf{y}^{(l)} = \sum_{k=1}^{K} \mathbf{H}_k^{(0)} \mathbf{w}_k \sqrt{P_k} s_k^{(l)} + \sum_{k=1}^{K} \mathbf{H}_k^{(1)} \mathbf{w}_k \sqrt{P_k} s_k^{(l-1)} + \mathbf{n}^{(l)} \tag{12.7}$$

where $\mathbf{H}_k^{(1)}$ is defined by

$$\mathbf{H}_k^{(1)} \triangleq \begin{pmatrix} 0 \cdots & 0 & h_{k,L_k-1} & \cdots & h_{k,1} \\ \vdots \ddots & \ddots & \ddots & \ddots & \vdots \\ \vdots \ddots & \ddots & \ddots & \ddots & h_{k,L_k-1} \\ \vdots \; \vdots & \ddots & \ddots & \ddots & 0 \\ \vdots \ddots & \cdots & \ddots & \ddots & \vdots \\ 0 \cdots & \cdots & \cdots & \cdots & 0 \end{pmatrix}.$$

If $\max_k(L_k)$ is small compared to N, as N grows large, it is rather intuitive that the term due to inter-symbol interference can be neglected, when estimating the performances of the CDMA decoders. This is because the matrices $\mathbf{H}_k^{(1)}$ are filled with zeros but in the upper $\frac{1}{2}L_k(L_k-1)$ elements. This number is of order $o(N)$ and the elements are uniformly bounded with increasing N. Informally, the following therefore ought to be somewhat correct for large N and K

$$\mathbf{y}^{(l)} \simeq \sum_{k=1}^{K} \mathbf{H}_k^{(0)} \mathbf{w}_k \sqrt{P_k} s_k^{(l)} + \mathbf{n}^{(l)}.$$

We could now work with this model and evaluate the performance of the different decoding modes, which will feature in this case the Gram matrix of a random matrix with independent columns $\mathbf{w}_k$ left-multiplied by $\mathbf{H}_k^{(0)}$ for $k \in \{1, \ldots, K\}$. Therefore, it will be possible to invoke Theorem 6.12 to compute in particular the performance of the matched-filter, MMSE and optimal uplink decoders. Nonetheless, the resulting final expressions will be given as a function of the matrices $\mathbf{H}_k^{(0)}$, rather than as a function of the entries $h_{k,j}$. Instead, we will continue the successive model approximations by modifying further $\mathbf{H}_k^{(0)}$ into a more convenient matrix form, asymptotically equivalent to $\mathbf{H}_k^{(0)}$.

From the same line of reasoning as previously, we can indeed guess that filling the triangle of $\frac{1}{2}L_k(L_k - 1)$ upper right entries of $\mathbf{H}_k^{(0)}$ with bounded elements should not alter the final result in the large N, K limit. We may therefore replace $\mathbf{H}_k^{(0)}$ by $\mathbf{H} \triangleq \mathbf{H}_k^{(0)} + \mathbf{H}_k^{(1)}$, leading to

$$\mathbf{y}^{(l)} \simeq \sum_{k=1}^{K} \mathbf{H}_k \mathbf{w}_k \sqrt{P_k} s_k^{(l)} + \mathbf{n}^{(l)}. \tag{12.8}$$

The right-hand side of (12.8) is more interesting to study, as $\mathbf{H}_k$ is a circulant matrix

$$\mathbf{H}_k \triangleq \begin{pmatrix} h_{k,0} & 0 & \cdots & h_{k,L_k-1} & \cdots & h_{k,1} \\ \vdots & \ddots & \ddots & \ddots & \ddots & \vdots \\ h_{k,L_k-1} & \ddots & h_{k,0} & \ddots & \ddots & h_{k,L_k-1} \\ 0 & h_{k,L_k-1} & \ddots & h_{k,0} & \ddots & \vdots \\ \vdots & \ddots & \ddots & \ddots & \ddots & 0 \\ 0 & \cdots & 0 & h_{k,L_k-1} & \cdots & h_{k,0} \end{pmatrix}$$

which can be written under the form $\mathbf{H}_k = \mathbf{F}_N^{\mathsf{H}} \mathbf{D}_k \mathbf{F}_N$, with $\mathbf{F}_N$ the discrete Fourier transform matrix of order N with entries $F_{N,ab} = e^{-2\pi i \frac{(a-1)(b-1)}{N}}$. Moreover, the diagonal entries of $\mathbf{D}_k$ are the discrete Fourier transform coefficients of the first column of $\mathbf{H}_k$ [Gray, 2006], i.e. with $d_{ab} \triangleq D_{a,bb}$

$$d_{ab} = \sum_{n=0}^{L_a-1} h_{a,n} e^{-2\pi i \frac{bn}{N}}.$$

All $\mathbf{H}_k$ matrices are therefore diagonalizable in a common eigenvector basis and we have that the right-hand side of (12.8), multiplied on the left by $\mathbf{F}_N$, reads:

$$\mathbf{z}^{(l)} \triangleq \sum_{k=1}^{K} \mathbf{D}_k \tilde{\mathbf{w}}_k \sqrt{P_k} s_k^{(l)} + \tilde{\mathbf{n}}^{(l)} \tag{12.9}$$

with $\tilde{\mathbf{w}}_k \triangleq \mathbf{F}_N \mathbf{w}_k$ and $\tilde{\mathbf{n}}^{(l)} \triangleq \mathbf{F}_N \mathbf{n}^{(l)}$.

In order to simplify the problem, we now need to make the strong assumption that the vectors $\mathbf{w}_k$ have independent *Gaussian* entries. This ensures that $\tilde{\mathbf{w}}_k$ also has i.i.d. entries (incidentally Gaussian).

We wish to study the performance of linear decoders for the model (12.7). To prove that it is equivalent to work with (12.7) or with (12.9), as N, K grow large, we need to prove that the difference between the figure of merit (say here, the SINR) for the model $\mathbf{y}^{(l)}$ and the figure of merit for the model $\mathbf{z}^{(l)}$ is asymptotically almost surely zero. This can be proved, see, e.g., [Chaufray *et al.*, 2004], using Szegö's theorem, the Markov inequality, Theorem 3.5, and the Borel–Cantelli lemma, Theorem 3.6, in a similar way as in the proof of Theorem 3.4 for instance. For this condition to hold, it suffices that $L/N \to 0$ as N grows large with L an upper bound on $L_1, \ldots, L_K$ for all K large, that the $\mathbf{H}_k$ matrices are bounded in spectral norm, and that there exists $a < b$ such that $0 < a < P_k < b < \infty$ for all k, uniformly on K.

Indeed, the condition on the equivalence of Toeplitz and circulant matrices is formulated rigorously by Szegö's theorem, given below.

Theorem 12.1 (Theorem 4.2 of [Gray, 2006]). *Let $\ldots, t_{-2}, t_{-1}, t_0, t_1, t_2, \ldots$ be a summable sequence of real numbers, i.e. such that*

$$\sum_{k=-\infty}^{\infty} |t_k| < \infty.$$

Denote $\mathbf{T}_N \in \mathbb{C}^{N \times N}$ the Toeplitz matrix with kth column the vector $(t_{-k+1}, \ldots, t_{N-k})^\mathsf{T}$. Then, denoting $\tau_{N,1}, \ldots, \tau_{N,N}$ the eigenvalues of $\mathbf{T}_N$, for any positive s

$$\lim_{N\to\infty} \frac{1}{N} \sum_{k=0}^{N-1} \tau_{N,k}^s = \frac{1}{2\pi} \int_0^{2\pi} f(\lambda) d\lambda$$

with

$$f(\lambda) \triangleq \sum_{k=-\infty}^{\infty} t_k e^{i\lambda}$$

the Fourier transform of $\ldots, t_{-2}, t_{-1}, t_0, t_1, t_2, \ldots$. In particular, if $t_k = 0$ for $k < 0$ and $k > K$ for some constant K, then the series is finite and then absolutely summable, and the l.s.d. of $\mathbf{T}_N$ is the l.s.d. of the circulant matrices with first column $(t_0, \ldots, t_{K-1})^\mathsf{T}$.

From now on, we claim that model (12.9) is equivalent to model (12.7) for the studies to come, in the sense that we can work either with (12.9) or with (12.7) and will end up with the same asymptotic performance results, but for a set of $\mathbf{w}_1, \ldots, \mathbf{w}_K$ of probability one. Equation (12.9) can be written more compactly as

$$\mathbf{z}^{(l)} = \mathbf{X}\mathbf{P}^{\frac{1}{2}}\mathbf{s} + \tilde{\mathbf{n}}^{(l)}$$

where the (i,j)th entry X_{ij} of $\mathbf{X} \in \mathbb{C}^{N \times K}$ has zero mean, $\mathrm{E}[|X_{ij}|^2] = |d_{ji}|^2$ and the elements X_{ij}/d_{ji} are identically distributed. This is the situation of a channel model with variance profile. This type of model was first studied in Theorem 3.14, when the matrix of the d_{ij} has a limiting spectrum, and then in Theorem 6.14 in terms of a deterministic equivalent, in a more general case.

We consider successively the performance of the matched-filter, the MMSE decoder, and the optimal receiver in the frequency selective case.

12.2.2.1 Matched-filter

The matched-filter here consists, for user k, in filtering $\mathbf{z}^{(l)}$ by the kth column of $\mathbf{X}$ as $\mathbf{x}_k^{\mathsf{H}}\mathbf{z}^{(l)}$. From the previous derivation, we have that the SINR $\gamma_k^{(\mathrm{MF})}$ at the output of the matched-filter reads:

$$\gamma_k^{(\mathrm{MF})} = \frac{P_k|\mathbf{x}_k^{\mathsf{H}}\mathbf{x}_k|^2}{\mathbf{x}_k^{\mathsf{H}}\left(\mathbf{X}\mathbf{P}\mathbf{X}^{\mathsf{H}} - P_k\mathbf{x}_k\mathbf{x}_k^{\mathsf{H}} + \sigma^2\mathbf{I}_N\right)\mathbf{x}_k}. \tag{12.10}$$

The $\mathbf{x}_k$ are defined as $\mathbf{x}_k = \mathbf{D}_k\tilde{\mathbf{w}}_k$, where $\tilde{\mathbf{w}}_k$ has i.i.d. entries of zero mean and variance $1/N$, independent of the $\mathbf{D}_k$. The trace lemma therefore ensures that

$$\tilde{\mathbf{w}}_k^{\mathsf{H}}\mathbf{D}_k^{\mathsf{H}}\mathbf{D}_k\tilde{\mathbf{w}}_k - \frac{1}{N}\operatorname{tr}\mathbf{D}_k^{\mathsf{H}}\mathbf{D}_k \xrightarrow{\text{a.s.}} 0$$

where the trace can be rewritten

$$\frac{1}{N}\operatorname{tr}\mathbf{D}_k^{\mathsf{H}}\mathbf{D}_k = \frac{1}{N}\sum_{i=1}^{N}|d_{k,i}|^2.$$

As for the denominator of (12.10), notice that the inner matrix has entries independent of $\mathbf{x}_k$, which we would ideally like to be of almost sure uniformly bounded spectral norm, so that the trace lemma, Lemma 14.2, can operate as before, i.e.

$$\mathbf{x}_k^{\mathsf{H}}\left(\mathbf{X}\mathbf{P}\mathbf{X}^{\mathsf{H}} - P_k\mathbf{x}_k\mathbf{x}_k^{\mathsf{H}} + \sigma^2\mathbf{I}_N\right)\mathbf{x}_k = \tilde{\mathbf{w}}_k^{\mathsf{H}}\mathbf{D}_k^{\mathsf{H}}\left(\mathbf{X}\mathbf{P}\mathbf{X}^{\mathsf{H}} - P_k\mathbf{x}_k\mathbf{x}_k^{\mathsf{H}} + \sigma^2\mathbf{I}_N\right)\mathbf{D}_k\tilde{\mathbf{w}}_k$$

would satisfy

$$\mathbf{x}_k^{\mathsf{H}}\left(\mathbf{X}\mathbf{P}\mathbf{X}^{\mathsf{H}} - P_k\mathbf{x}_k\mathbf{x}_k^{\mathsf{H}} + \sigma^2\mathbf{I}_N\right)\mathbf{x}_k - \frac{1}{N}\operatorname{tr}\mathbf{D}_k\mathbf{D}_k^{\mathsf{H}}\left(\mathbf{X}\mathbf{P}\mathbf{X}^{\mathsf{H}} + \sigma^2\mathbf{I}_N\right) \xrightarrow{\text{a.s.}} 0.$$

However, although extensive Monte Carlo simulations suggest that $\|\mathbf{X}\mathbf{P}\mathbf{X}^{\mathsf{H}}\|$ indeed is uniformly bounded almost surely, it is not proved to this day that this holds true. For the rest of this section, we therefore mainly conjecture this result.

From the definition of $\mathbf{X}$, we then have

$$\frac{1}{N}\operatorname{tr}\mathbf{D}_k\mathbf{D}_k^{\mathsf{H}}\left(\mathbf{X}\mathbf{P}\mathbf{X}^{\mathsf{H}} + \sigma^2\mathbf{I}_N\right) - \frac{1}{N^2}\sum_{n=1}^{N}|d_{k,n}|^2\left[\sigma^2 + \sum_{1\leq i\leq K}P_i|d_{i,n}|^2\right] \xrightarrow{\text{a.s.}} 0. \tag{12.11}$$

And we finally have

$$\gamma_k^{(\mathrm{MF})} - \frac{P_k\left(\frac{1}{N}\sum_{n=1}^{N}|d_{k,n}|^2\right)^2}{\frac{1}{N^2}\sum_{n=1}^{N}\sum_{i=1}^{K}P_i|d_{k,n}|^2|d_{i,n}|^2 + \sigma^2\frac{1}{N}\sum_{n=1}^{N}|d_{k,n}|^2} \xrightarrow{\text{a.s.}} 0$$

where the deterministic equivalent is now clearly dependent on k.

The spectral efficiency $C_{\rm MF}(\sigma^2)$ in that case satisfies

$$C_{\rm MF}(\sigma^2) - \frac{1}{N}\sum_{k=1}^{K}\log_2\left(1+\frac{P_k\left(\frac{1}{N}\sum_{n=1}^{N}|d_{k,n}|^2\right)^2}{\frac{1}{N^2}\sum_{n=1}^{N}\sum_{i=1}^{K}P_i|d_{k,n}|^2|d_{i,n}|^2+\sigma^2\frac{1}{N}\sum_{n=1}^{N}|d_{k,n}|^2}\right)\xrightarrow{\text{a.s.}}0.$$

Contrary to previous sections, we will not derive any limiting result when the random $h_1,\ldots,h_L$ variables have a given distribution. Indeed, in general, for growing N, there does not exist a limit to $C_{\rm MF}(\sigma^2)$. Instead, we can assume, as is often done, that the doubly infinite array $\{|d_{k,n}|^2,\ k\geq 1,\ n\geq 1\}$ converges, in the sense that there exists a function $p(x,y)$ for $(x,y)\in[0,1]\times[0,1]$ such that

$$d_{k,n}-\int_{\frac{k-1}{N}}^{\frac{k}{N}}\int_{\frac{n-1}{N}}^{\frac{n}{N}}p(x,y)dxdy\to 0. \tag{12.12}$$

This is convenient to obtain a deterministic limit of $C_{\rm MF}(\sigma^2)$. This is the assumption taken for instance in [Tulino *et al.*, 2005] in the case of multi-carrier random CDMA transmissions. Assume then that the $d_{k,n}$ are such that they converge to $p(x,y)$, in the sense of (12.12). The value $p(x,y)dxdy$ represents the square of the absolute channel fading coefficient experienced by dx terminal users indexed by the integers k such that $x-dx/2<k/K<x+dx/2$, at normalized frequency indexes n such that $y-dy/2<n/N<y+dy/2$. Therefore, we will denote $p(x,y)=|h(x,y)|^2$ for any convenient $h(x,y)$. We will further denote $P(x)$ a function such that user k satisfying $x-dx/2<k/K<x+dx/2$ consumes power $P(x)-dP(x)<P_k<P(x)+dP(x)$.

In this scenario, we have that the matched-filter spectral efficiency satisfies

$$C_{\rm MF}(\sigma^2)\xrightarrow{\text{a.s.}} c\int_0^1\log_2\left(1+\frac{P(\kappa)\left(\int_0^1|h(\kappa,f)|^2df\right)^2}{c\int_0^1\int_0^1P(\kappa')|h(\kappa,f)|^2|h(\kappa',f)|^2df d\kappa'+\sigma^2\int_0^1|h(\kappa,f)|^2df}\right)d\kappa.$$

12.2.2.2 MMSE decoder

We now consider the SINR minimizer MMSE decoder, that, in our context, filters the receiver input $\mathbf{z}^{(l)}$ as

$$P_k\mathbf{x}_k^{\sf H}\left(\mathbf{XPX}^{\sf H}+\sigma^2\mathbf{I}_N\right)^{-1}\mathbf{z}^{(l)}.$$

The SINR $\gamma_k^{(\rm MMSE)}$ relative to user k reads here

$$\gamma_k^{(\rm MMSE)}=P_k\mathbf{x}_k^{\sf H}\left(\sum_{\substack{1\leq i\leq K\\ i\neq k}}P_i\mathbf{x}_i\mathbf{x}_i^{\sf H}+\sigma^2\mathbf{I}_N\right)^{-1}\mathbf{x}_k.$$

The inner part of the inverse matrix is independent of the entries of $\mathbf{x}_k$. Since $\mathbf{x}_k = \mathbf{D}_k \tilde{\mathbf{w}}_k$ with $\tilde{\mathbf{w}}_k$ a vector of i.i.d. entries with variance $1/N$, we have, similarly as before, that

$$\gamma_k^{(\text{MMSE})} - \frac{P_k}{N} \operatorname{tr} \mathbf{D}_k \mathbf{D}_k^{\mathsf{H}} \left(\mathbf{X}\mathbf{P}\mathbf{X}^{\mathsf{H}} + \sigma^2 \mathbf{I}_N \right)^{-1} \xrightarrow{\text{a.s.}} 0$$

as N, n grow large. Now notice that $\mathbf{X}\mathbf{P}^{\frac{1}{2}}$ is still a matrix with independent entries and variance profile $\{P_j \sigma_{ij}^2\}$, $1 \leq i \leq N$, $1 \leq j \leq n$. From Theorem 6.10, it turns out that the trace on the left-hand side satisfies

$$\frac{P_k}{N} \operatorname{tr} \mathbf{D}_k \mathbf{D}_k^{\mathsf{H}} \left(\mathbf{X}\mathbf{P}\mathbf{X}^{\mathsf{H}} + \sigma^2 \mathbf{I}_N \right)^{-1} - e_k(-\sigma^2) \xrightarrow{\text{a.s.}} 0$$

where $e_k(z)$, $z \in \mathbb{C} \setminus \mathbb{R}^+$, is defined as the unique Stieltjes transform that satisfies

$$e_k(z) = -\frac{1}{z} \frac{1}{N} \sum_{n=1}^{N} \frac{P_k |d_{kn}|^2}{1 + \frac{K}{N} \bar{e}_n(z)}$$

where $\bar{e}_n(z)$ is given by:

$$\bar{e}_n(z) = -\frac{1}{z} \frac{1}{K} \sum_{i=1}^{K} \frac{P_i |d_{in}|^2}{1 + e_i(z)}.$$

Note that the factor K/N is placed in the denominator of the term $e_k(z)$ here instead of the denominator of the term $\bar{e}_k(z)$, contrary to the initial statement of Theorem 6.10. This is due to the fact that $\mathbf{X}$ is here an $N \times K$ matrix with entries of variance $|d_{ij}|^2/N$ and not $|d_{ij}|^2/K$. Particular care must therefore be taken here when propagating the term K/N in the formula of Theorem 6.10.

We conclude that the spectral efficiency $C_{(\text{MMSE})}(\sigma^2)$ for the MMSE decoder in that case satisfies

$$C_{(\text{MMSE})}(\sigma^2) - \frac{1}{N} \sum_{k=1}^{K} \log_2 \left(1 + e_k(-\sigma^2) \right) \xrightarrow{\text{a.s.}} 0.$$

Similar to the MF case, there does not exist a straightforward limit to $C_{(\text{MMSE})}(\sigma^2)$ for practical distributions of $h_1, \ldots, h_L$. However, if the user-frequency channel decay $|d_{k,n}|^2$ converges to a density $|h(\kappa, f)|^2$ for users indexed by $k \simeq \kappa K$ and for normalized frequencies $n \simeq fN$, and that users within $d\kappa$ of κ have power $P(\kappa)$, then $C_{(\text{MMSE})}(\sigma^2)$ has a deterministic almost sure limit, given by:

$$C_{(\text{MMSE})}(\sigma^2) \xrightarrow{\text{a.s.}} c \int_0^1 \log_2 \left(1 + e(\kappa) \right) d\kappa$$

for $e(\kappa)$ the function defined as a solution to the differential equations

$$e(\kappa) = \frac{1}{\sigma^2} \int_0^1 \frac{P(\kappa) |h(\kappa, f)|^2}{1 + c\bar{e}(f)} df$$

$$\bar{e}(f) = \frac{1}{\sigma^2} \int_0^1 \frac{P(\kappa) |h(\kappa, f)|^2}{1 + e(\kappa)} d\kappa. \tag{12.13}$$

12.2.2.3 Optimal decoder

The spectral efficiency $C_{\rm opt}(\sigma^2)$ of the frequency selective uplink CDMA reads:

$$C_{\rm opt}(\sigma^2) = \frac{1}{N}\log\det\left(\mathbf{I}_N + \frac{1}{\sigma^2}\mathbf{XPX}^{\sf H}\right).$$

A deterministic equivalent for $C_{\rm opt}(\sigma^2)$ is then provided by extending Theorem 6.11 with the conjectured asymptotic boundedness of $\mathbf{XPX}^{\sf H}$ by a straightforward application of the dominated convergence theorem, Theorem 6.3. Namely, we have:

$$C_{\rm opt}(\sigma^2) - \left[\frac{1}{N}\sum_{n=1}^{N}\log_2\left(1+\frac{K}{N}\bar{e}_n(-\sigma^2)\right) + \frac{1}{N}\sum_{k=1}^{K}\log_2\left(1+e_k(-\sigma^2)\right)\right.$$
$$\left. - \frac{\log_2(e)}{\sigma^2}\frac{1}{N^2}\sum_{\substack{1\leq n\leq N\\ 1\leq k\leq K}}\frac{|d_{kn}|^2}{\left(1+\frac{K}{N}\bar{e}_n(-\sigma^2)\right)\left(1+e_k(-\sigma^2)\right)}\right] \xrightarrow{\rm a.s.} 0$$

where the $\bar{e}_n(-\sigma^2)$ and $e_k(-\sigma^2)$ are defined as in the previous MMSE case. The conjectured result is however known to hold in expectation by applying directly the result of Theorem 6.11.

As for the linear decoders, if $|d_{k,n}|^2$ has a density limit $|h(\kappa,f)|^2$, then $C_{\rm opt}(\sigma^2)$ converges almost surely as follows.

$$C_{\rm opt}(\sigma^2) \xrightarrow{\rm a.s.} \left[\int_0^1 \log_2\left(1+c\bar{e}(f)\right)df + c\int_0^1\log_2\left(1+e(\kappa)\right)d\kappa\right.$$
$$\left. - \frac{\log_2(e)}{\sigma^2}c\int_0^1\int_0^1\frac{|h(\kappa,f)|^2}{\left(1+c\bar{e}(f)\right)\left(1+e(\kappa)\right)}d\kappa df\right]$$

where the functions e and $\bar{e}$ are solutions of (12.13).

In Figure 12.5, we compare the performance of random CDMA detectors as a function of the channel frequency selectivity, for different ratios K/N. We successively assume that the channel is Rayleigh fading, of length $L=1$, $L=2$, and $L=8$, the coefficients h_l being i.i.d. Gaussian of variance $1/L$. For simulation purposes, we consider a single realization, i.e. we do not average realizations, of the channel condition for an $N=512$ random CDMA transmission. We observe various behaviors of the detectors against frequency selectivity. In particular, the optimal decoder clearly benefits from channel diversity. Nonetheless, although this is not represented here, no further gain is obtained for $L>8$; therefore, there exists a diversity threshold above which the uplink data rate does not increase. The matched-filter follows the same trend, as it benefits as well from frequency diversity. The case of the MMSE decoder is more intriguing, as the latter benefits from frequency selectivity only for ratios K/N lower than one, i.e. for less users than code length, and suffers from channel frequency selectivity for $K/N \ll 1$. In our application example, $K/N \leq 1$ is the most likely assumption, in order for the CDMA codes to be almost orthogonal.

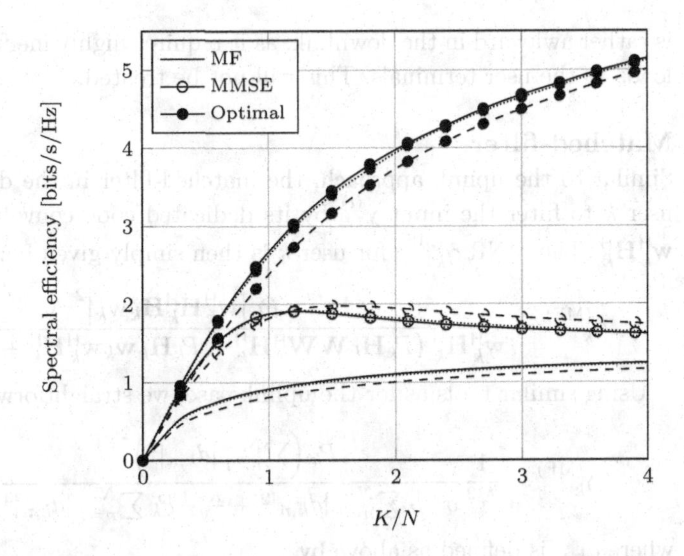

Figure 12.5 Spectral efficiency of random CDMA decoders, for different ratios K/N, SNR=10 dB, Rayleigh frequency selective fading channels. Deterministic equivalents for the matched-filter, the MMSE decoder, and the optimal decoder; $N = 512$, $L = 1$ in dashed lines, $L = 4$ in dotted lines, $L = 8$ in plain lines.

12.2.3 Random CDMA in downlink frequency selective channels

We now consider the downlink CDMA setting, where the base station issues data for the K terminals. Instead of characterizing the complete rate region of the broadcast channel, we focus on the capacity achieved by a specific terminal $k \in \{1, \ldots, K\}$. In the case of frequency selective transmissions, the communication model can be written at time l as

$$\mathbf{y}_k^{(l)} = \sqrt{P_k}\mathbf{H}_k^{(0)}\mathbf{W}\mathbf{s}^{(l)} + \sqrt{P_k}\mathbf{H}_k^{(1)}\mathbf{W}\mathbf{s}^{(l-1)} + \mathbf{n}_k^{(l)} \tag{12.14}$$

with $\mathbf{y}_k^{(l)} \in \mathbb{C}^N$ the N-chip signal received by terminal k at time l, $\mathbf{W} = [\mathbf{w}_1, \ldots, \mathbf{w}_K] \in \mathbb{C}^{N\times K}$, where $\mathbf{w}_i$ is now the downlink random CDMA code intended for user i, $\mathbf{s}^{(l)} = [s_1^{(l)}, \ldots, s_K^{(l)}]^\mathsf{T} \in \mathbb{C}^K$ with $s_i^{(l)}$ the signal intended for user i at time l, P_k is the mean transmit power of user k, $\mathbf{H}_k^{(0)} \in \mathbb{C}^{N\times N}$ is the Topelitz matrix corresponding to the frequency selective channel from the base station to user k, $\mathbf{H}_k^{(1)} \in \mathbb{C}^{N\times N}$ the upper-triangular matrix that takes into account the inter-symbol interference, and $\mathbf{n}_k^{(l)} \in \mathbb{C}^N$ is the N-dimensional noise received by terminal k at time l.

Similar to the uplink case, we can consider for (12.14) the approximated model

$$\mathbf{y}_k^{(l)} \simeq \sqrt{P_k}\mathbf{H}_k\mathbf{W}\mathbf{s}^{(l)} + \mathbf{n}_k^{(l)} \tag{12.15}$$

where $\mathbf{H}_k \in \mathbb{C}^{N\times K}$ is the circulant matrix equivalent to $\mathbf{H}_k^{(0)}$. It can indeed be shown, see, e.g., [Debbah *et al.*, 2003a], that the asymptotic SINR sought for are the same in both models. We study here the two receive linear decoders that are the matched-filter and the MMSE decoder. The optimal joint decoding strategy is rather awkward in the downlink, as it requires highly inefficient computational loads at the user terminals. This will not be treated.

12.2.3.1 Matched-filter

Similar to the uplink approach, the matched-filter in the downlink consists for user k to filter the input $\mathbf{y}_k^{(l)}$ by its dedicated code convoluted by the channel $\mathbf{w}_k^{\mathsf{H}}\mathbf{H}_k^{\mathsf{H}}$. The SINR $\gamma_k^{(\mathrm{MF})}$ for user k is then simply given by:

$$\gamma_k^{(\mathrm{MF})} = \frac{P_k \left|\mathbf{w}_k^{\mathsf{H}}\mathbf{H}_k^{\mathsf{H}}\mathbf{H}_k\mathbf{w}_k\right|^2}{\mathbf{w}_k^{\mathsf{H}}\mathbf{H}_k^{\mathsf{H}}\left(P_k\mathbf{H}_k\mathbf{W}\mathbf{W}^{\mathsf{H}}\mathbf{H}_k^{\mathsf{H}} - P_k\mathbf{H}_k\mathbf{w}_k\mathbf{w}_k^{\mathsf{H}}\mathbf{H}_k^{\mathsf{H}} + \sigma^2\mathbf{I}_N\right)\mathbf{H}_k\mathbf{w}_k}.$$

Using similar tools as for the uplink case, we straightforwardly have that

$$\gamma_k^{(\mathrm{MF})} - \frac{1}{N^2}\frac{P_k\left(\sum_{n=1}^N |d_{k,n}|^2\right)^2}{\sigma^2\frac{1}{N}\sum_{n=1}^N |d_{k,n}|^2 + \frac{K-1}{N^2}P_k\sum_{n=1}^N |d_{k,n}|^4} \xrightarrow{\text{a.s.}} 0 \tag{12.16}$$

where $d_{k,n}$ is defined as above by

$$d_{k,n} \triangleq \sum_{l=0}^{L_k-1} h_{k,l}e^{-2\pi i\frac{nl}{N}}.$$

From this expression, we then have that the sum rate C_{MF} achieved by the broadcast channel satisfies

$$C_{\mathrm{MF}}(\sigma^2) - \frac{1}{N}\sum_{k=1}^K \log_2\left(1 + \frac{P_k\left(\frac{1}{N}\sum_{n=1}^N |d_{k,n}|^2\right)^2}{\sigma^2\frac{1}{N}\sum_{n=1}^N |d_{k,n}|^2 + \frac{K-1}{N^2}P_k\sum_{n=1}^N |d_{k,n}|^4}\right) \xrightarrow{\text{a.s.}} 0.$$

When $P_1, \ldots, P_K$ have a limiting density $P(\kappa)$, the $|d_{k,n}|$ have a limiting density $|h(\kappa, f)|$, and $K/N \to c$, then asymptotically

$$C_{\mathrm{MF}}(\sigma^2) \xrightarrow{\text{a.s.}} c\int_0^1 \log_2\left[1 + P(\kappa)\frac{\left(\int_0^1 |h(\kappa,f)|^2 df\right)^2}{\sigma^2\int_0^1 |h(\kappa,f)|^2 df + c\int_0^1 |h(\kappa,f)|^4 df} d\kappa\right].$$

12.2.3.2 MMSE decoder

For the more advanced MMSE decoder, i.e. the linear decoder that consists in retrieving the transmit symbols from the product $\mathbf{w}_k^{\mathsf{H}}\mathbf{H}_k^{\mathsf{H}}\left(P_k\mathbf{H}_k\mathbf{W}\mathbf{W}^{\mathsf{H}}\mathbf{H}_k^{\mathsf{H}} + \sigma^2\mathbf{I}_N\right)^{-1}\mathbf{y}_k^{(l)}$, the SINR $\gamma_k^{(\mathrm{MMSE})}$ for the data intended for user k is explicitly given by:

$$\gamma_k^{(\mathrm{MMSE})} = P_k\mathbf{w}_k^{\mathsf{H}}\mathbf{H}_k^{\mathsf{H}}\left(P_k\mathbf{H}_k\mathbf{W}\mathbf{W}^{\mathsf{H}}\mathbf{H}_k^{\mathsf{H}} - P_k\mathbf{H}_k\mathbf{w}_k\mathbf{w}_k^{\mathsf{H}}\mathbf{H}_k^{\mathsf{H}} + \sigma^2\mathbf{I}_N\right)^{-1}\mathbf{H}_k\mathbf{w}_k.$$

As in the uplink case, the central matrix is independent of $\mathbf{w}_k$ and therefore the trace lemma ensures that the right-hand side expression is close to the normalized trace of the central matrix, which is its Stieltjes transform at point $-\sigma^2$. We therefore use again the deterministic equivalent of Theorem 6.1 to obtain

$$\gamma_k^{(\mathrm{MMSE})} - e_k(-\sigma^2) \xrightarrow{\text{a.s.}} 0$$

where $e_k(-\sigma^2)$ is the unique real positive solution to the equation in e

$$e = \frac{1}{N}\sum_{n=1}^{N} \frac{P_k|d_{k,n}|^2}{\frac{1}{1+e}\frac{K-1}{N}P_k|d_{k,n}|^2 + \sigma^2}. \tag{12.17}$$

The resulting achievable sum rate C_{MMSE} for the MMSE decoded broadcast channel is then such that

$$C_{\text{MMSE}}(\sigma^2) - \frac{1}{N}\sum_{k=1}^{K} \log_2\left(1 + e_k(-\sigma^2)\right) \xrightarrow{\text{a.s.}} 0.$$

When the channel has a limiting space-frequency power density $|h(\kappa, f)|^2$ and the users within $d\kappa$ of κ have inverse path loss $P(\kappa)$, the sum rate has a deterministic limit given by:

$$C_{\text{MMSE}}(\sigma^2) \xrightarrow{\text{a.s.}} c\int_0^1 \log_2\left(1 + e(\kappa)\right)$$

where the function $e(\kappa)$ satisfies

$$e(\kappa) = \int_0^1 \frac{P(\kappa)|h(\kappa,f)|^2}{\frac{c}{1+e(\kappa)}P(\kappa)|h(\kappa,f)|^2 + \sigma^2} df.$$

Before moving to the study of orthogonal CDMA transmissions, let us recall [Poor and Verdú, 1997] that the multiple access interference incurred by the non-orthogonal users can be considered roughly Gaussian in the large dimensional system limit. As a consequence, the bit error rate BER induced, e.g. by the MMSE decoder for user k and for QPSK modulation, is of order

$$\text{BER} \simeq Q\left(\sqrt{\gamma_k^{(\text{MMSE})}}\right)$$

with Q the Gaussian Q-function, defined by

$$Q(x) \triangleq \frac{1}{\sqrt{2\pi}}\int_x^\infty e^{-\frac{t^2}{2}} dt.$$

We can therefore give an approximation of the average bit error rate in the downlink decoding. This is provided in Figure 12.6, where it can be seen that, for small ratios K/N, the deterministic approximation of the bit error rate is very accurate even for not too large K and N. In contrast, for larger K/N ratios, large K, N are demanded for the deterministic approximation to be accurate.

Note additionally that asymptotic gaussianity of the SINR at the output of the MMSE receiver for all cases above can be also proved, although this is not detailed here, see, e.g., [Guo *et al.*, 2002; Tse and Zeitouni, 2000]. Also, extensions of the above results to the multiple antenna case were studied in [Bai and Silverstein, 2007; Hanly and Tse, 2001] as well as to asynchronous random CDMA in [Cottatellucci *et al.*, 2010a,b; Hwang, 2007; Mantravadi and Veeravalli, 2002].

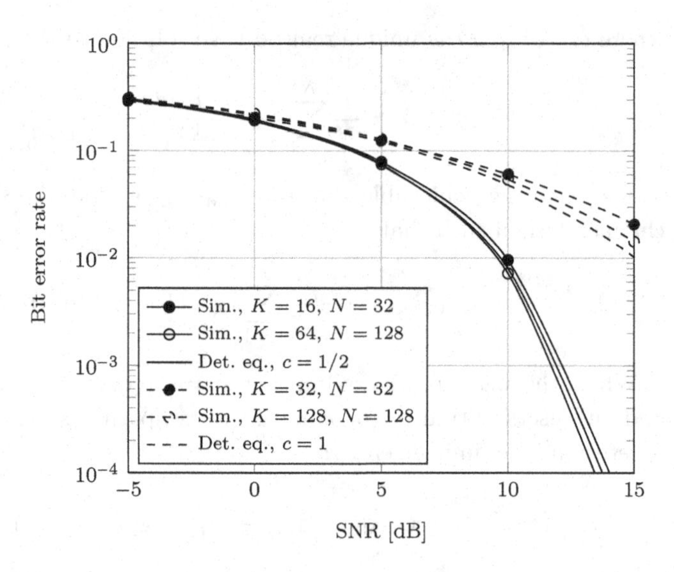

Figure 12.6 Bit error rate achieved by random CDMA decoders in the downlink, AWGN channel. Comparison between simulations (sim.) and deterministic equivalents (det. eq.) for the MMSE decoder, $K = 16$ users, $N = 32$ chips per code.

12.3 Performance of orthogonal CDMA technologies

The initial incentive for using CDMA schemes in multi-user wireless communications is based on the idea that orthogonality between users can be brought about by codes, instead of separating user transmissions by using time division or frequency division multiplexing. This is all the more convenient when the codes are perfectly orthogonal and the communication channel is frequency flat. In this scenario, the signals received either in the uplink by the base station or in the downlink by the users are perfectly orthogonal. If the channel is frequency selective, the code orthogonality is lost so that orthogonal codes are not much better than random codes, as will clearly appear in the following. Moreover, it is important to recall that orthogonality is preserved on the sole condition that all codes are sent and received simultaneously. In the uplink, this imposes all users to be synchronous in their transmissions and that the delay incurred by the difference of wave travel distance between the user closest and the user furthest to the base station is small enough.

For all these reasons, in days when CDMA technologies are no longer used exclusively for communications over narrowband channels, the question is posed of whether orthogonal CDMA is preferable to random CDMA. This section provides the orthogonal version of the results derived in the previous section for i.i.d. codes. It will be shown that orthogonal codes perform always better than random codes in terms of achievable rates, although the difference becomes marginal as the channel frequency selectivity increases.

Similar to the previous section, we start by the study of orthogonal CDMA in the uplink.

12.3.1 Orthogonal CDMA in uplink frequency flat channels

We start with the transmission model (12.1) where now $\mathbf{w}_1, \ldots, \mathbf{w}_K$ are $K \leq N$ columns of a Haar matrix, and $\mathbf{W} = [\mathbf{w}_1, \ldots, \mathbf{w}_K]$ is such that $\mathbf{W}^\mathsf{H}\mathbf{W} = \mathbf{I}_K$. We define $\mathbf{H}$ and $\mathbf{P}$ as before.

Consider the matched-filter for which the SINR $\bar{\gamma}_k^{(\mathrm{MF})}$ for user k is given by:

$$\begin{aligned}\bar{\gamma}_k^{(\mathrm{MF})} &= \frac{P_k|h_k|^2\mathbf{w}_k^\mathsf{H}\mathbf{w}_k}{\mathbf{w}_k^\mathsf{H}\left(\mathbf{WHPH}^\mathsf{H}\mathbf{W}^\mathsf{H} - P_k|h_k|^2\mathbf{w}_k\mathbf{w}_k^\mathsf{H} + \sigma^2\mathbf{I}_N\right)\mathbf{w}_k} \\ &= \frac{P_k|h_k|^2\mathbf{w}_k^\mathsf{H}\mathbf{w}_k}{\sigma^2}\end{aligned}$$

which unfolds from the code orthogonality.

Since we have, for all N

$$\mathbf{w}_i^\mathsf{H}\mathbf{w}_j = \delta_i^j$$

with δ_i^j the Kronecker delta, we have simply

$$\bar{\gamma}_k^{(\mathrm{MF})} = \frac{P_k|h_k|^2}{\sigma^2}$$

and the achievable sum rate $\bar{C}_{(\mathrm{MF})}$ satisfies

$$\bar{C}_{(\mathrm{MF})} = \frac{1}{N}\sum_{k=1}^{K}\log_2\left(1 + \frac{P_k|h_k|^2}{\sigma^2}\right).$$

This is the best we can get as this also corresponds to the capacity of both the MMSE and the optimal joint decoder.

12.3.2 Orthogonal CDMA in uplink frequency selective channels

We now move to the frequency selective model (12.8).

12.3.2.1 Matched-filter

The SINR $\bar{\gamma}_k^{(\mathrm{MF})}$ for the signal originating from user k reads as before

$$\bar{\gamma}_k^{(\mathrm{MF})} = \frac{P_k\left|\mathbf{x}_k^\mathsf{H}\mathbf{x}_k\right|^2}{\mathbf{x}_k^\mathsf{H}\left(\mathbf{XPX}^\mathsf{H} - P_k\mathbf{x}_k\mathbf{x}_k^\mathsf{H} + \sigma^2\mathbf{I}_N\right)\mathbf{x}_k}$$

where $\mathbf{x}_k = \mathbf{D}_k\mathbf{F}_N\mathbf{w}_k$, $\mathbf{D}_k$ being diagonal defined as previously, and $\mathbf{F}_N$ is the Fourier transform matrix. Since $\mathbf{W}$ is formed of columns of a Haar matrix, $\mathbf{F}_N\mathbf{W}$ is still Haar distributed.

From the trace lemma, Corollary 6.3, we have again that the numerator satisfies

$$\mathbf{w}_k^{\mathsf{H}}\mathbf{F}_N^{\mathsf{H}}\mathbf{D}_k^{\mathsf{H}}\mathbf{D}_k\mathbf{F}_N\mathbf{w}_k - \frac{1}{N}\operatorname{tr}\mathbf{D}_k^{\mathsf{H}}\mathbf{D}_k \xrightarrow{\text{a.s.}} 0$$

as long as the inner matrix has almost surely uniformly bounded spectral norm. The second term in the left-hand side is simply $\frac{1}{N}\sum_{n=1}^{N}|d_{kn}|^2$ with d_{kn} the nth diagonal entry of $\mathbf{D}_k$. To handle the denominator, we need the following result.

Lemma 12.1 ([Bonneau *et al.*, 2005]). *Let $\mathbf{W} = [\mathbf{w}_1, \ldots, \mathbf{w}_K] \in \mathbb{C}^{N\times K}$, $K \leq N$, be K columns of a Haar random unitary matrix, and $\mathbf{A} \in \mathbb{C}^{N\times K}$ be independent of $\mathbf{W}$ and have uniformly bounded spectral norm. Denote $\mathbf{X} = [\mathbf{x}_1, \ldots, \mathbf{x}_K]$ the matrix with (i,j)th entry $w_{ij}a_{ij}$. Then, for $k \in \{1, \ldots, K\}$*

$$\mathbf{x}_k^{\mathsf{H}}\mathbf{X}\mathbf{X}^{\mathsf{H}}\mathbf{x}_k - \left[\frac{1}{N^2}\sum_{n=1}^{N}\sum_{j\neq k}|a_{nk}|^2|a_{nj}|^2 - \frac{1}{N^3}\sum_{j\neq k}\left|\sum_{n=1}^{N}a_{kn}a_{jn}^*\right|^2\right] \xrightarrow{\text{a.s.}} 0.$$

Remark 12.1. Compared to the case where $\mathbf{w}_i$ has i.i.d. entries (see, e.g., (12.11)), observe that the i.i.d. and Haar cases only differ by the additional second term in brackets. Observe also that $\mathbf{w}_k^{\mathsf{H}}\mathbf{X}\mathbf{X}^{\mathsf{H}}\mathbf{w}_k$ is necessarily asymptotically smaller when $\mathbf{W}$ is Haar than if $\mathbf{W}$ has i.i.d. entries. Therefore, the interference term at the output of the matched-filter is asymptotically smaller and the resulting SINR larger. Note also that the almost sure convergence is easy to verify compared to the (only conjectured) i.i.d. counterpart, since the eigenvalues of unitary matrices are all of unit norm.

Applying Lemma 12.1 to the problem at hand, we finally have that the difference between $\bar{\gamma}_k^{(\mathrm{MF})}$ and

$$\frac{P_k\left(\frac{1}{N}\sum_{n=1}^{N}|d_{k,n}|^2\right)^2}{\sum_{i\neq k}\sum_{n=1}^{N}\frac{P_i}{N^2}|d_{k,n}|^2|d_{i,n}|^2 - \sum_{i\neq k}\frac{P_i}{N^3}\left|\sum_{n=1}^{N}d_{k,n}d_{i,n}^*\right|^2 + \frac{\sigma^2}{N}\sum_{n=1}^{N}|d_{k,n}|^2}$$

is asymptotically equal to zero in the large N, K limit, almost surely. The resulting deterministic equivalent for the capacity unfolds directly.

To this day, a convenient deterministic equivalent for the capacity of the MMSE decoder in the frequency selective class has not been proposed, although non-convenient forms can be obtained using similar derivations as in the first steps of the proof of Theorem 6.17. This is because the communication model involves random Haar matrices with a variance profile, which are more involved to study.

12.3.3 Orthogonal CDMA in downlink frequency selective channels

We now consider the downlink model (12.15) and, again, turn $\mathbf{W}$ into K columns of an $N \times N$ Haar matrix.

12.3.3.1 Matched-filter

The SINR $\bar{\gamma}_k^{(\mathrm{MF})}$ for user k at the output of the matched-filter is given by:

$$\bar{\gamma}_k^{(\mathrm{MF})} = \frac{P_k \left|\mathbf{w}_k^{\mathsf{H}}\mathbf{H}_k^{\mathsf{H}}\mathbf{H}_k\mathbf{w}_k\right|^2}{\mathbf{w}_k^{\mathsf{H}}\mathbf{H}_k^{\mathsf{H}}\left(P_k\mathbf{H}_k\mathbf{W}\mathbf{W}^{\mathsf{H}}\mathbf{H}_k^{\mathsf{H}} - P_k\mathbf{H}_k\mathbf{w}_k\mathbf{w}_k^{\mathsf{H}}\mathbf{H}_k^{\mathsf{H}} + \sigma^2\mathbf{I}_N\right)\mathbf{H}_k\mathbf{w}_k}.$$

Since $\mathbf{H}_k$ is a circulant matrix and $\mathbf{W}$ is unitarily invariant, by writing $\mathbf{F}_N\mathbf{H}_k\mathbf{F}_N^{\mathsf{H}} = \mathbf{D}_k$ and $\tilde{\mathbf{w}}_k = \mathbf{F}_N\mathbf{w}_k$, $\tilde{\mathbf{W}} = [\tilde{\mathbf{w}}_1, \ldots, \tilde{\mathbf{w}}_K]$ ($\tilde{\mathbf{W}}$ is still unitary and unitarily invariant), $\mathbf{D}_k$ is diagonal and the SINR reads:

$$\bar{\gamma}_k^{(\mathrm{MF})} = \frac{P_k \left|\tilde{\mathbf{w}}_k^{\mathsf{H}}\mathbf{D}_k^{\mathsf{H}}\mathbf{D}_k\tilde{\mathbf{w}}_k\right|^2}{\tilde{\mathbf{w}}_k^{\mathsf{H}}\mathbf{D}_k^{\mathsf{H}}\left(P_k\mathbf{D}_k\tilde{\mathbf{W}}\tilde{\mathbf{W}}^{\mathsf{H}}\mathbf{D}_k^{\mathsf{H}} - P_k\mathbf{D}_k\tilde{\mathbf{w}}_k\tilde{\mathbf{w}}_k^{\mathsf{H}}\mathbf{D}_k^{\mathsf{H}} + \sigma^2\mathbf{I}_N\right)\mathbf{D}_k\tilde{\mathbf{w}}_k}.$$

The numerator is as usual such that

$$\tilde{\mathbf{w}}_k^{\mathsf{H}}\mathbf{D}_k^{\mathsf{H}}\mathbf{D}_k\tilde{\mathbf{w}}_k - \frac{1}{N}\sum_{i=1}^{N}|d_{k,n}|^2 \xrightarrow{\text{a.s.}} 0.$$

As for the denominator, we invoke once more Lemma 12.1 with a variance profile a_{ij}^2 with a_{ij} constant over j. We therefore obtain

$$\tilde{\mathbf{w}}_k^{\mathsf{H}}\mathbf{D}_k^{\mathsf{H}}\left(\mathbf{D}_k\tilde{\mathbf{W}}\tilde{\mathbf{W}}^{\mathsf{H}}\mathbf{D}_k^{\mathsf{H}} - \mathbf{D}_k\tilde{\mathbf{w}}_k\tilde{\mathbf{w}}_k^{\mathsf{H}}\mathbf{D}_k^{\mathsf{H}}\right)\mathbf{D}_k\tilde{\mathbf{w}}_k$$
$$- \left[\frac{K-1}{N^2}\sum_{n=1}^{N}|d_{k,n}|^4 - \frac{K-1}{N^3}\left(\sum_{n=1}^{N}|d_{k,n}|^2\right)^2\right] \xrightarrow{\text{a.s.}} 0.$$

A deterministic equivalent for the SINR therefore unfolds as

$$\bar{\gamma}_k^{(\mathrm{MF})} - \frac{P_k\left(\frac{1}{N}\sum_{i=1}^{N}|d_{k,n}|^2\right)^2}{\frac{P_k(K-1)}{N^2}\sum_{n=1}^{N}|d_{k,n}|^4 - \frac{P_k(K-1)}{N^3}\left(\sum_{n=1}^{N}|d_{k,n}|^2\right)^2 + \frac{\sigma^2}{N}\sum_{i=1}^{N}|d_{k,n}|^2}$$
$$\xrightarrow{\text{a.s.}} 0.$$

Compared to Equation (12.16), observe that the term in the denominator is necessarily inferior to the term in the denominator of the deterministic equivalent of the SINR in (12.16), while the numerator is unchanged. As a consequence, at least asymptotically, the performance of the orthogonal CDMA transmission is better than that of the random CDMA scheme. Notice also, from the boundedness assumption on the $|d_{n,k}|$, that

$$\frac{K-1}{N^3}\left(\sum_{n=1}^{N}|d_{k,n}|^2\right)^2 \leq \frac{K-1}{N}\max_n|d_{k,n}|^4$$

the right-hand side of which is of order $O(K/N)$. The difference between the i.i.d. and orthogonal CDMA performance is then marginal for small K.

This expression has an explicit limit if $|d_{k,n}|$ and P_k have limiting densities $|h(\kappa, f)|$ and $P(\kappa)$, respectively, and $K/N \to c$. Precisely, we have:

$$\bar{\gamma}_k^{(\text{MF})} \xrightarrow{\text{a.s.}}$$

$$\frac{P(\kappa)\left(\int_0^1 |h(\kappa,f)|^2 df\right)^2}{P(\kappa)c\int_0^1 |h(\kappa,f)|^4 df - P(\kappa)c\int_0^1 \left(i\int_0^1 \sum_{n=1}^N |h(\kappa,f)|^2 df\right)^2 + \sigma^2 \int_0^1 |h(\kappa,f)|^2 df}.$$

The resulting limit of the deterministic equivalent of the capacity $\bar{C}_{(\text{MF})}$ unfolds directly.

12.3.3.2 MMSE decoder

The MMSE decoder leads to the SINR $\bar{\gamma}_k^{(\text{MMSE})}$ of the form

$$\bar{\gamma}_k^{(\text{MMSE})} = P_k \mathbf{w}_k^{\mathsf{H}} \mathbf{H}_k^{\mathsf{H}} \left(P_k \mathbf{H}_k \mathbf{W}\mathbf{W}^{\mathsf{H}} \mathbf{H}_k^{\mathsf{H}} - P_k \mathbf{H}_k \mathbf{w}_k \mathbf{w}_k^{\mathsf{H}} \mathbf{H}_k^{\mathsf{H}} + \sigma^2 \mathbf{I}_N\right)^{-1} \mathbf{H}_k \mathbf{w}_k.$$

A direct application of Theorem 6.19 when the sum of Gram matrices is taken over a single term leads to

$$\bar{\gamma}_k^{(\text{MMSE})} - \frac{P_k e_k}{1 - e_k \bar{e}_k}$$

where e_k and $\bar{e}_k$ satisfy the implicit equation

$$\bar{e}_k = \frac{K}{N} \frac{P_k}{1 + P_k e_k - \bar{e}_k e_k}$$

$$e_k = \frac{1}{N} \sum_{n=1}^N \frac{|d_{k,n}|^2}{\bar{e}_k |d_{k,n}|^2 + \sigma^2}.$$

In [Debbah *et al.*, 2003a,b], a free probability approach is used to derive the above deterministic equivalent. The precise result of [Debbah *et al.*, 2003a] is that

$$\bar{\gamma}_k^{(\text{MMSE})} - \eta_k \xrightarrow{\text{a.s.}} 0$$

where η_k is the unique positive solution to

$$\frac{\eta_k}{\eta_k + 1} = \frac{1}{N} \sum_{n=1}^N \frac{P_k |d_{n,k}|^2}{cP_k |d_{n,k}|^2 + \sigma^2 (1-c)\eta_k + \sigma^2}.$$

It can be shown that both expressions are consistent, i.e. that $\eta_k = \frac{P_k e_k}{1 - e_k \bar{e}_k}$, by writing

$$\frac{P_k e_k}{1 - e_k \bar{e}_k}\left(\frac{P_k e_k}{1 - e_k \bar{e}_k} + 1\right)^{-1} = \frac{1}{N} \sum_{n=1}^N \frac{P_k |d_{n,k}|^2}{cP_k |d_{n,k}|^2 + \sigma^2 (1 - e_k \bar{e}_k + P_k e_k)}$$

$$= \frac{1}{N} \sum_{n=1}^N \frac{P_k |d_{n,k}|^2}{cP_k |d_{n,k}|^2 + \sigma^2 (1 + (1-c)\frac{P_k e_k}{1 - e_k \bar{e}_k})}$$

where the second equality comes from the observation that

$$\begin{aligned}1-(1-c)\frac{P_k e_k}{1-e_k\bar{e}_k} &= \frac{(1-e_k\bar{e}_k+P_k e_k)-cP_k e_k}{1-e_k\bar{e}_k}\\ &= \frac{(1-e_k\bar{e}_k+P_k e_k)-e_k\bar{e}_k(1-e_k\bar{e}_k+P_k e_k)}{1-e_k\bar{e}_k}\\ &= 1-e_k\bar{e}_k+P_k e_k.\end{aligned}$$

This notation in terms of η_k is rather convenient, as it can be directly compared to the i.i.d. CDMA scenario of Equation (12.17). In both cases, denote η_k the deterministic equivalent for the SINR of user k. In the i.i.d. case, we found out that the solution for

$$\frac{\eta_k}{\eta_k+1} = \frac{1}{N}\sum_{n=1}^{N}\frac{P_k|d_{k,n}|^2}{cP_k|d_{k,n}|^2+\sigma^2(1+\eta_k)} \tag{12.18}$$

is such a deterministic equivalent, while in the orthogonal case, we now have

$$\frac{\eta_k}{\eta_k+1} = \frac{1}{N}\sum_{n=1}^{N}\frac{P_k|d_{n,k}|^2}{cP_k|d_{n,k}|^2+\sigma^2(1+\eta_k)-c\eta_k\sigma^2}. \tag{12.19}$$

Since $\frac{\eta_k}{\eta_k+1}=1-\frac{1}{\eta_k+1}$, clearly η_k is larger in the orthogonal CDMA setting than in its i.i.d. counterpart. Notice that $0\leq\eta_k\leq\frac{1}{\sigma^2}$ in both cases, and η_k is therefore uniformly bounded. As a consequence, as $K/N\to 0$, $\frac{\eta_k}{\eta_k+1}$ in both the orthogonal and the i.i.d. scenarios converge to the same value, which is similar to the matched-filter case.

We then have the following expression of a deterministic equivalent for $\bar{C}_{(\mathrm{MMSE})}$

$$\bar{C}_{(\mathrm{MMSE})}-\frac{1}{N}\sum_{k=1}^{K}\log_2(1+\eta_k)\xrightarrow{\text{a.s.}}0.$$

If $|d_{k,n}|$ and P_k have limiting densities $|h(\kappa,f)|$ and $P(\kappa)$, respectively, and $K/N\to c$, this leads to

$$\bar{C}_{(\mathrm{MMSE})}\xrightarrow{\text{a.s.}}c\int_0^1\log_2(1+P(\kappa)\eta(\kappa))\,d\kappa$$

where the function $\eta(\kappa)$ is solution to

$$\frac{\eta(\kappa)}{\eta(\kappa)+1}=\int\frac{P(\kappa)|h(\kappa,f)|^2}{cP(\kappa)|h(\kappa,f)|^2+\sigma^2\left(1+(1-c)\eta(\kappa)\right)}df.$$

In what follows, we compare the performance of the random CDMA and orthogonal CDMA in the downlink. We will specifically evidence the gain brought by orthogonal CDMA precoders when the channel frequency selectivity is not too strong. In Figure 12.7, we consider the case when the multi-path channel is composed of a single tap, and we depict the spectral efficiency for both orthogonal and i.i.d. codes as the ratio K/N varies (note that $K/N\leq 1$ necessarily in the

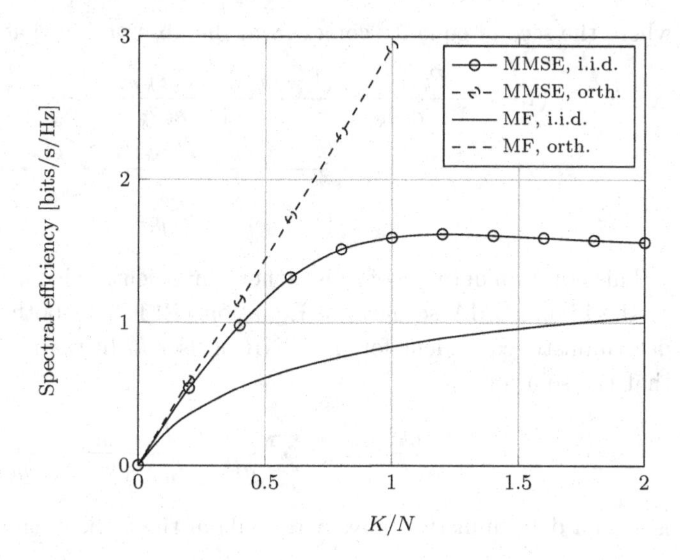

Figure 12.7 Spectral efficiency of random and orthogonal CDMA decoders, for different ratios K/N, $K = 512$, SNR=10 dB, Rayleigh frequency selective fading channels $L = 1$, in the downlink.

orthogonal case). That is, we consider that the orthogonal codes received by the users are perfectly orthogonal. The SNR is set to 10 dB, and the number of receivers taken for simulation is $K = 512$. In this case, we observe indeed that both the matched-filter and the MMSE filter for the orthogonal codes perform the same (as no inter-code interference is present), while the linear filters for the i.i.d. codes are highly suboptimal, as predicted. Then, in Figure 12.8, we consider the scenario of an $L = 8$-tap multi-path channel, which now shows a large advantage of the MMSE filter compared to the matched-filter, both for i.i.d. codes and for orthogonal codes. Nonetheless, in spite of the channel convolution effect, the spectral efficiency achieved by orthogonal codes is still largely superior to that achieved by random codes. This is different from the uplink case, where different channels affect the different codes. Here, from the point of view of the receiver, the channel convolution effect affects identically all user codes, therefore limiting the orthogonality reduction due to frequency selectivity and thus impacting the spectral efficiency in a limited manner.

This completes the chapter on CDMA technologies. We now move to the study of multiple antenna communications when the number of antennas on either communication side is large.

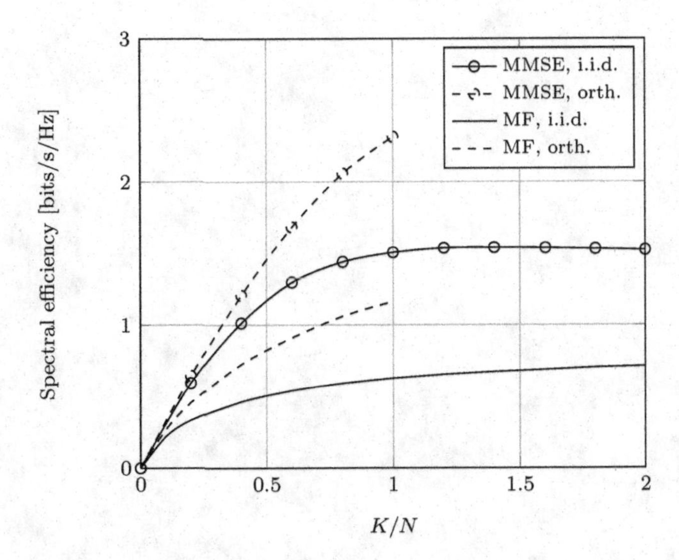

Figure 12.8 Spectral efficiency of random and orthogonal CDMA decoders, for different ratios K/N, $K = 512$, SNR=10 dB, Rayleigh frequency selective fading channels, $L = 8$, in the downlink.

13 Performance of multiple antenna systems

In this section, we study the second most investigated application of random matrix theory to wireless communications, namely multiple antenna systems, first introduced and motivated by the pioneering works of Telatar [Telatar, 1995] and Foschini [Foschini and Gans, 1998]. While large dimensional system analysis is easily defensible in CDMA networks, which typically allow for a large number of users with large orthogonal or random codes, it is not so for multiple antenna communications. Indeed, when it comes to applying approximated results provided by random matrix theory analysis, we expect that the typical system dimensions are of order ten to a thousand. However, for multiple input multiple output (MIMO) setups, the system dimensions can be of order 4, or even 2. Asymptotic results for such systems are then of minor interest. However, it will turn out in some specific scenarios that the difference between the *ergodic* capacity for multiple antenna systems and their respective deterministic equivalents is sometimes of order $O(1/N)$, N being the typical system dimension. The per-receive antenna rate, which is of interest for studying the cost and gain of bringing additional antennas on finite size devices, can therefore be approximated within $O(1/N^2)$. This is a rather convenient rate, even for small N. In fact, as will be observed through simulations, the accuracy of the deterministic equivalents is often even better.

13.1 Quasi-static MIMO fading channels

We hereafter recall the foundations of multiple antenna communications. We first assume a simple point-to-point communication between a transmitter equipped with n_t antennas and a receiver equipped with n_r antennas. The communication channel is assumed linear, frequency flat, and is modeled at any instant by the matrix $\mathbf{H} \in \mathbb{C}^{n_r \times n_t}$, with (i, j) entry h_{ij}. At time t, the transmitter emits the data vector $\mathbf{x}^{(t)} \in \mathbb{C}^{n_t}$ through $\mathbf{H}$, which is corrupted by additive white Gaussian noise $\sigma \mathbf{n}^{(t)} \in \mathbb{C}^{n_r}$ with entries of variance σ^2 and received as $\mathbf{y}^{(t)} \in \mathbb{C}^{n_r}$. We therefore have the classical linear transmission model

$$\mathbf{y}^{(t)} = \mathbf{H}\mathbf{x}^{(t)} + \sigma \mathbf{n}^{(t)}.$$

Assuming $\mathbf{H}$ constant over a long time period, compared to the transmission period of a given data block, and that $\mathbf{H}$ is perfectly known to the transmitter, the capacity $C^{(n_r,n_t)}$ of the flat fading MIMO point-to-point Gaussian channel is given by:

$$C^{(n_r,n_t)}(\sigma^2) = \max_{\substack{\mathbf{P} \\ \operatorname{tr}\mathbf{P} \leq P}} \mathcal{I}^{(n_r,n_t)}(\sigma^2; \mathbf{P})$$

with $\mathcal{I}^{(n_r,n_t)}(\sigma^2; \mathbf{P})$ the mutual information

$$\mathcal{I}^{(n_r,n_t)}(\sigma^2; \mathbf{P}) \triangleq \log_2 \det \left(\mathbf{I}_{n_r} + \frac{1}{\sigma^2} \mathbf{H}\mathbf{P}\mathbf{H}^{\mathsf{H}} \right) \tag{13.1}$$

where $\mathbf{P} \in \mathbb{C}^{n_t \times n_t}$ is the covariance matrix $\mathbf{P} \triangleq \mathrm{E}[\mathbf{x}^{(t)}\mathbf{x}^{(t)\mathsf{H}}]$ of the transmitted data and P is the maximal power allowed for transmission. If the channel is indeed constant, $C^{(n_r,n_t)}$ is determined by finding the matrix $\mathbf{P}$ under trace constraint $\operatorname{tr}\mathbf{P} \leq P$ such that the log determinant is maximized. That is, we need to determine $\mathbf{P}$ such that

$$\det \left(\mathbf{I}_{n_r} + \frac{1}{\sigma^2} \mathbf{H}\mathbf{P}\mathbf{H}^{\mathsf{H}} \right) = \det \left(\mathbf{I}_{n_t} + \frac{1}{\sigma^2} \mathbf{H}^{\mathsf{H}}\mathbf{H}\mathbf{P} \right) = \det \left(\mathbf{I}_{n_t} + \frac{1}{\sigma^2} \mathbf{P}\mathbf{H}^{\mathsf{H}}\mathbf{H} \right)$$

is maximal. Since $\mathbf{H}^{\mathsf{H}}\mathbf{H}$ is Hermitian and non-negative definite, we can write $\mathbf{H}^{\mathsf{H}}\mathbf{H} = \mathbf{U}\mathbf{\Lambda}\mathbf{U}^{\mathsf{H}}$ with $\mathbf{\Lambda} = \operatorname{diag}(l_1, \ldots, l_{n_t})$ diagonal non-negative and $\mathbf{U} \in \mathbb{C}^{n_t \times n_t}$ unitary. The precoding matrix $\mathbf{P}$ can then be rewritten $\mathbf{P} = \mathbf{U}^{\mathsf{H}}\mathbf{Q}\mathbf{U}$, with $\operatorname{tr}\mathbf{Q} = P$, and $\mathbf{Q}$ is still Hermitian non-negative definite. Denote q_{ij} the entry (i,j) of $\mathbf{Q}$. The maximization problem under constraint is therefore equivalent to maximizing

$$\det \left(\mathbf{I}_{n_t} + \frac{1}{\sigma^2} \mathbf{\Lambda}^{\frac{1}{2}}\mathbf{Q}\mathbf{\Lambda}^{\frac{1}{2}} \right)$$

under constraint $\operatorname{tr}\mathbf{Q} = P$. From Hadamard inequality [Telatar, 1999], we have that

$$\det \left(\mathbf{I}_{n_t} + \frac{1}{\sigma^2} \mathbf{\Lambda}^{\frac{1}{2}}\mathbf{Q}\mathbf{\Lambda}^{\frac{1}{2}} \right) \leq \prod_{i=1}^{\min(n_t,n_r)} \left(1 + \frac{1}{\sigma^2} l_i q_{ii} \right) \tag{13.2}$$

with equality if and only if $\mathbf{Q}$ is diagonal. For given q_{ii} values for all i, the Hadamard inequality implies that the maximizing $\mathbf{Q}$ must be diagonal, and therefore the eigenvectors of the capacity maximizing $\mathbf{P}$ must be aligned to those of $\mathbf{H}$. Maximizing the right-hand side of (13.2) under trace constraint is then easily shown through Lagrange multipliers to be found by the water-filling algorithm as

$$q_{ii} = \left(\mu - \frac{\sigma^2}{l_i} \right)^+$$

where μ is such that $\sum_{i=1}^{\min(n_t,n_r)} q_{ii} = P$, and $q_{ii} = 0$ for $i > \min(n_t, n_r)$.

The capacity $C^{(n_r,n_t)}$ found above, when $\mathbf{H}$ is constant over a long time (sufficiently long to be considered infinite) and $\mathbf{H}$ is known at the transmitter side, will be further referred to as the *quasi-static channel capacity.* That is, it concerns the scenario when $\mathbf{H}$ is a fading channel static over a long time, compared to the data transmission duration.

13.2 Time-varying Rayleigh channels

However, in mobile communications, it is often the case that $\mathbf{H}$ is varying fast, and often too fast for the *transmitter* to get to know the transmission environment perfectly prior to transmission. For simplicity here, we assume that some channel information is nonetheless emitted by the transmitter in the direction of the receiver prior to proper communication so that the *receiver* is at all times fully aware of $\mathbf{H}$. We also assume that the feedback effort is negligible in terms of consumed bit rate (think of it as being performed on an adjacent control channel). In this case, the computation of the mutual information between the transmitter and the receiver therefore assumes that the exact value of $\mathbf{H}$ is unknown to the transmitter, although the joint probability distribution $P_{\mathbf{H}}(\mathbf{H})$ of $\mathbf{H}$ is known (or at least that some statistical information about $\mathbf{H}$ has been gathered). The computation of Shannon's capacity $C_{\text{ergodic}}^{(n_r,n_t)}$ in this case reads:

$$C_{\text{ergodic}}^{(n_r,n_t)}(\sigma^2) \triangleq \max_{\substack{\mathbf{P} \\ \operatorname{tr}\mathbf{P}\leq P}} \int \log_2 \det\left(\mathbf{I}_{n_r} + \frac{1}{\sigma^2}\mathbf{H}\mathbf{P}\mathbf{H}^{\mathsf{H}}\right) dP_{\mathbf{H}}(\mathbf{H})$$

which is the maximization over $\mathbf{P}$ of the expectation over $\mathbf{H}$ of the mutual information $\mathcal{I}^{(n_r,n_t)}(\sigma^2;\mathbf{P})$ given in (13.1).

This capacity is usually referred to as the *ergodic capacity.* Indeed, we assume that $\mathbf{H}$ is drawn from an ergodic process, that is a process whose probability distribution can be deduced from successive observations. Determining the exact value of $C_{\text{ergodic}}^{(n_r,n_t)}$ in this case is more involved as an integral has to be solved and maximized over $\mathbf{P}$.

We now recall the early result from Telatar [Telatar, 1995, 1999] on the ergodic capacity of multiple antenna flat Rayleigh fading channels. Telatar assumes the now well spread i.i.d. Gaussian model, i.e. $\mathbf{H}$ has Gaussian i.i.d. entries of zero mean and variance $1/n_t$. This assumption amounts to assuming that the physical channel between the transmitter side and the receiver side is filled with numerous scatterers and that there exists no line-of-sight component. The choice of letting the entries of $\mathbf{H}$ have variances proportional to $1/n_t$ changes the power constraint into $\frac{1}{n_t}\operatorname{tr}\mathbf{P} \leq P$, which will turn out to be often more convenient for practical calculus.

13.2.1 Small dimensional analysis

When $\mathbf{H}$ is i.i.d. Gaussian, it is unitarily invariant so that the ergodic capacity for the channel $\mathbf{HU}$, for $\mathbf{U} \in \mathbb{C}^{n_t \times n_t}$ unitary, is identical to the ergodic capacity for $\mathbf{H}$ itself. The optimal precoding matrix $\mathbf{P}$ can therefore be considered diagonal (non-negative definite) with no generality restriction. Denote $\Pi^{(n_t)}$ the set of permutation matrices of size $n_t \times n_t$, whose cardinality is $(n_t!)$. Note [Telatar, 1999], by the concavity of $\log_2 \det(\mathbf{I}_{n_t} + \frac{1}{\sigma^2}\mathbf{HPH}^{\mathsf{H}})$ seen as a function of $\mathbf{P}$, that the matrix $\mathbf{Q} \triangleq \frac{1}{n_t!}\sum_{\mathbf{\Pi} \in \Pi^{(n_t)}} \mathbf{\Pi P \Pi}^{\mathsf{H}}$ is such that

$$\log_2 \det\left(\mathbf{I}_{n_t} + \frac{1}{\sigma^2}\mathbf{HQH}^{\mathsf{H}}\right) \geq \frac{1}{n_t!}\sum_{\mathbf{\Pi} \in \Pi^{(n_t)}} \log_2 \det\left(\mathbf{I}_{n_t} + \frac{1}{\sigma^2}\mathbf{H\Pi P \Pi}^{\mathsf{H}}\mathbf{H}^{\mathsf{H}}\right)$$
$$= \log_2 \det\left(\mathbf{I}_{n_t} + \frac{1}{\sigma^2}\mathbf{HPH}^{\mathsf{H}}\right).$$

This follows from Jensen's inequality. Since $\mathbf{P}$ was arbitrary, $\mathbf{Q}$ maximizes the capacity. But now notice that $\mathbf{Q}$ is, by construction, necessarily a multiple of the identity matrix. With the power constraint, we therefore have $\mathbf{Q} = \mathbf{I}_{n_t}$.

It therefore remains to evaluate the ergodic capacity as

$$C_{\text{ergodic}}^{(n_r,n_t)}(\sigma^2) = \int \log_2 \det\left(\mathbf{I}_{n_r} + \frac{1}{\sigma^2}\mathbf{HH}^{\mathsf{H}}\right) dP_{\mathbf{H}}(\mathbf{H})$$

where $P_{\mathbf{H}}$ is the density of an $(n_r \times n_t)$-variate Gaussian variable with entries of zero mean and variance $1/n_t$. To this purpose, we first diagonalize $\mathbf{HH}^{\mathsf{H}}$ and write

$$C_{\text{ergodic}}^{(n_r,n_t)}(\sigma^2) = n_r \int_0^{\infty} \log_2\left(1 + \frac{\lambda}{n_t\sigma^2}\right) p_\lambda(\lambda) d\lambda$$

where p_λ is the marginal eigenvalue distribution of the null Wishart matrix $n_t\mathbf{HH}^{\mathsf{H}}$. Remember now that this is exactly stated in Theorem 2.3. Hence, we have that

$$C_{\text{ergodic}}^{(n_r,n_t)}(\sigma^2) = \int_0^{\infty} \log_2\left(1 + \frac{\lambda}{n_t\sigma^2}\right) \sum_{k=0}^{m-1} \frac{n_r k!}{(k+n-m)!}[L_k^{n-m}(\lambda)]^2 \lambda^{n-m} e^{-\lambda} d\lambda \tag{13.3}$$

with $m = \min(n_r, n_t)$, $n = \max(n_r, n_t)$, and L_i^j are the Laguerre polynomials.

This important result is however difficult to generalize to more involved channel conditions, e.g. by introducing side correlations or a variance profile to the channel matrix. We will therefore quickly move to large dimensional analysis, where many results can be found to approximate the capacity of point-to-point MIMO communications in various channel conditions, through deterministic equivalents.

13.2.2 Large dimensional analysis

From a large dimensional point of view, the ergodic capacity evaluation for the Rayleigh i.i.d. channel is a simple application of the Marčenko–Pastur law. We have that the per-receive antenna capacity $\frac{1}{n_r}C^{(n_r,n_t)}_{\text{ergodic}}$ satisfies

$$\frac{1}{n_r}C^{(n_r,n_t)}_{\text{ergodic}}(\sigma^2) \to \int_0^\infty \log_2\left(1+\frac{x}{\sigma^2}\right)\frac{\sqrt{(x-(1-\sqrt{c})^2)((1+\sqrt{c}-x)^2-x)}}{2\pi c x}dx$$

as (n_t, n_r) grow large with asymptotic ratio $n_r/n_t \to c$, $0 < c < \infty$. This result was already available in Telatar's pioneering article [Telatar, 1999]. In fact, we even have that the quasi-static mutual information for $\mathbf{P} = \mathbf{I}_{n_t}$ converges almost surely to the right-hand side value. Since the channels $\mathbf{H}$ for which this is not the case lie in a space of zero measure, this implies that the convergence holds surely in expectation. Now, an explicit expression for the above integral can be derived. It suffices here to apply, e.g. Theorem 6.1 or Theorem 6.8 to the extremely simple case where the channel matrix has no correlation. In that case, the equations leading to the deterministic equivalent for the Stieltjes transform are explicit and we finally have the more interesting result

$$\frac{1}{n_r}C^{(n_r,n_t)}_{\text{ergodic}}(\sigma^2) - \left[\log_2\left(1+\frac{1}{\sigma^2(1+\frac{n_r}{n_t}\delta)}\right)+\frac{n_t}{n_r}\log_2\left(1+\frac{n_r}{n_t}\delta\right)+\log_2(e)\left[\sigma^2\delta-1\right]\right] \to 0$$

where δ is the positive solution to

$$\delta = \left(\frac{1}{1+\frac{n_r}{n_t}\delta}+\sigma^2\right)^{-1}$$

which is explicitly given by:

$$\delta = \frac{1}{2}\left[\frac{1}{\sigma^2}\left(1-\frac{n_t}{n_r}\right)-\frac{n_t}{n_r}+\sqrt{\left(\frac{1}{\sigma^2}\left(1-\frac{n_t}{n_r}\right)-\frac{n_t}{n_r}\right)^2+4\frac{n_t}{n_r\sigma^2}}\right].$$

Also, since $\mathbf{H}$ has Gaussian entries, by invoking, e.g. Theorem 6.8, it is known that the convergence is as fast as $O(1/n_t^2)$. Therefore, we also have that

$$C^{(n_r,n_t)}_{\text{ergodic}}(\sigma^2) - \left[n_r\log_2\left(1+\frac{1}{\sigma^2(1+\frac{n_r}{n_t}\delta)}\right)+n_t\log_2\left(1+\frac{n_r}{n_t}\delta\right)+n_r\log_2(e)\left[\sigma^2\delta-1\right]\right] = O(1/n_t). \tag{13.4}$$

In Table 13.1, we evaluate the absolute difference between $C^{(n_r,n_t)}(\sigma^2)$ (from Equation (13.3)) and its deterministic equivalent (given in Equation (13.4)), relative to $C^{(n_r,n_t)}(\sigma^2)$, for $(n_t, n_r) \in \{1,\ldots,8\}^2$. The SNR is 10 dB. We observe that, even for very small values of n_t and n_r, the relative difference does not

n_r, n_t	1	2	3	4	5	6	7	8
1	0.0630	0.0129	0.0051	0.0027	0.0016	0.0011	0.0008	0.0006
2	0.0116	0.0185	0.0072	0.0035	0.0020	0.0013	0.0009	0.0007
3	0.0039	0.0072	0.0080	0.0044	0.0025	0.0016	0.0011	0.0008
4	0.0019	0.0032	0.0046	0.0044	0.0029	0.0019	0.0013	0.0009
5	0.0011	0.0017	0.0025	0.0030	0.0028	0.0020	0.0014	0.0010
6	0.0007	0.0010	0.0015	0.0019	0.0021	0.0019	0.0015	0.0011
7	0.0005	0.0007	0.0009	0.0012	0.0015	0.0015	0.0014	0.0011
8	0.0003	0.0005	0.0006	0.0008	0.0010	0.0012	0.0012	0.0011

Table 13.1. Relative difference between true ergodic $n_r \times n_t$ MIMO capacity and associated deterministic equivalent.

exceed 6% and is of order 0.5% for n_t and n_r of order 4. This simple example motivates the use of large dimensional analysis to approximate the real capacity of even small dimensional systems. We will see in this chapter that this trend can be extended to more general models, although particular care has to be taken for some degenerated cases. It is especially of interest to determine the transmit precoders that maximize the capacity of MIMO communications under strong antenna correlations at both communication ends. It will be shown that the capacity in this corner case can still be approximated using deterministic equivalents, although fast convergence of the deterministic equivalents cannot be ensured and the resulting estimators can therefore be very inaccurate. In this case, theory requires that very large system dimensions be assumed to obtain acceptable results. Nonetheless, simulations still suggest, apart from very degenerated models, where, e.g. both transmit and receive sides have rank-1 correlation profiles, that the deterministic equivalents are still very accurate for small system dimensions.

Note additionally that random matrix theory, in addition to providing consistent estimates for the ergodic capacity, also ensures that, with probability one, as the system dimension grows large, the instantaneous mutual information of a given realization of a Rayleigh distributed channel is within $o(1)$ of the deterministic equivalent (this assumes however that only statistical channel state information is available at the transmitter). This unveils some sort of deterministic behavior for the achievable data rates as the number of antennas grows, which can be thought of as a channel hardening effect [Hochwald *et al.*, 2004], i.e. as the system dimensions grow large, the variance of the quasi-static capacity is significantly reduced.

13.2.3 Outage capacity

For practical finite dimensional quasi-static Rayleigh fading channel realizations, whose realizations are unknown beforehand, the value of the ergodic capacity, that can be only seen as an a priori "expected" capacity, is not a proper measure of the truly achievable transmission data rate. In fact, if the Rayleigh fading channel realization is unknown, the largest rate to which we can ensure data is transmitted reliably is in fact null. Indeed, for every given positive transmission rate, there exists a non-zero probability that the channel realization has a lesser capacity. As this statement is obviously not convenient, it is often preferable to consider the so-called *q-outage capacity* defined as the largest transmission data rate that is achievable at least a fraction q of the time. That is, for a random channel $\mathbf{H}$ with realizations $\mathbf{H}(\omega)$ and instantaneous capacity $C^{(n_t,n_r)}(\sigma^2;\omega)$, $\omega \in \Omega$, the q-outage capacity $C^{(n_t,n_r)}_{\text{outage}}(\sigma^2;q)$ is defined as

$$\begin{aligned} C^{(n_t,n_r)}_{\text{outage}}(\sigma^2;q) &= \sup_{R\geq 0}\left\{P\left(\left\{\omega,\ C^{(n_t,n_r)}(\sigma^2;\omega) > R\right\}\right) \leq q\right\} \\ &= \sup_{R\geq 0}\left\{P\left(C^{(n_t,n_r)}(\sigma^2) > R\right)\right\} \end{aligned} \tag{13.5}$$

with $C^{(n_t,n_r)}(\sigma^2)$ seen here as a random variable of $\mathbf{H}$.

It is often difficult to characterize fully the outage capacity under perfect channel knowledge at the transmitter, since the transmit precoding policy is different for each channel realization. In the following, we will in general refer to the outage capacity as the outage rate obtained under deterministic (often uniform) power allocation at the transmitter. In this scenario, we have instead the outage mutual information $\mathcal{I}^{(n_t,n_r)}_{\text{outage}}(\sigma^2;\mathbf{P};q)$, for a specific precoding matrix $\mathbf{P}$

$$\mathcal{I}^{(n_t,n_r)}_{\text{outage}}(\sigma^2;\mathbf{P};q) = \sup_{R\geq 0}\left\{P\left(\mathcal{I}^{(n_t,n_r)}(\sigma^2;\mathbf{P}) > R\right) \leq q\right\}. \tag{13.6}$$

To determine the outage capacity of a given communication channel, it suffices to be able to determine the complete probability distribution of $C^{(n_t,n_r)}(\sigma^2)$ for varying $\mathbf{H}$, if we consider the outage capacity definition (13.5), or the probability distribution of $\mathcal{I}^{(n_t,n_r)}(\sigma^2;\mathbf{P})$ for a given precoder $\mathbf{P}$ if we consider the definition (13.6). For the latter, with $\mathbf{P} = \mathbf{I}_{n_t}$, for the Rayleigh fading MIMO channel $\mathbf{H} \in \mathbb{C}^{n_r \times n_t}$, it suffices to describe the distribution of

$$\log_2\left(\mathbf{I}_N + \frac{1}{\sigma^2}\mathbf{H}\mathbf{H}^{\mathsf{H}}\right) = \sum_{k=1}^{n_r}\log_2\left(1 + \frac{\lambda_k}{\sigma^2}\right)$$

for the random variables $\mathbf{H}$, with $\lambda_1, \ldots, \lambda_{n_r}$ the eigenvalues of $\mathbf{H}\mathbf{H}^{\mathsf{H}}$.

Interestingly, as both n_t and n_r grow large, the distribution of the zero mean random variable $\mathcal{I}^{(n_t,n_r)}(\sigma^2;\mathbf{P}) - \mathrm{E}[\mathcal{I}^{(n_t,n_r)}(\sigma^2;\mathbf{P})]$ turns out to be asymptotically Gaussian [Kamath *et al.*, 2002]. This result is in fact a straightforward consequence of the central limit Theorem 3.17. Indeed, consider Theorem 3.17 in the case when the central $\mathbf{T}_N$ matrix is identity. Taking the uniform precoder $\mathbf{P} = \mathbf{I}_{n_t}$ and letting f_1 be defined as $f_1(x) = \log_2\left(1 + \frac{x}{\sigma^2}\right)$, clearly continuous on some closed set around the limiting support of the

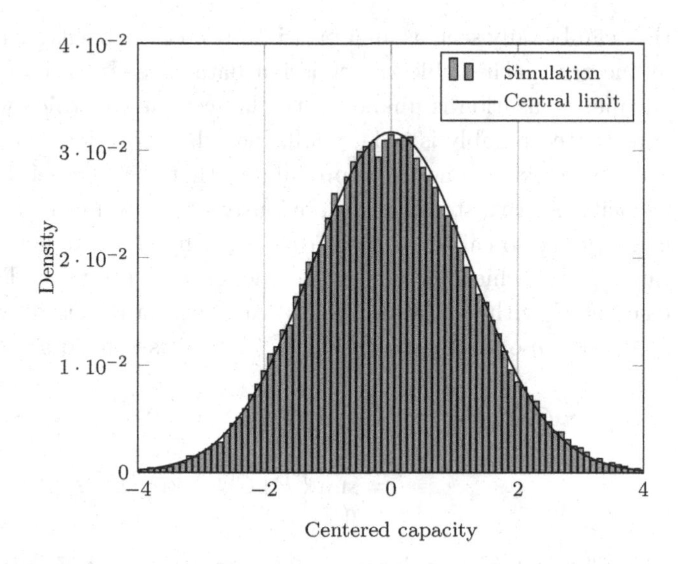

Figure 13.1 Simulated $\mathcal{I}^{(n_t,n_r)}(\sigma^2; \mathbf{I}_{n_t}) - \mathrm{E}[\mathcal{I}^{(n_t,n_r)}(\sigma^2; \mathbf{I}_{n_t})]$ against central limit, $\sigma^2 = -10$ dB.

Marčenko–Pastur law F, we obtain that

$$n_r \int \log_2\left(1 + \frac{\lambda}{\sigma^2}\right)\left(dF^{\mathbf{H}\mathbf{H}^{\mathsf{H}}}(\lambda) - dF(\lambda)\right) \Rightarrow X$$

with X a random real Gaussian variable with zero mean and variance computed from (3.26) as being equal to

$$\mathrm{E}\left[X^2\right] = -\log\left(1 - \frac{\sigma^4}{16c}\left(\sqrt{\frac{(1+\sqrt{c})^2}{\sigma^2} + 1} - \sqrt{\frac{(1-\sqrt{c})^2}{\sigma^2} + 1}\right)^4\right).$$

This result is also a direct application of Theorem 3.18 for Gaussian distributed random entries of $\mathbf{X}_N$.

13.3 Correlated frequency flat fading channels

Although the above analysis has the advantage to predict the potential capacity gains brought by multiple antenna communications, i.i.d. Rayleigh fading channel links are usually too strong an assumption to model practical communication channels. In particular, the multiplexing gain of order $\min(n_r, n_t)$ announced by Telatar [Telatar, 1999] and Foschini [Foschini and Gans, 1998] relies on the often unrealistic supposition that the channel links are frequency flat and have a multi-variate independent zero mean Gaussian distribution. In practical communication channels, this model faces strong limitations. We look at these limitations from a receiver point of view, although the same reasoning can be performed on the transmitter side.

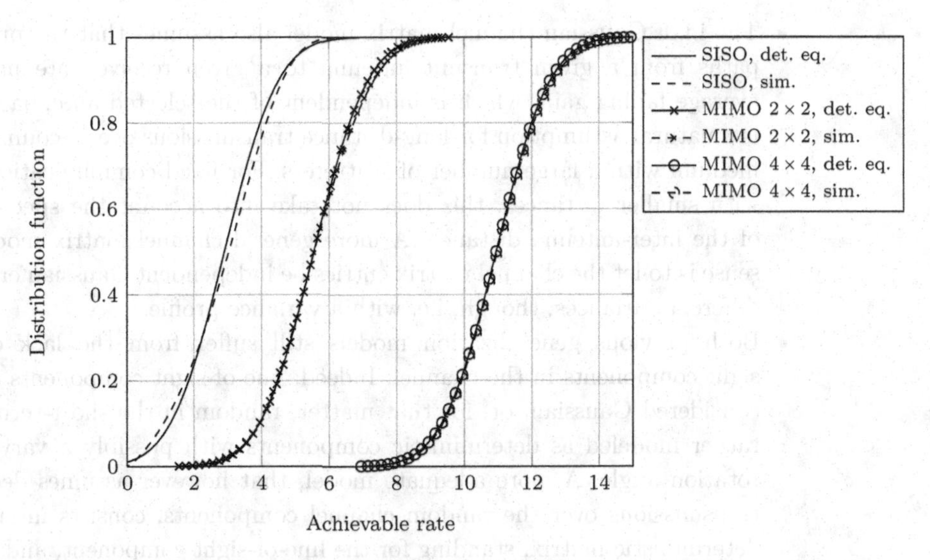

Figure 13.2 Distribution function of $C^{(n_t,n_r)}(\sigma^2)$, $\sigma^2 = 0.1$, for different values of n_t, n_r, and comparison against deterministic equivalents.

- To ensure conditional independence (with respect to the transmitted data) of the waveforms received by two distinct antennas but emerging from a single source, the propagation environments must be assumed decorrelated in some sense. Roughly speaking, two incoming waveforms can be stated independent if they propagate along different paths in the communication medium. This physically constrains the distance between receive antennas to be of an order larger than the transmission wavelength. Introducing a specific model of channel correlation, it can be in particular shown that increasing the number of antennas to infinity on finite size devices leads to a physically fundamental rate saturation, see, e.g., [Couillet *et al.*, 2008; Pollock *et al.*, 2003]. To model more realistic channel matrices, statistical correlation between transmit antennas and receive antennas must therefore be taken into account. The most largely spread channel model which accounts for both transmit and receive signal correlations is the Kronecker channel model. This model assumes that the transmitted signals are first emitted from a correlated source (e.g. close transmit antennas, privileged direction of wave transmission, etc.), then propagate through a largely scattered environment, which acts as a random i.i.d. linear filter decorrelated from transmit and receive parts, to be finally received on a correlated antenna array. Again, the correlation at the receiver is due either to the fact that receive antennas are so close, or that the solid angle of direction of arrival is so thin, that all incoming signals are essentially the same on all antennas. Note that this model, although largely spread, has been criticized and claimed unrealistic to some extent by field measurements in, e.g., [Ozcelik *et al.*, 2003; Weichselberger *et al.*, 2006].

- The i.i.d. Gaussian channel matrix model also assumes that the propagation paths from a given transmit antenna to a given receive antenna have an average fading gain, which is independent of the selected antenna pair. This is a natural assumption for long-distance transmissions over a communication medium with a large number of scatterers. For fixed communication systems with smaller distances, this does not take into account the specific impact of the inter-antenna distance. A more general channel matrix model in this sense is to let the channel matrix entries be independent Gaussian entries with different variances, though, i.e. with a variance profile.
- Both previous generalization models still suffer from the lack of line-of-sight components in the channel. Indeed, line-of-sight components cannot be considered Gaussian, or, for that matter, random in the short-term, but are rather modeled as deterministic components with possibly a varying phase rotation angle. A more adequate model, that however assumes decorrelated transmissions over the random channel components, consists in summing a deterministic matrix, standing for the line-of-sight component, and a random matrix with Gaussian entries and a variance profile. To account for the relative importance of the line-of-sight component and the random part, both deterministic and random matrices are scaled accordingly. This model is referred to as the Rician model.
- Although this section only deals with frequency flat fading channels, we recall that wideband communication channels, i.e. communication channels that span over a large range of frequencies (the adjective "large" qualifying the fact that the transmission bandwidth is several times larger than the channel coherence bandwidth) induce frequency dependent channel matrices. Therefore, in wideband transmissions, channel models cannot be simply represented as a single matrix $\mathbf{H} \in \mathbb{C}^{n_r \times n_t}$ at any given instant, but rather as a matrix-valued continuous function $\mathbf{H}(f) \in \mathbb{C}^{n_r \times n_t}$, with $f \in [-W/2, W/2]$ the communication bandwidth. This motivates in particular communication schemes such as OFDM, which practically exploit these frequency properties. This is however not the subject of the current section, which will be given deeper considerations in Section 13.5.

In this section, we discuss the case of communication channels with correlation patterns both at the transmitter and the receiver and a very scattered medium in between; i.e. we consider here the Kronecker channel model. The results to be introduced were initially derived in, e.g., [Debbah and Müller, 2003] using tools from free probability theory. We will see in Chapter 18 that the Kronecker model has deep information-theoretic grounds, in the sense that it constitutes the *least informative* channel model when statistical correlation matrices at both communication sides are a priori known by the system modeler.

Letting $\mathbf{H} \in \mathbb{C}^{n_r \times n_t}$ be a narrowband Kronecker communication channel matrix, we will write

$$\mathbf{H} = \mathbf{R}^{\frac{1}{2}} \mathbf{X} \mathbf{T}^{\frac{1}{2}}$$

where $\mathbf{X} \in \mathbb{C}^{n_r \times n_t}$ is a random matrix with independent Gaussian entries of zero mean and variance $1/n_t$, which models the rich scattering environment, $\mathbf{R}^{\frac{1}{2}} \in \mathbb{C}^{n_r \times n_r}$ is a non-negative definite Hermitian square root of the non-negative definite receive correlation matrix $\mathbf{R}$, and $\mathbf{T}^{\frac{1}{2}} \in \mathbb{C}^{n_t \times n_t}$ is a non-negative definite Hermitian square root of the non-negative definite receive correlation matrix $\mathbf{T}$. Note that the fact that $\mathbf{X}$ has Gaussian i.i.d. entries allows us to assume without loss of generality that both $\mathbf{R}$ and $\mathbf{T}$ are diagonal matrices (a remark which no longer holds in multi-user scenarios). The achievable quasi-static bit rate $C^{(n_t,n_r)}$ for the additive white Gaussian noise channel under this medium filtering matrix model reads:

$$C^{(n_t,n_r)}(\sigma^2) = \sup_{\substack{\mathbf{P} \\ \frac{1}{n_t} \operatorname{tr} \mathbf{P} \leq P}} \mathcal{I}^{(n_r,n_t)}(\sigma^2; \mathbf{P})$$

where we define as before the mutual information $\mathcal{I}^{(n_r,n_t)}(\sigma^2; \mathbf{P})$ as

$$\begin{aligned}
\mathcal{I}^{(n_t,n_r)}(\sigma^2; \mathbf{P}) &\triangleq \log_2 \det \left(\mathbf{I}_{n_r} + \frac{1}{\sigma^2} \mathbf{H} \mathbf{P} \mathbf{H}^{\mathsf{H}} \right) \\
&= \log_2 \det \left(\mathbf{I}_{n_r} + \frac{1}{\sigma^2} \mathbf{R}^{\frac{1}{2}} \mathbf{X} \mathbf{T}^{\frac{1}{2}} \mathbf{P} \mathbf{T}^{\frac{1}{2}} \mathbf{X}^{\mathsf{H}} \mathbf{R}^{\frac{1}{2}} \right)
\end{aligned}$$

where σ^2 is the variance of the individual i.i.d. receive noise vector entries. The corresponding *ergodic* capacity takes the same form but with an additional expectation in front of the log determinant, which is taken over the random $\mathbf{X}$ matrices.

Evaluating the ergodic capacity and the corresponding optimal covariance matrix in closed-form is however rather involved. This has been partially solved in [Hanlen and Grant, 2003] where an exact expression of the ergodic mutual information is given as follows.

Theorem 13.1 (Theorem 2 in [Hanlen and Grant, 2003]). *Let* $\mathbf{H} = \mathbf{R}^{\frac{1}{2}} \mathbf{X} \mathbf{T}^{\frac{1}{2}}$, *with* $\mathbf{R} \in \mathbb{C}^{n_r \times n_r}$ *and* $\mathbf{T} \in \mathbb{C}^{n_t \times n_t}$ *deterministic and* $\mathbf{X} \in \mathbb{C}^{n_r \times n_t}$ *random with i.i.d. Gaussian entries of zero mean and variance* $1/n_t$. *Then, denoting*

$$\mathcal{I}^{(n_t,n_r)}(\sigma^2; \mathbf{P}) = \log_2 \det \left(\mathbf{I}_{n_r} + \frac{1}{\sigma^2} \mathbf{H} \mathbf{P} \mathbf{H}^{\mathsf{H}} \right)$$

we have:

$$\begin{aligned}
\mathrm{E}\left[\mathcal{I}^{(n_t,n_r)}(\sigma^2; \mathbf{P}) \right] &= \frac{\det(\mathbf{R})^{-n_r} \det(\mathbf{T}\mathbf{P})^{-n_t}}{\prod_{i=1}^{m} (M-i)! \prod_{i=1}^{m} (m-i)!} \int_{\{\mathbf{\Lambda} > 0\}} {}_0F_0 \left(-\mathbf{R}^{-1}, \mathbf{\Lambda}, \mathbf{T}^{-1} \right) \\
&\quad \times \prod_{i=1}^{m} \lambda_i^{M-m} \prod_{i<j}^{m} (\lambda_i - \lambda_j)^2 \sum_{i=1}^{m} \log_2 \left(1 + \frac{\lambda_i}{\sigma^2 n_t} \right) d\mathbf{\Lambda}
\end{aligned}$$

with $m = \min(n_r, n_t)$, $M = \max(n_r, n_t)$, $\mathbf{\Lambda} = \operatorname{diag}(\lambda_1, \ldots, \lambda_m)$, ${}_0F_0$ *defined in Equation (2.2) and the integral is taken over the set of m-dimensional vectors with positive entries.*

Moreover, assuming $\mathbf{R}$ *and* $\mathbf{T}$ *diagonal without loss of generality, the matrix* $\mathbf{P}$ *that maximizes* $\mathrm{E}[\mathcal{I}^{(n_t,n_r)}(\sigma^2;\mathbf{P})]$ *under constraint* $\frac{1}{n_t}\operatorname{tr}\mathbf{P} \leq P$ *is the diagonal matrix* $\mathbf{P}^\star = \operatorname{diag}(p_1^\star,\ldots,p_{n_t}^\star)$ *such that, for all* $k \in \{1,\ldots,n_t\}$

$$\left(\mathrm{E}\left[(\mathbf{I}_{n_t} + \mathbf{H}^{\mathsf{H}}\mathbf{H}\mathbf{P}^\star)^{-1}\mathbf{H}^{\mathsf{H}}\mathbf{H}\right]\right)_{kk} = \mu, \quad \text{if } p_k^\star > 0$$
$$\left(\mathrm{E}\left[(\mathbf{I}_{n_t} + \mathbf{H}^{\mathsf{H}}\mathbf{H}\mathbf{P}^\star)^{-1}\mathbf{H}^{\mathsf{H}}\mathbf{H}\right]\right)_{kk} < \mu, \quad \text{if } p_k^\star = 0$$

for some μ *set to satisfy the power constraint* $\frac{1}{n_t}\operatorname{tr}\mathbf{P}^\star \leq P$.

These results, although exact, are difficult to use in practice. We will see in the next section that large dimensional random matrix analysis brings several interesting features to the study of MIMO Kronecker channels.

- It first allows us to provide a deterministic equivalent for the quasi-static mutual information of typical channels, assuming channel independent transmit data precoding. The informal "typical" adjective here suggests that such channels belong to a high probability subset of all possible channel realizations. Indeed, we have already discussed the fact that, for non-deterministic channel models, the capacity is ill-defined and is actually zero in the current Kronecker scenario. Instead, we will say that we provide a deterministic equivalent for the achievable rate of all highly probable deterministic channels, with deterministic transmit precoding. The reason why transmit precoding cannot be optimized will be made clear.
- It then allows us to obtain a deterministic equivalent of the ergodic capacity of correlated MIMO communications. As we will see, the transmit covariance matrix which maximizes the deterministic equivalent can be derived. The mutual information evaluated at this precoder can further be proved to be asymptotically close to the exact ergodic capacity. The main advantage of this approach is that, compared to the results of Theorem 13.1, deterministic equivalents provide much simpler and more elegant solutions to the ergodic capacity characterization.

Nonetheless, before going further, we address some key limitations of the Kronecker channel model. The major shortcoming of the Kronecker model lies in the assumption of a rich scattered environment. To ensure that a large number of antennas can be used on either communication side, the number of such scatterers must be of an order larger than the product between the number of transmit and received antennas, so as to generate diverse propagation paths for all transmit–receive antenna pairs. This assumption is typically not met in an outdoor environment. Also, no line-of-sight component is allowed for in the Kronecker model, nor does there exist a correlation between transmit and receive antennas. This is also restrictive in short range communications with few propagation paths.

To provide a deterministic equivalent for the capacity of the Kronecker channel model, we will use a simplified version of Corollary 6.1 of Theorem 6.4 and

Theorem 6.8. Both deterministic equivalents will obviously be consistent and in fact equal, although the theorems rely on different underlying assumptions. Thanks to these assumptions, different conclusions will be drawn in terms of the applicability to specific channel conditions. Note that these results were initially derived in [Tulino *et al.*, 2003] using Girko's result, Theorem 3.14, and were also derived later using tools borrowed from physics in [Sengupta and Mitra, 2006].

13.3.1 Communication in strongly correlated channels

We first recall Corollary 6.1 in the case of a single transmitter, i.e. $K = 1$. Assume, as above, a Gaussian channel with channel matrix $\mathbf{H}$ and additive noise variance σ^2. Let $\mathbf{H} = \mathbf{R}^{\frac{1}{2}}\mathbf{X}\mathbf{T}^{\frac{1}{2}} \in \mathbb{C}^{n_r \times n_t}$, with $\mathbf{X} \in \mathbb{C}^{n_r \times n_t}$ composed of Gaussian i.i.d. entries of zero mean and variance $1/n_t$, $\mathbf{R}^{\frac{1}{2}} \in \mathbb{C}^{n_r \times n_r}$ and $\mathbf{T}^{\frac{1}{2}} \in \mathbb{C}^{n_t \times n_t}$ be Hermitian non-negative definite, such that $F^{\mathbf{R}}$ and $F^{\mathbf{T}}$ form a tight sequence, as the dimensions n_r, n_r grow large, and the sequence n_t/n_r is uniformly bounded from below, away from $a > 0$, and from above, away from $b < \infty$. Also assume that there exists $\alpha > 0$ and a sequence s_{n_r}, such that, for all n_r

$$\max(t_{s_{n_r}+1}, r_{s_{n_r}+1}) \leq \alpha$$

where r_i and t_i denote the ith ordered eigenvalue of $\mathbf{R}$ and $\mathbf{T}$, respectively, and, denoting b_{n_r} an upper-bound on the spectral norm of $\mathbf{T}$ and $\mathbf{R}$ and β some real, such that $\beta > (b/a)(1+\sqrt{a})^2$, assume that $a_{n_r} = b_{n_r}^2 \beta$ satisfies

$$s_{n_r} \log_2(1 + a_{n_r}\sigma^{-2}) = o(n_r). \tag{13.7}$$

Then, for large n_r, n_t, the Shannon transform of $\mathbf{H}\mathbf{H}^{\mathsf{H}}$, given by:

$$\mathcal{V}_{\mathbf{H}\mathbf{H}^{\mathsf{H}}}(\sigma^{-2}) = \frac{1}{n_r} \log_2 \det(\mathbf{I}_{n_r} + \frac{1}{\sigma^2}\mathbf{H}\mathbf{H}^{\mathsf{H}})$$

satisfies

$$\mathcal{V}_{\mathbf{H}\mathbf{H}^{\mathsf{H}}}(\sigma^{-2}) - \mathcal{V}_{n_r}(\sigma^{-2}) \xrightarrow{\text{a.s.}} 0$$

where $\mathcal{V}_{n_r}(\sigma^{-2})$ satisfies

$$\mathcal{V}_{n_r}(\sigma^{-2}) = \frac{1}{n_r}\sum_{k=1}^{n_r} \log_2\left(1 + \bar{\delta} r_k\right) + \frac{1}{n_r}\sum_{k=1}^{n_t} \log_2\left(1 + \frac{n_r}{n_t}\delta t_k\right) - \sigma^2 \log_2(e)\delta\bar{\delta}$$

with δ and $\bar{\delta}$ the unique positive solutions to the fixed-point equations

$$\delta = \frac{1}{\sigma^2 n_r}\sum_{k=1}^{n_r} \frac{r_k}{1 + \bar{\delta} r_k}$$

$$\bar{\delta} = \frac{1}{\sigma^2 n_t}\sum_{k=1}^{n_t} \frac{t_k}{1 + \frac{n_r}{n_t}\delta t_k}.$$

This result provides a deterministic equivalent for the Shannon transform of $\mathbf{HH}^{\mathsf{H}}$, which is, in our information-theoretic context, a deterministic equivalent for the per-receive antenna mutual information between the multiple antenna transmitter and the multiple antenna receiver when uniform power allocation is used across transmit antennas. We need now to understand the extent of the applicability of the previous result, which consists in understanding exactly the underlying assumptions made on $\mathbf{T}$ and $\mathbf{R}$. Indeed, it is of particular importance to be able to study the capacity of MIMO systems when the correlation matrices $\mathbf{T}$ and $\mathbf{R}$ have very ill-conditioned profiles. It may seem in particular that the deterministic equivalents may not be valid if $\mathbf{T}$ and $\mathbf{R}$ are composed of a few very large eigenvalues and all remaining eigenvalues are close to zero. This intuition turns out not to be correct.

Remember that tightness, which is commonly defined as a probability theory notion, qualifies a sequence of distribution functions $F_1, F_2, \ldots$ (let us assume $F(x) = 0$ for $x < 0$), such that, for all $\varepsilon > 0$, there exists $M > 0$, such that

$$F_k(M) > 1 - \varepsilon$$

for all k. This is often thought of as the probability theory equivalent to boundedness. Indeed, a sequence $x_1, x_2, \ldots$ of real positive scalars is bounded if there exists M such that $x_k < M$ for all k. Here, the parameter ε allows for some event leakage towards infinity, although the probability of such events is increasingly small. Therefore, no positive mass can leak to infinity. In our setting, $F^{\mathbf{T}}$ and $F^{\mathbf{R}}$ do not really form tight probability distributions in the classical probabilistic meaning of boundedness as they are deterministic distribution functions rather than random distribution functions. This does not however affect the mathematical derivations to come.

Since $\mathbf{T}$ and $\mathbf{R}$ are correlation matrices, for a proper definition of the signal-to-noise ratio $1/\sigma^2$, we assume that they are constrained by $\frac{1}{n_t} \operatorname{tr} \mathbf{T} = 1$ and $\frac{1}{n_r} \operatorname{tr} \mathbf{R} = 1$. That is, we do not allow the power transmitted or received to grow as n_t, n_r grow. The trace constraint is set to one for obvious convenience. Note that it is classical to let all diagonal entries of $\mathbf{T}$ and $\mathbf{R}$ equal one, as we generally assume that every individual antenna on either communication side has the same physical properties. This assumption would no longer be valid under a channel with variance profile, for which the fading link h_{ij} between transmit antenna i and receive antenna j has different variances for different (i, j) pairs.

Observe that, because of the constraint $\frac{1}{n_r} \operatorname{tr} \mathbf{R} = 1$, the sequence $\{F^{\mathbf{R}}\}$ (for growing n_r) is necessarily tight. Indeed, given $\varepsilon > 0$, take $M = 2/\varepsilon$; $n_r[1 - F^{\mathbf{R}}(M)]$ is the number of eigenvalues in $\mathbf{R}$ larger than $2/\varepsilon$, which is necessarily smaller than or equal to $n_r\varepsilon/2$ from the trace constraint, leading to $1 - F^{\mathbf{R}}(M) \leq \varepsilon/2$ and then $F^{\mathbf{R}}(M) \geq 1 - \varepsilon/2 > 1 - \varepsilon$. The same naturally holds for matrix $\mathbf{T}$. Now the condition regarding the smallest eigenvalues of $\mathbf{R}$ and $\mathbf{T}$ (those less than α) requires a stronger assumption on the correlation matrices. Under the trace constraint, this requires that there exists $\alpha > 0$, such that the number of eigenvalues in $\mathbf{R}$ greater than α is of order $o(n_r/\log n_r)$. This may not always be the case, as we presently show with a counter-example. Take $n_r = 2^p + 1$ and the eigenvalues of $\mathbf{R}$ to be

$$2^{p-1}, \underbrace{p, \ldots, p}_{\frac{2^{p-1}}{p}}, \underbrace{0, \ldots, 0}_{2^p - \frac{2^{p-1}}{p}}.$$

The largest eigenvalue is of order n_r so that a_{n_r} is of order n_r^2, and the number s_{n_r} of eigenvalues larger than any $\alpha > 0$ for n_r large is of order $\frac{2^{p-1}}{p} \sim \frac{n_r}{\log(n_r)}$. Therefore $s_{n_r} \log(1 + a_{n_r}/x) = O(n_r)$ here. Nonetheless, most conventional models for $\mathbf{R}$ and $\mathbf{T}$, even when showing strong correlation properties, satisfy the assumptions of Equation (13.7). We mention in particular the following examples:

- if all $\mathbf{R}$ and $\mathbf{T}$ have uniformly bounded spectral norm, then there exists $\alpha > 0$ such that all eigenvalues of $\mathbf{R}$ and $\mathbf{T}$ are less then α for all n_r. This implies $s_{n_r} = 0$ for all n_r and therefore the condition is trivially satisfied. Our model is therefore compatible with loosely correlated antenna structures;
- when antennas are on the opposite densely packed on a volume limited device, the correlation matrices $\mathbf{R}$ and $\mathbf{T}$ tend to be asymptotically of finite rank, see, e.g., [Pollock *et al.*, 2003] in the case of a dense circular array. That is, for any given $\alpha > 0$, the number s_{n_r} of eigenvalues greater than α is finite for all large n_r, while a_{n_r} is of order ${n_r}^2$. This implies $s_{n_r} \log(1 + a_{n_r}/x) = O(\log n_r) = o(n_r)$.
- for one-, two- or three-dimensional antenna arrays with neighbors separated by half the wavelength as discussed by Moustakas *et al.* in [Moustakas *et al.*, 2000], the correlation figure corresponds to $O(n_r)$ eigenvalues of order of magnitude $O(1)$ for one-dimensional arrays, $O(\sqrt{n_r})$ large eigenvalues of order $O(\sqrt{n_r})$ for two-dimensional arrays or $O({n_r}^{\frac{2}{3}})$ large eigenvalues of order $O({n_r}^{\frac{1}{3}})$ for three-dimensional arrays, the remaining eigenvalues being close to zero. In the p-dimensional scenario, we can approximate s_{n_r} by ${n_r}^{\frac{p-1}{p}}$ and a_{n_r} by ${n_r}^{\frac{2}{p}}$, and we have:

$$s_{n_r} \log(1 + a_{n_r}/x) \sim {n_r}^{\frac{p-1}{p}} \log n_r = o(n_r).$$

As a consequence, a wide scope of antenna correlation models enter our deterministic equivalent framework, which comes again at the price of a slower theoretical convergence of the difference $\mathcal{V}_{\mathbf{HH}^{\mathsf{H}}}(\sigma^{-2}) - \mathcal{V}_{n_r}(\sigma^{-2})$.

When $\mathbf{T}$ is changed into $\mathbf{T}^{\frac{1}{2}}\mathbf{P}\mathbf{T}^{\frac{1}{2}}$, $\mathbf{P} \in \mathbb{C}^{n_t \times n_t}$ standing for the transmit power policy with constraint $\frac{1}{n_t} \operatorname{tr} \mathbf{P} \leq 1$ to accept power allocation in the system model, it is still valid that $\{F^{\mathbf{T}^{\frac{1}{2}}\mathbf{P}\mathbf{T}^{\frac{1}{2}}}\}$ forms a tight sequence and that the condition on the smallest eigenvalues of $\mathbf{T}^{\frac{1}{2}}\mathbf{P}\mathbf{T}^{\frac{1}{2}}$ is fulfilled for all matrices satisfying the mild assumption (13.7). Indeed, let $\mathbf{T}$ satisfy the trace constraint, then for $\varepsilon > 0$ such that $n_r\varepsilon \in \mathbb{N}$, we can choose M such that we have $1 - F^{\mathbf{T}}(\sqrt{M}) < \varepsilon/2$ and $1 - F^{\mathbf{P}}(\sqrt{M}) < \varepsilon/2$ for all n_t; since the smallest $n_t\varepsilon/2 + 1$ eigenvalues of both $\mathbf{T}$ and $\mathbf{P}$ are less than $\sqrt{M}$, at least the smallest $n_t\varepsilon + 1$ eigenvalues of

TP are less than M, hence $1 - F^{\mathbf{TP}}(M) < \varepsilon$ and $\{F^{\mathbf{TP}}\}$ is tight. Once again, the condition on the smallest eigenvalues can be satisfied for a vast majority of $\mathbf{T}^{\frac{1}{2}}\mathbf{P}\mathbf{T}^{\frac{1}{2}}$ matrices from the same argument.

The analysis above makes it possible to provide deterministic equivalents for the per-antenna mutual information of even strongly correlated antenna patterns. Assuming a quasi-static channel with imposed deterministic power allocation policy $\mathbf{P}$ at the transmitter (assume, e.g. short time data transmission over a typical quasi-static channel such that the transmitter does not have the luxury to estimate the propagation environment and chooses $\mathbf{P}$ in a deterministic manner), the mutual information $\mathcal{I}^{(n_t,n_r)}(\sigma^2;\mathbf{P})$ for this precoder satisfies

$$\frac{1}{n_r}\mathcal{I}^{(n_t,n_r)}(\sigma^2;\mathbf{P}) - \left[\frac{1}{n_r}\log_2\det\left(\mathbf{I}_{n_t} + \frac{n_r}{n_t}\delta\mathbf{T}^{\frac{1}{2}}\mathbf{P}\mathbf{T}^{\frac{1}{2}}\right) + \frac{1}{n_r}\sum_{k=1}^{n_r}\log_2\left(1 + \bar{\delta}r_k\right) - \sigma^2\log_2(e)\delta\bar{\delta}\right] \xrightarrow{\text{a.s.}} 0$$

with δ and $\bar{\delta}$ the unique positive solutions to

$$\delta = \frac{1}{\sigma^2 n_r}\sum_{k=1}^{n_r}\frac{r_k}{1+\bar{\delta}r_k},$$

$$\bar{\delta} = \frac{1}{\sigma^2 n_t}\operatorname{tr}\mathbf{T}^{\frac{1}{2}}\mathbf{P}\mathbf{T}^{\frac{1}{2}}\left(\mathbf{I}_{n_t} + \frac{n_r}{n_t}\delta\mathbf{T}^{\frac{1}{2}}\mathbf{P}\mathbf{T}^{\frac{1}{2}}\right)$$

and this is valid for all (but a restricted set of) choices of $\mathbf{T}$, $\mathbf{P}$, and $\mathbf{R}$ matrices.

This property of letting $\mathbf{T}$ and $\mathbf{R}$ have eigenvalues of order $O(n_r)$ is crucial to model the communication properties of very correlated antenna arrays, although extreme care must be taken in the degenerated case when both $\mathbf{T}$ and $\mathbf{R}$ have very few large eigenvalues and the remainder of their eigenvalues are close to zero. In such a scenario, extensive simulations suggest that the convergence of the deterministic equivalent is extremely slow, to the point that even the per-antenna capacity of a 1000×1000 MIMO channel is not well approximated by the deterministic equivalent. Also, due to strong correlation, it may turn out that the true per-antenna capacity decreases rather fast to zero, with growing n_r, n_t, while the difference between true per-antenna capacity and deterministic equivalent is only slowly decreasing. This may lead to the very unpleasant consequence that the *relative* difference grows to infinity while the effective difference goes to zero, but at a slow rate. On the other hand, extensive simulations also suggest that when only one of $\mathbf{T}$ and $\mathbf{R}$ is very ill-conditioned, the other having not too few large eigenvalues, deterministic equivalents are very accurate even for small dimensions.

When it comes to determining the optimal transmit data precoding matrix, we seek for the optimal matrix $\mathbf{P}$, $\mathbf{H}$ being fully known at the transmitter, such that the quasi-static mutual information is maximized. We might think that determining the precoding matrix which maximizes the deterministic equivalent of the quasi-static mutual information can provide at least an insight into the quasi-static capacity. This is however an incorrect reasoning, as the optimal precoding matrix, and therefore the system capacity, depend explicitly on the entries of $\mathbf{X}$. But then, the assumptions of Theorem 6.4 clearly state that the correlation matrices are deterministic or, as could be shown, are random but at least independent of the entries of $\mathbf{X}$. The deterministic equivalent of the mutual information can therefore not be extended to a deterministic equivalent of the quasi-static capacity. As an intuitive example, consider an i.i.d. Gaussian channel $\mathbf{H}$. The eigenvalue distribution of $\mathbf{H}\mathbf{H}^{\mathsf{H}}$ is, with high probability, close to the Marčenko–Pastur for sufficiently large dimensions. From our finite dimension analysis of the multiple antenna quasi-static capacity, water-filling over the eigenvalues of the Marčenko–Pastur law must be applied to maximize the capacity, so that strong communication modes receive much more power than strongly faded modes. However, it is clear, by symmetry, that the deterministic equivalent of the quasi-static mutual information is maximized under equal power allocation, which leads to a smaller rate.

13.3.2 Ergodic capacity in strongly correlated channels

Assuming $\mathbf{T}$ and $\mathbf{R}$ satisfy the conditions of Theorem 6.4, the space of matrices $\mathbf{X}$ over which the deterministic equivalent of the per-antenna quasi-static mutual information is an asymptotically accurate estimator has probability one. As a consequence, by integrating the per-antenna capacity over the space of Rayleigh matrices $\mathbf{X}$ and applying straightforwardly the dominated convergence theorem, Theorem 6.3, for all deterministic precoders $\mathbf{P}$, we have that the per-antenna ergodic mutual information is well approximated by the deterministic equivalent of the per-antenna quasi-static mutual information for this precoder.

It is now of particular interest to provide a deterministic equivalent for the per-antenna ergodic *capacity*, i.e. for unconstrained choice of a deterministic transmit data precoding matrix. Contrary to the quasi-static scenario, as the matrix $\mathbf{X}$ is unknown to the transmitter, the optimal precoding matrix is now chosen independently of $\mathbf{X}$. As a consequence, it seems possible to provide a deterministic equivalent for

$$\frac{1}{n_r}C_{\text{ergodic}}^{(n_t,n_r)}(\sigma^2) \triangleq \sup_{\substack{\mathbf{P}\\ \frac{1}{n_t}\operatorname{tr}\mathbf{P}=1}} \frac{1}{n_r}\mathrm{E}[\mathcal{I}^{(n_t,n_r)}(\sigma^2;\mathbf{P})]$$

with

$$\mathcal{I}^{(n_t,n_r)}(\sigma^2;\mathbf{P}) \triangleq \log_2\det\left(\mathbf{I}_{n_r} + \frac{1}{\sigma^2}\mathbf{H}\mathbf{P}\mathbf{H}^{\mathsf{H}}\right).$$

This is however not so obvious as $\mathbf{P}$ is allowed here to span over all matrices with constrained trace, which, as we saw, is a larger set than the set of matrices that satisfy the constraint (13.7) on their smallest eigenvalues. If we consider in the argument of the supremum only precoding matrices satisfying the assumption

(13.7), the per-antenna unconstrained ergodic capacity reads:

$$\frac{1}{n_r} C_{\text{ergodic}}^{(n_t,n_r)}(\sigma^2) = \sup_{\substack{\mathbf{P} \\ \frac{1}{n_t}\operatorname{tr}\mathbf{P}=1 \\ q_{s_{n_r}+1}\leq\alpha}} \frac{1}{n_r}\mathrm{E}[\mathcal{I}^{(n_t,n_r)}(\sigma^2;\mathbf{P})]$$

for some constant $\alpha > 0$, with $q_1, \ldots, q_{n_t}$ the eigenvalues of $\mathbf{T}^{\frac{1}{2}}\mathbf{P}\mathbf{T}^{\frac{1}{2}}$ and $s_1, s_2, \ldots$ any sequence which satisfies (13.7) (with $\mathbf{T}$ replaced by $\mathbf{T}^{\frac{1}{2}}\mathbf{P}\mathbf{T}^{\frac{1}{2}}$). We will assume in what follows that (13.7) holds for all precoding matrices considered. It is then possible to provide a deterministic equivalent for $\frac{1}{n_r}C_{\text{ergodic}}^{(n_t,n_r)}(\sigma^2)$ as it is possible to provide a deterministic equivalent for the right-hand side term for every deterministic $\mathbf{P}$. In particular, assume $\mathbf{P}^\star$ is a precoder that achieves $\frac{1}{n_r}C_{\text{ergodic}}^{(n_t,n_r)}$ and $\mathbf{P}^\circ$ is a precoder that maximizes its deterministic equivalent. We will denote $\frac{1}{n_r}\mathcal{I}^{(n_t,n_r)\circ}(\sigma^2;\mathbf{P})$ the value of the deterministic equivalent of the mutual information for precoder $\mathbf{P}$. In particular

$$\frac{1}{n_r}\mathcal{I}^{(n_t,n_r)\circ}(\sigma^2;\mathbf{P}) = \frac{1}{n_r}\sum_{k=1}^{n_r}\log_2\left(1+\bar{\delta}(\mathbf{P})r_k\right) + \frac{1}{n_r}\log_2\det\left[\mathbf{I}_{n_t} + \delta(\mathbf{P})\mathbf{T}^{\frac{1}{2}}\mathbf{P}\mathbf{T}^{\frac{1}{2}}\right] - \sigma^2\log_2(e)\delta(\mathbf{P})\bar{\delta}(\mathbf{P})$$

with $\delta(\mathbf{P})$ and $\bar{\delta}(\mathbf{P})$ the unique positive solutions to the equations in $(\delta, \bar{\delta})$

$$\delta = \frac{1}{\sigma^2 n_r}\sum_{k=1}^{n_r}\frac{r_k}{1+\bar{\delta}r_k},$$
$$\bar{\delta} = \frac{1}{\sigma^2 n_t}\operatorname{tr}\mathbf{T}^{\frac{1}{2}}\mathbf{P}\mathbf{T}^{\frac{1}{2}}\left(\mathbf{I}_{n_t} + \frac{n_r}{n_t}\delta\mathbf{T}^{\frac{1}{2}}\mathbf{P}\mathbf{T}^{\frac{1}{2}}\right). \tag{13.8}$$

We then have

$$\begin{aligned}
&\frac{1}{n_r}C_{\text{ergodic}}^{(n_t,n_r)}(\sigma^2) - \frac{1}{n_r}\mathcal{I}^{(n_t,n_r)\circ}(\sigma^2;\mathbf{P}^\circ) \\
&= \frac{1}{n_r}\mathrm{E}[\mathcal{I}^{(n_t,n_r)}(\sigma^2;\mathbf{P}^\star)] - \frac{1}{n_r}\mathcal{I}^{(n_t,n_r)\circ}(\sigma^2;\mathbf{P}^\circ) \\
&= \frac{1}{n_r}\left(\mathrm{E}[\mathcal{I}^{(n_t,n_r)}(\sigma^2;\mathbf{P}^\star)] - \mathcal{I}^{(n_t,n_r)\circ}(\sigma^2,\mathbf{P}^\star)\right) \\
&+ \frac{1}{n_r}\left(\mathcal{I}^{(n_t,n_r)\circ}(\sigma^2;\mathbf{P}^\star) - \mathcal{I}^{(n_t,n_r)\circ}(\sigma^2;\mathbf{P}^\circ)\right) \\
&= \frac{1}{n_r}\left(\mathrm{E}[\mathcal{I}^{(n_t,n_r)}(\sigma^2;\mathbf{P}^\star)] - \mathrm{E}[\mathcal{I}^{(n_t,n_r)}(\sigma^2,\mathbf{P}^\circ)]\right) \\
&+ \frac{1}{n_r}\left(\mathrm{E}[\mathcal{I}^{(n_t,n_r)}(\sigma^2,\mathbf{P}^\circ)] - \mathcal{I}^{(n_t,n_r)\circ}(\sigma^2;\mathbf{P}^\circ)\right).
\end{aligned}$$

In the second equality, as n_t, n_r grow large, the first term goes to zero, while the second is clearly negative by definition of $\mathbf{P}^\circ$, so that asymptotically $\frac{1}{n_r}(C_{\text{ergodic}}^{(n_t,n_r)}(\sigma^2) - \mathcal{I}^{(n_t,n_r)\circ}(\sigma^2;\mathbf{P}^\circ))$ is negative. In the third equality, the first term is clearly positive by definition of $\mathbf{P}^\star$ while the second term goes to zero, so that asymptotically $\frac{1}{n_r}(C_{\text{ergodic}}^{(n_t,n_r)}(\sigma^2) - \mathcal{I}^{(n_t,n_r)\circ}(\sigma^2;\mathbf{P}^\circ))$ is positive.

Therefore $\frac{1}{n_r}(C_{\text{ergodic}}^{(n_t,n_r)}(\sigma^2) - \mathcal{I}^{(n_t,n_r)\circ}(\sigma^2;\mathbf{P}^\circ)) \xrightarrow{\text{a.s.}} 0$, and the maximum value of the deterministic equivalent provides a deterministic equivalent for the ergodic capacity. This however does not say yet whether $\mathbf{P}^\circ$ is close to the optimal precoder $\mathbf{P}^\star$ itself. To verify this fact, we merely need to see that both the ergodic capacity and its deterministic equivalent, seen as functions of the precoder $\mathbf{P}$, are strictly concave functions, so that they both have a unique maximum; this can be proved without difficulty and therefore $\mathbf{P}^\star$ coincides with $\mathbf{P}^\circ$ asymptotically.

The analysis above has the strong advantage to be valid for all practical channel conditions that follow the Kronecker model, even strongly correlated channels. However, it also comes along with some limitations. The strongest limitation is that the rate of convergence of the deterministic equivalent of the per-antenna capacity is only ensured to be of order $o(1)$. This indicates that an approximation of the capacity falls within $o(n_t)$, which might be good enough (although not very satisfying) if the capacity scales as $O(n_t)$. In the particular case where an increase in the number of antennas on both communication ends increases correlation in some sense, the capacity no longer scales linearly with n_t, and the deterministic equivalent is of no practical use. The main two reasons why only $o(1)$ convergence could be proved is that the proof of Theorem 6.1 is (i) not restricted to Gaussian entries for $\mathbf{X}$, and (ii) assumes tight $\{F^{\mathbf{T}}\}$, $\{F^{\mathbf{R}}\}$ sequences. These two assumptions are shown, through truncation steps in the proof, to be equivalent to assuming that $\mathbf{T}$, $\mathbf{R}$, and $\mathbf{X}$ have entries bounded by $\log(n_t)$, therefore growing to infinity, but not too fast.

In the following, the same result is discussed but under tighter assumptions on the $\mathbf{T}$, $\mathbf{R}$, and $\mathbf{X}$ matrices. It will then be shown that a much faster convergence rate of the deterministic equivalent for the per-antenna capacity can be derived.

13.3.3 Ergodic capacity in weakly correlated channels

We now turn to the analysis provided in [Dupuy and Loubaton, 2009, 2010; Moustakas and Simon, 2007], leading in particular to Theorem 6.8. Obviously, the deterministic equivalent for the case when $K = 1$ is the same as the one from Theorem 6.4 for $K = 1$. However, the underlying conditions are now slightly different. In particular, large eigenvalues in $\mathbf{R}$ and $\mathbf{T}$ are no longer allowed as we now assume that both $\mathbf{R}$ and $\mathbf{T}$ have uniformly bounded spectral norm. Also, the proof derived in Theorem 6.8 explicitly takes into account the fact that $\mathbf{X}$ has Gaussian entries. In this scenario, it is shown that the ergodic capacity itself $\mathrm{E}[\mathcal{I}^{(n_t,n_r)}]$, and not simply the per-antenna ergodic mutual information, is well approximated by the above deterministic equivalent. Constraining uniform power allocation across the transmit antennas, we have:

$$\mathrm{E}[\mathcal{I}^{(n_t,n_r)}(\sigma^2;\mathbf{I}_{n_t})] - \left[\sum_{k=1}^{n_r}\log_2\left(1+\bar{\delta}r_k\right) + \sum_{k=1}^{n_t}\log_2\left(1+\frac{n_r}{n_t}\delta t_k\right) - n_r\sigma^2\log_2(e)\delta\bar{\delta}\right] = O(1/n_r).$$

The capacity maximizing transmit precoder $\mathbf{P}$ is shown similarly as before to coincide asymptotically with the transmit precoder $\mathbf{P}$ that maximizes the deterministic equivalent. We need however to constrain the matrices $\mathbf{P}$ to lie within a set of $(n_t \times n_t)$ Hermitian non-negative matrices with uniformly bounded spectral norm. Nonetheless, in their thorough analysis of the Rician model [Dumont *et al.*, 2010; Hachem *et al.*, 2007, 2008b], which we will discuss in the following section, Hachem *et al.* go much further and show explicitly that, under the assumption of uniformly bounded spectral norms for $\mathbf{T}$ and $\mathbf{R}$, the assumption that $\mathbf{P}$ lies within a set of matrices with uniformly bounded spectral norm is justified. The proof of this result is however somewhat cumbersome and is not further discussed.

In the following section, we turn to the proper evaluation of the capacity maximizing transmit precoder.

13.3.4 Capacity maximizing precoder

As was presented in the above sections, the capacity maximizing precoder $\mathbf{P}^\star$ is close to $\mathbf{P}^\circ$, the precoder that maximizes the deterministic equivalent of the true capacity, for large system dimensions. We will therefore determine $\mathbf{P}^\circ$ instead of $\mathbf{P}^\star$. Note nonetheless that [Hanlen and Grant, 2003] proves that the Gaussian input distribution is still optimal in this correlated setting and provides an iterative water-filling algorithm to obtain $\mathbf{P}^\star$, although the formulas involved are highly non-trivial. Since we do not deal with asymptotic considerations here, there is no need to restrict the definition domain of the non-negative Hermitian $\mathbf{P}$ matrices.

By definition

$$\mathbf{P}^\circ = \arg\max_{\mathbf{P}} \left[\sum_{k=1}^{n_r} \log_2\left(1+\bar{\delta} r_k\right) + \log_2\det\left[\mathbf{I}_{n_t} + \frac{n_r}{n_t}\delta \mathbf{T}^{\frac{1}{2}}\mathbf{P}\mathbf{T}^{\frac{1}{2}}\right] - n_r\sigma^2\log_2(e)\delta\bar{\delta}\right]$$

where δ and $\bar{\delta}$ are here function of $\mathbf{P}$, defined as the unique positive solutions to (13.8).

In order to simplify the differentiation of the deterministic equivalent along $\mathbf{P}$, which is made difficult due to the interconnection between δ, $\bar{\delta}$, and $\mathbf{P}$, we will use the differentiation chain rule. For this, we first denote V the function

$$V : (\Delta, \bar{\Delta}, \mathbf{P}) \mapsto \sum_{k=1}^{n_r} \log_2\left(1+\bar{\Delta} r_k\right) + \log_2\det\left(\mathbf{I}_{n_t} + \frac{n_r}{n_t}\Delta \mathbf{T}^{\frac{1}{2}}\mathbf{P}\mathbf{T}^{\frac{1}{2}}\right) - n_r\sigma^2\log_2(e)\Delta\bar{\Delta}.$$

That is, we define V as a function of the *independent* dummy parameters Δ, $\bar{\Delta}$, and $\mathbf{P}$. The function V is therefore a deterministic equivalent of the capacity only for restricted choices of Δ, $\bar{\Delta}$, and $\mathbf{P}$, i.e. $\Delta = \delta$, $\bar{\Delta} = \bar{\delta}$ that satisfy the implicit Equations (13.8). We have that

$$\frac{\partial V}{\partial \Delta}(\Delta, \bar{\Delta}, \mathbf{P}) = \log_2(e)\left[\frac{n_r}{n_t}\operatorname{tr}\mathbf{T}^{\frac{1}{2}}\mathbf{P}\mathbf{T}^{\frac{1}{2}}\left(\mathbf{I}_{n_t} + \frac{n_r}{n_t}\Delta\mathbf{T}^{\frac{1}{2}}\mathbf{P}\mathbf{T}^{\frac{1}{2}}\right)^{-1} - n_r\sigma^2\bar{\Delta}\right]$$

$$\frac{\partial V}{\partial \bar{\Delta}}(\Delta, \bar{\Delta}, \mathbf{P}) = \log_2(e)\left[\sum_{k=1}^{n_r}\frac{r_k}{1+\bar{\Delta}r_k} - n_r\sigma^2\Delta\right].$$

Observe now that, for δ and $\bar{\delta}$ the solutions of (13.8) (for any given $\mathbf{P}$), we have also by definition that

$$\frac{n_r}{n_t}\operatorname{tr}\mathbf{T}^{\frac{1}{2}}\mathbf{P}\mathbf{T}^{\frac{1}{2}}\left(\mathbf{I}_{n_t} + \frac{n_r}{n_t}\delta\mathbf{T}^{\frac{1}{2}}\mathbf{P}\mathbf{T}^{\frac{1}{2}}\right)^{-1} - n_r\sigma^2\bar{\delta} = 0$$

$$\sum_{k=1}^{n_r}\frac{r_k}{1+\bar{\delta}r_k} - n_r\sigma^2\delta = 0.$$

As a consequence

$$\frac{\partial V}{\partial \Delta}(\delta, \bar{\delta}, \mathbf{P}) = 0$$

$$\frac{\partial V}{\partial \bar{\Delta}}(\delta, \bar{\delta}, \mathbf{P}) = 0$$

and then:

$$\frac{\partial V}{\partial \Delta}(\delta, \bar{\delta}, \mathbf{P})\frac{\partial \Delta}{\partial \mathbf{P}}(\delta, \bar{\delta}, \mathbf{P}) + \frac{\partial V}{\partial \bar{\Delta}}(\delta, \bar{\delta}, \mathbf{P})\frac{\partial \bar{\Delta}}{\partial \mathbf{P}}(\delta, \bar{\delta}, \mathbf{P}) + \frac{\partial V}{\partial \mathbf{P}}(\delta, \bar{\delta}, \mathbf{P}) = \frac{\partial V}{\partial \mathbf{P}}(\delta, \bar{\delta}, \mathbf{P})$$

as all first terms vanish. But, from the differentiation chain rule, this expression coincides with the derivative of the deterministic equivalent of the mutual information along $\mathbf{P}$. For $\mathbf{P} = \mathbf{P}^\circ$, this derivative is zero by definition of $\mathbf{P}^\circ$. Therefore, setting the derivative along $\mathbf{P}$ of the deterministic equivalent of the mutual information to zero is equivalent to writing

$$\frac{\partial V}{\partial \mathbf{P}}(\delta(\mathbf{P}^\circ), \bar{\delta}(\mathbf{P}^\circ), \mathbf{P}^\circ) = 0$$

for $(\delta(\mathbf{P}^\circ), \bar{\delta}(\mathbf{P}^\circ))$ the solution of (13.8) with $\mathbf{P} = \mathbf{P}^\circ$. This is equivalent to

$$\frac{\partial}{\partial \mathbf{P}}\log_2\det\left(\mathbf{I}_{n_t} + \delta(\mathbf{P}^\circ)\mathbf{T}^{\frac{1}{2}}\mathbf{P}\mathbf{T}^{\frac{1}{2}}\right) = 0$$

which reduces to a water-filling problem for given $\delta(\mathbf{P}^\circ)$. Denoting $\mathbf{T} = \mathbf{U}\mathbf{\Lambda}\mathbf{U}^{\mathsf{H}}$ the spectral decomposition of $\mathbf{T}$ for some unitary matrix $\mathbf{U}$, $\mathbf{P}^\circ$ is defined as $\mathbf{P}^\circ = \mathbf{U}\mathbf{Q}^\circ\mathbf{U}^{\mathsf{H}}$, with $\mathbf{Q}^\circ$ diagonal with ith diagonal entry q_i° given by:

$$q_i^\circ = \left(\mu - \frac{1}{\delta(\mathbf{P}^\circ)t_i}\right)^+$$

μ being set so that $\frac{1}{n_t}\sum_{k=1}^{n_t} q_k^\circ = P$.

Obviously, the question is now to determine $\delta(\mathbf{P}^\circ)$. For this, the iterative water-filling algorithm of Table 13.2 is proposed.

Define $\eta > 0$ the convergence threshold and $l \geq 0$ the iteration step.
At step $l = 0$, for $k \in \{1, \ldots, n_t\}$, set $q_k^0 = P$
while $\max_k\{|q_k^l - q_k^{l-1}|\} > \eta$ **do**
  Define $(\delta^{l+1}, \bar{\delta}^{l+1})$ as the unique pair of positive solutions to (13.8)
  for $\mathbf{P} = \mathbf{U}\mathbf{Q}^l\mathbf{U}^\mathsf{H}$, $\mathbf{Q}^l = \mathrm{diag}(q_1^l, \ldots, q_{n_t}^l)$
  for $i \in \{1 \ldots, n_t\}$ **do**
    Set $q_i^{l+1} = \left(\mu - \frac{1}{ce^{l+1}t_i}\right)^+$, with μ such that $\frac{1}{n_t}\operatorname{tr}\mathbf{Q}^l = P$
  end for
  assign $l \leftarrow l + 1$
end while

Table 13.2. Iterative water-filling algorithm for the Kronecker channel model.

In [Dumont *et al.*, 2010], it is shown that, if the iterative water-filling algorithm does converge, then the iterated $\mathbf{Q}^l$ matrices of the algorithm in Table 13.2 necessarily converge towards $\mathbf{Q}^\circ$. To this day, though, no proof of the absolute or conditional convergence of this water-filling algorithm has been provided.

To conclude this section on correlated MIMO transmissions, we present simulation results for $\mathbf{R}$ and $\mathbf{T}$ modeled as a Jakes' correlation matrix, i.e. the (i,j)th entry of $\mathbf{R}$ or $\mathbf{T}$ equals $\mathrm{J}_0(2\pi d_{ij}/\lambda)$, with λ the transmission wavelength and d_{ij} the distance between antenna i and antenna j (on either of the two communication sides). We assume the antennas are distributed along a horizontal linear array and numbered in order (say, from left to right), so that $\mathbf{T}$ and $\mathbf{R}$ are simply Toeplitz matrices based on the vector $(1, \mathrm{J}_0(2\pi d/\lambda), \ldots, \mathrm{J}_0(2\pi(n_t - 1)d/\lambda)$, with d the distance between neighboring antennas. The eigenvalues of $\mathbf{R}$ and $\mathbf{T}$ for $n_t = 4$ are provided in Table 13.3 for different ratios d/λ. Jakes' model arises from the assumption that the antenna array under study transmits or receives waveforms of wavelength λ isotropically in the three-dimensional space, which is a satisfying assumption under no additional channel constraint.

In Figure 13.3, we depict simulation results of the MIMO mutual information with equal power allocation at the transmitter as well as the MIMO capacity (i.e. with optimal power allocation), and compare these results to the deterministic equivalents derived in this section. We assume 4×4 MIMO communication with inter-antenna spacing $d = 0.1\lambda$ and $d = \lambda$. It turns out that, even for strongly correlated channels, the mutual information with uniform power allocation is very well approximated for all SNR values, while a slight mismatch appears for the optimal power allocation policy. Although this is not appearing in this graph, we mention that the mismatch gets more acute for higher SNR values. This is consistent with the results derived in this section.

From an information-theoretic point of view, observe that the gains achieved by optimal power allocation in weakly correlated channels are very marginal, as the water-filling algorithm distributes power almost evenly on the channel eigenmodes, while the gains brought about by optimal power allocation in

Correlation factor	Eigenvalues of $\mathbf{T}, \mathbf{R}$			
$d = 0.1\lambda$	0.0	0.0	0.1	3.9
$d = \lambda$	0.3	1.0	1.4	1.5
$d = 10\lambda$	0.8	1.0	1.1	1.1

Table 13.3. Eigenvalues of correlation matrices for $n_t = n_r = 4$, under different correlations.

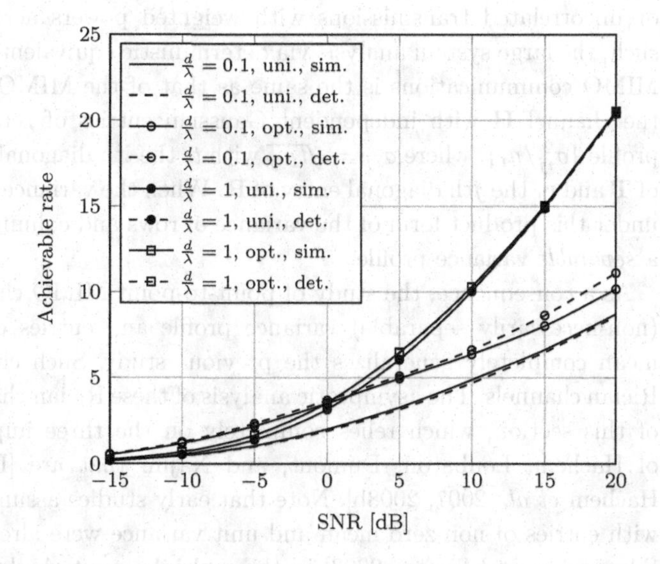

Figure 13.3 Ergodic capacity from simulation (sim.) and deterministic equivalent (det.) of the Jakes' correlated 4×4 MIMO, for SNR varying from -15 dB to 20 dB, for different values of $\frac{d}{\lambda}$, time-varying channels, uniform (uni.) and optimal (opt.) power allocation.

strongly correlated channels are much more relevant, as the water-filling algorithm manages to pour much of the power onto the stronger eigenmodes. Also, the difference between the mutual information for uniform and optimal power allocations reduces as the SNR grows. This is due to the fact that, for high SNR, the contribution to the capacity of every channel mode become sensibly the same, and therefore, from concavity arguments, it is optimal to evenly distribute the available transmit power along these modes. In contrast, for low SNR, $\log(1 + |h_k|^2\sigma^{-2})$ is close to $|h_k|^2\sigma^{-2}$, where $|h_k|^2$ denotes the kth eigenvalue of $\mathbf{H}\mathbf{H}^{\mathsf{H}}$, and therefore it makes sense to pour more power on the larger $|h_k|^2$ eigenvalues. Also notice, as described in [Goldsmith *et al.*, 2003] that, for low SNR, statistical correlation is in fact beneficial, as the capacity is larger than in the weak correlation case. This is here due to the fact that strong correlation comes along with high power modes (see Table 13.3), in which all the available power can be poured to increase the transmission bit rate.

13.4 Rician flat fading channels

Note that in the previous section, although we used Theorem 6.4 and Theorem 6.8 that provide in their general form deterministic equivalents that are functions of the eigenvectors of the underlying random matrices, all results derived so far are only functions of the eigenvalues of the correlation matrices $\mathbf{T}$ and $\mathbf{R}$. That is, we could have considered $\mathbf{T}$ and $\mathbf{R}$ diagonal. It is therefore equivalent for the point-to-point MIMO analysis to consider doubly correlated transmissions or uncorrelated transmissions with weighted powers across the antennas. As such, the large system analysis via deterministic equivalents of doubly correlated MIMO communications is the same as that of the MIMO communication over the channel $\mathbf{H}$ with independent Gaussian entries of zero mean and variance profile $\{\sigma_{ij}^2/n_t\}$, where $\sigma_{ij} = \sqrt{t_i r_j}$ with t_i the ith diagonal entry (or eigenvalue) of $\mathbf{T}$ and r_j the jth diagonal entry of $\mathbf{R}$. When the variance profile can be written under this product form of the variance of rows and columns, we say that $\mathbf{H}$ has a *separable* variance profile.

As a consequence, the study of point-to-point MIMO channels with a general (non-necessarily separable) variance profile and entries of non-necessary zero mean completely generalizes the previous study. Such channels are known as Rician channels. The asymptotic analysis of these Rician channels is the objective of this section, which relies completely on the three important contributions of Hachem, Loubaton, Dumont, and Najim that are [Dumont *et al.*, 2010; Hachem *et al.*, 2007, 2008b]. Note that early studies assuming channel matrices with entries of non-zero mean and unit variance were already provided in, e.g., [Moustakas and Simon, 2003] in the multiple input single output (MISO) case and [Cottatellucci and Debbah, 2004a,b] in the MIMO case.

Before moving to the study of the ergodic capacity, i.e. the study of the results from [Hachem *et al.*, 2007], it is important to remind that Rician channel models do *not* generalize Kronecker channel models. What we stated before is exactly that, for the capacity analysis of *point-to-point MIMO communications*, Kronecker channels can be substituted by Rician channels with zero mean and separable variance profiles. However, we will see in Chapter 14 that, for *multi-user* communications, this remark does not hold in general, unless all users have identical or at least co-diagonalizable channel correlation matrices. The coming analysis of Rician channel models is therefore not sufficient to treat the most general multi-user communications, which will be addressed in Chapter 14.

13.4.1 Quasi-static mutual information and ergodic capacity

Consider the point-to-point communication between an n_t-antenna transmitter and an n_r-antenna receiver. The communication channel is denoted by the random matrix $\mathbf{H} \in \mathbb{C}^{n_r \times n_t}$ which is modeled as Rician, i.e. the entries of $\mathbf{H}$ are Gaussian, independent and the (i,j)th entry h_{ij} has mean a_{ij} and variance

σ_{ij}^2/n_t, for $1 \leq i \leq n_r$ and $1 \leq j \leq n_t$. We further denote $\mathbf{A} \in \mathbb{C}^{n_r \times n_t}$ the matrix with (i,j)th entry a_{ij}, $\mathbf{\Sigma}_j \in \mathbb{R}^{n_r \times n_r}$ the diagonal matrix of ith entry σ_{ij}, and $\bar{\mathbf{\Sigma}}_i \in \mathbb{R}^{n_t \times n_t}$ the diagonal matrix of jth entry σ_{ij}. Assume, as in the previous sections, that the filtered transmit signal is corrupted by additive white Gaussian noise of variance σ^2 on all receive antennas. Here we assume that the σ_{ij} are uniformly bounded with respect to the system dimensions n_t and n_r.

Assuming equal power allocation at the transmitter, the mutual information for the per-antenna quasi-static case $\mathcal{I}^{(n_t,n_r)}(\sigma^2; \mathbf{I}_{n_t})$ reads:

$$\begin{aligned}\mathcal{I}^{(n_t,n_r)}(\sigma^2; \mathbf{I}_{n_t}) &= \log_2 \det \left(\mathbf{I}_{n_r} + \frac{1}{\sigma^2} \mathbf{H}\mathbf{H}^{\mathsf{H}} \right) \\ &= \log_2 \det \left(\mathbf{I}_{n_r} + \frac{1}{\sigma^2} \left(\mathbf{R}^{\frac{1}{2}} \mathbf{X} \mathbf{T}^{\frac{1}{2}} + \mathbf{A} \right) \left(\mathbf{R}^{\frac{1}{2}} \mathbf{X} \mathbf{T}^{\frac{1}{2}} + \mathbf{A} \right)^{\mathsf{H}} \right).\end{aligned}$$

From Theorem 6.14, we have that

$$\begin{aligned}&\frac{1}{n_t} \mathcal{I}^{(n_t,n_r)}(\sigma^2; \mathbf{I}_{n_t}) - \left[\frac{1}{n_r} \log_2 \det \left[\frac{1}{\sigma^2} \mathbf{\Psi}^{-1} + \mathbf{A} \bar{\mathbf{\Psi}} \mathbf{A}^{\mathsf{T}} \right] + \frac{1}{n_r} \log_2 \det \left(\frac{1}{\sigma^2} \bar{\mathbf{\Psi}}^{-1} \right) \right. \\ &\qquad \left. - \frac{\log_2(e) \sigma^2}{n_t n_r} \sum_{i,j} \sigma_{ij}^2 v_i \bar{v}_j \right] \xrightarrow{\text{a.s.}} 0 \qquad (13.9)\end{aligned}$$

where we defined $\mathbf{\Psi} \in \mathbb{C}^{n_r \times n_r}$ the diagonal matrix with ith entry ψ_i, $\bar{\mathbf{\Psi}} \in \mathbb{C}^{n_r \times n_r}$ the diagonal matrix with jth entry $\bar{\psi}_j$, with ψ_i and $\bar{\psi}_j$, $1 \leq i \leq n_r$, $1 \leq j \leq n_t$, the unique solutions of

$$\psi_i = \frac{1}{\sigma^2} \left[1 + \frac{1}{n_r} \operatorname{tr} \bar{\mathbf{\Sigma}}_i^2 \left(\bar{\mathbf{\Psi}}^{-1} + \sigma^2 \mathbf{A}^{\mathsf{T}} \mathbf{\Psi} \mathbf{A} \right)^{-1} \right]^{-1}$$

$$\bar{\psi}_j = \frac{1}{\sigma^2} \left[1 + \frac{1}{n_r} \operatorname{tr} \mathbf{\Sigma}_j^2 \left(\mathbf{\Psi}^{-1} + \sigma^2 \mathbf{A} \bar{\mathbf{\Psi}} \mathbf{A}^{\mathsf{T}} \right)^{-1} \right]^{-1}$$

which are Stieltjes transforms of distribution functions, while v_i and $\bar{v}_j$ are defined as the ith diagonal entry of $\left(\bar{\mathbf{\Psi}}^{-1} + \sigma^2 \mathbf{A}^{\mathsf{T}} \mathbf{\Psi} \mathbf{A} \right)^{-1}$ and the jth diagonal entry of $\left(\mathbf{\Psi}^{-1} + \sigma^2 \mathbf{A} \bar{\mathbf{\Psi}} \mathbf{A}^{\mathsf{T}} \right)^{-1}$, respectively.

Also, due to the Gaussian assumption (the general statement is based on some moment assumptions), [Hachem *et al.*, 2008b] ensures that the ergodic mutual information $\mathcal{I}^{(n_t,n_r)}(\sigma^2; \mathbf{I}_{n_t})$ under uniform transmit power allocation has deterministic equivalent

$$\begin{aligned}&\mathrm{E}[\mathcal{I}^{(n_t,n_r)}(\sigma^2; \mathbf{I}_{n_t})] \\ &\quad - \left[\log_2 \det \left(\frac{1}{\sigma^2} \mathbf{\Psi}^{-1} + \mathbf{A} \bar{\mathbf{\Psi}} \mathbf{A}^{\mathsf{T}} \right) + \log_2 \det \left(\frac{1}{\sigma^2} \bar{\mathbf{\Psi}}^{-1} \right) - \frac{\sigma^2 \log_2(e)}{n_t} \sum_{i,j} \sigma_{ij}^2 v_i \bar{v}_j \right] \\ &= O(1/n_t). \qquad (13.10)\end{aligned}$$

As we have already mentioned in the beginning of this section, Equations (13.9) and (13.10) can be verified to completely generalize the uncorrelated channel case, the Kronecker channel case, and the case of a channel with separable variance profile.

13.4.2 Capacity maximizing power allocation

In [Dumont *et al.*, 2010], the authors assume the above Rician channel model, however restricted to a separable variance profile, i.e. for all pairs (i, j), σ_{ij} can be written as a product $r_i t_j$. We then slightly alter the previous model setting by considering the channel model

$$\mathbf{H} = \mathbf{R}^{\frac{1}{2}}\mathbf{X}\mathbf{T}^{\frac{1}{2}}\mathbf{P}^{\frac{1}{2}} + \mathbf{A}\mathbf{P}^{\frac{1}{2}}$$

with $\mathbf{R} \in \mathbb{R}^{n_r \times n_r}$ diagonal with ith diagonal entry r_i, $\mathbf{T} \in \mathbb{R}^{n_t \times n_t}$ diagonal with jth diagonal entry t_j, and $\mathbf{P}$ Hermitian non-negative, the transmit precoding matrix. Note, as previously mentioned, that taking $\mathbf{R}$ and $\mathbf{T}$ to be diagonal does not restrict generality. Indeed, the mutual information for this channel model reads:

$$\begin{aligned}
&\mathcal{I}^{(n_t,n_r)}(\sigma^2;\mathbf{P}) \\
&= \log_2 \det\left(\mathbf{I}_{n_r} + \frac{1}{\sigma^2}\left(\mathbf{R}^{\frac{1}{2}}\mathbf{X}\mathbf{T}^{\frac{1}{2}}\mathbf{P}^{\frac{1}{2}} + \mathbf{A}\mathbf{P}^{\frac{1}{2}}\right)\left(\mathbf{R}^{\frac{1}{2}}\mathbf{X}\mathbf{T}^{\frac{1}{2}}\mathbf{P}^{\frac{1}{2}} + \mathbf{A}\mathbf{P}^{\frac{1}{2}}\right)^{\mathsf{H}}\right) \\
&= \log_2 \det\left(\mathbf{I}_{n_r} + \frac{1}{\sigma^2}\left(\mathbf{U}\mathbf{R}^{\frac{1}{2}}\mathbf{U}^{\mathsf{H}}\mathbf{X}\mathbf{V}\mathbf{T}^{\frac{1}{2}}\mathbf{V}^{\mathsf{H}}\mathbf{V}\mathbf{P}^{\frac{1}{2}}\mathbf{V}^{\mathsf{H}} + \mathbf{U}\mathbf{A}\mathbf{V}^{\mathsf{H}}\mathbf{V}\mathbf{P}^{\frac{1}{2}}\mathbf{V}^{\mathsf{H}}\right)\right. \\
&\qquad \left. \times \left(\mathbf{U}\mathbf{R}^{\frac{1}{2}}\mathbf{U}^{\mathsf{H}}\mathbf{X}\mathbf{V}\mathbf{T}^{\frac{1}{2}}\mathbf{V}^{\mathsf{H}}\mathbf{V}\mathbf{P}^{\frac{1}{2}}\mathbf{V}^{\mathsf{H}} + \mathbf{U}\mathbf{A}\mathbf{V}^{\mathsf{H}}\mathbf{V}\mathbf{P}^{\frac{1}{2}}\mathbf{V}^{\mathsf{H}}\right)^{\mathsf{H}}\right)
\end{aligned}$$

for any unitary matrices $\mathbf{U} \in \mathbb{C}^{n_t \times n_t}$ and $\mathbf{V} \in \mathbb{C}^{n_r \times n_r}$. The matrix $\mathbf{X}$ being Gaussian, its distribution is not altered by the left- and right-unitary products by $\mathbf{U}^{\mathsf{H}}$ and $\mathbf{V}$, respectively. Also, when addressing the power allocation problem, optimization can be carried out equally on $\mathbf{P}$ or $\mathbf{V}\mathbf{P}\mathbf{V}^{\mathsf{H}}$. Therefore, instead of the diagonal $\mathbf{R}$ and $\mathbf{T}$ covariance matrices, we could have considered the Kronecker channel with non-necessarily diagonal correlation matrices $\mathbf{U}\mathbf{R}\mathbf{U}^{\mathsf{H}}$ at the receiver and $\mathbf{V}\mathbf{T}\mathbf{V}^{\mathsf{H}}$ at the transmitter and the line-of-sight component matrix $\mathbf{U}\mathbf{A}\mathbf{V}^{\mathsf{H}}$.

The ergodic capacity optimizing power allocation matrix, under power constraint $\frac{1}{n_t}\operatorname{tr}\mathbf{P} \leq P$, is then shown in [Dumont *et al.*, 2010] to be determined as the solution of an iterative water-filling algorithm. Specifically, call $\mathbf{P}^\circ$ the optimal power allocation policy for the deterministic equivalent of the ergodic capacity. Using a similar approach as for the Kronecker channel case, it is shown that maximizing the deterministic equivalent of $\mathcal{I}^{(n_t,n_r)}(\sigma^2;\mathbf{P})$ is equivalent to maximizing the function

$$V : (\delta, \bar{\delta}, \mathbf{P}) \mapsto \log_2 \det\left(\mathbf{I}_{n_t} + \mathbf{P}\mathbf{G}(\delta, \bar{\delta})\right) + \log_2 \det\left(\mathbf{I}_{n_r} + \bar{\delta}\mathbf{R}\right) - \log_2(e) n_t \sigma^2 \delta\bar{\delta}$$

over $\mathbf{P}$ such that $\frac{1}{n_t}\operatorname{tr}\mathbf{P} = P$, where $\mathbf{G}$ is the deterministic matrix

$$\mathbf{G} = \delta\mathbf{T} + \frac{1}{\sigma^2}\mathbf{A}^{\mathsf{T}}\left(\mathbf{I}_{n_r} + \bar{\delta}\mathbf{R}\right)^{-1}\mathbf{A} \tag{13.11}$$

Define $\eta > 0$ the convergence threshold and $l \geq 0$ the iteration step.
At step $l = 0$, for $k \in \{1, \ldots, n_t\}$, set $q_k^0 = P$. At step $l \geq 1$,
while $\max_k\{|q_k^l - q_k^{l-1}|\} > \eta$ **do**
  Define $(\delta^{l+1}, \bar{\delta}^{l+1})$ as the unique pair of positive solutions to (13.12) for $\mathbf{P} = \mathbf{W}\mathbf{Q}^l\mathbf{W}^\mathsf{H}$, $\mathbf{Q}^l = \operatorname{diag}(q_1^l, \ldots, q_{n_t}^l)$ and $\mathbf{W}$ the matrix such that $\mathbf{G}$, given in (13.11) with $\delta = \delta^l$, $\bar{\delta} = \bar{\delta}^l$ has spectral decomposition $\mathbf{G} = \mathbf{W}\mathbf{\Lambda}\mathbf{W}^\mathsf{H}$, $\mathbf{\Lambda} = \operatorname{diag}(\lambda_1, \ldots, \lambda_{n_t})$
  for $i \in \{1 \ldots, n_t\}$ **do**
    Set $q_i^{l+1} = \left(\mu - \frac{1}{\lambda_i}\right)^+$, with μ such that $\frac{1}{n_t}\operatorname{tr}\mathbf{Q}^l = P$
  end for
  assign $l \leftarrow l + 1$
end while

Table 13.4. Iterative water-filling algorithm for the Rician channel model with separable variance profile.

in the particular case when $\delta = \delta(\mathbf{P}^\circ)$ and $\bar{\delta} = \bar{\delta}(\mathbf{P}^\circ)$, the latter two scalars being the unique solutions to the fixed-point equation

$$\delta = \frac{1}{\sigma^2 n_t}\operatorname{tr}\mathbf{R}\left((\mathbf{I}_{n_r} + \bar{\delta}\mathbf{R}) + \frac{1}{\sigma^2}\mathbf{A}\mathbf{P}^{\circ\frac{1}{2}}(\mathbf{I}_{n_t} + \delta\mathbf{P}^{\circ\frac{1}{2}}\mathbf{T}\mathbf{P}^{\circ\frac{1}{2}})^{-1}\mathbf{P}^{\circ\frac{1}{2}}\mathbf{A}^\mathsf{T}\right)^{-1}$$

$$\bar{\delta} = \frac{1}{\sigma^2 n_t}\operatorname{tr}\mathbf{P}^{\circ\frac{1}{2}}\mathbf{T}\mathbf{P}^{\circ\frac{1}{2}}\left((\mathbf{I}_{n_t} + \delta\mathbf{P}^{\circ\frac{1}{2}}\mathbf{T}\mathbf{P}^{\circ\frac{1}{2}}) + \frac{1}{\sigma^2}\mathbf{P}^{\circ\frac{1}{2}}\mathbf{A}^\mathsf{T}(\mathbf{I}_{n_r} + \bar{\delta}\mathbf{R})^{-1}\mathbf{A}\mathbf{P}^{\circ\frac{1}{2}}\right)^{-1} \tag{13.12}$$

which are Stieltjes transforms of distribution functions.

Similar to the Kronecker case, the iterative water-filling algorithm of Table 13.4 solves the power-optimization problem.

It is common to consider a parameter κ, the Rician factor, that accounts for the relative importance of the line-of-sight component $\mathbf{A}$ relative to the varying random part $\mathbf{R}^{\frac{1}{2}}\mathbf{X}\mathbf{T}^{\frac{1}{2}}$. To incorporate the Rician factor in the present model, we assume that $\frac{1}{n_t}\operatorname{tr}\mathbf{A}\mathbf{A}^\mathsf{T} = \frac{\kappa}{\kappa+1}$ and that $\mathbf{R}$ is constrained to satisfy $\frac{1}{n_r}\operatorname{tr}\mathbf{R} = \frac{1}{\kappa+1}$. Therefore, the larger the parameter κ the more important the line-of-sight contribution. In Figure 13.4, we consider the 4×4 MIMO Rician channel where $\mathbf{A}$ has entries equal to $\sqrt{\frac{1}{n_t}\frac{\kappa}{\kappa+1}}$, and $\mathbf{T}$, $\mathbf{R}$ are modeled as Jakes' correlation matrices for a linear antenna array with distance d between consecutive antennas. The transmission wavelength is denoted λ. We wish to study the relative impact of the line-of-sight component $\mathbf{A}$ on the ergodic capacity, and on the power allocation policy. Therefore, we consider first mildly correlated antennas at both communication ends with inter-antenna distance $d = \lambda$. We give the channel performance both for $\kappa = 1$ and for $\kappa = 100$.

We observe again, already for the 4×4 MIMO case, a very accurate match between the approximated and simulated ergodic capacity expressions. We first

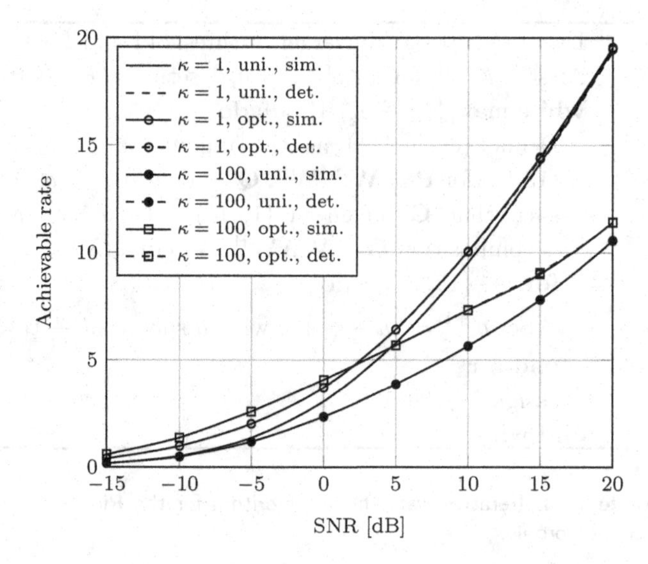

Figure 13.4 Ergodic capacity from simulation (sim.) and deterministic equivalent (det.) of the Jakes' correlated 4×4 MIMO model with line-of-sight component, linear antenna arrays with $\frac{d}{\lambda} = 1$ are considered at both communication ends. The SNR varies from -15 dB to 20 dB, and the Rician factor κ is chosen to be $\kappa = 1$ and $\kappa = 100$. Uniform (uni.) and optimal (opt.) power allocations are considered.

see that the capacity for κ small is larger than the capacity for κ large at high SNR, which is due to the limited multiplexing gain offered by the strongly line-of-sight channel. Note also that, for $\kappa = 1$, there is little room for high capacity gain by proceeding to optimal power allocation, while important gains can be achieved when $\kappa = 100$. For asymptotically large κ, the optimal power allocation policy requires that all power be poured on the unique non-zero eigenmode of the channel matrix. From the trace constraint on the precoding matrix, this entails a SNR gain of up to $\log_{10}(n_t) \simeq 6$ dB (already observed in the zero dB-10 dB region). For low SNR regimes, it is therefore again preferable to seek correlation, embodied here by the line-of-sight component. In contrast, for medium to high SNR regimes, correlation is better avoided to fully benefit from the channel multiplexing gain. The latter is always equal to $\min(n_t, n_r)$ for all finite κ, although extremely large SNR conditions might be necessary for this gain to be observable.

13.4.3 Outage mutual information

We complete this section by the results from [Hachem *et al.*, 2008b] on a central limit theorem for the capacity of the Rician channel $\mathbf{H}$, in the particular case when the channel coefficients h_{ij} have zero mean, i.e. when $\mathbf{A} = 0$. Note that the simple one-sided correlation case was treated earlier in [Debbah and R. Müller,

2003], based on a direct application of the central limit Theorem 3.17. Further, assume either of the following situations:

- the variance profile $\{\sigma_{ij}^2/n_t\}$ is separable in the sense that $\sigma_{ij} = \sqrt{r_i t_j}$ and the transmitter precodes its signals with matrix $\mathbf{P}$. In that case, we denote $\mathbf{T} = \mathrm{diag}(t_1, \ldots, t_{n_t})$ and $\{\sigma_{ij}'^2/n_t\}$ the variance profile with $\sigma'_{ij} = i\sqrt{r_i t'_j}$ with $t'_1, \ldots, t'_{n_t}$ the eigenvalues of $\mathbf{T}^{\frac{1}{2}}\mathbf{P}\mathbf{T}^{\frac{1}{2}}$.
- both $\mathbf{T}$ and the precoding matrix $\mathbf{P} = \mathrm{diag}(p_1, \ldots, p_{n_t})$ are diagonal. In this case, $\sigma'_{ij} = \sqrt{r_i t'_j}$, where $t'_j = \sigma_{i,j}^2 p_j$.

The result is summarized in Theorem 6.21, which states under the current channel conditions that, for a deterministic precoder $\mathbf{P}$ satisfying one of the above conditions, the mutual information $\mathcal{I}^{(n_t,n_r)}(\sigma^2; \mathbf{P})$ asymptotically varies around its mean $\mathrm{E}[\mathcal{I}^{(n_t,n_r)}(\sigma^2; \mathbf{P})]$ as a zero mean Gaussian random variable with, for each n_r, a variance close to $\theta_{n_r}^2$, defined as

$$\theta_{n_r}^2 = -\log\det\left(\mathbf{I}_{n_t} - \mathbf{J}_{n_t}\right)$$

where $\mathbf{J}_{n_t}$ is the matrix with (i,j)th entry $J_{ij}^{n_t}$, defined as

$$J_{ij}^{n_t} = \frac{1}{n_t} \frac{\frac{1}{n_t}\sum_{k=1}^{n_r} \sigma_{ki}'^2 \sigma_{kj}'^2 t_k^2}{\left(1 + \frac{1}{n_t}\sum_{k=1}^{n_r} \sigma_{ki}'^{\,2} t_k\right)^2}$$

with $t_1, \ldots, t_{n_r}$, defined as the unique solutions of

$$t_i = \left(\sigma^2 + \frac{1}{n_t}\sum_{j=1}^{n_t} \frac{\sigma_{ij}'^2}{1 + \frac{1}{n_t}\sum_{l=1}^{n_r} \sigma_{lj}'^2 t_l}\right)^{-1}$$

which are Stieltjes transforms of distribution functions when seen as functions of the variable $-\sigma^2$.

We therefore deduce a theoretical approximation of the outage mutual information $\mathcal{I}_{\mathrm{outage}}^{(n_t,n_r)}(\sigma^2; \mathbf{P}; q)$, for large n_t and n_r and deterministic precoder $\mathbf{P}$, defined by

$$\mathcal{I}_{\mathrm{outage}}^{(n_t,n_r)}(\sigma^2; \mathbf{P}; q) = \sup_{R \geq 0}\left\{P\left(\mathcal{I}^{(n_t,n_r)}(\sigma^2; \mathbf{P}) > R\right) \leq q\right\}$$

with $\mathcal{I}^{(n_t,n_r)}(\sigma^2; \mathbf{P})$ the quasi-static mutual information for the deterministic precoder $\mathbf{P}$. This is:

$$\mathcal{I}_{\mathrm{outage}}^{(n_t,n_r)}(\sigma^2; \mathbf{P}; q) \simeq Q^{-1}\left(\frac{q}{\theta_{n_r}}\right)$$

with $Q(x)$ the Gaussian Q-function, where the inverse is taken with respect to the conjugation.

In Figure 13.5, we provide the curves of the theoretical distribution function of the mutual information for the precoder $\mathbf{P} = \mathbf{I}_{n_t}$. The assumptions taken in Figure 13.5 are those of a Kronecker channel with transmit correlation matrices

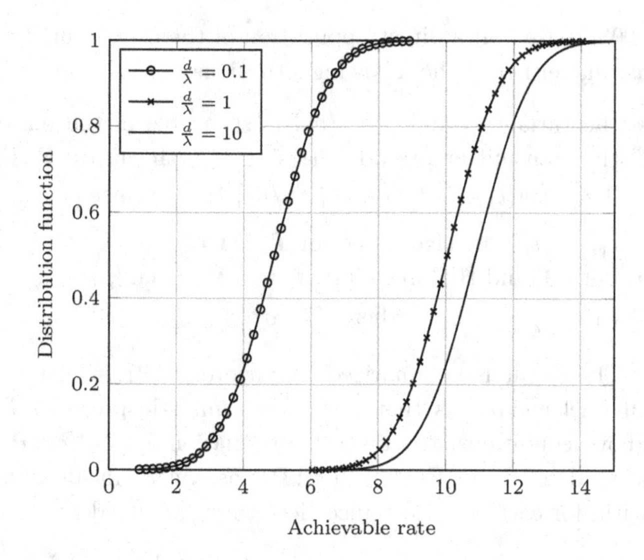

Figure 13.5 Deterministic equivalent of the outage capacity for the Jakes' correlated 4×4 MIMO channel model, linear antenna arrays with $\frac{d}{\lambda} = 0.1$, $\frac{d}{\lambda} = 1$ and $\frac{d}{\lambda} = 10$. The SNR is set to 10 dB. Uniform power allocation is considered.

$\mathbf{R} \in \mathbb{C}^{n_r \times n_r}$ and $\mathbf{T} \in \mathbb{C}^{n_t \times n_t}$ modeled along Jakes' correlation model for a linear antenna array with inter-antenna distance d (both at the transmit and receive ends) and transmission wavelength λ. We depict the outage performance of the MIMO Rician channel for different values of d/λ.

This concludes this section on the performance of point-to-point communications over MIMO Rician channels. We recall that the most general Rician setup generalizes all previously mentioned models. Nonetheless, the subject has not been fully covered as neither the optimal power allocation policy for a Rician channel with non-separable variance profile, nor the outage performance of a Rician channel with line-of-sight component have been addressed.

In what follows, we generalize multiple antenna point-to-point communications towards another direction, by introducing frequency selectivity. This will provide a first example of a channel model for which the MIMO Rician study developed in this section is not sufficient to provide a theoretical analysis of the communication performance.

13.5 Frequency selective channels

Due to the additional data processing effort required by multiple antenna communications compared to single antenna transmissions and to the non-negligible correlation arising in multiple antenna systems, MIMO technologies are mostly used as a solution to *further increase* the achievable data rate of existing wireless broadband communication networks, but are not used as a *substitute* for large bandwidth communications. Therefore, practical MIMO communication networks usually come along with large transmission bandwidths. It is therefore often unrealistic to assume narrowband MIMO transmission as we have done up to now. This section is dedicated to frequency selective transmissions, for which the channel coherence bandwidth is assumed smaller than the typical transmission bandwidth, or equivalently for which strong multi-path components convey signal energy.

Consider an L-path MIMO channel between an n_t-antenna transmitter and an n_r-antenna receiver, modeled as a sequence of L matrices $\{\mathbf{H}_1, \ldots, \mathbf{H}_L\}$, $\mathbf{H}_l \in \mathbb{C}^{n_r \times n_t}$. That is, each link from transmit antenna i and receive antenna j is a multi-path scalar channel with ordered path gains given by $(h_1(j,i), \ldots, h_L(j,i))$. In the frequency domain, the channel transfer matrix is modeled as a random matrix process $\mathcal{H}(f) \in \mathbb{C}^{n_r \times n_t}$, for every frequency $f \in [-W/2, W/2]$, defined as

$$\mathcal{H}(f) = \sum_{k=1}^{L} \mathbf{H}_k e^{-2\pi i k \frac{f}{W}}$$

with W the two-sided baseband communication bandwidth. The resulting (frequency normalized) quasi-static capacity $C^{(n_t,n_r)}$ of the frequency selective channel $\mathcal{H}(f)$, for a communication in additive white Gaussian noise and under power constraint P, reads:

$$C^{(n_t,n_r)}(\sigma^2) = \sup_{\{\mathbf{P}(f)\}} \frac{1}{W} \int_{-W/2}^{W/2} \log_2 \det \left(\mathbf{I}_{n_r} + \frac{1}{\sigma^2} \mathcal{H}(f) \mathbf{P}(f) \mathcal{H}^{\mathsf{H}}(f) \right) df$$

with $\mathbf{P}(f) \in \mathbb{C}^{n_t \times n_t}$, $f \in [-W/2, W/2]$, a matrix-valued function modeling the precoding matrix to be applied at all frequencies, with maximum mean power P (per Hertz) and therefore submitted to the power constraint

$$\frac{1}{n_t W} \int_f \operatorname{tr} \mathbf{P}(f) df \leq P. \tag{13.13}$$

According to the definition of $\mathcal{H}(f)$, the capacity $C^{(n_t,n_r)}(\sigma^2)$ also reads:

$$C^{(n_t,n_r)}(\sigma^2) = \sup_{\{\mathbf{P}(f)\}}$$
$$\frac{1}{W} \int_{-W/2}^{W/2} \log_2 \det \left(\mathbf{I}_{n_r} + \frac{1}{\sigma^2} \left[\sum_{k=1}^{L} \mathbf{H}_k e^{-2\pi i k \frac{f}{W}} \right] \mathbf{P}(f) \left[\sum_{k=1}^{L} \mathbf{H}_k^{\mathsf{H}} e^{2\pi i k \frac{f}{W}} \right] \right) df$$

under the trace constraint (13.13) on $\mathbf{P}(f)$.

We then assume that every channel $\mathbf{H}_k$, $k \in \{1, \ldots, L\}$, is modeled as a Kronecker channel, with non-negative definite left-correlation matrix $\mathbf{R}_k \in \mathbb{C}^{n_r \times n_r}$ and non-negative definite right-correlation matrix $\mathbf{T}_k \in \mathbb{C}^{n_t \times n_t}$. That is, $\mathbf{H}_k$ can be expressed as

$$\mathbf{H}_k = \mathbf{R}_k^{\frac{1}{2}} \mathbf{X}_k \mathbf{T}_k^{\frac{1}{2}}$$

where $\mathbf{X}_k \in \mathbb{C}^{n_r \times n_t}$ has i.i.d. Gaussian entries of zero mean and variance $1/n_t$. We also impose that the matrices $\mathbf{T}_k$ and $\mathbf{R}_k$, for all k, have uniformly bounded spectral norms and that the $\mathbf{X}_k$ are independent.

Although this is not strictly proven in the literature (Theorem 6.8), we infer that, for all fixed f and all deterministic choices of $\{\mathbf{P}(f)\}$ (with possibly some mild assumptions on the extreme eigenvalues), it is possible to provide a deterministic equivalent of the random quantity

$$\frac{1}{n_r}\log_2 \det\left(\mathbf{I}_{n_r} + \frac{1}{\sigma^2}\left[\sum_{k=1}^{L}\mathbf{H}_k e^{-2\pi i k \frac{f}{W}}\right]\mathbf{P}(f)\left[\sum_{k=1}^{L}\mathbf{H}_k^{\mathsf{H}} e^{2\pi i k \frac{f}{W}}\right]\right)$$

using the implicit equations of Theorem 6.8. The latter indeed only provides the convergence of such a deterministic equivalent in the mean. Integrating this deterministic equivalent over $f \in [-W/2, W/2]$ (and possibly averaging over W) would then lead to a straightforward deterministic equivalent for the per-receive-antenna quasi-static capacity (or its per-frequency version). Note that Theorem 6.4 and Theorem 6.8 are very similar in nature, so that the latter must be extensible to the quasi-static case, using tools from the proof of the former. Similar to previous sections, it will however not be possible to derive the matrix process $\{\mathbf{P}(f)\}$ which maximizes the capacity, as was performed for instance in [Tse and Hanly, 1998] in the single antenna (multi-user) case. We mention that [Scaglione, 2002] provides an explicit expression of the characteristic function of the above mutual information in the small dimensional setting.

13.5.1 Ergodic capacity

For the technical reasons explained above and also because this is a more telling measure of performance, we only consider the ergodic capacity of the frequency selective MIMO channel. Note that this frequency selective ergodic capacity $C_{\mathrm{ergodic}}^{(n_t,n_r)}(\sigma^2)$ reads:

$$\begin{aligned}
&C_{\mathrm{ergodic}}^{(n_t,n_r)}(\sigma^2)\\
&= \sup_{\{\mathbf{P}(f)\}} \frac{1}{W}\int_{-W/2}^{W/2} \mathrm{E}\left[\log_2\det\left(\mathbf{I}_{n_r} + \frac{1}{\sigma^2}\left[\sum_{k=1}^{L}\mathbf{H}_k\right]\mathbf{P}(f)\left[\sum_{k=1}^{L}\mathbf{H}_k^{\mathsf{H}}\right]\right)\right] df\\
&= \sup_{\mathbf{P}(0)} \mathrm{E}\left[\log_2\det\left(\mathbf{I}_{n_r} + \frac{1}{\sigma^2}\left[\sum_{k=1}^{L}\mathbf{H}_k\right]\mathbf{P}(0)\left[\sum_{k=1}^{L}\mathbf{H}_k^{\mathsf{H}}\right]\right)\right]
\end{aligned}$$

where in the second equality we discarded the terms $e^{-2\pi i k \frac{f}{W}}$ since it is equivalent to take the expectation over $\mathbf{X}_k$ or over $\mathbf{X}_k e^{-2\pi i k \frac{f}{W}}$, for all $f \in [-W/2, W/2]$ (since both matrices have the same joint Gaussian entry distribution). Therefore, on average, all frequencies are alike and the current problem reduces to finding a deterministic equivalent for the single frequency case. Also, it is obvious from convexity arguments that there is no reason to distribute the power P unevenly along the different frequencies. Therefore, the power optimization can be simply operated over a single frequency and the supremum can be taken over the single precoding matrix $\mathbf{P}(0)$. The new power constraint is therefore:

$$\frac{1}{n_t}\operatorname{tr}\mathbf{P}(0) \leq P.$$

For ease of read, from now on, we denote $\mathbf{P} \triangleq \mathbf{P}(0)$.

For all deterministic choices of precoding matrices $\mathbf{P}$, the ergodic mutual information $\mathrm{E}\mathcal{I}^{(n_t,n_r)}(\sigma^2;\mathbf{P})$ has a deterministic equivalent, given by Theorem 6.8, such that

$$\mathrm{E}[\mathcal{I}^{(n_t,n_r)}(\sigma^2;\mathbf{P})] - \left[\log_2\det\left(\mathbf{I}_{n_r} + \sum_{k=1}^{L}\bar{\delta}_k\mathbf{R}_k\right) + \log_2\det\left(\mathbf{I}_{n_t} + \sum_{k=1}^{L}\delta_k\mathbf{T}_k^{\frac{1}{2}}\mathbf{P}\mathbf{T}_k^{\frac{1}{2}}\right)\right.$$
$$\left. - n_r\log_2(e)\sigma^2\sum_{k=1}^{L}\bar{\delta}_k\delta_k\right] = O\left(\frac{1}{N}\right)$$

where $\bar{\delta}_k$ and δ_k, $k \in \{1,\ldots,L\}$, are defined as the unique positive solutions of

$$\delta_i = \frac{1}{n_r\sigma^2}\operatorname{tr}\mathbf{R}_i\left(\mathbf{I}_{n_r} + \sum_{k=1}^{L}\bar{\delta}_k\mathbf{R}_k\right)$$
$$\bar{\delta}_i = \frac{1}{n_r\sigma^2}\operatorname{tr}\mathbf{T}_i^{\frac{1}{2}}\mathbf{P}\mathbf{T}_i^{\frac{1}{2}}\left(\mathbf{I}_{n_t} + \sum_{k=1}^{L}\delta_k\mathbf{T}_k^{\frac{1}{2}}\mathbf{P}\mathbf{T}_k^{\frac{1}{2}}\right). \tag{13.14}$$

13.5.2 Capacity maximizing power allocation

Based on the standard methods evoked so far, the authors in [Dupuy and Loubaton, 2010] prove that the optimal power allocation strategy is to perform a standard water-filling procedure on the matrix

$$\sum_{k=1}^{L}\delta_k(\mathbf{P}^\circ)\mathbf{T}_k$$

where we define $\mathbf{P}^\circ$ as the precoding matrix that maximizes the deterministic equivalent of the ergodic mutual information, and we denote $\delta_k(\mathbf{P}^\circ)$ the (unique positive) solution of the system of Equations (13.14) when $\mathbf{P} = \mathbf{P}^\circ$.

Denote $\mathbf{U}\boldsymbol{\Lambda}\mathbf{U}^{\mathsf{H}}$ the spectral decomposition of $\sum_{k=1}^{L}\delta_k(\mathbf{P}^\circ)\mathbf{T}_k$, with $\mathbf{U}$ unitary and $\boldsymbol{\Lambda}$ a diagonal matrix with diagonal entries $\lambda_1,\ldots,\lambda_{n_t}$. We have that $\mathbf{P}^\circ$ asymptotically well approximates the ergodic mutual information maximizing precoder, and is given by:

$$\mathbf{P}^\circ = \mathbf{U}\mathbf{Q}^\circ\mathbf{U}^{\mathsf{H}}$$

where $\mathbf{Q}^\circ$ is diagonal with diagonal entries $q_1^\circ,\ldots,q_{n_t}^\circ$ defined by

$$q_k^\circ = \left(\mu - \frac{1}{\lambda_k}\right)^+$$

Define $\eta > 0$ the convergence threshold and $l \geq 0$ the iteration step.
At step $l = 0$, for $k \in \{1, \ldots, n_t\}$, set $q_k^0 = P$. At step $l \geq 1$,
while $\max_k\{|q_k^l - q_k^{l-1}|\} > \eta$ **do**
Define $(\delta^{l+1}, \bar{\delta}^{l+1})$ as the unique pair of positive solutions to (13.14) for $\mathbf{P} = \mathbf{U}\mathbf{Q}^l\mathbf{U}^{\mathsf{H}}$, $\mathbf{Q}^l = \mathrm{diag}(q_1^l, \ldots, q_{n_t}^l)$ and $\mathbf{U}$ the matrix such that $\sum_{k=1}^L \delta_k^l \mathbf{T}_k$ has spectral decomposition $\mathbf{U}\mathbf{\Lambda}\mathbf{U}^{\mathsf{H}}$, $\mathbf{\Lambda} = \mathrm{diag}(\lambda_1, \ldots, \lambda_{n_t})$
for $i \in \{1 \ldots, n_t\}$ **do**
Set $q_i^{l+1} = \left(\mu - \frac{1}{\lambda_i}\right)^+$, with μ such that $\frac{1}{n_t} \operatorname{tr} \mathbf{Q}^l = P$
end for
assign $l \leftarrow l + 1$
end while

Table 13.5. Iterative water-filling algorithm for the frequency selective channel.

where μ is set such that $\frac{1}{n_t}\sum_{k=1}^{n_t} q_k^\circ = P$. Similar to the previous sections, it is shown in [Dupuy and Loubaton, 2010] that, upon convergence, the precoder $\mathbf{P}$ in the iterative water-filling algorithm of Table 13.5 converges to $\mathbf{P}^\circ$.

For simulations, we consider a signal correlation model, which we will often use in the next chapter. Since all $\mathbf{R}_k$ and $\mathbf{T}_k$ matrices, $k \in \{1, \ldots, L\}$, account for the specific correlation pattern for every individual delayed propagation path on the antenna array, taking a standard Jakes' model for the $\mathbf{R}_k$ and $\mathbf{T}_k$ matrices would lead to $\mathbf{T}_1 = \ldots = \mathbf{T}_L$ and $\mathbf{R}_1 = \ldots = \mathbf{R}_L$. This is due to the fact that Jakes' model only accounts for inter-antenna distance, which is a physical device parameter independent of the propagation path. A further generalization of Jakes' model consists in taking into consideration the solid angle of energy departure or arrival. We will refer to this model as the *generalized Jakes' model.* Precisely, and for simplicity, we assume that the signal energy arrives (or departs) uniformly from angles in the vertical direction, but from a restricted angle in the horizontal direction, spanning from a minimum angle $\theta_{\min}$ to a maximum angle $\theta_{\max}$. Assuming linear antenna arrays at both communication ends and following the derivation of Jakes' model, the (n, m)th entry of, say, matrix $\mathbf{R}_1$ is given by:

$$R_{1_{nm}} = \int_{\theta_{\min}^{(\mathbf{R}_1)}}^{\theta_{\max}^{(\mathbf{R}_1)}} \exp\left(2\pi i |n - m| \frac{d^{\mathrm{R}}}{\lambda} \cos(\theta)\right) d\theta$$

where d^{R} is the distance between consecutive receive antennas, λ the signal wavelength, $\theta_{\min}^{(\mathbf{R}_1)}$ the minimum horizontal angle of energy arrival, and $\theta_{\max}^{(\mathbf{R}_1)}$ the maximum angle of energy arrival.

In Figure 13.6, we consider a two-path 4×4 frequency selective MIMO channel. We assume that antenna j, on either communication side, emits or receives data from a solid angle spanning the horizontal plane from the minimum angle $\theta_{\min}^{(\mathbf{R}_j)} = \theta_{\min}^{(\mathbf{T}_j)} = (j-1)\frac{\pi}{L}$ to the maximum angle $\theta_{\max}^{(\mathbf{R}_j)} = \theta_{\max}^{(\mathbf{T}_j)} = \Delta + (j -$

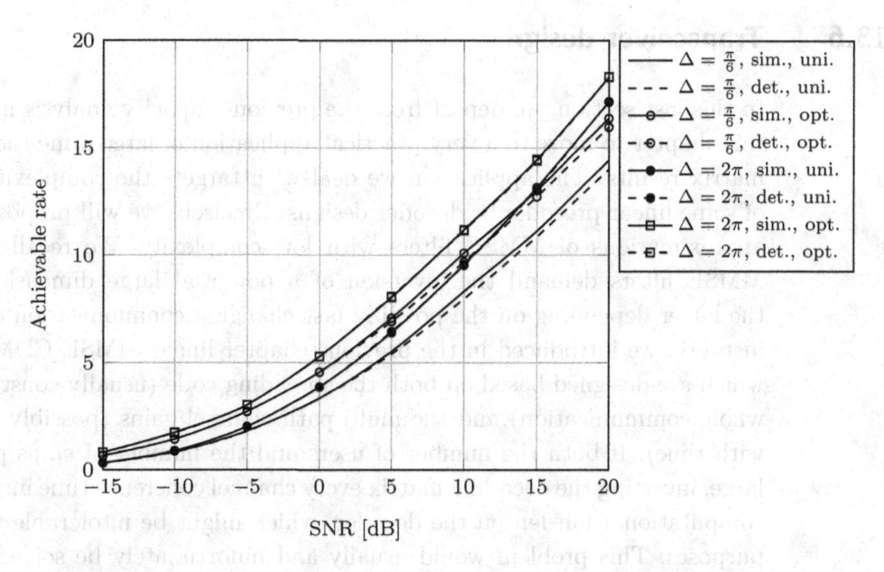

Figure 13.6 Ergodic capacity of the frequency selective 4×4 MIMO channel. Linear antenna arrays on each side, with correlation matrices modeled according to the generalized Jakes' model. Angle spreads in the horizontal direction set to $\Delta = 2\pi$ or $\Delta = \pi/6$. Comparison between simulations (sim.) and deterministic equivalent (det.), for uniform power allocation (uni.) and optimal power allocation (opt.).

$1)\frac{\pi}{L}$. We also assume an inter-antenna distance of $d^{\mathrm{R}} = \lambda$ at the receiver side and $d^{\mathrm{T}} = 0.1\lambda$ at the transmitter side. We take successively $\Delta = 2\pi$ and $\Delta = \frac{\pi}{6}$. We observe that the achievable rate is heavily impacted by the choice of a restricted angle of aperture at both transmission sides. This is because transmit and receive correlations increase with smaller antenna aperture, an effect which it is therefore essential to take into account.

We complete this chapter on single-user multiple antenna communications with more applied considerations on suboptimal transmitter and receiver design. From a practical point of view, be it for CDMA or MIMO technologies, achieving channel capacity requires in general heavy computational methods, the cost of which may be prohibitive for small communication devices. In place for optimal precoders and decoders, we have already seen several instances of linear precoders and decoders, such as the matched-filter and the MMSE filter. The subject of the next section is to study intermediary precoders, performing better than the matched-filter, less than the MMSE filter, but with adjustable complexity that can be made simple and efficient thanks to large dimensional random matrix results.

13.6 Transceiver design

In this last section, we depart from the previous capacity analysis introduced in this chapter to move to a very practical application of large dimensional random matrix results. The application we deal with targets the complexity reduction of some linear precoder or decoder designs. Precisely, we will propose successive approximations of MMSE filters with low complexity. We recall indeed that MMSE filters demand the inversion of a potential large dimensional matrix, the latter depending on the possibly fast changing communication channel. For instance, we introduced in the previous chapter linear MMSE CDMA decoders which are designed based on both the spreading code (usually constant over the whole communication) and the multi-path channel gains (possibly varying fast with time). If both the number of users and the number of chips per code are large, inverting the decoding matrix every channel coherence time imposes a large computational burden on the decoder, which might be intolerable for practical purposes. This problem would usually and unfortunately be solved by turning to less complex decoders, such as the classical matched-filter. Thanks to large dimensional random matrix theory, we will realize that most of the complexity involved in large matrix inverses can be fairly reduced by writing the matrix inverse as a finite weighted sum of matrices and by approximating the weights in this sum (which carry most of the computational burden) by deterministic equivalents.

We will address in this section the question of optimal low complex MMSE decoder design. The results mentioned below are initially due to Müller and Verdú in [Müller and Verdú, 2001] and are further developed in the work of Cottatellucci *et al.* in, e.g., [Cottatellucci and Müller, 2002, 2005; Cottatellucci *et al.*, 2004], Loubaton *et al.* [Loubaton and Hachem, 2003], and Hoydis *et al.* [Hoydis *et al.*, 2011c].

Consider the following communication channel

$$\mathbf{y} = \mathbf{H}\mathbf{x} + \sigma\mathbf{w}$$

with $\mathbf{x} = [x_1, \ldots, x_n]^\mathsf{T} \in \mathbb{C}^n$ some transmitted vectorial data of dimension n, assumed to have zero mean and covariance matrix $\mathbf{I}_n$, $\mathbf{w} \in \mathbb{C}^N$ is an additive Gaussian noise vector of zero mean and covariance $\mathbf{I}_N$, $\mathbf{y} = [y_1, \ldots, y_N]^\mathsf{T} \in \mathbb{C}^N$ is the received signal, and $\mathbf{H} \in \mathbb{C}^{N \times n}$ is the multi-dimensional communication channel.

Under these model assumptions, irrespective of the communication scenario under study (e.g. MIMO, CDMA), the minimum mean square error decoder output $\hat{\mathbf{x}}$ for the vector $\mathbf{x}$ reads:

$$\begin{aligned}\hat{\mathbf{x}} &= \left(\mathbf{H}^\mathsf{H}\mathbf{H} + \sigma^2\mathbf{I}_n\right)^{-1}\mathbf{H}^\mathsf{H}\mathbf{y} \\ &= \mathbf{H}^\mathsf{H}\left(\mathbf{H}\mathbf{H}^\mathsf{H} + \sigma^2\mathbf{I}_N\right)^{-1}\mathbf{y}.\end{aligned} \tag{13.15}$$

For practical applications, recovering $\hat{\mathbf{x}}$ therefore requires the inversion of the potentially large $(\mathbf{H}\mathbf{H}^{\mathsf{H}}+\sigma^2\mathbf{I}_N)^{-1}$ matrix. This inverted matrix has to be evaluated every time the channel matrix $\mathbf{H}$ changes. It unfolds that a high computational effort is required at the receiver to numerically evaluate such matrices. In some situations, where the computational burden at the receiver is an important constraint (e.g. impacting directly the battery consumption in cellular phones), this effort might be unbearable and we may have to resort to low complexity and less efficient detectors, such as the matched-filter $\mathbf{H}^{\mathsf{H}}$.

Now, from the Cayley-Hamilton theorem, any matrix is a (matrix-valued) root of its characteristic polynomial. That is, denoting $P(x)$ the characteristic polynomial of $\mathbf{H}^{\mathsf{H}}\mathbf{H}+\sigma^2\mathbf{I}_n$, i.e.

$$P(x)=\det\left(\mathbf{H}^{\mathsf{H}}\mathbf{H}+\sigma^2\mathbf{I}_n-x\mathbf{I}_n\right)$$

it is clear that $P(\mathbf{H}^{\mathsf{H}}\mathbf{H}+\sigma^2\mathbf{I}_n)=0$. Since the determinant above can be written as a polynomial of x of maximum degree n, $P(x)$ expresses as

$$P(x)=\sum_{i=0}^{n} a_i x^i \tag{13.16}$$

for some coefficients $a_0,\ldots,a_n$ to determine.

From (13.16), we then have

$$0=P(\mathbf{H}^{\mathsf{H}}\mathbf{H}+\sigma^2\mathbf{I}_n)=\sum_{i=0}^{n} a_i\left(\mathbf{H}^{\mathsf{H}}\mathbf{H}+\sigma^2\mathbf{I}_n\right)^i$$

from which

$$-a_0=\sum_{i=1}^{n} a_i\left(\mathbf{H}^{\mathsf{H}}\mathbf{H}+\sigma^2\mathbf{I}_n\right)^i.$$

Multiplying both sides by $(\mathbf{H}^{\mathsf{H}}\mathbf{H}+\sigma^2\mathbf{I}_n)^{-1}$, this becomes

$$\begin{aligned}\left(\mathbf{H}^{\mathsf{H}}\mathbf{H}+\sigma^2\mathbf{I}_n\right)^{-1}&=-\sum_{i=1}^{n}\frac{a_i}{a_0}\left(\mathbf{H}^{\mathsf{H}}\mathbf{H}+\sigma^2\mathbf{I}_n\right)^{i-1}\\&=-\sum_{i=1}^{n}\frac{a_i}{a_0}\sum_{j=0}^{i-1}\binom{i-1}{j}\sigma^{2(i-j-1)}\left(\mathbf{H}^{\mathsf{H}}\mathbf{H}\right)^{i-1}\end{aligned}$$

and therefore (13.15) can be rewritten under the form

$$\hat{\mathbf{x}}=\sum_{i=0}^{n-1} b_i\left(\mathbf{H}^{\mathsf{H}}\mathbf{H}\right)^i\mathbf{H}^{\mathsf{H}}\mathbf{y}$$

with $b_{i-1}=-\frac{a_i}{a_0}\sum_{j=0}^{i-1}\binom{i-1}{j}\sigma^{2(i-j-1)}$.

Obviously, the effort required to compute $(\mathbf{H}^{\mathsf{H}}\mathbf{H}+\sigma^2\mathbf{I}_n)^{-1}$ is equivalent to the effort required to compute the above sum. Nonetheless, it will appear that the b_i above can be expressed as a function of the trace of successive powers of

$\mathbf{H}^{\mathsf{H}}\mathbf{H} + \sigma^2\mathbf{I}_n$, which, in some cases, can be very well approximated thanks to random matrix theory approaches, *prior to communication.*

Nonetheless, even if $b_0, \ldots, b_{n-1}$ can be easily approximated, to evaluate $\hat{\mathbf{x}}$, we still have to take successive powers of $\mathbf{H}^{\mathsf{H}}\mathbf{H}$, which may still be computationally intense. On the other hand, the matched-filter performance may be so weak that it cannot be used either. Instead, it is always possible to use an estimator that both performs better than the matched-filter and is less computationally demanding than the MMSE filter. The idea here consists in substituting the MMSE or MF filters by a polynomial filter of order m, with $m \leq n-1$. The most natural decoder that comes to mind is a polynomially truncated version of the above decoder, i.e.

$$\sum_{i=0}^{m-1} b_i \left(\mathbf{H}^{\mathsf{H}}\mathbf{H}\right)^i \mathbf{H}^{\mathsf{H}}$$

for some $m \leq n$.

As often, though, sharp truncations do not lead to very efficient results. Instead, we may consider the polynomial precoder of order m in the variables $\mathbf{H}^{\mathsf{H}}\mathbf{H}$, which minimizes the mean square decoding error, among all such polynomial precoders. That is, we now seek for a precoder

$$\sum_{i=0}^{m-1} b_i^{(m)} \left(\mathbf{H}^{\mathsf{H}}\mathbf{H}\right)^i \mathbf{H}^{\mathsf{H}}$$

for some coefficients $b_k^{(m)}$ defined by

$$\mathbf{b}^{(m)} = (b_0^{(m)}, \ldots, b_{m-1}^{(m)}) = \arg \min_{(\beta_0,\ldots,\beta_{m-1})} \mathrm{E}\left[\left|\mathbf{x} - \sum_{i=0}^{m} \beta_i \left(\mathbf{H}^{\mathsf{H}}\mathbf{H}\right)^i \mathbf{H}^{\mathsf{H}}\mathbf{y}\right|^2\right].$$

Note that the $b_k^{(n)} = b_k$ are the coefficients of the MMSE decoder, while $b_1^{(1)}$ is the coefficient of the matched-filter. Obviously, for given m, m' with $m < m'$

$$\begin{aligned} &\arg \min_{(\beta_0,\ldots,\beta_{m'-1})} \mathrm{E}\left[\left|\mathbf{x} - \sum_{i=0}^{m'} \beta_i \left(\mathbf{H}^{\mathsf{H}}\mathbf{H}\right)^i \mathbf{H}^{\mathsf{H}}\mathbf{y}\right|^2\right] \\ &\leq \arg \min_{(\beta_0,\ldots,\beta_{m-1})} \mathrm{E}\left[\left|\mathbf{x} - \sum_{i=0}^{m} \beta_i \left(\mathbf{H}^{\mathsf{H}}\mathbf{H}\right)^i \mathbf{H}^{\mathsf{H}}\mathbf{y}\right|^2\right]. \end{aligned}$$

This is because equality is already achieved by taking $\beta_m = \ldots = \beta_{m'-1} = 0$ in the left-hand side argument. Therefore, with m ranging from 1 to n, we move gradually from the simplest and inexpensive matched-filter to the most elaborate but computationally demanding MMSE filter.

The above weights $\mathbf{b}^{(m)}$ are actually known and are provided explicitly by Moshavi *et al.* in [Moshavi *et al.*, 1996]. These are given by:

$$\mathbf{b}^{(m)} = \mathbf{\Phi}^{-1}\phi$$

where $\mathbf{\Phi} \in \mathbb{C}^{m \times m}$ and $\boldsymbol{\phi} \in \mathbb{C}^{(m)}$ depend only on the trace of the successive powers of $\mathbf{H}^{\mathsf{H}}\mathbf{H}$. Denoting Φ_{ij} the (i,j)th entry of $\mathbf{\Phi}$ and ϕ_i the ith entry of $\boldsymbol{\phi}$, we explicitly have

$$\Phi_{ij} = \frac{1}{n}\operatorname{tr}\left(\mathbf{H}^{\mathsf{H}}\mathbf{H}\right)^{i+j} + \sigma^2 \frac{1}{n}\operatorname{tr}\left(\mathbf{H}^{\mathsf{H}}\mathbf{H}\right)^{i+j-1}$$

$$\phi_i = \frac{1}{n}\operatorname{tr}\left(\mathbf{H}^{\mathsf{H}}\mathbf{H}\right)^{i}.$$

But then, from all limiting results on large dimensional random matrices introduced in Part I, either under the analytical Stieltjes transform approach or under the free probability approach, it is possible to approximate the entries of $\mathbf{\Phi}$ and $\boldsymbol{\phi}$ and to obtain deterministic equivalents for $b_0^{(m)}, \ldots, b_{m-1}^{(m)}$, for a large set of random matrix models for $\mathbf{H}$.

13.6.1 Channel matrix model with i.i.d. entries

Typically, for $\mathbf{H}$ with independent entries of zero mean, variance $1/N$, and finite $2+\varepsilon$ order moment, for some positive ε, from Theorem 2.14

$$\frac{1}{n}\operatorname{tr}\left(\mathbf{H}^{\mathsf{H}}\mathbf{H}\right)^{i} \xrightarrow{\text{a.s.}} \frac{1}{i}\sum_{k=0}^{i-1}\binom{i}{k}\binom{i}{k+1}c^k$$

as $N, n \to \infty$ with $n/N \to c$.

In place for the optimal MSE minimizing truncated polynomial decoders, we may then use the order m detector

$$\sum_{i=0}^{m-1} \tilde{b}_i^{(m)} \left(\mathbf{H}^{\mathsf{H}}\mathbf{H}\right)^{i} \mathbf{H}^{\mathsf{H}}$$

where $\tilde{\mathbf{b}}^{(m)} = (\tilde{b}_0^{(m)}, \ldots, \tilde{b}_{m-1}^{(m)})$ is defined as

$$\tilde{\mathbf{b}}^{(m)} = \tilde{\mathbf{\Phi}}^{-1}\tilde{\boldsymbol{\phi}}$$

where the entries of $\tilde{\mathbf{\Phi}}$ and $\tilde{\boldsymbol{\phi}}$ are the almost sure limit of $\mathbf{\Phi}$ and $\boldsymbol{\phi}$, respectively, as N, n grow to infinity with limiting ratio $n/N \to c$.

These suboptimal weights are provided, up to a scaling factor over all $\tilde{b}_i^{(m)}$, in Table 13.6.

Obviously, the partial MMSE detectors derived from asymptotic results differ from the exact partial MMSE detectors. They significantly differ if N, n are not large, therefore impacting the decoding performance. In Figure 13.7 and Figure 13.8, we depict the simulated bit error rate performance of partial MMSE detectors, using the weights $\mathbf{b}^{(m)}$ defined in this section, along with the bit error rate performance of the suboptimal detectors with weights $\tilde{\mathbf{b}}^{(m)}$. Comparison is made between both approaches, when $N = n = 4$ or $N = n = 64$ and $\mathbf{H}$ has independent Gaussian entries of zero mean and variance $1/N$. Observe that, for these small values of N and n, the large dimensional approximations $\tilde{\mathbf{b}}^{(m)}$ of $\mathbf{b}^{(m)}$ are far from accurate. Note in particular that the approximated MMSE detector

$N=1$	$\tilde{b}_0^{(1)}=1$
$N=2$	$\tilde{b}_0^{(2)}=\sigma^2-2(1+c)$ $\tilde{b}_1^{(2)}=1$
$N=3$	$\tilde{b}_0^{(3)}=3(1+c^2)-3\sigma^2(1+c)+\sigma^4+4c$ $\tilde{b}_1^{(3)}=\sigma^2-3(1+c)$ $\tilde{b}_2^{(3)}=1$
$N=4$	$\tilde{b}_0^{(4)}=6\sigma^2(1+c^2)+9\sigma^2c-4(1+c^3)-4\sigma^4(1+c)+\sigma^6-6c(1+c)$ $\tilde{b}_1^{(4)}=-6(1+c^2)+4\sigma^2(1+c)-\sigma^4-9c$ $\tilde{b}_2^{(4)}=\sigma^2-4(1+c)$ $\tilde{b}_3^{(4)}=1$

Table 13.6. Deterministic equivalents of the weights for the (MSE optimal) partial MMSE filters.

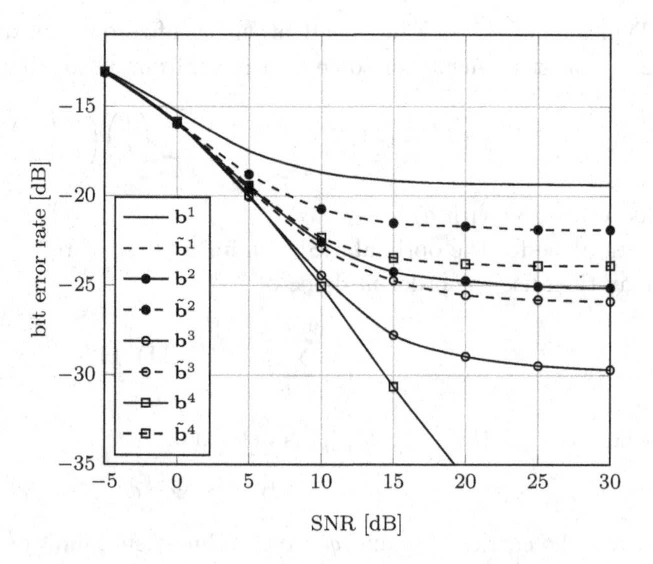

Figure 13.7 Bit error rate performance of partial MMSE filters, for exact weights $\mathbf{b}^{(m)}$ and approximated weights $\tilde{\mathbf{b}}^{(m)}$, $\mathbf{H}\in\mathbb{C}^{N\times n}$ has i.i.d. Gaussian entries, $N=n=4$.

($m=4$) is extremely badly approximated for these small values of N and n. For higher N and n, the decoders based on the approximated $\tilde{\mathbf{b}}^{(m)}$ perform very accurately for small m, as observed in Figure 13.8.

13.6.2 Channel matrix model with generalized variance profile

In the scenario when $\mathbf{H}\in\mathbb{C}^{N\times n}$ can be written under the very general form $\mathbf{H}=[\mathbf{h}_1,\ldots,\mathbf{h}_n]$, with $\mathbf{h}_i=\mathbf{R}_i^{\frac{1}{2}}\mathbf{x}_i$, with $\mathbf{R}_i\in\mathbb{C}^{N\times N}$ and $\mathbf{x}_i\in\mathbb{C}^N$, with i.i.d. entries of zero mean and variance $1/n$, we have from Theorem 6.13, for every m,

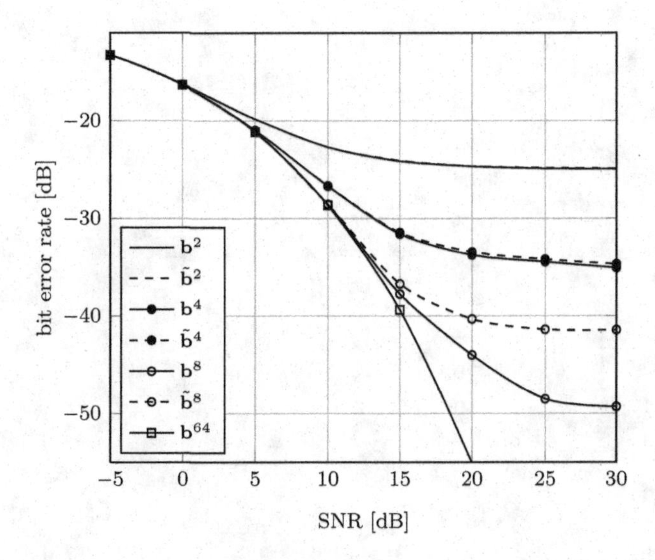

Figure 13.8 Bit error rate performance of partial MMSE filters, for exact weights $\mathbf{b}^{(m)}$ and approximated weights $\tilde{\mathbf{b}}^{(m)}$, $\mathbf{H} \in \mathbb{C}^{N \times n}$ has i.i.d. Gaussian entries, $N = n = 64$.

that $\tilde{\mathbf{b}}^{(m)} = \tilde{\mathbf{\Phi}}^{-1}\tilde{\phi}$, where

$$\tilde{\Phi}_{i,j} = M_{i+j} + \sigma^2 M_{i+j-1}$$
$$\phi_i = M_i$$

where the M_i are defined recursively in Theorem 6.13.

Other polynomial detector models for, e.g. downlink CDMA frequency selected channels, have been studied in [Hachem, 2004]. This concludes this chapter on point-to-point, or single-user, MIMO communications. In the following chapter, we extend some of the previous results to the scenario of multiple users possibly communicating with multiple antennas.

14 Rate performance in multiple access and broadcast channels

In this chapter, we consider both multiple access channels (MAC), which assume a certain number of users competing for the access to (i.e. to transmit data to) a single resource, and broadcast channels (BC), which assume the opposite scenario where a single transmitter multicasts data to multiple receivers.

The performance of multi-user communications can no longer be assessed from a single *capacity* parameter, as was the case for point-to-point communications. In a K-user MAC, we must evaluate what vectors $(R_1, \ldots, R_K)$ of rates, R_i being the data rate transmitted by user i, are achievable, in the sense that simultaneous reliable decoding of all data streams is possible at the receiver. Now, similar to single-user communications, where all rates R less than the capacity C are achievable and therefore define a rate set $\mathcal{R} = \{R, R \leq C\}$, for the multiple access channel, we define the multi-dimensional *MAC rate region* as the set $\mathcal{R}_{\text{MAC}}$ of all vectors $(R_1, \ldots, R_K)$ such that reliable decoding is possible at the receiver if users $1, \ldots, K$ transmit, respectively, at rate $R_1, \ldots, R_K$. Similarly, for the broadcast channel, we define the *BC rate region* $\mathcal{R}_{\text{BC}}$ as the (closed) set of all vectors $(R_1, \ldots, R_K)$, R_i being now the information data rate received by user i, such that every user can reliably decode its data. We further define the *rate region boundary*, either in the MAC or BC case, as the topological boundary of the rate region. These rate regions can be defined either in the quasi-static or in the ergodic sense. That is, the rate regions may assume perfect or only statistical channel state information at the transmitters. This is particularly convenient in the MAC, where in general perfect channel state information of all users' channel links is hardly accessible to each transmitter and where imperfect channel state information does not dramatically impact the achievable rates. In contrast, imperfect channel state information at the transmitters in the BC results in suboptimal beamforming strategies and thus high interference at all receivers, therefore reducing the rates achievable under perfect channel knowledge.

In the MAC, either with perfect or imperfect channel information at the transmitter, it is known that the boundary of the rate region can be achieved if the receiver performs MMSE decoding and successive interference cancellation (MMSE-SIC) of the input data streams. That is, the receiver decodes the strongest signal first using MMSE decoding, removes the signal contribution from the input data, then decodes the second to strongest signal, etc. until decoding the weakest signal. As for the BC with perfect channel information at the transmitter, it took researchers a long time to figure out a *precoding* strategy which achieves the boundary of the BC rate region; note here that the major processing effort is shifted to the transmitter. The main results were found almost simultaneously in the following articles [Caire and Shamai, 2003; Viswanath *et al.*, 2003; Viswanathan and Venkatesan, 2003; Weingarten *et al.*, 2006; Yu and Cioffi, 2004]. One of these boundary achieving codes (and for that matter, the only one we know of so far) is the so-called dirty-paper coding (DPC) algorithm [Costa, 1983]. The strategy of the DPC algorithm is to encode the data sequentially at the transmission in such a way that the interference created at every receiver and treated as noise by the latter allows for reliable data decoding. The approach is sometimes referred to as *successive encoding*, in duality reference with the successive decoding approach for the MAC. The DPC precoder is therefore non-linear and is to this day too complex to be implemented in practical communication systems. However, it has been shown in the information theory literature, see, e.g., [Caire and Shamai, 2003; Peel *et al.*, 2005; Wiesel *et al.*, 2008; Yoo and Goldsmith, 2006] that suboptimal *linear* precoders can achieve a large portion of the BC rate region while featuring low computational complexity. Thus, much research has recently focused on linear precoding strategies.

It is often not convenient though to derive complete achievable rate regions, especially for communications with a large number of users. Instead, *sum rate capacity* is often considered as a relevant performance metric, which corresponds to the maximally achievable sum $R_1 + \ldots + R_K$, with $(R_1, \ldots, R_K)$ elements of the achievable MAC or BC rate regions. Other metrics are sometimes used in place of the sum rate capacity, which allow for more user fairness. In particular, maximizing the minimum rate is a strategy that avoids leaving users with bad channel conditions with zero rate, therefore improving fairness among users. In this chapter, we will however only discuss sum rate maximization.

The next section is dedicated to the evaluation, through deterministic equivalents or large dimensional system limits, of the rate region or sum rate of MAC, BC, linearly precoded BC, etc. We assume a multi-user communication wireless network composed of K users, either transmitters (in the MAC) or receivers (in the BC), communicating with a multiple antenna access point or base station. User k, $k \in \{1, \ldots, K\}$ is equipped with n_k antennas, while the access point is equipped with N antennas. Since users and base stations are either transmitting or receiving data, we no longer use the notations n_t and n_r to avoid confusion.

We first consider the case of linearly precoded broadcast channels.

14.1 Broadcast channels with linear precoders

In this section, we consider the downlink communication of the N-antenna base station towards K single antenna receivers, i.e. $n_k = 1$ for all $k \in \{1, \ldots, K\}$, which is a common assumption in current broadcast channels, although studies regarding multiple antenna receivers have also been addressed, see, e.g., [Christensen *et al.*, 2008]. We further assume $N \geq K$. In some situations, we may need to further restrict this condition to $c_N \triangleq N/K > 1 + \varepsilon$ for some $\varepsilon > 0$ (as we will need to grow N and K large with specific rates). We denote by $\mathbf{h}_k^{\mathsf{H}} \in \mathbb{C}^{1\times N}$ the MISO channel from the base station to user k. At time t, denoting $s_k^{(t)}$ the signal intended to user k of zero mean and unit variance, $\sigma w_k^{(t)}$ an additive white Gaussian noise with zero mean and variance σ^2, and $y_k^{(t)}$ the signal received by user k, the transmission model reads:

$$y_k^{(t)} = \mathbf{h}_k^{\mathsf{H}} \sum_{j=1}^{K} \mathbf{g}_j s_j^{(t)} + \sigma w_k^{(t)}$$

where $\mathbf{g}_j \in \mathbb{C}^N$ denotes the linear vector precoder, also referred to as *beamforming vector* or *beamformer*, of user j. Gathering the transmit data into a vector $\mathbf{s}^{(t)} = (s_1^{(t)}, \ldots, s_K^{(t)})^{\mathsf{T}} \in \mathbb{C}^K$, the additive thermal noise into a vector $\mathbf{w}^{(t)} = (w_1^{(t)}, \ldots, w_K^{(t)})^{\mathsf{T}} \in \mathbb{C}^K$, the data received at the antenna array into a vector $\mathbf{y}^{(t)} = (y_1^{(t)}, \ldots, y_K^{(t)})^{\mathsf{T}} \in \mathbb{C}^K$, the beamforming vectors into a matrix $\mathbf{G} = [\mathbf{g}_1, \ldots, \mathbf{g}_K] \in \mathbb{C}^{N\times K}$, and the channel vectors into a matrix $\mathbf{H} = [\mathbf{h}_1, \ldots, \mathbf{h}_K]^{\mathsf{H}} \in \mathbb{C}^{K\times N}$, we have the compact transmission model

$$\mathbf{y}^{(t)} = \mathbf{H}\mathbf{G}\mathbf{s}^{(t)} + \sigma \mathbf{w}^{(t)}$$

where $\mathbf{G}$ must satisfy the power constraint

$$\operatorname{tr}(\mathrm{E}[\mathbf{G}\mathbf{s}^{(t)}\mathbf{s}^{(t)\mathsf{H}}\mathbf{G}^{\mathsf{H}}]) = \operatorname{tr}(\mathbf{G}\mathbf{G}^{\mathsf{H}}) \leq P \tag{14.1}$$

assuming $\mathrm{E}[\mathbf{s}^{(t)}\mathbf{s}^{(t)\mathsf{H}}] = \mathbf{I}_K$, for some available transmit power P at the base station.

When necessary, we will denote $\mathbf{z}^{(t)} \in \mathbb{C}^N$ the vector

$$\mathbf{z}^{(t)} \triangleq \mathbf{G}\mathbf{s}^{(t)}$$

of data transmitted from the antenna array of the base station at time t.

Due to its practical and analytical simplicity, this linear precoding model is very attractive. Most research in linear precoders has focused to this day both on analyzing the performance of some *ad-hoc* precoders and on determining the optimal linear precoders. Optimality is often taken with respect to sum rate maximization (or sometimes, for fairness reasons, with respect to maximization of the minimum user rate). In general, though, the rate maximizing linear precoder has no explicit form. Several iterative algorithms have been proposed in [Christensen *et al.*, 2008; Shi *et al.*, 2008] to come up with a sum rate optimal precoder, but no global convergence has yet been proved. Still, these iterative algorithms have a high computational complexity which motivates the use of further suboptimal linear transmit filters, by imposing more structure into the filter design.

In order to maximize the achievable sum rate, a first straightforward technique is to precode the transmit data by the inverse, or Moore–Penrose pseudo-inverse, of the channel matrix $\mathbf{H}$. This scheme is usually referred to as channel inversion (CI) or zero-forcing (ZF) precoding [Caire and Shamai, 2003]. That is, the ZF precoder $\mathbf{G}_{\text{zf}}$ reads:

$$\mathbf{G}_{\text{zf}} = \frac{\xi}{\sqrt{N}} \mathbf{H}^{\mathsf{H}} (\mathbf{H}\mathbf{H}^{\mathsf{H}})^{-1}$$

where ξ is set to fulfill some transmit power constraint (14.1).

The authors in [Hochwald and Vishwanath, 2002; Viswanathan and Venkatesan, 2003] carry out a large system analysis assuming that the number of transmit antennas N at the base station as well as the number of users K grow large while their ratio $c_N = N/K$ remains bounded. It is shown in [Hochwald and Vishwanath, 2002] that, for $c_N > 1 + \varepsilon$, uniformly on N, the achievable sum rate for ZF precoding has a multiplexing gain of K, which is identical to the optimal DPC-achieving multiplexing gain. The work in [Peel *et al.*, 2005] extends the analysis in [Hochwald and Vishwanath, 2002] to the case $K = N$ and shows that the sum rate of ZF saturates as K grows large. The authors in [Peel *et al.*, 2005] counter this problem by introducing a regularization term α into the inverse of the channel matrix. The precoder obtained is referred to as regularized zero-forcing (RZF) or regularized channel inversion, is denoted $\mathbf{G}_{\text{rzf}}$, and is given explicitly by

$$\mathbf{G}_{\text{rzf}} = \frac{\xi}{\sqrt{N}} \mathbf{H}^{\mathsf{H}} (\mathbf{H}\mathbf{H}^{\mathsf{H}} + \alpha \mathbf{I}_N)^{-1} \tag{14.2}$$

with ξ defined again to satisfy some transmit power constraint.

Under the assumption of large K and for any unitarily invariant channel distribution, [Peel *et al.*, 2005] derives the regularization term $\alpha^\star$ that maximizes the signal-to-interference plus noise ratio. It has been observed that the optimal RZF precoder proposed in [Peel *et al.*, 2005] is very similar to the transmit filters derived under the minimum mean square error (MMSE) criterion at every user, i.e. with $\alpha = \sigma^2$ [Joham *et al.*, 2002], and become identical in the large K limit.

Based on the tools developed in Part I, we provide in this section deterministic approximations of the achievable sum rate under ZF and RZF precoding, and determine the optimal $\alpha^\star$ parameter in terms of the sum rate as well as other interesting optimization measures discussed below. We also give an overview of the potential applications of random matrix theory for linearly precoded broadcast channels, under the most general channel conditions. In particular, consider that:

- the signal transmitted at the base station antenna array is correlated. That is, for every user k, $k \in \{1, \ldots, K\}$, $\mathbf{h}_k$ can be written under the form

$$\mathbf{h}_k = \mathbf{T}_k^{\frac{1}{2}} \mathbf{x}_k$$

where $\mathbf{x}_k \in \mathbb{C}^N$ has i.i.d. Gaussian entries of zero mean and variance $1/N$, and $\mathbf{T}_k^{\frac{1}{2}}$ is a non-negative definite square root of the transmit correlation matrix $\mathbf{T}_k \in \mathbb{C}^{N \times N}$ with respect to user k;

- the different users are assumed to have individual long-term path-losses $r_1, \ldots, r_K$. This allows us to further model $\mathbf{h}_k$ as

$$\mathbf{h}_k = \sqrt{r_k}\mathbf{T}_k^{\frac{1}{2}}\mathbf{x}_k.$$

The letter 'r' is chosen to indicate the channel fading seen at the *receiver*;

- the channel state information (CSI) at the transmitter side is assumed imperfect. That is, $\mathbf{H}$ is not completely known at the transmitter. Only an estimate $\hat{\mathbf{H}}$ is supposed to be available at the transmitter. We model $\hat{\mathbf{H}} = [\hat{\mathbf{h}}_1, \ldots, \hat{\mathbf{h}}_K]^{\mathsf{H}}$ with

$$\hat{\mathbf{h}}_j = \sqrt{1 - \tau_j^2}\mathbf{h}_j + \tau_j \tilde{\mathbf{h}}_j$$

with τ_j some parameter roughly indicating the accuracy of the jth channel estimate, and $\tilde{\mathbf{H}} = [\tilde{\mathbf{h}}_1, \ldots, \tilde{\mathbf{h}}_K]^{\mathsf{H}}$ as being the random matrix of channel errors with properties to be defined later. Note that a similar imperfect channel state analysis framework for the single-user MIMO case has been introduced in [Hassibi and Hochwald, 2003].

These channel conditions are rather realistic and include as particular cases the initial results found in the aforementioned literature contributions and others, e.g., [Ding *et al.*, 2007; Hochwald and Vishwanath, 2002; Jindal, 2006; Peel *et al.*, 2005]. An illustration of the linearly precoded MISO (or vector) broadcast channel is provided in Figure 14.1. The following development is heavily based on the work [Wagner *et al.*, 2011].

14.1.1 System model

Consider the transmission model described above. For simplicity here, we will assume that $\mathbf{T}_1 = \ldots = \mathbf{T}_K \triangleq \mathbf{T}$, i.e. the correlation at the base station of all users' channel vectors is identical. Field measurements [Kaltenberger *et al.*, 2009] suggest that this assumption is too strong and not fully realistic. As will be further discussed in subsequent sections, signal correlation at the transmitter does not only arise from close antenna spacing, but also from the different solid angles of signal departure. It could be argued though that the scenario where all users experience equal transmit covariance matrices represents a worst case scenario, as it reduces multi-user diversity. If not fully realistic, the current assumption on $\mathbf{T}$ is therefore still an interesting hypothesis.

Similar to [Chuah *et al.*, 2002; Shi *et al.*, 2008; Tulino and Verdú, 2005], we therefore denote

$$\mathbf{H} = \mathbf{R}^{\frac{1}{2}}\mathbf{X}\mathbf{T}^{\frac{1}{2}}$$

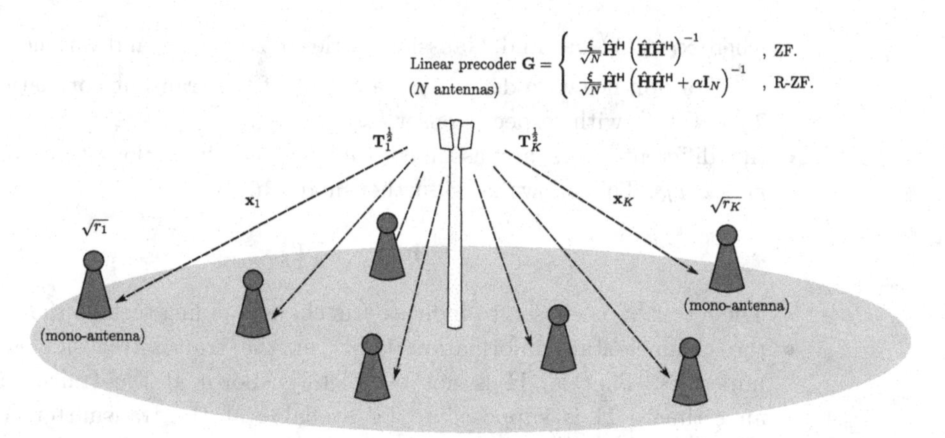

Figure 14.1 Linearly precoded vector broadcast channel, composed of K users and a base station. Each user is mono-antenna, and the base station is equipped with N antennas. The channel between the base station and user k is $\mathbf{h}_k = \sqrt{r_k}\mathbf{T}_k^{\frac{1}{2}}\mathbf{x}_k$, an approximation $\hat{\mathbf{h}}_k$ of which is available at the base station. The linear precoder is denoted $\mathbf{G}$ and is based on the channel matrix estimate $\hat{\mathbf{H}} = [\hat{\mathbf{h}}_1, \ldots, \hat{\mathbf{h}}_K]^{\mathsf{H}}$.

where $\mathbf{X} \in \mathbb{C}^{K \times N}$ has i.i.d. entries of zero mean and variance $1/N$, $\mathbf{T} \in \mathbb{C}^{N \times N}$ is the non-negative definite correlation matrix at the transmitter with eigenvalues $t_1, \ldots, t_N$, ordered as $t_1 \leq \ldots \leq t_N$, and $\mathbf{R} = \operatorname{diag}(r_1, \ldots, r_K)$, with entries ordered as $r_1 \leq \ldots \leq r_K$, contains the user channel gains, i.e. the inverse user path losses. We further assume that there exist $a_-, a_+ > 0$ such that

$$a_- < t_1 \leq t_N < a_+, \tag{14.3}$$

$$a_- < r_1 \leq r_K < a_+ \tag{14.4}$$

uniformly on N and K. That is, (14.3) assumes that the correlation between transmit antennas does not increase as the number of antennas increases. For practical systems, this is equivalent to requesting that neighboring antennas are spaced sufficiently apart. Equation (14.4) assumes that the users are not too close to the base station but also not too far away. This is a realistic assumption, as distant users would be served by neighboring base stations. Those requirements, although rather realistic, are obviously not mandatory for practical systems. However, they are required for the mathematical derivations of the present study. Further note that the Kronecker model assumes that the receivers are spaced sufficiently apart and are therefore spatially uncorrelated, an assumption which could also be argued against in some specific scenarios.

Besides, we suppose that only $\hat{\mathbf{H}} \in \mathbb{C}^{K \times N}$, an imperfect estimate of the true channel matrix $\mathbf{H}$, is available at the transmitter. The channel gain matrix $\mathbf{R}$ as well as the transmit correlation $\mathbf{T}$ are assumed to be slowly varying compared to the fast fading component $\mathbf{X}$ and are assumed to be perfectly known to the transmitter. We model $\hat{\mathbf{H}}$ as

$$\hat{\mathbf{H}} = \mathbf{R}^{\frac{1}{2}}\hat{\mathbf{X}}\mathbf{T}^{\frac{1}{2}} \tag{14.5}$$

with imperfect short-term statistics $\hat{\mathbf{X}} = [\hat{\mathbf{x}}_1, \ldots, \hat{\mathbf{x}}_K]^{\mathsf{H}}$ given by:

$$\hat{\mathbf{x}}_k = \sqrt{1 - \tau_k^2}\mathbf{x}_k + \tau_k \mathbf{q}_k$$

where we defined $\mathbf{Q} = [\mathbf{q}_1, \ldots, \mathbf{q}_K]^{\mathsf{H}} \in \mathbb{C}^{K \times N}$ the matrix of channel estimation errors containing i.i.d. entries of zero mean and variance $1/N$, and $\tau_k \in [0,1]$ the distortion in the channel estimate $\mathbf{h}_k$ of user k. We assume that the τ_k are perfectly known at the transmitter. However, as shown in [Dabbagh and Love, 2008], an approximated knowledge of τ_k will not lead to a severe performance degradation of the system. Furthermore, we suppose that $\mathbf{X}$ and $\mathbf{Q}$ are mutually independent as well as independent of the symbol vector $\mathbf{s}^{(t)}$ and noise vector $\mathbf{w}^{(t)}$. A similar model for the case of imperfect channel state information at the transmitter has been used in, e.g., [Dabbagh and Love, 2008; Hutter *et al.*, 2000; Yoo and Goldsmith, 2006].

We then define the average SNR ρ as $\rho \triangleq P/\sigma^2$. The received symbol $y_k^{(t)}$ of user k at time t is given by:

$$y_k^{(t)} = \mathbf{h}_k^{\mathsf{H}} \mathbf{g}_k s_k^{(t)} + \sum_{\substack{1 \leq i \leq K \\ i \neq k}} \mathbf{h}_k^{\mathsf{H}} \mathbf{g}_i s_i^{(t)} + \sigma w_k^{(t)}$$

where we recall that $\mathbf{h}_k^{\mathsf{H}} \in \mathbb{C}^N$ denotes the kth row of $\mathbf{H}$.

The SINR γ_k of user k reads:

$$\gamma_k = \frac{|\mathbf{h}_k^{\mathsf{H}} \mathbf{g}_k|^2}{\sum_{\substack{j=1 \\ j \neq k}}^{N} |\mathbf{h}_k^{\mathsf{H}} \mathbf{g}_j|^2 + \frac{\sigma^2}{N}}. \tag{14.6}$$

The system sum rate R_{sum} is defined as

$$R_{\text{sum}} = \sum_{k=1}^{K} \log_2(1 + \gamma_k)$$

evaluated in bits/s/Hz.

The objective of the next section is to provide a deterministic equivalent for the γ_k under RZF precoding.

14.1.2 Deterministic equivalent of the SINR

A deterministic equivalent for the SINR under RZF precoding is given as follows.

Theorem 14.1. *Let $\gamma_{\text{rzf},k}$ be the SINR of user k under RZF precoding, i.e. $\gamma_{\text{rzf},k}$ is given by* (14.6), *with* $\mathbf{G}$ *given by* (14.6) *and ξ such that the power constraint* (14.1) *is fulfilled. Then*

$$\gamma_{\text{rzf},k} - \bar{\gamma}_{\text{rzf},k} \xrightarrow{\text{a.s.}} 0$$

as $N, K \to \infty$, such that $1 \leq N/K \leq C$ for some $C > 1$, and where $\bar{\gamma}_{\mathrm{rzf},k}$ is given by:

$$\bar{\gamma}_{\mathrm{rzf},k} = \frac{r_k^2(1-\tau_k^2)\bar{m}^2}{r_k\bar{\Upsilon}\left(1-\tau_k^2\left[1-(1+r_k\bar{m})^2\right]\right)+\frac{\bar{\Psi}}{\rho}(1+r_k\bar{m})^2}$$

with

$$\bar{m} = \frac{1}{N}\operatorname{tr}\mathbf{T}\left(\alpha\mathbf{I}_N+\phi\mathbf{T}\right)^{-1}, \tag{14.7}$$

$$\bar{\Psi} = \frac{\frac{1}{N}\operatorname{tr}\mathbf{R}\left(\mathbf{I}_K+\bar{m}\mathbf{R}\right)^{-2}\frac{1}{N}\operatorname{tr}\mathbf{T}\left(\alpha\mathbf{I}_N+\phi\mathbf{T}\right)^{-2}}{1-\frac{1}{N}\operatorname{tr}\mathbf{R}^2\left(\mathbf{I}_K+\bar{m}\mathbf{R}\right)^{-2}\frac{1}{N}\operatorname{tr}\mathbf{T}^2\left(\alpha\mathbf{I}_N+\phi\mathbf{T}\right)^{-2}} \tag{14.8}$$

$$\bar{\Upsilon} = \frac{\frac{1}{N}\operatorname{tr}\mathbf{R}\left(\mathbf{I}_K+\bar{m}\mathbf{R}\right)^{-2}\frac{1}{N}\operatorname{tr}\mathbf{T}^2\left(\alpha\mathbf{I}_N+\phi\mathbf{T}\right)^{-2}}{1-\frac{1}{N}\operatorname{tr}\mathbf{R}^2\left(\mathbf{I}_K+\bar{m}\mathbf{R}\right)^{-2}\frac{1}{N}\operatorname{tr}\mathbf{T}^2\left(\alpha\mathbf{I}_N+\phi\mathbf{T}\right)^{-2}} \tag{14.9}$$

where ϕ is the unique real positive solution of

$$\phi = \frac{1}{N}\operatorname{tr}\mathbf{R}\left(\mathbf{I}_K+\mathbf{R}\frac{1}{N}\operatorname{tr}\mathbf{T}\left(\alpha\mathbf{I}_N+\phi\mathbf{T}\right)^{-1}\right)^{-1}. \tag{14.10}$$

Moreover, define $\phi_0 = 1/\alpha$, and for $k \geq 1$

$$\phi_k = \frac{1}{N}\operatorname{tr}\mathbf{R}\left(\mathbf{I}_K+\mathbf{R}\frac{1}{K}\operatorname{tr}\mathbf{T}\left(\alpha\mathbf{I}_N+\phi_{k-1}\mathbf{T}\right)^{-1}\right)^{-1}.$$

Then $\phi = \lim_{k\to\infty}\phi_k$.

We subsequently prove the above result by providing deterministic equivalents for ξ, then for the power of the signal of interest (numerator of the SINR) and finally for the interference power (denominator of the SINR). Throughout this proof, we will mainly use Theorem 6.12.

For convenience, we will constantly use the notation $m_{\mathbf{B}_N,\mathbf{Q}_N}(z)$ for a random matrix $\mathbf{B}_N \in \mathbb{C}^{N\times N}$ and a deterministic matrix $\mathbf{Q}_N \in \mathbb{C}^{N\times N}$ to represent

$$m_{\mathbf{B}_N,\mathbf{Q}_N}(z) \triangleq \frac{1}{N}\operatorname{tr}\mathbf{Q}_N\left(\mathbf{B}_N - z\mathbf{I}_N\right)^{-1}.$$

Also, we will denote $m_{\mathbf{B}_N}(z) \triangleq m_{\mathbf{B}_N,\mathbf{I}_N}(z)$. The character m is chosen here because all $m_{\mathbf{B}_N,\mathbf{Q}_N}(z)$ considered in the following will turn out to be Stieltjes transforms of finite measures on $\mathbb{R}^+$.

From the sum power constraint (14.1), we obtain

$$\xi^2 = \frac{P}{\frac{1}{N}\operatorname{tr}\left[\hat{\mathbf{H}}^{\mathsf{H}}\hat{\mathbf{H}}\left(\hat{\mathbf{H}}^{\mathsf{H}}\hat{\mathbf{H}}+\alpha\mathbf{I}_N\right)^{-2}\right]} = \frac{P}{m_{\hat{\mathbf{H}}^{\mathsf{H}}\hat{\mathbf{H}}}(-\alpha)-\alpha m'_{\hat{\mathbf{H}}^{\mathsf{H}}\hat{\mathbf{H}}}(-\alpha)} \triangleq \frac{P}{\Psi}$$

where the last equation follows from the decomposition

$$\hat{\mathbf{H}}^{\mathsf{H}}\hat{\mathbf{H}}(\hat{\mathbf{H}}^{\mathsf{H}}\hat{\mathbf{H}}+\alpha\mathbf{I}_M)^{-2} = (\hat{\mathbf{H}}^{\mathsf{H}}\hat{\mathbf{H}}+\alpha\mathbf{I}_N)^{-1} - \alpha(\hat{\mathbf{H}}^{\mathsf{H}}\hat{\mathbf{H}}+\alpha\mathbf{I}_N)^{-2}$$

and we define

$$\Psi \triangleq m_{\hat{\mathbf{H}}^{\mathsf{H}}\hat{\mathbf{H}}}(-\alpha) - \alpha m'_{\hat{\mathbf{H}}^{\mathsf{H}}\hat{\mathbf{H}}}(-\alpha).$$

The received symbol $y_k^{(t)}$ of user k at time t is given by:

$$y_k^{(t)} = \xi \mathbf{h}_k^{\mathsf{H}} \hat{\mathbf{W}} \hat{\mathbf{h}}_k s_k^{(t)} + \xi \sum_{\substack{i=1 \\ i \neq k}}^{K} \mathbf{h}_k^{\mathsf{H}} \hat{\mathbf{W}} \hat{\mathbf{h}}_i s_i^{(t)} + \sigma w_k^{(t)}$$

where $\hat{\mathbf{W}} \triangleq (\hat{\mathbf{H}}^{\mathsf{H}}\hat{\mathbf{H}} + \alpha \mathbf{I}_N)^{-1}$ and $\hat{\mathbf{h}}_k^{\mathsf{H}}$ is the kth row of $\hat{\mathbf{H}}$. The SINR $\gamma_{\mathrm{rzf},k}$ of user k can be written under the form

$$\gamma_{\mathrm{rzf},k} = \frac{|\mathbf{h}_k^{\mathsf{H}} \hat{\mathbf{W}} \hat{\mathbf{h}}_k|^2}{\mathbf{h}_k^{\mathsf{H}} \hat{\mathbf{W}} \hat{\mathbf{H}}_{(k)}^{\mathsf{H}} \hat{\mathbf{H}}_{(k)} \hat{\mathbf{W}} \mathbf{h}_k + \frac{1}{\rho}\Psi}$$

where $\hat{\mathbf{H}}_{(k)}^{\mathsf{H}} = [\hat{\mathbf{h}}_1, \ldots, \hat{\mathbf{h}}_{k-1}, \hat{\mathbf{h}}_{k+1}, \ldots, \hat{\mathbf{h}}_K] \in \mathbb{C}^{N \times (K-1)}$. Note that this SINR expression implicitly assumes that the receiver is perfectly aware of both the vector channel $\mathbf{h}_k$ *and* the phase of $\mathbf{h}_k^{\mathsf{H}} \hat{\mathbf{W}} \hat{\mathbf{h}}_k$ which rotates the transmitted signal. This requires to assume the existence of a dedicated training sequence for the receivers.

We will proceed by successively deriving deterministic equivalent expressions for Ψ, for the signal power $|\mathbf{h}_k^{\mathsf{H}} \hat{\mathbf{W}} \hat{\mathbf{h}}_k|^2$ and for the power of the interference $\mathbf{h}_k^{\mathsf{H}} \hat{\mathbf{W}} \hat{\mathbf{H}}_{(k)}^{\mathsf{H}} \hat{\mathbf{H}}_{(k)} \hat{\mathbf{W}} \mathbf{h}_k$.

We first consider the power regularization term Ψ. From Theorem 6.12, $m_{\hat{\mathbf{H}}^{\mathsf{H}}\hat{\mathbf{H}}}(-\alpha)$ is close to $\bar{m}_{\hat{\mathbf{H}}^{\mathsf{H}}\hat{\mathbf{H}}}(-\alpha)$ given as:

$$\bar{m}_{\hat{\mathbf{H}}^{\mathsf{H}}\hat{\mathbf{H}}}(-\alpha) = \frac{1}{N} \operatorname{tr} (\alpha \mathbf{I}_N + \phi \mathbf{T})^{-1} \tag{14.11}$$

where ϕ is defined in (14.10).

Remark now, and for further purposes, that $\bar{m}$ in (14.7) is a deterministic equivalent of the Stieltjes transform $m_{\mathbf{A}}(z)$ of

$$\mathbf{A} \triangleq \hat{\mathbf{X}}^{\mathsf{H}} \mathbf{R} \hat{\mathbf{X}} + \alpha \mathbf{T}^{-1}$$

evaluated at $z = 0$, i.e.

$$m_{\mathbf{A}}(0) - \bar{m} \xrightarrow{\text{a.s.}} 0. \tag{14.12}$$

Note that it is uncommon to evaluate Stieltjes transforms at $z = 0$. This is valid here, since we assumed in (14.3) that $1/t_N > 1/a_+$ and then the smallest eigenvalue of $\mathbf{A}$ is strictly greater than $1/(2a_+) > 0$, uniformly on N. Therefore, $m_{\mathbf{A}}(0)$ is well defined.

Since the deterministic equivalent of the Stieltjes transform of $\hat{\mathbf{H}}^{\mathsf{H}}\hat{\mathbf{H}}$ is itself the Stieltjes transform of a probability distribution, it is an analytic function outside the real positive half-line. The dominated convergence theorem, Theorem 6.3, ensures that the derivative $\bar{m}'_{\hat{\mathbf{H}}^{\mathsf{H}}\hat{\mathbf{H}}}(z)$ of $\bar{m}_{\hat{\mathbf{H}}^{\mathsf{H}}\hat{\mathbf{H}}}(z)$ is a deterministic equivalent of

$m'_{\hat{\mathbf{H}}^{\mathsf{H}}\hat{\mathbf{H}}}(z)$, i.e.

$$m'_{\hat{\mathbf{H}}^{\mathsf{H}}\hat{\mathbf{H}}}(z) - \bar{m}'_{\hat{\mathbf{H}}^{\mathsf{H}}\hat{\mathbf{H}}}(z) \xrightarrow{\text{a.s.}} 0.$$

After differentiation of (14.11) and standard algebraic manipulations, we obtain

$$\Psi - \bar{\Psi} \xrightarrow{\text{a.s.}} 0$$

as $N, K \to \infty$, where

$$\bar{\Psi} \triangleq \bar{m}_{\hat{\mathbf{H}}^{\mathsf{H}}\hat{\mathbf{H}}}(-\alpha) - \alpha \bar{m}'_{\hat{\mathbf{H}}^{\mathsf{H}}\hat{\mathbf{H}}}(-\alpha)$$

which is explicitly given by (14.8).

We now derive a deterministic equivalent for the signal power.

Applying Lemma 6.2 to $\hat{\mathbf{h}}_k^{\mathsf{H}}\hat{\mathbf{W}} = \hat{\mathbf{h}}_k^{\mathsf{H}}(\hat{\mathbf{H}}_{(k)}^{\mathsf{H}}\hat{\mathbf{H}}_{(k)} + \alpha\mathbf{I}_N + \hat{\mathbf{h}}_k\hat{\mathbf{h}}_k^{\mathsf{H}})^{-1}$, we have:

$$\hat{\mathbf{h}}_k^{\mathsf{H}}\hat{\mathbf{W}}\mathbf{h}_k = \frac{\hat{\mathbf{h}}_k^{\mathsf{H}}\left(\hat{\mathbf{H}}_{(k)}^{\mathsf{H}}\hat{\mathbf{H}}_{(k)} + \alpha\mathbf{I}_N\right)^{-1}\mathbf{h}_k}{1 + \hat{\mathbf{h}}_k^{\mathsf{H}}\left(\hat{\mathbf{H}}_{(k)}^{\mathsf{H}}\hat{\mathbf{H}}_{(k)} + \alpha\mathbf{I}_N\right)^{-1}\hat{\mathbf{h}}_k}.$$

Together with $\hat{\mathbf{h}}_k^{\mathsf{H}} = \sqrt{r_k}\left(\sqrt{1-\tau_k^2}\mathbf{x}_k^{\mathsf{H}} + \tau_k\mathbf{q}_k^{\mathsf{H}}\right)\mathbf{T}^{\frac{1}{2}}$, we obtain

$$\hat{\mathbf{h}}_k^{\mathsf{H}}\hat{\mathbf{W}}\mathbf{h}_k = \frac{\sqrt{1-\tau_k^2}\, r_k\mathbf{x}_k^{\mathsf{H}}\mathbf{A}_{(k)}^{-1}\mathbf{x}_k}{1 + r_k\hat{\mathbf{x}}_k^{\mathsf{H}}\mathbf{A}_{(k)}^{-1}\hat{\mathbf{x}}_k} + \frac{\tau_k r_k\mathbf{q}_k^{\mathsf{H}}\mathbf{A}_{(k)}^{-1}\mathbf{x}_k}{1 + r_k\hat{\mathbf{x}}_k^{\mathsf{H}}\mathbf{A}_{(k)}^{-1}\hat{\mathbf{x}}_k}$$

with $\mathbf{A}_{(k)} = \hat{\mathbf{X}}_{(k)}^{\mathsf{H}}\mathbf{R}_{(k)}\hat{\mathbf{X}}_{(k)} + \alpha\mathbf{T}^{-1}$ for $\hat{\mathbf{X}}_{(k)}^{\mathsf{H}} = [\hat{\mathbf{x}}_1, \ldots, \hat{\mathbf{x}}_{k-1}, \hat{\mathbf{x}}_{k+1}, \ldots, \hat{\mathbf{x}}_K]$, $\hat{\mathbf{x}}_n$ being the nth row of $\hat{\mathbf{X}}$, and $\mathbf{R}_{(k)} = \operatorname{diag}(r_1, \ldots, r_{k-1}, r_{k+1} \ldots r_K)$. Since both $\hat{\mathbf{x}}_k$ and $\mathbf{x}_k$ have i.i.d. entries of variance $1/N$ and are independent of $\mathbf{A}_{(k)}$, while $\mathbf{A}_{(k)}$ has uniformly bounded spectral norm since $t_1 > a_-$, we invoke the trace lemma, Theorem 3.4, and obtain

$$\mathbf{x}_k^{\mathsf{H}}\mathbf{A}_{(k)}^{-1}\mathbf{x}_k - \frac{1}{N}\operatorname{tr}\mathbf{A}_{(k)}^{-1} \xrightarrow{\text{a.s.}} 0$$
$$\hat{\mathbf{x}}_k^{\mathsf{H}}\mathbf{A}_{(k)}^{-1}\hat{\mathbf{x}}_k - \frac{1}{N}\operatorname{tr}\mathbf{A}_{(k)}^{-1} \xrightarrow{\text{a.s.}} 0.$$

Similarly, as $\mathbf{q}_k$ and $\mathbf{x}_k$ are independent, from Theorem 3.7

$$\mathbf{q}_k^{\mathsf{H}}\mathbf{A}_{(k)}^{-1}\mathbf{x}_k \xrightarrow{\text{a.s.}} 0.$$

Consequently, since $(1 + r_k\hat{\mathbf{x}}_k^{\mathsf{H}}\mathbf{A}_{(k)}^{-1}\hat{\mathbf{x}}_k)$ is bounded away from zero, we obtain

$$\hat{\mathbf{h}}_k^{\mathsf{H}}\hat{\mathbf{W}}\mathbf{h}_k - \sqrt{1-\tau_k^2}\frac{r_k\frac{1}{N}\operatorname{tr}\mathbf{A}_{(k)}^{-1}}{1 + r_k\frac{1}{N}\operatorname{tr}\mathbf{A}_{(k)}^{-1}} \xrightarrow{\text{a.s.}} 0. \tag{14.13}$$

To move from $\mathbf{A}_{(k)}$ to $\mathbf{A}$ in the previous equation, we will invoke the rank-1 perturbation lemma, Theorem 3.9. First, rewrite $\mathbf{A}_{(k)}^{-1}$ as

$$\mathbf{A}_{(k)}^{-1} = \left(\hat{\mathbf{X}}_{(k)}^{\mathsf{H}}\mathbf{R}_{(k)}\hat{\mathbf{X}}_{(k)} + \alpha\mathbf{T}^{-1}\right)^{-1}$$
$$= \left(\left[\hat{\mathbf{X}}_{(k)}^{\mathsf{H}}\mathbf{R}_{(k)}\hat{\mathbf{X}}_{(k)} + \alpha\mathbf{T}^{-1} - \alpha\frac{1}{2a_+}\mathbf{I}_N\right] + \alpha\frac{1}{2a_+}\mathbf{I}_N\right)^{-1}.$$

Since $1/t_N > 1/a_+$ uniformly on N, notice that the matrix in brackets on the right-hand side is still non-negative definite. Thus we can apply the rank-1 perturbation lemma to this matrix and the scalar $\alpha\frac{1}{2a_+} > 0$, and we obtain

$$\frac{1}{N}\operatorname{tr}\left[\left(\hat{\mathbf{X}}_{(k)}^{\mathsf{H}}\mathbf{R}_{(k)}\hat{\mathbf{X}}_{(k)} + \alpha\mathbf{T}^{-1}\right)^{-1}\right]$$
$$-\frac{1}{N}\operatorname{tr}\left[\left(\hat{\mathbf{X}}_{(k)}^{\mathsf{H}}\mathbf{R}_{(k)}\hat{\mathbf{X}}_{(k)} + r_k\hat{\mathbf{x}}_k\hat{\mathbf{x}}_k^{\mathsf{H}} + \alpha\mathbf{T}^{-1}\right)^{-1}\right] \xrightarrow{\text{a.s.}} 0.$$

Therefore

$$\frac{1}{N}\operatorname{tr}\mathbf{A}_{(k)}^{-1} - \frac{1}{N}\operatorname{tr}\mathbf{A}^{-1} \to 0 \tag{14.14}$$

where we remind that $\mathbf{A} = \hat{\mathbf{X}}^{\mathsf{H}}\mathbf{R}\hat{\mathbf{X}} + \alpha\mathbf{T}^{-1}$.

Thus, (14.14) and (14.12) together imply

$$\frac{1}{N}\operatorname{tr}\mathbf{A}_{(k)}^{-1} - \bar{m} \xrightarrow{\text{a.s.}} 0. \tag{14.15}$$

Finally, (14.13) takes the form

$$\hat{\mathbf{h}}_k^{\mathsf{H}}\hat{\mathbf{W}}\mathbf{h}_k - \sqrt{1-\tau_k^2}\frac{r_k\bar{m}}{1+r_k\bar{m}} \xrightarrow{\text{a.s.}} 0.$$

This establishes a deterministic equivalent for the square root of the signal power, which directly gives a deterministic equivalent for the signal power by taking the square of each term in the difference. Alternatively, similar to the more elaborated proofs of, e.g. Theorem 6.1, we could have worked directly on successive order moments of the difference between the signal power and its deterministic equivalent in order to apply the Markov inequality and the Borel–Cantelli lemma. A convenient result in this approach, that can come in handy for similar calculus, is given as follows [Hoydis *et al.*, 2010, Lemma 3].

Theorem 14.2. *Let p be an integer, $\mathbf{x}_1, \mathbf{x}_2, \ldots$, with $\mathbf{x}_N \in \mathbb{C}^N$, be a sequence of random vectors with i.i.d. entries of zero mean, variance $1/N$, and $12(p-1)$ moment of order $O(1/N^{6(p-1)})$, and let $\mathbf{A}_1, \mathbf{A}_2, \ldots$, with $\mathbf{A}_N \in \mathbb{C}^{N\times N}$, be matrices independent of $\mathbf{x}_N$. Then there exists C_p, a constant function of p only such that*

$$\mathrm{E}\left[\left|(\mathbf{x}_N^{\mathsf{H}}\mathbf{A}\mathbf{x}_N)^p - \left(\frac{1}{N}\operatorname{tr}\mathbf{A}_N\right)^p\right|^3\right] \le \frac{1}{N^{\frac{3}{2}}}C_p\|\mathbf{A}_N\|^{3p}.$$

This implies in particular that, if $\|\mathbf{A}_N\|$ is uniformly bounded

$$(\mathbf{x}_N^{\mathsf{H}}\mathbf{A}\mathbf{x}_N)^p - \left(\frac{1}{N}\operatorname{tr}\mathbf{A}_N\right)^p \xrightarrow{\text{a.s.}} 0.$$

We finally address the more involved question of the interference power. With $\hat{\mathbf{W}} = \mathbf{T}^{-\frac{1}{2}}\mathbf{A}^{-1}\mathbf{T}^{-\frac{1}{2}}$, the interference power can be written as

$$\mathbf{h}_k^{\mathsf{H}}\hat{\mathbf{W}}\hat{\mathbf{H}}_{(k)}^{\mathsf{H}}\hat{\mathbf{H}}_{(k)}\hat{\mathbf{W}}\mathbf{h}_k = r_k\mathbf{x}_k^{\mathsf{H}}\mathbf{A}^{-1}\hat{\mathbf{X}}_{(k)}^{\mathsf{H}}\mathbf{R}_{(k)}\hat{\mathbf{X}}_{(k)}\mathbf{A}^{-1}\mathbf{x}_k. \tag{14.16}$$

Denote $c_0 = (1-\tau_k^2)r_k$, $c_1 = \tau_k^2 r_k$ and $c_2 = \tau_k\sqrt{1-\tau_k^2}r_k$, then:

$$\mathbf{A} = \mathbf{A}_{(k)} + c_0\mathbf{x}_k\mathbf{x}_k^{\mathsf{H}} + c_1\mathbf{q}_k\mathbf{q}_k^{\mathsf{H}} + c_2\mathbf{x}_k\mathbf{q}_k^{\mathsf{H}} + c_2\mathbf{q}_k\mathbf{x}_k^{\mathsf{H}}.$$

In order to eliminate the dependence between $\mathbf{x}_k$ and $\mathbf{A}$ in (14.16), we rewrite (14.16) as

$$\begin{aligned}
&\mathbf{h}_k^{\mathsf{H}}\hat{\mathbf{W}}\hat{\mathbf{H}}_{(k)}^{\mathsf{H}}\hat{\mathbf{H}}_{(k)}\hat{\mathbf{W}}\mathbf{h}_k \\
&= r_k\mathbf{x}_k^{\mathsf{H}}\mathbf{A}_{(k)}^{-1}\hat{\mathbf{X}}_{(k)}^{\mathsf{H}}\mathbf{R}_{(k)}\hat{\mathbf{X}}_{(k)}\mathbf{A}^{-1}\mathbf{x}_k + r_k\mathbf{x}_k^{\mathsf{H}}\left[\mathbf{A}^{-1} - \mathbf{A}_{(k)}^{-1}\right]\hat{\mathbf{X}}_{(k)}^{\mathsf{H}}\mathbf{R}_{(k)}\hat{\mathbf{X}}_{(k)}\mathbf{A}^{-1}\mathbf{x}_k.
\end{aligned} \tag{14.17}$$

Applying the resolvent identity, Lemma 6.1, to the term in brackets in (14.17) and together with $\mathbf{A}_{(k)}^{-1}\hat{\mathbf{X}}_{(k)}^{\mathsf{H}}\mathbf{R}_{(k)}\hat{\mathbf{X}}_{(k)} = \mathbf{I}_N - \alpha\mathbf{A}_{(k)}^{-1}\mathbf{T}^{-1}$, (14.17) takes the form

$$\begin{aligned}
\mathbf{h}_k^{\mathsf{H}}\hat{\mathbf{W}}\hat{\mathbf{H}}_{(k)}^{\mathsf{H}}\hat{\mathbf{H}}_{(k)}\hat{\mathbf{W}}\mathbf{h}_k &= r_k\mathbf{x}_k^{\mathsf{H}}\mathbf{A}^{-1}\mathbf{x}_k - \alpha r_k\mathbf{x}_k^{\mathsf{H}}\mathbf{A}_{(k)}^{-1}\mathbf{T}^{-1}\mathbf{A}^{-1}\mathbf{x}_k \\
&\quad - c_0 r_k\mathbf{x}_k^{\mathsf{H}}\mathbf{A}^{-1}\mathbf{x}_k\left(\mathbf{x}_k^{\mathsf{H}}\mathbf{A}^{-1}\mathbf{x}_k - \alpha\mathbf{x}_k^{\mathsf{H}}\mathbf{A}_{(k)}^{-1}\mathbf{T}^{-1}\mathbf{A}^{-1}\mathbf{x}_k\right) \\
&\quad - c_1 r_k\mathbf{x}_k^{\mathsf{H}}\mathbf{A}^{-1}\mathbf{q}_k\left(\mathbf{q}_k^{\mathsf{H}}\mathbf{A}^{-1}\mathbf{x}_k - \alpha\mathbf{q}_k^{\mathsf{H}}\mathbf{A}_{(k)}^{-1}\mathbf{T}^{-1}\mathbf{A}^{-1}\mathbf{x}_k\right) \\
&\quad - c_2 r_k\mathbf{x}_k^{\mathsf{H}}\mathbf{A}^{-1}\mathbf{x}_k\left(\mathbf{q}_k^{\mathsf{H}}\mathbf{A}^{-1}\mathbf{x}_k - \alpha\mathbf{q}_k^{\mathsf{H}}\mathbf{A}_{(k)}^{-1}\mathbf{T}^{-1}\mathbf{A}^{-1}\mathbf{x}_k\right) \\
&\quad - c_2 r_k\mathbf{x}_k^{\mathsf{H}}\mathbf{A}^{-1}\mathbf{q}_k\left(\mathbf{x}_k^{\mathsf{H}}\mathbf{A}^{-1}\mathbf{x}_k - \alpha\mathbf{x}_k^{\mathsf{H}}\mathbf{A}_{(k)}^{-1}\mathbf{T}^{-1}\mathbf{A}^{-1}\mathbf{x}_k\right).
\end{aligned} \tag{14.18}$$

To find a deterministic equivalent for all of all the terms in (14.18), we need the following lemma, which is an extension of Theorem 3.4.

Lemma 14.1. *Let $\mathbf{U}_N, \mathbf{V}_N \in \mathbb{C}^{N\times N}$ be invertible and of uniformly bounded spectral norm. Let $\mathbf{x}_N, \mathbf{y}_N \in \mathbb{C}^N$ have i.i.d. complex entries of zero mean, variance $1/N$, and finite eighth order moment and be mutually independent as well as independent of $\mathbf{U}_N, \mathbf{V}_N$. Define $c_0, c_1, c_2 > 0$ such that $c_0c_1 - c_2^2 \geq 0$ and let $u \triangleq \frac{1}{N}\operatorname{tr}\mathbf{V}_N^{-1}$ and $u' \triangleq \frac{1}{N}\operatorname{tr}\mathbf{U}_N\mathbf{V}_N^{-1}$. Then we have, as $N \to \infty$*

$$\begin{aligned}
&\mathbf{x}_N^{\mathsf{H}}\mathbf{U}_N\left(\mathbf{V}_N + c_0\mathbf{x}_N\mathbf{x}_N^{\mathsf{H}} + c_1\mathbf{y}_N\mathbf{y}_N^{\mathsf{H}} + c_2\mathbf{x}_N\mathbf{y}_N^{\mathsf{H}} + c_2\mathbf{y}_N\mathbf{x}_N^{\mathsf{H}}\right)^{-1}\mathbf{x}_N \\
&\quad - \frac{u'(1+c_1u)}{(c_0c_1 - c_2^2)u^2 + (c_0+c_1)u + 1} \xrightarrow{\text{a.s.}} 0.
\end{aligned}$$

Furthermore

$$\mathbf{x}_N^{\mathsf{H}}\mathbf{U}_N\left(\mathbf{V}_N + c_0\mathbf{x}_N\mathbf{x}_N^{\mathsf{H}} + c_1\mathbf{y}_N\mathbf{y}_N^{\mathsf{H}} + c_2\mathbf{x}_N\mathbf{y}_N^{\mathsf{H}} + c_2\mathbf{y}_N\mathbf{x}_N^{\mathsf{H}}\right)^{-1}\mathbf{y}_N - \frac{-c_2uu'}{(c_0c_1 - c_2^2)u^2 + (c_0+c_1)u + 1} \xrightarrow{\text{a.s.}} 0.$$

The proof of Lemma 14.1 is just a matter of algebraic considerations, fully detailed in [Wagner *et al.*, 2011].

Denote $u \triangleq \frac{1}{N}\operatorname{tr}\mathbf{A}_{(k)}^{-1}$ and $u' \triangleq \frac{1}{N}\operatorname{tr}\mathbf{T}^{-1}\mathbf{A}_{(k)}^{-2}$. Note that, in the present scenario, $c_0c_1 = c_2^2$ and thus $(c_0c_1 - c_2^2)u^2 + (c_0+c_1)u + 1$ reduces to $1 + r_ku$. Applying Lemma 14.1 to each of the terms in (14.18), we obtain

$$\mathbf{h}_k^{\mathsf{H}}\hat{\mathbf{W}}\hat{\mathbf{H}}_{(k)}^{\mathsf{H}}\hat{\mathbf{H}}_{(k)}\hat{\mathbf{W}}\mathbf{h}_k - \left[\frac{r_k(1+c_1u)(u-\alpha u')}{1+r_ku} - \frac{r_kc_0u(u-\alpha u')}{(1+r_ku)^2}\right] \xrightarrow{\text{a.s.}} 0 \tag{14.19}$$

where the first term in brackets stems from the first line in (14.18) and the second term in brackets arises from the last four lines in (14.18). Replacing c_0 and c_1 by $(1-\tau_k^2)r_k$ and $\tau_k^2 r_k$, respectively, and after some algebraic manipulation, (14.19) takes the form

$$\mathbf{h}_k^{\mathsf{H}}\hat{\mathbf{W}}\hat{\mathbf{H}}_{(k)}^{\mathsf{H}}\hat{\mathbf{H}}_{(k)}\hat{\mathbf{W}}\mathbf{h}_k - \frac{r_k(u-\alpha u')\left[1 - \tau_k^2\left(1 - (1+r_ku)^2\right)\right]}{(1+r_ku)^2} \xrightarrow{\text{a.s.}} 0.$$

Since a rank-1 perturbation has no impact on $\frac{1}{N}\operatorname{tr}\mathbf{A}^{-1}$ for $N \to \infty$, we also have

$$u - m_{\mathbf{A}}(0) \xrightarrow{\text{a.s.}} 0$$
$$u' - m_{\mathbf{A}^2,\mathbf{T}^{-1}}(0) \xrightarrow{\text{a.s.}} 0.$$

Note again that $m_{\mathbf{A}^2,\mathbf{T}^{-1}}(0)$ is well defined since $t_1 > a_-$ is bounded away from zero. Denote

$$\Upsilon = m_{\mathbf{A}}(0) - \alpha m'_{\mathbf{A},\mathbf{T}^{-1}}(0)$$

and observe that

$$m_{\mathbf{A}^2,\mathbf{T}^{-1}}(0) = m'_{\mathbf{A}\mathbf{T}^{-1}}(0).$$

Furthermore, we have $m_{\mathbf{A}^2,\mathbf{T}^{-1}}(0) - \bar{m}'_{\mathbf{A},\mathbf{T}^{-1}}(0) \xrightarrow{\text{a.s.}} 0$, where $\bar{m}'_{\mathbf{A},\mathbf{T}^{-1}}(0)$ is obtained from Theorem 6.12. Similar to the derivations of $\bar{\Psi}$, we then obtain $\Upsilon - \bar{\Upsilon} \xrightarrow{\text{a.s.}} 0$, as N, K grow large with $\bar{\Upsilon}$ given by:

$$\bar{\Upsilon} = \bar{m}_{\mathbf{A}}(0) - \alpha\bar{m}'_{\mathbf{A},\mathbf{T}^{-1}}(0)$$

whose explicit form is given by (14.9). Substituting u and u' by their respective deterministic equivalent expressions $\bar{m}$ and $\bar{m}'_{\mathbf{A}\mathbf{T}^{-1}}(0)$, respectively, we obtain

$$\mathbf{h}_k^{\mathsf{H}}\hat{\mathbf{W}}\hat{\mathbf{H}}_{(k)}^{\mathsf{H}}\hat{\mathbf{H}}_{(k)}\hat{\mathbf{W}}\mathbf{h}_k - \frac{r_k\bar{\Upsilon}(1 - \tau_k^2[1-(1+r_k\bar{m})^2])}{(1+r_k\bar{m})^2} \xrightarrow{\text{a.s.}} 0.$$

This completes the first part of the proof. Now, the convergence of the sequence $\phi_1, \phi_2, \ldots$ to ϕ is a direct consequence of a trivial extension of the last statement in Theorem 6.12.

The next question we address is the precise characterization of the sum rate maximizing α parameter, call it $\alpha^\star$. Indeed, contrary to the MMSE precoding approach, that would consist in setting $\alpha = \sigma^2$, due to inter-user interference, it is not clear which α parameter is optimal in the sum rate viewpoint. Since this is a difficult problem, instead we determine here $\bar{\alpha}^\star$, the parameter α which maximizes the *deterministic equivalent* of the sum rate. This parameter can be shown to asymptotically well approximate the optimal sum rate maximizing α. As it turns out, though, $\bar{\alpha}^\star$ cannot be expressed explicitly and is the only solution to an implicit equation of the system parameters. However, in more symmetric scenarios, $\bar{\alpha}^\star$ takes an explicit form. This is discussed subsequently.

14.1.3 Optimal regularized zero-forcing precoding

Define $\bar{R}_{\mathrm{rzf}}$ to be

$$\bar{R}_{\mathrm{rzf}} = \sum_{k=1}^{K} \log_2\left(1 + \bar{\gamma}_{\mathrm{rzf},k}\right)$$

where the parameter α, implicitly present in the term $\bar{\gamma}_{\mathrm{rzf},k}$ is chosen to be the positive real that maximizes $\bar{R}_{\mathrm{rzf}}$, i.e. $\alpha = \bar{\alpha}^\star$, with $\bar{\alpha}^\star$ defined as

$$\bar{\alpha}^\star = \arg\max_{\alpha>0} \bar{R}_{\mathrm{rzf}}. \tag{14.20}$$

Under the channel model with imperfect CSIT (14.5), $\bar{\alpha}^\star$ is a positive solution of the implicit equation

$$\sum_{k=1}^{K} \frac{\partial \bar{\gamma}_{\mathrm{rzf},k}}{\partial \alpha} \frac{1}{1 + \bar{\gamma}_{\mathrm{rzf},k}} = 0.$$

The implicit equation above is not convex in α so that the solution needs to be computed via a one-dimensional line search. The RZF precoder with optimal regularization parameter $\bar{\alpha}^\star$ will be called RZF-O in the remainder of this section. For homogeneous networks, i.e. $\mathbf{R} = \mathbf{I}_K$, the user channels $\mathbf{h}_k$ are statistically equivalent and it is reasonable to assume that the distortions τ_k^2 of the CSIT $\hat{\mathbf{h}}_k$ are identical for all users, i.e. $\tau_1 = \ldots = \tau_K \triangleq \tau$. Under the additional assumption of uncorrelated transmit antennas, i.e. $\mathbf{T} = \mathbf{I}_N$, the solution to (14.20) has a closed-form and leads to the asymptotically optimal precoder given by:

$$\bar{\alpha}^\star = \left(\frac{1 + \tau^2 \rho}{1 - \tau^2}\right) \frac{K}{N} \frac{1}{\rho}.$$

To prove this, observe that $\bar{\Psi} = \bar{\Upsilon}$ and $\bar{m} = \bar{m}_{\hat{\mathbf{X}}^{\mathsf{H}}\hat{\mathbf{X}}}(-\alpha)$ is the Stieltjes transform of the Marčenko–Pastur law at $z = -\alpha$. The SINR and its derivative then take an explicit form, whose extrema can be computed.

Note that, for $\tau > 0$ and at asymptotically high SNR, the above regularization term $\bar{\alpha}^\star$ satisfies

$$\lim_{\rho\to\infty} \bar{\alpha}^\star = \frac{\tau^2}{1-\tau^2}\frac{K}{N}.$$

Thus, for asymptotically high SNR, RZF-O does *not* converge to the ZF precoding, in the sense that $\bar{\alpha}^\star$ does not converge to zero. This is to be opposed to the case when perfect channel state information at the transmitter is assumed where $\alpha^\star$ tends to zero.

Under this scenario and with $\alpha = \bar{\alpha}^\star$, the SINR $\bar{\gamma}_{\mathrm{rzf},k}$ is now independent of the index k and takes now the simplified form

$$\bar{\gamma}_{\mathrm{rzf},k} = \frac{\omega}{2}\rho\left(\frac{N}{K}-1\right) + \frac{\chi}{2} - \frac{1}{2}$$

where ω and χ are given by:

$$\omega = \frac{1-\tau^2}{1+\tau^2\rho}$$

$$\chi = \sqrt{\left(\frac{N}{K}-1\right)^2 \omega^2\rho^2 + 2\left(\frac{N}{K}+1\right)\omega\rho + 1}.$$

Further comments on the above are provided in [Wagner *et al.*, 2011]. We now move to ZF precoding, whose performance can be studied by having the regularization parameter of RZF tend to zero.

14.1.4 Zero-forcing precoding

For $\alpha = 0$, the RZF precoding matrix reduces to the ZF precoding matrix $\mathbf{G}_{\mathrm{zf}}$, which we recall is defined as

$$\mathbf{G}_{\mathrm{zf}} = \frac{\xi}{\sqrt{N}}\hat{\mathbf{H}}^{\mathsf{H}}\left(\hat{\mathbf{H}}\hat{\mathbf{H}}^{\mathsf{H}}\right)^{-1}$$

where ξ is a scaling factor to fulfill the power constraint (14.1).

To derive a deterministic equivalent of the SINR of ZF precoding, we cannot assume $N = K$ and apply the same techniques as for RZF, since by removing a row of $\hat{\mathbf{H}}$, the matrix $\hat{\mathbf{H}}^{\mathsf{H}}\hat{\mathbf{H}}$ becomes singular. Therefore, we adopt a different strategy and derive a deterministic equivalent $\bar{\gamma}_{\mathrm{zf},k}$ for the SINR of ZF for user k under the additional assumption that, uniformly on K, N, $\frac{N}{K} > 1 + \varepsilon$, for some $\varepsilon > 0$. The approximated SINR $\bar{\gamma}_{\mathrm{zf},k}$ is then given by:

$$\bar{\gamma}_{\mathrm{zf},k} = \lim_{\alpha\to 0} \bar{\gamma}_{\mathrm{rzf},k}.$$

The result is summarized in the following theorem.

Theorem 14.3. *Let $N/K > 1+\varepsilon$ for some $\varepsilon > 0$ and $\bar{\gamma}_{\mathrm{zf},k}$ be the SINR of user k for the ZF precoder. Then*

$$\gamma_{\mathrm{zf},k} - \bar{\gamma}_{\mathrm{zf},k} \xrightarrow{\text{a.s.}} 0$$

as $N, K \to \infty$, with uniformly bounded ratio, where $\bar{\gamma}_{\mathrm{zf},k}$ is given by:

$$\bar{\gamma}_{\mathrm{zf},k} = \frac{1-\tau_k^2}{r_k \tau_k^2 \bar{\Upsilon} + \frac{\bar{\Psi}}{\rho}} \tag{14.21}$$

with

$$\begin{aligned} \bar{\Psi} &= \frac{1}{\phi}\frac{1}{N}\operatorname{tr}\mathbf{R}^{-1} \\ \bar{\Upsilon} &= \frac{\psi}{\frac{N}{K}\phi^2 - \psi}\frac{1}{K}\operatorname{tr}\mathbf{R}^{-1} \\ \psi &= \frac{1}{N}\operatorname{tr}\mathbf{T}^2\left(\mathbf{I}_N + \frac{K}{N}\frac{1}{\phi}\mathbf{T}\right)^{-2} \end{aligned} \tag{14.22}$$

where ϕ is the unique solution of

$$\phi = \frac{1}{N}\operatorname{tr}\mathbf{T}\left(\mathbf{I}_N + \frac{K}{N}\frac{1}{\phi}\mathbf{T}\right)^{-1}. \tag{14.23}$$

Moreover, $\phi = \lim_{k\to\infty}\phi_k$, where $\phi_0 = 1$ and, for $k \geq 1$

$$\phi_k = \frac{1}{N}\operatorname{tr}\mathbf{T}\left(\mathbf{I}_N + \frac{K}{N}\frac{1}{\phi_{k-1}}\mathbf{T}\right)^{-1}.$$

Note, that by Jensen's inequality $\psi \geq \phi^2$ with equality if $\mathbf{T} = \mathbf{I}_N$. In the simpler case when $\mathbf{T} = \mathbf{I}_N$, $\mathbf{R} = \mathbf{I}_K$ and $\tau_1 = \ldots = \tau_K \triangleq \tau$, we have the following deterministic equivalent for the SINR under ZF precoding.

$$\bar{\gamma}_{\mathrm{zf},k} = \frac{1-\tau^2}{\tau^2 + \frac{1}{\rho}}\left(\frac{N}{K} - 1\right).$$

We hereafter prove Theorem 14.3.

Recall the terms in the SINR of RZF that depend on α, i.e. $m_{\mathbf{A}}$, Ψ, and Υ

$$\begin{aligned} m_{\mathbf{A}} &= \frac{1}{N}\operatorname{tr}\mathbf{TF} \\ \Psi &= \frac{1}{N}\operatorname{tr}\left(\mathbf{F} - \alpha\mathbf{F}^2\right) \\ \Upsilon &= \frac{1}{N}\operatorname{tr}\mathbf{T}\left(\mathbf{F} - \alpha\mathbf{F}^2\right) \end{aligned}$$

where we introduced the notation

$$\mathbf{F} = \left(\hat{\mathbf{H}}^{\mathsf{H}}\hat{\mathbf{H}} + \alpha\mathbf{I}_N\right)^{-1}.$$

In order to take the limit $\alpha \to 0$ of Ψ and Υ, we apply the resolvent lemma, Lemma 6.1, to $\mathbf{F}$ and we obtain

$$\mathbf{F} - \alpha\mathbf{F}^2 = \hat{\mathbf{H}}^{\mathsf{H}}\left(\hat{\mathbf{H}}\hat{\mathbf{H}}^{\mathsf{H}} + \alpha\mathbf{I}_N\right)^{-2}\hat{\mathbf{H}}.$$

Since $\hat{\mathbf{H}}\hat{\mathbf{H}}^{\mathsf{H}}$ is non-singular with probability one for all large N, K, we can take the limit $\alpha \to 0$ of Ψ and Υ in the RZF case, for all such $\hat{\mathbf{H}}\hat{\mathbf{H}}^{\mathsf{H}}$ and large enough N, K. This redefines Ψ and Υ as

$$\Psi = \frac{1}{N}\operatorname{tr}\left(\hat{\mathbf{H}}\hat{\mathbf{H}}^{\mathsf{H}}\right)^{-1}$$
$$\Upsilon = \frac{1}{N}\operatorname{tr}\hat{\mathbf{H}}\mathbf{T}\hat{\mathbf{H}}^{\mathsf{H}}\left(\hat{\mathbf{H}}\hat{\mathbf{H}}^{\mathsf{H}}\right)^{-2}.$$

Note that it is necessary to assume that $N/K > 1+\varepsilon$ uniformly for some $\varepsilon > 0$ to ensure that the maximum eigenvalue of matrix $(\hat{\mathbf{H}}\hat{\mathbf{H}}^{\mathsf{H}})^{-1}$ is uniformly bounded for all large N. Since $m_{\mathbf{A}}$ grows with α decreasing to zero as $O(1/\alpha)$, we have:

$$\gamma_{\mathrm{zf},k} = \lim_{\alpha\to 0}\gamma_{\mathrm{rzf},k} = \frac{1-\tau_k^2}{r_k\tau_k^2\Upsilon + \frac{\Psi}{\rho}}. \tag{14.24}$$

Now we derive deterministic equivalents $\bar{\Psi}$ and $\bar{\Upsilon}$ for Ψ and Υ, respectively. Theorem 6.12 can be directly applied to find $\bar{\Psi}$ as

$$\bar{\Psi} = \frac{1}{N}\frac{1}{\phi}\operatorname{tr}\mathbf{R}^{-1}$$

where ϕ is defined in (14.23).

To determine $\bar{\Upsilon}$, note that we can diagonalize $\mathbf{T}$ as $\mathbf{T} = \mathbf{U}\operatorname{diag}(t_1,\ldots,t_N)\mathbf{U}^{\mathsf{H}}$, where $\mathbf{U}$ is a unitary matrix, and still have i.i.d. elements in the kth column $\hat{\mathbf{x}}'_k$ of $\hat{\mathbf{X}}\mathbf{U}$. Denoting $\mathbf{C} = \hat{\mathbf{H}}\hat{\mathbf{H}}^{\mathsf{H}}$, $\mathbf{C}_{(k)} = \hat{\mathbf{H}}_{[k]}\hat{\mathbf{H}}_{[k]}^{\mathsf{H}} - t_k\mathbf{R}^{\frac{1}{2}}\hat{\mathbf{x}}'_k\hat{\mathbf{x}}'^{\mathsf{H}}_k\mathbf{R}^{\frac{1}{2}}$ and applying the usual matrix inversion lemma twice, we can write

$$\Upsilon = \frac{1}{N}\sum_{k=1}^{N} t_k^2 \frac{\hat{\mathbf{x}}'^{\mathsf{H}}_k\mathbf{R}^{\frac{1}{2}}\mathbf{C}_{(k)}^{-2}\mathbf{R}^{\frac{1}{2}}\hat{\mathbf{x}}'_k}{(1+t_k\hat{\mathbf{x}}'^{\mathsf{H}}_k\mathbf{R}^{\frac{1}{2}}\mathbf{C}_{(k)}^{-1}\mathbf{R}^{\frac{1}{2}}\hat{\mathbf{x}}'_k)^2}.$$

Notice here that $\mathbf{C}_{(k)}^{-1}$ does not have uniformly bounded spectral norm. The trace lemma, Theorem 3.4, can therefore not be applied straightforwardly here. However, since the ratio N/K is taken uniformly greater than $1+\varepsilon$ for some $\varepsilon > 0$, $\mathbf{C}_{(k)}^{-1}$ has almost surely bounded spectral norm for all large N (this unfolds from Theorem 7.1 and from standard matrix norm inequalities, reminding us that $0 < a < t_k, r_k < b < \infty$). This is in fact sufficient for a similar trace lemma to hold.

Lemma 14.2. *Let $\mathbf{A}_1, \mathbf{A}_2, \ldots$, with $\mathbf{A}_N \in \mathbb{C}^{N\times N}$, be a series of random matrices generated by the probability space $(\Omega, \mathcal{F}, P)$ such that, for $\omega \in A \subset \Omega$, with $P(A) = 1$, $\|\mathbf{A}_N(\omega)\| < K(\omega) < \infty$, uniformly on N. Let $\mathbf{x}_1, \mathbf{x}_2, \ldots$ be random vectors of i.i.d. entries such that the entries of $\mathbf{x}_N \in \mathbb{C}^N$ have zero mean,* *variance* $1/N$, *and eighth order moment of order* $O(1/N^4)$, *independent of* $\mathbf{A}_N$. *Then*

$$\mathbf{x}_N^{\mathsf{H}}\mathbf{A}_N\mathbf{x}_N - \frac{1}{N}\operatorname{tr}\mathbf{A}_N \xrightarrow{\text{a.s.}} 0$$

as $N \to \infty$.

Proof. The proof unfolds from a direct application of the Tonelli theorem, Theorem 3.16. Denoting $(X, \mathcal{X}, P_X)$ the probability space that generates the series $\mathbf{x}_1, \mathbf{x}_2, \ldots$, we have that, for every $\omega \in A$ (i.e. for every realization $\mathbf{A}_1(\omega), \mathbf{A}_2(\omega), \ldots$ with $\omega \in A$), the trace lemma, Theorem 3.4, holds true. Now, from Theorem 3.16, the space B of couples $(x, \omega) \in X \times \Omega$ for which the trace lemma holds satisfies

$$\int_{X\times\Omega} 1_B(x,\omega)dP_{X\times\Omega}(x,\omega) = \int_\Omega\int_X 1_B(x,\omega)dP_X(x)dP(\omega).$$

If $\omega \in A$, then $1_B(x,\omega) = 1$ on a subset of X of probability one. The inner integral therefore equals one whenever $\omega \in A$. As for the outer integral, since $P(A) = 1$, it also equals one, and the result is proved. □

Moreover, the rank-1 perturbation lemma, Theorem 3.9, no longer ensures that the normalized trace of $\mathbf{C}_{(k)}^{-1}$ and $\mathbf{C}^{-1}$ are asymptotically equal. In fact, following the same line of argument as above, we also have a generalized rank-1 perturbation lemma, which now holds only almost surely.

Lemma 14.3. *Let* $\mathbf{A}_1, \mathbf{A}_2, \ldots$, *with* $\mathbf{A}_N \in \mathbb{C}^{N\times N}$, *be deterministic with uniformly bounded spectral norm and* $\mathbf{B}_1, \mathbf{B}_2, \ldots$, *with* $\mathbf{B}_N \in \mathbb{C}^{N\times N}$, *be random Hermitian, with eigenvalues* $\lambda_1^{\mathbf{B}_N} \leq \ldots \leq \lambda_N^{\mathbf{B}_N}$ *such that, with probability one, there exist* $\varepsilon > 0$ *for which* $\lambda_1^{\mathbf{B}_N} > \varepsilon$ *for all large* N. *Then for* $\mathbf{v} \in \mathbb{C}^N$

$$\frac{1}{N}\operatorname{tr}\mathbf{A}_N\mathbf{B}_N^{-1} - \frac{1}{N}\operatorname{tr}\mathbf{A}_N(\mathbf{B}_N + \mathbf{v}\mathbf{v}^{\mathsf{H}})^{-1} \xrightarrow{\text{a.s.}} 0$$

as $N \to \infty$, *where* $\mathbf{B}_N^{-1}$ *and* $(\mathbf{B}_N + \mathbf{v}\mathbf{v}^{\mathsf{H}})^{-1}$ *exist with probability one.*

Proof. The proof unfolds similarly as above, with some particular care to be taken. Call B the set of probability one in question and take $\omega \in B$. The smallest eigenvalue of $\mathbf{B}_N(\omega)$ is greater than $\varepsilon(\omega)$ for all large N. Therefore, for such an N, first of all $\mathbf{B}_N(\omega)$ and $\mathbf{B}_N(\omega) + \mathbf{v}\mathbf{v}^{\mathsf{H}}$ are invertible and, taking $z = -\varepsilon(\omega)/2$, we can write

$$\frac{1}{N}\operatorname{tr}\mathbf{A}_N\mathbf{B}_N(\omega)^{-1} = \frac{1}{N}\operatorname{tr}\mathbf{A}_N\left(\left[\mathbf{B}_N(\omega) - \frac{\varepsilon(\omega)}{2}\mathbf{I}_N\right] + \frac{\varepsilon(\omega)}{2}\mathbf{I}_N\right)^{-1}$$

and

$$\frac{1}{N}\operatorname{tr}\mathbf{A}_N(\mathbf{B}_N(\omega)+\mathbf{v}\mathbf{v}^{\mathsf{H}})^{-1}$$
$$=\frac{1}{N}\operatorname{tr}\mathbf{A}_N\left(\left[\mathbf{B}_N(\omega)+\mathbf{v}\mathbf{v}^{\mathsf{H}}-\frac{\varepsilon(\omega)}{2}\mathbf{I}_N\right]+\frac{\varepsilon(\omega)}{2}\mathbf{I}_N\right)^{-1}.$$

With these notations, $\mathbf{B}_N(\omega)-\frac{\varepsilon(\omega)}{2}\mathbf{I}_N$ and $\mathbf{B}_N(\omega)+\mathbf{v}\mathbf{v}^{\mathsf{H}}-\frac{\varepsilon(\omega)}{2}\mathbf{I}_N$ are still nonnegative definite for all N. Therefore, the rank-1 perturbation lemma, Theorem 3.9, can be applied for this ω. But then, from the Tonelli theorem, in the space that generates $(\mathbf{B}_1,\mathbf{B}_2,\ldots)$, the subspace where the rank-1 perturbation lemma applies has probability one, which is what needed to be proved. □

Applying the above trace lemma and rank-1 perturbation lemma, we obtain

$$\Upsilon-\frac{1}{N}\operatorname{tr}\mathbf{R}\mathbf{C}^{-2}\frac{1}{N}\sum_{k=1}^{N}\frac{t_k^2}{(1+t_k\frac{1}{N}\operatorname{tr}\mathbf{R}\mathbf{C}^{-1})^2}\xrightarrow{\text{a.s.}}0.$$

To determine a deterministic equivalent $\hat{m}_{\mathbf{C},\mathbf{\Lambda}}(0)$ for $m_{\mathbf{C},\mathbf{\Lambda}}(0)=\frac{1}{K}\operatorname{tr}\mathbf{R}\mathbf{C}^{-1}$, we apply Theorem 6.12 again (noticing that there is once more no continuity issue in point $z=0$). For $\frac{1}{K}\operatorname{tr}\mathbf{R}\mathbf{C}^{-2}$, we have:

$$\frac{1}{K}\operatorname{tr}\mathbf{R}\mathbf{C}^{-2}=m_{\mathbf{C}^2,\mathbf{R}}(z)=m'_{\mathbf{C},\mathbf{R}}(0).$$

The derivative of $\bar{m}_{\mathbf{C},\mathbf{R}}(0)$ being a deterministic equivalent of $m'_{\mathbf{C},\mathbf{R}}(0)$, we have $\bar{\Upsilon}$, given as:

$$\bar{\Upsilon}=\frac{\psi}{\frac{N}{K}\phi^2-\psi}\frac{1}{K}\operatorname{tr}\mathbf{R}^{-1}$$

where ψ and ϕ are defined in (14.22) and (14.23), respectively. Finally, we obtain (14.21) by substituting Ψ and Υ in (14.24) by their respective deterministic equivalents, which completes the proof.

The above results allow for interesting characterizations of the linearly precoded broadcast channels. Some of these are provided below.

14.1.5 Applications

An interesting characterization of the performance of RZF-O derived above for imperfect channel state information at the transmitter is to evaluate the difference ΔR between the sum rate achieved under RZF-O and the sum rate achieved when perfect channel information is assumed. For $N=K$, from the deterministic equivalents obtained above, this is close to $\Delta\bar{R}$ which, in homogeneous channel conditions, takes the following convenient form

$$\Delta\bar{R}=\log_2\left(\frac{1+\sqrt{1+4\rho}}{1+\sqrt{1+4\rho\omega}}\right)$$

with

$$\omega = \frac{1-\tau^2}{1+\tau^2\rho}.$$

An interesting question is to determine how much feedback is required in the uplink for the sum rate between the perfect CSIT and imperfect CSIT case to be no larger than $\log_2 b$. That is, how much distortion τ^2 is maximally allowed to ensure a rate gap of $\log_2 b$. Under the above simplified channel conditions, with $N = K$

$$\tau^2 = \frac{1+4\rho-\frac{\omega^2}{b^2}}{3+\frac{\omega^2}{b^2}} \frac{1}{\rho}.$$

In particular, $\tau^2 \simeq b^2 - 1$ in the large ρ regime. This further allows us to evaluate the optimal number of training bits required to maintain a given rate loss. Then, taking into account the additional rate loss incurred by the process of channel training, the optimal sum rate under the channel training constraint can be made explicit. This is further discussed in [Wagner *et al.*, 2011].

A second question of interest lies in the optimal ratio N/K between the number of transmit antennas and the number of users that maximizes the sum rate per transmit antenna. This allows us to determine a figure of merit for the efficiency of every single antenna. In the uncorrelated channel conditions above, from the deterministic equivalents, we find the optimal ratio c to be given by

$$c = \left(1 - \frac{\tau^2 + \frac{1}{\rho}}{1-\tau^2}\right)\left(1 + \frac{1}{\mathcal{W}\left(\frac{1}{e}\left[\frac{1-\tau^2}{\tau^2+\frac{1}{\rho}} - 1\right]\right)}\right)$$

with $\mathcal{W}(x)$ the Lamber-W function, defined as the solution to $\mathcal{W}(x) = xe^{\mathcal{W}(x)}$, unique for $x \in [-\frac{1}{e}, \infty)$.

In Figure 14.2, the sum rate performance of RZF and ZF under imperfect channel state information at the transmitter are depicted and compared. We assume here that $N = K$, that the users are uniformly distributed around the transmitter, and that there exists no channel correlation at the transmitter. The channel estimation distortion equals $\tau^2 = 0.1$. Observe that a significant performance gain is achieved when the imperfect channel estimation parameter is taken into account, while, as expected, the performance of both RZF assuming perfect channel information and ZF converge asymptotically to the same limit.

In Figure 14.3, the performance of ZF precoding under different channel assumptions are compared against their respective deterministic equivalents. The exact channel conditions assumed here follow from those introduced in [Wagner *et al.*, 2011] and are omitted here. We only mention that homogeneous channel conditions, i.e. with $\mathbf{T} = \mathbf{I}_N$ and $\mathbf{R} = \mathbf{I}_K$, are compared against the case when correlation at the transmitter emerges from a compact array of antennas on a limited volume, with inter-antenna spacing equal to half the transmit wavelength, and the case when users have different path losses, following

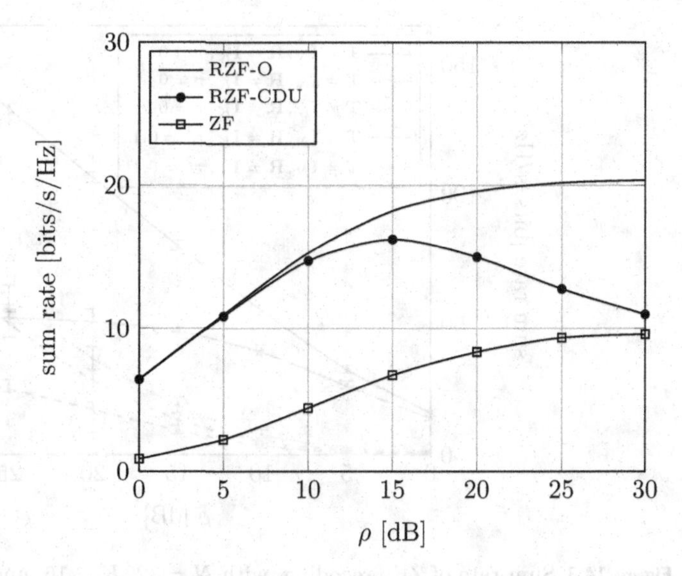

Figure 14.2 Ergodic sum rate of regularized zero-forcing with optimal α (RZF-O), α taken as if $\tau = 0$, i.e. for channel distortion unaware transmitter (RZF-CDU) and zero-forcing (ZF), $\mathbf{T} = \mathbf{I}_N$, $\mathbf{R} = \mathbf{I}_K$, $N = K$, $\tau^2 = 0.1$.

the COST231 Hata urban propagation model. We also take $\tau_1 = \ldots = \tau_K \triangleq \tau$, with $\tau = 0$ or $\tau^2 = 0.1$, alternatively. The number of users is $K = 16$ and the number of transmit antennas $N = 32$. The lines drawn are the deterministic equivalents, while the dots and error bars are the averaged sum rate evaluated from simulations and the standard deviation, respectively. Observe that the deterministic equivalents, already for these not too large system dimensions, fall within one standard deviation of the simulated sum rates and are in most situations very close approximations of the mean sum rates.

Similar considerations of optimal training time in large networks, but in the context of multi-cell uplink models are also provided in [Hoydis *et al.*, 2011d], also relying on deterministic equivalent techniques. This concludes this section on linearly precoded broadcast channels. In the subsequent section, we address the information-theoretic, although less practical, question of the characterization of the overall rate region of the dual multiple access channel.

14.2 Rate region of MIMO multiple access channels

We consider the generic model of an N-antenna access point (or base station) communicating with K users. User k, $k \in \{1, \ldots, K\}$, is equipped with n_k antennas. Contrary to the study developed in Section 14.1, we do not restrict receiver k to be equipped with a single antenna. To establish large

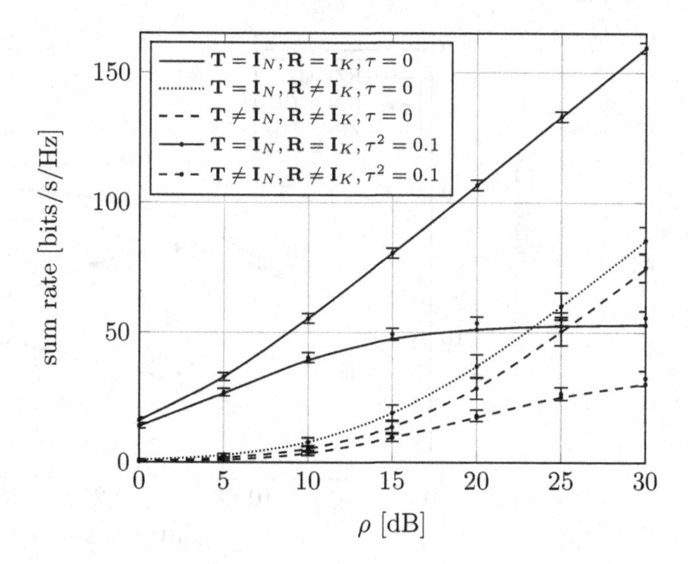

Figure 14.3 Sum rate of ZF precoding, with $N = 32$, $K = 16$, under different channel conditions. Uncorrelated transmit antennas ($\mathbf{T} = \mathbf{I}_N$) or volume limited transmit device with inter-antenna spacing of half the wavelength ($\mathbf{T} \neq \mathbf{I}_N$), equal path losses ($\mathbf{R} = \mathbf{I}_K$) or path losses based on modified COST231 Hata urban model ($\mathbf{T} \neq \mathbf{I}_K$), $\tau = 0$ or $\tau^2 = 0.1$. Simulation results are indicated by circle marks with error bars indicating one standard deviation in each direction.

dimensional matrix results, we will consider here that N and $n \triangleq \sum_{i=1}^K n_k$ are commensurable. That is, we will consider a large system analysis where:

- either a large number K of users, each equipped with few antennas, communicate with an access point, equipped with a large number N of antennas;
- either a small number K of users, each equipped with a large number of antennas, communicate with an access point, equipped also with a large number of antennas.

The channel between the base station and user k is modeled by the matrix $\mathbf{H}_k \in \mathbb{C}^{n_k \times N}$. In the previous section, where $n_k = 1$ for all k, $\mathbf{H}_k$ was denoted by $\mathbf{h}_k^{\mathsf{H}}$, the Hermitian sign being chosen for readability to avoid working with row vectors. With similar notations as in Section 14.1, the downlink channel model for user k at time t reads:

$$\mathbf{y}_k^{(t)} = \mathbf{H}_k \mathbf{z}^{(t)} + \sigma \mathbf{w}_k^{(t)}$$

where $\mathbf{z}^{(t)} \in \mathbb{C}^N$ is the transmit data vector and the receive symbols $\mathbf{y}_k^{(t)} \in \mathbb{C}^{n_k}$ and additive Gaussian noise $\mathbf{w}_k^{(t)} \in \mathbb{C}^{n_k}$ are now vector valued. Contrary to the linear precoding approach, the relation between the effectively transmitted $\mathbf{z}^{(t)}$ and the intended symbols $\mathbf{s}^{(t)} = (s_1^{(t)}, \ldots, s_K^{(t)})^{\mathsf{T}}$ is not necessarily linear. We also

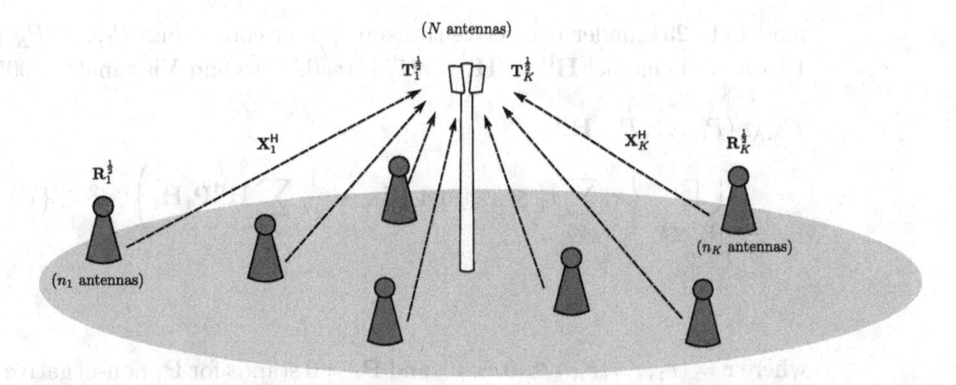

Figure 14.4 Multiple access MIMO channel, composed of K users and a base station. User k is equipped with n_k antennas, and the base station with N antennas. The channel between user k and the base station is $\mathbf{H}_k^{\mathsf{H}} = \mathbf{T}_k^{\frac{1}{2}}\mathbf{X}_k^{\mathsf{H}}\mathbf{R}_k^{\mathsf{H}}$.

denote $\mathbf{P}$ the transmit covariance matrix $\mathbf{P} = \mathrm{E}[\mathbf{z}^{(t)}\mathbf{z}^{(t)\mathsf{H}}]$, assumed independent of t, which satisfies the power constraint $\frac{1}{N}\operatorname{tr}\mathbf{P} = P$.

Denoting equivalently $\mathbf{z}_k^{(t)} \in \mathbb{C}^{n_k}$ the signal transmitted in the uplink (MAC) by user k, such that $\mathrm{E}[\mathbf{z}_k^{(t)}\mathbf{z}_k^{(t)\mathsf{H}}] = \mathbf{P}_k$, $\frac{1}{n_k}\operatorname{tr}\mathbf{P}_k \leq P_k$, $\mathbf{y}^{(t)}$ and $\mathbf{w}^{(t)}$ the signal and noise received by the base station, respectively, and assuming perfect channel reciprocity in the downlink and the uplink, we have the uplink transmission model

$$\mathbf{y}^{(t)} = \sum_{k=1}^{K} \mathbf{H}_k^{\mathsf{H}}\mathbf{z}_k^{(t)} + \mathbf{w}^{(t)}. \tag{14.25}$$

Similar to the linearly precoded case, we assume that $\mathbf{H}_k$, $k \in \{1, \ldots, K\}$, is modeled as Kronecker, i.e.

$$\mathbf{H}_k \triangleq \mathbf{R}_k^{\frac{1}{2}}\mathbf{X}_k\mathbf{T}_k^{\frac{1}{2}} \tag{14.26}$$

where $\mathbf{X}_k \in \mathbb{C}^{n_k \times N}$ has i.i.d. Gaussian entries of zero mean and variance $1/n_k$, $\mathbf{T}_k \in \mathbb{C}^{N \times N}$ is the Hermitian non-negative definite channel correlation matrix at the base station with respect to user k, and $\mathbf{R}_k \in \mathbb{C}^{n_k \times n_k}$ is the Hermitian non-negative definite channel correlation matrix at user k.

In the following, we study the MAC rate regions for quasi-static channels. We will then consider the ergodic rate region for time-varying MAC. An illustrative representation of a cellular uplink MAC channel as introduced above is provided in Figure 14.4.

14.2.1 MAC rate region in quasi-static channels

We start by assuming that the channels $\mathbf{H}_1, \ldots, \mathbf{H}_K$ are random realizations of the Kronecker channel model (14.26), considered constant over the observation period. The MIMO MAC rate region $C_{\mathrm{MAC}}(P_1, \ldots, P_K; \mathbf{H}^{\mathsf{H}})$ for the quasi-static model (14.25), under respective transmit power constraints $P_1, \ldots, P_K$ for users 1 to K and channel $\mathbf{H}^{\mathsf{H}} \triangleq [\mathbf{H}_1^{\mathsf{H}} \ldots \mathbf{H}_K^{\mathsf{H}}]$, reads [Tse and Viswanath, 2005]

$$C_{\mathrm{MAC}}(P_1, \ldots, P_K; \mathbf{H}^{\mathsf{H}}) = \bigcup_{\substack{\frac{1}{n_i}\operatorname{tr}(\mathbf{P}_i) \le P_i \\ \mathbf{P}_i \ge 0 \\ i=1,\ldots,K}} \left\{ \mathbf{r}; \sum_{i \in \mathcal{S}} r_i \le \log_2 \det \left(\mathbf{I}_N + \frac{1}{\sigma^2} \sum_{i \in \mathcal{S}} \mathbf{H}_i^{\mathsf{H}} \mathbf{P}_i \mathbf{H}_i \right), \forall \mathcal{S} \subset \{1, \ldots, K\} \right\} \tag{14.27}$$

where $\mathbf{r} = (r_1, \ldots, r_K) \in [0, \infty)^K$ and $\mathbf{P}_i \ge 0$ stands for $\mathbf{P}_i$ non-negative definite. That is, the set of achievable rate vectors $(r_1, \ldots, r_K)$ is such that the sum of the rates of any subset $\mathcal{S} = \{i_1, \ldots, i_{|\mathcal{S}|}\}$ is less than a classical log determinant expression for all possible precoders $\mathbf{P}_{i_1}, \ldots, \mathbf{P}_{i_{|\mathcal{S}|}}$.

Consider such a subset $\mathcal{S} = \{i_1, \ldots, i_{|\mathcal{S}|}\}$ of $\{1, \ldots, K\}$ and a set $\mathbf{P}_{i_1}, \ldots, \mathbf{P}_{i_{|\mathcal{S}|}}$ of deterministic precoders, i.e. precoders chosen independently of the particular realizations of the $\mathbf{H}_1, \ldots, \mathbf{H}_K$ matrices (although possibly taken as a function of the $\mathbf{R}_k$ and $\mathbf{T}_k$ correlation matrices).

At this point, depending on the underlying model assumptions, it is possible to apply either Corollary 6.1 of Theorem 6.4 or Remark 6.5 of Theorem 6.12. Although we mentioned in Remark 6.5 that it is highly probable that both theorems hold for the most general model hypotheses, we presently state which result can be applied to which situation based on the mathematical results available in the literature.

- If the $\mathbf{R}_k$ and $\mathbf{T}_k$ matrices are only constrained to have uniformly bounded normalized trace and the number of users K is small compared to the number of antennas n_k per user and the number of antennas at the transmitter, then Corollary 6.1 of Theorem 6.4 states (under some mild additional assumptions recalled in Chapter 13) that the per-antenna normalized log determinant expressions of (14.27) can be given a deterministic equivalent. This case allows us to assume very correlated antenna arrays, which is rather convenient for practical purposes.
- If the $\mathbf{R}_k$ and $\mathbf{T}_k$ matrices have uniformly bounded spectral norm as their dimensions grow, i.e. in practice if the largest eigenvalues of $\mathbf{R}_k$ or $\mathbf{T}_k$ are much smaller than the matrix size, and the total number of user antennas $n \triangleq \sum_{k=1}^K n_k$ is of the same order as the number N of antennas at the base station, then the hypotheses of Theorem 6.12 hold and a deterministic equivalent for the total capacity can be provided.

The first setting is more general in the sense that more general antenna correlation profiles can be used, while the second case is more general in the sense that the users' antennas can be distributed in a more heterogeneous way. Since we wish first to determine the rate region of MAC, though, we need, for all subset $\mathcal{S} \subset \{1, \ldots, K\}$, that N and $\sum_{i \in \mathcal{S}} n_i$ grow large simultaneously. This imposes in particular that all n_i grow large simultaneously with N. Later we will determine the achievable sum rate, for which only the subset $\mathcal{S} = \{1, \ldots, K\}$ will be considered. For the latter, Theorem 6.12 will be more interesting, as it can assume a large number of users with few antennas, which is a far more realistic assumption in practice.

For rate region analysis, consider then that all n_i are of the same order of dimension as N and that K is small in comparison. From Theorem 6.4, we have immediately that

$$\frac{1}{N}\log_2\det\left(\mathbf{I}_N + \frac{1}{\sigma^2}\sum_{i\in\mathcal{S}}\mathbf{H}_i^{\mathsf{H}}\mathbf{P}_i\mathbf{H}_i\right) - \left[\frac{1}{N}\log_2\det\left(\mathbf{I}_N + \sum_{k\in\mathcal{S}}\bar{e}_k\mathbf{T}_k\right)\right.$$
$$\left. - \log_2(e)\sigma^2\sum_{k\in\mathcal{S}}\bar{e}_k e_k + \frac{1}{N}\sum_{k\in\mathcal{S}}\log_2\det\left(\mathbf{I}_{n_k} + c_k e_k \mathbf{R}_k^{\frac{1}{2}}\mathbf{P}_k\mathbf{R}_k^{\frac{1}{2}}\right)\right] \xrightarrow{\text{a.s.}} 0$$

where $c_k \triangleq N/n_k$ and $e_{i_1}, \ldots, e_{i_{|\mathcal{S}|}}, \bar{e}_{i_1}, \ldots, \bar{e}_{i_{|\mathcal{S}|}}$ are the only positive solutions to

$$e_i = \frac{1}{\sigma^2 N}\operatorname{tr}\mathbf{T}_i\left(\mathbf{I}_N + \sum_{k\in\mathcal{S}}\bar{e}_k\mathbf{T}_k\right)^{-1}$$
$$\bar{e}_i = \frac{1}{\sigma^2 n_i}\operatorname{tr}\mathbf{R}_i^{\frac{1}{2}}\mathbf{P}_i\mathbf{R}_i^{\frac{1}{2}}\left(\mathbf{I}_{n_i} + c_i e_i \mathbf{R}_i^{\frac{1}{2}}\mathbf{P}_i\mathbf{R}_i^{\frac{1}{2}}\right)^{-1}. \tag{14.28}$$

This therefore provides a deterministic equivalent for the points in the rate region corresponding to *deterministic power allocation strategies*, i.e. power allocation strategies that do not depend on the $\mathbf{X}_k$ matrices. That is, not all points in the rate region can be associated with a deterministic equivalent (especially not the points on the rate region boundary) but only those points for which a deterministic power allocation is assumed.

Note now that we can similarly provide a deterministic equivalent to every point in the rate region of the quasi-static broadcast channel corresponding to a deterministic power allocation policy. As recalled earlier, the boundaries of the rate region $C_{\text{BC}}(P; \mathbf{H})$ of the broadcast channel have been recently shown [Weingarten *et al.*, 2006] to be achieved by dirty paper coding (DPC). For a transmit power constraint P over the compound channel $\mathbf{H}$, it is shown by MAC-BC duality that [Viswanath *et al.*, 2003]

$$C_{\text{BC}}(P; \mathbf{H}) = \bigcup_{\substack{P_1, \ldots, P_K \\ \sum_{k=1}^K P_k \leq P}} C_{\text{MAC}}(P_1, \ldots, P_K; \mathbf{H}^{\mathsf{H}}).$$

Therefore, from the deterministic equivalent formula above, we can also determine a portion of the BC rate region: that portion corresponding to the deterministic precoders. However, note that this last result has a rather limited interest. Indeed, channel-independent precoders in quasi-static BC inherently perform poorly compared to precoders adapted to the propagation channel, such as the optimal DPC precoder or the linear ZF and RZF precoders. This is because BC communications come along with potentially strong inter-user interference, which is only mitigated through adequate beamforming strategies. Deterministic precoders are incapable of providing efficient inter-user interference reduction and are therefore rarely considered in the literature.

Simulation results are provided in Figure 14.5 in which we assume a two-user MAC scenario. Each user is equipped with $n_1 = n_2$ antennas, where $n_1 = 8$ or $n_1 = 16$, while the base station is equipped with $N = n_1 = n_2$ antennas. The antenna array is linear with inter-antenna distance d^{R}/λ set to 0.5 or 0.1 at the users, and $d^{\mathrm{T}}/\lambda = 10$ at the base station. We further assume that the effectively transmitted energy propagates from a solid angle of $\pi/6$ on either communication side, with different propagation directions, and therefore consider the generalized Jakes' model for the $\mathbf{R}_k$ and $\mathbf{T}_k$ matrices. Specifically, we assume that user 2 sees the signal arriving at angle zero rad, and user 1 sees the signal arriving at angle π rad. We further assume uniform power allocation at the transmission. From the figure, we observe that the deterministic equivalent plot is centered somewhat around the mean value of the rates achieved for different channel realizations. As such, it provides a rather rough estimate of the instantaneous multiple access mutual information. It is nonetheless necessary to have at least 16 antennas on either side for the deterministic equivalent to be effectively useful. In terms of information-theoretical observations, note that a large proportion of the achievable rates is lost by increasing the antenna correlation. Also, as already observed in the single-user MIMO case, increasing the number of antennas in strongly correlated channels reduces the efficiency of every individual antenna.

As largely discussed above, it is in fact of limited interest to study the performance of quasi-static MAC and BC channels through large dimensional analysis, in a similar way to the single-user case, in the sense that optimal power allocation cannot be performed and the deterministic equivalent only provides a rough estimate of the effective rates achieved with high probability for a small number of antennas. When it comes to ergodic mutual information, though, similar to the point-to-point MIMO scenario, large system analysis can provide optimal power allocation policies and very accurate capacity approximations for small system dimensions.

14.2.2 Ergodic MAC rate region

Consider now the situation where the K channels are changing too fast for the users to be able to adapt adequately their transmit powers, while having constant Kronecker-type statistics. In this case, the MAC rate region is defined as

$$C_{\mathrm{MAC}}^{(\mathrm{ergodic})}(P_1, \ldots, P_K; \mathbf{H}^{\mathsf{H}})$$
$$= \bigcup_{\{\mathbf{P}_i\}} \left\{ \mathbf{r}, \sum_{i \in \mathcal{S}} r_i \leq \mathrm{E}\left[\log_2 \det\left(\mathbf{I}_N + \frac{1}{\sigma^2}\sum_{i \in \mathcal{S}} \mathbf{H}_i^{\mathsf{H}}\mathbf{P}_i\mathbf{H}_i\right)\right], \forall \mathcal{S} \subset \{1, \ldots, K\} \right\}$$

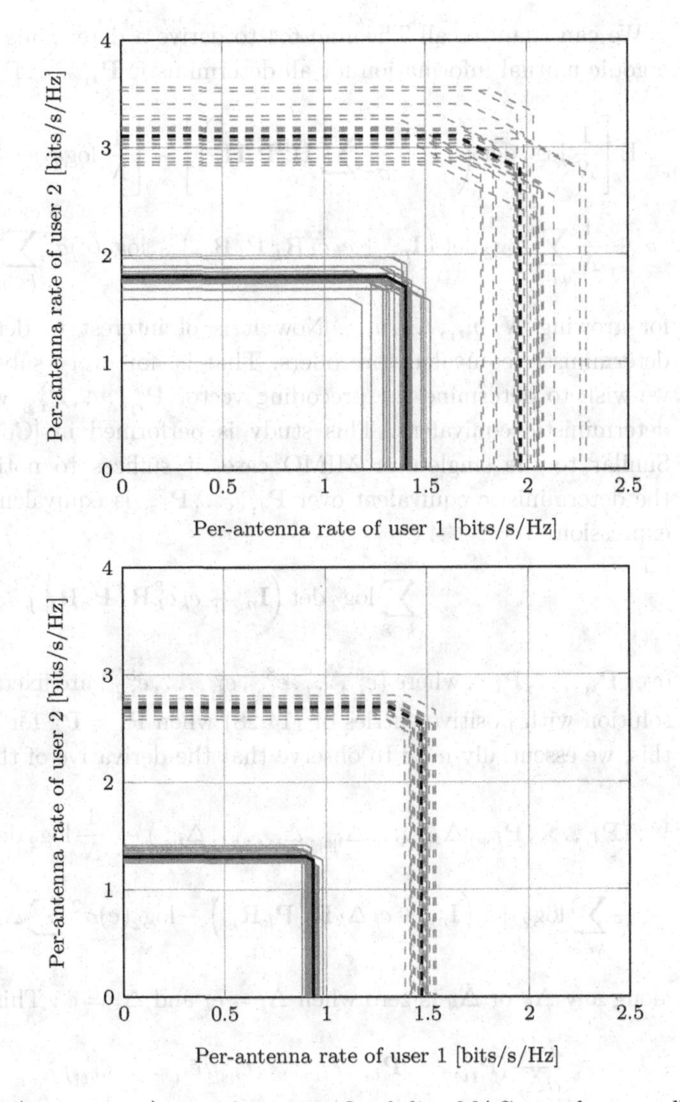

Figure 14.5 (Per-antenna) rate of two-user flat fading MAC, equal power allocation, for $N = 8$ (top), $N = 16$ (bottom) antennas at the base station, $n_1 = n_2 = N$ antennas at the transmitters, uniform linear antenna arrays, antenna spacing $\frac{d^{\mathrm{R}}}{\lambda} = 0.5$ (dashed) and $\frac{d^{\mathrm{R}}}{\lambda} = 0.1$ (solid) at the transmitters, $\frac{d^{\mathrm{T}}}{\lambda} = 10$ at the base station, $\mathrm{SNR} = 20$ dB. Deterministic equivalents are given in thick dark lines.

where the union is taken over all $\mathbf{P}_i$ non-negative definite such that

$$\frac{1}{n_i} \operatorname{tr}(\mathbf{P}_i) \leq P_i.$$

We can again recall Theorem 6.4 to derive a deterministic equivalent for the ergodic mutual information for all deterministic $\mathbf{P}_{i_1}, \ldots, \mathbf{P}_{i_{|\mathcal{S}|}}$ precoders, as

$$\mathrm{E}\left[\frac{1}{N}\log_2\det\left(\mathbf{I}_N + \frac{1}{\sigma^2}\sum_{i\in\mathcal{S}}\mathbf{H}_i^{\mathsf{H}}\mathbf{P}_i\mathbf{H}_i\right)\right] - \left[\frac{1}{N}\log_2\det\left(\mathbf{I}_N + \sum_{k\in\mathcal{S}}\bar{e}_k\mathbf{T}_k\right)\right.$$
$$\left. + \frac{1}{N}\sum_{k\in\mathcal{S}}\log_2\det\left(\mathbf{I}_{n_k} + c_k e_k \mathbf{R}_k^{\frac{1}{2}}\mathbf{P}_k\mathbf{R}_k^{\frac{1}{2}}\right) - \log_2(e)\sigma^2\sum_{k\in\mathcal{S}}\bar{e}_k e_k\right] \to 0$$

for growing N, $n_{i_1}, \ldots, n_{i_{|\mathcal{S}|}}$. Now it is of interest to determine the optimal deterministic equivalent precoders. That is, for every subset $\mathcal{S} = \{i_1, \ldots, i_{|\mathcal{S}|}\}$, we wish to determine the precoding vector $\mathbf{P}_{i_1}^{\mathcal{S}}, \ldots, \mathbf{P}_{i_{|\mathcal{S}|}}^{\mathcal{S}}$ which maximizes the deterministic equivalent. This study is performed in [Couillet *et al.*, 2011a]. Similar to the single-user MIMO case, it suffices to notice that maximizing the deterministic equivalent over $\mathbf{P}_{i_1}, \ldots, \mathbf{P}_{i_{|\mathcal{S}|}}$ is equivalent to maximizing the expression

$$\sum_{k\in\mathcal{S}}\log_2\det\left(\mathbf{I}_{n_k} + c_k e_k^{\mathcal{S}}\mathbf{R}_k^{\frac{1}{2}}\mathbf{P}_k\mathbf{R}_k^{\frac{1}{2}}\right)$$

over $\mathbf{P}_{i_1}, \ldots, \mathbf{P}_{i_{|\mathcal{S}|}}$, where $(e_{i_1}^{\mathcal{S}}, \ldots, e_{i_{|\mathcal{S}|}}^{\mathcal{S}}, \bar{e}_{i_1}^{\mathcal{S}}, \ldots, \bar{e}_{i_{|\mathcal{S}|}}^{\mathcal{S}})$ are fixed, equal to the unique solution with positive entries of (14.28) when $\mathbf{P}_i = \mathbf{P}_i^{\mathcal{S}}$ for all $i \in \mathcal{S}$. To observe this, we essentially need to observe that the derivative of the function

$$V : (\mathbf{P}_{i_1}, \ldots, \mathbf{P}_{i_{|\mathcal{S}|}}, \Delta_{i_1}, \ldots, \Delta_{i_{|\mathcal{S}|}}, \bar{\Delta}_{i_1}, \ldots, \bar{\Delta}_{i_{|\mathcal{S}|}}) \mapsto \frac{1}{N}\log_2\det\left(\mathbf{I}_N + \sum_{k\in\mathcal{S}}\bar{\Delta}_k\mathbf{T}_k\right)$$
$$+ \frac{1}{N}\sum_{k\in\mathcal{S}}\log_2\det\left(\mathbf{I}_{n_k} + c_k\Delta_k\mathbf{R}_k^{\frac{1}{2}}\mathbf{P}_k\mathbf{R}_k^{\frac{1}{2}}\right) - \log_2(e)\sigma^2\sum_{k\in\mathcal{S}}\bar{\Delta}_k\Delta_k$$

along any Δ_k or $\bar{\Delta}_k$ is zero when $\Delta_i = e_i$ and $\bar{\Delta}_i = \bar{e}_i$. This unfolds from

$$\begin{aligned}
&\frac{\partial V}{\partial\bar{\Delta}_k}(\mathbf{P}_{i_1}, \ldots, \mathbf{P}_{i_{|\mathcal{S}|}}, e_{i_1}, \ldots, e_{i_{|\mathcal{S}|}}, \bar{e}_{i_1}, \ldots, \bar{e}_{i_{|\mathcal{S}|}})\\
&= \log_2(e)\left[\frac{1}{N}\mathrm{tr}\left[\left(\mathbf{I}_N + \sum_{i\in\mathcal{S}}\bar{e}_i\mathbf{T}_i\right)^{-1}\mathbf{T}_k\right] - \sigma^2 e_k\right]\\
&\frac{\partial V}{\partial\Delta_k}(\mathbf{P}_{i_1}, \ldots, \mathbf{P}_{i_{|\mathcal{S}|}}, e_{i_1}, \ldots, e_{i_{|\mathcal{S}|}}, \bar{e}_{i_1}, \ldots, \bar{e}_{i_{|\mathcal{S}|}})\\
&= \log_2(e)\left[\frac{c_k}{N}\mathrm{tr}\left[\left(\mathbf{I} + c_k e_k\mathbf{R}_i^{\frac{1}{2}}\mathbf{P}_i\mathbf{R}_i^{\frac{1}{2}}\right)^{-1}\mathbf{R}_k^{\frac{1}{2}}\mathbf{P}_k\mathbf{R}_k^{\frac{1}{2}}\right] - \sigma^2\bar{e}_k\right]
\end{aligned}$$

both being null according to (14.28).

By the differentiation chain rule, the maximization of the log determinants over every $\mathbf{P}_i$ is therefore equivalent to the maximization of every term

$$\log_2\det\left(\mathbf{I}_{n_k} + c_k e_k^{\mathcal{S}}\mathbf{R}_k^{\frac{1}{2}}\mathbf{P}_k\mathbf{R}_k^{\frac{1}{2}}\right)$$

Define $\eta > 0$ the convergence threshold and $l \geq 0$ the iteration step.
At step $l = 0$, for $k \in \mathcal{S}$, $i \in \{1, \dots, n_k\}$, set $q_{k,i}^0 = P_k$. At step $l \geq 1$,
while $\max_{k,i}\{|q_{k,i}^l - q_{k,i}^{l-1}|\} > \eta$ **do**
  For $k \in \mathcal{S}$, define $(e_k^{l+1}, \bar{e}_k^{l+1})$ as the unique pair of positive solutions to (14.28) with, for all $j \in \mathcal{S}$, $\mathbf{P}_j = \mathbf{U}_j \mathbf{Q}_j^l \mathbf{U}_j^{\mathsf{H}}$, $\mathbf{Q}_j^l = \operatorname{diag}(q_{j,1}^l, \dots, q_{j,n_j}^l)$ and $\mathbf{U}_j$ the matrix such that $\mathbf{R}_j$ has spectral decomposition $\mathbf{U}_j \mathbf{\Lambda}_j \mathbf{U}_j^{\mathsf{H}}$, $\mathbf{\Lambda}_j = \operatorname{diag}(r_{j,1}, \dots, r_{j,n_j})$
  for $i \in \{1 \dots, n_k\}$ **do**
    Set $q_{k,i}^{l+1} = \left(\mu_k - \frac{1}{c_k e_k^{l+1} r_{k,i}}\right)^+$, with μ_k such that $\frac{1}{n_k} \operatorname{tr} \mathbf{Q}_k^l = P_k$
  end for
  assign $l \leftarrow l + 1$
end while

Table 14.1. Iterative water-filling algorithm for the determination of the MIMO MAC ergodic rate region boundary.

(remember that the power constraints over the $\mathbf{P}_i$ are independent). The maximum of the deterministic equivalent for the MAC ergodic mutual information is then found to be V evaluated at $e_k^{\mathcal{S}}$, $\bar{e}_k^{\mathcal{S}}$ and $\mathbf{P}_k^{\mathcal{S}}$ for all $k \in \mathcal{S}$. It unfolds that the capacity maximizing precoding matrices are given by a water-filling solution, as

$$\mathbf{P}_k^{\mathcal{S}} = \mathbf{U}_k \mathbf{Q}_k^{\mathcal{S}} \mathbf{U}_k^{\mathsf{H}}$$

where $\mathbf{U}_k \in \mathbb{C}^{n_k \times n_k}$ is the eigenvector matrix of the spectral decomposition of $\mathbf{R}_k$ as $\mathbf{R}_k = \mathbf{U}_k \operatorname{diag}(r_{k,1}, \dots, r_{k,n_k}) \mathbf{U}_k^{\mathsf{H}}$, and $\mathbf{Q}_k^{\mathcal{S}}$ is a diagonal matrix with ith diagonal entry $q_{ki}^{\mathcal{S}}$ given by:

$$q_{ki}^{\mathcal{S}} = \left(\mu_k - \frac{1}{c_k e_k^{\mathcal{S}} r_{k,i}}\right)^+$$

μ_k being set so that $\frac{1}{n_k}\sum_{i=1}^{n_k} q_{ki}^{\mathcal{S}} = P_k$, the maximum power allowed for user k. As usual, this can be determined from an iterative water-filling algorithm, provided that the latter converges. This is given in Table 14.1.

The performance of uniform and optimal power allocation strategies in the uplink ergodic MAC channel is provided in Figure 14.6 and Figure 14.7. As in the quasi-static case, the system comprises two users with $n_1 = n_2$ antennas, identical distance 0.5λ between consecutive antennas placed in linear arrays, and angle spread of energy arrival of $\pi/2$. User 1 sees the signal from an angle of zero rad, while user 1 sees the signal from an angle of π rad. In Figure 14.6, we observe, as was true already for the point-to-point MIMO case, that deterministic equivalents approximate very well the actual ergodic mutual information, for dimensions greater than or equal to four. It is then observed in Figure 14.7 that much data throughput can be gained by using optimal precoders at the user

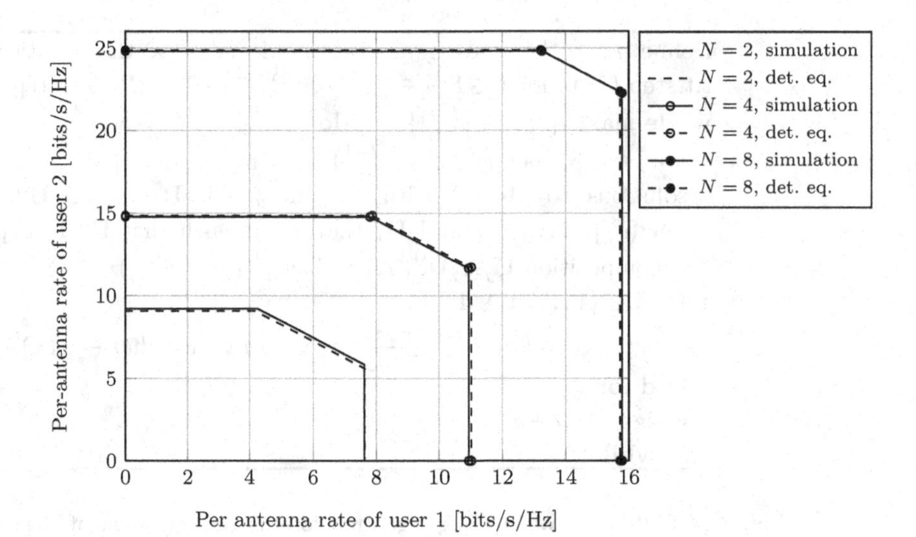

Figure 14.6 Ergodic rate region of two-user MAC, uniform power allocation, for $N = 2$, $N = 4$, and $N = 8$, $n_1 = n_2 = N$, uniform linear array model, antenna spacing at the users $\frac{d^R}{\lambda} = 0.5$, at the base station $\frac{d^T}{\lambda} = 10$. Comparison between simulations and deterministic equivalents (det. eq.).

terminals, especially on the rates of strongly correlated users. Notice also that in all previous performance plots, depending on the direction of energy arrival, a large difference in throughput can be achieved. This is more acute than in the single-user case where the resulting capacity is observed to be only slightly reduced by different propagation angles. Here, it seems that some users can either benefit or suffer greatly from the conditions met by other users.

We now turn to the specific study of the achievable ergodic sum rate.

14.2.3 Multi-user uplink sum rate capacity

As recalled earlier, when it comes to sum rate capacity, we only need to provide a deterministic equivalent for the log determinant expression

$$\mathrm{E}\left[\log_2 \det\left(\mathbf{I}_N + \frac{1}{\sigma^2}\sum_{i=1}^{K}\mathbf{H}_i^{\mathsf{H}}\mathbf{P}_i\mathbf{H}_i\right)\right]$$

for all deterministic $\mathbf{P}_1, \ldots, \mathbf{P}_K$. Obviously, this problem is even easier to treat than the previous case, as only one subset $\mathcal{S}$ of $\{1, \ldots, K\}$, namely $\mathcal{S} = \{1, \ldots, K\}$, has to be considered. As a consequence, the large dimensional constraint is just that both $n = \sum_{i=1}^{K} n_i$ and N are large and of similar dimension. This does no longer restrict any individual n_i to be large and of similar amplitude as N. Therefore, Theorem 6.12 will be used instead of Theorem 6.4.

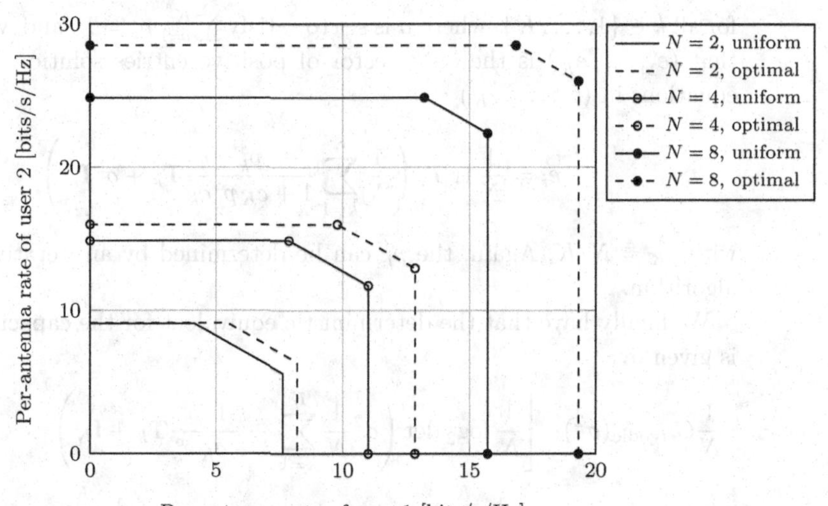

Figure 14.7 Deterministic equivalents for the ergodic rate region of two-user MAC, uniform power allocation against optimal power allocation, for $N = 2$, $N = 4$, and $N = 8$, $n_1 = n_2 = N$, uniform linear array model, antenna spacing at the users $\frac{d^R}{\lambda} = 0.5$, at the base station $\frac{d^T}{\lambda} = 10$.

We will treat a somewhat different problem in this section, which assumes that, instead of a per-user power constraint, all users can share a total energy budget to be used in the uplink. We also assume that the conditions of Theorem 6.12 now hold and we can consider, as in Section 14.1, that a large number of mono-antenna users share the channel to access a unique base station equipped with a large number of antennas. In this case, the deterministic equivalent developed in the previous section still holds true. Under transmit sum-power budget P, the power optimization problem does no longer take the form of a few optimal matrices but of a large number of scalars to be appropriately shared among the users. Recalling the notations of Section 14.1 and assuming perfect channel reciprocity, the ergodic MAC sum rate capacity $C_{\text{ergodic}}(\sigma^2)$ reads:

$$C_{\text{ergodic}}(\sigma^2) = \max_{\substack{p_1,\ldots,p_K \\ \sum_{i=1}^K p_i = P}} \mathrm{E}\left[\log_2 \det\left(\mathbf{I}_N + \frac{1}{\sigma^2}\sum_{i=1}^K p_i \mathbf{h}_i \mathbf{h}_i^{\mathsf{H}}\right)\right] \tag{14.29}$$

with $\mathbf{h}_i = \mathbf{T}_i^{\frac{1}{2}} \mathbf{x}_i \in \mathbb{C}^N$, $\mathbf{x}_i \in \mathbb{C}^N$ having i.i.d. Gaussian entries of zero mean and variance $1/N$. Remember that $\mathbf{T}_i$ stems for both the correlation at the base station and the path-loss component. In the MAC channel, $\mathbf{T}_i$ is now a *receive* correlation matrix.

The right-hand side of (14.29) is asymptotically maximized for $p_k = p_k^\star$ such that $p_k^\star - p_k^\circ \to 0$ as the system dimensions grow large, where p_k° is given by:

$$p_k^\circ = \left(\mu - \frac{K}{N}\frac{1}{e_k^\circ}\right)^+$$

for all $k \in \{1, \ldots, K\}$, where μ is set to satisfy $\sum_{i=1}^{K} p_i = P$, and where e_k° is such that $(e_1^\circ, \ldots, e_K^\circ)$ is the only vector of positive entries solution of the implicit equations in $(e_1, \ldots, e_K)$

$$e_i = \frac{1}{N} \operatorname{tr} \mathbf{T}_i \left(\frac{1}{N} \sum_{k=1}^{K} \frac{p_k^\circ}{1 + c_K p_k^\circ e_k} \mathbf{T}_k + \sigma^2 \mathbf{I}_N \right)^{-1}$$

with $c_K \triangleq N/K$. Again, the p_k° can be determined by an iterative water-filling algorithm.

We finally have that the deterministic equivalent for the capacity $C_{\text{ergodic}}(\sigma^2)$ is given by:

$$\frac{1}{N} C_{\text{ergodic}}(\sigma^2) - \left[\frac{1}{N} \log_2 \det \left(\sigma^2 \frac{1}{N} \sum_{k=1}^{K} \frac{1}{1 + c_K e_k^\circ} \mathbf{T}_k + \mathbf{I}_N \right) + \frac{1}{N} \sum_{k=1}^{K} \log_2 \left(1 + c_K e_k^\circ p_k^\circ\right) - \log_2(e) \frac{1}{K} \sum_{k=1}^{K} \frac{p_k^\circ e_k^\circ}{1 + c_K e_k^\circ p_k^\circ} \right] \to 0.$$

In Figure 14.8, the performance of the optimal power allocation scheme is depicted for correlation conditions similar to the previous scenarios, i.e. with correlation patterns at the base station accounting for inter-antenna distance and propagation direction, and different path loss exponents for the different users. It turns out numerically that, as in the single-user MIMO case, for low SNR, it is beneficial to reduce the number of transmitting users and allocate most of the available power to a few users, while this tendency fades away for higher SNR.

This closes this chapter on multi-user communications in both single antenna and multiple antenna regimes. In the next chapter, we extend the single-cell multiple antenna and multi-user framework to multi-cellular communications and relay communications.

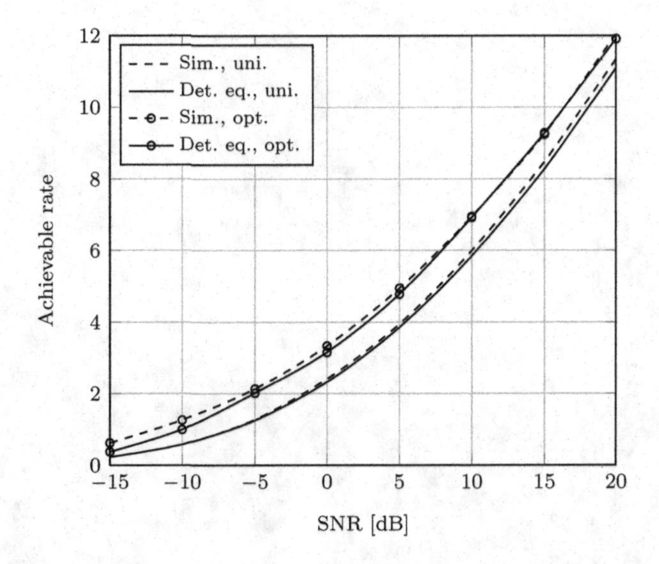

Figure 14.8 Ergodic MAC sum rate for an $N = 4$-antenna receiver and $K = 4$ mono-antenna transmitters under sum power constraint. Every user transmit signal has different correlation patterns at the receiver, and different path losses. Deterministic equivalents (det. eq.) against simulation (sim.), with uniform (uni.) or optimal (opt.) power allocation.

15 Performance of multi-cellular and relay networks

In this chapter, we move from single-cell considerations, with a central entity, e.g. access point or base station, to the wider multi-user multi-cell network point of view or the multi-hop relay point of view. For the former, we now consider that communications are performed in a cellular environment over a given shared communication resource (e.g. same frequency band, overlaying cell coverage), with possible inter-cell interference. This is a more realistic assumption to model practical communication networks than the isolated single-cell case with additive white Gaussian noise. Here, not only AWGN is affecting the different actors in the network but also cross-talk between adjacent base stations or cell-edge users. For the latter, we do not assume any cellular planning but merely consider multi-hop communications between relays which is of reduced use in commercial communication standards but of high interest to, e.g., *ad-hoc* military applications.

We start with the multi-cell viewpoint.

15.1 Performance of multi-cell networks

We consider the model of a multi-cellular network with overlaying coverage regions. In every cell, each user has a dedicated communication resource, supposed to be orthogonal to any other user's resource. For instance, we may assume that orthogonal frequency division multiple access (OFDMA) is used in every cell, so that users are orthogonally separated both in time and frequency. It is well-known that a major drawback of such networks is their being strongly interference limited since users located at cell edges may experience much interference from signals transmitted in adjacent cells. To mitigate or cancel this problem, a few solutions are considered, such as:

- ban the usage of certain frequencies in individual cells. This technique is referred to as *spatial frequency reuse* and consists precisely in making sure that two adjacent cells never use the same frequency band. This has the strong advantage that cross-talk in the same communication bandwidth can only come from remote cells. Nonetheless, this is a strong sacrifice in terms of available communication resources;

- create directional transmit signal beams. That is, at least in the downlink scenario, base stations use additional antennas to precode the transmit data in such a way that the information dedicated to the in-cell users is in the main lobe of a propagation beam, while the out-cell users are outside these lobes. This solution does not cancel adjacent cell interference but mitigates it strongly, although this requires additional antennas at the transmitter and more involved data processing.

An obvious solution to totally discard interference is to allow the base stations to cooperate. Precisely, if all network base stations are connected via a high rate backbone and are able to process simultaneously all uplink and downlink transmissions, then it is possible to remove the inter-cell interference completely by considering a single large cell instead, which is composed of multiple cooperative base stations. In this case, the network can be simply considered as a regular multi-access or broadcast channel, as those studied in Chapter 14. Nonetheless, while this idea of cooperative base stations has now made its way to standardization even for large range communications, e.g. in the 3GPP-LTE Advanced standard [Sesia *et al.*, 2009], this approach is both difficult to establish and of actually limited performance gain. Concerning the difficulty to put the system in place, we first mention the need to have a high rate backbone common to all base stations and a central processing unit so fast that joint decoding of all users can be performed under the network delay-limited constraints. As for the performance gain limitation, this mainly arises from the fact that, as the effective cell size increases, the central processing unit must be at all times aware of all communication channels and all intended data of all users. This imposes a large amount of synchronization and channel estimation information to be fed back to the central unit as the network size grows large. This therefore assumes that much time is spent by the network learning from its own structure. In fast mobile communication networks, where communication channels are changing fast, this is intolerable as too little time is then spent on effective communications.

This section is dedicated to the study of the intrinsic potential gains brought by cooperative cellular networks. As cooperative networks become large, large dimensional matrix theory can be used with high accuracy and allows us to obtain deterministic approximations of the system performance from which optimization can be conducted, such as determining the capacity maximizing feedback time, the capacity maximizing number of cooperative cells, etc. For simplicity, though, we assume perfect channel knowledge in the following in order to derive the potential gains obtained by unrealistic genie-aided multi-cellular cooperation.

Let us consider a K-cell network sharing a narrowband frequency resource, used by exactly one user per cell. Therefore, we can assume here that each cell is composed of a single-user. We call user k the user of cell k, for $k \in \{1, \ldots, K\}$. We can consider two interference limited scenarios, namely:

- the downlink scenario, where multiple base stations concurrently use the resource of all users, generating inter-cell interference to all users;
- the uplink scenario, where every base station suffers from inter-cell interference generated by the data transmitted by users in other cells.

We will consider here only the uplink scenario. We assume that base station k is equipped with N_k transmit antennas, and that user k is equipped with n_k receive antennas. Similar to previous chapters, we denote $n \triangleq \sum_{k=1}^{K} n_k$ and $N \triangleq \sum_{k=1}^{K} N_k$. We also denote $\mathbf{H}_{k,i} \in \mathbb{C}^{N_i \times n_k}$ the uplink multiple antenna channel between user k and base station i. We consider that $\mathbf{H}_{k,i}$ is modeled as a Kronecker channel, under the following form

$$\mathbf{H}_{k,i} = \mathbf{R}_{k,i}^{\frac{1}{2}} \mathbf{X}_{k,i} \mathbf{T}_{k,i}^{\frac{1}{2}}$$

for non-negative $\mathbf{R}_{k,i}$ with eigenvalues $r_{k,1}, \ldots, r_{k,N}$, non-negative $\mathbf{T}_{k,i}$ with eigenvalues $t_{k,i,1}, \ldots, t_{k,i,n_k}$, and $\mathbf{X}_{k,i}$ with i.i.d. Gaussian entries of zero mean and variance $1/n_k$. As usual, we can assume the $\mathbf{T}_{k,i}$ diagonal without loss of generality.

Assume that a given subset $\mathcal{C} \subset \{1, \ldots, K\}$ of cells cooperates and is interfered with by the remaining users from the complementary set $\mathcal{C}^c = \{1, \ldots, K\} \setminus \mathcal{C}$. Denote $N_{\mathcal{C}} = \sum_{k \in \mathcal{C}} N_k$, $N_{\mathcal{C}^c} = \sum_{k \in \mathcal{C}^c} N_k$, and $\mathbf{H}_{k,\mathcal{C}} \in \mathbb{C}^{N_{\mathcal{C}} \times n_k}$ the channel $[\mathbf{H}_{k,i_1}^{\mathsf{H}}, \ldots, \mathbf{H}_{k,i_{|\mathcal{C}|}}^{\mathsf{H}}]^{\mathsf{H}}$, with $\{i_1, \ldots, i_{|\mathcal{C}|}\} = \mathcal{C}$. The channel $\mathbf{H}_{k,\mathcal{C}}$ is the joint channel between user k and the cooperative large base station composed of the $|\mathcal{C}|$ subordinate base stations indexed by elements of $\mathcal{C}$. Assuming uniform power allocation across the antennas of the transmit users, the ergodic sum rate $\mathrm{E}[\mathcal{I}(\mathcal{C};\sigma^2)]$ of the multiple access channel between the $|\mathcal{C}|$ users and the cooperative base stations, considered as a unique receiver, under AWGN of variance σ^2 reads:

$$\begin{aligned}
&\mathrm{E}[\mathcal{I}(\mathcal{C};\sigma^2)] \\
&= \mathrm{E}\left[\log_2 \det\left(\mathbf{I}_{N_{\mathcal{C}}} + \left[\sum_{k \in \mathcal{C}} \mathbf{H}_{k,\mathcal{C}} \mathbf{H}_{k,\mathcal{C}}^{\mathsf{H}}\right]\left[\sigma^2 \mathbf{I}_{N_{\mathcal{C}}} + \sum_{k \in \mathcal{C}^c} \mathbf{H}_{k,\mathcal{C}} \mathbf{H}_{k,\mathcal{C}}^{\mathsf{H}}\right]^{-1}\right)\right] \\
&= \mathrm{E}\left[\log_2 \det\left(\mathbf{I}_{N_{\mathcal{C}}} + \frac{1}{\sigma^2} \sum_{k=1}^{K} \mathbf{H}_{k,\mathcal{C}} \mathbf{H}_{k,\mathcal{C}}^{\mathsf{H}}\right)\right] \\
&- \mathrm{E}\left[\log_2 \det\left(\mathbf{I}_{N_{\mathcal{C}}} + \frac{1}{\sigma^2} \sum_{k \in \mathcal{C}^c} \mathbf{H}_{k,\mathcal{C}} \mathbf{H}_{k,\mathcal{C}}^{\mathsf{H}}\right)\right].
\end{aligned}$$

Under this second form, it unfolds that a deterministic equivalent of $\mathrm{E}[\mathcal{I}(\mathcal{C};\sigma^2)]$ can be found as the difference of two deterministic equivalents, similar to Chapter

14. Indeed, the jth column $\mathbf{h}_{k,\mathcal{C},j} \in \mathbb{C}^{N_\mathcal{C}}$ of $\mathbf{H}_{k,\mathcal{C}}$ can be written under the form

$$\mathbf{h}_{k,\mathcal{C},j} = \begin{pmatrix} \mathbf{h}_{k,i_1,j} \\ \vdots \\ \mathbf{h}_{k,i_{|\mathcal{C}|},j} \end{pmatrix} = \begin{pmatrix} \sqrt{t_{k,i_1,j}}\mathbf{R}_{k,i_1}^{\frac{1}{2}} & \cdots & 0 \\ \vdots & \ddots & \vdots \\ 0 & \cdots & \sqrt{t_{k,i_{|\mathcal{C}|},j}}\mathbf{R}_{k,i_{|\mathcal{C}|}}^{\frac{1}{2}} \end{pmatrix} \begin{pmatrix} \mathbf{x}_{k,i_1,j} \\ \vdots \\ \mathbf{x}_{k,i_{|\mathcal{C}|},j} \end{pmatrix}$$

with $\mathbf{x}_{k,i,j}$ the jth column of $\mathbf{X}_{k,i}$, which has independent Gaussian entries of zero mean and variance $1/n_k$. We will denote $\bar{\mathbf{R}}_{k,\mathcal{C},j}$ the block diagonal matrix

$$\bar{\mathbf{R}}_{k,\mathcal{C},j} \triangleq \begin{pmatrix} t_{k,i_1,j}\mathbf{R}_{k,i_1} & \cdots & 0 \\ \vdots & \ddots & \vdots \\ 0 & \cdots & t_{k,i_{|\mathcal{C}|},j}\mathbf{R}_{k,i_{|\mathcal{C}|}} \end{pmatrix}.$$

With these notations, the random matrix $\mathbf{H}_{k,\mathcal{C}}\mathbf{H}_{k,\mathcal{C}}^{\mathsf{H}}$ therefore follows the model of Theorem 6.12 with correlation matrices $\bar{\mathbf{R}}_{k,\mathcal{C},j}$ for all $j \in \{1, \ldots, n_k\}$ and the normalized ergodic sum rate

$$\frac{1}{N_\mathcal{C}} \mathrm{E}[\mathcal{I}(\mathcal{C};\sigma^2)] = \mathrm{E}\left[\frac{1}{N_\mathcal{C}} \log_2 \det\left(\mathbf{I}_{N_\mathcal{C}} + \frac{1}{\sigma^2}\sum_{k=1}^{K}\sum_{j_k=1}^{n_k} \mathbf{h}_{k,\mathcal{C},j_k}\mathbf{h}_{k,\mathcal{C},j_k}^{\mathsf{H}}\right)\right]$$
$$- \mathrm{E}\left[\frac{1}{N_\mathcal{C}} \log_2 \det\left(\mathbf{I}_{N_\mathcal{C}} + \frac{1}{\sigma^2}\sum_{k\in\mathcal{C}^c}\sum_{j_k=1}^{n_k} \mathbf{h}_{k,\mathcal{C},j_k}\mathbf{h}_{k,\mathcal{C},j_k}^{\mathsf{H}}\right)\right]$$

has a deterministic equivalent that can be split into the difference of the following expression (i)

$$\frac{1}{N_\mathcal{C}} \log_2 \det\left(\sigma^2 \frac{1}{n}\sum_{k=1}^{K}\sum_{j_k=1}^{n_k} \frac{1}{1+c_N e_{k,j_k}^{\mathcal{K}}}\bar{\mathbf{R}}_{k,\mathcal{C},j_k} + \mathbf{I}\right)$$
$$+ \frac{1}{N_\mathcal{C}}\sum_{k=1}^{K}\sum_{j_k=1}^{n_k}\log_2\left(1+c_N e_{k,j_k}^{\mathcal{K}}\right) - \frac{1}{n}\sum_{k=1}^{K}\log_2(e)\sum_{j_k=1}^{n_k}\frac{e_{k,j_k}^{\mathcal{K}}}{1+c_N e_{k,j_k}}$$

with $c_N \triangleq N/n$ and $e_{k,j_k}^{\mathcal{K}}$, $\mathcal{K} \triangleq \{1, \ldots, K\}$, defined as the only all positive solutions of

$$e_{k,j_k}^{\mathcal{K}} = \frac{1}{N_\mathcal{C}} \operatorname{tr} \bar{\mathbf{R}}_{k,\mathcal{C},j_k}\left(\frac{1}{n}\sum_{k'=1}^{K}\sum_{j_{k'}=1}^{n_{k'}}\frac{1}{1+c_N e_{k',j_{k'}}^{\mathcal{K}}}\bar{\mathbf{R}}_{k',\mathcal{C},j_{k'}} + \sigma^2\mathbf{I}\right)^{-1} \tag{15.1}$$

and of this second term (ii)

$$\frac{1}{N_\mathcal{C}} \log_2 \det\left(\sigma^2 \frac{1}{n}\sum_{k\in\mathcal{C}^c}\sum_{j_k=1}^{n_k} \frac{1}{1+c_N e_{k,j_k}^{\mathcal{C}^c}}\bar{\mathbf{R}}_{k,\mathcal{C},j_k} + \mathbf{I}\right)$$
$$+ \frac{1}{N_\mathcal{C}}\sum_{k\in\mathcal{C}^c}\sum_{j_k=1}^{n_k}\log_2\left(1+c_N e_{k,j_k}^{\mathcal{C}^c}\right) - \frac{1}{n}\sum_{k\in\mathcal{C}^c}\sum_{j_k=1}^{n_k}\log_2(e)\frac{e_{k,j_k}^{\mathcal{C}^c}}{1+c_N e_{k,j_k}^{\mathcal{C}^c}}$$

with $e_{k,j_k}^{\mathcal{C}^c}$ the only all positive solutions of

$$e_{k,j_k}^{\mathcal{C}^c} = \frac{1}{N_{\mathcal{C}}} \operatorname{tr} \bar{\mathbf{R}}_{k,\mathcal{C},j_k} \left(\frac{1}{n} \sum_{k' \in \mathcal{C}^c} \sum_{j_{k'}=1}^{n_{k'}} \frac{1}{1 + c_N e_{k',j_{k'}}^{\mathcal{C}^c}} \bar{\mathbf{R}}_{k',\mathcal{C},j_{k'}} + \sigma^2 \mathbf{I} \right)^{-1}.$$

Although these expressions are all the more general under our model assumptions, simplifications are required to obtain tractable expressions. We will consider two situations. First, that of a two-cell network with different interference powers, and second that of a large network with mono-antenna devices. The former allows us to discuss the interest of cooperation in a simple two-cell network as a function of the different channel conditions affecting capacity: overall interference level, number of available antennas, transmit and receive correlation, etc. The latter allows us to determine how many base stations are required to be connected in order for the overall system capacity not to be dramatically impaired. On top of these discussions, the problem of optimal channel state information feedback time can also be considered when the channel coherence time varies. It is then possible, for practical networks, to have a rough overview on the optimal number of base stations to inter-connect under mobility constraints. This last point is not presented here.

15.1.1 Two-cell network

In this first example, we consider a two-cell network for which we provide the theoretical sum rate capacities achieved when:

- both cells cooperate in the sense that they proceed to joint decoding of the users operating at the same frequency;
- cells do not cooperate and are therefore interference limited.

When base stations do not cooperate, it is a classical assumption that the channels from user k to base station k, $k \in \{1, 2\}$, are not shared among the adjacent cells. An optimal power allocation policy can therefore not be clearly defined. However, in the cooperative case, power allocation can be clearly performed, under either sum power constraint or individual power constraint for each user. We will consider the latter, more realistic, choice. Also, we make the strong necessary assumption that, at the transmitter side, the effective energy is transmitted isotropically, i.e. there is no privileged direction of propagation. This assumption is in fact less restrictive when the mobile transmitter is located in a very scattered environment, such as in the house or in the street while the receive base station is located far away. This assumption would however not hold for base stations, which are typically located in a poorly scattered environment. Following Jakes' model for the transmit correlation $\mathbf{T}_{k,i}$ matrices, we will therefore assume that $\mathbf{T}_{k,1} = \mathbf{T}_{k,2} \triangleq \mathbf{T}_k$, $k \in \{1, 2\}$. We are now in the situation of the previous study for $\mathcal{K} = \{1, 2\}$. The subset $\mathcal{C} \subset \mathcal{K}$ of simultaneously decoded users is alternatively $\{1\}$ and $\{2\}$ in the non-cooperative case, and $\{1,2\}$ in the cooperative case.

The real difficulty lies in determining the optimal power allocation policy for the cooperative case. Denoting $\mathbf{P}_k^\circ \in \mathbb{C}^{n_k \times n_k}$ this optimal power allocation policy for user k in the cooperative case, we have in fact trivially that $\mathbf{P}_k^\circ = \mathbf{U}_k \mathbf{Q}_k^\circ \mathbf{U}_k^\mathsf{H}$ with $\mathbf{U}_k$ the eigenvector basis of $\mathbf{T}_k$, and $\mathbf{Q}_k^\circ = \operatorname{diag}(q_{k,1}^\circ, \ldots, q_{k,n_k}^\circ)$, with $q_{k,i}^\circ$ given by:

$$q_{k,i}^\circ = \left(\mu_k - \frac{1}{1 + c_N e_{k,j_k}^{\mathcal{K}\circ}} \right)^+$$

where μ_k is set to satisfy $\frac{1}{n_k} \sum_{i=1}^{n_k} q_{k,i}^\circ = P_k$, P_k being the power allowed for transmission to user k, and $e_{k,j_k}^{\mathcal{K}\circ}$ is the solution to the fixed-point Equation (15.1) in $e_{k,j_k}^{\mathcal{K}}$, with $e_{k,j_k}^{\mathcal{K}}$ replaced by $e_{k,j_k}^{\mathcal{K}} p_{k,j}^\circ$ (present now in the expression of $\bar{\mathbf{R}}_{k,\mathcal{K},j}$). The only difference compared to the multi-user mono-antenna MAC model is that multiple power constraints are imposed. This is nonetheless not hard to solve as the maximization is equivalent to the maximization of

$$\sum_{k \in \mathcal{K}} \sum_{j_k=1}^{n_k} \log_2 \left(1 + c_N e_{k,j_k}^{\mathcal{K}} p_{k,j}^\circ\right).$$

As usual, the entries of $\mathbf{P}_k$ can be determined from an appropriate iterative water-filling algorithm. In the following, we provide performance results for the two-cell network. In Figure 15.1 and Figure 15.2, we consider the case of two users, equipped with four antennas each, and two base stations, also equipped with four antennas each. In Figure 15.1, both transmit and receive sides are loosely correlated, i.e. the inter-antenna spacing is large and isotropic transmissions are assumed. In Figure 15.2, the same scenario is considered, although, at the transmitting users, strong signal correlation is present (the distance between successive antennas is taken to be half the wavelength). We observe that strong correlation at the transmitter side reduces the achievable optimal sum rate, although in this case optimal power allocation policy helps to recovere part of the lost capacity, notably for low SNR regimes. However, for strongly correlated transmitters, it turns out that the performance of single-user decoding in every cell is not strongly affected and is even beneficial for high SNR. This is mostly due to the uniform power allocation policy adopted in the interference limited case, which could be greatly improved.

In the following, we address the problem of the uplink communication in a large dimensional network, somewhat following the Wyner model [Wyner, 1994] with one user per cell and only one transmitting antenna on either communication device.

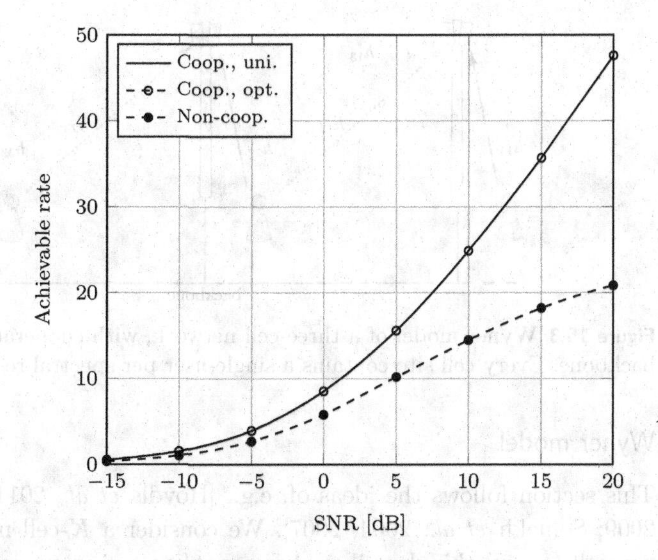

Figure 15.1 Sum rate capacity of two-cell network with two users per cell. Comparison between cooperative MAC scenario (coop.) for uniform (uni.) and optimal (opt.) power allocation, and interference limited scenario (non-coop.). Loosely correlated signals at both communication ends. Interference power 0.5.

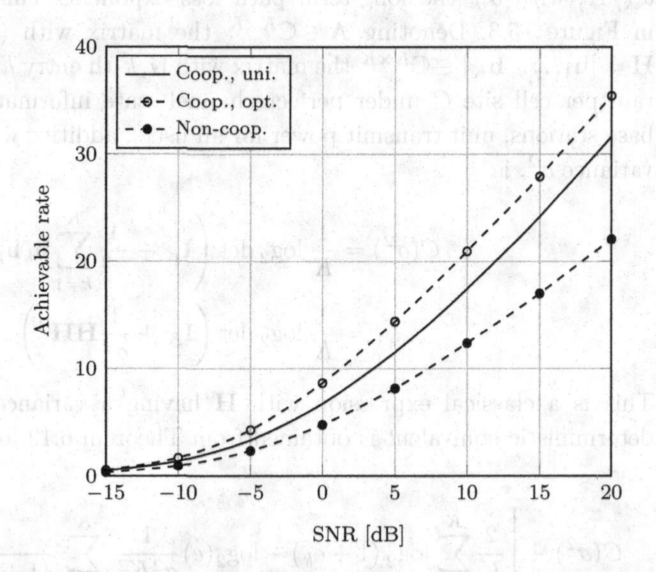

Figure 15.2 Sum rate capacity of two-cell network with two users per cell. Comparison between cooperative MAC scenario (coop.) for uniform (uni.) and optimal (opt.) power allocation, and interference limited scenario (non-coop.). Loosely correlated signals at the receiver end, strongly correlated signals at the transmission end. Interference power 0.5.

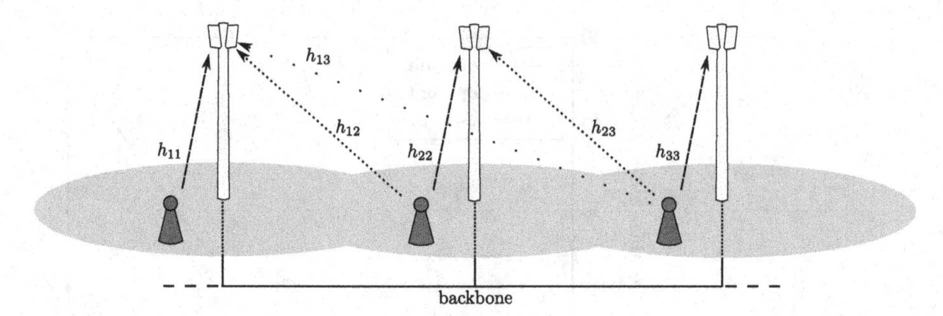

Figure 15.3 Wyner model of a three-cell network, with cooperation via a central backbone. Every cell site contains a single-user per spectral resource.

15.1.2 Wyner model

This section follows the ideas of, e.g., [Hoydis *et al.*, 2011d; Levy and Shamai, 2009; Somekh *et al.*, 2004, 2007]. We consider a K-cell network with one user per cell (again, this describes the case when a single resource is analyzed) and one antenna per user. We further assume that the base stations of all cells are connected via an infinite capacity backbone, so that inter-cell communications are possible at a high transmission rate. The link h_{ik} between user k and base station i is assumed to be random Gaussian with zero mean and variance a_{ik}/K, with a_{ik} the long-term path loss exponent. This scenario is depicted in Figure 15.3. Denoting $\mathbf{A} \in \mathbb{C}^{K \times K}$ the matrix with (i, k)th entry a_{ik} and $\mathbf{H} = [\mathbf{h}_1, \ldots, \mathbf{h}_K] \in \mathbb{C}^{K \times K}$ the matrix with (i, k)th entry h_{ik}, the achievable sum rate per cell site C under perfect channel state information at the connected base stations, unit transmit power for all users, additive white Gaussian noise of variance σ^2, is

$$
\begin{aligned}
C(\sigma^2) &= \frac{1}{K} \log_2 \det \left(\mathbf{I}_K + \frac{1}{\sigma^2} \sum_{k=1}^{K} \mathbf{h}_k \mathbf{h}_k^{\mathsf{H}} \right) \\
&= \frac{1}{K} \log_2 \det \left(\mathbf{I}_K + \frac{1}{\sigma^2} \mathbf{H}\mathbf{H}^{\mathsf{H}} \right).
\end{aligned}
$$

This is a classical expression with $\mathbf{H}$ having a variance profile, for which a deterministic equivalent is obtained from Theorem 6.12 for instance.

$$
C(\sigma^2) - \left[\frac{2}{K} \sum_{k=1}^{K} \log_2 (1 + e_k) - \log_2(e) \frac{1}{\sigma^2 K^2} \sum_{i,k=1}^{K} \frac{a_{k,i}}{(1 + e_i)(1 + e_k)} \right] \xrightarrow{\text{a.s.}} 0
$$

where $e_1, \ldots, e_K$ are the only all positive solutions to

$$
e_i = \frac{1}{\sigma^2} \frac{1}{K} \sum_{k=1}^{K} \frac{a_{k,i}}{1 + e_k}. \tag{15.2}
$$

We consider the (admittedly unrealistic) situation when the cells are uniformly distributed on a linear array, following the Wyner model [Wyner, 1994]. The cells are numbered in order, following the linear structure. The path loss from user k to cell n has variance $\alpha^{|n-k|}$, for some fixed parameter α. As such, $\mathbf{A}$ is explicitly given by:

$$\mathbf{A} = \begin{pmatrix} 1 & \alpha & \alpha^2 & \cdots & \alpha^{K-1} \\ \alpha & 1 & \alpha & \cdots & \alpha^{K-2} \\ \vdots & \ddots & \ddots & \ddots & \vdots \\ \alpha^{K-2} & \cdots & \alpha & 1 & \alpha \\ \alpha^{K-1} & \cdots & \alpha^2 & \alpha & 1 \end{pmatrix}.$$

In the case where α is much smaller than one, we can approximate the Toeplitz matrix $\mathbf{A}$ by a matrix with all terms α^n replaced by zeros, for all n greater than some integer L, of order $O(K)$ but strictly less than $K/2$. In this case, the resulting Toeplitz matrix is a band matrix, which in turn can be approximated for large dimensions by an equivalent circulant matrix $\bar{\mathbf{A}}$, whose entries are the same as $\mathbf{A}$ in the $2L$ main diagonals, and which is made circulant by filling the upper right and lower left matrix corners accordingly. This must nonetheless be performed with extreme care, following the conditions of Szegö's theorem, Theorem 12.1. The matrix $\bar{\mathbf{A}}$, in addition to being circulant, has the property to have all sums in rows and columns equal. This matrix is referred to as a *doubly regular* variance profile matrix. In this particular case, it is shown in [Tulino and Verdú, 2005] that the asymptotic eigenvalue distribution of the matrix

$$\frac{1}{\frac{1}{K}\sum_{k=1}^{K}\bar{a}_{ki}}\mathbf{H}\mathbf{H}^{\mathsf{H}}$$

with $\bar{a}_{ki}$ the entry (k,i) of $\bar{\mathbf{A}}$, is the Marčenko–Pastur law. To observe this fact, it suffices to notice that the system of Equations (15.2), with a_{ij} replaced by $\bar{a}_{ij}$, is solved for $e_1 = \ldots = e_K = f$, with f defined as the only Stieltjes transform solution to

$$f = \frac{1}{\sigma^2}\frac{1}{K}\sum_{k=1}^{K}\frac{\bar{a}_{k,i}}{1+f}$$

for all i. This is easily verified as $\sum_{k=1}^{K}\bar{a}_{k,i}$ is constant irrespective of the choice of the row i. Now, f is also the solution of (15.2) if all $\bar{a}_{ki}$ were taken to be identical, all equal to $\frac{1}{K}\sum_{k=1}^{K}\bar{a}_{ki}$ for any i, in which case $\mathbf{H}$ would be a matrix with i.i.d. entries of zero mean and variance $\frac{1}{K}\sum_{k=1}^{K}\bar{a}_{ki}$. Hence the result.

We therefore finally have, analogously to the uncorrelated antenna case in Chapter 13, that C can be approximated explicitly as

$$C(\sigma^2) - \left[\log_2\left(1+\delta+\frac{s}{\sigma^2}\right) + \log_2(e)\left[\frac{\sigma^2}{s}\delta - 1\right]\right] \xrightarrow{\text{a.s.}} 0$$

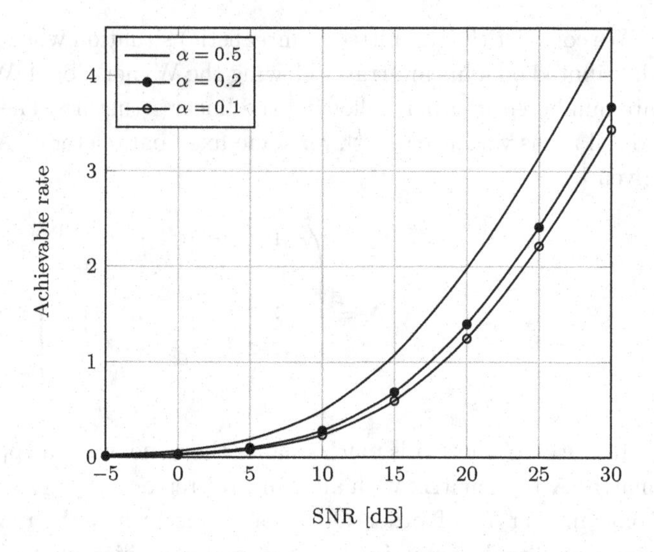

Figure 15.4 Multi-cell site capacity performance for $K = 64$ cell sites distributed on a linear array. Path loss decay parameter $\alpha = 0.5$, $\alpha = 0.2$, $\alpha = 0.1$.

where δ is given by:

$$\delta = \frac{1}{2}\left[\sqrt{1 + \frac{4s}{\sigma^2}} - 1\right]$$

and where we denoted $s \triangleq \frac{1}{K}\sum_{k=1}^{K} \bar{a}_{k1}$.

In Figure 15.4, we provide the deterministic equivalents for the capacity C for different values of α, and for $K = 64$. The observed gain in capacity achieved by inter-cell cooperation when cells are rather close encourages a shift towards inter-cell cooperation, although in this particular example we did not consider at all the inherent problems of such communications: base station synchronization, cost of simultaneous data decoding, and more importantly feedback cost necessary for the joint base station to learn the various (possibly fast varying) communication channels.

In the next section, we turn to the analysis of multi-hop multiple antenna communications assuming the number of antennas per relay is large.

15.2 Multi-hop communications

Relay communications are often seen as a practical solution to extend the coverage of multi-cellular networks to remote areas. In this context, the main interest is focused on two-hop relaying, where the relays are forwarding directly the information from the source to the receiver. In the simple scenario where the relay only *amplifies and forwards* the data of the source, the receiver captures identical signals, with possibly some delay, so that the overall system can be roughly modeled as a multiple antenna channel. However, information theory requires that more intelligent processing be done at the relay than enhancing and forwarding both received signal and noise. In particular, *decoding, re-encoding, and forwarding* the information at the relay is a better approach in terms of achievable rate, but is much more difficult to analyze. This becomes even more difficult as the number of simultaneously transmitting relays increases. Simple models are therefore often called for when studying these scenarios. Large dimensional random matrix theory is particularly helpful here. Increasing the number of relays in the multi-hop model has different applications, especially for *ad-hoc* networks, e.g. intended for military field operations.

We will study in this section the scenario of multiple hops in a relay network assuming a linear array composed of a source, $K-1$ relays, and a destination. Each hop between relay pairs or source to first relay will be assumed noise-free for simplicity, while the last hop will experience additive noise. Each communication entity will be assumed to be equipped with a large number of antennas and to receive signals only from the last backward relay or source; we therefore assume the simple scenario where distant relays do not interfere with each other. At each hop, the communication strategy is to re-encode the receive data by a linear precoder in a hybrid manner between the amplify and forward and the decode and forward strategies. The study of the achievable rates in this setup will therefore naturally call for the analysis of successive independent channel matrix products. Therefore, tools such as the S-transform, and in particular Theorem 4.7, will be extremely useful here.

The following relies heavily on the work of Müller [Müller, 2002] on the capacity of large dimensional product channels and on the work of Fawaz *et al.* [Fawaz *et al.*, 2011].

15.2.1 Multi-hop model

Consider a multi-hop relaying system with a source, $K-1$ relays, and a destination. The source is equipped with N_0 antennas, the destination with N_K antennas and the kth relay level with N_k antennas. We assume that the noise power is negligible at all relays, while at the destination, at all times l, an additive Gaussian noise vector $\mathbf{w}^{(l)} \in \mathbb{C}^{N_K}$ with zero mean and covariance $\sigma^2 \mathbf{I}_{N_K}$ is received. In effect, the simplifying assumption of noise-free relays is made to have a white aggregate noise at the destination and consequently more tractable derivations. Note that several works have implicitly used a similar noise-free relay assumption by assuming that the noise at the destination of a multiple antenna multi-hop relay network is white. In [Yang and Belfiore, 2008], the authors prove that in an amplify and forward multi-hop relay system the resulting colored noise at the destination can be well approximated by white noise in the high SNR regime. In terms of practical relevance, the mutual information expression derived in the case of noise-free relays can be seen as an upper-bound for the case of noisy relays.

We further assume that the channel matrix $\mathbf{H}_k$ at hop $k \in \{1, \ldots, K\}$ (hop k is from relay $k-1$ to relay k, with relay zero the source and relay K the destination) is a random realization of a Kronecker channel model

$$\mathbf{H}_k = \mathbf{R}_k^{\frac{1}{2}} \mathbf{X}_k \mathbf{T}_k^{\frac{1}{2}}$$

where $\mathbf{T}_k \in \mathbb{C}^{N_{k-1}}$ and $\mathbf{R}_k \in \mathbb{C}^{N_k}$ are the transmit and receive correlation matrices, respectively, $\mathbf{X}_k \in \mathbb{C}^{N_k \times N_{k-1}}$ has i.i.d. Gaussian entries of zero mean and variance a_k/N_{k-1}, where $a_k = d_k^{-\beta}$ represents the long-term path loss attenuation with β and d_k the path loss exponent and the length of the kth hop, respectively.

Note that, by adapting the correlation matrix structures, the Kronecker model can be used to model relay-clustering. Given a total number of antennas N_k at relaying level k, instead of considering that the relaying level consists of a single relay equipped with N_k antennas, we can consider that a relaying level contains r_k relays equipped with (N_k/r_k) antennas each. Clustering has a direct impact on the structure of correlation matrices; when the N_k antennas at level k are distributed among several relays, correlation matrices become block diagonal matrices, whose blocks represent the correlation between antennas at a relay, while antennas at different relays sufficiently separated in space are supposed uncorrelated. If each relaying level k contains N_k relays equipped with a single antenna each, we fall back to the case of uncorrelated fading with correlation matrices equal to identity.

Within one channel coherence block, the signal transmitted by the N_0 source antennas at time l is given by the vector $\mathbf{s}_0^{(l)} = \mathbf{P}_0 \mathbf{y}_0^{(l-1)}$, where $\mathbf{P}_0$ is the source precoding matrix and $\mathbf{y}_0^{(l)}$ is a random vector with zero mean and covariance $\mathrm{E}[\mathbf{y}_0^{(l)} \mathbf{y}_0^{(l)\mathsf{H}}] = \mathbf{I}_{N_0}$. This implies $\mathrm{E}[\mathbf{s}_0^{(l)} \mathbf{s}_0^{(l)\mathsf{H}}] = \mathbf{P}_0 \mathbf{P}_0^{\mathsf{H}}$. Assuming that relays work in full-duplex mode, at time l the relay at level k uses a precoding matrix $\mathbf{P}_k$ to linearly precode its received signal $\mathbf{y}_k^{(l-1)} = \mathbf{H}_k \mathbf{s}_{k-1}^{(l-1)}$ and to form its transmitted signal (we remind that these channels are noise-free)

$$\mathbf{s}_k^{(l)} = \mathbf{P}_k \mathbf{y}_k^{(l-1)}.$$

The precoding matrices $\mathbf{P}_k$, $k \in \{0, \ldots, K-1\}$, at source and relays are moreover subject to the per-node power constraints

$$\frac{1}{N_k} \operatorname{tr}(\mathrm{E}[\mathbf{s}_k^{(l)} \mathbf{s}_k^{(l)\mathsf{H}}]) \leq P_k. \tag{15.3}$$

This scenario is depicted in Figure 15.5.

It should be noticed that choosing diagonal precoding matrices would reduce the above scheme to the simpler amplify and forward relaying strategy. Also, the proposed linear precoding relaying technique is adapted for high SNR regimes, but not for low SNR regimes. In the low SNR regime, the performance of the relay system is imposed by the noise figure and linear precoding performs poorly because power is wasted on forwarding noise; other relaying strategies such as decode and forward are more appropriate in this case, see, e.g., [Fawaz

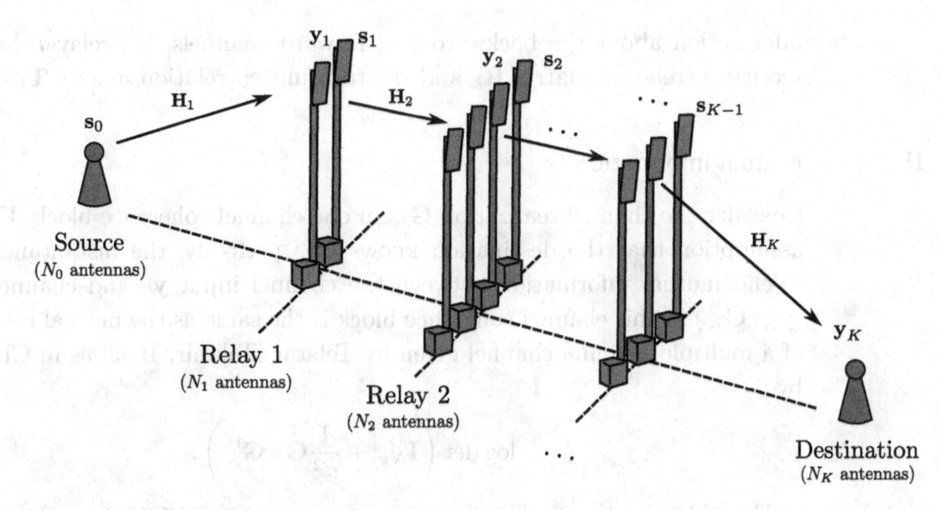

Figure 15.5 Linear relay network.

and Médard, 2010; Maric and Yates, 2010]. On the contrary, in the high SNR regime, linear precoding techniques such as amplify and forward perform well [Borade *et al.*, 2007; Maric *et al.*, 2010]. Finally, from a practical point of view, limited channel knowledge and simple linear precoding techniques at relays are particularly relevant for systems where relays have limited processing capabilities.

The signal received at the destination at time l is given by:

$$\begin{aligned}\mathbf{y}_K^{(l)} &= \mathbf{H}_K\mathbf{P}_{K-1}\mathbf{H}_{K-1}\mathbf{P}_{K-2}\ldots\mathbf{H}_2\mathbf{P}_1\mathbf{H}_1\mathbf{P}_0\mathbf{y}_0^{(l-K)} + \mathbf{w}^{(l)}\\ &= \mathbf{G}_K\mathbf{y}_0^{(l-K)} + \mathbf{w}^{(l)}\end{aligned}$$

where the end-to-end equivalent channel $\mathbf{G}_K$ is given by:

$$\begin{aligned}\mathbf{G}_K &\triangleq \mathbf{H}_K\mathbf{P}_{K-1}\mathbf{H}_{K-1}\mathbf{P}_{K-2}\ldots\mathbf{H}_2\mathbf{P}_1\mathbf{H}_1\mathbf{P}_0\\ &= \mathbf{R}_K^{\frac{1}{2}}\mathbf{X}_K\mathbf{T}_K^{\frac{1}{2}}\mathbf{P}_{K-1}\mathbf{R}_{K-1}^{\frac{1}{2}}\mathbf{X}_{K-1}\mathbf{T}_{K-1}^{\frac{1}{2}}\mathbf{P}_{K-2}\ldots\mathbf{R}_2^{\frac{1}{2}}\mathbf{X}_2\mathbf{T}_2^{\frac{1}{2}}\mathbf{P}_1\mathbf{R}_1^{\frac{1}{2}}\mathbf{X}_1\mathbf{T}_1^{\frac{1}{2}}\mathbf{P}_0.\end{aligned}$$

For clarity in what follows, let us introduce the notations

$$\begin{aligned}\mathbf{M}_0 &= \mathbf{T}_1^{\frac{1}{2}}\mathbf{P}_0\\ \mathbf{M}_k &= \mathbf{T}_{k+1}^{\frac{1}{2}}\mathbf{P}_k\mathbf{R}_k^{\frac{1}{2}},\ k\in\{1,\ldots,K-1\}\\ \mathbf{M}_K &= \mathbf{R}_K^{\frac{1}{2}}.\end{aligned}$$

Then $\mathbf{G}_K$ can be rewritten as

$$\mathbf{G}_K = \mathbf{M}_K\mathbf{X}_K\mathbf{M}_{K-1}\mathbf{X}_{K-1}\ldots\mathbf{M}_2\mathbf{X}_2\mathbf{M}_1\mathbf{X}_1\mathbf{M}_0.$$

In what follows, we will always assume that the destination knows perfectly the channel $\mathbf{G}_K$ at all times in order to decode the source data. We may additionally assume that the source and the relays have statistical channel state information about the backward and forward channels, i.e. relay k knows the receive correlation matrix $\mathbf{R}_k$ and the transmit correlation matrix $\mathbf{T}_{k+1}$.

15.2.2 Mutual information

Consider the channel realization $\mathbf{G}_K$ in one channel coherence block. Under the assumption that the destination knows $\mathbf{G}_K$ perfectly, the instantaneous end-to-end mutual information between the channel input $\mathbf{y}_0$ and channel output $(\mathbf{y}_K, \mathbf{G}_K)$ in this channel coherence block is the same as the mutual information of a multiple antenna channel given by Telatar [Telatar, 1999] as in Chapter 13 by

$$\log\det\left(\mathbf{I}_{N_K} + \frac{1}{\sigma^2}\mathbf{G}_K\mathbf{G}_K^{\mathsf{H}}\right).$$

The end-to-end mutual information averaged over multiple channel realizations is in turn given by:

$$\mathrm{E}_{(\mathbf{X}_k)}\left[\log\det\left(\mathbf{I}_{N_K} + \frac{1}{\sigma^2}\mathbf{G}_K\mathbf{G}_K^{\mathsf{H}}\right)\right]$$

where the expectation is taken over the joint realization of the $K-1$ random matrices $\mathbf{X}_k$, $k \in \{1, \ldots, K-1\}$.

As in Chapter 13, we will not be able to optimize the instantaneous mutual information, but will rather focus on optimizing the average mutual information, when the different relays have at least statistical information about the channels $\mathbf{H}_k$. Under adequate assumptions on the various channel state information known at the relays, we will therefore try to find the precoders $\mathbf{P}_k$ that maximize the end-to-end mutual information subject to power constraints (15.3). This will give us the end-to-end average mutual information

$$C(\sigma^2) \triangleq \sup_{\substack{\{\mathbf{P}_k\} \\ \frac{1}{N_k}\operatorname{tr}(\mathrm{E}[\mathbf{s}_k\mathbf{s}_k^{\mathsf{H}}]) \leq P_k}} \mathrm{E}_{(\mathbf{X}_k)}\left[\log\det\left(\mathbf{I}_{N_K} + \frac{1}{\sigma^2}\mathbf{G}_K\mathbf{G}_K^{\mathsf{H}}\right)\right].$$

Note that the average mutual information above does not necessarily represent the channel capacity in the Shannon sense here. In the next section, we will derive a limiting result for the mutual information using tools from free probability theory.

15.2.3 Large dimensional analysis

In this section, we consider the instantaneous mutual information per source antenna between the source and the destination and derive its asymptotic value as $N_0, N_1, \ldots, N_K$ grow large at a similar rate. We obtain the following result.

Theorem 15.1. *For the system described above, and under the assumption that the destination knows $\mathbf{G}_K$ at all times, that $\frac{N_k}{N_K} \to c_k$, $0 < c_k < \infty$, and $\mathbf{M}_k^{\mathsf{H}}\mathbf{M}_k$*

has an almost sure l.s.d. F_k *and has uniformly bounded spectral norm along growing* N_k*, for all* k*, then the normalized (per source antenna) instantaneous mutual information*

$$I \triangleq \frac{1}{N_0} \log\det\left(\mathbf{I}_{N_K} + \frac{1}{\sigma^2}\mathbf{G}_K\mathbf{G}_K^{\mathsf{H}}\right)$$

converges almost surely to

$$I^{\infty} = \sum_{k=0}^{K} \frac{c_i}{c_0} \int \log\left(1 + \frac{1}{\sigma^2}\frac{a_{k+1}}{c_k} h_k^K t\right) dF_k(t) - K\frac{\sigma^2}{c_0}\prod_{k=0}^{K} h_k$$

where $a_{K+1} = 1$ *by convention and* $h_0, h_1, \ldots, h_K$ *are the solutions of*

$$\prod_{j=0}^{K} h_j = \int \frac{a_{k+1} h_k^K t}{1 + \frac{1}{\sigma^2}\frac{a_{k+1}}{c_k} h_k^K t} dF_k(t).$$

We give hereafter the main steps of the proof.

Proof. The proof of Theorem 15.1 consists of the following steps:

- *Step 1.* We first obtain an expression for the limiting S-transform $S_G(z)$ of $\mathbf{G}_K\mathbf{G}_K^{\mathsf{H}}$ using the fact that the matrix $\mathbf{G}_K$ is composed of products of asymptotically almost everywhere free matrices.
- *Step 2.* We then use $S_G(z)$ to determine the limiting ψ-transform $\psi_G(z)$ of $\mathbf{G}_K\mathbf{G}_K^{\mathsf{H}}$, which we recall is closely related to the Stieltjes transform of $\mathbf{G}_K\mathbf{G}_K^{\mathsf{H}}$, see Definition 3.6.
- *Step 3.* We finally use the relation between the ψ-transform and the Shannon transform to complete the derivation.

Step 1.
We show first the following result. As all $N_k \to \infty$ with the same rate, the S-transform of $\mathbf{G}_K\mathbf{G}_K^{\mathsf{H}}$ converges almost surely to $S_G(z)$, given by:

$$S_G(z) = S_{F_K}(z) \prod_{k=1}^{K} \frac{c_{k-1}}{a_k} \frac{1}{(z + c_{k-1})} S_{F_{k-1}}\left(\frac{z}{c_{k-1}}\right). \tag{15.4}$$

The proof is done by induction. First, we prove the case $K = 1$. Note that

$$\mathbf{G}_1\mathbf{G}_1^{\mathsf{H}} = \mathbf{M}_1\mathbf{X}_1\mathbf{M}_0\mathbf{M}_0^{\mathsf{H}}\mathbf{X}_1^{\mathsf{H}}\mathbf{M}_1^{\mathsf{H}}$$

and therefore, denoting systematically $S_{\mathbf{Z}}^{\infty}(z)$ the limiting almost sure S-transform of the random Hermitian matrix $\mathbf{Z}$ as the dimensions grow to infinity

$$S_G(z) = S^{\infty}_{\mathbf{X}_1\mathbf{M}_0\mathbf{M}_0^{\mathsf{H}}\mathbf{X}_1^{\mathsf{H}}\mathbf{M}_1^{\mathsf{H}}\mathbf{M}_1}(z)$$

thanks to the S-transform matrix exchange identity, Lemma 4.3. Then, from the asymptotic freeness almost everywhere of Wishart matrices and deterministic matrices, Theorem 4.5, and the S-transform product relation, Theorem 4.7, we further have

$$S_G(z) = S^{\infty}_{\mathbf{X}_1\mathbf{M}_0\mathbf{M}_0^{\mathsf{H}}\mathbf{X}_1^{\mathsf{H}}}(z) S_{F_1}(z).$$

Using again Lemma 4.3, we exchange the order in matrix $\mathbf{X}_1\mathbf{M}_0\mathbf{M}_0^{\mathsf{H}}\mathbf{X}_1^{\mathsf{H}}$ to obtain

$$S_G(z) = \frac{z+1}{z+\frac{N_0}{N_1}} S^{\infty}_{\mathbf{M}_0\mathbf{M}_0^{\mathsf{H}}\mathbf{X}_1^{\mathsf{H}}\mathbf{X}_1}\left(z\frac{N_1}{N_0}\right) S_{F_1}(z).$$

A second application of Theorem 4.7 gives

$$S_G(z) = \frac{z+1}{z+\frac{N_0}{N_1}} S_{F_0}\left(z\frac{N_1}{N_0}\right) S^{\infty}_{\mathbf{X}_1^{\mathsf{H}}\mathbf{X}_1}\left(z\frac{N_1}{N_0}\right) S_{F_1}(z)$$

where we recognize the S-transform of (a scaled version of) the Marčenko–Pastur law. Applying both Lemma 4.2 and Theorem 4.8, we obtain

$$S_G(z) = \frac{z+1}{z+\frac{N_0}{N_1}} S_{F_0}\left(z\frac{N_1}{N_0}\right) \frac{1}{a_1} \frac{1}{z\frac{N_1}{N_0}+\frac{N_1}{N_0}} S_{F_1}(z)$$

and finally

$$S_G(z) = S_{F_1}(z) \frac{c_0}{a_1} \frac{1}{z+c_0} S_{F_0}\left(\frac{z}{c_0}\right)$$

which proves the case $K = 1$.

Now, we need to prove that, if the result holds for $K = q$, it also holds for $K = q + 1$. Note that

$$\mathbf{G}_{q+1}\mathbf{G}_{q+1}^{\mathsf{H}} = \mathbf{M}_{q+1}\mathbf{X}_{q+1}\mathbf{M}_q\mathbf{X}_q \ldots \mathbf{M}_1\mathbf{X}_1\mathbf{M}_0\mathbf{M}_0^{\mathsf{H}}\mathbf{X}_1^{\mathsf{H}}\mathbf{M}_1^{\mathsf{H}} \ldots \mathbf{X}_q^{\mathsf{H}}\mathbf{M}_q^{\mathsf{H}}\mathbf{X}_{q+1}^{\mathsf{H}}\mathbf{M}_{q+1}^{\mathsf{H}}.$$

Therefore

$$S_{\mathbf{G}_{q+1}\mathbf{G}_{q+1}^{\mathsf{H}}}(z) = S_{\mathbf{X}_{q+1}\mathbf{M}_q \ldots \mathbf{M}_q^{\mathsf{H}}\mathbf{X}_{q+1}^{\mathsf{H}}\mathbf{M}_{q+1}^{\mathsf{H}}\mathbf{M}_{q+1}}(z).$$

We use once more the same approach and theorems as above to obtain successively, with $K = q + 1$

$$\begin{aligned}
S_G(z) &= S^{\infty}_{\mathbf{X}_{q+1}\ldots\mathbf{X}_{q+1}^{\mathsf{H}}}(z) S_{\mathbf{M}_{q+1}^{\mathsf{H}}\mathbf{M}_{q+1}}(z) \\
&= \frac{z+1}{z+\frac{N_q}{N_{q+1}}} S^{\infty}_{\mathbf{M}_q\ldots\mathbf{M}_q^{\mathsf{H}}\mathbf{X}_{q+1}^{\mathsf{H}}\mathbf{X}_{q+1}}\left(z\frac{N_{q+1}}{N_q}\right) S_{F_{q+1}}(z) \\
&= \frac{z+1}{z+\frac{N_q}{N_{q+1}}} S_{F_q}\left(z\frac{N_{q+1}}{N_q}\right) S^{\infty}_{\mathbf{X}_{q+1}^{\mathsf{H}}\mathbf{X}_{q+1}}\left(z\frac{N_{q+1}}{N_q}\right) S_{F_{q+1}}(z).
\end{aligned}$$

As above, developing the expression of the S-transform of the Marčenko–Pastur law, we then obtain

$$S_G(z) = \frac{z+1}{z+\frac{N_q}{N_{q+1}}} S_{F_q}\left(z\frac{N_{q+1}}{N_q}\right)\left[\prod_{i=1}^{q}\frac{1}{a_i}\frac{N_{i-1}}{N_q}\frac{1}{z\frac{N_{q+1}}{N_q}+\frac{N_{i-1}}{N_q}} S_{F_{i-1}}\left(z\frac{N_{q+1}}{N_{i-1}}\right)\right]$$
$$\times \frac{1}{a_{q+1}}\frac{1}{\frac{N_{q+1}}{N_q}+z\frac{N_{q+1}}{N_q}} S_{F_{q+1}}(z)$$

which further simplifies as

$$\begin{aligned} S_G(z) &= \frac{z+1}{z+\frac{N_q}{N_{q+1}}} S_{F_{q+1}}(z)\frac{1}{a_{q+1}}\frac{N_q}{N_{q+1}}\frac{1}{z+1} S_{F_q}\left(z\frac{N_{q+1}}{N_q}\right) \\ &\quad\times \prod_{i=1}^{q}\frac{\frac{N_{i-1}}{N_{q+1}}}{a_i}\frac{1}{z+\frac{N_{i-1}}{N_{q+1}}} S_{F_{i-1}}\left(z\frac{N_{q+1}}{N_{i-1}}\right) \\ &= S_{F_{q+1}}(z)\prod_{i=1}^{q+1}\frac{1}{a_i}\frac{N_{i-1}}{N_{q+1}}\frac{1}{z+\frac{N_{i-1}}{N_{q+1}}} S_{F_{i-1}}\left(z\frac{N_{q+1}}{N_{i-1}}\right) \\ &= S_{F_{q+1}}(z)\prod_{i=1}^{q+1}\frac{c_{i-1}}{a_i}\frac{1}{(z+c_{i-1})} S_{F_{i-1}}\left(\frac{z}{c_{i-1}}\right) \end{aligned}$$

which is the intended result.

Step 2.
We now prove the following. Denoting $a_{K+1} = 1$, for $s \in \mathbb{C}\setminus\mathbb{R}^+$, we have:

$$s(\psi_G(s))^K = \prod_{i=0}^{K}\frac{c_i}{a_{i+1}}\psi_{F_i}^{-1}\left(\frac{\psi_G(s)}{c_i}\right). \tag{15.5}$$

This unfolds first from (15.4), by multiplying each side by $z/(z+1)$

$$\frac{z}{z+1}S_G(z) = \frac{z}{z+1}S_{F_K}(z)\prod_{k=1}^{K}\frac{c_{k-1}}{a_k}\frac{1}{(z+c_{k-1})} S_{F_{k-1}}\left(\frac{z}{c_{k-1}}\right)$$

where the last right-hand side term can be rewritten

$$S_{F_{k-1}}\left(\frac{z}{c_{k-1}}\right) = \frac{1+\frac{z}{c_{k-1}}}{\frac{z}{c_{k-1}}}\frac{\frac{z}{c_{k-1}}}{1+\frac{z}{c_{k-1}}} S_{F_{k-1}}\left(\frac{z}{c_{k-1}}\right).$$

Using the free probability definition of the S-transform in relation to the ψ-transform, Theorem 4.3, we obtain

$$\psi_G^{-1}(z) = \frac{1}{z^K}\psi_{F_K}^{-1}(z)\prod_{i=1}^{K}\frac{c_{i-1}}{a_i}\psi_{F_{i-1}}^{-1}\left(\frac{z}{c_{i-1}}\right)$$

or equivalently

$$\psi_G^{-1}(z) = \frac{1}{z^K} \prod_{i=0}^{K} \frac{c_i}{a_{i+1}} \psi_{F_i}^{-1}\left(\frac{z}{c_i}\right).$$

Substituting $z = \psi_G(s)$ gives the result.

Step 3.

We subsequently show that, as $N_0, N_1, \ldots, N_K$ go to infinity, the derivative of the almost surely limiting instantaneous mutual information I^∞ is given by:

$$\frac{dI^\infty}{d(\sigma^{-2})} = \frac{1}{c_0} \prod_{i=0}^{K} h_i$$

where $h_0, h_1, \ldots, h_K$ are the solutions to the following

$$\prod_{j=0}^{K} h_j = c_i \int \frac{h_i^K t}{\frac{c_i}{a_i+1} + \frac{1}{\sigma^2} h_i^K t} dF_i(t).$$

First, we note that

$$I = \frac{1}{N_0} \log\det\left(\mathbf{I}_{N_K} + \frac{1}{\sigma^2} \mathbf{G}_K \mathbf{G}_K^{\mathsf{H}}\right) \xrightarrow{\text{a.s.}} \frac{1}{c_0} \int \log\left(1 + \frac{1}{\sigma^2} t\right) dG(t)$$

the convergence result being valid here because the largest eigenvalue of $\mathbf{G}_K\mathbf{G}_K^{\mathsf{H}}$ is almost surely bounded as the system dimensions grow large. This is due to the fact that the deterministic matrices $\mathbf{M}_k\mathbf{M}_k^{\mathsf{H}}$ are uniformly bounded in spectral norm and that the largest eigenvalue of $\mathbf{X}_k\mathbf{X}_k^{\mathsf{H}}$ is almost surely bounded for all N_k by Theorem 7.1; therefore, the largest eigenvalue of $\mathbf{G}_K\mathbf{G}_K^{\mathsf{H}}$, which is smaller than or equal to the product of all these largest eigenvalues is almost surely uniformly bounded. The dominated convergence theorem, Theorem 6.3, then ensures the convergence.

Now, we also have the relation

$$\begin{aligned} \frac{dI^\infty}{d(\sigma^{-2})} &= \frac{1}{c_0} \int \frac{t}{1 + \frac{1}{\sigma^2} t} dF_G(t) \\ &= -\frac{\sigma^2}{c_0} \psi_G(-\sigma^{-2}). \end{aligned} \tag{15.6}$$

Let us denote

$$\tau = \psi_G(-\sigma^{-2})$$
$$g_i = \psi_{F_i}^{-1}\left(\frac{t}{c_i}\right).$$

From (15.6), we have:

$$\tau = -\frac{c_0}{\sigma^2} \frac{dI^\infty}{d(\sigma^{-2})}.$$

Substituting $s = -\sigma^{-2}$ in (15.5) and using τ and g_i above, it follows that

$$-\frac{\tau^K}{\sigma^2} = \prod_{i=0}^{K} \frac{c_i}{a_{i+1}} g_i.$$

Using the definition of the ψ-transform, this is finally

$$\tau = c_i \int \frac{g_i t}{1 - g_i t} dF_i(t).$$

These last three equations together give

$$\left(-\frac{1}{\sigma^2}\right)^{K+1} \left(c_0 \frac{dI^\infty}{d(\sigma^{-2})}\right)^K = \prod_{i=0}^{K} \frac{c_i}{a_{i+1}} g_i$$

and

$$-\frac{1}{\sigma^2}\left(c_0 \frac{dI^\infty}{d(\sigma^{-2})}\right) = c_i \int \frac{g_i t}{1 - g_i t} dF_i(t).$$

Defining now

$$h_i \triangleq \left(\frac{c_i}{a_{i+1}}\right)^{\frac{1}{K}} (-g_i \sigma^2)^{\frac{1}{K}}$$

we have

$$c_0 \frac{dI^\infty}{d(\sigma^{-2})} = \prod_{i=0}^{K} h_i. \tag{15.7}$$

Using again these last three equations, we obtain

$$-\frac{1}{\sigma^2} \prod_{j=0}^{K} h_j = c_i \int \frac{-\frac{1}{\sigma^2} h_i^K \frac{a_{i+1}}{c_i} t}{1 - (-\frac{1}{\sigma^2}) h_i^K \frac{a_{i+1}}{c_i} t} dF_i(t)$$

or equivalently

$$\prod_{j=0}^{N} h_j = c_i \int \frac{h_i^K t}{\frac{c_i}{a_{i+1}} + \frac{1}{\sigma^2} h_i^K t} dF_i(t).$$

This, along with Equation (15.7), gives the result.

We finally need to prove that the final result is indeed a pre-derivative of $dI^\infty / d\sigma^{-2}$ that also verifies $\lim_{\sigma^2 \to \infty} I^\infty = 0$, i.e. the mutual information is zero for null SNR. This unfolds, as usual, by differentiating the final result. In particular, we successively obtain

$$\begin{aligned}
&\sum_{i=0}^{K} c_i \int \frac{t\left(h_i^K + \frac{K}{\sigma^2} h_i^{K-1} h_i'\right)}{\frac{c_i}{a_{i+1}}\left(1 + \frac{a_{i+1}}{\sigma^2 c_i} h_i^K t\right)} dF_i(t) - K \prod_{i=0}^{K} h_i - \frac{K}{\sigma^2}\left(\sum_{i=0}^{K} h_i' \prod_{j \neq i} h_j\right) \\
&= \sum_{i=0}^{K} c_i \int \frac{t h_i^K dF_i(t)}{\frac{c_i}{a_{i+1}} + \frac{1}{\sigma^2} h_i^K t} + \frac{K}{\sigma^2} \sum_{i=0}^{K} \frac{h_i'}{h_i} c_i \int \frac{t h_i^K dF_i(t)}{\frac{c_i}{a_{i+1}} + \frac{1}{\sigma^2} h_i^K t} - K \prod_{i=0}^{K} h_i \\
&- \frac{K}{\sigma^2}\left(\sum_{i=0}^{K} \frac{h_i'}{h_i} \prod_{j=0}^{K} h_j\right) \\
&= \sum_{i=0}^{K} \prod_{j=0}^{K} h_j + \frac{K}{\sigma^2}\left(\sum_{i=0}^{K} \frac{h_i'}{h_i} \prod_{j=0}^{K} h_j\right) - K \prod_{i=0}^{K} h_i - \frac{K}{\sigma^2}\left(\sum_{i=0}^{K} \frac{h_i'}{h_i} \prod_{j=0}^{K} h_j\right) \\
&= (K+1) \prod_{j=0}^{K} h_j - K \prod_{j=0}^{K} h_j \\
&= \prod_{j=0}^{K} h_j
\end{aligned}$$

where $h_i' \triangleq \frac{dh_i}{d(\sigma^{-2})}$. This completes the proof. □

Theorem 15.1 holds as usual for any arbitrary set of precoding matrices $\mathbf{P}_k$, $k \in \{0, \ldots, K-1\}$ such that $\mathbf{M}_k^{\mathsf{H}} \mathbf{M}_k$ has uniformly bounded spectral norm.

15.2.4 Optimal transmission strategy

In this section, we analyze the optimal linear precoding strategies $\mathbf{P}_k$, $k \in \{0, \ldots, K-1\}$, at the source and the relays that allow us to maximize the average mutual information. We characterize the optimal transmit directions determined by the singular vectors of the precoding matrices at source and relays, for a system with finite dimensions $N_0, \ldots, N_K$ before considering large dimensional limits.

The main result of this section is given by the following theorem.

Theorem 15.2. *For each $i \in \{1, \ldots, K\}$, denote $\mathbf{T}_i = \mathbf{U}_{t,i} \mathbf{\Lambda}_{t,i} \mathbf{U}_{t,i}^{\mathsf{H}}$ and $\mathbf{R}_i = \mathbf{U}_{r,i} \mathbf{\Lambda}_{r,i} \mathbf{U}_{r,i}^{\mathsf{H}}$ the spectral decompositions of the correlation matrices $\mathbf{T}_i$ and $\mathbf{R}_i$, where $\mathbf{U}_{t,i}$ and $\mathbf{U}_{r,i}$ are unitary and $\mathbf{\Lambda}_{t,i}$ and $\mathbf{\Lambda}_{r,i}$ are diagonal, with their respective eigenvalues ordered in non-increasing order. Then, assuming the destination knows $\mathbf{G}_K$ at all times and that the source and intermediary relays have local statistical information about the backward and forward channels, the optimal linear precoding matrices that maximize the average mutual information* *under power constraints* (15.3) *can be written as*

$$\mathbf{P}_0 = \mathbf{U}_{t,1}\mathbf{\Lambda}_{P_0}$$
$$\mathbf{P}_i = \mathbf{U}_{t,i+1}\mathbf{\Lambda}_{P_i}\mathbf{U}_{r,i}^{\mathsf{H}}, \ i \in \{1,\ldots,K-1\}$$

where $\mathbf{\Lambda}_{P_i}$ *are diagonal matrices with non-negative real diagonal elements.*

We do not provide here the proof of Theorem 15.2 that can be found in [Fawaz *et al.*, 2011, Appendix C]. Theorem 15.2 indicates that the power maximization can then be divided into two phases: the alignment of the eigenvectors on the one hand and the search for the optimal eigenvalues (entries of $\mathbf{\Lambda}_{t,i}$ and $\mathbf{\Lambda}_{r,i}$) on the other hand.

We now apply the above result to two specific multi-hop communication scenarios. In these scenarios, a multi-hop multiple antenna system as above is considered and the asymptotic mutual information is developed in the uncorrelated and exponential correlation cases, respectively.

15.2.4.1 Uncorrelated multi-hop MIMO

In this example, we consider an uncorrelated multi-hop MIMO system, i.e. all correlation matrices are equal to the identity matrix.

Before analyzing this scenario, we mention the following finite dimensional result, which can be proved rather easily by induction on i.

$$\operatorname{tr}(\mathrm{E}[\mathbf{s}_i\mathbf{s}_i^{\mathsf{H}}]) = a_i \operatorname{tr}(\mathbf{P}_i\mathbf{R}_i\mathbf{P}_i^{\mathsf{H}}) \prod_{k=0}^{i-1} \frac{a_k}{N_k} \operatorname{tr}(\mathbf{T}_{k+1}\mathbf{P}_k\mathbf{P}_k^{\mathsf{H}}). \tag{15.8}$$

By Theorem 15.2, in the uncorrelated case ($\mathbf{R}_k$ and $\mathbf{T}_k$ taken identity) the optimal precoding matrices should be diagonal. Assuming equal power allocation at source and relays, the precoding matrices are of the form $\mathbf{P}_k = \alpha_k\mathbf{I}_{N_k}$, where α_k is real positive and chosen to satisfy the power constraints. Using the expression (15.8), it can be shown by induction on k that the coefficients α_k in the uncorrelated case are necessarily given by:

$$\alpha_0 = \sqrt{P_0}$$
$$\alpha_i = \sqrt{\frac{P_i}{a_i P_{i-1}}}, \ i \in \{1,\ldots,K-1\}$$
$$\alpha_K = 1. \tag{15.9}$$

Then the asymptotic mutual information for the uncorrelated multi-hop MIMO system with equal power allocation is given by:

$$I^\infty = \sum_{i=0}^{K} \frac{c_i}{c_0} \log\left(1 + \frac{h_i^K a_{i+1}\alpha_i^2}{\sigma^2 c_i}\right) - K\frac{\sigma^2}{c_0}\prod_{i=0}^{K} h_i \tag{15.10}$$

where $h_0, h_1, \ldots, h_K$ are the solutions of the system of $K+1$ multivariate polynomial equations

$$\prod_{j=0}^{K} h_j = \frac{h_i^K \alpha_i^2 a_{i+1}}{1 + \frac{h_i^K a_{i+1} \alpha_i^2}{\sigma^2 c_i}}.$$

15.2.4.2 Exponentially correlated multi-hop MIMO

In this second example, the asymptotic mutual information is developed in the case of exponential correlation matrices and precoding matrices with singular vectors as in Theorem 15.2. Similar to Jakes' model, exponential correlation matrices are a common model of correlation, particularly for uniform linear antenna array, see, e.g., [Loyka, 2001; Martin and Ottersten, 2004; Oestges *et al.*, 2008].

We assume that the relay at level i is equipped with a uniform linear antenna array of length L_i, characterized by its antenna spacing $l_i = L_i/N_i$ and its characteristic distances $\Delta_{t,i}$ and $\Delta_{r,i}$ proportional to transmit and receive spatial coherences, respectively. Then the receive and transmit correlation matrices at relaying level i can, respectively, be modeled by the following Hermitian Toeplitz matrices

$$\mathbf{R}_i = \begin{pmatrix} 1 & r_i & r_i^2 & \ldots & r_i^{N_i-1} \\ r_i & 1 & \ddots & \ddots & \vdots \\ r_i^2 & \ddots & \ddots & \ddots & r_i^2 \\ \vdots & \ddots & \ddots & 1 & r_i \\ r_i^{N_i-1} & \ldots & r_i^2 & r_i & 1 \end{pmatrix}$$

and

$$\mathbf{T}_{i+1} = \begin{pmatrix} 1 & t_{i+1} & t_{i+1}^2 & \ldots & t_{i+1}^{N_i-1} \\ t_{i+1} & 1 & \ddots & \ddots & \vdots \\ t_{i+1}^2 & \ddots & \ddots & \ddots & t_{i+1}^2 \\ \vdots & \ddots & \ddots & 1 & t_{i+1} \\ t_{i+1}^{N_i-1} & \ldots & t_{i+1}^2 & t_{i+1} & 1 \end{pmatrix}$$

where the antenna correlation at receive and transmit sides read $r_i = e^{-\frac{l_i}{\Delta_{r,i}}}$ and $t_{i+1} = e^{-\frac{l_i}{\Delta_{t,i}}}$, respectively. It can be verified that these Toeplitz matrices are of Wiener-class thanks to Szegö's theorem, Theorem 12.1. We therefore know also from Theorem 12.1 the limiting eigenvalue distribution of those deterministic matrices as their dimensions grow large. We further assume here equal power allocation over the optimal directions, i.e. the singular values of $\mathbf{P}_i$ are chosen to be all equal: $\mathbf{\Lambda}_{P_i} = \alpha_i \mathbf{I}_{N_i}$, where α_i is real positive and chosen to satisfy the power constraint (15.3). Equal power allocation may not be the optimal power allocation scheme, but it is considered in this example for simplicity.

Using the power constraint expression for general correlation models (15.8) and considering precoding matrices $\mathbf{P}_i = \mathbf{U}_{r,i}^{\mathsf{H}}(\alpha_i \mathbf{I}_{N_i})\mathbf{U}_{t,i+1}$, with $\mathbf{U}_{r,i}$ unitary such that $\mathbf{R}_i = \mathbf{U}_{r,i}\mathbf{\Lambda}_{r,i}\mathbf{U}_{r,i}^{\mathsf{H}}$ with $\mathbf{\Lambda}_{r,i}$ diagonal and $\mathbf{U}_{t,i}$ such that $\mathbf{T}_i = \mathbf{U}_{t,i}\mathbf{\Lambda}_{t,i}\mathbf{U}_{t,i}^{\mathsf{H}}$ with $\mathbf{\Lambda}_{t,i}$ diagonal, following the conditions of Theorem 15.2 with equal singular values α_i, we can show by induction on i that the coefficients α_i respecting the power constraints for any correlation model are now given by:

$$\alpha_0 = \sqrt{P_0}$$

$$\alpha_i = \sqrt{\frac{P_i}{a_i P_{i-1}} \frac{\operatorname{tr}(\mathbf{\Lambda}_{r,i-1})}{\operatorname{tr}(\mathbf{\Lambda}_{r,i})} \frac{k_i}{\operatorname{tr}(\mathbf{\Lambda}_{t,i}\mathbf{\Lambda}_{r,i-1})}}, \; i \in \{1, \ldots, K-1\}$$

$$\alpha_K = 1.$$

Applying the exponential correlation model to the above relations and making the dimensions of the system grow large, it can be shown that in the asymptotic regime, the α_i respecting the power constraint for the exponentially correlated system converge to the same value, given in (15.9), as for the uncorrelated system. It can then be shown that the asymptotic mutual information in this scenario is given by:

$$I^\infty = \sum_{i=0}^{K} \frac{c_i}{c_0 \pi^2} \int_{t=-\infty}^{\infty} \int_{u=-\infty}^{\infty} g(t, u; \mathbf{h}) dt du - K \frac{\sigma^2}{c_0} \prod_{i=0}^{K} h_i$$

where, denoting $\mathbf{h} = (h_0, \ldots, h_K)$

$$g(t, u; \mathbf{h}) \triangleq \frac{1}{1+t^2} \frac{1}{1+u^2} \log\left(1 + \rho_{r,i}\rho_{t,i+1} \frac{h_i^K a_{i+1}\alpha_i^2}{\sigma^2 c_i} \frac{(1+t^2)}{(\rho_{r,i}^2 + t^2)} \frac{(1+u^2)}{(\rho_{t,i+1}^2 + u^2)}\right)$$

and $h_0, h_1, \ldots, h_K$ are the solutions of the system

$$\prod_{j=0}^{K} h_j = \frac{2}{\pi} \frac{h_i^K a_{i+1}\alpha_i^2}{\sqrt{\rho_{r,i}\rho_{t,i+1} + \frac{h_i^K a_{i+1}\alpha_i^2}{\sigma^2 c_i}} \sqrt{\frac{1}{\rho_{r,i}\rho_{t,i+1}} + \frac{h_i^K a_{i+1}\alpha_i^2}{\sigma^2 c_i}}} F\left(\frac{\pi}{2}, \sqrt{m_i}\right)$$

with $F(\theta, x)$ the incomplete elliptic integral of the first kind given by:

$$F(\theta, x) = \int_0^\theta \frac{1}{\sqrt{1 - x^2 \sin^2(t)}} dt$$

and, for all $i \in \{0, K\}$

$$\rho_{r,i} = \frac{1 - r_i}{1 + r_i}$$

$$\rho_{t,i+1} = \frac{1 - t_{i+1}}{1 + t_{i+1}}$$

$$m_i = 1 - \frac{\left(\frac{\rho_{t,i+1}}{\rho_{r,i}} + \frac{h_i^K a_{i+1}\alpha_i^2}{\sigma^2 c_i}\right)\left(\frac{\rho_{r,i}}{\rho_{t,i+1}} + \frac{h_i^K a_{i+1}\alpha_i^2}{\sigma^2 c_i}\right)}{\left(\frac{1}{\rho_{r,i}\rho_{t,i+1}} + \frac{h_i^K a_{i+1}\alpha_i^2}{\sigma^2 c_i}\right)\left(\rho_{r,i}\rho_{t,i+1} + \frac{h_i^K a_{i+1}\alpha_i^2}{\sigma^2 c_i}\right)}$$

with the convention $r_0 = t_{N+1} = 0$. The details of these results can be found in [Fawaz *et al.*, 2011].

This concludes this chapter on the performance of multi-cellular and relay communication systems. In the subsequent chapters, the signal processing problems of source detection, separation, and statistical inference for large dimensional systems are addressed.

16 Detection

In this chapter, we now address a quite different problem than the performance evaluation of data transmissions in large dimensional communication channel models. The present chapter, along with Chapter 17, deals with practical signal processing techniques to solve problems involving (possibly large dimensional) random matrix models. Specifically in this chapter, we will first address the question of signal sensing using multi-dimensional sensor arrays.

16.1 Cognitive radios and sensor networks

A renewed motivation for large dimensional signal sensing has been recently triggered by the *cognitive radio* incentive, which, according to some, may be thought of as the next information-theoretic revolution after the original work of Shannon [Shannon, 1948] and the introduction of multiple antenna systems by Foshini [Foschini and Gans, 1998] and Telatar [Telatar, 1999]. In addition to the theoretical expression of the point-to-point noisy channel capacity in [Shannon, 1948], Shannon made us realize that, in order to achieve high rate of information transfer, increasing the transmission bandwidth is largely preferred over increasing the transmission power. Therefore, to ensure high rate communications with a finite power budget, we have to consider frequency multiplexing. This constituted the first and most important revolution in modern telecommunications and most notably wireless communications, which led today to an almost complete saturation of all possible transmission frequencies. By "all possible," we mean those frequencies that can efficiently carry information (high frequencies tend to be rapidly attenuated when propagating in the atmosphere) and be adequately processed by analog and digital devices (again, high frequency carriers require expensive and sometimes even physically infeasible radio front-ends). Foschini and Telatar brought forward the idea of multiplexing the information, not only through orthogonal frequency bands, but also in the space domain, by using spatially orthogonal propagation paths in multi-dimensional channels. As we saw in the previous chapter, though, this assumed orthogonality only holds for fairly unrealistic communication channels (very scattered propagation environments filled with objects the size of the transmission wavelength, very distant transmit and receive antennas, etc.). Also, the multiple antenna multiplexing gain is only apparent for high signal-to-noise ratios, which is inconsistent with most contemporary interference limited cellular networks. We also discussed in Chapter 15 the impracticality of large cooperative networks which require a huge load of channel state information feedback. The cognitive radio incentive, initiated with the work of Mitola [Mitola III and Maguire Jr, 1999], follows the same idea of *communication resource harvesting*. That is, cognitive radios intend to communicate not by exploiting the over-used frequency domain, or by exploiting the over-used space domain, but by exploiting so-called spectrum *holes*, jointly in time, space, and frequency. The basic idea is that, while the time, frequency, and spatial domains are over-used in the sense that telecommunication service providers have already bought all frequencies and have already placed base stations, access points, and relays to cover most areas, the effectively delivered communication service is largely discontinuous. That is, the telecommunication networks do not operate constantly in all frequency bands, at all times, and in all places at the maximum of their deliverable capacities. Multiple situations typically arise.

- A licensed network is under-used over a given period of time. This is typically the case during night-time, when little large range telecommunication service is provided.
- The frequency band exploited by a network is left free of use or the delivered content is not of interest to potential receivers. This is the case of broadcast television, whose frequency multiplexed channels are not all used simultaneously at any given space location.
- A licensed network is not used locally. This arises whenever no close user is found in a given space area, where a licensed network is operating.
- A licensed network is used simultaneously on all resource dimensions, but the users' service request induces transmission rates below the channel capacity. This arises when for instance a wideband CDMA network is used for a single-user voice call. The single-user is clearly capable, in the downlink, of decoding the CDMA stream with few errors, even if it were slightly interfered with by overlaying communications, since the CDMA code redundancy induces resistence against interfering with data streams.

The concept of cognitive radios covers a very large framework, not clearly unified to this day, though, which intends to reuse spectrum left-overs (or holes), four examples of which were given above. A cognitive radio network can be described as an autonomous network overlaying one or many existing legacy networks. While the established networks have dedicated bandwidths and spatial planning to operate, cognitive radios are not using any licensed resource, be it in space, time, or frequency. Cognitive radios are however free to use any licensed spectrum, as long as by doing so they do not dramatically interfere with the licensed networks. That is, they are able to reuse the spectrum left unused by so-called *primary networks* whenever possible, while generating minimum harm to the on-going communications. Considering the four examples above, a cognitive radio network, also called *secondary network*, could operate, respectively:

- on a given time–frequency–space resource when no communication is found to take place in the licensed frequencies for a certain amount of time;
- if the delivered data content is not locally used, by overlaying the (now interfering) network transmissions;
- if the delivered data content is intended for some user but the cognitive radio is aware that this user is sufficiently far away, by again overlaying the locally unused spectrum;
- by intentionally interfering with the established network but using a sufficiently low transmit power that still allows the licensed user to decode its own data.

Now, what makes the secondary network *cognitive* is that all the aforementioned ways of action require constant awareness of the operations taking place in the licensed networks. Indeed, as it is an absolute necessity not to interfere with the licensed users, some sort of dynamic monitoring, or information feedback, is required for the secondary network to abide by the rules. Since secondary networks are assumed to minimally impact on the networks in place, it is a conventional assumption to consider that the licensed networks do not proactively deliver network information to the cognitive radio. It is even conventional to assume that the licensed networks are completely oblivious of the existence of potential interferers. Therefore, legacy telecommunication networks need not be restructured in order to face the interference of the new secondary networks. As a consequence, all the burden is placed on the cognitive radio to *learn* about its own environment. This is relatively easy when dealing with surrounding base stations and other fixed transmitters, as much data can be exploited in the long-term, but this is not so for mobile users. Service providers sometimes do not transmit at all (apart from pilot data), in which case secondary networks can detect a spectrum hole and exploit it. However, the real gain of cognitive radios does not come solely from benefiting from completely unused access points, but rather from benefiting from overlaying on-going communications while not affecting the licensed users. A classical example is that of a mobile phone network, which can cover an area as large as a few kilometers. In day-time, it is uncommon for a given base station never to be in use (for CDMA transmissions, remember that this means that the whole spectrum is then used at once), but it is also uncommon that the users communicating with this base station are always located close to a secondary network. A cognitive radio can always overlay the data transmitted by an operating base station if the user, located somewhere in a large area, is not found to be anywhere close to the cognitive network. For in-house cognitive radios, such as femto-cells in closed access (see, e.g., [Calin *et al.*, 2010; Chandrasekhar *et al.*, 2009; Claussen *et al.*, 2008], it can even be assumed that overlaying communication can take place almost continuously, as long as no user inside the house or in neighboring houses establishes a communication with this network.

The question of whether an active user is to be found in the vicinity of a cognitive radio is therefore of prior importance to establish reliable overlaying communications in cognitive radios. For that, a cognitive radio needs to be able to *sense* neighboring users in active transmissions. This can be performed by simple energy detection, as in the original work from Urkowitz [Urkowitz, 1967]. However, energy detection is meant for single antenna transmitters and receivers under additive white Gaussian noise conditions and does therefore not take into account the possibility of joint processing at the sensor network level in MIMO fading channel conditions. In this chapter, we will investigate the various approaches brought by random matrix theory to perform signal detection as reliably as possible. We will first investigate the generalization of the Urkowitz approach to multiple sources and multiple receivers under a small dimensional random matrix approach. The rather involved result we will present will then motivate large dimensional random matrix analysis. Most notably, approaches that require minimum a priori knowledge of the environment will be studied from a large dimensional perspective. Indeed, it must be assumed that the cognitive radio is completely unaware even of the expected received signal-to-noise ratio in a given frequency band, as it exactly intends to decide whether only noise or informative signals are received within this band.

Before getting into random matrix applications, let us model the signal sensing problem.

16.2 System model

We consider a communication network composed of K transmitting sources, e.g. this can be either a K-antenna transmitter or K single antenna (not necessarily uncorrelated) information sources, and a receiver composed of N sensors, be they the uncorrelated antennas of a single terminal or a mesh of scattered sensors, similar to the system model exploited in, e.g., [Cabric *et al.*, 2006; Ghasemi and Sousa, 2005, 2007; Mishra *et al.*, 2006; Sun *et al.*, 2007a,b; Wang *et al.*, 2010; Zhang and Letaief, 2008]. To enhance the multiple antenna (MIMO) model analogy, the set of sources and the set of sensors will be collectively referred to as *the transmitter* and *the receiver*, respectively. The communication channel between the transmitter and the receiver is modeled by the matrix $\mathbf{H} \in \mathbb{C}^{N \times K}$, with (i,j)th entry h_{ij}. If at time l the transmitter emits data, those are denoted by the K-dimensional vector $\mathbf{x}^{(l)} = (x_1^{(l)}, \ldots, x_K^{(l)})^\mathsf{T} \in \mathbb{C}^K$ and are assumed independent across time. The additive white Gaussian noise at the receiver is modeled, at time l, by the vector $\sigma\mathbf{w}^{(l)} = \sigma(w_1^{(l)}, \ldots, w_N^{(l)})^\mathsf{T} \in \mathbb{C}^N$, where σ^2 denotes the variance of the noise vector entries, again assumed independent across time. Without generality restriction, we consider in the following zero

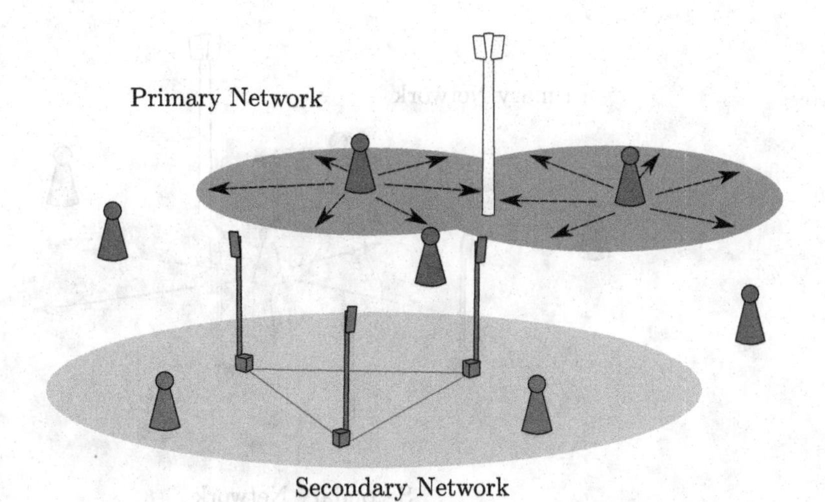

Figure 16.1 A cognitive radio network under hypothesis $\mathcal{H}_0$, i.e. no close user is transmitting during the exploration period.

mean and unit variance of the entries of both $\mathbf{w}^{(l)}$ and $\mathbf{x}^{(l)}$, i.e. $\mathrm{E}[|w_i^{(l)}|^2] = 1$, $\mathrm{E}[|x_i^{(l)}|^2] = 1$ for all i. We then denote $\mathbf{y}^{(l)} = (y_1^{(l)}, \ldots, y_N^{(l)})^\mathsf{T}$ the N-dimensional data received at time l. Assuming the channel coherence time is at least as long as M sampling periods, we finally denote $\mathbf{Y} = [\mathbf{y}^{(1)}, \ldots, \mathbf{y}^{(M)}] \in \mathbb{C}^{N \times M}$ the matrix of the concatenated receive i.i.d. vectors.

Depending on whether the transmitter emits informative signals, we consider the following hypotheses.

- $\mathcal{H}_0$. Only background noise is received.
- $\mathcal{H}_1$. Informative signals plus background noise are received.

Both scenarios of cognitive radio networks under hypotheses $\mathcal{H}_0$ and $\mathcal{H}_1$ are depicted in Figure 16.1 and Figure 16.2, respectively. Figure 16.1 illustrates the case when users neighboring the secondary network are not transmitting, while Figure 16.2 illustrates the opposite situation when a neighboring user is found to transmit in the frequency resource under exploration.

Therefore, under condition $\mathcal{H}_0$, we have the model

$$\mathbf{Y} = \sigma \mathbf{W}$$

with $\mathbf{W} = [\mathbf{w}^{(1)}, \ldots, \mathbf{w}^{(M)}] \in \mathbb{C}^{N \times M}$ and under condition $\mathcal{H}_1$

$$\mathbf{Y} = \begin{pmatrix} \mathbf{H} & \sigma \mathbf{I}_N \end{pmatrix} \begin{pmatrix} \mathbf{X} \\ \mathbf{W} \end{pmatrix} \tag{16.1}$$

with $\mathbf{X} = [\mathbf{x}^{(1)}, \ldots, \mathbf{x}^{(M)}] \in \mathbb{C}^{N \times M}$.

Under this hypothesis, we further denote $\mathbf{\Sigma}$ the covariance matrix of $\mathbf{y}^{(1)}$

$$\mathbf{\Sigma} = \mathrm{E}[\mathbf{y}^{(1)}\mathbf{y}^{(1)\mathsf{H}}] = \mathbf{H}\mathbf{H}^\mathsf{H} + \sigma^2 \mathbf{I}_N = \mathbf{U}\mathbf{G}\mathbf{U}^\mathsf{H}$$

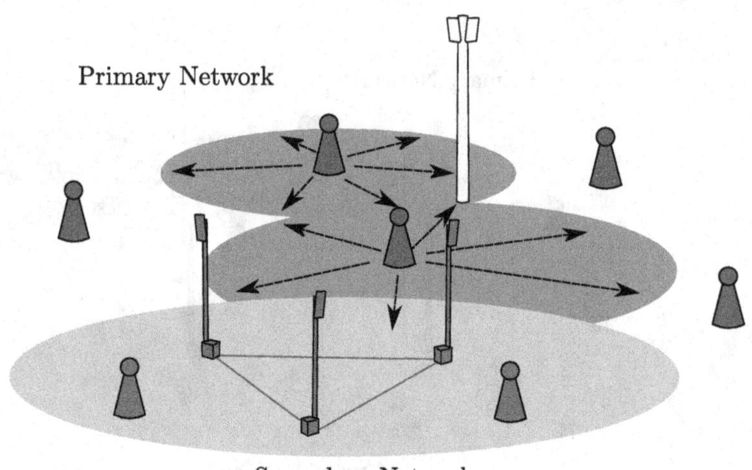

Figure 16.2 A cognitive radio network under hypothesis $\mathcal{H}_1$, i.e. at least one close user is transmitting during the exploration period.

where $\mathbf{G} = \operatorname{diag}\left(\nu_1 + \sigma^2, \ldots, \nu_N + \sigma^2\right) \in \mathbb{R}^{N \times N}$, with $\{\nu_1, \ldots, \nu_N\}$ the eigenvalues of $\mathbf{H}\mathbf{H}^{\mathsf{H}}$ and $\mathbf{U} \in \mathbb{C}^{N \times N}$ a certain unitary matrix.

The receiver is entitled to decide whether the primary users are transmitting informative signals or not. That is, the receiver is required to *test* the hypothesis $\mathcal{H}_0$ against the hypothesis $\mathcal{H}_1$. The receiver is however considered to have very limited information about the transmission channel and is in particular not necessarily aware of the exact number K of sources and of the signal-to-noise ratio. For this reason, following the maximum entropy principle [Jaynes, 1957a,b], we seek a probabilistic model for the unknown variables, which is both (i) consistent with the little accessible prior information available to the sensor network and (ii) has maximal entropy over the set of densities that validate (i). Maximum entropy considerations, which we do not develop here, are further discussed in Chapter 18, as they are at the core of the channel models developed in this chapter. We therefore admit for the time being that the entropy maximizing probability distribution of a random vector, the knowledge about which is limited to its population covariance matrix, is a multivariate Gaussian distribution with zero mean and covariance matrix the known population covariance matrix. If the population covariance matrix is unknown but is known to be of unit trace, then the entropy maximizing distribution is now multivariate independent Gaussian with zero mean and normalized identity covariance matrix. Therefore, if the channel matrix $\mathbf{H}$ is only known to satisfy, as is often the case in the short-term, $\mathrm{E}[\frac{1}{N} \operatorname{tr} \mathbf{H}\mathbf{H}^{\mathsf{H}}] = E$, with E the total power carried through the channel, the maximum entropy principle states that the entries h_{ij} should be modeled as independent and all Gaussian distributed with zero mean and variance E/K. For the same reason, both noise $w_i^{(l)}$ and signal $x_i^{(l)}$ entries are taken independent Gaussian with zero mean and variance $\mathrm{E}[|w_i^{(l)}|^2] = 1$, $\mathrm{E}[|x_i^{(l)}|^2] = 1$. Obviously, the above scalings depend on the definition of the signal-to-noise ratio.

Now that the model is properly defined, we turn to the question of testing hypothesis $\mathcal{H}_0$ against hypothesis $\mathcal{H}_1$. The idea is to decide, based on the available prior information and upon observation of $\mathbf{Y}$, whether $\mathcal{H}_0$ is more likely than $\mathcal{H}_1$. Instead of exploiting structural features of the signal, such as cyclostationarity as in, e.g., [Enserink and Cochran, 1994; Gardner, 1991; Kim and Shin, 2008], we consider here the optimal Neyman–Pearson decision test. This is what we study first in the following (Section 16.3) under different prior information on all relevant system parameters. We will realize that the optimal Neyman–Pearson test, be it explicitly derivable for the model under study, leads nonetheless to very involved formulations, which cannot flexibly be extended to more involved system models. We will therefore turn to simpler suboptimal tests, whose behavior can be controlled based on large dimensional analysis. This is dealt with in Section 16.4.

16.3 Neyman–Pearson criterion

The Neyman–Pearson criterion [Poor, 1994; Vantrees, 1968] for the receiver to establish whether an informative signal was transmitted is based on the ratio

$$C(\mathbf{Y}) = \frac{P_{\mathcal{H}_1|\mathbf{Y}}(\mathbf{Y})}{P_{\mathcal{H}_0|\mathbf{Y}}(\mathbf{Y})} \tag{16.2}$$

where, following the conventions of Chapter 2, $P_{\mathcal{H}_i|\mathbf{Y}}(\mathbf{Y})$ is the probability of the event $\mathcal{H}_i$ conditioned on the observation $\mathbf{Y}$. For a given receive space–time matrix $\mathbf{Y}$, if $C(\mathbf{Y}) > 1$, then the odds are that an informative signal was transmitted, while if $C(\mathbf{Y}) < 1$, it is more likely that no informative signal was transmitted and therefore only background noise was captured. To ensure a low probability of false alarms (or false positives), i.e. the probability of declaring a pure noise sample to carry an informative signal, a certain threshold ξ is generally set such that, when $C(\mathbf{Y}) > \xi$, the receiver declares an informative signal was sent, while when $C(\mathbf{Y}) < \xi$, the receiver declares that no informative signal was sent. The question of what ratio ξ to be set to ensure a given maximally acceptable false alarm rate will not be treated in the following. We will however provide an explicit expression of (16.2) for the aforementioned model, and will compare its performance to that achieved by classical detectors. The results provided in this section are borrowed from [Couillet and Debbah, 2010a].

Thanks to Bayes' rule, (16.2) becomes

$$C(\mathbf{Y}) = \frac{P_{\mathcal{H}_1} \cdot P_{\mathbf{Y}|\mathcal{H}_1}(\mathbf{Y})}{P_{\mathcal{H}_0} \cdot P_{\mathbf{Y}|\mathcal{H}_0}(\mathbf{Y})}$$

with $P_{\mathcal{H}_i}$ the a priori probability for hypothesis $\mathcal{H}_i$ to hold. We suppose that no side information allows the receiver to consider that $\mathcal{H}_1$ is more or less probable than $\mathcal{H}_0$, and therefore set $P_{\mathcal{H}_1} = P_{\mathcal{H}_0} = \frac{1}{2}$, so that

$$C(\mathbf{Y}) = \frac{P_{\mathbf{Y}|\mathcal{H}_1}(\mathbf{Y})}{P_{\mathbf{Y}|\mathcal{H}_0}(\mathbf{Y})} \tag{16.3}$$

reduces to a maximum likelihood ratio.

In the next section, we derive closed-form expressions for $C(\mathbf{Y})$ under the hypotheses that the values of K and the SNR, that we define as $1/\sigma^2$, are either perfectly or only partially known at the receiver.

16.3.1 Known signal and noise variances

16.3.1.1 Derivation of $P_{\mathbf{Y}|\mathcal{H}_i}$ in the SIMO case

We first analyze the situation where the noise power σ^2 and the number K of signal sources are known to the receiver. We also assume in this first scenario that $K = 1$. Since it is a common assumption that the number of available samples at the receiver is larger than the number of sensors, we further consider that $M > N$ and $N \geq 2$ (the case $N = 1$ is already known to be solved by the classical energy detector [Kostylev, 2002]).

Likelihood under $\mathcal{H}_0$.

In this first scenario, the noise entries $w_i^{(l)}$ are Gaussian and independent. The probability density of $\mathbf{Y}$, that can be seen as a random vector with NM entries, is then an NM multivariate uncorrelated complex Gaussian with covariance matrix $\sigma^2 \mathbf{I}_{NM}$

$$P_{\mathbf{Y}|\mathcal{H}_0}(\mathbf{Y}) = \frac{1}{(\pi\sigma^2)^{NM}} e^{-\frac{1}{\sigma^2} \operatorname{tr} \mathbf{Y}\mathbf{Y}^\mathsf{H}}. \tag{16.4}$$

Denoting $\boldsymbol{\lambda} = (\lambda_1, \ldots, \lambda_N)^\mathsf{T}$ the eigenvalues of $\mathbf{Y}\mathbf{Y}^\mathsf{H}$, (16.4) only depends on $\sum_{i=1}^N \lambda_i$ as follows.

$$P_{\mathbf{Y}|\mathcal{H}_0}(\mathbf{Y}) = \frac{1}{(\pi\sigma^2)^{NM}} e^{-\frac{1}{\sigma^2} \sum_{i=1}^N \lambda_i}.$$

Likelihood under $\mathcal{H}_1$.

Under the information plus noise hypothesis $\mathcal{H}_1$, the problem is more involved. The entries of the channel matrix $\mathbf{H}$ were previously modeled as jointly uncorrelated Gaussian, with $\mathrm{E}[|h_{ij}|^2] = E/K$. From now on, for simplicity, we take $E = 1$ without loss of generality. Therefore, since here $K = 1$, $\mathbf{H} \in \mathbb{C}^{N \times 1}$ and $\boldsymbol{\Sigma} = \mathbf{H}\mathbf{H}^\mathsf{H} + \sigma^2 \mathbf{I}_N$ has $N - 1$ eigenvalues $g_2 = \ldots = g_N$ equal to σ^2 and another distinct eigenvalue $g_1 = \nu_1 + \sigma^2 = (\sum_{i=1}^N |h_{i1}|^2) + \sigma^2$. Since the $|h_{i1}|^2$ are the sum of two Gaussian independent variables of zero mean and variance $\frac{1}{2}$ (the real and imaginary parts of h_{ij}), $2(g_1 - \sigma^2)$ is a χ^2_{2N} distribution. Hence, the unordered eigenvalue distribution of $\mathbf{\Sigma}$, defined on $[\sigma^2, \infty)^N$, reads:

$$P_{\mathbf{G}}(\mathbf{G}) = \frac{1}{N}(g_1 - \sigma^2)^{N-1} \frac{e^{-(g_1-\sigma^2)}}{(N-1)!} \prod_{i=2}^{N} \delta(g_i - \sigma^2).$$

From the model $\mathcal{H}_1$, $\mathbf{Y}$ is distributed as correlated Gaussian, as follows.

$$P_{\mathbf{Y}|\mathbf{\Sigma},I_1}(\mathbf{Y}, \mathbf{\Sigma}) = \frac{1}{\pi^{MN} \det(\mathbf{G})^M} e^{-\operatorname{tr}(\mathbf{Y}\mathbf{Y}^{\mathsf{H}}\mathbf{U}\mathbf{G}^{-1}\mathbf{U}^{\mathsf{H}})}$$

where I_k denotes the prior information at the receiver "$\mathcal{H}_1$ and $K = k$." This additional notation is very conventional for Bayesian probabilists, as it helps remind that all derived probability expressions are the outcomes of a so-called *plausible reasoning* based on the prior information available at the system modeler.

Since the channel $\mathbf{H}$ is unknown, we need to integrate out all possible channels for the transmission model under $\mathcal{H}_1$ over the probability space of $N \times K$ matrices with Gaussian i.i.d. distribution. From the unitarily invariance of Gaussian i.i.d. random matrices, this is equivalent to integrating out all possible covariance matrices $\mathbf{\Sigma}$ over the space of such non-negative definite Hermitian matrices, as follows.

$$P_{\mathbf{Y}|\mathcal{H}_1}(\mathbf{Y}) = \int_{\mathbf{\Sigma}} P_{\mathbf{Y}|\mathbf{\Sigma},\mathcal{H}_1}(\mathbf{Y}, \mathbf{\Sigma}) P_{\mathbf{\Sigma}}(\mathbf{\Sigma}) d\mathbf{\Sigma}.$$

Eventually, after complete integration calculus given in the proof below, the Neyman–Pearson decision ratio (16.2) for the single input multiple output channel takes an explicit expression, given by the following theorem.

Theorem 16.1. *The Neyman–Pearson test ratio $C_{\mathbf{Y}}(\mathbf{Y})$ for the presence of an informative signal when the receiver knows $K = 1$, the signal power $E = 1$, and the noise power σ^2, reads:*

$$C_{\mathbf{Y}}(\mathbf{Y}) = \frac{1}{N} \sum_{l=1}^{N} \frac{\sigma^{2(N+M-1)} e^{\sigma^2 + \frac{\lambda_l}{\sigma^2}}}{\prod_{\substack{i=1 \\ i \neq l}}^{N} (\lambda_l - \lambda_i)} J_{N-M-1}(\sigma^2, \lambda_l), \tag{16.5}$$

with $\lambda_1, \ldots, \lambda_N$ the eigenvalues of $\mathbf{Y}\mathbf{Y}^{\mathsf{H}}$ and where

$$J_k(x, y) \triangleq \int_x^{\infty} t^k e^{-t-\frac{y}{t}} dt = 2y^{\frac{k+1}{2}} K_{-k-1}(2\sqrt{y}) - \int_0^x t^k e^{-t-\frac{y}{t}} dt. \tag{16.6}$$

The proof of Theorem 16.1 is provided below. Among the interesting features of (16.5), note that the Neyman–Pearson test does only depend on the eigenvalues of $\mathbf{Y}\mathbf{Y}^{\mathsf{H}}$. This suggests that the eigenvectors of $\mathbf{Y}\mathbf{Y}^{\mathsf{H}}$ do not provide any information regarding the presence of an informative signal. The essential reason is that, both under $\mathcal{H}_0$ and $\mathcal{H}_1$, the eigenvectors of $\mathbf{Y}$ are isotropically distributed on the unit N-dimensional complex sphere due to the Gaussian assumptions made here. As such, a given realization of the eigenvectors of $\mathbf{Y}$ does indeed not carry any relevant information to the hypothesis test. The Gaussian assumption for $\mathbf{H}$ brought by the maximum entropy principle, or as a matter of fact for any unitarily invariant distribution assumption for $\mathbf{H}$, is therefore essential here. Note however that (16.5) is not reduced to a function of the sum $\sum_i \lambda_i$ of the eigenvalues, as suggested by the classical energy detector.

On the practical side, note that the integral $J_k(x,y)$ does not take a closed-form expression, but for $x = 0$, see, e.g., pp. 561 of [Gradshteyn and Ryzhik, 2000]. This is rather inconvenient for practical purposes, since $J_k(x,y)$ must either be evaluated every time, or be tabulated. It is also difficult to get any insight on the performance of such a detector for different values of σ^2, N, and K. We provide hereafter a proof of Theorem 16.1, in which classical multi-dimensional integration techniques are required. In particular, the tools introduced in Section 2.1, such as the important Harish–Chandra formula, Theorem 2.4, will be shown to be key ingredients of the derivation.

Proof. We start by writing the probability $P_{\mathbf{Y}|I_1}(\mathbf{Y})$ as the marginal probability of $P_{\mathbf{Y},\mathbf{\Sigma},I_1}$ after integration along all possible $\mathbf{\Sigma}$. This is:

$$P_{\mathbf{Y}|I_1}(\mathbf{Y}) = \int_{\mathcal{S}(\sigma^2)} P_{\mathbf{Y}|\mathbf{\Sigma},I_1}(\mathbf{Y},\mathbf{\Sigma}) P_{\mathbf{\Sigma}}(\mathbf{\Sigma}) d\mathbf{\Sigma}$$

with $\mathcal{S}(\sigma^2) \subset \mathbb{C}^{N\times N}$ the cone of positive definite complex matrices with smallest $N-1$ eigenvalues equal to σ^2.

We now consider the one-to-one mapping

$$\begin{aligned} B : (\mathcal{U}(N)/T) \times (\sigma^2, \infty) &\to \mathcal{S}(\sigma^2) \\ (\mathbf{U}, g_1) &\mapsto \mathbf{\Sigma} = \mathbf{U}\mathbf{G}\mathbf{U}^{\mathsf{H}} \end{aligned}$$

where

$$\mathbf{G} = \begin{pmatrix} g_1 & 0 \\ 0 & \sigma^2 \mathbf{I}_{N-1} \end{pmatrix}$$

and where $\mathcal{U}(N)/T$ is the space of unitary matrices of $N \times N$ with first column composed of real positive entries. More information on this mapping is provided in [Hiai and Petz, 2006], which is reused later in Chapter 18. From variable change calculus, see, e.g., [Billingsley, 1995], we have

$$\begin{aligned} &P_{\mathbf{Y}|I_1}(\mathbf{Y}) \\ &= \int_{(\mathcal{U}(N)/T)\times(\sigma^2,\infty)} P_{\mathbf{Y}|B(\mathbf{U},g_1)}(\mathbf{Y},\mathbf{U},g_1) P_{B(\mathbf{U},g_1)}(\mathbf{U},g_1) \det(\mathbf{J}(B)) d\mathbf{U} dg_1 \end{aligned}$$

with $\mathbf{J}(B)$ the Jacobian matrix of B.

Notice now that $\mathbf{\Sigma} - \sigma^2 \mathbf{I}_N = \mathbf{H}\mathbf{H}^{\mathsf{H}}$ is a Wishart matrix. The density of its entries is therefore invariant by left- and right-product by unitary matrices. The eigenvectors of $\mathbf{\Sigma}$ are as a consequence uniformly distributed over $\mathcal{U}(N)$, the space of complex unitary matrices of size $N \times N$. Moreover, the eigenvalue distribution of $\mathbf{\Sigma}$ is independent of the matrix $\mathbf{U}$. From these observations, we conclude that the joint density

$$P_{(\mathbf{U},g_1)}(\mathbf{U},g_1) = P_{B(\mathbf{U},g_1)}(\mathbf{U},g_1)\det(\mathbf{J}(B))$$

can be written under the product form

$$P_{(\mathbf{U},g_1)}(\mathbf{U},g_1) = P_{\mathbf{U}}(\mathbf{U})P_{g_1}(g_1).$$

As in Chapter 2, we assume that $d\mathbf{U}$ is the Haar measure with density $P_{\mathbf{U}}(\mathbf{U}) = 1$. We can therefore write

$$P_{\mathbf{Y}|I_1}(\mathbf{Y}) = \int_{\mathcal{U}(N)\times(\sigma^2,\infty)} P_{\mathbf{Y}|\mathbf{\Sigma},\mathcal{H}_1}(\mathbf{Y},\mathbf{\Sigma})P_{g_1}(g_1)d\mathbf{U}dg_1.$$

The latter can further be equated to

$$P_{\mathbf{Y}|I_1}(\mathbf{Y}) = \int_{\mathcal{U}(N)\times(\sigma^2,\infty)} \frac{e^{-\operatorname{tr}(\mathbf{Y}\mathbf{Y}^{\mathsf{H}}\mathbf{U}\mathbf{G}^{-1}\mathbf{U}^{\mathsf{H}})}}{\pi^{NM}\det(\mathbf{G})^M}(g_1-\sigma^2)^{N-1}\frac{e^{-(g_1-\sigma^2)}}{N!}d\mathbf{U}dg_1.$$

The next step is to use the Harish–Chandra identity provided in Theorem 2.4. Denoting $\Delta(\mathbf{Z})$ the Vandermonde determinant of matrix $\mathbf{Z} \in \mathbb{C}^{N\times N}$ with eigenvalues $z_1 \leq \ldots \leq z_N$

$$\Delta(\mathbf{Z}) \triangleq \prod_{i>j}(z_i - z_j) \tag{16.7}$$

the likelihood $P_{\mathbf{Y}|I_1}(\mathbf{Y})$ further develops as

$$P_{\mathbf{Y}|I_1}(\mathbf{Y}) = \lim_{g_2,\ldots,g_N\to\sigma^2} \frac{e^{\sigma^2}(-1)^{\frac{N(N-1)}{2}}\prod_{j=1}^{N-1}j!}{\pi^{MN}\sigma^{2M(N-1)}N!}$$
$$\times\int_{\sigma^2}^{+\infty}(g_1-\sigma^2)^{N-1}e^{-g_1}\frac{1}{g_1^M}\frac{\det\left(\left\{e^{-\frac{\lambda_i}{g_j}}\right\}\right)}{\Delta(\mathbf{Y}\mathbf{Y}^{\mathsf{H}})\Delta(\mathbf{G}^{-1})}dg_1.$$

Now, noticing that $\Delta(\mathbf{G}^{-1}) = (-1)^{N(N+3)/2}\frac{\Delta(\mathbf{G})}{\det(\mathbf{G})^{N-1}}$, this is also

$$P_{\mathbf{Y}|I_1}(\mathbf{Y})$$
$$= \lim_{g_2,\ldots,g_N\to\sigma^2}\frac{\pi^{-MN}e^{\sigma^2}\prod_{j=1}^{N-1}j!}{N!\sigma^{2(N-1)(M-N+1)}}\int_{\sigma^2}^{+\infty}\frac{(g_1-\sigma^2)^{N-1}e^{-g_1}}{g_1^{M-N+1}}\frac{\det\left(\left\{e^{-\frac{\lambda_i}{g_j}}\right\}\right)}{\Delta(\mathbf{Y}\mathbf{Y}^{\mathsf{H}})\Delta(\mathbf{G})}dg_1$$

in which we remind that $\lambda_1,\ldots,\lambda_N$ are the eigenvalues of $\mathbf{Y}\mathbf{Y}^{\mathsf{H}}$. Note the trick of replacing the known values of $g_2,\ldots,g_N$ by limits of scalars converging to these known values, which dodges the problem of improper ratios. To derive the explicit limits, we then proceed as follows.

Denoting $\mathbf{y} = (\gamma_1,\ldots,\gamma_{N-1},\gamma_N) = (g_2,\ldots,g_N,g_1)$ and defining the functions

$$f(x_i,\gamma_j) \triangleq e^{-\frac{x_i}{\gamma_j}}$$
$$f_i(\gamma_j) \triangleq f(x_i,\gamma_j)$$

we then have from Theorem 2.9

$$\lim_{g_2,\ldots,g_N\to\sigma^2} \frac{\det\left(\left\{e^{-\frac{\lambda_i}{g_j}}\right\}_{\substack{1\le i\le N\\1\le j\le N}}\right)}{\Delta(\mathbf{YY}^{\mathsf{H}})\Delta(\mathbf{G})}$$
$$= \lim_{\substack{\gamma_1,\ldots,\gamma_{N-1}\to\sigma^2\\ \gamma_M\to g_1}} (-1)^{N-1}\frac{\det\left(\{f_i(\lambda_j)\}_{i,j}\right)}{\Delta(\mathbf{YY}^{\mathsf{H}})\Delta(\mathbf{G})}$$
$$= (-1)^{N-1}\frac{\det\left[f_i(\sigma^2),\ f_i'(\sigma^2),\ldots,\ f^{(N-2)}(\sigma^2),\ f_i(g_1)\right]}{\prod_{i<j}(\lambda_i-\lambda_j)(g_1-\sigma^2)^{N-1}\prod_{j=1}^{N-2} j!}.$$

The change of variables led to a switch of one column and explains the $(-1)^{N-1}$ factor appearing when computing the resulting determinant. The partial derivatives of f along the second variable is

$$\left(\frac{\partial}{\partial\gamma^k}f\right)_{k\ge1}(a,b) = \sum_{m=1}^{k}\frac{(-1)^{k+m}}{b^{m+k}}\binom{m}{k}\frac{(k-1)!}{(m-1)!}a^m e^{-\frac{a}{b}}$$
$$\triangleq \kappa_k(a,b)e^{-\frac{a}{b}}.$$

Back to the full expression of $P_{\mathbf{Y}|\mathcal{H}_1}(\mathbf{Y})$, we then have

$$P_{\mathbf{Y}|I_1}(\mathbf{Y})$$
$$= \frac{e^{\sigma^2}\sigma^{2(N-1)(N-M-1)}}{N\pi^{MN}}$$
$$\times\int_{\sigma^2}^{+\infty}(-1)^{N-1}g_1^{N-M-1}e^{-g_1}\frac{\det\left[f_i(\sigma^2),f_i'(\sigma^2),\ldots,f^{(N-2)}(\sigma^2),f_i(g_1)\right]}{\prod_{i<j}(\lambda_i-\lambda_j)}dg_1$$
$$= \frac{e^{\sigma^2}\sigma^{2(N-1)(N-M-1)}}{N\pi^{MN}\prod_{i<j}(\lambda_i-\lambda_j)}$$
$$\times\int_{\sigma^2}^{+\infty}(-1)^{N-1}g_1^{N-M-1}e^{-g_1}\det\left[\begin{array}{c|c|c} e^{-\frac{x_1}{\sigma^2}} & & e^{-\frac{\lambda_1}{g_1}}\\ \vdots & \left(\kappa_j(\lambda_i,\sigma^2)e^{-\frac{\lambda_i}{\sigma^2}}\right)_{\substack{1\le i\le N\\1\le j\le N-2}} & \vdots\\ e^{-\frac{x_N}{\sigma^2}} & & e^{-\frac{\lambda_N}{g_1}}\end{array}\right]dg_1.$$

Before going further, we need the following result, often required in the calculus of marginal eigenvalue distributions for Gaussian matrices.

Lemma 16.1. *For any family $\{a_1,\ldots,a_N\}\in\mathbb{R}^N$, $N\ge2$, and for any $b\in\mathbb{R}^*$*

$$\det\left[\begin{array}{c|c} 1 & \\ \vdots & (\kappa_j(a_i,b))_{\substack{1\le i\le N\\1\le j\le N-1}}\\ 1 & \end{array}\right] = \frac{1}{b^{N(N-1)}}\prod_{i<j}(a_j-a_i).$$

This identity follows from the observation that column k of the matrix above is a polynomial of order k. Since summations of linear combinations of the columns do not affect the determinant, each polynomial can be replaced by the monomial of highest order, i.e. $b^{-2(k-1)}a_i^k$ in row i. Extracting the product $1 \cdot b^{-2} \cdots b^{-2(N-1)} = b^{-(N-1)N}$ from the determinant, what remains is the determinant of a Vandermonde matrix based on the vector $a_1, \ldots, a_N$.

By factorizing every row of the matrix by $e^{-\frac{\lambda_i}{\sigma^2}}$ and developing the determinant on the last column, we obtain

$$\begin{aligned}
&P_{\mathbf{Y}|I_1}(\mathbf{Y}) \\
&= \frac{e^{\sigma^2}\sigma^{2(N-1)(N-M-1)}}{N\pi^{MN}\prod_{i<j}(\lambda_i - \lambda_j)} \\
&\times \int_{\sigma^2}^{+\infty} g_1^{N-M-1} e^{-g_1 - \frac{\sum_{i=1}^N \lambda_i}{\sigma^2}} \sum_{l=1}^{N} \frac{(-1)^{2N+l-1} e^{\lambda_l\left(\frac{1}{g_1} - \frac{1}{\sigma^2}\right)}}{\sigma^{2(N-1)(N-2)}} \prod_{\substack{i<j \\ i\neq l \\ j\neq l}} (\lambda_i - \lambda_j) dg_1 \\
&= \frac{e^{\sigma^2 - \frac{1}{\sigma^2}\sum_{i=1}^N \lambda_i}}{N\pi^{MN}\sigma^{2(N-1)(M-1)}} \sum_{l=1}^{N} (-1)^{l-1} \int_{\sigma^2}^{+\infty} \frac{g_1^{N-M-1} e^{-g_1} e^{-\lambda_l\left(\frac{1}{g_1} - \frac{1}{\sigma^2}\right)}}{\prod_{i<l}(\lambda_i - \lambda_l)\prod_{i>l}(\lambda_l - \lambda_i)} dg_1 \\
&= \frac{e^{\sigma^2 - \frac{1}{\sigma^2}\sum_{i=1}^N \lambda_i}}{N\pi^{MN}\sigma^{2(N-1)(M-1)}} \sum_{l=1}^{N} \frac{e^{\frac{\lambda_l}{\sigma^2}}}{\prod_{\substack{i=1 \\ i\neq l}}^{N}(\lambda_l - \lambda_i)} \int_{\sigma^2}^{+\infty} g_1^{N-M-1} e^{-\left(g_1 + \frac{\lambda_l}{g_1}\right)} dg_1
\end{aligned}$$

which finally gives

$$P_{\mathbf{Y}|I_1}(\mathbf{Y}) = \frac{e^{\sigma^2 - \frac{1}{\sigma^2}\sum_{i=1}^N \lambda_i}}{N\pi^{MN}\sigma^{2(N-1)(M-1)}} \sum_{l=1}^{N} \frac{e^{\frac{\lambda_l}{\sigma^2}}}{\prod_{\substack{i=1 \\ i\neq l}}^{N}(\lambda_l - \lambda_i)} J_{N-M-1}(\sigma^2, \lambda_l)$$

where

$$J_k(x,y) = \int_x^{+\infty} t^k e^{-t-\frac{y}{t}} dt = 2y^{\frac{k+1}{2}} K_{-k-1}(2\sqrt{y}) - \int_0^x t^k e^{-t-\frac{y}{t}} dt$$

and K_n denotes the modified Bessel function of the second kind. □

We now turn to the more general case where $K \geq 1$, which unfolds similarly.

16.3.1.2 Multi-source case

In the generalized multi-source configuration, where $K \geq 1$, the likelihood $P_{\mathbf{Y}|\mathcal{H}_0}$ remains unchanged and therefore the previous expression for $K = 1$ is still correct. For the subsequent derivations, we only treat the situation where $K \leq N$ but the case $K > N$ is a rather similar extension.

In this scenario, $\mathbf{H} \in \mathbb{C}^{N \times K}$ is now a random matrix (instead of a vector) with i.i.d. zero mean Gaussian entries. The variance of every row is $\mathrm{E}[\sum_{j=1}^K |h_{ij}|^2] = 1$. Therefore $K\mathbf{H}\mathbf{H}^{\mathsf{H}}$ is distributed as a null Wishart matrix. Hence, observing that $\mathbf{\Sigma} - \sigma^2\mathbf{I}_N$ is the diagonal matrix of the eigenvalues of $\mathbf{H}\mathbf{H}^{\mathsf{H}}$

$$\mathbf{\Sigma} = \mathbf{U} \cdot \mathrm{diag}(\nu_1 + \sigma^2, \ldots, \nu_K + \sigma^2, \sigma^2, \ldots, \sigma^2) \cdot \mathbf{U}^{\mathsf{H}} \tag{16.8}$$

for some unitary matrix $\mathbf{U} \in \mathbb{C}^{N\times N}$ and with $\nu_1, \ldots, \nu_K$ the eigenvalues of $\mathbf{H}^\mathsf{H}\mathbf{H}$, the unordered eigenvalue density of $\mathbf{G}$ unfolds from Theorem 2.3

$$P_{\mathbf{G}}(\mathbf{G}) = \frac{(N-K)!K^{KN}}{N!} \prod_{i=1}^{K} e^{-K\sum_{i=1}^{K}(g_i-\sigma^2)} \frac{(g_i-\sigma^2)_+^{N-K}}{(K-i)!(N-i)!} \prod_{i<j}^{K} (g_i - g_j)^2. \tag{16.9}$$

From the Equations (16.8) and (16.9) above, it is possible to extend Theorem 16.1 to the multi-source scenario, using similar techniques as for the proof of Theorem 16.1, which we do not further develop here, but can be found in [Couillet and Debbah, 2010a]. This extended result is provided below.

Theorem 16.2. *The Neyman–Pearson test ratio $C_{\mathbf{Y}}(\mathbf{Y})$ for the presence of informative signals when the receiver perfectly knows the number K $(K \leq N)$ of signal sources, the source power $E = 1$, and the noise power σ^2, reads:*

$$C_{\mathbf{Y}}(\mathbf{Y}) = \frac{\sigma^{2K(N+M-K)}(N-K)!e^{K^2\sigma^2}}{N!K^{(K-1-2M)K/2}\prod_{j=1}^{K-1} j!} \sum_{\mathbf{a}\subset[1,N]} \frac{e^{\frac{\sum_{i=1}^{K}\lambda_{a_i}}{\sigma^2}}}{\prod_{a_i}\prod_{\substack{j\neq a_1\\ \cdots \\ j\neq a_i}}(\lambda_{a_i}-\lambda_j)}$$

$$\times \sum_{\mathbf{b}\in\mathcal{P}(K)} (-1)^{\mathrm{sgn}(\mathbf{b})+K} \prod_{l=1}^{K} J_{N-M-2+b_l}(K\sigma^2, K\lambda_{a_l})$$

with $\mathcal{P}(K)$ the ensemble of permutations of $\{1,\ldots,K\}$, $\mathbf{b} = (b_1,\ldots,b_K)$ and $\mathrm{sgn}(\mathbf{b})$ *the signature of the permutation* $\mathbf{b}$. *The function J_k is defined as in Theorem 16.1.*

Observe again that $C_{\mathbf{Y}}(\mathbf{Y})$ is a function of the empirical eigenvalues $\lambda_1, \ldots, \lambda_N$ of $\mathbf{Y}\mathbf{Y}^\mathsf{H}$ only. In the following, we extend the current signal detector to the more realistic situations where K, E, and σ^2 are not a priori known to the receiver.

16.3.2 Unknown signal and noise variances

Efficient signal detection when the noise variance is unknown is highly desirable [Tandra and Sahai, 2005]. Indeed, as recalled earlier, if the noise and signal variances were exactly known, some prior noise detection mechanism would be required. The difficulty here is handily avoided thanks to *ad-hoc* methods that are asymptotically independent of the noise variance, as in, e.g., [Cardoso *et al.*, 2008; Zeng and Liang, 2009], or more theoretical, although suboptimal, approaches as in [Bianchi *et al.*, 2011], which will be discussed when dealing with large dimensional random matrix considerations.

In the following, we consider the general case when the knowledge about the signal and noise variances can range from a total absence of information to a perfect knowledge, and will represent this knowledge under the form of a prior probability distribution, as per classical Bayesian derivation. It might happen in particular that the receiver has no knowledge whatsoever on the values of the noise power and the expected signal power, but obviously knows that these powers are positive values. When such a situation arises, the unknown parameter must be assigned a so-called *uninformative prior*, such as the widely spread Jeffreys prior [Jeffreys, 1946]. Assigning uninformative priors of variables defined in a continuum is however, still to this day, a controverted issue of the maximum entropy principle [Caticha, 2001]. The classical uninformative priors considered in the literature are (i) the uniform prior, i.e. every two positive values for the signal/noise power are equi-probable, which experiences problems of scaling invariance thoroughly discussed in [Jaynes, 2003], and (ii) the aforementioned Jeffreys prior [Jeffreys, 1946], i.e. the prior distribution for the variance parameter σ^2 takes the form $\sigma^{-2\beta}$ for any deterministic choice of positive β, which is invariant under scaling but is not fully attractive as it requires a subjective choice of β.

In the case where the signal power E is known to be contained between E_- and E_+ (for the time being, we had considered $E = 1$), and the noise power σ^2 is known at least to be bounded by σ_-^2 and σ_+^2, we will consider the "desirable" assumption of uniform prior

$$P_E(E) = \frac{1}{E_+ - E_-}$$
$$P_{\sigma^2}(\sigma^2) = \frac{1}{\sigma_+^2 - \sigma_-^2}.$$

Denoting I'_k the event "$\mathcal{H}_1, K = k, E_- \leq E \leq E_+$ and $\sigma_-^2 \leq \sigma^2 \leq \sigma_+^2$," this leads to the updated decisions of the form

$$C_{\mathbf{Y}}(\mathbf{Y}) = \frac{1}{E_+ - E_-} \frac{\int_{\sigma_-^2}^{\sigma_+^2} \int_{E_-}^{E^+} P_{\mathbf{Y}|\sigma^2, I'_K}(\mathbf{Y}, \sigma^2, E) d\sigma^2 dE}{\int_{\sigma_-^2}^{\sigma_+^2} P_{\mathbf{Y}|\sigma^2, \mathcal{H}_0}(\mathbf{Y}, \sigma^2) d\sigma^2} \tag{16.10}$$

where $P_{\mathbf{Y}|\sigma^2, I'_K}(\mathbf{Y}, \sigma^2, E)$ is obtained as previously by assuming a transmit power E, instead of 1. Precisely, it suffices to consider the density of $E\mathbf{Y}$ with σ^2 changed into σ^2/E.

The computational difficulty raised by the integrals $J_k(x, y)$ does not allow for any satisfying closed-form expression for (16.10) so that only numerical integrations can be performed at this point.

16.3.3 Unknown number of sources

In practical cases, the number of transmitting sources is only known to be finite and discrete. If only an upper bound $K_{\max}$ on K is known, a uniform prior is assigned to K. The probability distribution of $\mathbf{Y}$ under hypothesis $I_0 \triangleq$"σ^2

known, $1 \leq K \leq K_{\max}$ unknown," reads:

$$P_{\mathbf{Y}|I_0}(\mathbf{Y}) = \sum_{i=1}^{K_{\max}} P_{\mathbf{Y}|``K=i", I_0}(\mathbf{Y}) \cdot P_{``K=i"|I_0}$$
$$= \frac{1}{K_{\max}} \sum_{i=1}^{K_{\max}} P_{\mathbf{Y}|``K=i", I_0}(\mathbf{Y})$$

which does not meet any computational difficulty.

Assuming again equal probability for the hypotheses $\mathcal{H}_0$ and $\mathcal{H}_1$, this leads to the decision ratio

$$C_{\mathbf{Y}}(\mathbf{Y}) = \frac{1}{K_{\max}} \sum_{i=1}^{K_{\max}} \frac{P_{\mathbf{Y}|``K=i", I_0}(\mathbf{Y})}{P_{\mathbf{Y}|\mathcal{H}_0}(\mathbf{Y})}.$$

Note now that it is possible to make a decision test on the number of sources itself in a rather straightforward extension of the previous formula. Indeed, given a space–time matrix realization $\mathbf{Y}$, the probability for the number of transmit antennas to be i is from Bayes' rule

$$P_{``K=i"|\mathbf{Y}}(\mathbf{Y})$$
$$= \frac{P_{\mathbf{Y}|``K=i"}(\mathbf{Y}) P_{``K=i"}}{\sum_{j=0}^{K_{\max}} P_{\mathbf{Y}|``K=j"}(\mathbf{Y}) P_{``K=j"}}$$
$$= \begin{cases} P_{\mathbf{Y}|\mathcal{H}_0}(\mathbf{Y}) \left[P_{\mathbf{Y}|\mathcal{H}_0}(\mathbf{Y}) + \frac{1}{K_{\max}} \sum_{j=1}^{K_{\max}} P_{\mathbf{Y}|``K=j"}(\mathbf{Y}) \right]^{-1} & , i = 0 \\ \frac{1}{K_{\max}} P_{\mathbf{Y}|``K=i"}(\mathbf{Y}) \left[P_{\mathbf{Y}|\mathcal{H}_0}(\mathbf{Y}) + \frac{1}{K_{\max}} \sum_{j=1}^{K_{\max}} P_{\mathbf{Y}|``K=j"}(\mathbf{Y}) \right]^{-1} & , i \geq 1 \end{cases}$$

where all the quantities of interest here were derived in previous sections. The multiple hypothesis test on K is then based on a comparison of the *odds* $O(``K = i")$ for the events "$K = i$," for all $i \in \{0, \ldots, K_{\max}\}$. Under Bayesian terminology, we remind that the odds for the event "$K = i$" are defined as

$$O(``K = i") = \frac{P_{``K=i"|\mathbf{Y}}(\mathbf{Y})}{\sum_{\substack{j=0 \\ j \neq i}}^{K_{\max}} P_{``K=j"|\mathbf{Y}}(\mathbf{Y})}.$$

In the current scenario, these odds express as

$$O(``K = i")$$
$$= \begin{cases} P_{\mathbf{Y}|\mathcal{H}_0}(\mathbf{Y}) \left[\frac{1}{K_{\max}} \sum_{j=1}^{K_{\max}} P_{\mathbf{Y}|``K=j"}(\mathbf{Y}) \right]^{-1} & , i = 0 \\ \frac{1}{K_{\max}} P_{\mathbf{Y}|``K=i"}(\mathbf{Y}) \left[P_{\mathbf{Y}|\mathcal{H}_0}(\mathbf{Y}) + \frac{1}{K_{\max}} \sum_{j \neq i} P_{\mathbf{Y}|``K=j"}(\mathbf{Y}) \right]^{-1} & , i \geq 1. \end{cases}$$

We now provide a few simulation results that confirm the optimality of the Neyman–Pearson test for the channel model under study, i.e. with i.i.d. Gaussian channel, signal, and noise. We also provide simulation results when these assumptions are not met, in particular when a line-of-sight component is present in the channel and when the signal samples are drawn from a quadrature phase shift-keying (QPSK) constellation.

First, we provide in Figure 16.3 the simulated plots of the false alarm and correct detection rates obtained for the Neyman–Pearson test derived in Theorem 16.1 when $K = 1$, with respect to the decision threshold above which correct detection is claimed. To avoid trivial scenarios, we consider a rather low SNR of -3 dB, and $N = 4$ receivers capturing only $M = 8$ signal instances. The channel conditions are assumed to match the conditions required by the maximum entropy model, i.e. channel, signal, and noise are all i.i.d. Gaussian. Note that such conditions are desirable when fast decisions are demanded. In a cognitive radio setup, secondary networks are expected to be capable of very fast and reliable signal detection, in order to be able to optimally exploit spectrum opportunities [Hoyhtya *et al.*, 2007]. This is often referred to as the *exploration* versus *exploitation* trade-off, which balances the time spent exploring for available resources with high enough detection reliability and the time spent exploiting the available spectrum resources. Observe in Figure 16.3 that the false alarm rate curve shows a steep drop around $C_{\mathbf{Y}}(\mathbf{Y}) = 1$ (or zero dB). This however comes along with a drop, although not so steep, of the correct detection rate. A classical way to assess the performance of various detection tests is to evaluate how much correct detection rate is achieved for a given fixed tolerable false alarm rate. Comparison of correct detection rates for given false alarm rates is obtained in the so-called receiver operating characteristic (ROC) curve. The ROC curve of the Neyman–Pearson test against that of the energy detector is provided in Figure 16.4 under the channel model conditions, for $N = 4$, $M = 8$, and $\sigma^2 = -3$ dBm as above, with the transmit power E equal to zero dBm, this last information being either perfectly known or only known to belong to $[-10 \text{ dBm}, 10 \text{ dBm}]$. We only focus on a section of the curve which corresponds to low false alarm rates (FAR), which is a classical assumption. We recall that the energy detector consists in summing up λ_1 to λ_N, the eigenvalues of $\mathbf{Y}\mathbf{Y}^{\mathsf{H}}$ (or equivalently taking the trace of $\mathbf{Y}\mathbf{Y}^{\mathsf{H}}$) and comparing it against some deterministic threshold. The larger the sum the more we expect the presence of an informative signal in the received signal. Observe that the Neyman–Pearson test is effectively superior in correct detection rate than the legacy energy detector, with up to 10% detection gain for low false alarm rates.

We then test the robustness of the Neyman–Pearson test by altering the effective transmit channel model. We specifically consider that a line-of-sight component of amplitude one fourth of the mean channel energy is present. This is modeled by letting the effective channel matrix $\mathbf{H}$ be $\mathbf{H} = \sqrt{1-\alpha^2}\mathbf{Z} + \alpha\mathbf{A}$, where $\mathbf{Z} \in \mathbb{C}^{N \times K}$ has i.i.d. Gaussian entries of variance $1/K$ and $\mathbf{A} \in \mathbb{C}^{N \times K}$ has all entries equal to $1/K$. This is depicted in Figure 16.5 with $\alpha^2 = \frac{1}{4}$. We observe once more that the Neyman–Pearson test performs better than the power detector, especially at low SNR. It therefore appears to be quite robust to alterations in the system model such as the existence of a line-of-sight component, although this was obviously not a design purpose.

In Figure 16.6, we now vary the SNR range, and evaluate the correct detection rates under different false alarm rate constraints, for the Gaussian i.i.d. signal

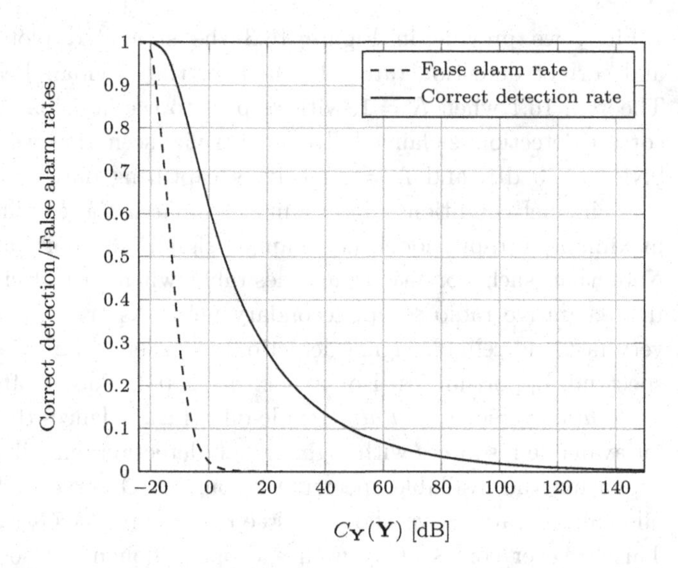

Figure 16.3 Neyman–Pearson test performance in single-source scenario. Correct detection rates and false alarm rates for $K = 1$, $N = 4$, $M = 8$, SNR $= -3$ dB.

and channel model. This graph confirms the previous observation that the stronger the false alarm request, the more efficient the Neyman–Pearson test comparatively with the energy detection approach. Note in particular that as much as 10% of correct detection can again be gained in the low SNR regime and for a tolerable FAR of 10^{-3}.

Finally, in Figure 16.7, we provide the ROC curve performance for the multi-source scheme, when K ranges from $K = 1$ to $K = 3$, still under the Gaussian i.i.d. system model. We observe notably that, as the number of sources increases, the energy detector closes in the performance gap observed in the single source case. This arises both from a performance decrease of the Neyman–Pearson test, which can be interpreted from the fact that the more the unknown variables (there are more unknown channel links) the less reliable the noise-versus-information comparative test, and from a performance increase of the power detector, which can be interpreted as a channel hardening effect (the more the channel links the less the received signal variance).

A more interesting problem though is to assume that the noise variance σ^2 is not a priori known at the receiver end since the receiver is entitled to determine whether noise or informative signals are received without knowing the noise statistics in the first place. We have already seen that the Neyman–Pearson test approach leads to a multi-dimensional integral form, which is difficult to further simplify. Practical systems however call for low complex implementation [Cabric *et al.*, 2004]. We therefore turn to alternative approaches bearing ideas in the large dimensional random matrix field to cover this particularly interesting case. It will turn out that very simple tests can be determined for the scenario where the noise variance is not known to the receiver, and theoretical derivations of the

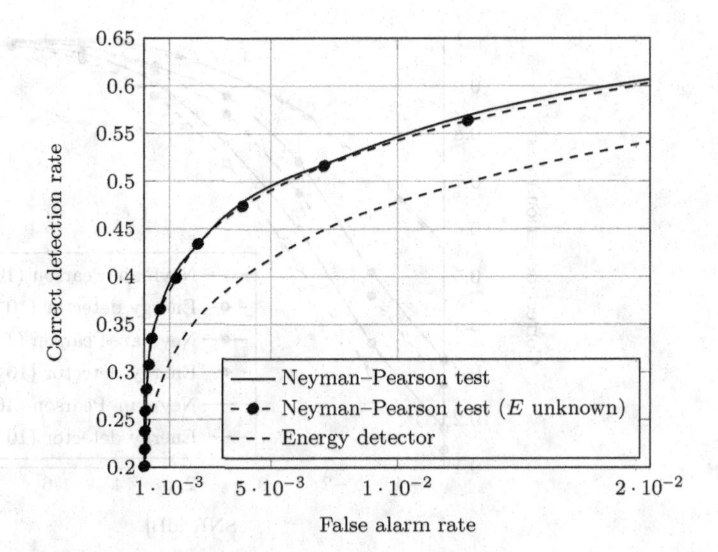

Figure 16.4 ROC curve for single-source detection, $K = 1$, $N = 4$, $M = 8$, SNR = −3 dB, FAR range of practical interest, with signal power $E = 0$ dBm, either known or unknown at the receiver.

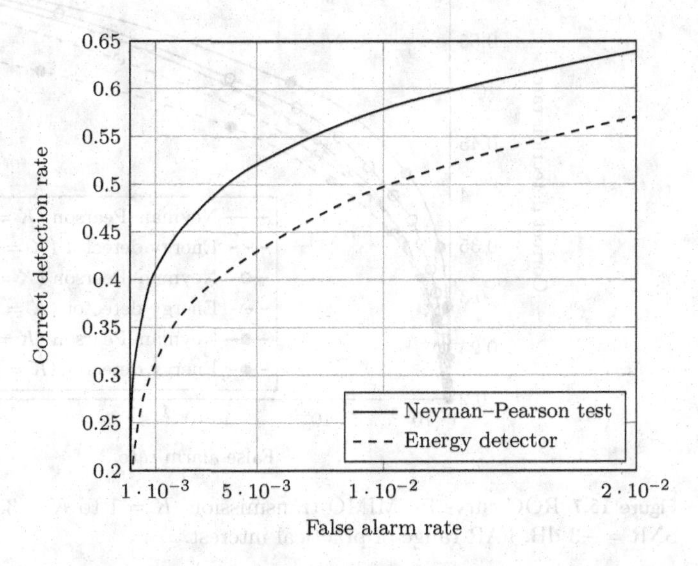

Figure 16.5 ROC curve for single-source detection, $K = 1$, $N = 4$, $M = 8$, SNR = −3 dB, FAR range of practical interest, under Rician channel with line-of-sight component of amplitude 1/4 and QPSK modulated input signals.

correct detection rate against the false alarm rate can be performed. This is the subject of the subsequent section.

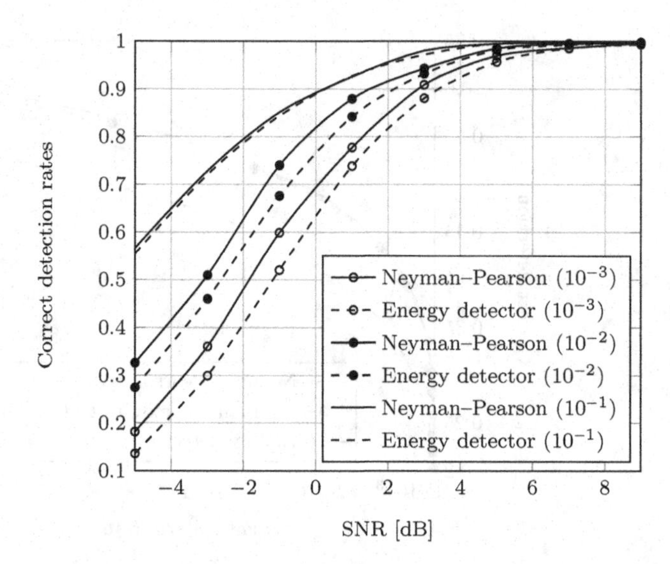

Figure 16.6 Correct detection rates under different FAR constraints (in parentheses in the legend) and for different SNR levels, $K = 1$, $N = 4$, $M = 8$.

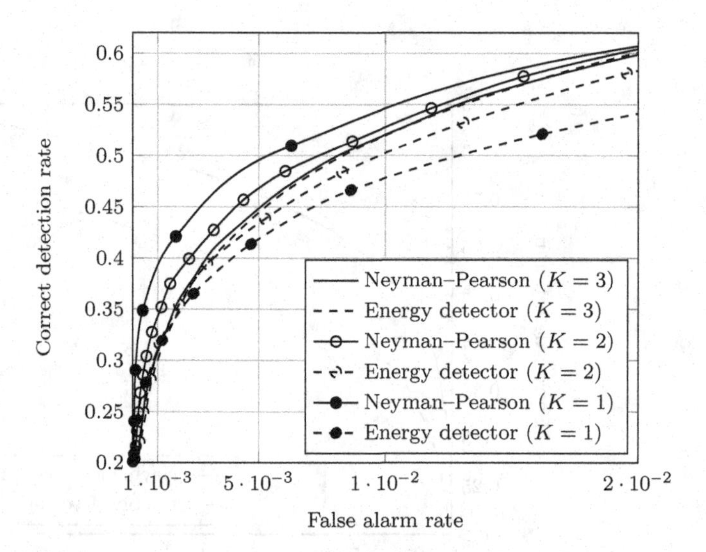

Figure 16.7 ROC curve for MIMO transmission, $K = 1$ to $K = 3$, $N = 4$, $M = 8$, SNR $= -3$ dB. FAR range of practical interest.

16.4 Alternative signal sensing approaches

The major results of interest in the large dimensional random matrix field for signal detection are those regarding the position of extreme eigenvalues of a sample covariance matrix. The first idea we will discuss, namely the condition number test, arises from the simple observation that, under hypothesis $\mathcal{H}_0$, not only should the empirical eigenvalue distribution of $\mathbf{YY}^{\mathsf{H}}$ be close to the Marčenko–Pastur law, but also should the largest eigenvalue of $\mathbf{YY}^{\mathsf{H}}$ be close to the rightmost end of the Marčenko–Pastur law support, as both the number of sensors and the number of available time samples grow large. If an informative signal is present in the observation $\mathbf{Y}$, we expect instead the largest eigenvalue of $\mathbf{YY}^{\mathsf{H}}$ to be found sufficiently far away from the Marčenko–Pastur law support.

The methods proposed below therefore heavily rely on Bai and Silverstein's Theorem 7.1 and its various extensions and corollaries, e.g. Theorem 9.8.

16.4.1 Condition number method

The first method we introduce is an *ad-hoc* approach based on the observation that in the large dimensional regime, as both N and M grow large, the ratio between the largest and the smallest eigenvalue of $\frac{1}{M}\mathbf{YY}^{\mathsf{H}}$, often referred to as the condition number of $\frac{1}{M}\mathbf{YY}^{\mathsf{H}}$, converges almost surely to a deterministic value. Ordering the eigenvalues $\lambda_1, \ldots, \lambda_N$ of $\mathbf{YY}^{\mathsf{H}}$ as $\lambda_1 \geq \ldots \geq \lambda_N$, under hypothesis $\mathcal{H}_0$, this convergence reads:

$$\frac{\lambda_1}{\lambda_N} \xrightarrow{\text{a.s.}} \frac{\sigma^2 (1+\sqrt{c})^2}{\sigma^2 (1-\sqrt{c})^2} = \frac{(1+\sqrt{c})^2}{(1-\sqrt{c})^2}$$

with c defined as the limiting ratio $c \triangleq \lim_{N\to\infty} N/M$. This is an immediate consequence of Theorem 7.1 and Theorem 9.8. This ratio is seen no longer to depend on the specific value of the noise variance σ^2. Under hypothesis $\mathcal{H}_1$, notice that the model (16.1) is related to a spiked model, as the population covariance matrix of $\frac{1}{M}\mathbf{YY}^{\mathsf{H}}$ is formed of $N-K$ eigenvalues equal to σ^2 and K other eigenvalues strictly superior to σ^2 and all different with probability one. In the particular case when $K=1$, all eigenvalues of $\mathrm{E}[\mathbf{y}^{(1)}\mathbf{y}^{(1)\mathsf{H}}]$ equal σ^2 but the largest which equals $\sigma^2 + \sum_{i=1}^N |h_{i1}|^2$. As previously, call $g_1 \triangleq \sigma^2 + \sum_{i=1}^N |h_{i1}|^2$. Similar to the previous section, let us consider the $K=1$ scenario. We still assume that $M > N$, i.e. that more time samples are collected than there are sensors. From Theorem 9.1, we then have that, as M and N grow large with limiting ratio $c \triangleq \lim \frac{N}{M}$ and such that $\frac{g_1}{\sigma^2} - 1 \to \rho$, if $\rho > \sqrt{c}$

$$\frac{\lambda_1}{M} \xrightarrow{\text{a.s.}} (1+\rho)\left(1+\frac{c}{\rho}\right) \triangleq \lambda_{\text{sp}}$$

and

$$\frac{\lambda_N}{M} \xrightarrow{\text{a.s.}} \sigma^2 (1-\sqrt{c})^2$$

while if $\rho < \sqrt{c}$

$$\frac{\lambda_1}{M} \xrightarrow{\text{a.s.}} \sigma^2 (1+\sqrt{c})^2$$

and

$$\frac{\lambda_N}{M} \xrightarrow{\text{a.s.}} \sigma^2 \left(1 - \sqrt{c}\right)^2.$$

Thus, under the condition that M is large enough to ensure that $g_1 > 1 + \sqrt{c}$, it is asymptotically possible to detect the presence of informative signals, without explicit knowledge of σ^2. To this end, we may compare the ratio λ_1/λ_N to the value

$$\left(\frac{1+\sqrt{c}}{1-\sqrt{c}}\right)^2$$

corresponding to the asymptotically expected ratio under $\mathcal{H}_0$. This defines a new test, rather empirical, which consists in considering a threshold around the ratio $\left(\frac{1+\sqrt{c}}{1-\sqrt{c}}\right)^2$ and of deciding for hypothesis $\mathcal{H}_1$ whenever λ_1/λ_N exceeds this value, or $\mathcal{H}_0$ otherwise.

The condition number approach is interesting, although it is totally empirical. In the following section, we will derive the generalized likelihood ratio test (GLRT). Although suboptimal from a Bayesian point of view, this test will be shown through simulations to perform much more accurately than the present condition number test and in fact very close to the optimal Bayesian test. It will in particular appear that the intuitive choice of λ_1/λ_N as a decision variable was not so appropriate and that the appropriate choice (at least, the choice that is appropriate in the GLRT approach) is in fact $\lambda_1/(\frac{1}{N}\operatorname{tr}(\mathbf{Y}\mathbf{Y}^{\mathsf{H}}))$.

16.4.2 Generalized likelihood ratio test

As we concluded in Section 16.3, it is rather difficult to exploit the final formula obtained in Theorem 16.1, let alone its generalized form of Theorem 16.2. This is the reason why a different approach is taken in this section. Instead of considering the optimal Neyman–Pearson test, which is nothing more than a likelihood ratio test when $P_{\mathcal{H}_0} = P_{\mathcal{H}_1}$, we consider the suboptimal generalized likelihood ratio test, which is based on the calculus of the ratio $C_{\text{GLRT}}(\mathbf{Y})$ below

$$C_{\text{GLRT}}(\mathbf{Y}) = \frac{\sup_{\mathbf{H},\sigma^2} P_{\mathbf{Y}|\mathbf{H},\sigma^2,\mathcal{H}_1}(\mathbf{Y})}{\sup_{\sigma^2} P_{\mathbf{Y}|\sigma^2,\mathcal{H}_0}(\mathbf{Y})}.$$

This test differs from the likelihood ratio test (or Neyman–Pearson test) by the introduction of the $\sup_{\mathbf{H},\sigma^2}$ in the numerator and the $\sup_{\sigma^2}$ in the denominator. That is, among all possible $\mathbf{H}$ and σ^2 that are tested against the observation $\mathbf{Y}$, we consider only the most probable $(\mathbf{H}, \sigma^2)$ pair in the calculus of the numerator and the most probable σ^2 in the calculus of the denominator. This is a rather appropriate approach whenever $\mathbf{Y}$ carries much information about the possible $\mathbf{H}$ and σ^2, but a rather hazardous one when a large extent of $(\mathbf{H}, \sigma^2)$ pairs can account for the observation $\mathbf{Y}$, most of these being discarded by taking the supremum.

The explicit calculus of $C_{\mathrm{GLRT}}(\mathbf{Y})$ is rather classical and not new. It is particularly based on, e.g., [Anderson, 1963] and [Wax and Kailath, 1985]. The complete calculus, which we do not recall here, leads to the following result.

Theorem 16.3. *Call T_M the ratio*

$$T_M = \frac{\lambda_1}{\frac{1}{N}\operatorname{tr}\mathbf{Y}\mathbf{Y}^{\mathsf{H}}}.$$

Then the generalized likelihood ratio $C_{\mathrm{GLRT}}(\mathbf{Y})$ is given by:

$$C_{\mathrm{GLRT}}(\mathbf{Y}) = \left(1 - \frac{1}{N}\right)^{(1-N)M} T_M^{-M}\left(1 - \frac{T_M}{N}\right)^{(1-N)M}.$$

The function

$$\phi_N : t \mapsto \left(1 - \frac{1}{N}\right)^{(1-N)M} t^{-M}\left(1 - \frac{t}{N}\right)^{(1-N)M}$$

turns out to be increasing on $(1, N)$. Since T_M lies in this interval with probability one, $C_{\mathrm{GLRT}}(\mathbf{Y}) > \xi_N$ is equivalent to $T_M > \phi_N^{-1}(\xi_N)$, with ϕ_N^{-1} the local inverse of ϕ_N on $(1, N)$. Therefore, setting $C_{\mathrm{GLRT}}(\mathbf{Y})$ above a threshold ξ_N is equivalent to setting the ratio T_M of the largest eigenvalue to the trace of $\mathbf{Y}\mathbf{Y}^{\mathsf{H}}$ above $\phi_N^{-1}(\xi_N)$. Theorem 16.3 therefore provides a close alternative approach to the condition number criterion, which we remind was linked on the ratio of the largest to the smallest eigenvalues of $\mathbf{Y}\mathbf{Y}^{\mathsf{H}}$ instead.

For practical purposes, it is fundamental to be able to set adequately the decision threshold over which $\mathcal{H}_1$ is preferred to $\mathcal{H}_0$. To this end, we need to be able to derive, e.g., for all desired false alarm rates $\alpha \in (0, 1)$, the threshold γ_M such that $\phi_N^{-1}(\gamma_M) = \alpha$. Then, for such γ_M, the so-called *power of the test*, i.e. the correct detection rate for fixed false alarm rate, needs be evaluated to assess the performance of the test. It turns out that the probability of missing the detection of $\mathcal{H}_1$ (i.e. the complementary to the power of the test) can be shown to be approximately exponential with growing M. That is:

$$P_{T_M|\mathcal{H}_1}(x) \simeq e^{-MI(x)}$$

where, in large deviation theory [Dembo and Zeitouni, 2009], the function $I(x)$ is called the rate function associated with T_M. An interesting figure of performance to characterize the probability of missing detection under $\mathcal{H}_1$ is therefore given by the asymptotic value

$$I_\infty(\alpha) \triangleq \lim_{M\to\infty} -\frac{1}{M}\log\int_0^{\gamma_M} P_{T_M|\mathcal{H}_1}(x)dx$$

which depends on α through the fact that $\phi_N^{-1}(\gamma_M) = \alpha$.

16.4.3 Test power and error exponents

As was mentioned earlier, the explicit computation of a decision threshold and of error exponents for the optimal Neyman–Pearson test is prohibitive due to the expression taken by $C_{\mathbf{Y}}$, not to mention the expression when the signal-to-noise ratio is a priori unknown. However, we will presently show that, for the GLRT approach it is possible to derive both the optimal threshold and the error exponent for increasingly large system dimensions N and M (growing simultaneously large). For the condition number approach, the optimal threshold cannot be derived due to an eigenvalue independence assumption, which is not proved to this day, although it is possible to derive an expression for the error exponent for this threshold. In fact, it will turn out that the error exponents are independent of the choice of the detection threshold.

We first consider the GLRT test. The optimal threshold γ_M for a fixed false alarm rate α can be expressed as follows.

Theorem 16.4. *For fixed false alarm rate $\alpha \in (0,1)$, the power of the generalized likelihood ratio test is maximum for the threshold*

$$\gamma_M = \left(1 + \sqrt{\frac{N}{M}}\right)^2 + \frac{b_M}{M^{\frac{2}{3}}}\zeta_M$$

for some sequence ζ_M such that ζ_M converges to $(\bar{F}^+)^{-1}(\alpha)$, with $\bar{F}^+$ the complementary Tracy–Widom law defined as $\bar{F}^+(x) = 1 - F^+(x)$ and where $b_M \triangleq \left(1 + \sqrt{\frac{N}{M}}\right)^{\frac{4}{3}} \left(\frac{N}{M}\right)^{-\frac{1}{6}}$.

Moreover the false alarm rate of the GLRT with threshold

$$\gamma_M = \left(1 + \sqrt{\frac{N}{M}}\right)^2 + \frac{b_M}{M^{\frac{2}{3}}}(\bar{F}^+)^{-1}(\alpha)$$

converges to α, and more generally

$$\int_{\gamma}^{\infty} P_{T_M|\mathcal{H}_0}(x) - \bar{F}^+\left(\frac{M^{\frac{2}{3}}(\gamma - (1 + \sqrt{\frac{N}{M}})^2)}{b_M}\right) \to 0.$$

Let us assume $\sigma = 1$ without loss of generality. To prove this result, we first need to remember from Theorem 9.5 that the largest eigenvalue λ_1 of $\mathbf{Y}\mathbf{Y}^{\mathsf{H}}$ is related to the Tracy–Widom distribution as follows.

$$M^{\frac{2}{3}}\left(\frac{\frac{\lambda_1}{M} - \left(1 + \frac{N}{M}\right)^2}{b_M}\right) \Rightarrow X^+$$

as N, M grow large, where X^+ is distributed according to the Tracy–Widom distribution F^+. At the same time, $\frac{1}{MN} \operatorname{tr} \mathbf{Y}\mathbf{Y}^{\mathsf{H}} \xrightarrow{\text{a.s.}} 1$. Therefore, from the

Slutsky theorem, Theorem 8.12, we have that

$$\tilde{T}_M \triangleq M^{\frac{2}{3}} \left(\frac{T_M - (1 + \frac{N}{M})^2}{b_M} \right) \Rightarrow X^+.$$

Denoting $F_M(x) \triangleq \int_x^\infty P_{\tilde{T}_M|\mathcal{H}_0}(x)dx$, the distribution function of $\tilde{T}_M$ under $\mathcal{H}_0$, it can in fact be shown that the convergence of F_M towards F^+ is in fact uniform over $\mathbb{R}$. By definition of α, we have that

$$1 - F_M \left(M^{\frac{2}{3}} \left(\frac{\gamma_M - (1 + \frac{N}{M})^2}{b_M} \right) \right) = \alpha.$$

From the convergence of F_M to F^+, we then have that

$$\bar{F}^+ \left(M^{\frac{2}{3}} \left(\frac{\gamma_M - (1 + \frac{N}{M})^2}{b_M} \right) \right) \to \alpha$$

from which, taking the (continuous) inverse of $\bar{F}^+$, we have the first identity. The remaining expressions are merely due to the fact that $F_M(\tilde{T}_M) - F^+(\tilde{T}_M) \to 0$.

Deriving a similar threshold for the condition number test requires to prove the asymptotic independence of λ_1 and λ_N. This has been shown for the Gaussian unitary ensemble, Theorem 9.6, i.e. for the extreme eigenvalues of the semi-circle law, but to this day not for the extreme eigenvalues of the Marčenko–Pastur law.

After setting the decision threshold, it is of interest to evaluate the corresponding theoretical power test, i.e. the probability of correct detection. To this end, instead of an explicit expression of the theoretical test power, we assume large system dimensions and use tools from the theory of large deviations. Those tools come along with mathematical requirements, though, which are outside the scope of this book. We will therefore only state the main conclusions. For details, the reader is referred to [Bianchi *et al.*, 2011]. As stated in [Bianchi *et al.*, 2011], as the system dimensions grow large, it is sensible for us to reduce the acceptable false alarm rate. Instead of a mere expression of the test power for a given false alarm, we may be interested in the *error exponent curve*, which is the set of points (a, b), such that there exists sequences $\alpha_1, \alpha_2, \ldots$ such that $\lim_M -\frac{1}{M} \log \alpha_M = a$ and $I_\infty(a) = b$. We first address the case of the GLRT.

Theorem 16.5. *Assume, under* $\mathcal{H}_1$, *that* $\sum_{k=1}^N \frac{|h_k|^2}{\sigma^2}$ *converges to* ρ *as* N, M *grow large, and that* $N/M \to c$. *Then, for any* $\alpha \in (0, 1)$, $I_\infty(\alpha)$ *is well defined and reads:*

$$I_\infty(\alpha) = \begin{cases} I_\rho\left((1+\sqrt{c})^2\right), & \text{if } \rho > \sqrt{c} \\ 0 & \text{, otherwise} \end{cases}$$

where $I_\rho(x)$ is defined as

$$I_\rho(x) = \frac{x - \lambda_{sp}}{1+\rho} - (1-c)\log\left(\frac{x}{\lambda_{sp}}\right) - c(V^+(x) - V^+(\lambda_{sp})) + \Delta(x|[(1+\sqrt{c})^2, \infty))$$

with $\Delta(x|A)$ the function equal to zero is $x \in A$ or to infinity otherwise, and $V^+(x)$ the function defined by

$$V^+(x) = \log(x) + \frac{1}{c}\log(1 + cm(x)) + \log(1 + \underline{m}(x)) + xm(x)\underline{m}(x)$$

with $\underline{m}(x) = cm(x) - \frac{1-c}{x}$ and $m(x)$ the Stieltjes transform of the Marčenko–Pastur law with ratio c

$$m(z) = \frac{(1-z-c) + \sqrt{(1-z-c)^2 - 4cz}}{2cz}$$

(the branch of the square-root being such that m is a Stieltjes transform).
Moreover, the error exponent curve is described by the set of points

$$\left\{(I_0(x), I_\rho(x)),\ x \in \left((1+\sqrt{c})^2, \lambda_{sp}\right)\right\}$$

with I_0 defined as

$$I_0(x) = x - (1+\sqrt{c})^2 - (1-c)\log\left(\frac{x}{(1+\sqrt{c})^2}\right) - c[V(x) - V((1+\sqrt{c})^2)] + \Delta(x|[(1+\sqrt{c})^2, \infty)).$$

A similar result is obtained for the condition number test, namely:

Theorem 16.6. *For any false alarm rate $\alpha \in (0,1)$ and for each limiting ρ, the error exponent of the condition number test coincides with the error exponent for the GLRT. Moreover, the error exponent curve is given by the set of points*

$$\left\{(J_0(x), J_\rho(x)),\ x \in \left(\frac{(1+\sqrt{c})^2}{(1-\sqrt{c})^2}, \frac{\lambda_{sp}}{(1-\sqrt{c})^2}\right)\right\}$$

with

$$J_\rho(x) = \inf\left\{I_\rho(x_1) + I^-(x_2),\ \frac{x_1}{x_2} = x\right\}$$

$$J_0(x) = \inf\left\{I_0(x_1) + I^-(x_2),\ \frac{x_1}{x_2} = x\right\}$$

where I^- is defined as

$$I^-(x) = x - (1-\sqrt{c})^2 - (1-c)\log\left(\frac{x}{(1-\sqrt{c})^2}\right) - 2c[V^-(x) - V^-((1-\sqrt{c})^2)] + \Delta(x|(0, (1-\sqrt{c})^2])$$

and V^- *is given by:*

$$V^-(x) = \log(x) + \frac{1}{c}\log(1 + cm(x)) + \log(-1 - \underline{m}(x)) + xm(x)\underline{m}(x).$$

The ROC curve performance of the optimal Neyman–Pearson test with unknown SNR is compared against the condition number test and the GLRT in Figure 16.8. For the Neyman–Pearson test, we remind that the unknown SNR parameter must be integrated out, which assumes the need for a prior probability distribution for σ^2. We provide in Figure 16.8 two classical approaches, namely uniform distribution and Jeffreys prior with coefficient $\beta = 1$, i.e. $P_{\sigma^2}(\sigma^2) = \frac{1}{\sigma^2}$. The simulation conditions are as before with $K = 1$ transmit source, $N = 4$ receive sensors, $M = 8$ samples. The SNR is now set to zero dB in order to have non-trivial correct detection values. Observe that, as expected, the Neyman–Pearson test outperforms both the GLRT and condition number tests, either for uniform or Jeffreys prior. More surprising is the fact that the generalized likelihood ratio test largely outperforms the condition number test and performs rather close to the optimal Neyman–Pearson test. Therefore, the choice of the ratio between the largest eigenvalue and the normalized trace of the sample covariance matrix as a test comparison criterion is much more appropriate than the ratio between the largest eigenvalue and the smallest eigenvalue. Given the numerical complexity involved by the explicit computation of the Neyman–Pearson test, the GLRT can be considered as an interesting suboptimal substitute for this test when the signal and noise powers are a priori unknown.

It is also mentioned in [Bianchi *et al.*, 2011] that the error exponent curve of the GLRT dominates that of the condition number test in the sense that, for each (a, b) in the error exponent curve of the condition number test, there exists $b' > b$ such that (a, b') is in the error exponent curve of the GLRT. Therefore, from the above theorems, at least asymptotically, the GLRT always outperforms the condition number test. The practical simulations confirm this observation.

In this section, we mainly used Theorem 7.1 and Theorem 9.1 that basically state that, for large dimensional sample covariance matrices with population covariance eigenvalues converging to a single mass in 1, the largest eigenvalue is asymptotically found at the edge of the support of the Marčenko–Pastur law or outside this support, depending on whether a spike is found among the population eigenvalues. This allowed us to proceed to hypothesis tests discriminating both models with or without a spiked population eigenvalue. In the next section, we go further by considering more involved random matrix models for which not only hypothesis testing is performed but also statistical inference. That is, we will proceed to the estimation of system parameters using eigen-inference methods. This chapter will therefore require the tools developed in Chapter 7 and Chapter 8 of Part I.

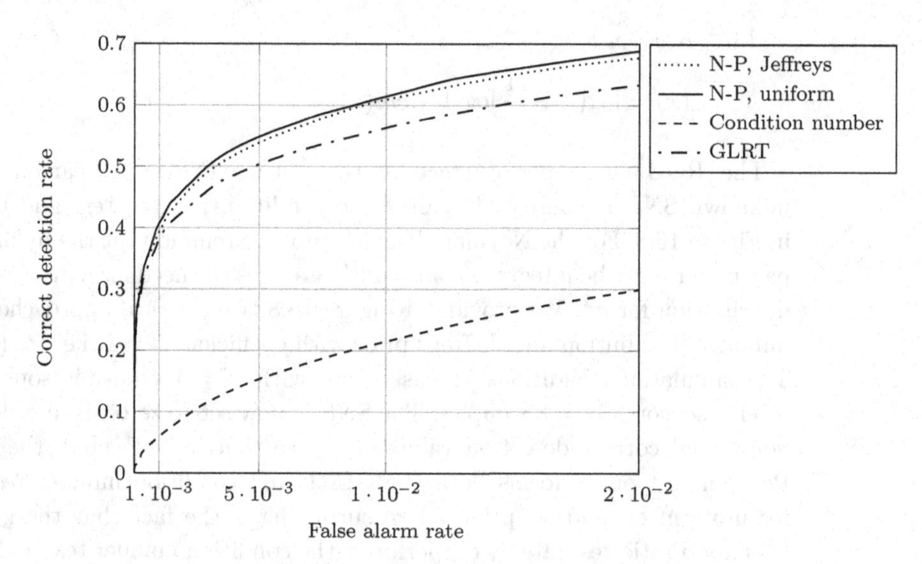

Figure 16.8 ROC curve for a priori unknown σ^2 of the Neyman–Pearson test (N-P), condition number method and GLRT, $K = 1$, $N = 4$, $M = 8$, $\mathsf{SNR} = 0$ dB. For the Neyman–Pearson test, both uniform and Jeffreys prior, with exponent $\beta = 1$, are provided.

17 Estimation

In this chapter, we consider the consistent estimation of system parameters involving random matrices with large dimensions. When it comes to estimation or statistical inference in signal processing, there often exists a large number of different methods proposed in the literature, most of which are usually based on a reference, simple, and robust method which has various limitations such as the Urkowitz's power detector [Urkowitz, 1967] that only assumes the additive white Gaussian noise (AWGN) model, or the multiple signal classification (MUSIC) algorithm [Schmidt, 1986] of Schmidt that suffers from undecidability issues when the signal to noise ratio reaches a critically low value. When performing statistical inference based on a limited number of large dimensional vector inputs, the main limitation is due to the fact that those legacy estimators are usually built under the assumption that the number of available observations is extremely large compared to the number of system parameters to identify. In modern signal processing applications, especially for large sensor networks, the estimators receive as inputs the M stacked N-dimensional observation vectors $\mathbf{Y} = [\mathbf{y}^{(1)}, \ldots, \mathbf{y}^{(M)}] \in \mathbb{C}^{N \times M}$ of some observation vectors $\mathbf{y}^{(m)} \in \mathbb{C}^N$ at time m, M and N being of similar size, or even sometimes M being much smaller than N. Novel estimators that can cope with this large population size limitation are therefore required in place of the historical estimators. In this chapter, we introduce such (N, M)-consistent estimators, which we recall are estimators which are asymptotically unbiased when both N and M grow large at a similar rate.

Since the significant advances in this field of research are rather new, only two main examples will be treated here. The first example is that of the consistent estimation of direction of arrivals (DoA) in linear sensor arrays (such as radars) [Kay, 1993; Scharf, 1991] when the number of sensors is of similar dimension as the number of available observations. The major works in this direction are [Mestre and Lagunas, 2008] and [Vallet *et al.*, 2010] for almost identical situations, involving nonetheless different system models. We will then move to the question of blind user sensing and power estimation. Specifically, we will consider a sensor array receiving simultaneous transmissions from multiple signal emitters, the objective being to estimate both the number of transmitters and the power used by each one of those. The latter is mainly based on [Couillet *et al.*, 2011c].

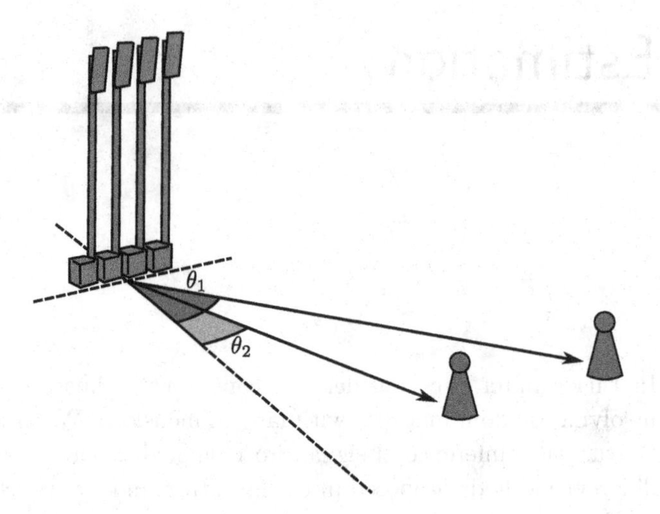

Figure 17.1 Two-user line-of-sight transmissions with different angles of arrival, θ_1 and θ_2.

17.1 Directions of arrival

In this section, we consider the problem of a sensor array impinged by multiple signals, each one of which comes from a given direction. This is depicted in Figure 17.1, where two signals transmitted simultaneously by two terminal users (positioned far away from the receiving end) are received with angles θ_1 and θ_2 at the sensor array. The objective here is to detect both the number of signal sources and the direction of arrival from each of these signals. This has natural applications in radar detection for instance, where multiple targets need to be localized. In general, thanks to the diversity offered by the sensor array, and the phase shifts in the signals impacting every antenna, it is possible to determine the angle of signal arrival from basic geometrical optics. In the following, we will recall the classical so-called multiple signal classification estimator (MUSIC) [Schmidt, 1986], which is suited for large streams of data and small dimensional sensor array as it can be proved to be a consistent estimator in this setting. However, it can be proved that the MUSIC technique is not consistent with increasing dimensions of both the number of sensors and the number of samples. To cope with this problem, a G-estimator is proposed, essentially based on Theorem 8.7. This recent estimator, developed in [Mestre and Lagunas, 2008] by Mestre and Lagunas, is based on the concept of G-estimation and is referred to as G-MUSIC. We first introduce the system model under consideration.

17.1.1 System model

We consider the communication setup between K signal sources (that would be, in the radar context, the reflected waveforms from detected targets) and N

receive sensors, $N > K$. Denote $x_k^{(t)}$ the signal issued by source k at time t. The received signals at time t, corrupted by the additive white Gaussian noise vector $\sigma\mathbf{w}^{(t)} \in \mathbb{C}^N$ with $\mathrm{E}[\mathbf{w}^{(t)}\mathbf{w}^{(t')}] = \delta_{t,t'}\mathbf{I}_N$, are gathered into the vector $\mathbf{y}^{(t)} \in \mathbb{C}^N$. We assume that the channel between the sources and the sensors creates only phase rotations, that essentially depend on the antenna array geometry. Other parameters such as known scattering effects might be taken into account as well. To be all the more general, we assume that the channel steering effect on signal $x_k^{(t)}$ for sensor i is modeled through the time invariant function $s_i(\theta)$ for $\theta = \theta_k$. As such, we characterize the transmission model at time t as

$$\mathbf{y}^{(t)} = \sum_{k=1}^{K} \mathbf{s}(\theta_k)x_k^{(t)} + \sigma\mathbf{w}^{(t)} \tag{17.1}$$

where $\mathbf{s}(\theta_k) = [s_1(\theta_k), \ldots, s_N(\theta_k)]^\mathsf{T}$. For simplicity, we assume that the vectors $\mathbf{s}(\theta_k)$ have unit Euclidean norm.

Suppose for the time being that $\mathbf{x}^{(t)} = [x_1^{(t)}, \ldots, x_K^{(t)}]^\mathsf{T} \in \mathbb{C}^K$ are i.i.d. along the time domain t and have zero mean and covariance matrix $\mathbf{P} \in \mathbb{C}^{K\times K}$. This assumption, which is not necessarily natural, will be discarded in Section 17.1.4. The vectors $\mathbf{y}^{(t)}$ are sampled M times, with M of the same order of magnitude as N, and are gathered into the matrix $\mathbf{Y} = [\mathbf{y}^{(1)}, \ldots, \mathbf{y}^{(M)}] \in \mathbb{C}^{N\times M}$. From the assumptions above, the columns of $\mathbf{Y}$ have zero mean and covariance $\mathbf{R}$, given by:

$$\mathbf{R} = \mathbf{S}(\Theta)\mathbf{P}\mathbf{S}(\Theta)^\mathsf{H} + \sigma^2\mathbf{I}_N$$

where $\mathbf{S}(\Theta) = [\mathbf{s}(\theta_1), \ldots, \mathbf{s}(\theta_K)] \in \mathbb{C}^{N\times K}$.

The DoA detection question amounts to estimating $\theta_1, \ldots, \theta_K$ based on $\mathbf{Y}$, knowing the steering vector function $\mathbf{s}(\theta) = [s_1(\theta), \ldots, s_N(\theta)]^\mathsf{T}$ for all θ. To this end, not only eigenvalues of $\frac{1}{M}\mathbf{Y}\mathbf{Y}^\mathsf{H}$ but also eigenvectors are necessary. This is why we will resort to the G-estimators introduced in Section 17.1.3. Before that, we discuss the classical subspace methods and the MUSIC approach.

17.1.2 The MUSIC approach

We denote $\lambda_1 \leq \ldots \leq \lambda_N$ the eigenvalues of $\mathbf{R}$ and $\mathbf{e}_1, \ldots, \mathbf{e}_N$ their corresponding eigenvectors. Similarly, we denote $\hat{\lambda}_1 \leq \ldots \leq \hat{\lambda}_N$ the eigenvalues of $\mathbf{R}_N \triangleq \frac{1}{M}\mathbf{Y}\mathbf{Y}^\mathsf{H}$, with respective eigenvectors $\hat{\mathbf{e}}_1, \ldots, \hat{\mathbf{e}}_N$. If some eigenvalue has multiplicity greater than one, the set of corresponding eigenvectors is taken to be any orthonormal basis of the associated eigenspace. From the assumption that the number of sensors N is greater than the number of transmit sources K, the last $N - K$ eigenvalues of $\mathbf{R}$ equal σ^2 and we can represent $\mathbf{R}$ under the form

$$\mathbf{R} = \begin{pmatrix}\mathbf{E}_W & \mathbf{E}_S\end{pmatrix} \begin{pmatrix}\sigma^2\mathbf{I}_{N-K} & \mathbf{0} \\ \mathbf{0} & \mathbf{\Lambda}_S\end{pmatrix} \begin{pmatrix}\mathbf{E}_W^\mathsf{H} \\ \mathbf{E}_S^\mathsf{H}\end{pmatrix}$$

with $\mathbf{\Lambda}_S = \text{diag}(\lambda_{N-K+1}, \ldots, \lambda_N)$, $\mathbf{E}_S = [\mathbf{e}_{N-K+1}, \ldots, \mathbf{e}_N]$ the so-called *signal space* and $\mathbf{E}_W = [\mathbf{e}_1, \ldots, \mathbf{e}_{N-K}]$ the so-called *noise space.*

The basic idea of the subspace approach, which is at the core of the MUSIC method, is to observe that any vector lying in the signal space is orthogonal to the noise space. This leads in particular to

$$\mathbf{E}_W^{\mathsf{H}}\mathbf{s}(\theta_k) = 0$$

for all $k \in \{1, \ldots, K\}$, which is equivalent to

$$\eta(\theta_k) \triangleq \mathbf{s}(\theta_k)\mathbf{E}_W\mathbf{E}_W^{\mathsf{H}}\mathbf{s}(\theta_k) = 0.$$

The idea behind the MUSIC approach is simple in that it suggests, according to the large M-dimension approach, that the covariance matrix $\mathbf{R}$ is well approximated by $\mathbf{R}_N$ as M grows to infinity. Therefore, denoting $\hat{\mathbf{E}}_W = [\hat{\mathbf{e}}_1, \ldots, \hat{\mathbf{e}}_{N-K}]$ the eigenvector space corresponding to the smallest eigenvalues of $\mathbf{R}_N$, the MUSIC estimator consists in retrieving the arguments θ which minimize the function

$$\hat{\eta}(\theta) \triangleq \mathbf{s}(\theta)^{\mathsf{H}}\hat{\mathbf{E}}_W\hat{\mathbf{E}}_W^{\mathsf{H}}\mathbf{s}(\theta).$$

Notice that it may not be possible for $\hat{\eta}(\theta)$ to be zero for any θ, so that by looking for minima in $\eta(\theta)$, we are not necessarily looking for roots. This approach is originally due to Schmidt in [Schmidt, 1986]. However, the finite number of available samples strongly affects the efficiency of the MUSIC algorithm. In order to come up with more efficient approaches, the subspace approach was further refined by taking into account the fact that, in addition to be orthogonal to the noise space, $\mathbf{s}(\theta_k)$ is aligned to the signal space $\mathbf{S}(\Theta)\mathbf{P}\mathbf{S}(\Theta)^{\mathsf{H}}$. One of the known examples is the so-called SSMUSIC approach due to McCloud and Scharf [McCloud and Scharf, 2002]. The approach considered in the SSMUSIC method is now to determine the local minima of the function

$$\hat{\eta}_{\text{SS}}(\theta) \triangleq \frac{\mathbf{s}(\theta)^{\mathsf{H}}\hat{\mathbf{E}}_W\hat{\mathbf{E}}_W^{\mathsf{H}}\mathbf{s}(\theta)}{\mathbf{s}(\theta)^{\mathsf{H}}\hat{\mathbf{E}}_S\left(\hat{\mathbf{\Lambda}}_S - \hat{\sigma}^2\mathbf{I}_K\right)^{-1}\hat{\mathbf{E}}_S^{\mathsf{H}}\mathbf{s}(\theta)}$$

where the denominator comes from $\left(\mathbf{S}(\Theta)\mathbf{P}\mathbf{S}(\Theta)^{\mathsf{H}}\right)^{-1} = \mathbf{E}_S\left(\mathbf{\Lambda}_S - \sigma^2\mathbf{I}_K\right)^{-1}\mathbf{E}_S^{\mathsf{H}}$ (when $\mathbf{S}(\Theta)\mathbf{P}\mathbf{S}(\Theta)^{\mathsf{H}}$ is not invertible, the same remark holds with the inverse sign replaced by the Moore–Penrose pseudo-inverse sign), with $\hat{\mathbf{\Lambda}}_S = \text{diag}(\hat{\lambda}_{N-K+1}, \ldots, \hat{\lambda}_N)$ and $\hat{\sigma}^2 = \frac{1}{N-K}\sum_{k=1}^{N-K}\hat{\lambda}_k$. The SSMUSIC technique was proved to outperform the MUSIC approach for finite M, as it has a higher resolution power to distinguish close angles of arrival [McCloud and Scharf, 2002].

However, even though it is proved to be better, this last approach is still not (N, M)-consistent. This fact, which we do not prove here, is the point made by Mestre in [Mestre, 2008a] and [Mestre and Lagunas, 2008]. Instead of this classical large dimensional M approach, we will assume that both N and M grow large at similar pace while K is kept constant, so that we can use the results from Chapter 8.

17.1.3 Large dimensional eigen-inference

The improved MUSIC estimator unfolds from a trivial application of Theorem 8.7. The cost function u introduced in Theorem 8.7 is simply replaced by the subspace cost function $\eta(\theta)$, defined by

$$\eta(\theta_k) = \mathbf{s}(\theta_k)\mathbf{E}_W\mathbf{E}_W^{\mathsf{H}}\mathbf{s}(\theta_k).$$

We therefore have the following improved MUSIC estimator, called by the authors in [Mestre and Lagunas, 2008] the G-MUSIC estimator.

Theorem 17.1 ([Mestre and Lagunas, 2008]). *Under the above conditions, we have:*

$$\eta(\theta) - \bar{\eta}(\theta) \xrightarrow{\text{a.s.}} 0$$

as N, M grow large with limiting ratio satisfying $0 < \lim N/M < \infty$, *where*

$$\bar{\eta}(\theta) = \mathbf{s}(\theta)^{\mathsf{H}} \left(\sum_{n=1}^{N} \phi(n)\hat{\mathbf{e}}_n\hat{\mathbf{e}}_n^{\mathsf{H}} \right) \mathbf{s}(\theta)$$

with $\phi(n)$ defined as

$$\phi(n) = \begin{cases} 1 + \sum_{k=N-K+1}^{N} \left(\frac{\hat{\lambda}_k}{\hat{\lambda}_n - \hat{\lambda}_k} - \frac{\hat{\mu}_k}{\hat{\lambda}_n - \hat{\mu}_k} \right), & n \leq N-K \\ -\sum_{k=1}^{N-K} \left(\frac{\hat{\lambda}_k}{\hat{\lambda}_n - \hat{\lambda}_k} - \frac{\hat{\mu}_k}{\hat{\lambda}_n - \hat{\mu}_k} \right) & , n > N-K \end{cases}$$

and with $\mu_1 \leq \ldots \leq \mu_N$ the eigenvalues of $\operatorname{diag}(\hat{\boldsymbol{\lambda}}) - \frac{1}{M}\sqrt{\hat{\boldsymbol{\lambda}}}\sqrt{\hat{\boldsymbol{\lambda}}}^{\mathsf{T}}$, *where we denoted* $\hat{\boldsymbol{\lambda}} = (\hat{\lambda}_1, \ldots, \hat{\lambda}_N)^{\mathsf{T}}$.

This derives naturally from Theorem 8.7 by noticing that the noise space $\mathbf{E}_W$ is the space of the smallest eigenvalue of $\mathbf{R}$ with multiplicity $N-K$, which is mapped to the space of the smallest $N-K$ eigenvalues of the empirical $\mathbf{R}_N$ to derive the consistent estimate.

It is also possible to derive an (N, M)-consistent estimate for the improved SSMUSIC method. The latter will be referred to as G-SSMUSIC This unfolds from a similar application of Theorem 8.7 and is given in [Mestre and Lagunas, 2008] under the following form.

Theorem 17.2 ([Mestre and Lagunas, 2008]). *Under the above conditions*

$$\eta_{\text{SS}}(\theta) - \bar{\eta}_{\text{SS}}(\theta) \xrightarrow{\text{a.s.}} 0$$

as N, M grow large with ratio uniformly bounded away from zero and infinity, where

$$\bar{\eta}_{\text{SS}}(\theta) = \frac{\bar{\eta}(\theta)}{\varepsilon\bar{\eta}(\theta) + \bar{\chi}(\theta)}$$

for any $\varepsilon \geq 0$ *that guarantees that the denominator does not vanish for all* θ*, where* $\bar{\eta}(\theta)$ *is given in Theorem 17.1 and* $\bar{\chi}(\theta)$ *is given by:*

$$\bar{\chi}(\theta) = \mathbf{s}(\theta)^{\mathsf{H}} \left(\sum_{n=1}^{N} \psi(n) \hat{\mathbf{e}}_n \hat{\mathbf{e}}_n^{\mathsf{H}} \right) \mathbf{s}(\theta)$$

with $\psi(n)$ *defined as*

$$\psi(n) = \begin{cases} \frac{1}{\hat{\sigma}^2} \sum_{k=N-K+1}^{N} \left(\frac{\hat{\lambda}_k}{\hat{\lambda}_n - \hat{\lambda}_k} - \frac{\hat{\nu}_k}{\hat{\lambda}_n - \hat{\nu}_k} \right) & , n \leq N-K \\ \frac{1}{\hat{\sigma}^2} \left(\sum_{k=0}^{N-K} \frac{\hat{\nu}_k}{\hat{\lambda}_n - \hat{\nu}_k} - \sum_{k=1}^{N-K} \frac{\hat{\lambda}_k}{\hat{\lambda}_n - \hat{\lambda}_k} \right) & , n > N-K \end{cases}$$

with $\mu_1 \leq \ldots \leq \mu_N$ *the eigenvalues of* $\mathrm{diag}(\hat{\boldsymbol{\lambda}}) - \frac{1}{M} \sqrt{\hat{\boldsymbol{\lambda}}} \sqrt{\hat{\boldsymbol{\lambda}}}^{\mathsf{T}}$, $\hat{\nu}_0 \leq \ldots \leq \hat{\nu}_N$ *the solutions of the equation in* ν

$$\nu = \hat{\sigma}^2 \left(1 - \frac{1}{M} \sum_{k=1}^{N} \frac{\hat{\lambda}_k}{\hat{\lambda}_k - \nu} \right)$$

and $\hat{\sigma}^2$ *is an* (N, M)*-consistent estimator for* σ^2*, given here by*

$$\hat{\sigma}^2 = \frac{M}{N-K} \sum_{k=1}^{N-K} \left(\hat{\lambda}_k - \hat{\mu}_k \right).$$

We hereafter provide one-shot realizations of the cost functions $\bar{\eta}(\theta)$ and $\bar{\eta}_{\mathrm{SS}}(\theta)$ for the different DoA estimation methods proposed above. We take the assumptions that $K = 3$ signal sources are emitting and that an array of $N = 20$ sensors is used to perform the statistical inference, that samples $M = 150$ times the incoming waveform. The angles of arrival are $10°$, $35°$, and $37°$, while the SNR is set to 10 dB. This situation is particularly interesting as two incoming waveforms are found with very close DoA. From the discussion above, we therefore hope that SSMUSIC would better resolve the two close angles. In fact, we will see that the G-estimators that are G-MUSIC and G-SSMUSIC are even more capable of discriminating between close angles. Figure 17.2 and Figure 17.3 provide the comparative performance plots of the MUSIC against G-MUSIC approaches, for θ ranging from $-45°$ to $45°$ in Figure 17.2 and for θ varying from $-33°$ to $-38°$ in Figure 17.3. Observe that, while the MUSIC approach is not able to resolve the two close DoA, the G-MUSIC technique clearly isolates two minima of $\bar{\eta}(\theta)$ around $35°$ and $37°$. Apart from that, both performance plots look alike. Similarly, the SSMUSIC and G-SSMUSIC cost functions for the same random realization as in Figure 17.2 and Figure 17.3 are provided in Figure 17.4 and Figure 17.5. Observe here that the SSMUSIC estimator is able to resolve both angles, although it is clearly not as efficient as the G-SSMUSIC estimator. Performance figures in terms of mean square error are found in [Mestre and Lagunas, 2008]. It is observed in particular by the authors that the improved estimators still do not solve the inherent problem of both MUSIC and SS-MUSIC estimators, which is that both perform very

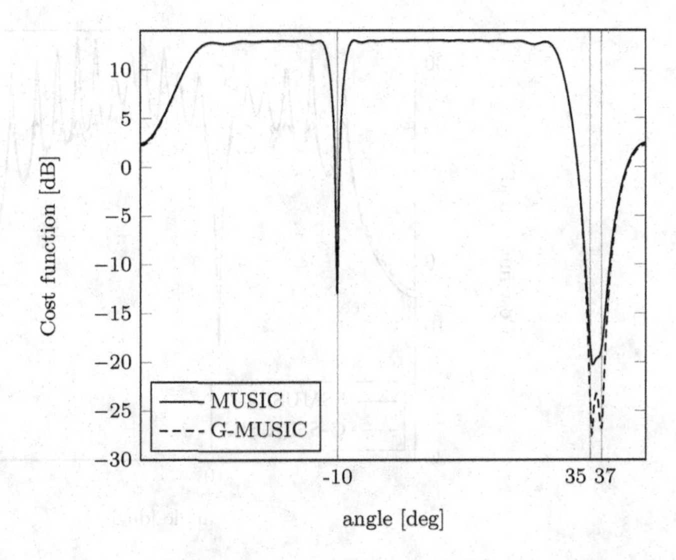

Figure 17.2 MUSIC against G-MUSIC for DoA detection of $K = 3$ signal sources, $N = 20$ sensors, $M = 150$ samples, SNR of 10 dB. Angles of arrival of 10°, 35°, and 37°.

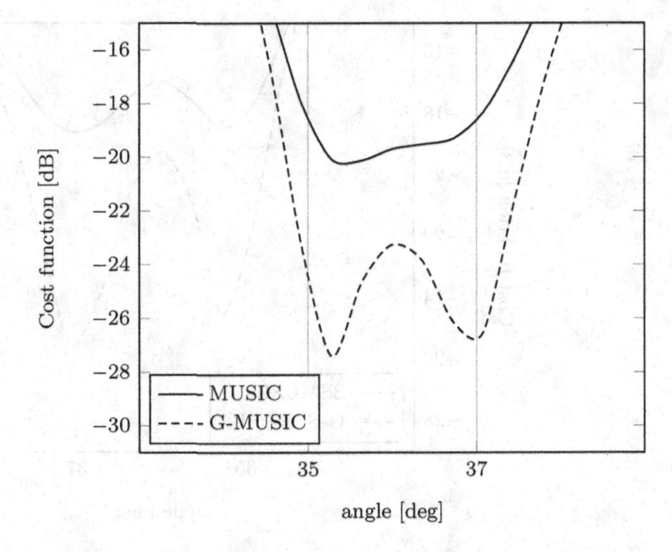

Figure 17.3 MUSIC against G-MUSIC for DoA detection of $K = 3$ signal sources, $N = 20$ sensors, $M = 150$ samples, SNR of 10 dB. Angles of arrival of 10°, 35°, and 37°.

badly in the low SNR regime. Nevertheless, the improved G-estimators manage to repel to a lower level the SNR limit for which performance decays significantly. The same performance behavior will also be observed in Section 17.2, where the performance of blind multi-source power estimators is discussed.

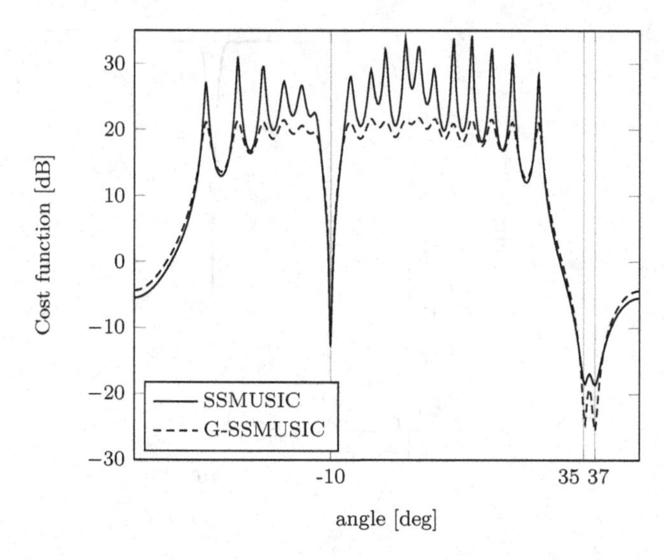

Figure 17.4 SSMUSIC against G-SSMUSIC for DoA detection of $K = 3$ signal sources, $N = 20$ sensors, $M = 150$ samples, SNR of 10 dB. Angles of arrival of 10°, 35°, and 37°.

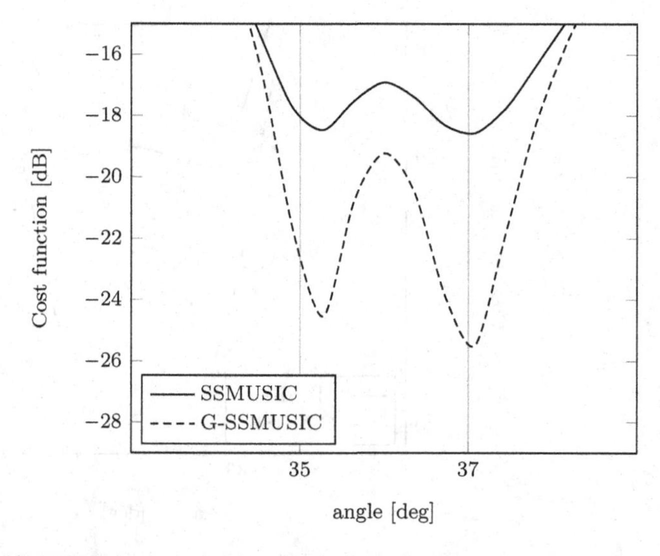

Figure 17.5 SSMUSIC against G-SSMUSIC for DoA detection of $K = 3$ signal sources, $N = 20$ sensors, $M = 150$ samples, SNR of 10 dB. Angles of arrival of 10°, 35°, and 37°.

Further work has been done on the DoA topic, especially in the case where, instead of i.i.d. samples, the sensors receive correlated data. These data can be assumed not to be known to the sensors, so that no specific random model can be applied. This is discussed in the next section.

17.1.4 The correlated signal case

We recall that in the previous section, we explicitly assumed that the vector of transmit signals are independent for successive samples, have zero mean, and have the same covariance matrix. The present section is merely an extension of Section 17.1.3 to the even more restrictive case when the transmit data structure is unknown to the receiving sensors. In this case, the sample covariance matrix model assumed in Section 17.1.3, which allowed us to use Theorem 8.7, is no longer available. This section mainly recalls the results of [Vallet *et al.*, 2010].

Remark 17.1. It must be noted that the authors of [Vallet *et al.*, 2010], instead of mentioning that the signal source is random with unknown distribution, state that the source is *deterministic but unknown* to the sensors. We prefer to say that the source, being unknown to the receiver, is therefore *random from the point of view of the receiver*, and not deterministic. However, to be able to use the tools hereafter, we will need to assume that, although unknown to the receiver, the particular realization of the random incoming data satisfies some important boundedness assumptions, known to the receiver.

The model (17.1) is still valid, i.e. the receive data vector $\mathbf{y}^{(t)}$ at time instant t reads:

$$\mathbf{y}^{(t)} = \sum_{k=1}^{K} \mathbf{s}(\theta_k) x_k^{(t)} + \sigma \mathbf{w}^{(t)}$$

where the vector $\mathbf{x}^{(t)} = [x_1^{(t)}, \ldots, x_K^{(t)}]^\mathsf{T} \in \mathbb{C}^K$ is no longer i.i.d. along the time index t. We therefore collect the M vector samples into the random matrix $\mathbf{X} = [\mathbf{x}^{(1)}, \ldots, \mathbf{x}^{(M)}] \in \mathbb{C}^{K \times M}$ and we obtain the receive model

$$\mathbf{Y} = \mathbf{S}\mathbf{X} + \sigma \mathbf{W}$$

with $\mathbf{W} = [\mathbf{w}^{(1)}, \ldots, \mathbf{w}^{(M)}] \in \mathbb{C}^{N \times M}$ a Gaussian matrix and $\mathbf{S} \in \mathbb{C}^{N \times K}$ the matrix with columns the K steering vectors, defined as previously. In Section 17.1.3, $\mathbf{X}$ was of the form $\mathbf{P}^{\frac{1}{2}}\mathbf{Z}$ with $\mathbf{Z} \in \mathbb{C}^{K \times M}$ filled with i.i.d. entries of zero mean and unit variance, and $\mathbf{W}$ was naturally filled with i.i.d. entries of zero mean and unit variance, so that $\mathbf{Y}$ took the form of a sample covariance matrix. This is no longer valid in this new scenario. The matrix $\mathbf{Y}$ can instead be considered as an information plus noise matrix, if we take the additional assumption that we can ensure $\|\mathbf{S}\mathbf{X}\mathbf{X}^\mathsf{H}\mathbf{S}\|$ uniformly bounded for all matrix sizes. Denoting $\mathbf{E}_W$ the noise subspace of $\mathbf{R} \triangleq \frac{1}{M}\mathbf{S}\mathbf{X}\mathbf{X}^\mathsf{H}\mathbf{S}$, our objective is now to estimate the cost function

$$\eta(\theta) = \mathbf{s}(\theta)^\mathsf{H} \mathbf{E}_W \mathbf{E}_W^\mathsf{H} \mathbf{s}(\theta).$$

Again, the traditional MUSIC approach replaces the noise subspace $\mathbf{E}_W$ by the empirical subspace $\hat{\mathbf{E}}_W$ composed of the eigenvectors corresponding to the $N - K$ smallest eigenvalues of $\frac{1}{M}\mathbf{Y}\mathbf{Y}^\mathsf{H}$. The resulting cost function therefore reads:

$$\hat{\eta}(\theta) = \mathbf{s}(\theta)^{\mathsf{H}}\hat{\mathbf{E}}_W\hat{\mathbf{E}}_W^{\mathsf{H}}\mathbf{s}(\theta)$$

and the MUSIC algorithm consists once more in finding the K deepest minima of $\hat{\eta}(\theta)$. Assuming this time that the noise variance σ^2 is known, as per the assumptions of [Vallet *et al.*, 2010], the improved MUSIC approach, call it once more the G-MUSIC technique, now for the information plus noise model, derives directly from Theorem 8.10.

Theorem 17.3 ([Vallet *et al.*, 2010]). *As N, M grow to infinity with limiting ratio satisfying* $0 < \lim N/M < \infty$

$$\eta(\theta) - \bar{\eta}(\theta) \xrightarrow{\text{a.s.}} 0$$

for all θ, where

$$\bar{\eta}(\theta) = \mathbf{s}(\theta)^{\mathsf{H}}\left(\sum_{k=1}^{N}\phi(k)\hat{\mathbf{e}}_k\hat{\mathbf{e}}_k^{\mathsf{H}}\right)\mathbf{s}(\theta)$$

with $\phi(k)$ defined as

$$\phi(k) = \begin{cases} 1 + \frac{\sigma^2}{M}\sum_{i=N-K+1}^{N}\frac{1}{\hat{\lambda}_i - \hat{\lambda}_k} + \frac{2\sigma^2}{M}\sum_{i=N-K+1}^{N}\frac{\hat{\lambda}_k}{\hat{\lambda}_i - \hat{\lambda}_k} \\ +\frac{\sigma^2(M-N)}{M}\sum_{i=N-K+1}^{N}\left(\frac{1}{\hat{\lambda}_i - \hat{\lambda}_k} - \frac{1}{\hat{\mu}_i - \hat{\lambda}_k}\right) & , k \leq N-K \\ \frac{\sigma^2}{M}\sum_{i=1}^{N-K}\frac{1}{\hat{\lambda}_i - \hat{\lambda}_k} - \frac{2\sigma^2}{M}\sum_{i=1}^{N-K}\frac{\hat{\lambda}_k}{\hat{\lambda}_i - \hat{\lambda}_k} \\ +\frac{\sigma^2(M-N)}{M}\sum_{i=1}^{N-K}\left(\frac{1}{\hat{\lambda}_i - \hat{\lambda}_k} - \frac{1}{\hat{\mu}_i - \hat{\lambda}_k}\right) & , k > N-K \end{cases}$$

and with $\hat{\mu}_1 \leq \ldots \leq \hat{\mu}_N$ the N roots of the equation in μ

$$1 + \frac{\sigma^2}{M}\sum_{k=1}^{N}\frac{1}{\hat{\lambda}_k - \mu}.$$

The performance of the G-MUSIC approach which assumes the information plus noise model against the previous G-MUSIC technique is compared for the one-shot random transmission in Figure 17.6. We observe that both estimators detect very accurately both directions of arrival. Incidentally, the information plus noise G-MUSIC shows slightly deeper minima than the sample covariance matrix G-MUSIC. This is only an outcome of the one-shot observation at hand and does not affect the average performance of both approaches, as shown more precisely in [Vallet *et al.*, 2010].

Note that Theorem 17.3 only proves the asymptotic consistency for the function $\bar{\eta}(\theta)$. The consistency of the angle estimator itself, which is the result of interest for practical applications, is derived in [Vallet *et al.*, 2011a]. The fluctuations of the estimator are then provided in [Mestre *et al.*, 2011]. We also mention that the same authors also proposed a G-MUSIC alternative relying on an additive spike model approach. Both limit and fluctuations are derived also for this technique, which turns out to perform worse than the approach presented

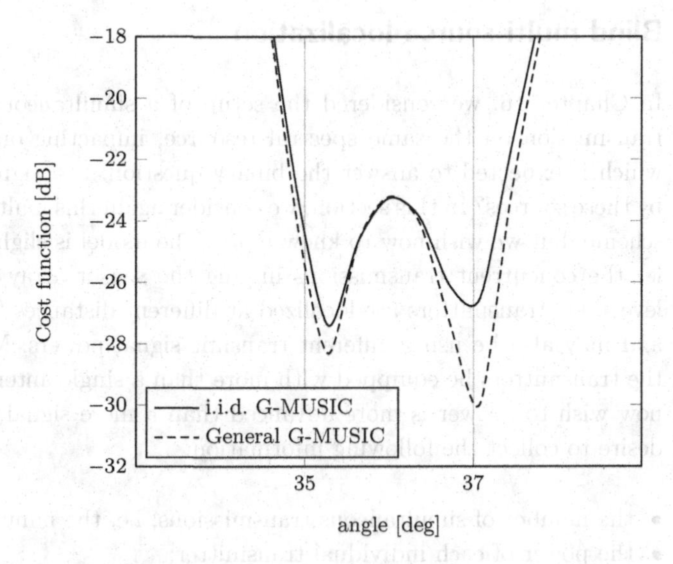

Figure 17.6 G-MUSIC tailored to i.i.d. samples (i.i.d. G-MUSIC) against unconditional G-MUSIC (General G-MUSIC) for DoA detection of $K = 3$ signal sources, $N = 20$ sensors, $M = 150$ samples, SNR of 10 dB. Angles of arrival of 10°, 35°, and 37°.

in this section. The initial results can be found in [Hachem *et al.*, 2011; Vallet *et al.*, 2011b].

This completes this section on DoA localization. We take the opportunity of the information plus noise study above to mention that similar G-estimators have been derived by the same authors for system models involving the information plus noise scenario. In particular, in [Vallet and Loubaton, 2009], a consistent estimator for the capacity of a MIMO channels $\mathbf{H}$ under additive white noise of variance σ^2, i.e. $\log\det(\mathbf{I}_N + \sigma^{-2}\mathbf{H}\mathbf{H}^{\mathsf{H}})$, is derived based on successive observations $\mathbf{y}^{(t)} = \mathbf{H}\mathbf{s}^{(t)} + \sigma\mathbf{w}^{(t)}$ with known pilot sequence $\mathbf{s}^{(1)}, \mathbf{s}^{(2)}, \ldots$ and with additive standard Gaussian noise vector $\mathbf{w}^{(t)}$.

In the following, we consider a similar inference problem, relative to the localization of multiple transmit sources, not from an angular point of view, but rather from a distance point of view. The main idea now is to consider a statistical model where hidden eigenvalues must be recovered that give information on the power transmitted by distinct signal sources. Similar to the DoA estimation, the problem of resolving transmissions of close power will be raised for which a complete analysis of the conditions of source resolution is performed. As such, we will treat the following section in more detail than the current DoA estimation section.

17.2 Blind multi-source localization

In Chapter 16, we considered the setup of a simultaneous multi-source signal transmission on the same spectral resource, impacting on an array of sensors which is expected to answer the binary question: is a signal being transmitted by these sources? In this section, we consider again this multi-source transmission scheme, but we wish now to know more. The model is slightly generalized as we let the concurrent transmissions imping the sensor array with different power levels, i.e. transmitters are localized at different distances from the sensor array and may also be using different transmit signal powers. Moreover, we now let the transmitters be equipped with more than a single antenna. The question we now wish to answer is more advanced than a mere signal sensing decision. We desire to collect the following information:

- the number of simultaneous transmissions, i.e. the number of active users;
- the power of each individual transmitter;
- the number of antennas of each transmitter.

The relative importance of the above pieces of information to the sensor array depends on the problem at hand. We will mainly discuss the problem of user localization in a cognitive radio setting. In the introduction of Chapter 16, we mentioned that cognitive radios, whose objective is to reuse licensed spectrum holes, basically work on a two-step mechanism as they successively need to explore the available spectrum for transmission opportunities and to exploit the spectrum found unused. Through the dual hypothesis test analyzed in Chapter 16 (presence or absence of on-going transmissions), a secondary network is capable of deciding with more or less accuracy whether a given spectral resource is free of use. It is however rather unusual that a spectrum resource be completely left unused within a sufficiently large network coverage area. Typically, a secondary network will sense that no transmission is on-going in a close neighborhood, as it may sense only very low power signals coming from remote transmitters. This situation will then be associated with the no-transmission $\mathcal{H}_0$ hypothesis and exploitation of the spectral resource under study will then be declared possible. How to optimally exploit the spectrum holes depends then on the maximally acceptable secondary transmit coverage area that lets the primary transmissions be free of interference. This question is however not fully answered by the dual hypothesis test.

This is where the question of estimating the power of on-going transmissions is of utmost importance. Obtaining a rough estimate of the total or mean power used by primary transmitters is a first step towards assessing the acceptable secondary transmit coverage area. But this is not the whole story. Indeed, if the secondary network is only able to state that a signal of cumulated power P is received, then the secondary network will dynamically adapt its transmit coverage area as follows:

- Assuming the sensed data are due to primary uplink transmissions by mobile users to a fixed network, the primary uplink frequency band will be reused in such a way that no primary user emitting with power P is interfered with by any transmission from the secondary network;
- if P is above a certain threshold, the cognitive radio will decide that neighboring primary cell sites are in use by primary users. Therefore, also downlink transmissions are not to be interfered with, so that the downlink spectrum is not considered a spectrum hole.

If the secondary network is able to do more than just overall power estimation, namely if it is capable of estimating both the number of concurrent simultaneous transmissions in a given spectral resource, call this number K, and the power of each individual source, call them $P_1, \ldots, P_K$ for source 1 to K, respectively, with $P_1 \leq \ldots \leq P_K$, then the secondary network can adapt its coverage area in a more accurate way as follows.

- Since the strongest transmitter has power P_K, the secondary cell coverage area can be set such that the primary user with power P_K is not interfered with. This will automatically induce that the other primary users are not interfered with (if it is further assumed that no power control is performed by the primary users). As an immediate consequence, the primary uplink transmission will be stated as reusable if P_K is not too large. Also, if P_K is so little that no primary user is expected to use primary downlink data sent by neighboring cells, also the downlink spectrum will be reused. In the case where multiple transmissions happen simultaneously, this strategy will turn out to be much more efficient than the estimation of the overall transmit power $P \triangleq \sum_{k=1}^{K} P_k$.
- Also, by measuring the transmit powers of multiple primary users within multiple distant secondary networks, information can be shared (via low speed links) among these networks so as to eventually pinpoint the precise location of the users. This brings even more information about the occupancy (and therefore the spectrum reusability) of each primary cell site. Moreover, it will turn out that most methods presented below show a strong limitation when it comes to resolving different users transmitting with almost equal power. Quite often, it is difficult to discriminate between the scenario of a single-user transmitting with power P or multiple transmitters with similar transmit powers, the sum of which being equal to P. Communications between distant secondary networks can therefore bring more information on the number of users with almost equal power. This eventually leads to the same performance gain as given in the previous point when it comes for the cognitive network to decide on the maximally acceptable coverage area.
- We also mentioned the need for estimating the number of transmit antennas per user. In fact, in the eigen-inference techniques presented below, the ability to estimate the number of antennas per user is an aftermath of the estimation algorithms. The interest of estimating the number of eigenvalues is linked

to the previous point, concerning the difficulty to differentiate between one user transmitting with strong power or many users transmitting with almost equal powers. If it is observed that power P is used by a device equipped with many more antennas than the communication protocol allows, then this should indicate to the sensors the presence of multiple transmitters with close transmit powers instead of a unique transmitter. This again allows for more precise inference on the system parameters.

Note from the discussion above that estimating P_K is in fact more important to the secondary network than estimating P_1, as P_K can by itself already provide a major piece of information concerning the largest coverage radius for secondary transmissions. When the additive noise variance is large, or when the number of available sensors is too small, inferring the smallest transmit powers is rather difficult. This is one of the reasons why eigen-inference methods that are capable of estimating a particular P_k are preferred over methods that jointly estimate the power distribution with masses in $P_1, \ldots, P_K$.

We hereafter introduce the general communication model discussed in the rest of this section. We will then derive eigen-inference techniques based on either exact small dimensional approaches, asymptotic free deconvolution approaches (as presented in Section 8.2), or the more involved but more efficient Stieltjes transform methods, relying on the theorems derived in Section 8.1.2.

17.2.1 System model

Consider a wireless (primary) network in which K entities are transmitting data simultaneously on the same frequency resource. Transmitter $k \in \{1, \ldots, K\}$ has transmission power P_k and is equipped with n_k antennas. We denote $n \triangleq \sum_{k=1}^{K} n_k$ the total number of transmit antennas within the primary network. Consider also a secondary network composed of a total of N, $N > n$, sensing devices (either N single antenna devices or multiple devices equipped with a total of N antennas); we refer to the N sensors collectively as *the receiver*. This scenario relates in particular to the configuration depicted in Figure 17.7.

To ensure that every sensor in the secondary network, e.g. in a closed-access femto-cell [Claussen *et al.*, 2008], roughly captures the same amount of energy from a given transmitter, we need to assume that all distances between a given transmitter and the individual sensors are alike. This is a realistic assumption for instance for an in-house femto-cell network, where all sensors lie in a restricted space and transmitters are found far away from the sensors. Denote $\mathbf{H}_k \in \mathbb{C}^{N \times n_k}$ the channel matrix between transmitter k and the receiver. We assume that the entries of $\sqrt{N}\mathbf{H}_k$ are i.i.d. with zero mean, unit variance, and finite fourth order moment. At time instant m, transmitter k emits the signal $\mathbf{x}_k^{(m)} \in \mathbb{C}^{n_k}$, with entries assumed to be independent, independent along m, k, identically distributed along m, and all have zero mean, unit variance, and finite fourth order moment (the $\mathbf{x}_k^{(m)}$ need not be identically distributed along k). Assume

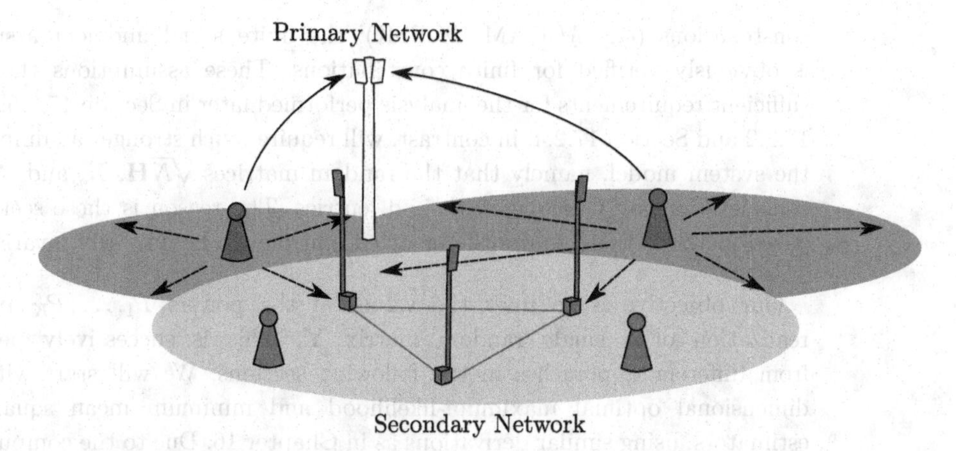

Figure 17.7 A cognitive radio network.

further that at time instant m the receive signal is impaired by additive white noise with entries of zero mean, variance σ^2, and finite fourth order moment on every sensor; we denote $\sigma\mathbf{w}^{(m)} \in \mathbb{C}^N$ the receive noise vector where the entries of $\mathbf{w}_k^{(m)}$ have unit variance. At time m, the receiver therefore senses the signal $\mathbf{y}^{(m)} \in \mathbb{C}^N$ defined as

$$\mathbf{y}^{(m)} = \sum_{k=1}^{K} \sqrt{P_k}\mathbf{H}_k\mathbf{x}_k^{(m)} + \sigma\mathbf{w}^{(m)}.$$

Assuming the channel fading coefficients are constant over at least M consecutive sampling periods, by concatenating M successive signal realizations into $\mathbf{Y} = [\mathbf{y}^{(1)}, \ldots, \mathbf{y}^{(M)}] \in \mathbb{C}^{N \times M}$, we have:

$$\mathbf{Y} = \sum_{k=1}^{K} \sqrt{P_k}\mathbf{H}_k\mathbf{X}_k + \sigma\mathbf{W}$$

where $\mathbf{X}_k = [\mathbf{x}_k^{(1)}, \ldots, \mathbf{x}_k^{(M)}] \in \mathbb{C}^{n_k \times M}$ and $\mathbf{W} = [\mathbf{w}^{(1)}, \ldots, \mathbf{w}^{(M)}] \in \mathbb{C}^{N \times M}$. This can be further rewritten as

$$\mathbf{Y} = \mathbf{H}\mathbf{P}^{\frac{1}{2}}\mathbf{X} + \sigma\mathbf{W} \tag{17.2}$$

where $\mathbf{P} \in \mathbb{R}^{n \times n}$ is diagonal with first n_1 entries P_1, subsequent n_2 entries P_2, etc. and last n_K entries P_K, $\mathbf{H} = [\mathbf{H}_1, \ldots, \mathbf{H}_K] \in \mathbb{C}^{N \times n}$ and $\mathbf{X} = [\mathbf{X}_1^\mathsf{T}, \ldots, \mathbf{X}_K^\mathsf{T}]^\mathsf{T} \in \mathbb{C}^{n \times M}$. By convention, we assume $P_1 \leq \ldots \leq P_K$.

Remark 17.2. The statement that $\sqrt{N}\mathbf{H}$, $\mathbf{X}$ and $\mathbf{W}$ have independent entries of finite fourth order moment is meant to provide as loose assumptions as possible on the channel, signal, and noise properties. In the simulations carried out later in this section, the entries of $\mathbf{H}$, $\mathbf{W}$ are taken Gaussian. Nonetheless, according to our assumptions, the entries of $\mathbf{X}$ need not be identically distributed, but may originate from a maximum of K distinct distributions. This translates the realistic assumption that different data sources may use different symbol constellations (e.g. M-QAM, M-PSK); the finite fourth moment assumption is obviously verified for finite constellations. These assumptions though are sufficient requirements for the analysis performed later in Section 17.2.5. Section 17.2.2 and Section 17.2.4, in contrast, will require much stronger assumptions on the system model, namely that the random matrices $\sqrt{N}\mathbf{H}$, $\mathbf{X}$, and $\mathbf{W}$ under consideration are Gaussian with i.i.d. entries. The reason is these sections are based methods that require the involved matrices to be unitarily invariant.

Our objective is to infer the values of the powers $P_1, \ldots, P_K$ from the realization of a single random matrix $\mathbf{Y}$. This is successively performed from different approaches in the following sections. We will start with small dimensional optimal maximum-likelihood and minimum mean square error estimators, using similar derivations as in Chapter 16. Due to the computational complexity of the method, we then consider large dimensional approaches. The fist of those is the conventional approach that assumes n small, N much larger than n, and M much larger than N. This will lead to a simple although largely biased estimation algorithm when tested in practical small dimensional scenarios. This algorithm will be corrected first by using moment approaches and specifically free deconvolution approaches, although this approach requires strong system assumptions and will be proved not to be very efficient, both with respect to performance and to computational effort. Finally, the latter will be further improved using Stieltjes transform approaches in the same spirit as in Section 17.1.

17.2.2 Small dimensional inference

This first approach consists in evaluating the exact distribution of the powers $P_1, \ldots, P_K$ given the observations $\mathbf{Y}$, modeled in (17.2), when $\mathbf{H}$, $\mathbf{X}$, and $\mathbf{W}$ are assumed Gaussian. Noticing that we can write $\mathbf{Y}$ under the unitarily invariant form

$$\mathbf{Y} = \begin{pmatrix} \mathbf{H}\mathbf{P}^{\frac{1}{2}} & \sigma \mathbf{I}_N \end{pmatrix} \begin{pmatrix} \mathbf{X} \\ \mathbf{W} \end{pmatrix} \tag{17.3}$$

the derivation unfolds similarly as that proposed in Chapter 16 with the noticeable exception that the matrix $N\mathbf{H}\mathbf{P}\mathbf{H}^{\mathsf{H}}$ has now a correlated Wishart distribution instead of an uncorrelated Wishart distribution. This makes the calculus somewhat more involved. We do not provide the successive steps of the full derivations that mimic those of Chapter 16 and that can be found in detail in [Couillet and Guillaud, 2011]. The final result is given as follows.

Theorem 17.4 ([Couillet and Guillaud, 2011]). *Assume $P_1, \ldots, P_K$ are all different and have multiplicity $n_1 = \ldots = n_K = 1$, hence $n = K$. Then, denoting*

$\boldsymbol{\lambda} = (\lambda_1, \ldots, \lambda_N)$ *the eigenvalues of* $\frac{1}{M}\mathbf{Y}\mathbf{Y}^{\mathsf{H}}$, $P_{\mathbf{Y}|P_1,\ldots,P_K}(\mathbf{Y})$ *reads:*

$$P_{\mathbf{Y}|P_1,\ldots,P_K}(\mathbf{Y}) = \frac{C(-1)^{Nn+1}e^{N\sigma^2\sum_{i=1}^{n}\frac{1}{P_i}}}{\sigma^{2(N-n)(M-n)}\prod_{i=1}^{n}P_i^{M-n+1}\Delta(\mathbf{P})}\sum_{\mathbf{a}\in\mathcal{S}_n^N}(-1)^{|\mathbf{a}|}\mathrm{sgn}(\mathbf{a})e^{-\frac{M}{\sigma^2}|\boldsymbol{\lambda}[\bar{\mathbf{a}}]|}$$
$$\times\frac{\Delta(\mathrm{diag}(\boldsymbol{\lambda}[\bar{\mathbf{a}}]))}{\Delta(\mathrm{diag}(\boldsymbol{\lambda}))}\sum_{\mathbf{b}\in\mathcal{S}_n}\mathrm{sgn}(\mathbf{b})\prod_{i=1}^{n}J_{N-M-1}\left(\frac{N\sigma^2}{P_{b_i}},\frac{NM\lambda_{a_i}}{P_{b_i}}\right)$$

where $\mathcal{S}_n^N$ *is the set of all permutations of* n*-subsets of* $\{1,\ldots,N\}$, $\mathcal{S}_n = \mathcal{S}_n^n$, $|\mathbf{x}| = \sum_i x_i$, $\bar{\mathbf{x}}$ *is the complementary of the set* $\mathbf{x}$, $\mathbf{x}[\mathbf{a}]$ *is the restriction of* $\mathbf{x}$ *to the indexes stored in the vector* $\mathbf{a}$, $\Delta(\mathbf{X})$ *is the Vandermonde determinant of the matrix* $\mathbf{X}$, *the constant* C *is given by:*

$$C = \frac{1}{\pi^{NM}n!}\frac{N^{n(M-\frac{n-1}{2})}}{M^{n(N-\frac{n+1}{2})}}$$

and $J_k(x,y)$ *is the integral form defined in* (16.6) *by*

$$J_k(x,y) = 2y^{\frac{k+1}{2}}K_{-k-1}(2\sqrt{y}) - \int_0^x u^k e^{-u-\frac{y}{u}}du.$$

The generalization of Theorem 17.4 to powers $P_1,\ldots,P_K$ of multiplicities greater than one is obtained by exploiting Theorem 2.9. The final result however takes a more involved form which we do not provide here.

From Theorem 17.4, we can derive the maximum likelihood (ML) estimator $\underline{\hat{P}}^{(\mathrm{ML})} = \hat{P}_1^{(\mathrm{ML})},\ldots,\hat{P}_K^{(\mathrm{ML})}$ of the joint $(P_1,\ldots,P_K)$ vector as

$$\underline{\hat{P}}^{(\mathrm{ML})} = \arg\max_{P_1,\ldots,P_K} P_{\mathbf{Y}|P_1,\ldots,P_K}(\mathbf{Y})$$

or the minimum mean square error (MMSE) estimator $\underline{\hat{P}}^{(\mathrm{MMSE})} = \hat{P}_1^{(\mathrm{MMSE})},\ldots,\hat{P}_K^{(\mathrm{MMSE})}$ as

$$\underline{\hat{P}}^{(\mathrm{MMSE})} = \int_{[0,\infty)^K}(P_1,\ldots,P_K)P_{P_1,\ldots,P_K|\mathbf{Y}}(P_1,\ldots,P_K)dP_1\ldots dP_K$$

with $P_{P_1,\ldots,P_K|\mathbf{Y}}(\mathbf{Y})P_{\mathbf{Y}}(\mathbf{Y}) = P_{\mathbf{Y}|P_1,\ldots,P_K}(\mathbf{Y})P_{P_1,\ldots,P_K}(P_1,\ldots,P_K)$. Under uniform a priori distribution of the powers, this is simply

$$\underline{\hat{P}}^{(\mathrm{MMSE})} = \frac{\int_{[0,\infty)^K}(P_1,\ldots,P_K)P_{\mathbf{Y}|P_1,\ldots,P_K}(P_1,\ldots,P_K)dP_1\ldots dP_K}{\int_{[0,\infty)^K}P_{\mathbf{Y}|P_1,\ldots,P_K}(P_1,\ldots,P_K)dP_1\ldots dP_K}.$$

However, both approaches are computationally complex, the complexity scaling exponentially with N in particular, and require multi-dimensional line searches on fine grids for proper evaluation, which also do not scale nicely with growing K. We will therefore no longer consider this optimal approach, which is only useful in providing lower bounds on the inference performance for small system dimensions. We now turn directly to alternative estimators using large dimensional random matrix theory.

17.2.3 Conventional large dimensional approach

The classical large dimensional approach assumes numerous sensors in order to have much diversity in the observation vectors, as well as an even larger number of observations so as to create an averaging effect on the incoming random data. In this situation, let us consider the system model (17.3) and now denote $\lambda_1 \leq \ldots \leq \lambda_N$ the *ordered* eigenvalues of $\frac{1}{M}\mathbf{Y}\mathbf{Y}^{\mathsf{H}}$ (the non-zero eigenvalues of which are almost surely different).

Appending $\mathbf{Y} \in \mathbb{C}^{N \times M}$ into the larger matrix $\underline{\mathbf{Y}} \in \mathbb{C}^{(N+n)\times M}$

$$\underline{\mathbf{Y}} = \begin{pmatrix} \mathbf{H}\mathbf{P}^{\frac{1}{2}} & \sigma \mathbf{I}_N \\ 0 & 0 \end{pmatrix} \begin{pmatrix} \mathbf{X} \\ \mathbf{W} \end{pmatrix}$$

we recognize that, conditional on $\mathbf{H}$, $\frac{1}{M}\underline{\mathbf{Y}\mathbf{Y}}^{\mathsf{H}}$ is a *sample covariance matrix*, for which the *population covariance matrix* is

$$\mathbf{T} \triangleq \begin{pmatrix} \mathbf{H}\mathbf{P}\mathbf{H}^{\mathsf{H}} + \sigma^2 \mathbf{I}_N & 0 \\ 0 & 0 \end{pmatrix}$$

and the random matrix

$$\begin{pmatrix} \mathbf{X} \\ \mathbf{W} \end{pmatrix}$$

has independent (non-necessarily identically distributed) entries of zero mean and unit variance. The population covariance matrix $\mathbf{T}$, whose upper left entries also form a matrix unitarily equivalent to a sample covariance matrix, clearly has an almost sure l.s.d. as N grows large for fixed or slowly growing n. Extending Theorem 3.13 and Theorem 9.1 to $c = 0$ and applying them twice (once for the population covariance matrix $\mathbf{T}$ and once for $\frac{1}{M}\underline{\mathbf{Y}\mathbf{Y}}^{\mathsf{H}}$), we finally have that, as $M, N, n \to \infty$ with $M/N \to \infty$ and $N/n \to \infty$, the distribution of the largest n eigenvalues of $\frac{1}{M}\mathbf{Y}\mathbf{Y}^{\mathsf{H}}$ is asymptotically almost surely composed of a mass $\sigma^2 + P_1$ of weight $\lim n_1/n$, a mass $\sigma^2 + P_2$ of weight $\lim n_2/n$, etc. and a mass $\sigma^2 + P_K$ of weight $\lim n_K/n$. As for the distribution of the smallest $N - n$ eigenvalues of $\frac{1}{M}\mathbf{Y}\mathbf{Y}^{\mathsf{H}}$, it converges to a single mass in σ^2.

If σ^2 is a priori known, a rather trivial estimator of P_k is then given by:

$$\frac{1}{n_k} \sum_{i \in \mathcal{N}_k} (\lambda_i - \sigma^2)$$

where we denoted $\mathcal{N}_k = \{N - \sum_{j=k}^{K} n_j + 1, \ldots, N - \sum_{j=k+1}^{K} n_j\}$ and we recall that $\lambda_1 \leq \ldots \leq \lambda_N$ are the ordered eigenvalues of $\frac{1}{M}\mathbf{Y}\mathbf{Y}^{\mathsf{H}}$.

This means in practice that P_K is asymptotically well approximated by the averaged value of the n_K largest eigenvalues of $\frac{1}{M}\mathbf{Y}\mathbf{Y}^{\mathsf{H}}$, P_{K-1} is well approximated by the averaged value of the n_{K-1} eigenvalues before that, etc. This also assumes that σ^2 is perfectly known at the receiver. If it were not, observe that the averaged value of the $N - n$ smallest eigenvalues of $\frac{1}{M}\mathbf{Y}\mathbf{Y}^{\mathsf{H}}$ is a consistent estimate for σ^2. This therefore leads to the second estimator $\hat{P}_k^{\infty}$ for

P_k, that will constitute our reference estimator

$$\hat{P}_k^{\infty} = \frac{1}{n_k} \sum_{i \in \mathcal{N}_k} (\lambda_i - \hat{\sigma}^2)$$

where

$$\hat{\sigma}^2 = \frac{1}{N-n} \sum_{i=1}^{N-n} \lambda_i.$$

Note that the estimation of P_k only relies on n_k contiguous eigenvalues of $\frac{1}{M}\mathbf{Y}\mathbf{Y}^{\mathsf{H}}$, which suggests that the other eigenvalues are asymptotically uncorrelated from these. It will turn out that the improved (n, N, M)-consistent estimator does take into account all eigenvalues for each k, in a certain manner.

As a reference example, we assume the scenario of three simultaneous transmissions with transmit powers P_1, P_2, and P_3 equal to 1/16, 1/4, and 1, respectively. We assume that each user possesses four transmit antennas, i.e. $K = 3$ and $n_1 = n_2 = n_3 = 4$. The receiver is an array of $N = 24$ sensors, that samples as many as 128 independent (and identically distributed) observations. The SNR is set to 20 dB. In this reference scenario, we assume that K, n_1, n_2, n_3 are known. The question of estimating these values will be discussed later in Section 17.2.6. In Figure 17.8 and Figure 17.9, the performance of the estimator $\hat{P}_k^{\infty}$ for k ranging from one to three is evaluated, for 1000 random realizations of Gaussian channels $\mathbf{H}$, Gaussian additive noise $\mathbf{W}$ and QPSK modulated user transmissions $\mathbf{X}$. This is gathered in Figure 17.8 under the form of an histogram of the estimated $\hat{P}_k^{\infty}$ in linear scale and in Figure 17.9 under the form of the distribution function of the marginal distribution of the $\hat{P}_k^{\infty}$ in logarithmic scale. While our analysis ensures consistency of the $\hat{P}_k^{\infty}$ estimates for extremely large M and very large N, we observe that, for not-too-large system dimensions, the $\hat{P}_k^{\infty}$ are very biased estimates of the true P_k powers. In particular here, both P_1 and P_2 are largely underestimated overall, while P_3 is clearly overestimated. Since the system dimensions under study are rather realistic in practical cognitive (secondary) networks, i.e. the number of sensors is not assumed extremely large and the number of observation samples is such that the exploration phase is short, this means that the estimator $\hat{P}_k^{\infty}$ is inappropriate to applications in cognitive radios. These performance figures naturally call for improved estimates. In particular, it will turn out that estimates that take into account the facts that M is not much larger than N and that N is not significantly larger than n will provide unbiased estimates in the large dimensional setting, which will be seen through simulations to be very accurate even for small system dimensions.

We start with a moment approach, which recalls the free probability and moment-based eigen-inference methods detailed in Section 8.2 of Chapter 8.

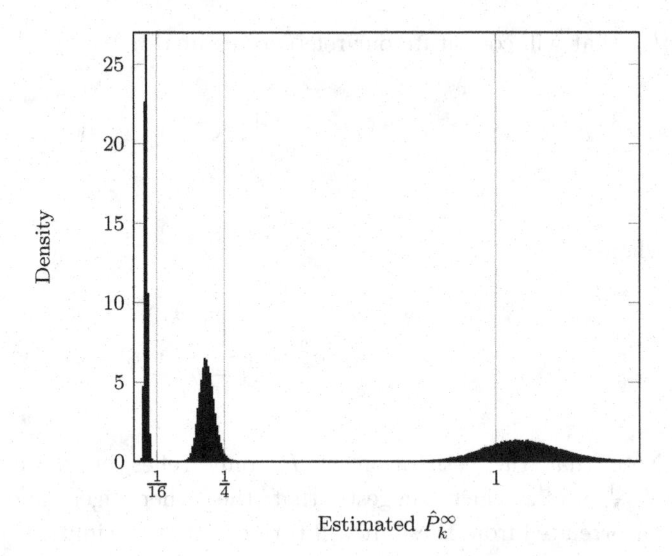

Figure 17.8 Histogram of the $\hat{P}_k^\infty$ for $k \in \{1, 2, 3\}$, $P_1 = 1/16$, $P_2 = 1/4$, $P_3 = 1$, $n_1 = n_2 = n_3 = 4$ antennas per user, $N = 24$ sensors, $M = 128$ samples and SNR = 20 dB.

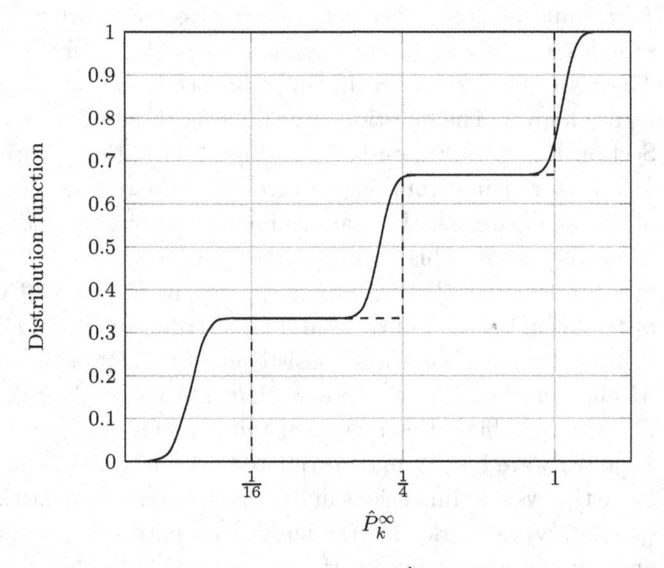

Figure 17.9 Distribution function of the estimator $\hat{P}_k^\infty$ for $k \in \{1, 2, 3\}$, $P_1 = 1/16$, $P_2 = 1/4$, $P_3 = 1$, $n_1 = n_2 = n_3 = 4$ antennas per user, $N = 24$ sensors, $M = 128$ samples and SNR = 20 dB. Optimum estimator shown in dashed lines.

17.2.4 Free deconvolution approach

To be able to proceed with free deconvolution, similar to Section 17.2.2, we will need to take some further assumptions on the system model at hand. Precisely, we will require the random matrices $\mathbf{H}$, $\mathbf{W}$, and $\mathbf{X}$ to be filled with Gaussian i.i.d. entries. That is, we no longer allow for arbitrary modulated transmissions such as QPSK or QAM (at least from a theoretical point of view). This is a key assumption to ensure asymptotic freeness of products and sums of random matrices. For the sake of diversity in the methods developed in this chapter, we no longer consider the compact model (17.3) for $\mathbf{Y}$ but instead we will see $\mathbf{Y}$ as an information plus noise matrix whose information matrix is random but has an almost surely l.s.d. We also assume that, as the system dimensions grow large, we have $M/N \to c$, for $k \in \{1,\ldots,K\}$, $N/n_k \to c_k$ and $N/n \to c_0$, with $0 < c, c_0, c_1, \ldots, c_K < \infty$.

We can write

$$\mathbf{B}_N \triangleq \frac{1}{M}\mathbf{Y}\mathbf{Y}^{\mathsf{H}} = \frac{1}{M}\left(\mathbf{H}\mathbf{P}^{\frac{1}{2}}\mathbf{X} + \sigma\mathbf{W}\right)\left(\mathbf{H}\mathbf{P}^{\frac{1}{2}}\mathbf{X} + \sigma\mathbf{W}\right)^{\mathsf{H}}$$

which is such that, as M, N, and n grow to infinity with positive limiting ratios, the e.s.d. of $\mathbf{H}\mathbf{P}^{\frac{1}{2}}\mathbf{X}\mathbf{X}^{\mathsf{H}}\mathbf{P}^{\frac{1}{2}}\mathbf{H}^{\mathsf{H}}$ converges weakly and almost surely to a limiting distribution function. This is ensured by iterating twice Theorem 3.13: a first time on the sample covariance matrix $\mathbf{P}^{\frac{1}{2}}\mathbf{H}^{\mathsf{H}}\mathbf{H}\mathbf{P}^{\frac{1}{2}}$ with population covariance matrix $\mathbf{P}$ (which we assume converges towards a l.s.d. composed of K masses in $P_1, \ldots, P_K$) and a second time on the (conditional) sample covariance matrix $\mathbf{H}\mathbf{P}^{\frac{1}{2}}\mathbf{X}\mathbf{X}^{\mathsf{H}}\mathbf{P}^{\frac{1}{2}}\mathbf{H}^{\mathsf{H}}$ with population covariance matrix $\mathbf{H}\mathbf{P}\mathbf{H}^{\mathsf{H}}$, that was proved in the first step to have an almost sure l.s.d.

In what follows, we will denote for readability $\mu_{\mathbf{Z}}^{\infty}$ the probability distribution associated with the l.s.d. of the Hermitian random matrix $\mathbf{Z}$. We can now refer to Theorem 4.9, which ensures under the system model above that the limiting distribution $\mu_{\mathbf{B}_N}^{\infty}$ of $\mathbf{B}_N$ (in the free probability terminology) can be expressed as a function of the limiting distribution $\mu^{\infty}_{\frac{1}{M}\mathbf{H}\mathbf{P}^{\frac{1}{2}}\mathbf{X}\mathbf{X}^{\mathsf{H}}\mathbf{P}^{\frac{1}{2}}\mathbf{H}^{\mathsf{H}}}$ of $\frac{1}{M}\mathbf{H}\mathbf{P}^{\frac{1}{2}}\mathbf{X}\mathbf{X}^{\mathsf{H}}\mathbf{P}^{\frac{1}{2}}\mathbf{H}^{\mathsf{H}}$. Using the free probability operators, this reads:

$$\mu^{\infty}_{\frac{1}{M}\mathbf{H}\mathbf{P}^{\frac{1}{2}}\mathbf{X}\mathbf{X}^{\mathsf{H}}\mathbf{P}^{\frac{1}{2}}\mathbf{H}^{\mathsf{H}}} = \left(\left(\mu_{\mathbf{B}_N}^{\infty} \boxslash \mu_{\frac{1}{c}}\right) \boxminus \delta_{\sigma^2}\right) \boxtimes \mu_{\frac{1}{c}}$$

where $\mu_{\frac{1}{c}}$ is the probability distribution of a random variable with distribution function the Marčenko–Pastur law with ratio $\frac{1}{c}$ and δ_{σ^2} the probability distribution of a single mass in σ^2. Remember that the above formula (through the convolution operators) translates by definition the fact that all moments of the left-hand side can be computed iteratively from the moments of the terms in the right-hand side. Since all eigenvalue distributions under study satisfy Carleman's condition, Theorem 5.1, this is equivalent to saying that the l.s.d. of $\frac{1}{M}\mathbf{H}\mathbf{P}^{\frac{1}{2}}\mathbf{X}\mathbf{X}^{\mathsf{H}}\mathbf{P}^{\frac{1}{2}}\mathbf{H}^{\mathsf{H}}$ is entirely defined through the l.s.d. of $\mu_{\mathbf{B}_N}^{\infty}$, a fact which is obvious from Theorem 3.13.

Instead of describing step by step the link between the moments of the l.s.d. of $\mathbf{B}_N$ and the moments of the l.s.d. of the deconvolved matrices, we perform deconvolution remembering that automated algorithms can provide us effortlessly with the final relations between the moments of the l.s.d. of $\mathbf{B}_N$ and the moments of the l.s.d. of $\mathbf{P}$, i.e. all the sums $\frac{1}{n}\sum_{k=1}^{K} n_k P_k^m$, for all integers m.

Before we move to the second deconvolution step, we rewrite $\frac{1}{M}\mathbf{H}\mathbf{P}^{\frac{1}{2}}\mathbf{X}\mathbf{X}^{\mathsf{H}}\mathbf{P}^{\frac{1}{2}}\mathbf{H}^{\mathsf{H}}$ under the form of a product of a scaled zero Wishart matrix with another matrix. This is:

$$\mu^{\infty}_{\frac{1}{M}\mathbf{P}^{\frac{1}{2}}\mathbf{H}^{\mathsf{H}}\mathbf{H}\mathbf{P}^{\frac{1}{2}}\mathbf{X}\mathbf{X}^{\mathsf{H}}} = c_0\mu^{\infty}_{\frac{1}{M}\mathbf{H}\mathbf{P}^{\frac{1}{2}}\mathbf{X}\mathbf{X}^{\mathsf{H}}\mathbf{P}^{\frac{1}{2}}\mathbf{H}^{\mathsf{H}}} + (1-c_0)\,\delta_0$$

with $\mu^{\infty}_{\frac{1}{M}\mathbf{P}^{\frac{1}{2}}\mathbf{H}^{\mathsf{H}}\mathbf{H}\mathbf{P}^{\frac{1}{2}}\mathbf{X}\mathbf{X}^{\mathsf{H}}}$ the l.s.d. of $\frac{1}{M}\mathbf{P}^{\frac{1}{2}}\mathbf{H}^{\mathsf{H}}\mathbf{H}\mathbf{P}^{\frac{1}{2}}\mathbf{X}\mathbf{X}^{\mathsf{H}}$.

In terms of moments, this introduces a scaling factor c_0 to all successive moments of the limiting distribution. Under this form, we can proceed to the second deconvolution step, which writes

$$\mu^{\infty}_{\mathbf{P}^{\frac{1}{2}}\mathbf{H}^{\mathsf{H}}\mathbf{H}\mathbf{P}^{\frac{1}{2}}} = \mu^{\infty}_{\frac{1}{M}\mathbf{P}^{\frac{1}{2}}\mathbf{H}^{\mathsf{H}}\mathbf{H}\mathbf{P}^{\frac{1}{2}}\mathbf{X}\mathbf{X}^{\mathsf{H}}} \boxslash \mu_{\frac{1}{cc_0}},$$

with $\mu^{\infty}_{\mathbf{P}^{\frac{1}{2}}\mathbf{H}^{\mathsf{H}}\mathbf{H}\mathbf{P}^{\frac{1}{2}}}$ the l.s.d. of $\mathbf{P}^{\frac{1}{2}}\mathbf{H}^{\mathsf{H}}\mathbf{H}\mathbf{P}^{\frac{1}{2}}$.

Note that the scaling factor $\frac{1}{M}$ disappeared due to the fact that $\mathbf{X}$ has entries of unit variance. With the same line of reasoning as before, we then write the resulting matrix under the form of a matrix product containing a Wishart matrix as a second factor. The step is rather immediate here as no additional mass in zero is required to be added

$$\mu^{\infty}_{\mathbf{P}\mathbf{H}^{\mathsf{H}}\mathbf{H}} = \mu^{\infty}_{\mathbf{P}^{\frac{1}{2}}\mathbf{H}^{\mathsf{H}}\mathbf{H}\mathbf{P}^{\frac{1}{2}}}$$

with $\mu^{\infty}_{\mathbf{P}\mathbf{H}^{\mathsf{H}}\mathbf{H}}$ the l.s.d. of $\mathbf{P}\mathbf{H}^{\mathsf{H}}\mathbf{H}$.

The final deconvolution step consists in removing the effect of the scaled Wishart matrix $\mathbf{H}^{\mathsf{H}}\mathbf{H}$. Incidentally, since $\mathbf{H}^{\mathsf{H}}$ has N columns and has entries of variance $1/N$, we finally have the simple expression

$$\mu^{\infty}_{\mathbf{P}} = \mu^{\infty}_{\mathbf{P}\mathbf{H}^{\mathsf{H}}\mathbf{H}} \boxslash \mu_{\frac{1}{c_0}}$$

where $\mu^{\infty}_{\mathbf{P}}$ is the probability distribution associated with the l.s.d. of $\mathbf{P}$, i.e. $\mu^{\infty}_{\mathbf{P}}$ is the probability distribution of a random variable with K masses in $P_1, \ldots, P_K$ with respective weights $\frac{c}{c_1}, \ldots, \frac{c}{c_K}$. This completes the free deconvolution steps. It is therefore possible, going algebraically or numerically through the successive deconvolution steps, to express all moments of the l.s.d. of $\mathbf{P}$ as a function of the moments of the almost sure l.s.d. of $\mathbf{B}_N$.

Remember now from Section 5.3 that it is possible to generalize further the concept of free deconvolution to finite dimensional random matrices, if we replace $\mu^{\infty}_{\mathbf{B}_N}$, $\mu^{\infty}_{\mathbf{P}}$ and the intermediate probability distributions introduced so far by the probability distributions of the *averaged e.s.d.*, instead of the l.s.d. That is, for any random matrix $\mathbf{X} \in \mathbb{C}^{N\times N}$ with compactly supported eigenvalue distribution for all N, similar to Chapter 4, we define $\mu^{N}_{\mathbf{X}}$ as the probability distribution with mth order moment $\mathrm{E}\left[\int x^m dF^{\mathbf{X}}(x)\right]$. Substituting the $\mu^{\infty}_{\mathbf{X}}$ by the $\mu^{N}_{\mathbf{X}}$ in the derivations above and changing the definitions of the free convolution operators accordingly, see, e.g., [Masucci *et al.*, 2011], we can finally derive the combinatorial expressions that link the moments of the eigenvalue distribution of $\mathbf{B}_N$ (seen here as a random matrix and not as a particular realization) to $\frac{1}{K}\sum_{k=1}^{K} P_k^m$, for all integer m.

We will not go into excruciating details as to how this expresses theoretically and will merely state the first three moment relations, whose output was generated automatically by a computer software, see, e.g., [Koev; Ryan, 2009a,b]. Denote

$$p_m \triangleq \sum_{k=1}^{K} \frac{n_k}{n} P_k^m$$

and

$$b_m \triangleq \frac{1}{N} \mathrm{E}\left[\operatorname{tr} \mathbf{B}_N^m\right]$$

where the expectation is taken over the joint realization of the random matrices $\mathbf{H}$, $\mathbf{X}$, and $\mathbf{W}$. In the case where $n_1 = \ldots = n_K$, the first moments p_m and b_m relate together as

$$\begin{aligned}
b_1 &= N^{-1} n p_1 + 1 \\
b_2 &= \left(N^{-2} M^{-1} n + N^{-1} n\right) p_2 + \left(N^{-2} n^2 + N^{-1} M^{-1} n^2\right) p_1^2 \\
&\quad + \left(2N^{-1} n + 2M^{-1} n\right) p_1 + \left(1 + N M^{-1}\right) \\
b_3 &= \left(3N^{-3} M^{-2} n + N^{-3} n + 6N^{-2} M^{-1} n + N^{-1} M^{-2} n + N^{-1} n\right) p_3 \\
&\quad + \left(6N^{-3} M^{-1} n^2 + 6N^{-2} M^{-2} n^2 + 3N^{-2} n^2 + 3N^{-1} M^{-1} n^2\right) p_2 p_1 \\
&\quad + \left(N^{-3} M^{-2} n^3 + N^{-3} n^3 + 3N^{-2} M^{-1} n^3 + N^{-1} M^{-2} n^3\right) p_1^3 \\
&\quad + \left(6N^{-2} M^{-1} n + 6N^{-1} M^{-2} n + 3N^{-1} n + 3M^{-1} n\right) p_2 \\
&\quad + \left(3N^{-2} M^{-2} n^2 + 3N^{-2} n^2 + 9N^{-1} M^{-1} n^2 + 3M^{-2} n^2\right) p_1^2 \\
&\quad + \left(3N^{-1} M^{-2} n + 3N^{-1} n + 9M^{-1} n + 3N M^{-2} n\right) p_1.
\end{aligned} \tag{17.4}$$

As a consequence, if L instances of the random matrices $\mathbf{Y}(\omega_1), \ldots, \mathbf{Y}(\omega_L)$ are available, and L grows large, then asymptotically the averaged moments of the e.s.d. of the $\frac{1}{M}\mathbf{Y}(\omega_i)\mathbf{Y}(\omega_i)^{\mathsf{H}}$ converge to the moments $b_1, b_2, \ldots$. This however requires that multiple realizations of $\mathbf{Y}$ matrices are indeed available, and that changes the conditions of the problem in the first place. Nonetheless, if effectively only one such matrix $\mathbf{Y}$ is available, it is possible to handily use this multi-instance approach by breaking down $\mathbf{Y}$ into several parts of smaller column dimension. That is, $\mathbf{Y}$ can be rewritten under the form $\mathbf{Y} = [\mathbf{Y}_1, \ldots, \mathbf{Y}_L]$, where $\mathbf{Y}_i \in \mathbb{C}^{N \times (M/L)}$, for some L which divides M. Note importantly that this approach is totally empirical and not equivalent to L independent realizations of the random $\mathbf{Y}$ matrix for all $\mathbf{Y}_i$, since the channel matrix $\mathbf{H}$ is kept identical for all realizations.

If large dimensions are assumed, then the terms that go to zero in the above relations must be discarded. Two different approaches can then be taken to use the moment approach. Either we assume large dimensions and keeps the realization of $\mathbf{Y}$ as is, or we may rewrite $\mathbf{Y}$ under the form of multiple submatrices and use the approximated averaged relations. In either case, the relations between consecutive moments must be dealt with carefully. We develop hereafter two estimation methods, already introduced in Chapter 8, namely the fast but inaccurate Newton–Girard approach and a computationally expensive method, which we will abusively call the maximum-likelihood approach.

17.2.4.1 Newton–Girard method

Let us work in the small system dimension regime and consider the scenario where the realization of $\mathbf{Y}$ is divided into L submatrices $\mathbf{Y}_1, \ldots, \mathbf{Y}_L$. The Newton–Girard approach consists in taking, for $m \in \{1, \ldots, K\}$, the estimate

$$\hat{b}_m = \frac{1}{NL} \sum_{l=1}^{L} \operatorname{tr}(\mathbf{Y}_l \mathbf{Y}_l^{\mathsf{H}})^m$$

of b_m for $m = 1, 2, \ldots$. Remember that those estimates are not L-consistent, since the random realization of $\mathbf{H}$ is the same for all $\mathbf{Y}_l$ (unless the time between successive observations of $\mathbf{Y}_i$ is long enough for the independence of the successive $\mathbf{H}$ matrices to hold). From $\hat{b}_m$, we may then successively take estimates of $p_1, \ldots, p_m$ by simply recursively inverting the formulas (17.4), with b_m replaced by $\hat{b}_m$. These estimates are denoted $\hat{p}_1, \ldots, \hat{p}_K$.

Newton–Girard formulas, see Section 5.2, allow us to recover estimates $\hat{P}_1^{(\mathrm{mom})}, \ldots, \hat{P}_K^{(\mathrm{mom})}$ of the transmit powers $P_1, \ldots, P_K$ by inverting the relations

$$\sum_{k=1}^{K} \frac{n_k}{n} \left(\hat{P}_k^{(\mathrm{mom})}\right)^m = \hat{p}_m$$

for $m \in \{1, \ldots, K\}$. For instance, in the case of our example where $K = 3$, $n_1 = n_2 = n_3$, we have that $\hat{P}_1^{(\mathrm{mom})}$, $\hat{P}_2^{(\mathrm{mom})}$, and $\hat{P}_3^{(\mathrm{mom})}$ are the roots of the polynomial in X

$$\left(-\frac{9}{2}\hat{p}_1^3 + \frac{9}{2}\hat{p}_1\hat{p}_2 - \hat{p}_3\right) X^2 + \left(\frac{9}{2}\hat{p}_1 - \frac{3}{2}\hat{p}_2\right) X + 1 = 0.$$

This method is simple and does not require a lot of computational resources, but fails in accuracy for several reasons discussed in Section 5.2 and which we presently recall. First, the free deconvolution approach aims at providing consistent estimates of the successive moments of the e.s.d. of $\mathbf{P}$, in order to obtain a good estimate on the e.s.d. of $\mathbf{P}$, while our current objective is rather to estimate some or all entries of $\mathbf{P}$ instead. Second, the Newton–Girard approach does not take into account the fact that the estimated random moments $\hat{b}_m$, and consequently the moments $\hat{p}_m$, do not all have the same variance around their means. Inverting the moment relations linking the b_m to the p_m by replacing the moments by their estimates assumes implicitly that all estimated moments equally well approximate the true values, which is in fact far from being correct. Finally, the roots of the polynomial that lead to the estimates $\hat{P}_k^{(\mathrm{mom})}$ are not ensured to be non-negative and worse not ensured to be real. Post-processing is then required to deal with such estimates. In the simulations below, we will

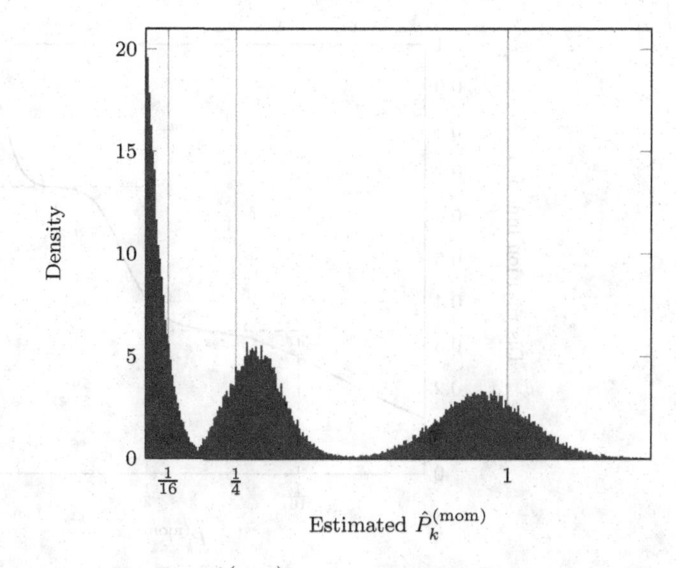

Figure 17.10 Histogram of the $\hat{P}_k^{(\mathrm{mom})}$ for $k \in \{1,2,3\}$, $P_1 = 1/16$, $P_2 = 1/4$, $P_3 = 1$, $n_1 = n_2 = n_3 = 4$ antennas per user, $N = 24$ sensors, $M = 128$ samples, and SNR = 20 dB.

simply discard the realizations leading to purely complex or real negative values, altering therefore the final result.

Figure 17.10 and Figure 17.11 provide the performance of the free deconvolution approach with Newton–Girard inversion for the same system model as before, i.e. three simultaneous transmissions, each user being equipped with four antennas, $N = 24$ sensors, $M = 128$ samples, and the SNR is 20 dB. We consider the case where $L = 1$, which is observed to perform overall similarly as the case where L is taken larger (with M scaled accordingly), with some minor differences. Notice that, although we now moved from a nested M- and N-consistent estimator to an (N, n, M)-consistent estimator of $P_1, \ldots, P_K$, the contestable Newton–Girard inversion has very awkward side effects, both in terms of bias for small dimensions, especially for the smallest powers, and in terms of variance of the estimate, which is also no match for the variance of the previous estimator when it comes to estimating very low powers. In terms of absolute mean square error, the conventional approach is therefore still better here. More elaborate post-processing methods are thus demanded to cope with the issues of the Newton–Girard inversion. This is what the subsequent section is devoted to.

17.2.4.2 ML and MMSE methods

The idea is now to consider the distribution of the $\hat{b}_m$ moment estimates and to take the estimated powers $(\hat{P}_1^{(\mathrm{mom,ML})}, \ldots, \hat{P}_K^{(\mathrm{mom,ML})})$ as the K-tuple that maximizes some reward function of the joint variable $\hat{b}_1, \ldots, \hat{b}_T$, for some integer T. A classical approach is to consider as a reward function the maximum

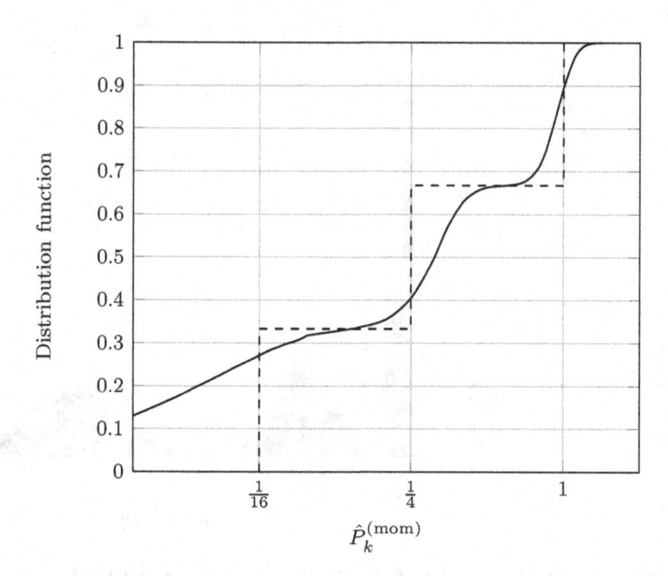

Figure 17.11 Distribution function of the estimator $\hat{P}_k^{(\mathrm{mom})}$ for $k \in \{1,2,3\}$, $P_1 = 1/16$, $P_2 = 1/4$, $P_3 = 1$, $n_1 = n_2 = n_3 = 4$ antennas per user, $N = 24$ sensors, $M = 128$, samples and SNR $= 20$ dB. Optimum estimator shown in dashed lines.

likelihood of $\hat{b}_1, \ldots, \hat{b}_T$ given $(\hat{P}_1^{(\mathrm{mom,ML})}, \ldots, \hat{P}_K^{(\mathrm{mom,ML})})$ or the minimum mean square error in the estimation of $\hat{b}_1, \ldots, \hat{b}_T$. This however implies that the joint probability distribution of $\hat{b}_1, \ldots, \hat{b}_T$ is known. Reminding that $\mathbf{B}_N$ can be written under the form of a covariance matrix with population covariance matrix whose e.s.d. converges almost surely to a limit distribution function, we are tempted to use Theorem 3.17 and to state that, for all finite T, the vector $N(\hat{b}_1 - \mathrm{E}[\hat{b}_1], \ldots, \hat{b}_T - \mathrm{E}[\hat{b}_T])$ converges in distribution and almost surely towards a T-variate Gaussian random variable. However, the assumptions of Theorem 3.17 do not let the sample covariance matrix be random, so that the central limit of the vector $N(\hat{b}_1 - \mathrm{E}[\hat{b}_1], \ldots, \hat{b}_T - \mathrm{E}[\hat{b}_T])$ cannot be stated but only conjectured. For the rest of the coming derivation, we will assume that the result does hold, as it may likely be the case. We therefore need to compute the covariance matrix of the vector $N(\hat{b}_1 - \mathrm{E}[\hat{b}_1], \ldots, \hat{b}_T - \mathrm{E}[\hat{b}_T])$, i.e. we need to compute all cross-moments

$$C_{ij}^N(P_1, \ldots, P_K) \triangleq N^2 \mathrm{E}\left[\left(\hat{b}_i - \mathrm{E}[\hat{b}_i]\right)\left(\hat{b}_j - \mathrm{E}[\hat{b}_j]\right)\right]$$

for $1 \leq i,j \leq T$, for some integer T, where the exponent N reminds the system dimension. Call $\mathbf{C}^N(P_1, \ldots, P_K) \in \mathbb{C}^{T \times T}$ the matrix with entries $C_{ij}^N(P_1, \ldots, P_K)$. As N grows large, $\mathbf{C}^N(P_1, \ldots, P_K)$ converges point-wise to some matrix, which we denote $\mathbf{C}^\infty(P_1, \ldots, P_K)$. The central limit theorem, be it valid in our situation, would therefore state that, asymptotically, the vector $N(\hat{b}_1 - \mathrm{E}[\hat{b}_1], \ldots, \hat{b}_T - \mathrm{E}[\hat{b}_T])$ is jointly Gaussian with zero mean and covariance matrix $\mathbf{C}^\infty(P_1, \ldots, P_K)$. We therefore determine the estimate vector

$\underline{P}^{(\mathrm{mom,ML})} \triangleq (\hat{P}_1^{(\mathrm{mom,ML})}, \ldots, \hat{P}_K^{(\mathrm{mom,ML})})$ as

$$\underline{P}^{(\mathrm{mom,ML})} = \arg \inf_{\substack{(\tilde{P}_1,\ldots,\tilde{P}_K) \\ \tilde{P}_i \geq 0}} (\hat{\mathbf{b}} - \mathrm{E}[\hat{\mathbf{b}}])\mathbf{C}^N(\tilde{P}_1, \ldots, \tilde{P}_K)^{-1}(\hat{\mathbf{b}} - \mathrm{E}[\hat{\mathbf{b}}])^\mathsf{T}$$

where $\hat{\mathbf{b}} = (\hat{b}_1, \ldots, \hat{b}_T)^\mathsf{T}$ and the expectations $\mathrm{E}[\hat{\mathbf{b}}]$ are conditioned with respect to $(\tilde{P}_1, \ldots, \tilde{P}_K)$ and can therefore be computed explicitly as described in the previous sections.

Similar to the maximum likelihood vector estimator $\underline{P}^{(\mathrm{mom,ML})}$, we can define the vector estimator $\underline{P}^{(\mathrm{mom,MMSE})} \triangleq (\hat{P}_1^{(\mathrm{mom,MMSE})}, \ldots, \hat{P}_K^{(\mathrm{mom,MMSE})})$, which realizes the minimum mean square error of $P_1, \ldots, P_K$ as

$$\underline{P}^{(\mathrm{mom,MMSE})} = \frac{1}{Z} \int_{\substack{\underline{P} d\underline{P} = (\tilde{P}_1,\ldots,\tilde{P}_K)^\mathsf{T} \\ \tilde{P}_m \geq 0}} \frac{\underline{P}}{\det \mathbf{C}^N(\tilde{P}_1, \ldots, \tilde{P}_K)} e^{-(\hat{\mathbf{b}} - \mathrm{E}[\hat{\mathbf{b}}])\mathbf{C}^N(\tilde{P}_1,\ldots,\tilde{P}_K)^{-1}(\hat{\mathbf{b}} - \mathrm{E}[\hat{\mathbf{b}}])^\mathsf{T}}$$

with Z a normalization factor.

Practically speaking, the problem is two-fold. First, either in the maximum-likelihood or in the minimum mean square error approach, a multi-dimensional line search is required. This is obviously extremely expensive compared to the Newton–Girard method. Simplification methods, such as iterative algorithms can be thought of, although they still require a lot of computations and are rarely ensured to converge to the global extremum sought for. The second problem is that the on-line or off-line computation of $\mathbf{C}^N(P_1, \ldots, P_K)$ is also extremely tedious. Note that the matrix $\mathbf{C}^N(P_1, \ldots, P_K)$ depends on the parameters K, M, N, $n_1, \ldots, n_K$, $P_1, \ldots, P_K$ and σ^2, and is of size $T \times T$. Based on combinatorial approaches, recent work has led to the possibility of a partly on-line, partly off-line computation of such matrices. Typically, it is intolerable that tables of $\mathbf{C}^N(P_1, \ldots, P_K)$ be kept in memory for multiple values of K, M, N, $n_1, \ldots, n_K$, σ^2 and $P_1, \ldots, P_K$. It is nonetheless acceptable that reference matrices be kept in memory so to fast compute on-line $\mathbf{C}^N(P_1, \ldots, P_K)$ for all K, M, N, $n_1, \ldots, n_K$, σ^2 and $P_1, \ldots, P_K$.

17.2.5 Analytic method

We finally introduce in this section the method based on G-estimation, which will be seen to have numerous advantages compared to the previous methods, although it is slightly handicapped by a cluster separability condition. The approach relies heavily on the recent techniques from Mestre, established in [Mestre, 2008b] and discussed at length in Chapter 8. This demands much more work than the combinatorial and rather automatic moment free deconvolution approach. Nevertheless, it appears that this approach can somewhat be reproduced for different models, as long as exact separation theorems, such as Theorem 7.2 or Theorem 7.8, are available.

The main strategy is the following:

- We first need to study the asymptotic spectrum of $\mathbf{B}_N \triangleq \frac{1}{M}\mathbf{Y}\mathbf{Y}^{\mathsf{H}}$, as all system dimensions (N, n, M) grow large (remember that K is fixed). For this, we will proceed by
  - determining the almost sure l.s.d. of $\mathbf{B}_N$ as all system dimensions grow large with finite limiting ratio. Practically, this will allow us to connect the asymptotic spectrum of $\mathbf{B}_N$ to the spectrum of $\mathbf{P}$;
  - studying the exact separation of the eigenvalues of $\mathbf{B}_N$ in clusters of eigenvalues. This is necessary first to determine whether the coming step of complex integration is possible and second to determine a well-chosen integration contour for the estimation of every P_k.
- Then, we will write P_k under the form of the complex integral of a functional of the spectrum of $\mathbf{P}$ over this well-chosen contour. Since the spectrum of $\mathbf{P}$ can be linked to that of $\mathbf{B}_N$ (at least asymptotically) through the previous step, a change of variable will allow us to rewrite P_k under the form of an integral of some functional of the l.s.d. of $\mathbf{B}_N$. This point is the key step in our derivation, where P_k is now connected to the observation matrix $\mathbf{Y}$ (although only in an asymptotic way).
- Finally, the estimate $\hat{P}_k$ of P_k will be computed from the previous step by replacing the l.s.d. of $\mathbf{B}_N$ by its e.s.d., i.e. by the truly observed eigenvalues of $\frac{1}{M}\mathbf{Y}\mathbf{Y}^{\mathsf{H}}$ in the expression relating P_k to the l.s.d. of $\mathbf{B}_N$.

We therefore divide this section into three subsections that analyze successively the almost sure l.s.d. of $\mathbf{B}_N$, then the conditions for cluster separation, and finally the actual calculus of the power estimator.

17.2.5.1 Limiting spectrum of $\mathbf{B}_N$

In this section, we prove the following result.

Theorem 17.5. *Let $\mathbf{B}_N = \frac{1}{M}\mathbf{Y}\mathbf{Y}^{\mathsf{H}}$, with $\mathbf{Y}$ defined as in (17.2). Then, for M, N, n growing large with limit ratios $M/N \to c$, $N/n_k \to c_k$, $0 < c, c_1, \ldots, c_K < \infty$, the e.s.d. $F^{\mathbf{B}_N}$ of $\mathbf{B}_N$ converges almost surely to the distribution function F, whose Stieltjes transform $m_F(z)$ satisfies, for $z \in \mathbb{C}^+$*

$$m_F(z) = c m_{\underline{F}}(z) + (c-1)\frac{1}{z} \tag{17.5}$$

where $m_{\underline{F}}(z)$ is the unique solution with positive imaginary part of the implicit equation in $m_{\underline{F}}$

$$\frac{1}{m_{\underline{F}}} = -\sigma^2 + \frac{1}{f} - \sum_{k=1}^{K} \frac{1}{c_k} \frac{P_k}{1 + P_k f} \tag{17.6}$$

in which we denoted f the value

$$f = (1-c)m_{\underline{F}} - czm_{\underline{F}}^2.$$

The rest of this section is dedicated to the proof of Theorem 17.5. First remember that the matrix $\mathbf{Y}$ in (17.2) can be extended into the larger sample covariance matrix $\underline{\mathbf{Y}} \in \mathbb{C}^{(N+n)\times M}$ (conditionally on $\mathbf{H}$)

$$\underline{\mathbf{Y}} = \begin{pmatrix} \mathbf{H}\mathbf{P}^{\frac{1}{2}} & \sigma \mathbf{I}_N \\ 0 & 0 \end{pmatrix} \begin{pmatrix} \mathbf{X} \\ \mathbf{W} \end{pmatrix}.$$

From Theorem 3.13, since $\mathbf{H}$ has independent entries with finite fourth order moment, we have that the e.s.d. of $\mathbf{H}\mathbf{P}\mathbf{H}^{\mathsf{H}}$ converges weakly and almost surely to a limit distribution G as $N, n_1, \ldots, n_K \to \infty$ with $N/n_k \to c_k > 0$. For $z \in \mathbb{C}^+$, the Stieltjes transform $m_G(z)$ of G is the unique solution with positive imaginary part of the equation in m_G

$$z = -\frac{1}{m_G} + \sum_{k=1}^{K} \frac{1}{c_k} \frac{P_k}{1 + P_k m_G}. \tag{17.7}$$

The almost sure convergence of the e.s.d. of $\mathbf{H}\mathbf{P}\mathbf{H}^{\mathsf{H}}$ ensures the almost sure convergence of the e.s.d. of $\begin{pmatrix} \mathbf{H}\mathbf{P}\mathbf{H}^{\mathsf{H}} + \sigma^2 \mathbf{I}_N & 0 \\ 0 & 0 \end{pmatrix}$. Since $m_G(z)$ evaluated at $z \in \mathbb{C}^+$ is the Stieltjes transform of the l.s.d. of $\mathbf{H}\mathbf{P}\mathbf{H}^{\mathsf{H}} + \sigma^2 \mathbf{I}_N$ evaluated at $z + \sigma^2$, adding n zero eigenvalues, we finally have that the e.s.d. of $\begin{pmatrix} \mathbf{H}\mathbf{P}\mathbf{H}^{\mathsf{H}} + \sigma^2 \mathbf{I}_N & 0 \\ 0 & 0 \end{pmatrix}$ tends almost surely to a distribution H whose Stieltjes transform $m_H(z)$ satisfies

$$m_H(z) = \frac{c_0}{1 + c_0} m_G(z - \sigma^2) - \frac{1}{1 + c_0} \frac{1}{z} \tag{17.8}$$

for $z \in \mathbb{C}^+$ and c_0 the limit of N/n, i.e. $c_0 = (c_1^{-1} + \ldots + c_K^{-1})^{-1}$.

As a consequence, the sample covariance matrix $\frac{1}{M}\underline{\mathbf{Y}}\underline{\mathbf{Y}}^{\mathsf{H}}$ has a population covariance matrix which is not deterministic but whose e.s.d. has an almost sure limit H for increasing dimensions. Since $\mathbf{X}$ and $\mathbf{W}$ have entries with finite fourth order moment, we can again apply Theorem 3.13 and we have that the e.s.d. of $\underline{\mathbf{B}}_N \triangleq \frac{1}{M}\underline{\mathbf{Y}}^{\mathsf{H}}\underline{\mathbf{Y}}$ converges almost surely to the limit $\underline{F}$ whose Stieltjes transform $m_{\underline{F}}(z)$ is the unique solution in $\mathbb{C}^+$ of the equation in $m_{\underline{F}}$

$$\begin{aligned} z &= -\frac{1}{m_{\underline{F}}} + \frac{1}{c}\left(1 + \frac{1}{c_0}\right) \int \frac{t}{1 + t m_{\underline{F}}} dH(t) \\ &= -\frac{1}{m_{\underline{F}}} + \frac{1 + \frac{1}{c_0}}{c m_{\underline{F}}} \left[1 - \frac{1}{m_{\underline{F}}} m_H\left(-\frac{1}{m_{\underline{F}}}\right)\right] \end{aligned} \tag{17.9}$$

for all $z \in \mathbb{C}^+$.

For $z \in \mathbb{C}^+$, $m_{\underline{F}}(z) \in \mathbb{C}^+$. Therefore $-1/m_{\underline{F}}(z) \in \mathbb{C}^+$ and we can evaluate (17.8) at $-1/m_{\underline{F}}(z)$. Combining (17.8) and (17.9), we then have

$$z = -\frac{1}{c} \frac{1}{m_{\underline{F}}(z)^2} m_G\left(-\frac{1}{m_{\underline{F}}(z)} - \sigma^2\right) + \left(\frac{1}{c} - 1\right) \frac{1}{m_{\underline{F}}(z)} \tag{17.10}$$

where, according to (17.7), $m_G(-1/m_{\underline{F}}(z) - \sigma^2)$ satisfies

$$\frac{1}{m_{\underline{F}}(z)} = -\sigma^2 + \frac{1}{m_G(-\frac{1}{m_{\underline{F}}(z)} - \sigma^2)} - \sum_{k=1}^{K} \frac{1}{c_k} \frac{P_k}{1 + P_k m_G(-\frac{1}{m_{\underline{F}}(z)} - \sigma^2)}. \tag{17.11}$$

Together with (17.10), denoting $f(z) = m_G(-\frac{1}{m_{\underline{F}}(z)} - \sigma^2) = (1-c)m_{\underline{F}}(z) - czm_{\underline{F}}(z)^2$, this is exactly (17.6).

Since the eigenvalues of the matrices $\mathbf{B}_N$ and $\underline{\mathbf{B}}_N$ only differ by $M - N$ zeros, we also have that the Stieltjes transform $m_F(z)$ of the l.s.d. of $\mathbf{B}_N$ satisfies

$$m_F(z) = cm_{\underline{F}}(z) + (c-1)\frac{1}{z}. \tag{17.12}$$

This completes the proof of Theorem 17.5. For further usage, notice here that (17.12) provides a simplified expression for $m_G(-1/m_{\underline{F}}(z) - \sigma^2)$. Indeed, we have:

$$m_G(-1/m_{\underline{F}}(z) - \sigma^2) = -zm_F(z)m_{\underline{F}}(z). \tag{17.13}$$

Therefore, the support of the (almost sure) l.s.d. F of $\mathbf{B}_N$ can be evaluated as follows: for any $z \in \mathbb{C}^+$, $m_F(z)$ is given by (17.5), in which $m_{\underline{F}}(z)$ solves (17.6); the inverse Stieltjes transform formula (3.2) then allows us to evaluate F from $m_F(z)$, for values of z spanning over the set $\{z = x + iy, x > 0\}$ and y small. This is depicted in Figure 17.12, where $\mathbf{P}$ has three distinct values $P_1 = 1$, $P_2 = 3$, $P_3 = 10$ and $n_1 = n_2 = n_3$, $N/n = 10$, $M/N = 10$, $\sigma^2 = 0.1$, as well as in Figure 17.13 for the same setup but with $P_3 = 5$.

Two remarks on Figure 17.12 and Figure 17.13 are of fundamental importance to the following. Similar to the study carried out in Chapter 7, it appears that the asymptotic l.s.d. F of $\mathbf{B}_N$ is compactly supported and divided into up to $K+1$ disjoint compact intervals, which we further refer to as *clusters*. Each cluster can be mapped onto one or many values in the set $\{\sigma^2, P_1, \ldots, P_K\}$. For instance, in Figure 17.13, the first cluster is mapped to σ^2, the second cluster to P_1, and the third cluster to the set $\{P_2, P_3\}$. Depending on the ratios c and c_0 and on the particular values taken by $P_1, \ldots, P_K$ and σ^2, these clusters are either disjoint compact intervals, as in Figure 17.12, or they may overlap to generate larger compact intervals, as in Figure 17.13. As is in fact required by the law of large numbers, for increasing c and c_0, the asymptotic spectrum tends to be divided into thinner and thinner clusters. The inference technique proposed hereafter relies on the separability of the clusters associated with each P_i and to σ^2. Precisely, to be able to derive a consistent estimate of the transmitted power P_k, the cluster associated with P_k in F, number it cluster k_F, must be distinct from the neighboring clusters $(k-1)_F$ and $(k+1)_F$, associated with P_{k-1} and P_{k+1}, respectively (when they exist), and also distinct from cluster 1 in F associated with σ^2. As such, in the scenario of Figure 17.13, our method will be able to provide a consistent estimate for P_1, but (so far) will not succeed

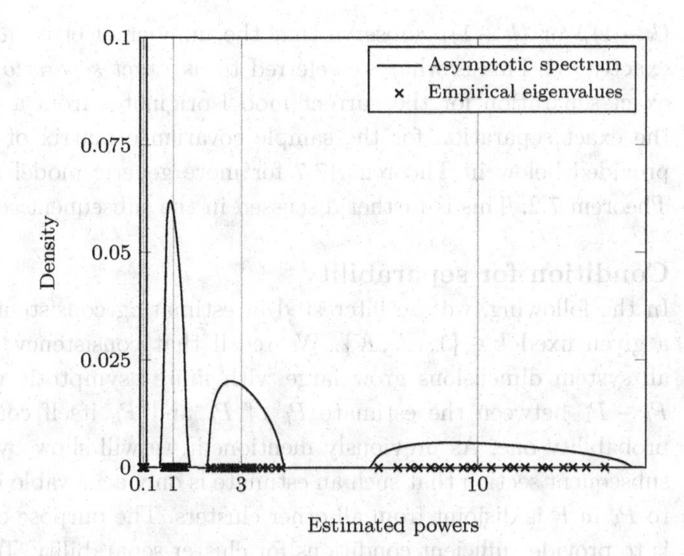

Figure 17.12 Empirical and asymptotic eigenvalue distribution of $\frac{1}{M}\mathbf{Y}\mathbf{Y}^{\mathsf{H}}$ when $\mathbf{P}$ has three distinct entries $P_1 = 1$, $P_2 = 3$, $P_3 = 10$, $n_1 = n_2 = n_3$, $c_0 = 10$, $c = 10$, $\sigma^2 = 0.1$. Empirical test: $n = 60$.

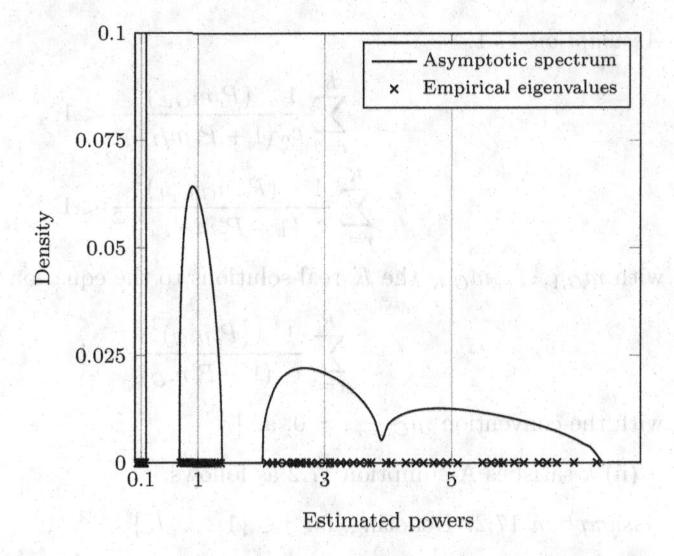

Figure 17.13 Empirical and asymptotic eigenvalue distribution of $\frac{1}{M}\mathbf{Y}\mathbf{Y}^{\mathsf{H}}$ when $\mathbf{P}$ has three distinct entries $P_1 = 1$, $P_2 = 3$, $P_3 = 5$, $n_1 = n_2 = n_3$, $c_0 = 10$, $c = 10$, $\sigma^2 = 0.1$. Empirical test: $n = 60$.

in providing a consistent estimate for either P_2 or P_3, since $2_F = 3_F$. We will see that a consistent estimate for $(P_2 + P_3)/2$ is accessible though. Secondly, notice that the empirical eigenvalues of $\mathbf{B}_N$ are all inside the asymptotic clusters and, most importantly, in the case where cluster k_F is distinct from either cluster 1,

$(k-1)_F$ or $(k+1)_F$, observe that the number of eigenvalues in cluster k_F is exactly n_k. This is what we referred to as *exact separation* in Chapter 7. The exact separation for the current model originates from a direct application of the exact separation for the sample covariance matrix of Theorem 7.2 and is provided below in Theorem 17.7 for more generic model assumptions than in Theorem 7.2. This is further discussed in the subsequent sections.

17.2.5.2 Condition for separability

In the following, we are interested in estimating consistently the power P_k for a given fixed $k \in \{1, \ldots, K\}$. We recall that consistency means here that, as all system dimensions grow large with finite asymptotic ratios, the difference $\hat{P}_k - P_k$ between the estimate $\hat{P}_k$ of P_k and P_k itself converges to zero with probability one. As previously mentioned, we will show by construction in the subsequent section that such an estimate is only achievable if the cluster mapped to P_k in F is disjoint from all other clusters. The purpose of the present section is to provide sufficient conditions for cluster separability. To ensure that cluster k_F (associated with P_k in F) is distinct from cluster 1 (associated with σ^2) and clusters i_F, $i \neq k$ (associated with all other P_i), we assume now and for the rest of this section that the following conditions are fulfilled:

(i) k satisfies Assumption 17.1, given as follows.

Assumption 17.1.

$$\sum_{r=1}^{K} \frac{1}{c_r} \frac{\left(P_r m_{G,k}\right)^2}{\left(1 + P_r m_{G,k}\right)^2} < 1, \tag{17.14}$$

$$\sum_{r=1}^{K} \frac{1}{c_r} \frac{\left(P_r m_{G,k+1}\right)^2}{\left(1 + P_r m_{G,k+1}\right)^2} < 1 \tag{17.15}$$

with $m_{G,1}, \ldots, m_{G,K}$ the K real solutions to the equation in m_G

$$\sum_{r=1}^{K} \frac{1}{c_r} \frac{\left(P_r m_G\right)^3}{\left(1 + P_r m_G\right)^3} = 1 \tag{17.16}$$

with the convention $m_{G,K+1} = 0$, and

(ii) k satisfies Assumption 17.2 as follows.

Assumption 17.2. Denoting, for $j \in \{1, \ldots, K\}$

$$j_G \triangleq \#\{i \leq j \mid i \text{ satisfies Assumption 17.1}\} \tag{17.17}$$

we have the two conditions

$$\frac{1-c_0}{c_0} \frac{(\sigma^2 m_{\underline{F},k_G})^2}{(1+\sigma^2 m_{\underline{F},k_G})^2} + \sum_{r=1}^{k_G-1} \frac{1}{c_r} \frac{(x_{G,r}^+ + \sigma^2)^2 m_{\underline{F},k_G}^2}{(1+(x_{G,r}^+ + \sigma^2) m_{\underline{F},k_G})^2}$$
$$+ \sum_{r=k_G}^{K_G} \frac{1}{c_r} \frac{(x_{G,r}^- + \sigma^2)^2 m_{\underline{F},k_G}^2}{(1+(x_{G,r}^- + \sigma^2) m_{\underline{F},k_G})^2} < c$$

and

$$\frac{1-c_0}{c_0}\frac{(\sigma^2 m_{\underline{F},k_G+1})^2}{(1+\sigma^2 m_{\underline{F},k_G+1})^2}+\sum_{r=1}^{k_G}\frac{1}{c_r}\frac{(x_{G,r}^+ +\sigma^2)^2 m_{\underline{F},k_G+1}^2}{(1+(x_{G,r}^+ +\sigma^2)m_{\underline{F},k_G+1})^2}$$
$$+\sum_{r=k_G+1}^{K_G}\frac{1}{c_r}\frac{(x_{G,r}^- +\sigma^2)^2 m_{\underline{F},k_G+1}^2}{(1+(x_{G,r}^- +\sigma^2)m_{\underline{F},k_G+1})^2}<c$$

where $x_{G,i}^-, x_{G,i}^+$, $i\in\{1,\ldots,K_G\}$, are defined by

$$x_{G,i}^- = -\frac{1}{m_{G,i}^-}+\sum_{r=1}^{K}\frac{1}{c_r}\frac{P_r}{1+P_r m_{G,i}^-} \tag{17.18}$$

$$x_{G,i}^+ = -\frac{1}{m_{G,i}^+}+\sum_{r=1}^{K}\frac{1}{c_r}\frac{P_r}{1+P_r m_{G,i}^+}, \tag{17.19}$$

with $m_{G,1}^-, m_{G,1}^+, \ldots, m_{G,K_G}^-, m_{G,K_G}^+$ the $2K_G$ real roots of (17.14) and $m_{\underline{F},j}$, $j\in\{1,\ldots,K_G+1\}$, the jth real root (in increasing order) of the equation in $m_{\underline{F}}$

$$\frac{1-c_0}{c_0}\frac{(\sigma^2 m_{\underline{F}})^3}{(1+\sigma^2 m_{\underline{F}})^3}+\sum_{r=1}^{j-1}\frac{1}{c_r}\frac{(x_{G,r}^+ +\sigma^2)^3 m_{\underline{F}}^3}{(1+(x_{G,r}^+ +\sigma^2)m_{\underline{F}})^3}$$
$$+\sum_{r=j}^{K_G}\frac{1}{c_r}\frac{(x_{G,r}^- +\sigma^2)^3 m_{\underline{F}}^3}{(1+(x_{G,r}^- +\sigma^2)m_{\underline{F}})^3}=c.$$

Although difficult to fathom at this point, the above assumptions will be clarified later. We give here a short intuitive explanation of the role of every condition. Assumption 17.1 is a necessary and sufficient condition for cluster k_G, that we define as the cluster associated with P_k in G (the l.s.d. of $\mathbf{HPH}^{\mathsf{H}}$), to be distinct from the clusters $(k-1)_G$ and $(k+1)_G$, associated with P_{k-1} and P_{k+1} in G, respectively. Note that we implicitly assume a unique mapping between the P_i and clusters in G; this statement will be made more rigorous in subsequent sections. Assumption 17.1 only deals with the inner $\mathbf{HPH}^{\mathsf{H}}$ covariance matrix properties and ensures specifically that the powers to be estimated differ sufficiently from one another for our method to be able to resolve them.

Assumption 17.2 deals with the complete $\mathbf{B}_N$ matrix model. It is however a non-necessary but sufficient condition for cluster k_F, associated with P_k in F, to be distinct from clusters $(k-1)_F$, $(k+1)_F$, and 1 (cluster 1 being associated with σ^2). The exact necessary and sufficient condition will be stated further in the next sections; however, the latter is not exploitable in practice and Assumption 17.2 will be shown to be an appropriate substitute. Assumption 17.2 is concerned with the value of c necessary to avoid:

(i) cluster k_G (associated with P_k in G) to further overlap the clusters k_G-1 and k_G+1 associated with P_{k-1} and P_{k+1},

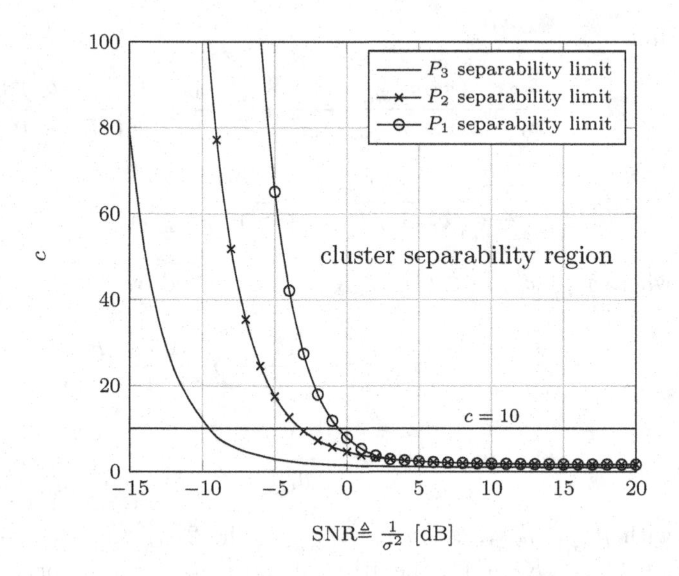

Figure 17.14 Limiting ratio c as a function of σ^2 to ensure consistent estimation of $P_1 = 1$, $P_2 = 3$ and $P_3 = 10$, $c_0 = 10$, $c_1 = c_2 = c_3$.

(ii) cluster 1 associated with σ^2 in F to merge with cluster k_F.

As will become evident in the next sections, when σ^2 is large, the tendency is for the cluster associated with σ^2 to become large and overlap the clusters associated with P_1, then P_2, etc. To counter this effect, we must increase c, i.e. take more signal samples. Figure 17.14 depicts the critical ratio c that satisfies Assumption 17.2 as a function of σ^2, in the case $K = 3$, $(P_1, P_2, P_3) = (1, 3, 10)$, $c_0 = 10$, $c_1 = c_2 = c_3$. Notice that, in the case $c = 10$, below $\sigma^2 \simeq 1$, it is possible to separate all clusters, which is compliant with Figure 17.12 where $\sigma^2 = 0.1$.

As a consequence, under the assumption (partly proved later) that our proposed method can only perform consistent power estimation when the cluster separability conditions are met, we have two first conclusions:

- if we want to increase the sensitivity of the estimator, i.e. to be able to separate two sources of close transmit powers, we need to increase the number of sensors (by increasing c_0);
- if we want to detect and reliably estimate power sources in a noise-limited environment, we need to increase the number of sensed samples (by increasing c).

In the subsequent section, we study the properties of the asymptotic spectrum of $\mathbf{HPH}^\mathsf{H}$ and $\underline{\mathbf{B}}_N$ in more detail. These properties will lead to an explanation for Assumptions 17.1 and 17.2. Under those assumptions, we will then derive the Stieltjes transform-based power estimator.

17.2.5.3 Multi-source power inference

In the following, we finally prove the main result of this section, which provides the G-estimator $\hat{P}_1, \ldots, \hat{P}_K$ of the transmit powers $P_1, \ldots, P_K$.

Theorem 17.6. *Let $\mathbf{B}_N \in \mathbb{C}^{N \times N}$ be defined as $\mathbf{B}_N = \frac{1}{M}\mathbf{Y}\mathbf{Y}^{\mathsf{H}}$ with $\mathbf{Y}$ defined as in (17.2), and $\boldsymbol{\lambda} = (\lambda_1, \ldots, \lambda_N)$, $\lambda_1 \leq \ldots \leq \lambda_N$, be the vector of the ordered eigenvalues of $\mathbf{B}_N$. Further, assume that the limiting ratios $c_0, c_1, \ldots, c_K$, c and $\mathbf{P}$ are such that Assumptions 17.1 and 17.2 are fulfilled for some $k \in \{1, \ldots, K\}$. Then, as N, n, M grow large, we have:*

$$\hat{P}_k - P_k \xrightarrow{\text{a.s.}} 0$$

where the estimate $\hat{P}_k$ is given by:

- *if $M \neq N$*

$$\hat{P}_k = \frac{NM}{n_k(M-N)} \sum_{i \in \mathcal{N}_k} (\eta_i - \mu_i)$$

- *if $M = N$*

$$\hat{P}_k = \frac{N}{n_k(N-n)} \sum_{i \in \mathcal{N}_k} \left(\sum_{j=1}^{N} \frac{\eta_i}{(\lambda_j - \eta_i)^2} \right)^{-1}$$

in which $\mathcal{N}_k = \{N - \sum_{i=k}^{K} n_i + 1, \ldots, N - \sum_{i=k+1}^{K} n_i\}$, $\eta_1 \leq \ldots \leq \eta_N$ are the ordered eigenvalues of the matrix $\operatorname{diag}(\boldsymbol{\lambda}) - \frac{1}{N}\sqrt{\boldsymbol{\lambda}}\sqrt{\boldsymbol{\lambda}}^{\mathsf{T}}$ and $\mu_1 \leq \ldots \leq \mu_N$ are the ordered eigenvalues of the matrix $\operatorname{diag}(\boldsymbol{\lambda}) - \frac{1}{M}\sqrt{\boldsymbol{\lambda}}\sqrt{\boldsymbol{\lambda}}^{\mathsf{T}}$.

Remark 17.3. We immediately notice that, if $N < n$, the powers $P_1, \ldots, P_l$, with l the largest integer such that $N - \sum_{i=l}^{K} n_i < 0$, cannot be estimated since clusters may be empty. The case $N \leq n$ turns out to be of no practical interest as clusters always merge and no consistent estimate of either P_i can be described.

The approach pursued to prove Theorem 17.6 relies strongly on the original idea of [Mestre, 2008a] which was detailed for the case of sample covariance matrices in Section 8.1.2 of Chapter 8. From Cauchy's integration formula, Theorem 8.5

$$\begin{aligned} P_k &= c_k \frac{1}{2\pi i} \oint_{\mathcal{C}_k} \frac{1}{c_k} \frac{\omega}{P_k - \omega} d\omega \\ &= c_k \frac{1}{2\pi i} \oint_{\mathcal{C}_k} \sum_{r=1}^{K} \frac{1}{c_r} \frac{\omega}{P_r - \omega} d\omega \end{aligned} \tag{17.20}$$

for any negatively oriented contour $\mathcal{C}_k \subset \mathbb{C}$, such that P_k is contained in the surface described by the contour, while for every $i \neq k$, P_i is outside this surface. The strategy unfolds as follows: we first propose a convenient integration contour $\mathcal{C}_k$ which is parametrized by a function of the Stieltjes transform $m_F(z)$ of the

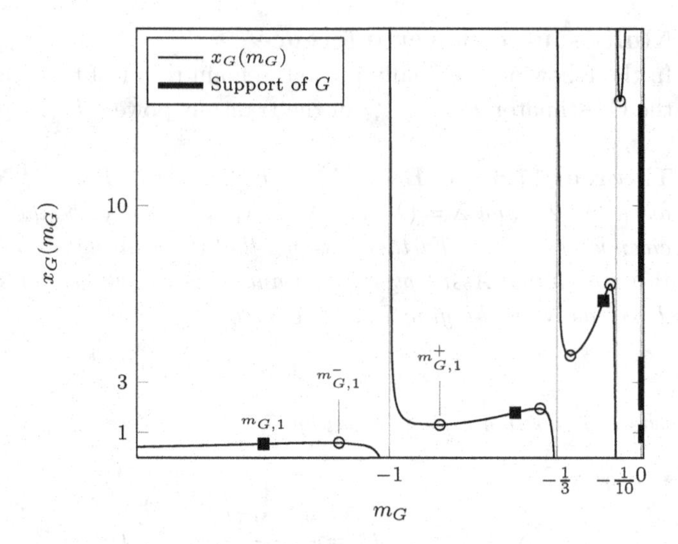

Figure 17.15 $x_G(m_G)$ for m_G real, $\mathbf{P}$ diagonal composed of three evenly weighted masses in 1, 3 and 10. Local extrema are marked in circles, inflexion points are marked in squares.

l.s.d. of $\mathbf{B}_N$; this is the technical part of the proof. We then proceed to a variable change in (17.20) to express P_k as a function of $m_F(z)$. We evaluate the complex integral resulting from replacing the limiting $m_F(z)$ in (17.20) by its empirical counterpart $m_{\mathbf{B}_N}(z) = \frac{1}{N}\operatorname{tr}(\mathbf{B}_N - z\mathbf{I}_N)^{-1}$. This new integral, whose value we name $\hat{P}_k$, is shown to be almost surely equal to P_k in the large N limit. It then suffices to evaluate $\hat{P}_k$, which is just a matter of residue calculus.

We start by determining the integration contour $\mathcal{C}_k$. For this, we first need to study the distributions G and F in more detail, following the study carried out in Chapter 7.

Properties of G and F.

First consider the matrix $\mathbf{HPH}^{\mathsf{H}}$, and let the function $x_G(m_G)$ be defined, for scalars $m_G \in \mathbb{R} \setminus \{0, -1/P_1, \ldots, -1/P_K\}$, by

$$x_G(m_G) = -\frac{1}{m_G} + \sum_{r=1}^{K} \frac{1}{c_r} \frac{P_r}{1 + P_r m_G}. \tag{17.21}$$

The function $x_G(m_G)$ is depicted in Figure 17.15 and Figure 17.16 for the cases where $c_0 = 10$, $c_1 = c_2 = c_3$ and (P_1, P_2, P_3) equal $(1, 3, 10)$ and $(1, 3, 5)$, respectively. As expected by Theorem 7.4, $x_G(m_G)$ is increasing for m_G such that $x_G(m_G)$ is outside the support of G. Note now that the function x_G presents asymptotes in the positions $-1/P_1, \ldots, -1/P_K$

$$\lim_{m_G \downarrow (-1/P_i)} x_G(m_G) = \infty$$

$$\lim_{m_G \uparrow (-1/P_i)} x_G(m_G) = -\infty$$

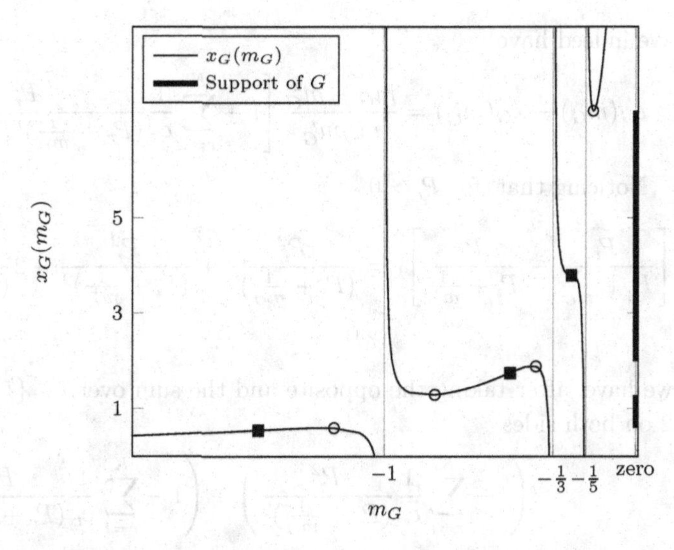

Figure 17.16 $x_G(m_G)$ for m_G real, $\mathbf{P}$ diagonal composed of three evenly weighted masses in 1, 3 and 5. Local extrema are marked in circles, inflexion points are marked in squares.

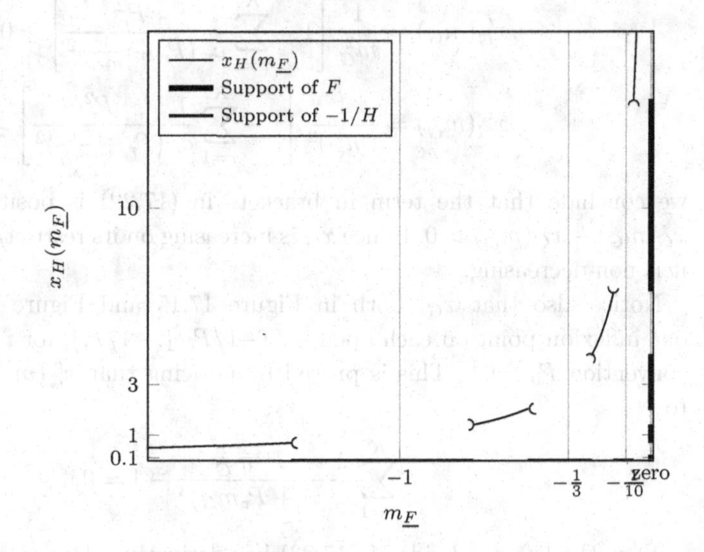

Figure 17.17 $x_{\underline{F}}(m_{\underline{F}})$ for $m_{\underline{F}}$ real, $\sigma^2 = 0.1$, $c = c_0 = 10$, $\mathbf{P}$ diagonal composed of three evenly weighted masses in 1, 3 and 10. The support of F is read on the right vertical axis.

and that $x_G(m_G) \to 0^+$ as $m_G \to -\infty$. Note also that, on its restriction to the set where it is non-decreasing, x_G is increasing. To prove this, let m_G and $m_G^\star$ be two distinct points such that $x_G(m_G) > 0$ and $x_G(m_G^\star) > 0$, and $m_G^\star < m_G < 0$.

We indeed have[1]

$$x_G(m_G) - x_G(m_G^\star) = \frac{m_G - m_G^\star}{m_G m_G^\star}\left[1 - \sum_{r=1}^{K} \frac{1}{c_r} \frac{P_r^2}{(P_r + \frac{1}{m_G})(P_r + \frac{1}{m_G^\star})}\right]. \quad (17.22)$$

Noticing that, for $P_i > 0$

$$\left[\frac{P_i}{P_i + \frac{1}{m_G}} - \frac{P_i}{P_i + \frac{1}{m_G^\star}}\right]^2 = \frac{P_i^2}{(P_i + \frac{1}{m_G})^2} + \frac{P_i^2}{(P_i + \frac{1}{m_G^\star})^2} - 2\frac{P_i^2}{(P_i + \frac{1}{m_G})(P_i + \frac{1}{m_G^\star})}$$
$$> 0$$

we have, after taking the opposite and the sum over $i \in \{1, \ldots, K\}$ and adding 2 on both sides

$$\left(1 - \sum_{r=1}^{K} \frac{1}{c_r} \frac{P_r^2}{(P_r + \frac{1}{m_G})^2}\right) + \left(1 - \sum_{r=1}^{K} \frac{1}{c_r} \frac{P_r^2}{(P_r + \frac{1}{m_G^\star})^2}\right)$$
$$< 2 - 2\sum_{r=1}^{K} \frac{1}{c_r} \frac{P_r^2}{(P_r + \frac{1}{m_G})(P_r + \frac{1}{m_G^\star})}.$$

Since we also have

$$x_G'(m_G) = \frac{1}{m_G^2}\left[1 - \sum_{r=1}^{K} \frac{1}{c_r} \frac{P_r^2}{(P_r + \frac{1}{m_G})^2}\right] \geq 0$$
$$x_G'(m_G^\star) = \frac{1}{(m_G^\star)^2}\left[1 - \sum_{r=1}^{K} \frac{1}{c_r} \frac{P_r^2}{(P_r + \frac{1}{m_G^\star})^2}\right] \geq 0$$

we conclude that the term in brackets in (17.22) is positive and then that $x_G(m_G) - x_G(m_G^\star) > 0$. Hence x_G is increasing on its restriction to the set where it is non-decreasing.

Notice also that x_G, both in Figure 17.15 and Figure 17.16, has exactly one inflexion point on each open set $(-1/P_{i-1}, -1/P_i)$, for $i \in \{1, \ldots, K\}$, with convention $P_0 = 0^+$. This is proved by noticing that $x_G''(m_G) = 0$ is equivalent to

$$\sum_{r=1}^{K} \frac{1}{c_r} \frac{P_r^3 m_G^3}{(1 + P_r m_G)^3} - 1 = 0. \quad (17.23)$$

Now, the left-hand side of (17.23) has derivative along m_G

$$3\sum_{r=1}^{K} \frac{1}{c_r} \frac{P_r^3 m_G^2}{(1 + P_r m_G)^4} \quad (17.24)$$

which is always positive. Notice that the left-hand side of (17.23) has asymptotes for $m_G = -1/P_i$ for all $i \in \{1, \ldots, K\}$ and has limits 0 as $m_G \to 0$ and $1/c_0 - 1$

[1] This proof is borrowed from the proof of [Mestre, 2008b], with different notations.

as $m_G \to -\infty$. If $c_0 > 1$, Equation (17.23) (and then $x''_G(m_G) = 0$) therefore has a unique solution in $(-1/P_{i-1}, -1/P_i)$ for all $i \in \{1, \ldots, K\}$. When x_G is increasing somewhere on $(-1/P_{i-1}, -1/P_i)$, the inflexion point, i.e. the solution of $x''_G(m_G) = 0$ in $(-1/P_{i-1}, -1/P_i)$, is necessarily found in the region where x_G increases. If $c_0 \leq 1$, the leftmost inflexion point may not exist.

From the discussion above and from Theorem 7.4 (and its corollaries discussed in Section 7.1.3), it is clear that the support of G is divided into $K_G \leq K$ compact subsets $[x^-_{G,i}, x^+_{G,i}]$, $i \in \{1, \ldots, K_G\}$. Also, if $c_0 > 1$, G has an additional mass in zero of probability $G(0) - G(0^-) = (c_0 - 1)/c_0$; this mass will not be counted as a cluster in G. Observe that every P_i can be uniquely mapped to a corresponding subset $[x^-_{G,j}, x^+_{G,j}]$ in the following fashion. The power P_1 is mapped onto the first cluster in G; we then have $1_G = 1$. Then the power P_2 is either mapped onto the second cluster in G if x_G increases in the subset $(-1/P_1, -1/P_2)$, which is equivalent to saying that $x'_G(m_{G,2}) > 0$ for $m_{G,2}$ the only solution to $x''_G(m_G) = 0$ in $(-1/P_1, -1/P_2)$; in this case, we have $2_G = 2$ and the clusters associated with P_1 and P_2 in G are distinct. Otherwise, if $x'_G(m_{G,2}) \leq 0$, P_2 is mapped onto the first cluster in F; in this case, $2_G = 1$. The latter scenario visually corresponds to the case when P_1 and P_2 engender "overlapping clusters." More generally, P_j, $j \in \{1, \ldots, K\}$, is uniquely mapped onto the cluster j_G such that

$$j_G = \#\{i \leq j \mid \min[x'_G(m_{G,i}), x'_G(m_{G,i+1})] > 0\}$$

with convention $m_{G,K+1} = 0$, which is exactly

$$j_G = \#\{i \leq j \mid i \text{ satisfies Assumption 17.1}\}$$

when $c_0 > 1$. If $c_0 \leq 1$, $m_{G,1}$, the zero of x''_G in $(-\infty, -1/P_1)$ may not exist. If $c_0 < 1$, we claim that P_1 cannot be evaluated (as was already observed in Remark 17.3). The special case when $c_0 = 1$ would require a restatement of Assumption 17.1 to handle the special case of P_1; this will however not be done, as it will turn out that Assumption 17.2 is violated for P_1 if $\sigma^2 > 0$, which we assume.

In the particular case of the power P_k of interest in Theorem 17.6, because of Assumption 17.1, $x'_G(m_{G,k}) > 0$. Therefore the index k_G of the cluster associated with P_k in G satisfies $k_G = (k-1)_G + 1$ (with convention $0_G = 0$). Also, from Assumption 17.1, $x'_G(m_{G,k+1}) > 0$. Therefore $(k+1)_G = k_G + 1$. In that case, we have that P_k is the only power mapped to cluster k_G in G, and then we have the required cluster separability condition.

We now proceed to the study of F, the almost sure limit spectrum distribution of $\mathbf{B}_N$. In the same way as previously, we have that the support of $\underline{F}$ is fully determined by the function $x_{\underline{F}}(m_{\underline{F}})$, defined for $m_{\underline{F}}$ real, such that $-1/m_{\underline{F}}$ lies outside the support of H, by

$$x_{\underline{F}}(m_{\underline{F}}) = -\frac{1}{m_{\underline{F}}} + \frac{1+c_0}{cc_0}\int \frac{t}{1+tm_{\underline{F}}} dH(t).$$

Figure 17.17 depicts the function $x_{\underline{F}}$ in the system conditions already used in Figure 17.12, i.e. $K = 3$, $P_1 = 1$, $P_2 = 3$, $P_3 = 10$, $c_1 = c_2 = c_3$, $c_0 = 10$, $c = 10$,

$\sigma^2 = 0.1$. Figure 17.17 has the peculiar behavior that it does not have asymptotes as in Figure 17.15 where the population eigenvalue distribution was discrete. As a consequence, our previous derivations cannot be straightforwardly adapted to derive the spectrum separability condition. If $c_0 > 1$, note also, although it is not appearing in the abscissa range of Figure 17.17, that there exist asymptotes in the position $m_{\underline{F}} = -1/\sigma^2$. This is due to the fact that $G(0) - G(0^-) > 0$, and therefore $H(\sigma^2) - H((\sigma^2)^-) > 0$. We assume $c_0 > 1$ until further notice.

Applying a second time Theorem 7.4, the support of $\underline{F}$ is complementary to the set of real non-negative x such that $x = x_{\underline{F}}(m_{\underline{F}})$ and $x'_{\underline{F}}(m_{\underline{F}}) > 0$ for a certain real $m_{\underline{F}}$, with $x'_{\underline{F}}(m_{\underline{F}})$ given by:

$$x'_{\underline{F}}(m_{\underline{F}}) = \frac{1}{m_{\underline{F}}^2} - \frac{1+c_0}{cc_0} \int \frac{t^2}{(1+tm_{\underline{F}})^2} dH(t).$$

Reminding that $H(t) = \frac{c_0}{c_0+1} G(t-\sigma^2) + \frac{1}{1+c_0}\delta(t)$, this can be rewritten

$$x'_{\underline{F}}(m_{\underline{F}}) = \frac{1}{m_{\underline{F}}^2} - \frac{1}{c} \int \frac{t^2}{(1+tm_{\underline{F}})^2} dG(t-\sigma^2). \tag{17.25}$$

It is still true that $x_{\underline{F}}(m_{\underline{F}})$, restricted to the set of $m_{\underline{F}}$ where $x'_{\underline{F}}(m_{\underline{F}}) \geq 0$, is increasing. As a consequence, it is still true also that each cluster of H can be mapped to a unique cluster of $\underline{F}$. It is then possible to iteratively map the power P_k onto cluster k_G in G, as previously described, and to further map cluster k_G in G (which is also cluster k_G in H) onto a unique cluster k_F in $\underline{F}$ (or equivalently in F).

Therefore, a necessary and sufficient condition for the separability of the cluster associated with P_k in $\underline{F}$ reads:

Assumption 17.3. There exist two distinct real values $m^{(l)}_{\underline{F},k_G} < m^{(r)}_{\underline{F},k_G}$ such that:

1. $x'_{\underline{F}}(m^{(l)}_{\underline{F},k_G}) > 0$, $x'_{\underline{F}}(m^{(r)}_{\underline{F},k_G}) > 0$
2. there exist $m^{(l)}_{G,k}, m^{(r)}_{G,k} \in \mathbb{R}$ such that

$$x_G(m^{(l)}_{G,k}) = -\frac{1}{m^{(l)}_{\underline{F},k_G}} - \sigma^2$$
$$x_G(m^{(r)}_{G,k}) = -\frac{1}{m^{(r)}_{\underline{F},k_G}} - \sigma^2$$

that satisfy:
a. $x'_G(m^{(l)}_{G,k}) > 0$, $x'_G(m^{(r)}_{G,k}) > 0$
b. and

$$P_{k-1} < -\frac{1}{m^{(l)}_{G,k}} < P_k < -\frac{1}{m^{(r)}_{G,k}} < P_{k+1} \tag{17.26}$$

with the convention $P_0 = 0^+$, $P_{K+1} = \infty$.

Assumption 17.3 states first that cluster k_G in G is distinct from clusters $(k-1)_G$ and $(k+1)_G$ (Item 2b), which is equivalent to Assumption 17.1, and second that $m_{\underline{F},k_G}^{(l)} \triangleq -1/(x_G(m_{G,k_G}^{(l)})+\sigma^2)$ and $m_{\underline{F},k_G}^{(r)} \triangleq -1/(x_G(m_{G,k_G}^{(r)})+\sigma^2)$ (which lie on either side of cluster k_G in H) have respective images $x_{k_F}^{(l)} \triangleq x_{\underline{F}}(m_{\underline{F},k_G}^{(l)})$ and $x_{k_F}^{(r)} \triangleq x_{\underline{F}}(m_{\underline{F},k_G}^{(r)})$ by $x_{\underline{F}}$, such that $x'_{\underline{F}}(m_{\underline{F},k_G}^{(l)}) > 0$ and $x'_{\underline{F}}(m_{\underline{F},k_G}^{(r)}) > 0$, i.e. $x_{k_F}^{(l)}$ and $x_{k_F}^{(r)}$ lie outside the support of $\underline{F}$, on either side of cluster k_F.

However, Assumption 17.3, be it a necessary and sufficient condition for the separability of cluster k_F, is difficult to exploit in practice. Indeed, it is not satisfactory to require the verification of the existence of such $m_{\underline{F},k_G}^{(l)}$ and $m_{\underline{F},k_G}^{(r)}$. More importantly, the computation of $x_{\underline{F}}$ requires to know H, which is only fully accessible through the non-convenient inverse Stieltjes transform formula

$$H(x) = \frac{1}{\pi} \lim_{y \to 0} \int_{-\infty}^{x} m_H(t+iy)dt. \tag{17.27}$$

Instead of Assumption 17.3, we derive here a sufficient condition for cluster separability in $\underline{F}$, which can be explicitly verified without resorting to involved Stieltjes transform inversion formulas. Note from the clustering of G into K_G clusters plus a mass at zero that (17.25) becomes

$$x'_{\underline{F}}(m_{\underline{F}}) = \frac{1}{m_{\underline{F}}^2} - \frac{1}{c} \sum_{r=1}^{K_G} \int_{x_{G,r}^-}^{x_{G,r}^+} \frac{t^2}{(1+tm_{\underline{F}})^2} dG(t-\sigma^2) - \frac{c_0 - 1}{c c_0} \frac{\sigma^4}{(1+\sigma^2 m_{\underline{F}})^2} \tag{17.28}$$

where we remind that $[x_{G,i}^-, x_{G,i}^+]$ is the support of cluster i in G, i.e. $x_{G,1}^-, x_{G,1}^+, \ldots, x_{G,K_G}^-, x_{G,K_G}^+$ are the images by x_G of the $2K_G$ real solutions to $x'_G(m_G) = 0$.

Observe now that the function $-t^2/(1+tm_{\underline{F}})^2$, found in the integrals of (17.28), has derivative along t

$$\left(-\frac{t^2}{(1+tm_{\underline{F}})^2}\right)' = -\frac{2t}{(1+tm_{\underline{F}})^4}(1+tm_{\underline{F}})$$

and is therefore strictly increasing when $m_{\underline{F}} < -1/t$ and strictly decreasing when $m_{\underline{F}} > -1/t$. For $m_{\underline{F}} \in (-1/(x_{G,i}^+ + \sigma^2), -1/(x_{G,i+1}^- + \sigma^2))$, we then have the inequality

$$x'_{\underline{F}}(m_{\underline{F}}) \geq \frac{1}{m_{\underline{F}}^2} - \frac{1}{c}\left(\sum_{r=1}^{i} \frac{(x_{G,r}^+ + \sigma^2)^2}{(1+(x_{G,r}^+ + \sigma^2)m_{\underline{F}})^2} \right.$$
$$\left. + \sum_{r=i+1}^{K_G} \frac{(x_{G,r}^- + \sigma^2)^2}{(1+(x_{G,r}^- + \sigma^2)m_{\underline{F}})^2} + \frac{c_0 - 1}{c_0} \frac{\sigma^4}{(1+\sigma^2 m_{\underline{F}})^2}\right). \tag{17.29}$$

Denote $f_i(m_{\underline{F}})$ the right-hand side of (17.29). Through the inequality (17.29), we then fall back on a finite sum expression as in the previous study of the support of G. In that case, we can exhibit a sufficient condition to ensure the separability of cluster k_F from the neighboring clusters. Specifically, we only need to verify that $f_{k_G-1}(m_{\underline{F},k_G}) > 0$, with $m_{\underline{F},k_G}$ the single solution to $f'_{k_G-1}(m_{\underline{F}}) = 0$ in the set $(-1/(x^+_{G,k_G-1}+\sigma^2), -1/(x^-_{G,k_G}+\sigma^2))$, and $f_{k_G}(m_{\underline{F},k_G+1}) > 0$, with $m_{\underline{F},k_G+1}$ the unique solution to $f'_{k_G}(m_{\underline{F}}) = 0$ in the set $(-1/(x^+_{G,k_G}+\sigma^2), -1/(x^-_{G,k_G+1}+\sigma^2))$. This is exactly what Assumption 17.2 states.

Remember now that we assumed in this section $c_0 > 1$. If $c_0 \leq 1$, then zero is in the support of H, and therefore the leftmost cluster in F, i.e. that attached to σ^2, is necessarily merged with that of P_1. This already discards the possibility of spectrum separation for P_1 and therefore P_1 cannot be estimated. It is therefore not necessary to update Assumption 17.1 for the particular case of P_1 when $c_0 = 1$.

Finally, Assumptions 17.1 and 17.2 ensure that $(k-1)_F < k_F < (k+1)_F$, $k_F \neq 1$, and there exists a constructive way to derive the mapping $k \mapsto k_F$. We are now in position to determine the contour $\mathcal{C}_k$.

Determination of $\mathcal{C}_k$.

From Assumption 17.2 and Theorem 7.4, there exist $x^{(l)}_{k_F}$ and $x^{(r)}_{k_F}$ outside the support of F, on either side of cluster k_F, such that $m_{\underline{F}}(z)$ has limits $m^{(l)}_{\underline{F},k_G} \triangleq m^\circ_{\underline{F}}(x^{(l)}_{k_F})$ and $m^{(r)}_{\underline{F},k_G} \triangleq m^\circ_{\underline{F}}(x^{(r)}_{k_F})$, as $z \to x^{(l)}_{k_F}$ and $z \to x^{(r)}_{k_F}$, respectively, with $m^\circ_{\underline{F}}$ the analytic extension of $m_{\underline{F}}$ in the points $x^{(l)}_{k_F} \in \mathbb{R}$ and $x^{(r)}_{k_F} \in \mathbb{R}$. These limits $m^{(l)}_{\underline{F},k_G}$ and $m^{(r)}_{\underline{F},k_G}$ are on either side of cluster k_G in the support of $-1/H$, and therefore $-1/m^{(l)}_{\underline{F},k_G} - \sigma^2$ and $-1/m^{(l)}_{\underline{F},k_G} - \sigma^2$ are on either side of cluster k_G in the support of G.

Consider any continuously differentiable complex path $\Gamma_{F,k}$ with endpoints $x^{(l)}_{k_F}$ and $x^{(r)}_{k_F}$, and interior points of positive imaginary part. We define the contour $\mathcal{C}_{F,k}$ as the union of $\Gamma_{F,k}$ oriented from $x^{(l)}_{k_F}$ to $x^{(r)}_{k_F}$ and its complex conjugate $\Gamma^*_{F,k}$ oriented backwards from $x^{(r)}_{k_F}$ to $x^{(l)}_{k_F}$. The contour $\mathcal{C}_{F,k}$ is clearly continuous and piecewise continuously differentiable. Also, the support of cluster k_F in $\underline{F}$ is completely inside $\mathcal{C}_{F,k}$, while the supports of the neighboring clusters are away from $\mathcal{C}_{F,k}$. The support of cluster k_G in H is then inside $-1/m_{\underline{F}}(\mathcal{C}_{F,k})$,[2] and therefore the support of cluster k_G in G is inside $\mathcal{C}_{G,k} \triangleq -1/m_{\underline{F}}(\mathcal{C}_{F,k}) - \sigma^2$. Since $m_{\underline{F}}$ is continuously differentiable on $\mathbb{C} \setminus \mathbb{R}$ (it is in fact holomorphic there [Silverstein and Choi, 1995]) and has limits in $x^{(l)}_{k_F}$ and $x^{(r)}_{k_F}$, $\mathcal{C}_{G,k}$ is also continuous and piecewise continuously differentiable. Going one more step in this process, we finally have that P_k is inside the contour $\mathcal{C}_k \triangleq -1/m_G(\mathcal{C}_{G,k})$, while P_i, for all $i \neq k$, is outside $\mathcal{C}_k$. Since m_G is also holomorphic on $\mathbb{C} \setminus \mathbb{R}$ and has limits in $-1/m^\circ_{\underline{F}}(x^{(l)}_{k_F}) - \sigma^2$ and $-1/m^\circ_{\underline{F}}(x^{(r)}_{k_F}) - \sigma^2$, $\mathcal{C}_k$ is a continuous and piecewise continuously differentiable complex path, which is sufficient to perform complex integration [Rudin, 1986].

[2] We slightly abuse notations here and should instead say that the support of cluster k_G in H is inside the contour described by the image by $-1/m_{\underline{F}}$ of the restriction to $\mathbb{C}^+$ and $\mathbb{C}^-$ of $\mathcal{C}_{F,k}$, continuously extended to $\mathbb{R}$ in the points $-1/m^{(l)}_{\underline{F},k_G}$ and $-1/m^{(r)}_{\underline{F},k_G}$.

Figure 17.18 depicts the contours $\mathcal{C}_1, \mathcal{C}_2, \mathcal{C}_3$ originating from circular integration contours $\mathcal{C}_{F,k}$ of diameter $[x_{k_F}^{(l)}, x_{k_F}^{(r)}]$, $k \in \{1,2,3\}$, for the case of Figure 17.12. The points $x_{k_F}^{(l)}$ and $x_{k_F}^{(r)}$ for $k_F \in \{1,2,3\}$ are taken to be $x_{k_F}^{(l)} = x_{\underline{F}}(m_{\underline{F},k_G})$, $x_{k_F}^{(r)} = x_{\underline{F}}(m_{\underline{F},k_G+1})$, with $m_{\underline{F},i}$ the real root of $f_i'(m_{\underline{F}})$ in $(-1/(x_{G,i-1}^+ + \sigma^2), -1/(x_{G,i}^- + \sigma^2))$ when $i \in \{1,2,3\}$, and we take the convention $m_{G,4} = -1/(15+\sigma^2)$.

Recall now that P_k was defined as

$$P_k = c_k \frac{1}{2\pi i} \oint_{\mathcal{C}_k} \sum_{r=1}^{K} \frac{1}{c_r} \frac{\omega}{P_r - \omega} d\omega.$$

With the variable change $\omega = -1/m_G(t)$, this becomes

$$\begin{aligned}
P_k &= \frac{c_k}{2\pi i} \oint_{\mathcal{C}_{G,k}} \sum_{r=1}^{K} \frac{1}{c_r} \frac{-1}{1+P_r m_G(t)} \frac{m_G'(t)}{m_G(t)^2} dt \\
&= \frac{c_k}{2\pi i} \oint_{\mathcal{C}_{G,k}} \left[m_G(t) \sum_{r=1}^{K} \frac{1}{c_r} \frac{P_r}{1+P_r m_G(t)} - \sum_{r=1}^{K} \frac{1}{c_r} \right] \frac{m_G'(t)}{m_G(t)^2} dt \\
&= \frac{c_k}{2\pi i} \oint_{\mathcal{C}_{G,k}} \left(m_G(t) \left[-\frac{1}{m_G(t)} + \sum_{r=1}^{K} \frac{1}{c_r} \frac{P_r}{1+P_r m_G(t)} \right] + \frac{c_0 - 1}{c_0} \right) \frac{m_G'(t)}{m_G(t)^2} dt.
\end{aligned}$$

From Equation (17.7), this simplifies into

$$P_k = \frac{c_k}{c_0} \frac{1}{2\pi i} \oint_{\mathcal{C}_{G,k}} (c_0 t m_G(t) + c_0 - 1) \frac{m_G'(t)}{m_G(t)^2} dt. \tag{17.30}$$

Using (17.10) and proceeding to the change of variable $t = -1/m_{\underline{F}}(z) - \sigma^2$, (17.30) becomes

$$P_k = \frac{c_k}{2\pi i} \oint_{\mathcal{C}_{F,k}} (1 + \sigma^2 m_{\underline{F}}(z)) \left[-\frac{1}{z m_{\underline{F}}(z)} - \frac{m_{\underline{F}}'(z)}{m_{\underline{F}}(z)^2} - \frac{m_F'(z)}{m_F(z) m_{\underline{F}}(z)} \right] dz. \tag{17.31}$$

This whole process of variable changes allows us to describe P_k as a function of $m_F(z)$, the Stieltjes transform of the almost sure limiting spectral distribution of $\mathbf{B}_N$, as $N \to \infty$. It then remains to exhibit a relation between P_k and the e.s.d. of $\mathbf{B}_N$ for finite N. This is to what the subsequent section is dedicated.

17.2.5.4 Evaluation of $\hat{P}_k$

Let us now define $m_{\mathbf{B}_N}(z)$ and $m_{\underline{\mathbf{B}}_N}(z)$ as the Stieltjes transforms of the empirical eigenvalue distributions of $\mathbf{B}_N$ and $\underline{\mathbf{B}}_N$, respectively, i.e.

$$m_{\mathbf{B}_N}(z) = \frac{1}{N} \sum_{i=1}^{N} \frac{1}{\lambda_i - z} \tag{17.32}$$

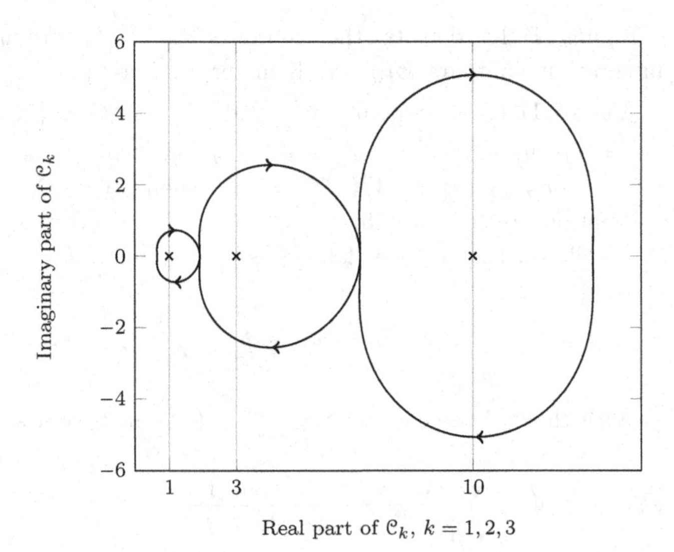

Figure 17.18 (Negatively oriented) integration contours $\mathcal{C}_1$, $\mathcal{C}_2$ and $\mathcal{C}_3$, for $c = 10$, $c_0 = 10$, $P_1 = 1$, $P_2 = 3$, $P_3 = 10$.

and

$$m_{\underline{\mathbf{B}}_N}(z) = \frac{N}{M} m_{\mathbf{B}_N}(z) - \frac{M-N}{M}\frac{1}{z}.$$

Instead of going further with (17.31), define $\hat{P}_k$, the "empirical counterpart" of P_k, as

$$\hat{P}_k = \frac{n}{n_k}\frac{1}{2\pi i}\oint_{\mathcal{C}_{F,k}} \frac{N}{n}\left(1 + \sigma^2 m_{\underline{\mathbf{B}}_N}(z)\right) \times \left[-\frac{1}{z m_{\underline{\mathbf{B}}_N}(z)} - \frac{m'_{\underline{\mathbf{B}}_N}(z)}{m_{\underline{\mathbf{B}}_N}(z)^2} - \frac{m'_{\mathbf{B}_N}(z)}{m_{\mathbf{B}_N}(z) m_{\underline{\mathbf{B}}_N}(z)}\right] dz. \quad (17.33)$$

The integrand can then be expanded into several terms, for which residue calculus can easily be performed. Denote first $\eta_1, \ldots, \eta_N$ the N real roots of $m_{\mathbf{B}_N}(z) = 0$ and $\mu_1, \ldots, \mu_N$ the N real roots of $m_{\underline{\mathbf{B}}_N}(z) = 0$. We identify three sets of possible poles for the aforementioned terms: (i) the set $\{\lambda_1, \ldots, \lambda_N\} \cap [x_{k_F}^{(l)}, x_{k_F}^{(r)}]$, (ii) the set $\{\eta_1, \ldots, \eta_N\} \cap [x_{k_F}^{(l)}, x_{k_F}^{(r)}]$, and (iii) the set $\{\mu_1, \ldots, \mu_N\} \cap [x_{k_F}^{(l)}, x_{k_F}^{(r)}]$. For $M \neq N$, the full calculus leads to

$$\hat{P}_k = \frac{NM}{n_k(M-N)}\left[\sum_{\substack{1 \le i \le N \\ x_{k_F}^{(l)} \le \eta_i \le x_{k_F}^{(r)}}} \eta_i - \sum_{\substack{1 \le i \le N \\ x_{k_F}^{(l)} \le \mu_i \le x_{k_F}^{(r)}}} \mu_i\right.$$

$$+\frac{N}{n_k}\left[\sum_{\substack{1\leq i\leq N \\ x_{k_F}^{(l)}\leq\eta_i\leq x_{k_F}^{(r)}}}\sigma^2-\sum_{\substack{1\leq i\leq N \\ x_{k_F}^{(l)}\leq\lambda_i\leq x_{k_F}^{(r)}}}\sigma^2+\sum_{\substack{1\leq i\leq N \\ x_{k_F}^{(l)}\leq\mu_i\leq x_{k_F}^{(r)}}}\sigma^2-\sum_{\substack{1\leq i\leq N \\ x_{k_F}^{(l)}\leq\lambda_i\leq x_{k_F}^{(r)}}}\sigma^2\right]. \tag{17.34}$$

We know from Theorem 17.5 that $m_{\mathbf{B}_N}(z)\xrightarrow{\text{a.s.}}m_F(z)$ and $m_{\underline{\mathbf{B}}_N}(z)\xrightarrow{\text{a.s.}}m_{\underline{F}}(z)$ as $N\to\infty$. Observing that the integrand in (17.33) is uniformly bounded on the compact $\mathcal{C}_{F,k}$, the dominated convergence theorem, Theorem 6.3, ensures $\hat{P}_k\xrightarrow{\text{a.s.}}P_k$.

To go further, we now need to determine which of $\lambda_1,\ldots,\lambda_N$, $\eta_1,\ldots,\eta_N$ and $\mu_1,\ldots,\mu_N$ lie inside $\mathcal{C}_{F,k}$. This requires a result of eigenvalue exact separation that extends Theorem 7.1 [Bai and Silverstein, 1998] and Theorem 7.2 [Bai and Silverstein, 1999], as follows.

Theorem 17.7. *Let $\mathbf{B}_n=\frac{1}{n}\mathbf{T}_n^{\frac{1}{2}}\mathbf{X}_n\mathbf{X}_n^{\mathsf{H}}\mathbf{T}_n^{\frac{1}{2}}\in\mathbb{C}^{p\times p}$, where we assume the following conditions:*

1. *$\mathbf{X}_n\in\mathbb{C}^{p\times n}$ has entries x_{ij}, $1\leq i\leq p$, $1\leq j\leq n$, extracted from a doubly infinite array $\{x_{ij}\}$ of independent variables, with zero mean and unit variance.*
2. *There exist K and a random variable X with finite fourth order moment such that, for any $x>0$*

$$\frac{1}{n_1n_2}\sum_{i\leq n_1,j\leq n_2}P(|x_{ij}|>x)\leq KP(|X|>x) \tag{17.35}$$

   for any n_1,n_2.
3. *There is a positive function $\psi(x)\uparrow\infty$ as $x\to\infty$, and $M>0$, such that*

$$\max_{ij}\mathrm{E}[|x_{ij}^2|\psi(|x_{ij}|)]\leq M. \tag{17.36}$$

4. *$p=p(n)$ with $c_n=p/n\to c>0$ as $n\to\infty$.*
5. *For each n, $\mathbf{T}_n\in\mathbb{C}^{p\times p}$ is Hermitian non-negative definite, independent of $\{x_{ij}\}$, satisfying $H_n\triangleq F^{\mathbf{T}_n}\Rightarrow H$, H a non-random probability distribution function, almost surely. $\mathbf{T}_n^{\frac{1}{2}}$ is any Hermitian square root of $\mathbf{T}_n$.*
6. *The spectral norm $\|\mathbf{T}_n\|$ of $\mathbf{T}_n$ is uniformly bounded in n almost surely.*
7. *Let $a,b>0$, non-random, be such that, with probability one, $[a,b]$ lies in an open interval outside the support of F^{c_n,H_n} for all large n, with $F^{y,G}$ defined to be the almost sure l.s.d. of $\frac{1}{n}\mathbf{X}_n^{\mathsf{H}}\mathbf{T}_n\mathbf{X}_n$ when $H=G$ and $c=y$.*

Denote $\lambda_1^{\mathbf{Y}}\geq\ldots\geq\lambda_p^{\mathbf{Y}}$ the ordered eigenvalues of the Hermitian matrix $\mathbf{Y}\in\mathbb{C}^{p\times p}$. Then, we have that:

1. *$P(\text{no eigenvalues of }\mathbf{B}_n\text{ appear in }[a,b]\text{ for all large }n)=1$.*
2. *If $c(1-H(0))>1$, then x_0, the smallest value in the support of $F^{c,H}$, is positive, and with probability one, $\lambda_n^{\mathbf{B}_n}\to x_0$ as $n\to\infty$.*

3. *If $c(1-H(0)) \leq 1$, or $c(1-H(0)) > 1$ but $[a,b]$ is not contained in $[0, x_0]$, then $m_{F^{c,H}}(a) < m_{F^{c,H}}(b) < 0$. Almost surely, there exists, for all n large, an index $i_n \geq 0$ such that $\lambda_{i_n}^{\mathbf{T}_n} > -1/m_{F^{c,H}}(b)$ and $\lambda_{i_n+1}^{\mathbf{T}_n} > -1/m_{F^{c,H}}(a)$ and we have:*

$$P(\lambda_{i_n}^{\mathbf{B}_n} > b \text{ and } \lambda_{i_n+1}^{\mathbf{B}_n} < a \text{ for all large } n) = 1.$$

Theorem 17.7 is proved in [Couillet *et al.*, 2011c]. This result is more general than Theorem 7.2, but the assumptions are so involved that we preferred to state Theorem 7.2 in Chapter 7 in its original form with i.i.d. entries in matrix $\mathbf{X}_n$.

To apply Theorem 17.7 to $\underline{\mathbf{B}}_N$ in our scenario, we need to ensure all assumptions are met. Only Items 2–6 need particular attention. In our scenario, the matrix $\mathbf{X}_n$ of Theorem 17.7 is $\left(\begin{smallmatrix}\mathbf{X}\\ \mathbf{W}\end{smallmatrix}\right)$, while $\mathbf{T}_n$ is $\mathbf{T} \triangleq \left(\begin{smallmatrix}\mathbf{HPH}^{\mathsf{H}}+\sigma^2\mathbf{I}_N & 0\\ 0 & 0\end{smallmatrix}\right)$. The latter has been proved to have almost sure l.s.d. H, so that Item 5 is verified. Also, from Theorem 7.1 upon which Theorem 17.7 is based, there exists a subset of probability one in the probability space that engenders the $\mathbf{T}$ over which, for n large enough, $\mathbf{T}$ has no eigenvalues in any closed set strictly outside the support of H; this ensures Item 6. Now, by construction, $\mathbf{X}$ and $\mathbf{W}$ have independent entries of zero mean, unit variance, fourth order moment and are composed of at most $K+1$ distinct distributions, irrespective of M. Denote $X_1, \ldots, X_d$, $d \leq K+1$, d random variables distributed as those distinct distributions. Letting $X = |X_1| + \ldots + |X_d|$, we have that

$$\begin{aligned}\frac{1}{n_1 n_2} \sum_{i \leq n_1, j \leq n_2} P(|z_{ij}| > x) &\leq P\left(\sum_{i=1}^{d} |X_i| > x\right)\\ &= P(|X| > x)\end{aligned}$$

where z_{ij} is the (i,j)th entry of $\left(\begin{smallmatrix}\mathbf{X}\\ \mathbf{W}\end{smallmatrix}\right)$. Since all X_i have finite order four moments, so does X, and Item 2 is verified. From the same argument, Item 3 follows with $\phi(x) = x^2$. Theorem 17.7 can then be applied to $\underline{\mathbf{B}}_N$.

The corollary of Theorem 17.7 applied to $\underline{\mathbf{B}}_N$ is that, with probability one, for N sufficiently large, there will be no eigenvalue of $\mathbf{B}_N$ (or $\underline{\mathbf{B}}_N$) outside the support of F, and the number of eigenvalues inside cluster k_F is exactly n_k. Since $\mathcal{C}_{F,k}$ encloses cluster k_F and is away from the other clusters, $\{\lambda_1, \ldots, \lambda_N\} \cap [x_{k_F}^{(l)}, x_{k_F}^{(r)}] = \{\lambda_i, i \in \mathcal{N}_k\}$ almost surely, for all large N. Also, for any $i \in \{1, \ldots, N\}$, it is easy to see from (17.32) that $m_{\mathbf{B}_N}(z) \to \infty$ when $z \uparrow \lambda_i$ and $m_{\mathbf{B}_N}(z) \to -\infty$ when $z \downarrow \lambda_i$. Therefore $m_{\mathbf{B}_N}(z)$ has at least one root in each interval $(\lambda_{i-1}, \lambda_i)$, with $\lambda_0 = 0$, hence $\mu_1 < \lambda_1 < \mu_2 < \ldots < \mu_N < \lambda_N$. This implies that, if k_0 is the index such that $\mathcal{C}_{F,k}$ contains exactly $\lambda_{k_0}, \ldots, \lambda_{k_0+(n_k-1)}$, then $\mathcal{C}_{F,k}$ also contains $\{\mu_{k_0+1}, \ldots, \mu_{k_0+(n_k-1)}\}$. The same result holds for $\eta_{k_0+1}, \ldots, \eta_{k_0+(n_k-1)}$. When the indexes exist, due to cluster separability, η_{k_0-1} and μ_{k_0-1} belong, for N large, to cluster $k_F - 1$. We are then left with determining whether μ_{k_0} and η_{k_0} are asymptotically found inside $\mathcal{C}_{F,k}$.

For this, we use the same approach as in Chapter 8 by noticing that, since zero is not included in $\mathcal{C}_k$, we have:

$$\frac{1}{2\pi i}\oint_{\mathcal{C}_k}\frac{1}{\omega}d\omega = 0.$$

Performing the same changes of variables as previously, we have:

$$\oint_{\mathcal{C}_{F,k}}\frac{-m_{\underline{F}}(z)m_F(z) - zm'_{\underline{F}}(z)m_F(z) - zm_{\underline{F}}(z)m'_F(z)}{z^2 m_{\underline{F}}(z)^2 m_F(z)^2}dz = 0. \tag{17.37}$$

For N large, the dominated convergence theorem, Theorem 6.3, ensures again that the left-hand side of the (17.37) is close to

$$\oint_{\mathcal{C}_{F,k}}\frac{-m_{\underline{\mathbf{B}}_N}(z)m_{\mathbf{B}_N}(z) - zm'_{\underline{\mathbf{B}}_N}(z)m_{\mathbf{B}_N}(z) - zm_{\underline{\mathbf{B}}_N}(z)m'_{\mathbf{B}_N}(z)}{z^2 m_{\underline{\mathbf{B}}_N}(z)^2 m_{\mathbf{B}_N}(z)^2}dz. \tag{17.38}$$

The residue calculus of (17.38) then leads to

$$\sum_{\substack{1\leq i\leq N\\ \lambda_i\in[x_{k_F}^{(l)},x_{k_F}^{(r)}]}} 2 - \sum_{\substack{1\leq i\leq N\\ \eta_i\in[x_{k_F}^{(l)},x_{k_F}^{(r)}]}} 1 - \sum_{\substack{1\leq i\leq N\\ \mu_i\in[x_{k_F}^{(l)},x_{k_F}^{(r)}]}} 1 \xrightarrow{\text{a.s.}} 0. \tag{17.39}$$

Since the cardinalities of $\{i, \eta_i \in [x_{k_F}^{(l)}, x_{k_F}^{(r)}]\}$ and $\{i, \mu_i \in [x_{k_F}^{(l)}, x_{k_F}^{(r)}]\}$ are at most n_k, (17.39) is satisfied only if both cardinalities equal n_k in the limit. As a consequence, $\mu_{k_0} \in [x_{k_F}^{(l)}, x_{k_F}^{(r)}]$ and $\eta_{k_0} \in [x_{k_F}^{(l)}, x_{k_F}^{(r)}]$ for all N large, almost surely. For N large, $N \neq M$, this allows us to simplify (17.34) into

$$\hat{P}_k = \frac{NM}{n_k(M-N)}\sum_{\substack{1\leq i\leq N\\ \lambda_i\in\mathcal{N}_k}}(\eta_i - \mu_i) \tag{17.40}$$

with probability one. The same reasoning holds for $M = N$. This is our final relation. It now remains to show that the η_i and the μ_i are the eigenvalues of $\mathrm{diag}(\boldsymbol{\lambda}) - \frac{1}{N}\sqrt{\boldsymbol{\lambda}}\sqrt{\boldsymbol{\lambda}}^\mathsf{T}$ and $\mathrm{diag}(\boldsymbol{\lambda}) - \frac{1}{M}\sqrt{\boldsymbol{\lambda}}\sqrt{\boldsymbol{\lambda}}^\mathsf{T}$, respectively. But this is merely an application of Lemma 8.1.

This concludes the elaborate proof of Theorem 17.6. We now turn to the proper evaluation of this last power inference method, for the two system models studied so far. The first system model, Scenario (a), has the following characteristics: $K = 3$ sources, $P_1 = 1$, $P_2 = 3$, and $P_3 = 10$, $N = 60$ sensors, $M = 600$ samples, and $n_1 = n_2 = n_3 = 2$ antennas per transmit source, while for the second system model, Scenario (b): $K = 3$ sources, $P_1 = 1/16$, $P_2 = 1/4$, $N = 24$ sensors, $M = 128$ samples, and $n_1 = n_2 = n_3 = 4$ antennas per transmit source. The histogram and distribution function of the estimated powers for Scenario (b) are depicted in Figure 17.19 and Figure 17.20. Observe that this last estimator seems rather unbiased and very precise for all three powers under study.

In all previous approaches to the problem of power inference, we have assumed to this point that the number of simultaneous transmissions is known and that the number of antennas used by every transmitter is known. In the moment deconvolution approach, this has to be assumed either when inverting the

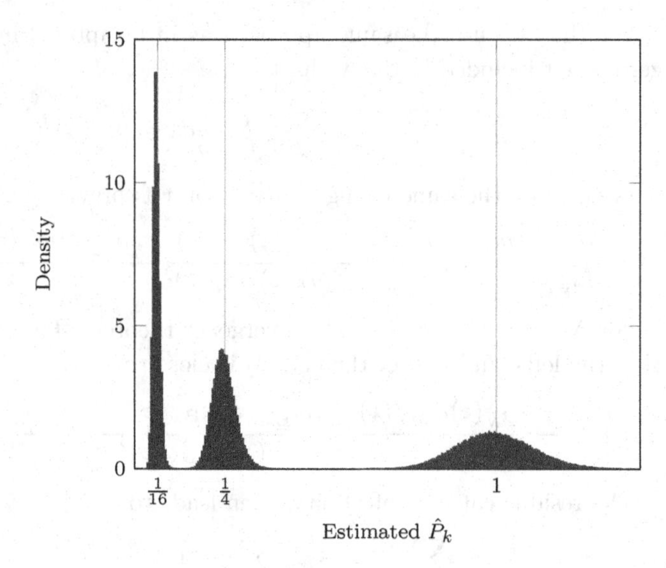

Figure 17.19 Histogram of the $\hat{P}_k$ for $k \in \{1, 2, 3\}$, $P_1 = 1/16$, $P_2 = 1/4$, $P_3 = 1$, $n_1 = n_2 = n_3 = 4$ antennas per user, $N = 24$ sensors, $M = 128$ samples, and SNR = 20 dB.

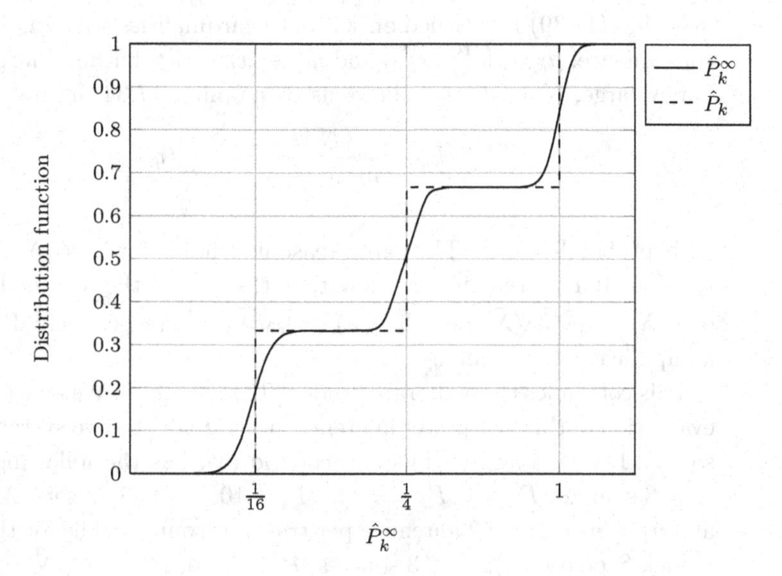

Figure 17.20 Distribution function of the estimator $\hat{P}_k$ for $k \in \{1, 2, 3\}$, $P_1 = 1/16$, $P_2 = 1/4$, $P_3 = 1$, $n_1 = n_2 = n_3 = 4$ antennas per user, $N = 24$ sensors, $M = 128$ samples, and SNR = 20 dB. Optimum estimator shown in dashed lines.

Newton–Girard formulas or when finding the maximum likelihood or minimum mean square error estimator for the transmit powers. As for the Stieltjes transform approach, this is required to determine which eigenvalues actually form a cluster. The same remark holds for the nest N- and M-consistent approach. It is therefore of prior importance to be first able to detect the number of simultaneous transmissions and the number of antennas per user. In the following, we will see that this is possible using *ad-hoc* tricks, although in most practical cases, more theoretical methods are required that are yet to be investigated.

17.2.6 Joint estimation of number of users, antennas and powers

It is obvious that the less is a priori known to the estimator, the less reliable estimation of the system parameters is possible. We discuss hereafter the problems linked to the absence of knowledge of some system parameters as well as what this entails from a cognitive radio point of view. Some further comments on the way to use the above estimators are also made.

- If both the number of transmit sources and the number of antennas per source are known prior to signal sensing, then all aforementioned methods will give more or less accurate estimates of the transmit powers. The accuracy depends in that case on whether transmit sources are sufficiently distinct from one another (depending on the cluster separability condition for Theorem 17.6) and on the efficiency of the algorithm used. From a cognitive radio viewpoint, that would mean that the secondary network is aware of the number of users exploiting a resource and of the number of antennas per user. It is in fact not necessary to know exactly how many users are currently transmitting, but only the maximum number of such users, as the sensing array would then always detect the maximum amount of users, some transmitting with null power. The assumption that the cognitive radio is aware of this maximal number of users per resource is therefore tenable. The assumption that the number of transmit antennas is known also makes sense if the primary communication protocols is known not to allow multiple antenna transmissions for instance. Note however that the overall performance in that case is rather degraded by the fact that single antenna transmissions do not provide much channel diversity. If this is so, it is reasonable for the sensing array to acquire more samples for different realizations of channel $\mathbf{H}$, which would take more time, or to be composed of numerous sensors, which might not be a realistic assumption.
- If the number of users is unknown, as discussed in the previous point, this might not be a dramatic issue on practical grounds if we can at least assume a maximal number of simultaneous transmissions. Typically, though, in a wideband CDMA network, a large number of users may simultaneously occupy a given frequency resource. If a cognitive radio is to operate on this frequency resource, it must then cope with the fact that a very large number of user transmit powers may need be estimated. Nonetheless, and rather fortunately, it is fairly untypical that all transmit users are found at the same location, close to the secondary network. The most remote users would in that case be hidden by thermal noise and the secondary network would then only need to deal with the closest users. Anyhow, if ever a large number of users is to be

found in the neighborhood of a cognitive radio, it is very unlikely that the frequency resource be reusable at all.

- If now the number of antennas per user is unknown, then more elaborate methods are demanded since this parameter is essential to all algorithms. Indeed, for both classical and Stieltjes transform approaches, we need to be able to distribute the empirical eigenvalues of $\frac{1}{M}\mathbf{Y}\mathbf{Y}^{\mathsf{H}}$ in several clusters, one for each source, the size of each cluster matching the number of antennas used by the transmitter. The same holds true for the exact inference method or the moment approach that both assume known power multiplicities. Among the methods to cope with this issue, we present below an *ad-hoc* suboptimal approach. We first assume for readability that we know the number K of transmit sources (taken large enough to cover all possible hypotheses), some having possibly a null number of transmit antenna. The approach consists in the following steps:

  1. we first identify a set of plausible hypotheses for $n_1, \ldots, n_K$. This can be performed by inferring clusters based on the spacing between consecutive eigenvalues: if the distance between neighboring eigenvalues is more than a threshold, then we add an entry for a possible cluster separation in the list of all possible positions of cluster separation. From this list, we create all possible K-dimensional vectors of eigenvalue clusters. Obviously, the choice of the threshold is critical to reduce the number of hypotheses to be tested;
  2. for each K-dimensional vector with assumed numbers of antennas $\hat{n}_1, \ldots, \hat{n}_K$, we use Theorem 17.6 in order to obtain estimates of the $\hat{P}_1, \ldots, \hat{P}_K$ (some being possibly null);
  3. based on these estimates, we compare the e.s.d. $F^{\mathbf{B}_N}$ of $\mathbf{B}_N$ to $\hat{F}$ defined as the l.s.d. of the matrix model $\hat{\mathbf{Y}} = \mathbf{H}\hat{\mathbf{P}}\mathbf{X} + \mathbf{W}$ with $\hat{\mathbf{P}}$ the diagonal matrix composed of $\hat{n}_1$ entries equal to $\hat{P}_1$, $\hat{n}_2$ entries equal to $\hat{P}_2$, etc. up to $\hat{n}_K$ entries equal to $\hat{P}_K$. The l.s.d. $\hat{F}$ is obtained from Theorem 17.5. The comparison can be performed based on different metrics. In the simulations carried hereafter, we consider as a metric the mean absolute difference between the Stieltjes transform of $F^{\mathbf{B}_N}$ and of $\hat{F}$ on the segment $[-1, -0.1]$.

  A more elaborate approach would consist in analyzing the second order statistics of $F^{\mathbf{B}_N}$, and therefore determining decision rules, such as hypothesis tests for every possible set $(K, n_1, \ldots, n_K)$.

Note that, when the number of antennas per user is unknown to the receiver and clusters can be clearly identified, another problem still occurs. Indeed, even if the clusters are perfectly disjoint, to this point in our study, the receiver has no choice but to assume that the cluster separability condition is always met and therefore that exactly as many users as visible clusters are indeed transmitting. If the condition is in fact not met, say the empirical eigenvalues corresponding to the p power values $P_i, \ldots, P_{i+(p-1)}$ are merged into a single cluster, i.e. with the notations of Section 17.2.5 $i_F = \ldots = (i+p-1)_F$, then applying the methods described above leads to an estimator of their mean $P_0 = \frac{1}{n_0}\sum_{k=0}^{p-1} n_{i+k}P_{i+k}$ with $n_0 = n_i + \ldots + n_{i+(p-1)}$ (since the integration contour encloses all power values or the link between moments and $P_1, \ldots, P_K$ takes into account the assumed eigenvalue multiplicities), instead of an estimator of their individual values. In this case, the receiver can therefore only declare that a given estimate $\hat{P}_0$ corresponds either to a single transmit source with dimension n_0 or to multiple transmit sources of cumulated dimension n_0 with average transmit power P_0, well approximated by $\hat{P}_0$. For practical blind detection purposes in cognitive radios, this leads the secondary network to infer a number of transmit entities that is less than the effective number of transmitters. In general, this would not have serious consequences on the decisions made by the secondary network but this might at least reduce the capabilities of the secondary network to optimally overlay the licensed spectrum. To go past this limitation, current investigations are performed to allow multiple eigenvalue estimations within a given cluster of eigenvalues. This can be performed by studying the second order statistics of the estimated powers.

17.2.7 Performance analysis

17.2.7.1 Method comparison

We first compare the conventional method against the Stieltjes transform approach for Scenario (a). Under the hypotheses of this scenario, the ratios c and c_0 equal 10, leading therefore the conventional detector to be close to unbiased. We therefore suspect that the normalized mean square error (NMSE) performance in the estimation of the powers for both detectors is alike. This is described in Figure 17.21, which suggests as predicted that in the high SNR regime (when cluster separability is reached) the conventional estimator performs similar to the Stieltjes transform method. However, it appears that a 3 dB gain is achieved by the Stieltjes transform method around the position where cluster separability is no longer satisfied. This is due to the fact that, when subsequent clusters tend to merge as σ^2 increases, the Stieltjes transform method manages to track the position of the powers P_k while the conventional method keeps assuming each P_k is located at the center of gravity of cluster k_F. This observation is very similar to that made in [Mestre and Lagunas, 2008] where the improved G-MUSIC estimator pushes further the SNR limit where the performance of the classical MUSIC estimator starts to decay significantly.

We now consider Scenario (b). We first compare the performance of the conventional, Stieltjes transform and moment estimators for a SNR of 20 dB. In order to compare techniques with similar computational complexity, we use Newton–Girard inversion formulas to retrieve the powers from the estimated moments. Figure 17.22 depicts the distribution function of the estimated powers in logarithmic scale. The Stieltjes transform method appears here to be very precise and seemingly unbiased. In contrast, the conventional method, with a

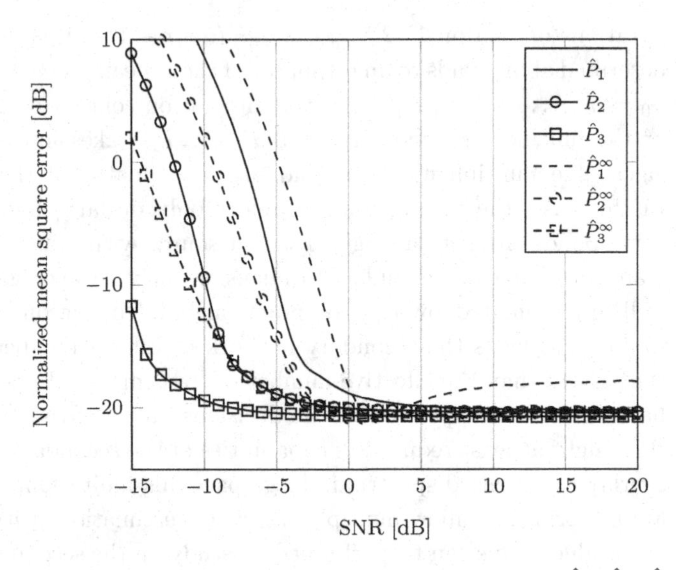

Figure 17.21 Normalized mean square error of individual powers $\hat{P}_1$, $\hat{P}_2$, $\hat{P}_3$, $P_1 = 1, P_2 = 3, P_3 = 10$, $n_1/n = n_2/n = n_3/n = 1/3$,$n/N = N/M = 1/10$, $n = 6$. Comparison between the conventional and Stieltjes transform approaches.

slightly smaller variance shows a large bias as was anticipated. As for the moment method, it shows rather accurate performance for the stronger estimated power, but proves very inaccurate for smaller powers. This follows from the inherent shortcomings of the moment method. The performance of the estimator $\hat{P}'_k$ will be commented on later.

We then focus on the estimate of the larger power P_3 and take now the SNR to range from -15 to 30 dB under the same conditions as previously and for the same estimators. The NMSE for the estimators of P_3 is depicted in Figure 17.23. The curve marked with squares will be commented on in the next section. As already observed in Figure 17.22, in the high SNR regime, the Stieltjes transform estimator outperforms both alternative methods. We also notice the SNR gain achieved by the Stieltjes transform approach with respect to the conventional method in the low SNR regime, as already observed in Figure 17.21. However, it now turns out that in this low SNR regime, the moment method is gaining ground and outperforms both cluster-based methods. This is due to the cluster separability condition, which is not a requirement for the moment approach. This indicates that much can be gained by the Stieltjes transform method in the low SNR regime if a more precise treatment of overlapping clusters is taken into account.

17.2.7.2 Joint estimation of K, n_k, P_k

So far, we have assumed that the number of users K and the numbers of antennas per user $n_1, \ldots, n_K$ were perfectly known. As discussed previously, this may not be a strong assumption if it is known in advance how many antennas are

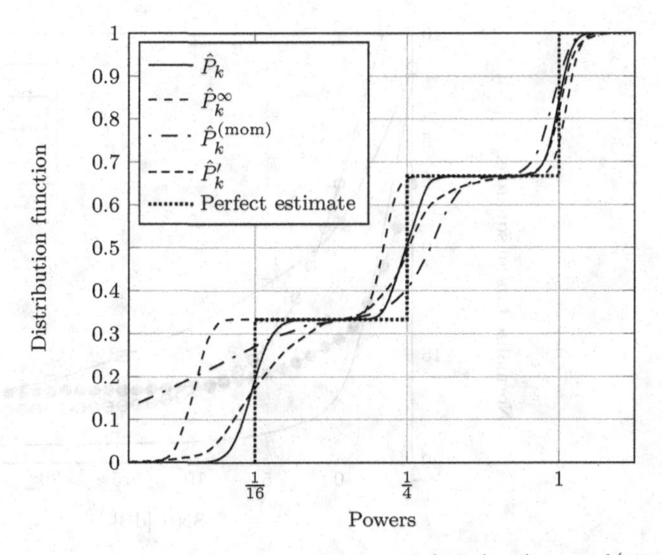

Figure 17.22 Distribution function of the estimators $\hat{P}_k^{\infty}$, $\hat{P}_k$, $\hat{P}'_k$ and $\hat{P}_k^{(\mathrm{mom})}$ for $k \in \{1, 2, 3\}$, $P_1 = 1/16$, $P_2 = 1/4$, $P_3 = 1$, $n_1 = n_2 = n_3 = 4$ antennas per user, $N = 24$ sensors, $M = 128$ samples and SNR $= 20$ dB. Optimum estimator shown in dotted line.

systematically used by every source or if another mechanism, such as in [Chung *et al.*, 2007], can provide this information. Nonetheless, these are in general strong assumptions. Based on the *ad-hoc* method described above, we therefore provide the performance of our novel Stieltjes transform method in the high SNR regime when only n is known; this assumption is less stringent since in the medium to high SNR regime we can easily decide which eigenvalues of $\mathbf{B}_N$ belong to the cluster associated with σ^2 and which eigenvalues do not. We denote $\hat{P}'_k$ the estimator of P_k when K and $n_1, \ldots, n_K$ are unknown. We assume for this estimator that all possible combinations of 1 to 3 clusters can be generated from the $n = 6$ observed eigenvalues in Scenario (a) and that all possible combinations of 1 to 3 clusters with even cluster size can be generated from the $n = 12$ eigenvalues of $\mathbf{B}_N$ in Scenario (b). For Scenario (a), the NMSE performance of the estimators $\hat{P}_k$ and $\hat{P}'_k$ is proposed in Figure 17.24 for the SNR ranging from 5 dB to 30 dB. For Scenario (b), the distribution function of the inferred $\hat{P}'_k$ is depicted in Figure 17.22, while the NMSE performance for the inference of P_3 is proposed in Figure 17.23; these are both compared against the conventional, moment, and Stieltjes transform estimators. We also indicate in Table 17.1 the percentage of correct estimations of the triplet (n_1, n_2, n_3) for both Scenario (a) and Scenario (b). In Scenario (a), this amounts to 12 such triplets that satisfy $n_k \geq 0$, $n_1 + n_2 + n_3 = 6$, while, in Scenario (b), this corresponds to 16 triplets that satisfy $n_k \in 2\mathbb{N}$, $n_1 + n_2 + n_3 = 12$. Observe that the noise variance, assumed to be known a priori in this case, plays an important role with respect to the statistical inference of the n_k. In Scenario (a), for a

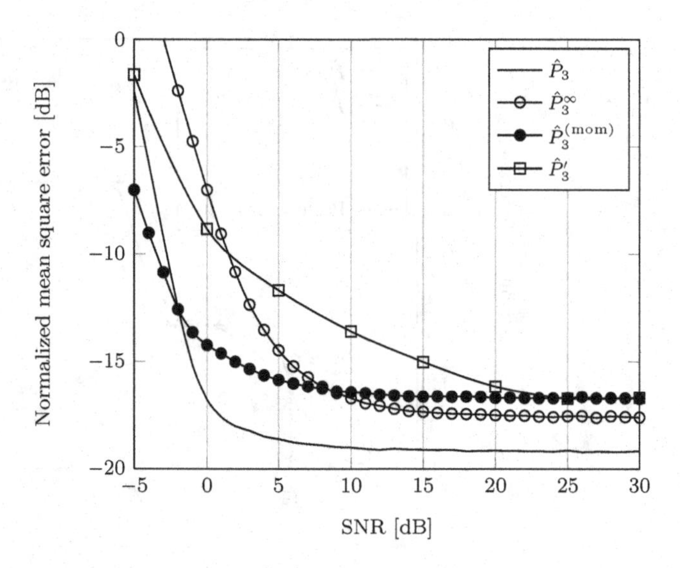

Figure 17.23 Normalized mean square error of largest estimated power P_3, $P_1 = 1/16, P_2 = 1/4, P_3 = 1$, $n_1 = n_2 = n_3 = 4$,$N = 24$, $M = 128$. Comparison between conventional, moment, and Stieltjes transform approaches.

SNR	RCI (a)	RCI (b)
5 dB	0.8473	0.1339
10 dB	0.9026	0.4798
15 dB	0.9872	0.4819
20 dB	0.9910	0.5122
25 dB	0.9892	0.5455
30 dB	0.9923	0.5490

Table 17.1. Rate of correct inference (RCI) of the triplet (n_1, n_2, n_3) for scenarios (a) and (b).

SNR greater than 15 dB, the correct hypothesis for the n_k is almost always taken and the performance of the estimator is similar to that of the optimal estimator. In Scenario (b), the detection of the exact cluster separation is less accurate and the performance for the inference of P_3 saturates at high SNR to −16 dB of NMSE, against −19 dB when the exact cluster separation is known. It therefore seems that, in the high SNR regime, the performance of the Stieltjes transform detector is loosely affected by the absence of knowledge about the cluster separation. This statement is also confirmed by the distribution function of $\hat{P}_k'$ in Figure 17.22, which still outperforms the conventional and moment methods. We underline again here that this is merely the result of an *ad-hoc* approach; this performance could be greatly improved if, e.g. more was known about the second order statistics of $F^{\mathbf{B}_N}$.

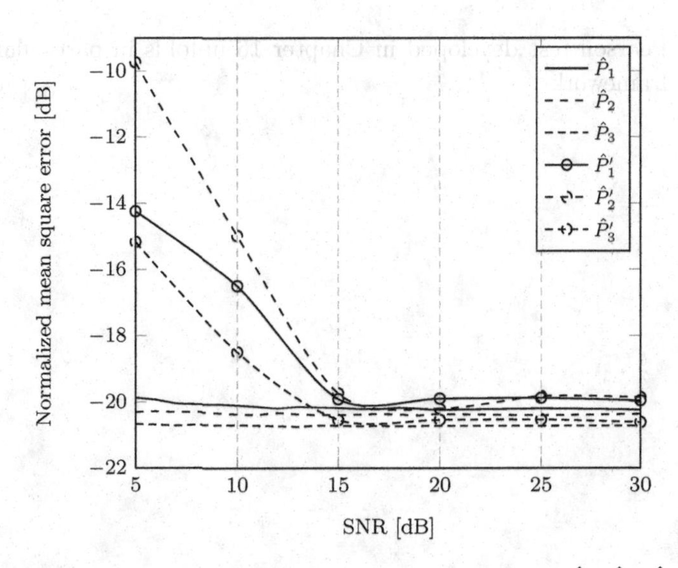

Figure 17.24 Normalized mean square error of individual powers $\hat{P}_1$, $\hat{P}_2$, $\hat{P}_3$ and $\hat{P}'_1$, $\hat{P}'_2$, $\hat{P}'_3$, $P_1 = 1, P_2 = 3, P_3 = 10$, $n_1/n = n_2/n = n_3/n = 1/3$,$n/N = N/M = 1/10$, $n = 6$, 10 000 simulation runs.

This concludes the present chapter on eigen-inference methods using large dimensional random matrices. Note that, to obtain the above estimators, a strong mathematical effort was put into the macroscopic analysis of the asymptotic spectra for rather involved random matrix models as well as in the microscopic analysis of the behavior of individual eigenvalues and eigenvectors. We believe that much more emphasis will be cast in the near future on G-estimation for other signal processing and wireless communication issues. The main limitation today to further develop multi-dimensional consistent estimators is that only few models have been carefully studied. In particular, we mentioned repeatedly in Part I the sample covariance matrix model and the spiked model, for which we have a large number of results concerning exact spectrum separation, limit distributions of the extreme eigenvalues in the Gaussian case, etc. When it comes to slightly more elaborate models, such as the information plus noise model, even the result on exact separation is yet unproved in the general i.i.d. case (with obviously some moment assumptions). Only the exact separation result for the information plus noise model with uncorrelated Gaussian noise is known. Nothing is yet known about limit distribution of the extreme eigenvalues. There is therefore a wide range of new results to come along with the deeper study of such random matrix models.

In the next chapter, we will deal with a more specific application of random matrices to the problem of channel modeling. This problem enters a much wider framework of statistical inference based on Bayesian probability theory and the maximum entropy principle [Jaynes, 1957a,b]. The introduction of the Neyman–

Pearson test developed in Chapter 16 unfolds in particular from this Bayesian framework.

18 System modeling

Channel modeling is a fundamental field of research, as all communication models that were used so far in this book as well as in the whole mobile communication literature are based on such models. These models are by definition meant to provide an adequate representation of real practical communication environments. As such, i.i.d. Gaussian channel models are meant to represent the most scattered environment possible where multiple waveforms reflecting on a large number of scattering objects add up non-coherently and independently on the receive antennas. From this point of view, the Gaussian assumption is due to a loose application of the law of large numbers. Due to the mobility of the communication devices in the propagation environment, the statistical disposition of scatterers is rather uniform, hence the i.i.d. property. This is basically the physical arguments for using Gaussian i.i.d. models, along with confirmation by field measurements. An alternative explanation for Gaussian channel models will be provided in this chapter, which accounts for a priori information available at the system modeler, rather than for physical interpretations.

However, the complexity of propagation environments call for more involved models. This is why the Kronecker model comes into play to adequately model correlations arising at either communication end, supposedly independent from one another. This is also why the Rician model is of common use, as it can take into account possible line-of-sight components in the channel, which can often be assumed of constant fading value for a rather long time period, compared to the fast varying scattering part of the channel. Those are channels usually considered in the scientific literature for their overall simplicity, but whose accuracy is somewhat disputed [Ozcelik *et al.*, 2003]. Nonetheless, the literature is also full of alternative models, mostly based on field tests. After fifty years of modern telecommunications, it is still unclear which models to value, which models are relevant and for what reasons. This chapter intends first to propose a joint information-theoretic framework for telecommunications that encompasses many topics such as channel modeling, source sensing, parameter estimation, etc. from a common probability-theoretic basis. This enables channel modeling to be seen, no longer as a remote empirical field of research, but rather as a component of a larger information-theoretic framework. More details about the current literature on channel modeling can be found in [Almers *et al.*, 2007]. The role of random matrix theory in this field will become clear when we address the question of finding the maximum entropy distribution of random matrix models.

Before we proceed with the channel modeling problem, explored through several articles by Debbah, Müller, and Guillaud mainly [de Lacerda Neto *et al.*, 2006; Debbah and Müller, 2005; Guillaud *et al.*, 2006], we need to quickly recall the theoretical foundation of Bayesian probability theory and the maximum entropy principle.

18.1 Introduction to Bayesian channel modeling

The ultimate objective of the field of channel modeling is to provide us with probabilistic models that are consistent with the *randomness observed in actual communication channels.* It therefore apparently makes sense to probe realistic channels in order to infer a general model by "sampling" the observed randomness. It also makes sense to come up with simple mathematical models that (i) select only the essential features of the channels, (ii) build whatever simple probability distribution around them, and (iii) can be compared to actual channel statistics to see if the unknown random part of the effective channel is well approximated by the model. In our opinion, though, these widely spread approaches suffer from a severe misconception of the word *randomness.* From a frequentist probability point of view, randomness reflects the unseizable character that makes particular events different every time they are observed. For instance, we may think of a coin toss as being a physical action ruled by chance, as if it were obeying no law of nature. This is however conceivably absurd when we consider the same toss coin played in slow motion in such a way that the observer can actually compute from the observed initial rotation speed, height, and other parameters, such as air friction, the exact outcome of the experiment in advance. Playing a toss coin in slow motion removes its random part. A more striking example, that does not require to modify the conditions of the experiment, is that of consecutive withdrawals of balls concealed in identical boxes with same initial content. Assume that we know prior to the first withdrawal from the first box that some balls are red, the others being blue. From the frequentist conception of randomness, the probability of drawing a red ball at first, second, or third withdrawal (the nth withdrawal is done from the nth box on the line) is identical since the boxes are all identical in content and the successive *physical events* are the same. But this is clearly untrue. Without any need for advanced mathematics, if we are told that the first box contains red and blue balls (but not the exact number), we must infer in total honesty that the probability of getting a blue ball is somewhere around one half. However, if after a hundred successive withdrawals, all selected balls turned out to be red, we would reasonably think that the probability for the hundred-first ball withdrawn to be blue is much lower than first anticipated.

From the above examples, it is more sensitive to see *randomness*, not as the result of unruled chance, but rather as the result of a lack of knowledge. The more we know about an event, the more confidence we have about the outcome of this event. This is the basis of Bayesian probability theory. The *confidence* factor here is what we wish to call the probability of the event. This conception of probabilities is completely anchored in our everyday life where our decisions are based on our knowledge and appreciation of the probability of possible events. In the same way, communication channels, be they often called the *environment*, as if we did not have any control over them, can be reduced to the knowledge we have about them. If we know the communication channels at all times, i.e. channels for which we often coin the phrase "perfect channel state information," then the channel is no longer conceived as random. If we do not have perfect channel state information at all times, though, then the channel is random in the sense that it is one realization of all possible channels consistent with the reduced knowledge we have on this channel.

Modeling channels under imperfect state information therefore consists in providing a probability distribution for all such channels consistent with this restricted information. From a *confidence* point of view, this means we must assign to each potential channel realization a degree of confidence in such a realization. This degree of confidence must be computed by taking into account only the information available, and by discarding as much as possible all supplementary unwanted hypotheses. It has been proved, successively by Cox [Cox, 1946], Shannon [Shannon, 1948], Jaynes [Jaynes, 1957a,b], and Shore and Johnson [Shore and Johnson, 1980] that, under a reasonable axiomatic definition of the information-theoretic key notion of *ignorance* (preferably referred to as *information content* by Shannon), the most *non-committal* way to attribute degrees of confidence of possible realizations of an event is to assign to it the probability distribution that has maximal entropy, among all probability distributions consistent with the prior knowledge. That is, the process of assigning degrees of confidence to a parameter x, given some information I, consists in the following recipe:

- Among all probability distributions for x, discard those that are inconsistent with I. For instance, if I contains information about the statistical mean of x, then all probability distributions that have different means must be discarded;
- among the remaining set $\mathcal{S}_I$ of such probability distributions, select the one which maximizes the entropy, i.e. calling $p^\star$ this probability distribution, we have:

$$p^\star \triangleq \arg\max_{p \in \mathcal{S}_I} \int p(x) \log(p(x)) dx.$$

This principle is referred to as the *maximum entropy principle* [Jaynes, 2003], which is widely spread in the signal processing community [Bretthorst, 1987], in econometrics [Zellner, 1971], engineering [Kapur, 1989], and in general science [Brillouin, 1962] in spite of several century-old philosophical divisions inside the community. Since such philosophical debates are nowhere near the target of this section, let alone the subject of this book, we claim from now on and without further justification that the maximum entropy principle is the most reliable tool to build statistical models for parameters regarding which limited information is available.

In the next section, we address the channel modeling question through a Bayesian and maximum entropy point of view.

18.2 Channel modeling under environmental uncertainty

Fast varying communication channels are systems for which a full parametrical description is lacking to the observer. Since these channels are changing too fast in time to be adequately tracked by communication devices without incurring too much information feedback, only limited inexpensive information is usually collected. For instance, the most trivial quantities that we can collect without effort on the successive channels is their empirical mean and their empirical variance (or covariance matrix in the case of multiple antenna communications). Assuming channel stationarity (at least in the wide sense), it is then possible to propose a consistent non-committal model for the channel at hand, by following the steps of the maximum entropy principle. In this particular case where mean and variance are known, it can in fact be shown that the channel model under consideration is exactly the Gaussian matrix channel. Some classical channel models such as the correlated Gaussian model and the Kronecker model were recovered in [Debbah and Müller, 2005] and [Guillaud *et al.*, 2006], while some new channel maximum entropy-consistent models were also derived. In the following, we detail the methodology used to come up with these models when statistical knowledge is available to the system modeler, similar to the ideas developed in [Franceschetti *et al.*, 2003]. In [Debbah and Müller, 2005], the maximum entropy principle is used also when deterministic knowledge is available at the system modeler. Both problems lead to mathematically different approaches, problems of the former type being rather easy to treat, while problems of latter type are usually not tractable. We only deal in this chapter with problems when statistical information about the channel is available.

Let us consider the multiple antenna wireless channel with n_t transmit and n_r receive antennas. We assume narrowband transmission so that the MIMO communication channels to be modeled are non-frequency selective. Let the complex scalar coefficient $h_{i,j}$ denote the channel attenuation between the transmit antenna j and the receive antenna i, $j \in \{1, \ldots, n_t\}$, $i \in \{1, \ldots, n_r\}$. Let $\mathbf{H}^{(t)}$ denote the $n_r \times n_t$ channel matrix at time instant t. We recall the general model for a time-varying flat-fading channel with additive noise

$$\mathbf{y}^{(t)} = \mathbf{H}^{(t)}\mathbf{x}^{(t)} + \mathbf{w}^{(t)} \tag{18.1}$$

where the noise vector $\mathbf{w}^{(t)} \in \mathbb{C}^{n_r}$ at time t is modeled as a complex circularly symmetric Gaussian random variable with i.i.d. coefficients (in compliance with maximum entropy requirements) and $\mathbf{x}^{(t)} \in \mathbb{C}^{n_t}$ denotes the transmit data vector at time t. In the following, we focus on the derivation of the fading characteristics of $\mathbf{H}^{(t)}$. When we are not concerned with the time-related properties of $\mathbf{H}^{(t)}$, we will drop the time index t, and refer to the channel realization $\mathbf{H}$ or equivalently to its vectorized notation $\mathbf{h} \triangleq \text{vec}(\mathbf{H}) = (h_{1,1}, \ldots, h_{n_r,1}, h_{1,2}, \ldots, h_{n_r,n_t})^\mathsf{T}$. Let us also denote $N \triangleq n_r n_t$ and map the antenna indices into $\{1 \ldots, N\}$; that is, $\mathbf{h} = (h_1, \ldots, h_N)^\mathsf{T}$.

18.2.1 Channel energy constraints

18.2.1.1 Average channel energy constraint

In this section, we recall the results of [Debbah and Müller, 2005] where an entropy-maximizing probability distribution is derived for the case where the average energy carried through a MIMO channel is known deterministically. This probability distribution is obtained by maximizing the entropy

$$\int_{\mathbb{C}^N} -\log(P_\mathbf{H}(\mathbf{H}))P_\mathbf{H}(\mathbf{H})d\mathbf{H}$$

under the only assumption that the channel has a finite average energy NE_0 and the normalization constraint associated with the definition of a probability density, i.e.

$$\int_{\mathbb{C}^N} \|\mathbf{H}\|_F^2 P_\mathbf{H}(\mathbf{H})d\mathbf{H} = NE_0 \tag{18.2}$$

with $\|\mathbf{H}\|_F$ the matrix Frobenius norm and

$$\int_{\mathbb{C}^N} P_\mathbf{H}(\mathbf{H})d\mathbf{H} = 1.$$

This is achieved through the method of Lagrange multipliers, by writing

$$L(P_\mathbf{H}) = -\int_{\mathbb{C}^N} \log(P_\mathbf{H}(\mathbf{H}))P_\mathbf{H}(\mathbf{H})d\mathbf{H} + \beta\left[1 - \int_{\mathbb{C}^N} P_\mathbf{H}(\mathbf{H})d\mathbf{H}\right] + \gamma\left[NE_0 - \int_{\mathbb{C}^N} \|\mathbf{H}\|_F^2 P_\mathbf{H}(\mathbf{H})d\mathbf{H}\right]$$

where we introduce the scalar Lagrange coefficients β and γ, and by taking the functional derivative [Fomin and Gelfand, 2000] with respect to $P_\mathbf{H}$ equal to zero

$$\frac{\delta L(P_\mathbf{H})}{\delta P_\mathbf{H}} = 0.$$

This functional derivative takes the form of an integral over $\mathbf{H}$, which is in particular identically null if the integrand is null for all $\mathbf{H}$. We pick this one solution and therefore write

$$-\log(P_\mathbf{H}(\mathbf{H})) - 1 - \beta - \gamma\|\mathbf{H}\|_F^2 = 0$$

for all $\mathbf{H}$.

The latter equation yields

$$P_{\mathbf{H}}(\mathbf{H}) = \exp\left(-(\beta+1) - \gamma \|\mathbf{H}\|_F^2\right)$$

and the normalization of this distribution according to (18.2) finally allows us to compute the coefficients β and γ. Observing in particular that $\beta = -1$ and $\gamma = \frac{1}{E_0}$ are consistent with the initial constraints, the final distribution is given by:

$$P_{\mathbf{H}|E_0}(\mathbf{H}) = \frac{1}{(\pi E_0)^N} \exp\left(-\sum_{i=1}^{N} \frac{|h_i|^2}{E_0}\right). \tag{18.3}$$

Interestingly, the distribution defined by (18.3) corresponds to a complex Gaussian random variable with independent fading coefficients, although neither Gaussianity nor independence were among the initial constraints. Via the maximum entropy principle, these properties are the consequence of the ignorance of the modeler regarding any constraint other than the total average energy NE_0.

18.2.1.2 Probabilistic average channel energy constraint

Let us now introduce a new model for situations where the channel model defined in the previous section applies locally in time but where E_0 cannot be expected to be constant, e.g. due to short-term shadowing. Therefore, let us replace E_0 in (18.3) by the random quantity E known only through its p.d.f. $P_E(E)$. In this case, the p.d.f. of the channel $\mathbf{H}$ can be obtained by marginalizing over E, as follows.

$$P_{\mathbf{H}}(\mathbf{H}) = \int_0^\infty P_{\mathbf{H},E}(\mathbf{H}, E)dE = \int_0^\infty P_{\mathbf{H}|E}(\mathbf{H})P_E(E)dE. \tag{18.4}$$

In order to determine the probability distribution P_E, let us find the maximum entropy distribution under the constraints:

- $0 \leq E \leq E_{\max}$, where $E_{\max}$ represents an absolute constraint on the power carried through the channel;
- the mean $E_0 \triangleq \int_0^{E_{\max}} EP_E(E)dE$ is known.

Applying once more the Lagrange multiplier method, we introduce the scalar unknowns β and γ, and maximize the functional

$$L(P_E) = -\int_0^{E_{\max}} \log(P_E(E))P_E(E)dE + \beta\left[\int_0^{E_{\max}} EP_E(E)dE - E_0\right]$$
$$+\gamma\left[\int_0^{E_{\max}} P_E(E)dE - 1\right].$$

Equating the derivative to zero

$$\frac{\partial L(P_E)}{\partial P_E} = 0$$

and picking the solution corresponding to taking all integrand terms of the resulting integral to be identically zero yields

$$P_E(E) = \exp\left(\beta E - 1 + \gamma\right)$$

and the Lagrange multipliers are finally eliminated by solving the normalization equations

$$\int_0^{E_{\max}} E \exp\left(\beta E - 1 + \gamma\right) dE = E_0$$

$$\int_0^{E_{\max}} \exp\left(\beta E - 1 + \gamma\right) dE = 1.$$

The Lagrangian multiplier $\beta < 0$ is then the solution to the implicit equation

$$E_{\max} e^{\beta E_{\max}} - \left(\frac{1}{\beta} + E_0\right)\left(e^{\beta E_{\max}} - 1\right) = 0 \tag{18.5}$$

and finally P_E is obtained as the truncated exponential law

$$P_E(E) = \frac{\beta}{\exp(\beta E_{\max}) - 1} e^{\beta E}$$

for $0 \leq E \leq E_{max}$ and $P_E(E) = 0$ elsewhere. Note that taking $E_{\max} = \infty$ in (18.5) yields $\beta = -\frac{1}{E_0}$ and the classical exponential law

$$P_E(E) = E_0 e^{-\frac{E}{E_0}}.$$

The final maximum entropy model for $P_{\mathbf{H}}$ is then:

$$P_{\mathbf{H}}(\mathbf{H}) = \int_0^{E_{\max}} \frac{1}{(\pi E)^N} \frac{\beta \exp(\beta E)}{\exp(\beta E_{\max}) - 1} \exp\left(-\sum_{i=1}^{N} \frac{|h_i|^2}{E}\right) dE.$$

18.2.1.3 Application to the single antenna channel

In order to illustrate the difference between the two situations presented so far, let us investigate the single input single output (SISO) case $n_t = n_r = 1$ where the channel is represented by a single complex scalar h. Furthermore, since the distribution is circularly symmetric, it is more convenient to consider the distribution of $r \triangleq |h|$. After the change of variables $h \triangleq r(\cos\theta + i\sin\theta)$, and marginalization over θ, (18.3) becomes

$$P_r(r) = \frac{2r}{E_0} e^{-\frac{r^2}{E_0}} \tag{18.6}$$

whereas (18.4) yields

$$P_r(r) = \int_0^{E_{\max}} \frac{\beta}{e^{\beta E_{\max}} - 1} \frac{2r}{E} e^{\beta E - \frac{r^2}{E}} dE. \tag{18.7}$$

Note that the integral always exists since $\beta < 0$. Figure 18.1 depicts the p.d.f. of r under known energy constraint ((18.6), with $E_0 = 1$) and the known energy distribution constraint ((18.7) is computed numerically, for $E_{\max} = 2$ and

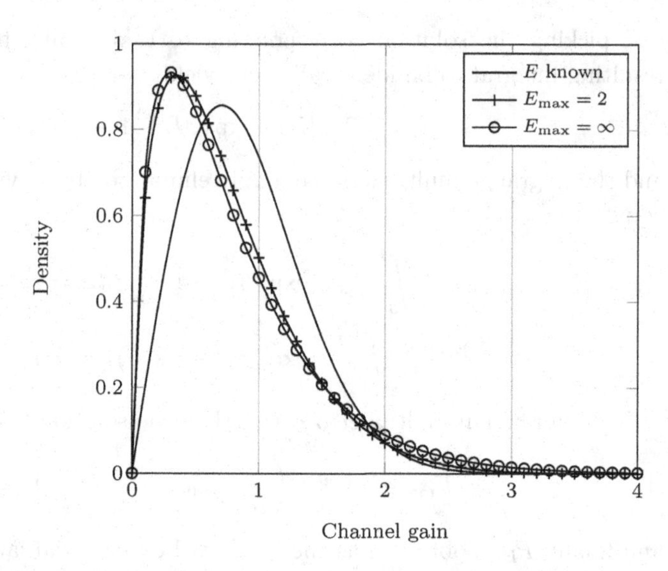

Figure 18.1 Amplitude distribution of the maximum entropy SISO channel models, for $E_0 = 1$, $E_{\max} \in \{1, 2, \infty\}$.

$E_{\max} = \infty$, taking $E_0 = 1$). Figure 18.2 depicts the distribution function of the corresponding instantaneous mutual information $C_{\mathrm{SISO}}(r) \triangleq \log_2(1 + \frac{1}{\sigma^2}r^2)$ for a signal-to-noise ratio $\frac{1}{\sigma^2}$ of 15 dB. The lowest range of the d.f. is of particular interest for wireless communications since it indicates the probability of a channel outage for a given transmission rate. The curves clearly show that the models corresponding to the unknown energy have a lower outage probability than the Gaussian channel model.

We now consider the more involved scenario of channels with known spatial correlation at either communication end. We will provide the proofs in their complete versions as they are instrumental to the general manipulation of Jacobian determinants, marginal eigenvalue distribution for small dimensional matrices, etc. and as they provide a few interesting tools to deal with matrix models with unitary invariance.

18.2.2 Spatial correlation models

In this section, we will incorporate several states of knowledge about the spatial correlation characteristics of the channel in the framework of maximum entropy modeling. We first study the case where the correlation matrix is deterministic and subsequently extend the result to an unknown covariance matrix.

18.2.2.1 Deterministic knowledge of the correlation matrix

In this section, we establish the maximum entropy distribution of $\mathbf{H}$ under the assumption that the covariance matrix $\mathbf{Q} \triangleq \int_{\mathbb{C}^N} \mathbf{h}\mathbf{h}^{\mathsf{H}} P_{\mathbf{H}|\mathbf{Q}}(\mathbf{H}) d\mathbf{H}$ is known, where $\mathbf{Q}$ is a $N \times N$ complex Hermitian matrix. Each component of the

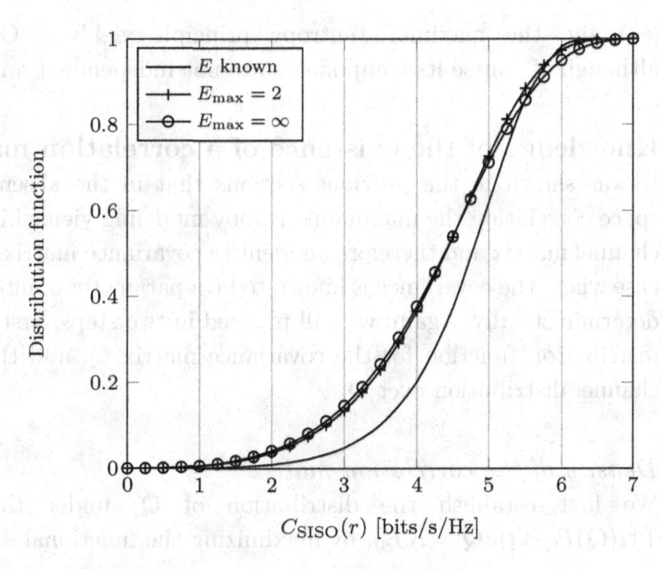

Figure 18.2 Mutual information distribution for maximum entropy SISO channel models, when $E_0 = 1$, $E_{\max} \in \{1, 2, \infty\}$, SNR of 15 dB.

covariance constraint represents an independent linear constraint of the form

$$\int_{\mathbb{C}^N} h_a h_b^* P_{\mathbf{H}|\mathbf{Q}}(\mathbf{H}) d\mathbf{H} = q_{a,b}$$

for $(a, b) \in \{1, \dots, N\}^2$. Note that this constraint makes any previous energy constraint redundant since $\int_{\mathbb{C}^N} \|\mathbf{H}\|_F^2 P_{\mathbf{H}|\mathbf{Q}}(\mathbf{H}) d\mathbf{H} = \operatorname{tr} \mathbf{Q}$. Proceeding along the lines of the method exposed previously, we introduce N^2 Lagrange coefficients $\alpha_{a,b}$, and maximize

$$L(P_{\mathbf{H}|\mathbf{Q}}) = \int_{\mathbb{C}^N} -\log(P_{\mathbf{H}|\mathbf{Q}}(\mathbf{H})) P_{\mathbf{H}|\mathbf{Q}}(\mathbf{H}) d\mathbf{H} + \beta \left[1 - \int_{\mathbb{C}^N} P_{\mathbf{H}|\mathbf{Q}}(\mathbf{H}) d\mathbf{H}\right]$$
$$+ \sum_{\substack{a \in \{1,\dots,N\} \\ b \in \{1,\dots,N\}}} \alpha_{a,b} \left[\int_{\mathbb{C}^N} h_a h_b^* P_{\mathbf{H}|\mathbf{Q}}(\mathbf{H}) d\mathbf{H} - q_{a,b}\right].$$

Denoting $\mathbf{A} \in \mathbb{C}^{N \times N}$ the matrix with (a, b) entry $\alpha_{a,b}$ and equating the derivative of the Lagrangian to zero, one solution satisfies

$$-\log(P_{\mathbf{H}|\mathbf{Q}}(\mathbf{H})) - 1 - \beta - \mathbf{h}^{\mathsf{T}} \mathbf{A} \mathbf{h}^* = 0. \tag{18.8}$$

Therefore we take

$$P_{\mathbf{H}|\mathbf{Q}}(\mathbf{H}) = \exp\left(-(\beta + 1) - \mathbf{h}^{\mathsf{T}} \mathbf{A} \mathbf{h}^*\right)$$

which leads, after elimination of the Lagrange coefficients through proper normalization, to

$$P_{\mathbf{H}|\mathbf{Q}}(\mathbf{H}, \mathbf{Q}) = \frac{1}{\det(\pi \mathbf{Q})} \exp\left(-\mathbf{h}^{\mathsf{H}} \mathbf{Q}^{-1} \mathbf{h}\right). \tag{18.9}$$

Again, the maximum entropy principle yields a Gaussian distribution, although of course its components are not independent anymore.

18.2.2.2 Knowledge of the existence of a correlation matrix

It was shown in the previous sections that in the absence of information on space correlation the maximum entropy modeling yields i.i.d. coefficients for the channel matrix and therefore an identity covariance matrix. We now consider the case where the covariance is known to be a parameter of interest but is not known deterministically. Again, we will proceed in two steps, first seeking a probability distribution function for the covariance matrix $\mathbf{Q}$, and then marginalizing the channel distribution over $\mathbf{Q}$.

Density of the correlation matrix.
We first establish the distribution of $\mathbf{Q}$, under the energy constraint $\int \operatorname{tr}(\mathbf{Q}) P_{\mathbf{Q}}(\mathbf{Q}) d\mathbf{Q} = NE_0$, by maximizing the functional

$$L(P_{\mathbf{Q}}) = \int_{\mathcal{S}} -\log(P_{\mathbf{Q}}(\mathbf{Q}))P_{\mathbf{Q}}(\mathbf{Q})d\mathbf{Q} + \beta\left[\int_{\mathcal{S}} P_{\mathbf{Q}}(\mathbf{Q})d\mathbf{Q} - 1\right] + \gamma\left[\int_{\mathcal{S}} \operatorname{tr}(\mathbf{Q})P_{\mathbf{Q}}(\mathbf{Q})d\mathbf{Q} - NE_0\right]. \tag{18.10}$$

Due to their structure, covariance matrices are restricted to the space $\mathcal{S}$ of $N \times N$ non-negative definite complex Hermitian matrices. Therefore, let us perform the variable change to the eigenvalues and eigenvectors space as was performed in Chapter 2 and in more detail in Chapter 16. Specifically, denote $\mathbf{\Lambda} \triangleq \operatorname{diag}(\lambda_1 \ldots \lambda_N)$ the diagonal matrix containing the eigenvalues $\lambda_1, \ldots, \lambda_N$ of $\mathbf{Q}$ and let $\mathbf{U}$ be the unitary matrix containing the associated eigenvectors, such that $\mathbf{Q} = \mathbf{U}\mathbf{\Lambda}\mathbf{U}^{\mathsf{H}}$.

We use the mapping between the space of complex $N \times N$ self-adjoint matrices (of which $\mathcal{S}$ is a subspace) and $\mathcal{U}(N)/T \times \mathbb{R}_{\leq}^{N}$, where $\mathcal{U}(N)/T$ denotes the space of unitary $N \times N$ matrices with first row composed of real non-negative entries, and $\mathbb{R}_{\leq}^{N}$ is the space of real N-tuples with non-decreasing components (see Lemma 4.4.6 of [Hiai and Petz, 2006]). The positive semidefinite property of the covariance matrices further restricts the components of $\mathbf{\Lambda}$ to non-negative values, and therefore $\mathcal{S}$ maps into $\mathcal{U}(N)/T \times \mathbb{R}_{\leq}^{+N}$.

Let us now define the function F over $\mathcal{U}(N)/T \times \mathbb{R}_{\leq}^{+N}$ as

$$F(\mathbf{U}, \mathbf{\Lambda}) \triangleq P_{\mathbf{Q}}(\mathbf{U}\mathbf{\Lambda}\mathbf{U}^{\mathsf{H}})$$

where $\mathbf{U} \in \mathcal{U}(N)/T$ and the ordered vector of diagonal entries of $\mathbf{\Lambda}$ lies in $\mathbb{R}_{\leq}^{+N}$. According to this mapping, (18.10) becomes

$$L(F) = \int_{\mathcal{U}(N)/T \times \mathbb{R}_{\leq}^{+N}} -\log(F(\mathbf{U}, \mathbf{\Lambda}))F(\mathbf{U}, \mathbf{\Lambda})K(\mathbf{\Lambda})d\mathbf{U}d\mathbf{\Lambda}$$

$$+\beta\left[\int_{\mathcal{U}(N)/T\times\mathbb{R}_{\leq}^{+N}} F(\mathbf{U},\mathbf{\Lambda})K(\mathbf{\Lambda})d\mathbf{U}d\mathbf{\Lambda}-1\right]$$
$$+\gamma\left[\int_{\mathcal{U}(N)/T\times\mathbb{R}_{\leq}^{+N}}\left(\sum_{i=1}^{N}\lambda_i\right)F(\mathbf{U},\mathbf{\Lambda})K(\mathbf{\Lambda})d\mathbf{U}d\mathbf{\Lambda}-NE_0\right] \quad (18.11)$$

where we introduced the corresponding Jacobian

$$K(\mathbf{\Lambda})\triangleq\frac{\pi^{N(N-1)/2}}{\prod_{j=1}^{N}j!}\prod_{i<j}(\lambda_i-\lambda_j)^2$$

and used $\operatorname{tr}\mathbf{Q}=\operatorname{tr}\mathbf{\Lambda}=\sum_{i=1}^{N}\lambda_i$. Maximizing the entropy of the distribution $P_{\mathbf{Q}}$ by taking $\frac{\partial L(F)}{\partial F}=0$ and equating all entries in the integrand to zero yields

$$-K(\mathbf{\Lambda})-K(\mathbf{\Lambda})\log(F(\mathbf{U},\mathbf{\Lambda}))+\beta K(\mathbf{\Lambda})+\gamma\left(\sum_{i=1}^{N}\lambda_i\right)K(\mathbf{\Lambda})=0.$$

Since $K(\mathbf{\Lambda})\neq 0$ except on a set of measure zero, this is equivalent to

$$F(\mathbf{U},\mathbf{\Lambda})=e^{\beta-1+\gamma\sum_{i=1}^{N}\lambda_i}. \quad (18.12)$$

Note that the distribution $F(\mathbf{U},\mathbf{\Lambda})K(\mathbf{\Lambda})$ does not explicitly depend on $\mathbf{U}$. This indicates that $\mathbf{U}$ is uniformly distributed, with constant density $P_{\mathbf{U}}=(2\pi)^N$ over $\mathcal{U}(N)/T$. Therefore, the joint density can be factored under the form $F(\mathbf{U},\mathbf{\Lambda})K(\mathbf{\Lambda})=P_{\mathbf{U}}P_{\mathbf{\Lambda}}(\mathbf{\Lambda})$ where the distribution of the eigenvalues over $\mathbb{R}_{\leq}^{+N}$ is

$$P_{\mathbf{\Lambda}}(\mathbf{\Lambda})=\frac{e^{\beta-1}}{P_{\mathbf{U}}}e^{\gamma\sum_{i=1\dots N}\lambda_i}\frac{\pi^{N(N-1)/2}}{\prod_{j=1}^{N}j!}\prod_{i<j}(\lambda_i-\lambda_j)^2. \quad (18.13)$$

At this point, notice that the form of (18.13) indicates that the order of the eigenvalues is immaterial. Therefore, for the sake of simplicity, we will now work with the p.d.f. $P'_{\mathbf{\Lambda}}(\mathbf{\Lambda})$ of the joint distribution of the *unordered* eigenvalues, defined over $\mathbb{R}^{+N}$. Note that its restriction to the set of the ordered eigenvalues is proportional to $P_{\mathbf{\Lambda}}(\mathbf{\Lambda})$. More precisely

$$P'_{\mathbf{\Lambda}}(\mathbf{\Lambda})=Ce^{\gamma\sum_{i=1\dots N}\lambda_i}\prod_{i<j}(\lambda_i-\lambda_j)^2 \quad (18.14)$$

where the value

$$C=\frac{e^{\beta-1}}{P_{\mathbf{U}}}\frac{\pi^{N(N-1)/2}}{N!\prod_{j=1}^{N}j!}$$

can be determined by solving the normalization equation for the probability distribution $P'_{\mathbf{\Lambda}}$

$$\int_{\mathbb{R}^{+N}}P'_{\mathbf{\Lambda}}(\mathbf{\Lambda})d\mathbf{\Lambda}=1$$

where we used the change of variables $x_i = -\gamma\lambda_i$ and the Selberg integral (see (17.6.5) of [Mehta, 2004]). Furthermore, $\frac{d\log(C)}{d(-\gamma)} = \frac{N^2}{-\gamma} = NE_0$, and we finally obtain the final expression of the eigenvalue distribution

$$P'_{\mathbf{\Lambda}}(\mathbf{\Lambda}) = \left(\frac{N}{E_0}\right)^{N^2} \prod_{n=1}^{N} \frac{1}{n!(n-1)!} e^{-\frac{N}{E_0}\sum_{i=1\ldots N}\lambda_i} \prod_{i<j}(\lambda_i - \lambda_j)^2. \qquad (18.15)$$

In order to obtain the final distribution of $\mathbf{Q}$, first note that since the order of the eigenvalues is immaterial, the restriction of $\mathbf{U}$ to $\mathcal{U}(N)/T$ is not necessary, and $\mathbf{Q}$ is distributed as $\mathbf{U\Lambda U}^{\mathsf{H}}$ where the distribution of $\mathbf{\Lambda}$ is given by (18.15) and $\mathbf{U}$ is a Haar matrix. Furthermore, note that (18.15) is a particular case of the density of the eigenvalues of a complex Wishart matrix, described in Chapter 2. We recall that the complex $N \times N$ Wishart matrix with K degrees of freedom and covariance $\mathbf{\Sigma}$, denoted $\mathcal{CW}_N(K, \mathbf{\Sigma})$, is the matrix $\mathbf{A} = \mathbf{BB}^{\mathsf{H}}$ where $\mathbf{B}$ is a $N \times K$ matrix whose columns are complex independent Gaussian vectors with covariance $\mathbf{\Sigma}$. Indeed, (18.15) describes the unordered eigenvalue density of a $\mathcal{CW}_N(N, \frac{E_0}{N}\mathbf{I}_N)$ matrix. Taking into account the isotropic property of the distribution of $\mathbf{U}$, we can conclude that $\mathbf{Q}$ itself is also a $\mathcal{CW}_N(N, \frac{E_0}{N}\mathbf{I}_N)$ Wishart matrix. A similar result with a slightly different constraint was obtained by Adhikari in [Adhikari, 2006] where it is shown that the entropy-maximizing distribution of a positive definite matrix with known mean $\mathbf{G}$ follows a Wishart distribution with $N+1$ degrees of freedom, more precisely the $\mathcal{CW}_N(N+1, \frac{\mathbf{G}}{N+1})$ distribution.

The isotropic property of the obtained Wishart distribution is a consequence of the fact that no spatial constraints were imposed on the correlation. The energy constraint imposed through the trace only affects the distribution of the eigenvalues of $\mathbf{Q}$.

We highlight the fact that the result is directly applicable to the case where the channel correlation is known to be separable between transmitter and receiver. In this case, the full correlation matrix $\mathbf{Q}$ is known to be the Kronecker product of the transmit $\mathbf{Q}_t$ and receive $\mathbf{Q}_r$ correlation matrices. This channel model is therefore the channel with separable variance profile, or equivalently the Kronecker model in the MIMO case. The stochastic nature of $\mathbf{Q}_t$ and $\mathbf{Q}_r$ is barely mentioned in the literature, since the correlation matrices are usually assumed to be measurable quantities associated with a particular antenna array shape and propagation environment. However, in situations where these are not known (for instance, if the array shape is not known at the time of the channel code design, or if the properties of the scattering environment cannot be determined), but the Kronecker model is assumed to hold, the above analysis suggests that the maximum entropy choice for the distribution of $\mathbf{Q}_t$ and $\mathbf{Q}_r$ is independent, complex Wishart distributions with, respectively, n_t and n_r degrees of freedom. A Kronecker channel representation is provided in Figure 18.3.

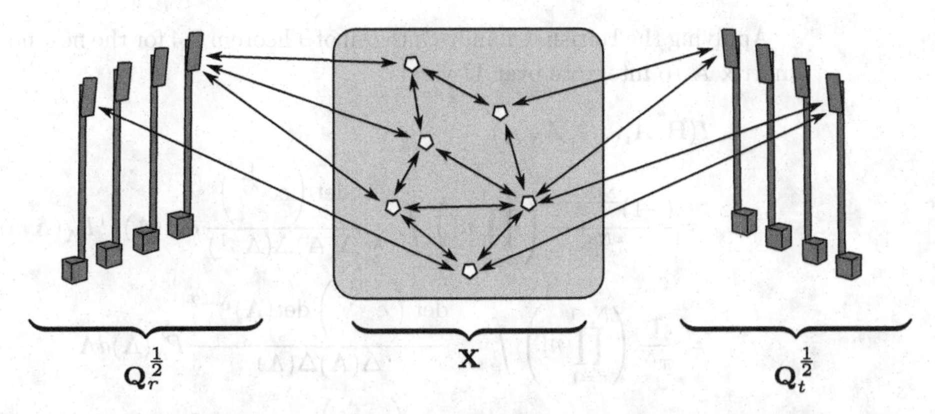

Figure 18.3 MIMO Kronecker channel representation, with $\mathbf{Q}_t \in \mathbb{C}^{n_t \times n_t}$ the transmit covariance matrix, $\mathbf{Q}_r \in \mathbb{C}^{n_r \times n_r}$ the receive correlation matrix and $\mathbf{X} \in \mathbb{C}^{n_r \times n_t}$ the i.i.d. Gaussian scattering matrix.

Marginalization over $\mathbf{Q}$.
The complete distribution of the correlated channel can be obtained by marginalizing out $\mathbf{Q}$, using its distribution as established in the previous paragraph. The distribution of $\mathbf{H}$ is obtained through

$$P_{\mathbf{H}}(\mathbf{H}) = \int_{\mathbb{S}} P_{\mathbf{H}|\mathbf{Q}}(\mathbf{H},\mathbf{Q})P_{\mathbf{Q}}(\mathbf{Q})d\mathbf{Q} = \int_{\mathcal{U}(N)\times\mathbb{R}^{+N}} P_{\mathbf{H}|\mathbf{Q}}(\mathbf{H},\mathbf{U},\mathbf{\Lambda})P'_{\mathbf{\Lambda}}(\mathbf{\Lambda})d\mathbf{U}d\mathbf{\Lambda}. \tag{18.16}$$

Let us rewrite the conditional probability density of (18.9) as

$$P_{\mathbf{H}|\mathbf{Q}}(\mathbf{h},\mathbf{U},\mathbf{\Lambda}) = \frac{1}{\pi^N \det(\mathbf{\Lambda})} e^{-\mathbf{h}^{\mathsf{H}}\mathbf{U}\mathbf{\Lambda}^{-1}\mathbf{U}^{\mathsf{H}}\mathbf{h}} = \frac{1}{\pi^N \det(\mathbf{\Lambda})} e^{-\operatorname{tr}(\mathbf{h}\mathbf{h}^{\mathsf{H}}\mathbf{U}\mathbf{\Lambda}^{-1}\mathbf{U}^{\mathsf{H}})}. \tag{18.17}$$

Using this expression in (18.16), we obtain

$$P_{\mathbf{H}}(\mathbf{H}) = \frac{1}{\pi^N}\int_{\mathbb{R}^{+N}}\int_{\mathcal{U}(N)} e^{-\operatorname{tr}(\mathbf{h}\mathbf{h}^{\mathsf{H}}\mathbf{U}\mathbf{\Lambda}^{-1}\mathbf{U}^{\mathsf{H}})}d\mathbf{U}\,\det(\mathbf{\Lambda})^{-1}P'_{\mathbf{\Lambda}}(\mathbf{\Lambda})d\mathbf{\Lambda}. \tag{18.18}$$

Now, similar to the proof of Theorem 16.1, let $\det(f(i,j))$ denote the determinant of a matrix with the (i,j)th element given by an arbitrary function $f(i,j)$. Let $\mathbf{A}$ be a Hermitian matrix which has its Nth eigenvalue A_N equal to $\mathbf{h}^{\mathsf{H}}\mathbf{h}$, and the others $A_1,\ldots,A_{N-1}$ are arbitrary, positive values that will eventually be set to zero. Letting

$$I(\mathbf{H},A_1,\ldots,A_{N-1}) = \frac{1}{\pi^N}\int_{\mathbb{R}^{+N}}\int_{\mathcal{U}(N)} e^{-\operatorname{tr}(\mathbf{A}\mathbf{U}\mathbf{\Lambda}^{-1}\mathbf{U}^{\mathsf{H}})}P_{\mathbf{U}}d\mathbf{U}\,\det(\mathbf{\Lambda})^{-1}P'_{\mathbf{\Lambda}}(\mathbf{\Lambda})d\mathbf{\Lambda}$$

the probability $P_{\mathbf{H}}(\mathbf{H})$ can be determined as the limit distribution when the first $N-1$ eigenvalues of $\mathbf{A}$ go to zero

$$P_{\mathbf{H}}(\mathbf{H}) = \lim_{A_1,\ldots,A_{N-1}\to 0} I(\mathbf{H},A_1,\ldots,A_{N-1}).$$

Applying the Harish–Chandra integral of Theorem 2.4 for the now non-singular matrix $\mathbf{A}$ to integrate over $\mathbf{U}$ yields

$$
\begin{aligned}
&I(\mathbf{H}, A_1, \ldots, A_{N-1}) \\
&= \frac{(-1)^{\frac{N(N-1)}{2}}}{\pi^N} \left(\prod_{n=1}^{N-1} n! \right) \int_{\mathbb{R}^{+N}} \frac{\det\left(e^{-\frac{A_i}{\lambda_j}}\right)}{\Delta(\mathbf{\Lambda})\Delta(\mathbf{\Lambda}^{-1})} \det(\mathbf{\Lambda})^{-1} P'_{\mathbf{\Lambda}}(\mathbf{\Lambda}) d\mathbf{\Lambda} \\
&= \frac{1}{\pi^N} \left(\prod_{n=1}^{N-1} n! \right) \int_{\mathbb{R}^{+N}} \frac{\det\left(e^{-\frac{A_i}{\lambda_j}}\right) \det(\mathbf{\Lambda})^{N-2}}{\Delta(\mathbf{\Lambda})\Delta(\mathbf{\Lambda})} P'_{\mathbf{\Lambda}}(\mathbf{\Lambda}) d\mathbf{\Lambda} \\
&= \frac{C}{\pi^N} \left(\prod_{n=1}^{N-1} n! \right) \int_{\mathbb{R}^{+N}} \frac{\det\left(e^{-\frac{A_i}{\lambda_j}}\right) \det(\mathbf{\Lambda})^{N-2} \Delta(\mathbf{\Lambda})}{\Delta(\mathbf{\Lambda})} e^{-\frac{N}{E_0}\operatorname{tr}(\mathbf{\Lambda})} d\mathbf{\Lambda}
\end{aligned}
$$

where we used the identity $\Delta(\mathbf{\Lambda}^{-1}) = \det(\frac{1}{\lambda_i}^{j-1}) = (-1)^{N(N+3)/2} \frac{\Delta(\mathbf{\Lambda})}{\det(\mathbf{\Lambda})^{N-1}}$.

Then, we decompose the determinant product using the classical expansion formula. That is, for an arbitrary $N \times N$ matrix $\mathbf{X} = (X_{ij})$

$$
\det(\mathbf{X}) = \sum_{\mathbf{a} \in \mathcal{S}_N} \operatorname{sgn}(\mathbf{a}) \prod_{n=1}^{N} X_{n,a_n} = \frac{1}{N!} \sum_{\mathbf{a},\mathbf{b} \in \mathcal{S}_N} \operatorname{sgn}(\mathbf{a}) \operatorname{sgn}(\mathbf{b}) \prod_{n=1}^{N} X_{a_n,b_n}
$$

where $\mathbf{a} = (a_1, \ldots, a_N)$, $\mathcal{S}_N$ denotes the set of all permutations of $\{1, \ldots, N\}$, and $\operatorname{sgn}(\mathbf{a})$ is the signature of the permutation $\mathbf{a}$. Using the first form of the determinant expansion, we obtain

$$
\begin{aligned}
\Delta(\mathbf{\Lambda}) \det\left(\left\{e^{-\frac{A_i}{\lambda_j}}\right\}\right) &= \det(\lambda_i^{j-1}) \det(e^{-\frac{A_j}{\lambda_i}}) \\
&= \sum_{\mathbf{a},\mathbf{b} \in \mathcal{S}_N^2} \operatorname{sgn}(\mathbf{a}) \operatorname{sgn}(\mathbf{b}) \prod_{n=1}^{N} \lambda_n^{a_n - 1} e^{-\frac{A_{b_n}}{\lambda_n}}.
\end{aligned} \tag{18.19}
$$

Note that in (18.19) we used the invariance of the second determinant by transposition in order to simplify subsequent derivations. Therefore

$$
\begin{aligned}
&I(\mathbf{H}, A_1, \ldots, A_{N-1}) \\
&= \frac{C}{\pi^N \Delta(\mathbf{A})} \left(\prod_{n=1}^{N-1} n! \right) \sum_{\mathbf{a},\mathbf{b} \in \mathcal{S}_N} \operatorname{sgn}(\mathbf{a}) \operatorname{sgn}(\mathbf{b}) \prod_{n=1}^{N} \int_{\mathbb{R}^+} \lambda_n^{N+a_n-3} e^{-\frac{A_{b_n}}{\lambda_n}} e^{-\frac{N}{E_0}\lambda_n} d\lambda_n \\
&= \frac{CN!}{\pi^N} \left(\prod_{n=1}^{N-1} n! \right) \frac{\det(f_i(A_j))}{\Delta(\mathbf{A})}
\end{aligned}
$$

where we let

$$
f_i(x) = \int_{\mathbb{R}^+} t^{N+i-3} e^{-\frac{x}{t}} e^{-\frac{N}{E_0} t} dt
$$

and we recognize the second form of the determinant expansion. In order to obtain the limit as $A_1, \ldots A_{N-1}$ go to zero, similar to the proof of Theorem 16.1, we apply Theorem 2.9 with $p = 1$, $N_1 = N - 1$ and $y_1 = 0$ since $\mathbf{A}$ has only one non-zero eigenvalue. This yields

$$\begin{aligned} P_{\mathbf{H}}(\mathbf{H}) &= \lim_{A_1, A_2, \ldots, A_{N-1} \to 0} I(\mathbf{H}, A_1, \ldots, A_{N-1}) \\ &= \frac{(-\gamma)^{N^2}}{\pi^N x_N^{N-1}} \prod_{n=1}^{N-1} [n!(n-1)!]^{-1} \det \left[f_i(0), f_i'(0), \ldots, f_i^{(N-2)}(0), f_i(x_N) \right]. \end{aligned} \tag{18.20}$$

At this point, it becomes obvious from (18.20) that the probability of $\mathbf{H}$ depends only on its norm (recall that $x_N = \mathbf{h}^{\mathsf{H}}\mathbf{h}$ by definition of $\mathbf{A}$). The distribution of $\mathbf{h}$ is isotropic, and is completely determined by the p.d.f. $P_{\mathbf{h}^{\mathsf{H}}\mathbf{h}}(x)$ of having $\mathbf{h}$ such that $\mathbf{h}^{\mathsf{H}}\mathbf{h} = x$.

Thus, for given x, $\mathbf{h}$ is uniformly distributed over $\mathbb{S}^{N-1}(x) = \{\mathbf{h}, \mathbf{h}^{\mathsf{H}}\mathbf{h} = x\}$, the complex hypersphere of radius $\sqrt{x}$ centered on zero. Its volume is $V_N(x) = \frac{\pi^N x^N}{N!}$, and its surface is $S_N(x) = \frac{dV_N(x)}{dx} = \frac{\pi^N x^{N-1}}{(N-1)!}$. Therefore, we can write the p.d.f. of x as

$$\begin{aligned} P_{\mathbf{h}^{\mathsf{H}}\mathbf{h}}(x) &= \int_{\mathbb{S}^{N-1}(x)} P_{\mathbf{H}}(\mathbf{h}) d\mathbf{h} \\ &= \frac{(-\gamma)^{N^2}}{(N-1)!} \prod_{n=1}^{N-1} [n!(n-1)!]^{-1} \det \left[f_i(0), f_i'(0), \ldots, f_i^{(N-2)}(0), f_i(x) \right]. \end{aligned}$$

In order to simplify the expression of the successive derivatives of f_i, it is useful to identify the Bessel K-function, Section 8.432 of [Gradshteyn and Ryzhik, 2000], and to replace it by its infinite sum expansion, Section 8.446 of [Gradshteyn and Ryzhik, 2000].

$$\begin{aligned} f_i(x) &= 2 \left(\sqrt{\frac{x}{-\gamma}} \right)^{i+N-2} K_{i+N-2}(2\sqrt{-\gamma x}) \\ &= (-\gamma)^{-i-N+2} \left[\sum_{k=0}^{i+N-3} (-1)^k \frac{(i+N-3-k)!}{k!} (-\gamma x)^k + (-1)^{i+N-1} \right. \\ &\quad \left. \times \sum_{k=0}^{+\infty} \frac{(-\gamma x)^{i+N-2+k}}{k!(i+N-2+k)!} \left(\log(-\gamma x) - \psi(k+1) - \psi(i+N-1+k) \right) \right]. \end{aligned}$$

Note that there is only one term in the sum with a non-zero pth derivative at zero. Therefore, the pth derivative of f_i at zero is simply (for $0 \leq p \leq N-2$)

$$f_i^{(p)}(0) = (-1)^{-i-N} \gamma^{p-i-N+2} (i+N-3-p)!. \tag{18.21}$$

Let us bring the last column to become the first, and expand the resulting determinant along its first column

$$\begin{aligned}
&\det\left[f_i^{(0)}(0),\ldots,f_i^{(N-2)}(0),f_i(x)\right]\\
&=(-1)^{N-1}\det\left[f_i(x),f_i^{(0)}(0),\ldots,f_i^{(N-2)}(0)\right]\\
&=(-1)^{N-1}\sum_{n=1}^{N}(-1)^{1+n}f_n(x)\det\left[\tilde{f}_{i,n}^{(0)}(0),\ldots,\tilde{f}_{i,n}^{(N-2)}(0)\right]
\end{aligned}$$

where $\tilde{f}_{i,n}^{(p)}(0)$ is the $N-1$ dimensional column obtained by removing the nth element from $f_i^{(p)}(0)$. Factorizing the $(-1)^p\gamma^{p-i-N+2}$ in the expression of $f_i^{(p)}(0)$ out of the determinant yields

$$\det\left[\tilde{f}_{i,n}^{(0)}(0),\ldots,\tilde{f}_{i,n}^{(N-2)}(0)\right]=(-1)^{n-N(N+1)/2}\gamma^{n-N^2+N-1}\det(\mathbf{G}^{(n)})$$

where the $N-1$ dimensional matrix $\mathbf{G}^{(n)}$ has (l,k) entry

$$\mathbf{G}_{l,k}^{(n)}=\Gamma(q_l^{(n)}+N-k-1)$$

where $\Gamma(i)=(i-1)!$ for i positive integer, and

$$q_l^{(n)}=\begin{cases}l & ,\ l\leq n-1,\\ l+1 & ,\ l\geq n.\end{cases}$$

Using the fact that $\Gamma(q_l^{(n)}+i)=q_l^{(n)}\Gamma(q_l^{(n)}+i-1)+(i-1)\Gamma(q_l^{(n)}+i-1)$, note that the kth column of $\mathbf{G}^{(n)}$ is

$$\mathbf{G}_{l,k}^{(n)}=q_l^{(n)}\Gamma(q_l^{(n)}+N-k-2)+(N-k-2)\mathbf{G}_{l,k+1}^{(n)}.$$

Since the second term is proportional to the $(k+1)$th column, it can be omitted without changing the value of the determinant. Applying this property to the first $N-2$ pairs of consecutive columns and repeating this process again to the first $N-2,\ldots,1$ pairs of columns, we obtain

$$\begin{aligned}
\det(\mathbf{G}^{(n)})&=\det\left[\Gamma(q_l^{(n)}+N-2),\ldots,\Gamma(q_l^{(n)}+2),\Gamma(q_l^{(n)}+1),\Gamma(q_l^{(n)})\right]\\
&=\det\left[q_l^{(n)}\Gamma(q_l^{(n)}+N-3),\ldots,q_l^{(n)}\Gamma(q_l^{(n)}+1),q_l^{(n)}\Gamma(q_l^{(n)}),\Gamma(q_l^{(n)})\right]\\
&=\det\left({q_l^{(n)}}^{N-1-k}\Gamma(q_l^{(n)})\right)\\
&=\frac{\prod_{i=1}^{N}\Gamma(i)}{\Gamma(n)}\det\left(\left\{{q_l^{(n)}}^{N-1-k}\right\}\right)\\
&=\frac{\prod_{i=1}^{N}\Gamma(i)}{\Gamma(n)}(-1)^{\frac{1}{2}(N-1)(N-2)}\det\left(\left\{{q_l^{(n)}}^{k-1}\right\}\right)
\end{aligned}$$

where the last two equalities are obtained, respectively, by factoring out the $\Gamma(q_l^{(n)})$ factors (common to all terms on the lth row) and inverting the order of the columns in order to get a proper Vandermonde structure. Finally, the determinant can be computed as

$$\begin{aligned}
\det\left(\left\{q_l^{(n)\,k-1}\right\}\right) &= \prod_{1\leq j<i\leq N-1}\left(q_i^{(n)}-q_j^{(n)}\right)\\
&= \prod_{i=1}^{n-2} i! \prod_{i=n}^{N-1} \frac{i!}{(i-n+1)!} \prod_{i=n+1}^{N-1}(i-n)!\\
&= \frac{\prod_{i=1}^{N-1} i!}{(n-1)!(N-n)!}.
\end{aligned}$$

Wrapping up the above derivations, we obtain successively

$$\det(\mathbf{G}^{(n)}) = \left(\prod_{i=1}^{N-1} i!\right)^2 (-1)^{(N-1)(N-2)/2} \frac{1}{[(n-1)!]^2 (N-n)!}$$

then:

$$\det\left[\tilde{f}_{i,n}^{(0)}(0),\ldots,\tilde{f}_{i,n}^{(N-2)}(0)\right] = \left(\prod_{i=1}^{N-1} i!\right)^2 \frac{(-1)^{n+1}\gamma^{n-N^2+N-1}}{[(n-1)!]^2 (N-n)!}$$

which gives

$$\begin{aligned}
&\det\left[f_i^{(0)}(0),\ldots,f_i^{(N-2)}(0),f_i(x)\right]\\
&= \sum_{n=1}^{N}(-1)^{1-N} f_n(x)\left(\prod_{i=1}^{N-1} i!\right)^2 \frac{\gamma^{n-N^2+N-1}}{[(n-1)!]^2 (N-n)!}.
\end{aligned}$$

Finally, we have:

$$P_{\mathbf{h}^{\mathsf{H}}\mathbf{h}}(x) = -\sum_{n=1}^{N} f_n(x) \frac{\gamma^{N+n-1}}{[(n-1)!]^2 (N-n)!}$$

where $\gamma = -\frac{N}{E_0}$. This leads to the maximum entropy distribution for $\mathbf{H}$, given by:

$$P_{\mathbf{H}}(\mathbf{H}) = -\frac{1}{\pi^N (\mathbf{h}^{\mathsf{H}}\mathbf{h})^{N-1}} \sum_{n=1}^{N} f_n(\mathbf{h}^{\mathsf{H}}\mathbf{h}) \frac{\gamma^{N+n-1}(N-1)!}{[(n-1)!]^2 (N-n)!}.$$

The corresponding p.d.f. is shown in Figure 18.4, as well as the p.d.f. of the instantaneous power of a Gaussian i.i.d. channel of the same size and mean power. As expected, the energy distribution of the proposed model is more spread out than the energy of a Gaussian i.i.d. channel.

Figure 18.5 shows the d.f. curves of the instantaneous mutual information achieved over the channel described in (18.1) by these two channel models. The proposed model differs in particular in the tails of the distribution: for instance, the 1% outage capacity is reduced from 8 to 7 bits/s/Hz with respect to the Gaussian i.i.d. model.

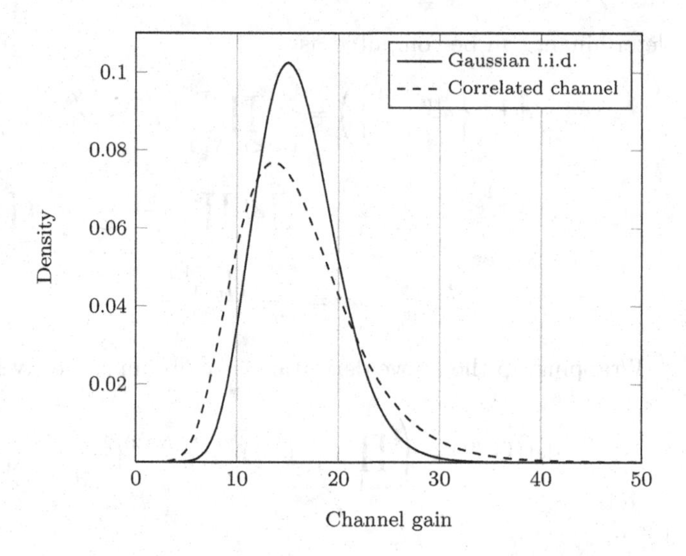

Figure 18.4 Amplitude distribution of the maximum entropy 4×4 MIMO channel models, with known identity correlation (Gaussian i.i.d.) or unknown correlation.

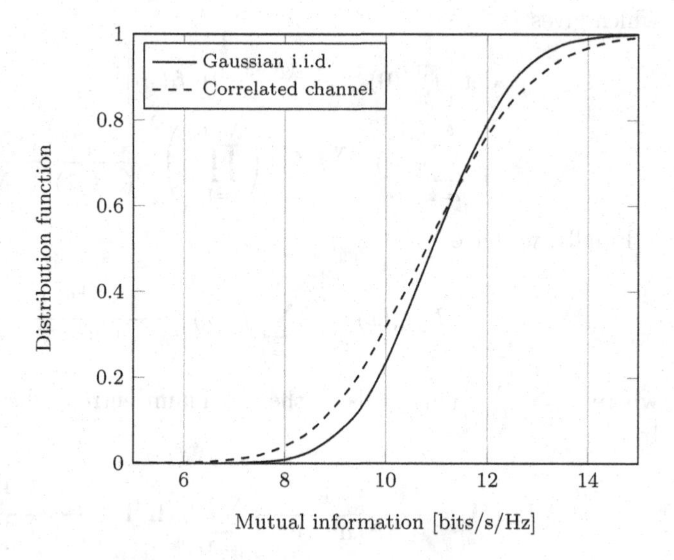

Figure 18.5 Mutual information distribution of the maximum entropy 4×4 MIMO channel models, with known identity correlation (Gaussian i.i.d.) or unknown correlation, SNR of 10 dB.

18.2.2.3 Limited-rank covariance matrix

In this section, we address the situation where the modeler takes into account the existence of a covariance matrix of rank $L < N$ (we assume that L is known). Such a situation arises in particular when the communication channel is a priori known not to offer numerous degrees of freedom, or when the MIMO antennas on

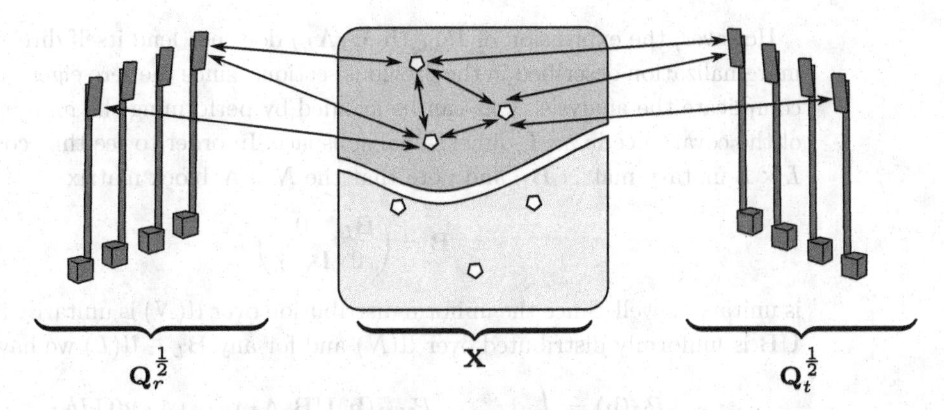

Figure 18.6 MIMO Kronecker channel representation with limited number of scatterers in the propagation environment, with $\mathbf{Q}_t \in \mathbb{C}^{n_t \times n_t \cdot}$ the (possibly rank-limited) transmit covariance matrix, $\mathbf{Q}_r \in \mathbb{C}^{n_r \times n_r}$ the (possibly rank-limited) receive correlation matrix and $\mathbf{X} \in \mathbb{C}^{n_r \times n_t}$ the i.i.d. Gaussian scattering matrix.

either communication side are known to be close enough for correlation to arise. Figure 18.6 depicts a Kronecker channel environment with limited diversity.

As in the full-rank case, we will use the spectral decomposition $\mathbf{Q} = \mathbf{U}\mathbf{\Lambda}\mathbf{U}^{\mathsf{H}}$ of the covariance matrix, with $\mathbf{\Lambda} = \text{diag}(\lambda_1, \ldots, \lambda_L, 0, \ldots, 0)$. Let us denote $\mathbf{\Lambda}_L = \text{diag}(\lambda_1, \ldots, \lambda_L)$. The maximum entropy probability density of $\mathbf{Q}$ with the extra rank constraint is unsurprisingly similar to that derived previously, with the difference that all the energy is carried by the first L eigenvalues, i.e. $\mathbf{U}$ is uniformly distributed over $\mathcal{U}(N)$, while

$$P_{\mathbf{\Lambda}_L}(\mathbf{\Lambda}_L) = \left(\frac{L^2}{NE_0}\right)^{L^2} \prod_{n=1}^{L} \frac{1}{n!(n-1)!} e^{-\frac{L^2}{NE_0}\sum_{i=1\ldots L}\lambda_i} \prod_{i<j\leq L} (\lambda_i - \lambda_j)^2. \quad (18.22)$$

However, the definition of the conditional probability density $P_{\mathbf{H}|\mathbf{Q}}(\mathbf{h}, \mathbf{U}, \mathbf{\Lambda})$ in (18.9) does not hold when $\mathbf{Q}$ is not full rank. The channel vector $\mathbf{h}$ becomes a degenerate Gaussian random variable. Its projection onto the L-dimensional subspace associated with the non-zero eigenvalues of $\mathbf{Q}$ follows a Gaussian law, whereas the probability of $\mathbf{h}$ being outside this subspace is zero. The conditional probability in (18.17) must therefore be rewritten as

$$P_{\mathbf{H}|\mathbf{Q}}(\mathbf{h}, \mathbf{U}, \mathbf{\Lambda}_L) = 1_{\text{span}(\mathbf{U}_{[L]})}(\mathbf{h}) \frac{1}{\pi^L \prod_{i=1}^{L} \lambda_i} e^{-\mathbf{h}^{\mathsf{H}}\mathbf{U}_{[L]}\mathbf{\Lambda}_L^{-1}\mathbf{U}_{[L]}{}^{\mathsf{H}}\mathbf{h}} \quad (18.23)$$

where $\mathbf{U}_{[L]}$ denotes the $N \times L$ matrix obtained by truncating the last $N - L$ columns of $\mathbf{U}$. The indicator function ensures that $P_{\mathbf{H}|\mathbf{Q}}(\mathbf{h}, \mathbf{U}, \mathbf{\Lambda})$ is zero for $\mathbf{h}$ outside of the column span of $\mathbf{U}_{[L]}$.

We need now to marginalize $\mathbf{U}$ and $\mathbf{\Lambda}$ in order to obtain the p.d.f. of $\mathbf{h}$.

$$P_{\mathbf{H}}(\mathbf{h}) = \int_{\mathcal{U}(N)\times\mathbb{R}^{+L}} P_{\mathbf{H}|\mathbf{Q}}(\mathbf{h}, \mathbf{U}, \mathbf{\Lambda}_L) P_{\mathbf{\Lambda}_L}(\mathbf{\Lambda}_L) d\mathbf{U} d\mathbf{\Lambda}_L.$$

However, the expression of $P_{\mathbf{H}|\mathbf{Q}}(\mathbf{h}, \mathbf{U}, \mathbf{\Lambda}_L)$ does not lend itself directly to the marginalization described in the previous sections, since the zero eigenvalues of $\mathbf{Q}$ complicate the analysis. This can be avoided by performing the marginalization of the covariance in an L-dimensional subspace. In order to see this, consider an $L \times L$ unitary matrix $\mathbf{B}_L$ and note that the $N \times N$ block matrix

$$\mathbf{B} = \begin{pmatrix} \mathbf{B}_L & 0 \\ 0 & \mathbf{I}_{N-L} \end{pmatrix}$$

is unitary as well. Since the uniform distribution over $\mathcal{U}(N)$ is unitarily invariant, $\mathbf{UB}$ is uniformly distributed over $\mathcal{U}(N)$ and for any $\mathbf{B}_L \in \mathcal{U}(L)$ we have:

$$P_{\mathbf{H}}(\mathbf{h}) = \int_{\mathcal{U}(N)\times\mathbb{R}^{+L}} P_{\mathbf{H}|\mathbf{Q}}(\mathbf{h}, \mathbf{UB}, \mathbf{\Lambda}_L) P_{\mathbf{\Lambda}_L}(\mathbf{\Lambda}_L) d\mathbf{U} d\mathbf{\Lambda}_L.$$

Furthermore, since $\int_{\mathcal{U}(L)} d\mathbf{B}_L = 1$

$$\begin{aligned} P_{\mathbf{H}}(\mathbf{h}) &= \int_{\mathcal{U}(L)} \int_{\mathcal{U}(N)\times\mathbb{R}^{+L}} P_{\mathbf{H}|\mathbf{Q}}(\mathbf{h}, \mathbf{UB}, \mathbf{\Lambda}_L) P_{\mathbf{\Lambda}_L}(\mathbf{\Lambda}_L) d\mathbf{U} d\mathbf{\Lambda}_L d\mathbf{B}_L \\ &= \int_{\mathbf{U}\in\mathcal{U}(N)} 1_{\text{span}(\mathbf{U}_{[L]})}(\mathbf{h}) P_{\mathbf{k}}(\mathbf{U}_{[L]}{}^{\mathsf{H}}\mathbf{h}) d\mathbf{U} \end{aligned} \tag{18.24}$$

where (18.24) is obtained by letting $\mathbf{k} = \mathbf{U}_{[L]}{}^{\mathsf{H}}\mathbf{h}$ and

$$P_{\mathbf{k}}(\mathbf{k}) = \int_{\mathcal{U}(L)\times\mathbb{R}^{+L}} \frac{1}{\pi^L \prod_{i=1}^L \lambda_i} e^{-\mathbf{k}^{\mathsf{H}}\mathbf{B}_L \mathbf{\Lambda}_L^{-1} \mathbf{B}_L^{\mathsf{H}} \mathbf{k}} P_{\mathbf{\Lambda}_L}(\mathbf{\Lambda}_L) d\mathbf{B}_L d\mathbf{\Lambda}_L. \tag{18.25}$$

We can then exploit the similarity of (18.25) and (18.18) and, by the same reasoning as in previous sections, conclude directly that $\mathbf{k}$ is isotropically distributed in $\mathcal{U}(L)$ and that its p.d.f. depends only on its Frobenius norm, following

$$P_{\mathbf{k}}(\mathbf{k}) = \frac{1}{S_L(\mathbf{k}^{\mathsf{H}}\mathbf{k})} P_x^{(L)}(\mathbf{k}^{\mathsf{H}}\mathbf{k})$$

where

$$P_x^{(L)}(x) = \frac{2}{x} \sum_{i=1}^{L} \left(-L\sqrt{\frac{x}{NE_0}}\right)^{L+i} K_{i+L-2}\left(2L\sqrt{\frac{x}{NE_0}}\right) \frac{1}{[(i-1)!]^2 (L-i)!}.$$

Finally, note that $\mathbf{h}^{\mathsf{H}}\mathbf{h} = \mathbf{k}^{\mathsf{H}}\mathbf{k}$, and that the marginalization over the random rotation that transforms $\mathbf{k}$ into $\mathbf{h}$ in (18.24) preserves the isotropic property of the distribution. Therefore

$$P_{\mathbf{H}}(\mathbf{h}) = \frac{1}{S_N(\mathbf{h}^{\mathsf{H}}\mathbf{h})} P_x^{(L)}(\mathbf{h}^{\mathsf{H}}\mathbf{h}).$$

Examples of the corresponding p.d.f. for $L \in \{1, 2, 4, 8, 12, 16\}$ are represented in Figure 18.7 for a 4×4 channel, together with the p.d.f. of the instantaneous power of a Gaussian i.i.d. channel of the same size and mean power. As expected, the energy distribution of the proposed maximum entropy model is more spread out than the energy of a Gaussian i.i.d. channel.

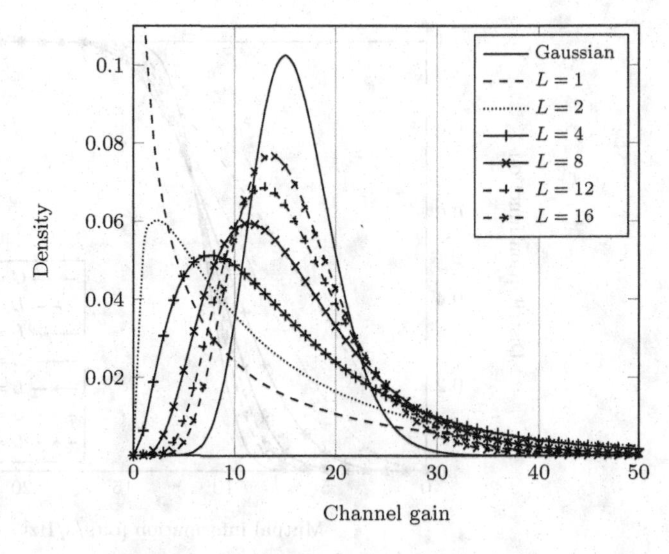

Figure 18.7 Amplitude distribution of the maximum entropy 4×4 MIMO channel models, for $E_0 = 1$, and limited degrees of freedom $L \in \{1, 2, 4, 8, 12, 16\}$.

The d.f. of the mutual information achieved over the limited-rank ($L < 16$) and full rank ($L = 16$) covariance maximum entropy channel at a signal-to-noise ratio of 10 dB is depicted in Figure 18.8 for various ranks L, together with the Gaussian i.i.d. channel. As already mentioned, the proposed model differs especially in the tails of the distribution. In particular, the outage capacity for low outage probability is greatly reduced with respect to the Gaussian i.i.d. channel model.

18.2.2.4 Discussion

It is important to understand the reason why maximum entropy channels are designed. It is of interest to characterize ergodic and outage capacities when very limited information is known about the channel as this can provide a figure of what mean or minimum transmission rates can be expected in a channel that is known to have limited degrees of freedom. Typically, MIMO communication channels with small devices embedded with multiple antennas tend to have strong correlation. Measuring correlation by a simple scalar number is however rather difficult and can be done through many approaches. Measuring correlation through the number of degrees of freedom left in the channel is one of those. The study above therefore helps us anticipate the outage performance of multiple antenna communications in more or less scattered environments.

Another very interesting feature of maximum entropy channel models is that they can be plugged into problems such as source sensing when the channel environment is known to enjoy some specific features. For instance, remember that in Chapter 16 we derived Neyman–Pearson tests based on the maximum entropy principle, in the sense that we assumed the communication channel

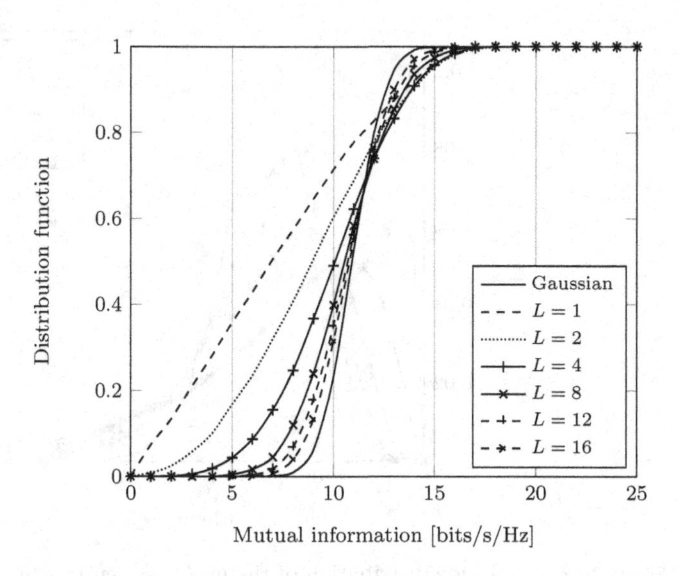

Figure 18.8 Mutual information distribution of the maximum entropy 4×4 MIMO channel models, with known identity correlation (Gaussian i.i.d.) or unknown correlation, SNR of 15 dB.

was only known to have signal-to-noise ratio $1/\sigma^2$, in which case we considered a Gaussian i.i.d. channel model (the choice of which is now confirmed by the analysis above). We then assumed the SNR was imperfectly known, so that we obtained an integral form over possible σ^2 of the Neyman–Pearson test. Alternatively, we could have assumed from the beginning that the channel variance was imperfectly known and used the expressions derived above of the distribution of the channel variance. Consistency of the maximum entropy principle, detailed in [Jaynes, 2003], would then ensure identical results at the end. Now, in the case when further information is known, such as the channel degrees of freedom are limited (for instance when a sensor network with multiple close antennas scans a low frequency resource), adaptive sensing strategies can be put in place that account for the expected channel correlation. In such scenarios, Neyman–Pearson tests can be more adequately designed than when assuming Gaussian i.i.d. propagation channels.

We believe that the maximum entropy approach, often used in signal processing questions, while less explored in wireless communications, can provide interesting solutions to problems dealing with too many unknown variables. Instead of relying on various *ad-hoc* approaches, the maximum entropy principle manages to provide an information-theoretic optimum solution to a large range of problems. It was in particular noticed in [Couillet *et al.*, 2010] that conventional minimum mean square error channel estimators enter the framework of maximum entropy channel estimation, when only the number of propagation paths in the frequency selective channel is a priori known. Then, for unknown channel delay spread, extensions of the classical minimum mean square error approach can be designed, whose increased complexity can then be further reduced based on suboptimal algorithms. This is the basic approach of maximum entropy solutions, which seek for optimal solutions prior to providing suboptimal implementations, instead of using simplified suboptimal models in the first place. Similarly, a maximum entropy optimal data-aided coarse frequency offset estimator for orthogonal frequency division multiplexing protocols is provided in [Couillet and Debbah, 2010b], along with a suboptimal iterative algorithm.

This completes this chapter on maximum entropy channel modeling.

19 Perspectives

Before concluding this book, we will briefly discuss open questions in random matrix theory. We will describe current research topics, results that are still unknown and that would be worth looking into. We also introduce briefly the replica methods, an alternative to the Stieltjes transform, and free probability methods proposed in this book which have been gaining a lot of interest lately, and will conclude with possible extensions of random matrix theory to a more elaborate time-varying (stochastic) mathematical framework.

19.1 From asymptotic results to finite dimensional studies

First, we recall the recent advances in random matrix theory, already introduced in this book, and which will be studied further in the near future. In Part I, we first explored limiting spectral distributions of some simple random matrix models. For instance, we characterized the l.s.d. of the sample covariance matrix model when the e.s.d. of the population covariance matrix converges weakly to some distribution function. We then observed that, for more practical system models, such as the sum of matrices with independent entries and left- and right-correlations, there may not exist a limiting spectral distribution, even when all deterministic matrices in the model do have a l.s.d. This led us to consider the deterministic equivalent approach instead of the l.s.d. approach. These deterministic equivalents have an outstanding advantage over l.s.d. and can be considered a major breakthrough for applied random matrix theory for the following reasons.

- They no longer require the existence of converging series of deterministic matrices in the model. For instance, in the sample covariance matrix case applied to the characterization of the sum rate capacity of a MISO broadcast channel with N users, we can freely assume that the distances of the users to the base station (which are modeled in the entries of the population covariance matrix) take any given values for all finite N. There is therefore no need to assume the existence of an unrealistic converging user-to-base station distance distribution as the number of users grows. This alleviates basic system modeling issues.

- More importantly, deterministic equivalents provide an approximation of the performance of such systems for all finite N, and not as N tends to infinity. Based on the previous example, we can imagine the case of a cellular MISO broadcast channel with users being successively connected to or disconnected from the base station. In this scenario, the analyzes based on l.s.d. or deterministic equivalents differ as follows.
  - with l.s.d. considerations, the sum rate for all finite N can be approximated by a single value corresponding to some functional of the l.s.d. of the sample covariance matrix when the population covariance matrix models the scenario of an increasingly high user density. The approximation here lies therefore in the fact that the reality does not fit the asymptotic model;
  - with deterministic equivalents, it is possible to derive an approximation of the system performance for every N, whatever the position of the active users. Therefore, even if the large N asymptotic performance (when all users are connected and their number grows to infinity) leads to a unique expression, the performances for all configurations of N users lead to various results. The inaccuracy here does not lie in the system model but rather in the inexactness of the finite N approximation, which is more convenient.
- Remember finally that, for more involved system models, limiting spectral distributions may not exist at all, and therefore the l.s.d. approach cannot be used any longer. This led quite a few authors to assume very unrealistic system models in order for a l.s.d. to exist, so to obtain exploitable results. From the theory of deterministic equivalents developed in Part I, this is unnecessary.

We wanted to insist on the considerations above a second time since random matrix theory applications to wireless communications are suffering from the false impression that the models designed assume an infinite number of antennas, an infinite number of users, etc. and that these models are so unrealistic that the results obtained are worthless. We hope that the reader now has a clear understanding that such criticism, totally acceptable ten years ago, are no longer justified. In Chapters 13–14, we derived approximated expressions of the capacity of multiple antenna systems (single-user MIMO, MIMO MAC, MIMO BC) and observed that the theoretical curves are indiscernible from the simulated curves, sometimes for N as small as 4. In this book, when dealing with practical applications, we systematically and purposely replaced most asymptotic results found in the literature by deterministic equivalents and replaced any mention of the term *asymptotic* or the phrase *infinite size matrices* by phrases such as *for all finite N* or *accurate as N grows large*.

We believe that much more is to be done regarding deterministic equivalents for more involved system models than those presented in this book. Such models are in particular demanded for the understanding of future cognitive radio networks as well as small-cell networks which will present more elaborate system conditions, such as cooperating base stations, short-range communications with numerous propagation paths, involved interference patterns, intense control data exchange, limited real-time channel state information, etc. With all these parameters taken into account, it is virtually impossible to assume large dimensional scenarios of converging matrix models. Deterministic equivalents can instead provide very precise system performance characterizations. It is important also to realize that, while the system models under study in the application chapters, Chapters 12–15, were sometimes very intricate, questions such as capacity optimization often resulted in very elegant and compact forms and come along with simple iterative algorithms, often with ensured convergence. It is therefore to be believed that even more complex models can still be provided with simple optimizations. It is important again to recall at this point that accurate mathematical derivations are fundamental to ensure in particular that the capacity maximizing algorithms do converge surely. We also mentioned that second order statistics of the performance of such models can also be well approximated in the large dimensional regime, with simple forms involving Jacobian matrices of the fundamental equations appearing systematically. Second order statistics provide further information on the outage capacity of these systems but also on the overall reliability of the deterministic equivalents. In heterogeneous primary-secondary networks where low probability of interference is a key problem to be considered, simple expressions of the second order statistics are of dramatic importance. As a consequence, in the near future, considerable effort will still need to be cast on deterministic equivalents and central limit theorems. Given the manpower demanded to treat the vastness of the recent small cell incentive, a systematic simplification of classical random matrix methods presented in this book will become a research priority.

From a more technical point of view, we also insist on the fact that the existence of a trace lemma for some matrix models is often sufficient to establish deterministic equivalents for involved system models. This was recently exemplified by the case of Haar matrices for which the trace lemma, Theorem 6.15, allows us to determine the approximated capacity of multi-cellular orthogonal CDMA setups with multi-path channels based on the Stieltjes and Shannon transforms provided in Theorem 6.17. Future multi-station short-range communication models with strong line-of-sight components may surely demand more exotic channel models than the simple models based on i.i.d. random matrices. We think in particular of Euclidean matrices [Bordenave, 2008] that can be used to model random grids of access points. If trace lemmas can be found for these models, it is likely that results similar to the i.i.d. case will be derived, for which systematic optimization methods and statistical analysis will have to be generated.

The methods for deterministic equivalents presented in this book therefore only pave the way for much more complex system model characterizations. Nonetheless, we also noticed that many random matrix models, more structured, are still beyond analytical reach, although combinatoric moment approaches are filling the gap. In particular, the characterization of the limiting spectrum of some random Vandermonde matrix models has known an increasing interest since tools from combinatorics can be used to efficiently derive all successive moments of the mean e.s.d. for all finite dimensions. As long as the random matrices under study exhibit symmetric structures, in the sense of rotational invariance, moments of the mean e.s.d. and of the l.s.d. can be usually characterized. For a complete distribution function to be analyzed, though, more advanced tools are demanded. As reminded earlier, the existence of a trace lemma makes the Stieltjes transform approach usable. However, for the random Vandermonde matrices, such a trace lemma is beyond one's knowledge to this day. Instead, it may be that other types of matrix transforms or possibly the Gaussian method, introduced briefly in Section 6.2 and which relies to some extent on moment calculus, can be used efficiently for these involved random matrix models. The Gaussian tools in particular, which are gaining more and more interest these days, may open new routes to future breakthroughs in the characterization of spectral distribution of structured matrices with invariance properties. Instead of a trace lemma, this approach essentially requires an appropriate integration by parts formula, Theorem 6.6, and a Nash–Poincaré inequality, Theorem 6.7, adapted to the structure of the random matrices under study.

In terms of estimation and detection, the increasing interest for cognitive radio networks, sensor networks with non-necessarily regular topology, and the need to share control information across large networks demand for more and more complex parameter estimators. Since these network structures typically have several large dimensions of similar sizes, the random matrix-based inference techniques developed in Chapter 8 are expected to be further developed in the future. From a purely mathematical viewpoint, these techniques are based on advanced results of limiting exact eigenvalue separation and central limit theorems. Such results are however known only to hold for a limited number of random matrix models and will need to be further extended in the future. Moreover, similar to central limit theorems for the deterministic equivalents of some functionals of the eigenvalues, central limit theorems for G-estimators are fundamental to characterize the sensitivity and the reliability of the parameter estimates. For practical cognitive radio and sensor network applications, such results will be importantly needed.

From a small dimensional viewpoint, we observed in Chapter 9 with the extreme eigenvalue distribution of spiked models, in Chapter 16 with detection tests for multi-dimensional systems, and in Chapter 18 with multi-dimensional channel modeling that small dimensional random matrix analysis is still a hot topic, which also demands for further generalizations. In fact, while random matrix theory started off with the study of small dimensional random matrices with the work of Wishart, it quickly moved to asymptotic considerations that have reached their glory in the past ten years, while today the study of the extreme eigenvalues and spiked models call for a revisit of the finite size considerations. We mentioned many times that an important field of random matrices, involving orthogonal polynomials and Fredholm determinants, has been providing numerous key results very recently for the characterization of limiting distributions and asymptotic independence of extreme eigenvalues. This topic, which originates from rather old works on the inner symmetry of unitarily invariant random matrices, e.g., [James, 1964], is still being thoroughly investigated these days. The tools required to study such models are very different from those proposed here and call for deeper mathematical considerations. A systematic simplification of these methods which should also generalize to more challenging random matrix models is also the key for future usage of these rather difficult tools in signal processing and wireless communications.

As we mention the democratization of some mathematical tools for random matrices, this book being a strong effort in this direction, we discuss briefly hereafter the method of statistical mechanics known as the *replica trick* or *replica method* which has not been mentioned so far but which has been generating lately an important wave of results in terms of l.s.d. and deterministic equivalents. Instead of introducing the specifics of this tool, which has the major drawback of relying on non-mathematically rigorous methods, we discuss its interaction with the methods used in this book.

19.2 The replica method

In addition to the approaches treated in Part I to determine deterministic equivalents, another approach, known as the *replica method*, is gaining ground in the wireless communication community. In a similar way as deterministic equivalent derivations based on the 'guessing' part of the Bai and Silverstein approach, this technique provides in general a first rapid hint on the expected solutions. The replica method does however not come along with appropriate mathematical tools to prove the accuracy of the derived solutions. More precisely, the replica derivations assume several mathematical properties of continuity and limit-integral interchange, which are assumed valid at first (in order to obtain the hypothetical solutions) but which are very challenging to prove. This tool therefore has to be used with extreme care. For a short introduction to the replica method, see, e.g., the appendix of [Müller, 2003].

The replica method is an approach borrowed from physics and especially from the field of statistical mechanics, see, e.g., [Mézard *et al.*, 1987; Nishimori, 2001]. It was then extensively used in the field of wireless communications, starting with the work of Tanaka [Tanaka, 2002] on maximum-likelihood CDMA detectors. The asymptotic behavior of numerous classical detectors were then derived using the replica method, e.g., [Guo and Verdú, 2002]. The replica method is used in statistical physics to evaluate thermodynamical entropies and free energy. In a wireless communication context, those are closely linked to the mutual information, i.e. the difference of receive signal and source signal entropies [Shannon, 1948], in the sense that free energy and mutual information only differ from an additive constant and a scalar factor. Replica methods have in particular proved to be very useful when determining central limit theorems for the e.s.d. of involved random matrix models. While classical deterministic equivalents and Stieltjes transform approaches used to fail to derive nice closed-form formulas for the asymptotic covariance matrices of functionals of the e.s.d., the replica method often conjectured that these covariance matrices take the form of Jacobian matrices. So far, all conjectures of the replica method going in this Jacobian direction turned out to be exact. For instance, the proof of Theorem 6.21 relies on martingale theory, similar to the proof of Theorem 3.18. With these tools alone, the limiting variance of the central limit often takes an unpleasant form, which is not obvious to relate to a Jacobian, although it effectively is asymptotically equivalent to a Jacobian. In this respect, replica methods have turned out to be extremely useful. However, some examples of calculus where the replica method fails have also been identified. Today, mathematicians are progressively trying to raise necessary and sufficient conditions for the replica method to be valid. That is, situations where the critical points of the replica derivations are valid or not are progressively being identified.

So far, however, the validity conditions are not sufficient for limiting laws and deterministic equivalents to be accurately proved using this method. We therefore see the replica method as a good opportunity for engineers and researchers to easily come up with would-be results that can then be accurately proved using classical random matrix techniques. We do not develop further this technique, though, as it requires the introduction of several additional tools, and we leave it to the reader to refer to alternative introductory articles.

We complete this chapter with the introduction of some ideas on the generalization of random matrix theory to continuous time random matrix theory and possible applications to wireless communications.

19.3 Towards time-varying random matrices

In addition to characterizing the capacity of wireless channels, it has always been of interest in wireless communications to study their time evolution. We have to this point gone successively through the following large dimensional network characterizations:

- the capacity, sum rate, or rate regions, that allow us to anticipate either the averaged achievable rate of quasi-static channels or the exact achievable rate of long coded sequences over very fast varying channels;
- the outage capacity, sum rate, or rate regions, which constitute a quality of service parameter relative to the rates achievable with high probability.

Now, since communication channels vary with time, starting from a given deterministic channel realization, it is possible to anticipate the rate evolution of this channel. Indeed, similar to the averaging effect arising when the matrix dimensions grow large, that turn random eigenvalue distributions into asymptotically deterministic quantities, solutions to implicit equations, the behavior of the time-varying random eigenvalues of a deterministic matrix affected by a Wiener process (better known as Brownian motion) can be deterministically characterized as the solution of differential equations. Although no publication related to these time-varying aspects for wireless communications has been produced so far (mostly because the tools are not mature enough), it is to be believed that random matrix theory for wireless communications may move on a more or less long-term basis towards *random matrix process theory* for wireless communications. Nonetheless, these random matrix processes are nothing new and have been the interest of several generations of mathematicians.

We hereafter introduce briefly the fundamental ideas, borrowed from a tutorial by Guionnet [Guionnet, 2006]. The initial interest of Guionnet is to derive the limiting spectral distribution of a non-central Wigner matrix with Gaussian entries, based on stochastic calculus. The result we will present here provides, under the form of the solution of a differential equation, the limiting eigenvalue distribution of such a random matrix affected by Brownian noise at all times $t > 0$.

We briefly introduce the notion of a Wigner matrix-valued Wiener process. A Wiener process W_t is defined as a stochastic process with the following properties:

- $W_0 = 0$
- W_t is a random variable, almost surely continuous over t
- for $s, t > 0$, $W_t - W_s$ is Gaussian with zero mean and variance $t - s$
- for $s_1 \leq t_1 < s_2 \leq t_2$, $W_{t_1} - W_{s_1}$ is independent of $W_{t_2} - W_{s_2}$.

This definition allows for the generation of random processes with independent increments. That is, if W_t is seen as the trajectory of a moving particle, the Wiener process assumptions ensure that the increment of the trajectory between two time instants is independent of the increments observed between any two instants in the past. This will be suitable in wireless communications to model the evolution of an unpredictable time-varying process such as the evolution of a time-varying channel matrix from a deterministically known matrix and an additional random time-varying innovation term; the later being conventionally modeled as Gaussian at time $t = 1$.

Instead of considering channel matrices, though, we restrict this introduction to Wigner matrices. We define the Wigner matrix-valued Wiener process as the time-varying matrix $\mathbf{X}_N(t) \in \mathbb{C}^{N \times N}$ with (m, n) entry $X_{N,mn}(t)$ given by:

$$X_{N,mn}(t) = \begin{cases} \frac{1}{\sqrt{2N}}(W_{m,n}(t) + iW'_{m,n}(t)) & , \ m < n \\ \frac{1}{\sqrt{N}} W_{m,m}(t) & , \ m = n \end{cases}$$

where $W_{m,n}(t)$ and $W'_{m,n}(t)$ are independent Wiener processes. As such, from the above definition, $\mathbf{X}_N(1)$ is a Gaussian Wigner matrix. We then define $\mathbf{Y}_N(t) \in \mathbb{C}^{N \times N}$ as

$$\mathbf{Y}_N(t) = \mathbf{Y}_N(0) + \mathbf{X}_N(t)$$

for some deterministic Hermitian matrix $\mathbf{Y}_N(0) \in \mathbb{C}^{N \times N}$.

We recognize that at time $t = 1$, $\mathbf{Y}_N(1) = \mathbf{Y}_N(0) + \mathbf{X}_N(1)$ is a Gaussian Wigner matrix with Gaussian independent entries of mean given by the entries of $\mathbf{Y}_N(0)$ and variance $1/N$. The current question though is to analyze the time evolution of the eigenvalue distribution of $\mathbf{Y}_N(t)$.

Denote $\boldsymbol{\lambda}^N(t) = (\lambda_1^N(t), \ldots, \lambda_N^N(t))$ the set of random eigenvalues of $\mathbf{Y}_N(t)$ and $F^{\mathbf{Y}_N(t)}$ the e.s.d. of $\mathbf{Y}_N$ at time t. The following result, due to Dyson [Dyson, 1962b], characterizes the time-varying e.s.d. $F^{\mathbf{Y}_N(t)}$ as the solution of a stochastic differential equation.

Theorem 19.1. *Let $\boldsymbol{\lambda}^N(0)$ be such that $\lambda_1^N(0) < \ldots < \lambda_N^N(0)$. Then $F^{\mathbf{Y}_N(t)}$ is the unique (weak) solution of the stochastic differential system*

$$d\lambda_i^N(t) = \frac{1}{\sqrt{N}} dV_t^i + \frac{1}{N} \sum_{j \neq i} \frac{1}{\lambda_i^N(t) - \lambda_j^N(t)} dt$$

with initial condition $\boldsymbol{\lambda}^N(0)$, such that $\lambda_1^N(t) < \ldots < \lambda_N^N(t)$, where $(V_t^1, \ldots, V_t^N)$ is an N-dimensional Wiener process.

This characterizes the distribution of $\boldsymbol{\lambda}^N(t)$ for all finite N. A large dimensional limit for such processes is then characterized by the following result, [Guionnet, 2006, Lemma 12.5].

Theorem 19.2. *Let $\boldsymbol{\lambda}^N(0) \in \mathbb{R}^N$ such that $F^{\mathbf{Y}_N(0)}$ converges weakly towards F_0 as N tends to infinity. Further, assume that*

$$\sup_N \int \log(1 + x^2) dF^{\mathbf{Y}_N(0)}(x) < \infty.$$

Then, for all $T > 0$, the measure-valued process $(F^{\mathbf{Y}_N(t)},\ t \in [0, T])$ converges almost surely in the set of distribution function-valued continuous functions defined on $[0, T]$ towards $(F_t,\ t \in [0, T])$, such that, for all $z \in \mathbb{C} \setminus \mathbb{R}$

$$m_{F_t}(z) = m_{F_0}(z) + \int_0^t m_{F_s}(z) m'_{F_s}(z) ds$$

where the derivative of the Stieltjes transform is taken along z.

This result generalizes the free additive convolution to time continuous processes. What this exactly states is that, as N grows large, $F^{\mathbf{Y}_N(t)}$ converges almost surely to some d.f. F_t, which is continuous along the time variable t. This indicates that, for large N, the eigenvalues of the time-evolving random matrix $\mathbf{Y}_N(t)$ follow a trajectory, whose Stieltjes transform satisfies the above differential equation.

It is to be believed that such time-varying considerations, along with the recent growing interest in mean field and mean field game theories [Bordenave *et al.*, 2005; Buchegger and Le Boudec, 2005; Le Boudec *et al.*, 2007; Sarafijanovic and

Le Boudec, 2005; Sharma *et al.*, 2006], may lead to the opening up of a new field of research in wireless communications. Indeed, mean field theory is dealing with the characterization of large dimensional time-varying systems for which asymptotic behavior are found to be solutions of stochastic differential equations. Mean field theory is in particular used in game-theoretic settings where numerous players, whose space distribution enjoys some symmetric structure, compete under some cost function constraint. Such characterizations are suitable for the study of the medium access control for future large dimensional networks, where the adjective *large* qualifies the number of users in the network. The time-varying aspects developed above may well turn in the end into a characterization of the time evolution of the physical layer for future large dimensional networks, where the adjective *large* now characterizes, e.g. the number of transmit antennas at the base station in a cellular broadcast channel or the number of users in a CDMA cell.

The possibility to study the time evolution of large dimensional networks, be it from a physical layer, medium access control layer, or network layer point of view, provides much more information than discrete time analysis in the sense that:

- as already mentioned, the classical static analysis brought by random matrix theory in wireless communications only allows us to anticipate the average performance and outage performance of a given communication system. That is, irrespective of the time instant t, quality of service figures such as averaged or minimally ensured data delivery rate can be derived. Starting from a data rate R_0 at time t_0, it is however not possible to anticipate the averaged rate or minimally ensured rate R_t at time $t > t_0$ (unless t is large enough for the knowledge of R_0 to become irrelevant). This can be performed by continuous time analysis though;
- dynamic system analysis also allows us to anticipate probabilities of chaotic situations such as system failure after a time $t > t_0$, with t_0 some initial time when the system is under control. We mention specifically the scenario of automatized systems with control, such as recent smart energy distribution networks in which information feedback is necessary to set the system as a whole in equilibrium. However, in wireless communications, as in any other communication system, feeding information back comes at a price and is preferably limited. This is why being able to anticipate the (possibly erratic) evolution of a system in free motion is necessary so to be able to decide when control has to take place;
- along the same line of thought as in the previous point, it is also important for system designers to take into account and anticipate the consequences of mobility within a large dimensional network. Indeed, in a network where users' mobility is governed by some time-varying stochastic process, the achievable transmission data rates depend strongly on channel state information exchanges within the network. In the discrete time random matrix

framework, deterministic considerations of the necessary information feedback is assumed, irrespective of the users' mobility (i.e. irrespective of the frequency to which information would effectively need to be fed back). The introduction of dynamics and user mobility distributions would allow for a disruptive study of such networks from a physical layer point of view. We mention for instance the long-standing problem of multi-cellular cooperation for future wireless communication networks, which takes time to appear in telecommunication standards since it is still unclear whether the extra feedback cost involved (i.e. the channel fading information of all base station-to-user links required to all base stations) is detrimental to the cooperation gain in terms of peak data rate. This demands the performance analysis of a typical network with users in motion according to some stochastic behavior, which again is not accessible to this day in the restrictive framework of random matrix theory.

This concludes this short chapter on the envisioned perspectives for random matrix theory applications in wireless communication networks. We end this book with a general conclusion of Part I and Part II.

20 Conclusion

Throughout this book, we tried to propose an up-to-date vision of the fundamental applications of random matrix theory to wireless communications. "Up-to-date" refers to the time when these lines were written. At the pace which the random matrix field evolves these days, the current book will be largely outdated when published. This is one of the two fundamental reasons why we thoroughly introduced the methods used to derive most of the results known to this day, as these technical approaches will take more time to be replaced by more powerful tools. The other, more important, reason why such an emphasis was made on these techniques, is that the wireless communication community is moving fast and is in perpetual need for new random matrix models for which mathematical research has no answer yet. Quite often, such an answer does not exist because it is either of no apparent interest to mathematicians or simply because too many of these problems are listed that cannot all be solved in a reasonable amount of time. But very often also, these problems can be directly addressed by non-mathematical experts. We desired this book to be both accessible in the sense that fast solutions to classical problems can be derived by wireless communication engineers and rigorous in some of the proofs so that precise proof techniques be known to whomever desires to derive mathematically sound information-theoretic results.

An important outcome of the current acceleration of the breakthroughs made in random matrix theory for wireless communications is the generalization of non-mathematically accurate methods, such as the replica method, introduced briefly in Chapter 19. From our point of view, thanks to the techniques developed in this book, it is also fairly simple to derive deterministic equivalents for the very same problems addressed by the replica approach. Nonetheless, the replica method is a very handy tool for fast results that may take time to obtain using conventional methods. This is confirmed for instance by the work of Moustakas *et al.* [Moustakas and Simon, 2007] on frequency selective MIMO channels, later proved accurate by Dupuy and Loubaton in an unpublished work, or by the work of Taricco [Taricco, 2008] on MIMO Rician models, largely generalized by the work of Hachem *et al.* [Hachem *et al.*, 2008b], or again by the work of Simon *et al.* [Simon *et al.*, 2006] generalized and accurately proved by Couillet *et al.* [Couillet *et al.*, 2011a] and further extended in [Wagner *et al.*, 2011] by Wagner *et al.*. As reminded in Chapter 19, replica methods also provide results that are not at all immediate using conventional tools, and constitute, as such, an important tool to be further developed. However, we intend this book to reinforce the idea to the reader that the difficult Stieltjes transform and Gaussian method tools of yesterday are now clearly understood and have moved to a simple and very accessible framework, no longer exclusive to a few mathematicians among the research community. We recall in particular that the deterministic equivalent method we referred to as *Bai and Silverstein's approach* in this book is rather simple and only requires practice. Once deterministic equivalents are *inferred* via the "guess-work" technique, accurate proofs can then be performed, which are usually based on very classical techniques. Moreover, as most results end up being solutions of implicit equations, it is important for practical purposes to be able to prove solution uniqueness and if possible sure convergence of some fixed-point algorithm to the solution. One of the reasons comes as follows: in the situation where we have to estimate some key system parameters (such as the optimal transmit covariance matrix in MIMO communications) and that these parameters are one of the solutions to an implicit equation, if sure convergence of some iterative algorithm towards this specific solution is proved, then the stability of the system under consideration is ensured.

Another source of debate within the random matrix community is the question of whether free probabilistic tools or more conventional random matrix tools must be used to solve problems dealing with large dimensional random matrices. Some time ago, problems related to Haar matrices were all approached using free probability tools since the R- and S-transforms are rather convenient for dealing with sums or products of these types of matrices. Nonetheless, as an equivalent trace lemma for Haar random matrices, Theorem 6.15, exists, it is also possible to treat models involving Haar matrices with the same tools as those used for i.i.d. matrices, see, e.g., Theorem 6.17. Moreover, free probability relies on stringent assumptions on the matrices involved in sums and products, starting with the eigenvalue boundedness assumption that can be somewhat extended using more conventional random matrix techniques. There is therefore no fundamental reason to prefer the exclusive usage of free probability theory, as the same derivations and much more are accessible through classical random matrix theory. Nonetheless, it is usually simpler, when possible, to exploit directly free probability theorems for involved sums and products of asymptotically free matrix families than to resort to complete random matrix derivations and convergence proofs, see, e.g., Section 15.2 on multi-hop communications. Since both fields are not orthogonal, it is therefore possible to use results from both of them to come up fast with results on more involved matrix models.

Regarding the latest contributions on signal sensing and parameter estimation, the studies provided in this book showed that, while important limitations (linked to spectrum clustering) restrict the use of recent Stieltjes transform-based techniques, these tools perform outstandingly better than any other moment-based approach and obviously better than original algorithms that only assume one large system dimension. Obtaining Stieltjes transform estimators requires work, but is not so difficult once the relation between the l.s.d. of the observed matrix model and the e.s.d. of the hidden parameter matrix is found. Proving that the estimator is indeed correct requires to ensure that exact separation of the eigenvalues in the observed matrix holds true. Nonetheless, as recalled many times, proving exact separation, already for the simple information plus noise model, is a very involved problem. Obtaining exact separation for more involved models is therefore an open question, still under investigation and still rather exclusive to pure mathematicians. To this day, for these complex models, intellectual honesty requires to step back to less precise combinatoric moment-based methods, which are also much easier to derive, and, as we have seen, can be often automatically obtained by computer software.

Moment approaches are also the only access we have to even more involved random matrix models, such as random Vandermonde or random Toeplitz matrices. Parameter estimation can then be performed for models involving Vandermonde matrices using the inaccurate moment approach, as no alternative technique is available yet. Moment approaches also have the property to be able to provide the exact moments of some functionals of small dimensional random matrices on average, as well as exact covariance matrices of the successive moments. Nonetheless, we also saw that functionals of random matrices as large as 4×4 matrices are often very accurately deterministically approximated using methods that assume large dimensions. The interest of tools that assume small dimensions is therefore often very limited.

Small dimensional random matrix theory need not be discarded, though, as exemplified in Chapter 16 and Chapter 18, where important results on multi-source detection and more generally optimum statistical inference through the maximum entropy principle were introduced. Even if such approaches often lead to very involved expressions, from which sometimes not much can be said, they always provide upper-bounds on alternative approaches which are fundamental to assess the performance of such alternative suboptimal methods. However, small dimensional techniques are very often restricted to simple problems that are very symmetrical in the sense that the matrices involved need to have pleasant invariance properties. The increasing complexity of large dimensional systems comes however in contradiction with this simplicity requirement. It is therefore believed that small dimensional random matrix theory will leave more and more room for the much better performing large dimensional random matrix theory for all applications.

Finally, we mention that the most recent field of study, for which new results appear at an increasing pace, is that of the limiting distribution and the large deviation of smallest and largest eigenvalues for different types of models, asymptotic independence within the spectrum of large dimensional matrices, etc. These new ideas, stated in this book under the form of a series of important results rely on powerful tools, which necessitate a lengthy mathematical introduction to Fredholm determinants or operator theory, which we briefly provided in Chapter 19. It is believed that the time will come when these tools will be made simpler and more accessible so that non-specialists can also benefit from these important results in the medium to long-term.

To conclude, we wish to insist once more that random matrix theory, which was ten years ago still in its infancy with techniques only exploitable by mathematicians of the field, has now become more popular, is better understood, and provides wireless telecommunication researchers with a large pool of useful and accessible tools. We now enter an era where the initial results on system performance evaluation are commonplace and where thrilling results have now to do with statistical inference in large dimensional inverse problems involving possibly time-varying random matrix processes.

References

T. B. Abdallah and M. Debbah. Downlink CDMA: to cell or not to cell. In *12th European Signal Processing Conference (EUSIPCO'04)*, pages 197–200, Vienna, Austria, September 2004.

S. Adhikari. A non-parametric approach for uncertainty quantification in elastodynamics. In *Proceedings of the 47th AIAA/ASME/ASCE/AHS/ASC Structures, Structural Dynamics and Materials Conference*, 2006.

D. Aktas, M. N. Bacha, J. S. Evans, and S. V. Hanly. Scaling results on the sum capacity of cellular networks with MIMO links. *IEEE Transactions on Information Theory*, 52(7):3264–3274, 2006.

I. F. Akyildiz, W. Y. Lee, M. C. Vuran, and S. Mohanty. Next generation/dynamic spectrum access/cognitive radio wireless networks: a survey. *Computer Networks Journal*, 50(13):2127–2159, 2006.

P. Almers, E. Bonek, A. Burr, N. Czink, M. Debbah, V. Degli-Esposti, H. Hofstetter, P. Kyosti, D. Laurenson, and G. Matz et al. Survey of channel and radio propagation models for wireless MIMO systems. *EURASIP Journal on Wireless Communications and Networking*, 2007.

G. W. Anderson, A. Guionnet, and O. Zeitouni. Lecture notes on random matrices, 2006. `www.mathematik.uni-muenchen.de/~lerdos/SS09/Random/randommatrix.pdf`. SAMSI, Lecture Notes.

G. W. Anderson, A. Guionnet, and O. Zeitouni. *An introduction to random matrices.* Cambridge University Press, 2010. ISBN 0521194520.

T. W. Anderson. The non-central Wishart distribution and certain problems of multivariate statistics. *The Annals of Mathematical Statistics*, 17(4):409–431, 1946.

T. W. Anderson. Asymptotic theory for principal component analysis. *Annals of Mathematical Statistics*, 34(1):122–148, March 1963.

L. Arnold. On the asymptotic distribution of the eigenvalues of random matrices. *Journal of Mathematics and Analytic Applications*, 20:262–268, 1967.

L. Arnold. On Wigner's semi-circle law for the eigenvalues of random matrices. *Probability Theory and Related Fields*, 19(3):191–198, September 1971.

L. Arnold, V. M. Gundlach, and L. Demetrius. Evolutionary formalism for products of positive random matrices. *The Annals of Applied Probability*, 4 (3):859–901, 1994.

Z. D. Bai. Circular law. *The Annals of Probability*, 25(1):494–529, 1997.

Z. D. Bai and J. W. Silverstein. No eigenvalues outside the support of the limiting spectral distribution of large dimensional sample covariance matrices. *The Annals of Probability*, 26(1):316–345, January 1998.

Z. D. Bai and J. W. Silverstein. Exact separation of eigenvalues of large dimensional sample covariance matrices. *The Annals of Probability*, 27(3): 1536–1555, 1999.

Z. D. Bai and J. W. Silverstein. CLT of linear spectral statistics of large dimensional sample covariance matrices. *The Annals of Probability*, 32(1A): 553–605, 2004.

Z. D. Bai and J. W. Silverstein. On the signal-to-interference-ratio of CDMA systems in wireless communications. *Annals of Applied Probability*, 17(1):81–101, 2007.

Z. D. Bai and J. W. Silverstein. *Spectral analysis of large dimensional random matrices.* Springer Series in Statistics, New York, NY, USA, second edition, 2009.

Z. D. Bai and J. F. Yao. Central limit theorems for eigenvalues in a spiked population model. *Annales de lInstitut Henri Poincaré-Probabilités et Statistiques*, 44(3):447–474, 2008a.

Z. D. Bai and J. F. Yao. Limit theorems for sample eigenvalues in a generalized spiked population model. 2008b. `http://arxiv.org/abs/0806.1141`.

J. Baik and J. W. Silverstein. Eigenvalues of large sample covariance matrices of spiked population models. *Journal of Multivariate Analysis*, 97(6):1382–1408, 2006.

J. Baik, G. Ben Arous, and S. Péché. Phase transition of the largest eigenvalue for non-null complex sample covariance matrices. *The Annals of Probability*, 33(5):1643–1697, 2005.

F. Benaych-Georges. Rectangular random matrices, related free entropy and free Fisher's information. *Journal of Operator Theory*, 62(2):371–419, 2009.

F. Benaych-Georges and R. Rao. The eigenvalues and eigenvectors of finite, low rank perturbations of large random matrices. *Advances in Mathematics*, 227 (1):494–521, 2011. ISSN 0001-8708.

F. Benaych-Georges, A. Guionnet, and M. Maida. Fluctuations of the extreme eigenvalues of finite rank deformations of random matrices. 2010. `http://arxiv.org/abs/1009.0145`.

H. Bercovici and V. Pata. The law of large numbers for free identically distributed random variables. *The Annals of Probability*, 24(1):453–465, 1996.

P. Bianchi, M. Debbah, and J. Najim. Asymptotic independence in the spectrum of the Gaussian Unitary Ensemble. *Electronic Communications in Probability*, 15:376–395, 2010. `http://arxiv.org/abs/0811.0979`.

P. Bianchi, J. Najim, M. Maida, and M. Debbah. Performance of some eigen-based hypothesis tests for collaborative sensing. *IEEE Transactions on Information Theory*, 57(4):2400–2419, 2011.

P. Biane. Free probability for probabilists. *Quantum Probability Communications*, 11:55–71, 2003.

E. Biglieri, G. Caire, and G. Tarico. CDMA system design through asymptotic analysis. *IEEE Transactions on Communications*, 48:1882–1896, November 2000.

E. Biglieri, G. Caire, G. Taricco, and E. Viterbo. How fading affects CDMA: an asymptotic analysis with linear receivers. *IEEE Journal on Selected Areas in Communications (Wireless Series)*, 19(2):191–201, 2001.

P. Billingsley. *Convergence of Probability Measures.* John Wiley and Sons, Inc., Hoboken, NJ, 1968.

P. Billingsley. *Probability and Measure.* John Wiley and Sons, Inc., Hoboken, NJ, third edition, 1995.

N. Bonneau, E. Altman, M. Debbah, and G. Caire. When to synchronize in uplink CDMA. In *Proceedings of IEEE International Symposium on Information Theory (ISIT'05)*, pages 337–341, 2005.

N. Bonneau, M. Debbah, and E. Altman. Wardrop equilibrium in CDMA networks. In *Workshop on Resource Allocation in Wireless Networks*, Limassol, Cyprus, 2007.

N. Bonneau, M. Debbah, E. Altman, and A. Hjørungnes. Non-atomic games for multi-user systems. *IEEE Journal on Selected Areas in Communications*, 26 (7):1047–1058, 2008.

S. Borade, L. Zheng, and R. Gallager. Amplify-and-forward in wireless relay networks: rate, diversity, and network size. *IEEE Transactions on Information Theory*, 53(10):3302–3318, 2007.

C. Bordenave. Eigenvalues of euclidean random matrices. *Random Structures and Algorithms*, 33(4):515–532, 2008.

C. Bordenave, D. McDonald, and A. Proutière. Random multi-access algorithms, a mean field analysis. In *Proceedings of IEEE Annual Allerton Conference on Communication, Control, and Computing (Allerton'05)*, Allerton, IL, USA, 2005.

G. L. Bretthorst. *Bayesian spectrum analysis and parameter estimation.* PhD thesis, Washington University, St. Louis, 1987.

L. Brillouin. *Science and Information Theory.* Academic Press, New York, second edition, 1962.

S. Buchegger and J. Y. Le Boudec. Self-policing mobile ad-hoc networks by reputation systems. *IEEE Communication Magazine*, 43(7):101–107, 2005.

D. Cabric, S. M. Mishra, and R. W. Brodersen. Implementation issues in spectrum sensing for cognitive radios. In *Proceedings of IEEE Conference Record of the Asilomar Conference on Signals, Systems, and Computers (ASILOMAR'04)*, pages 772–776, Pacific Grove, CA, USA, 2004.

D. Cabric, A. Tkachenko, and R. W. Brodersen. Spectrum sensing measurements of pilot, energy and collaborative detection. In *IEEE Military Communications Conference*, October 2006.

G. Caire and S. Shamai. On the achievable throughput of a multiantenna Gaussian broadcast channel. *IEEE Transactions on Information Theory*, 49 (7):1691–1706, 2003.

D. Calin, H. Claussen, and H. Uzunalioglu. On femto deployment architectures and macrocell offloading benefits in joint macro-femto deployments. *IEEE Transactions on Communications*, 48(1):26–32, January 2010.

L. S. Cardoso, M. Debbah, P. Bianchi, and J. Najim. Cooperative spectrum sensing using random matrix theory. In *IEEE Pervasive Computing (ISWPC'08)*, pages 334–338, Santorini, Greece, May 2008.

A. Caticha. Maximum entropy, fluctuations and priors. *Maximum Entropy and Bayesian Methods in Science and Engineering*, 568:94–106, 2001.

H. Chandra. Differential operators on a semi-simple Lie algebra. *American Journal of Mathematics*, 79:87–120, 1957.

V. Chandrasekhar, M. Kountouris, and J. G. Andrews. Coverage in multi-antenna two-tier networks. *IEEE Transactions on Wireless Communications*, 8(10):5314–5327, 2009.

J. M. Chaufray, W. Hachem, and P. Loubaton. Asymptotic analysis of optimum and sub-optimum CDMA downlink MMSE receivers. *IEEE Transactions on Information Theory*, 50(11):2620–2638, 2004.

S. S. Christensen, R. Agarwal, E. D. Carvalho, and J. M. Cioffi. Weighted sum-rate maximization using weighted MMSE for MIMO-BC beamforming design, Part I. *IEEE Transactions on Wireless Communications*, 7(12):4792–4799, 2008.

C. N. Chuah, D. N. C. Tse, J. M. Kahn, and R. A. Valenzuela. Capacity scaling in MIMO wireless systems under correlated fading. *IEEE Transactions on Information Theory*, 48(3):637–650, March 2002.

P. Chung, J. Böhme, C. Mecklenbraüker, and A. Hero. Detection of the number of signals using the Benjamini-Hochberg procedure. *IEEE Transactions on Signal Processing*, 55(6):2497–2508, 2007.

H. Claussen, L. T. Ho, and L. G. Samuel. An overview of the femtocell concept. *Bell Labs Technical Journal*, 13(1):221–245, May 2008.

M. H. M. Costa. Writing on dirty paper. *IEEE Transactions on Information Theory*, 29(3):439–441, 1983.

L. Cottatellucci and M. Debbah. The effect of line of sight on the asymptotic capacity of MIMO systems. In *Proceedings of IEEE International Symposium on Information Theory (ISIT'04)*, page 542, Chicago, USA, July 2004a.

L. Cottatellucci and M. Debbah. On the capacity of MIMO Rice channels. In *Proceedings of IEEE Annual Allerton Conference on Communication, Control, and Computing (Allerton'04)*, Allerton, IL, USA, October 2004b.

L. Cottatellucci and R. Müller. Asymptotic design and analysis of multistage detectors with unequal powers. In *Proceedings of IEEE Information Theory Workshop (ITW'02)*, pages 167–170, Bangalore, India, 2002.

L. Cottatellucci and R. Müller. A systematic approach to multistage detectors in multipath fading channels. *IEEE Transactions on Information Theory*, 51 (9):3146–3158, 2005.

L. Cottatellucci, R. Müller, and M. Debbah. Asymptotic design and analysis of linear detectors for asynchronous CDMA systems. In *Proceedings of IEEE*

International Symposium on Information Theory (ISIT'04), Chicago, IL, USA, 2004.

L. Cottatellucci, R. Müller, and M. Debbah. Asynchronous CDMA systems with random spreading – Part I: Fundamental limits. *IEEE Transactions on Information Theory*, 56(4):1477–1497, 2010a.

L. Cottatellucci, R. Müller, and M. Debbah. Asynchronous CDMA systems with random spreading – Part II: Design criteria. *IEEE Transactions on Information Theory*, 56(4):1498–1520, 2010b.

R. Couillet and M. Debbah. Free deconvolution for OFDM multicell SNR detection. In *Proceedings of IEEE International Symposium on Personal, Indoor and Mobile Radio Communications (PIMRC'08)*, Cannes, France, 2008.

R. Couillet and M. Debbah. Uplink capacity of self-organizing clustered orthogonal CDMA networks in flat fading channels. In *Proceedings of IEEE Information Theory Workshop*, Taormina, Sicily, 2009.

R. Couillet and M. Debbah. A Bayesian framework for collaborative multi-source signal detection. *IEEE Transactions on Signal Processing*, 58(10):5186–5195, October 2010a.

R. Couillet and M. Debbah. Information theoretic approach to synchronization: the OFDM carrier frequency offset example. In *Sixth Advanced International Conference on Telecommunications (AICT)*, Barcelona, Spain, 2010b.

R. Couillet and M. Guillaud. Performance of statistical inference methods for the energy estimation of multiple sources. In *Proceedings of IEEE Workshop on Statistical Signal Processing (SSP'11)*, Nice, France, 2011. To appear.

R. Couillet and W. Hachem. Local failure detection and diagnosis in large sensor networks. *IEEE Transactions on Information Theory*, 2011. Submitted for publication.

R. Couillet, S. Wagner, M. Debbah, and A. Silva. The space frontier: physical limits of multiple antenna information transfer. In *Workshop on Interdisciplinary Systems Approach in Performance Evaluation and Design of Computer and Communication Systems (Inter-Perf'08)*, Athens, Greece, 2008.

R. Couillet, A. Ancora, and M. Debbah. Bayesian foundations of channel estimation for smart radios. *Advances in Electronics and Telecommunications*, 1(1):41–49, 2010.

R. Couillet, M. Debbah, and J. W. Silverstein. A deterministic equivalent for the analysis of correlated MIMO multiple access channels. *IEEE Transactions on Information Theory*, 57(6):3493–3514, June 2011a.

R. Couillet, J. Hoydis, and M. Debbah. Deterministic equivalents for the analysis of unitary precoded systems. *IEEE Transactions on Information Theory*, 2011b. `ttp://arxiv.org/abs/1011.3717`. Submitted for publication.

R. Couillet, J. W. Silverstein, Z. D. Bai, and M. Debbah. Eigen-inference for energy estimation of multiple sources. *IEEE Transactions on Information Theory*, 57(4):2420–2439, 2011c.

R. T. Cox. Probability, frequency and reasonable expectation. *American Journal of Physics*, 14(1):1–13, 1946.

A. D. Dabbagh and D. J. Love. Multiple antenna MMSE based downlink precoding with quantized feedback or channel mismatch. *IEEE Transactions on Communications*, 56(11):1859–1868, 2008.

R. de Lacerda Neto, A. Menouni Hayar, M. Debbah, and B. H. Fleury. A maximum entropy approach to ultra-wideband channel modelling. *Proceedings of the 31st IEEE International Conference on Acoustics, Speech, and Signal Processing*, 2006.

M. Debbah and R. Müller. Impact of the power of the steering directions on the asymptotic capacity of MIMO channels. In *Proceedings of IEEE International Symposium on Signal Processing and Information Technology (ISSPIT'03)*, Darmstadt, Germany, December 2003.

M. Debbah and R. Müller. MIMO channel modelling and the principle of maximum entropy. *IEEE Transactions on Information Theory*, 51(5):1667–1690, 2005.

M. Debbah and R. Müller. Capacity complying MIMO channel models. In *Proceedings of IEEE Conference Record of the Asilomar Conference on Signals, Systems, and Computers (ASILOMAR'03)*, Pacific Grove, CA, USA, November 2003.

M. Debbah, W. Hachem, P. Loubaton, and M. de Courville. MMSE analysis of certain large isometric random precoded systems. *IEEE Transactions on Information Theory*, 49(5):1293–1311, May 2003a.

M. Debbah, P. Loubaton, and M. de Courville. The spectral efficiency of linear precoders. In *Proceedings of IEEE Information Theory Workshop (ITW'03)*, pages 90–93, Paris, France, March 2003b.

P. Deift. *Orthogonal Polynomials and Random Matrices: a Riemann-Hilbert Approach.* New York University Courant Institute of Mathematical Sciences, New York, NY, USA, 2000.

A. Dembo and O. Zeitouni. *Large Deviations Techniques and applications.* Springer Verlag, 2009.

P. Ding, D. J. Love, and M. D. Zoltowski. Multiple antenna broadcast channels with shape feedback and limited feedback, Part I. *IEEE Transactions on Signal Processing*, 55(7):3417–3428, 2007.

B. Dozier and J. W. Silverstein. On the empirical distribution of eigenvalues of large dimensional information plus noise-type matrices. *Journal of Multivariate Analysis*, 98(4):678–694, 2007a.

B. Dozier and J. W. Silverstein. Analysis of the limiting spectral distribution of large dimensional information-plus-noise type matrices. *Journal of Multivariate Analysis*, 98(6):1099–1122, 2007b.

J. Dumont, W. Hachem, S. Lasaulce, P. Loubaton, and J. Najim. On the capacity achieving covariance matrix for Rician MIMO channels: an asymptotic approach. *IEEE Transactions on Information Theory*, 56(3):1048–1069, 2010.

F. Dupuy and P. Loubaton. Mutual information of frequency selective MIMO systems: an asymptotic approach, 2009. http://www-syscom.univ-mlv.fr/~fdupuy/publications.php.

F. Dupuy and P. Loubaton. On the capacity achieving covariance matrix for frequency selective MIMO channels using the asymptotic approach. *IEEE Transactions on Information Theory*, 2010. http://arxiv.org/abs/1001.3102. To appear.

F. J. Dyson. Statistical theory of the energy levels of complex systems, Part II. *Journal of Mathematical Physics*, 3:157–165, January 1962a.

F. J. Dyson. A Brownian-motion model for the eigenvalues of a random matrix. *Journal of Mathematical Physics*, 3:1191–1198, 1962b.

S. Enserink and D. Cochran. A cyclostationary feature detector. In *Proceedings of IEEE Conference Record of the Asilomar Conference on Signals, Systems, and Computers (ASILOMAR'94)*, pages 806–810, Pacific Grove, CA, USA, 1994.

J. Evans and D. N. C. Tse. Large system performance of linear multiuser receivers in multipath fading channels. *IEEE Transactions on Information Theory*, 46 (6):2059–2078, 2000.

K. Fan. Maximum properties and inequalities for the eigenvalues of completely continuous operators. *Proceedings of the National Academy of Sciences of the United States of America*, 37(11):760–766, 1951.

J. Faraut. Random matrices and orthogonal polynomials. Lecture Notes, CIMPA School of Merida, 2006. www.math.jussieu.fr/~faraut/Merida.Notes.pdf.

N. Fawaz and M. Médard. On the non-coherent wideband multipath fading relay channel. In *Proceedings of IEEE International Symposium on Information Theory (ISIT'10)*, pages 679–683, Austin, Texas, USA, 2010.

N. Fawaz, K. Zarifi, M. Debbah, and D. Gesbert. Asymptotic capacity and optimal precoding in MIMO multi-hop relay networks. *IEEE Transactions on Information Theory*, 57(4):2050–2069, 2011. ISSN 0018-9448.

O. N. Feldheim and S. Sodin. A universality result for the smallest eigenvalues of certain sample covariance matrices. *Geometric And Functional Analysis*, 20 (1):88–123, 2010. ISSN 1016-443X.

R. A. Fisher. The sampling distribution of some statistics obtained from non-linear equations. *The Annals of Eugenics*, 9:238–249, 1939.

S. V. Fomin and I. M. Gelfand. *Calculus of Variations.* Prentice Hall, 2000.

G. J. Foschini and M. J. Gans. On limits of wireless communications in a fading environment when using multiple antennas. *Wireless Personal Communications*, 6(3):311–335, March 1998.

M. Franceschetti, S. Marano, and F. Palmieri. The role of entropy in wave propagation. In *Proceedings of IEEE International Symposium on Information Theory (ISIT'03)*, Yokohama, Japan, July 2003.

Y. V. Fyodorov. Introduction to the random matrix theory: Gaussian unitary ensemble and beyond. *Recent Perspectives in Random Matrix Theory and Number Theory*, 322:31–78, 2005.

W. A. Gardner. Exploitation of spectral redundancy in cyclostationary signals. *IEEE Signal Processing Magazine*, 8(2):14–36, 1991.

S. Geman. A limit theorem for the norm of random matrices. *The Annals of Probability*, 8(2):252–261, 1980.

A. Ghasemi and E. S. Sousa. Collaborative spectrum sensing for opportunistic access in fading environments. In *IEEE Proceedings of the International Symposium on Dynamic Spectrum Access Networks*, pages 131–136, 2005.

A. Ghasemi and E. S. Sousa. Spectrum sensing in cognitive radio networks: the cooperation-processing tradeoff. *Wireless Communications and Mobile Computing*, 7(9):1049–1060, 2007.

V. L. Girko. Ten years of general statistical analysis. `www.general-statistical-analysis.girko.freewebspace.com/chapter14.pdf`.

V. L. Girko. *Theory of Random Determinants.* Kluwer, Kluwer Academic Publishers, Dordrecht, The Netherlands, 1990.

M. A. Girshick. On the sampling theory of roots of determinantal equations. *The Annals of Math. Statistics*, 10:203–204, 1939.

J. Glimm and A. Jaffe. *Quantum Physics.* Springer, New York, NY, USA, 1981.

A. Goldsmith, S. A. Jafar, N. Jindal, and S. Vishwanath. Capacity limits of MIMO channels. *IEEE Journal on Selected Areas in Communications*, 21(5): 684–702, 2003.

I. S. Gradshteyn and I. M. Ryzhik. *Table of Integrals, Series and Products.* Academic Press, sixth edition, 2000.

A. J. Grant and P. D. Alexander. Random sequence multisets for synchronous code-division multiple-access channels. *IEEE Transactions on Information Theory*, 44(7):2832–2836, November 1998.

R. M. Gray. Toeplitz and circulant matrices: a review. *Foundations and Trends in Communications and Information Theory*, 2(3), 2006.

D. Gregoratti and X. Mestre. Random DS/CDMA for the amplify and forward relay channel. *IEEE Transactions on Wireless Communications*, 8(2):1017–1027, 2009.

D. Gregoratti, W. Hachem, and X. Mestre. Randomized isometric linear-dispersion space-time block coding for the DF relay channel. *IEEE Transactions on Signal Processing*, 2010. Submitted for publication.

M. Guillaud, M. Debbah, and A. L. Moustakas. Maximum entropy MIMO wireless channel models, 2006. `http://arxiv.org/abs/cs.IT/0612101`.

M. Guillaud, M. Debbah, and A. L. Moustakas. Modeling the multiple-antenna wireless channel using maximum entropy methods. In *International Workshop on Bayesian Inference and Maximum Entropy Methods in Science and Engineering (MaxEnt'07)*, pages 435–442, Saratoga Springs, NY, November 2007.

A. Guionnet. Large random matrices: lectures on macroscopic asymptotics. École d'Été de Probabilités de Saint-Flour XXXVI-2006, 2006. `www.umpa.ens-lyon.fr/~aguionne/cours.pdf`.

D. Guo and S. Verdú. Multiuser detection and statistical mechanics. *Communications, Information and Network Security*, pages 229–277, 2002.

D. Guo, S. Verdú, and L. K. Rasmussen. Asymptotic normality of linear multiuser receiver outputs. *IEEE Transactions on Information Theory*, 48 (12):3080–3095, December 2002.

U. Haagerup, H. Schultz, and S. Thorbjørnsen. A random matrix approach to the lack of projections in Cred*(F2). *Advances in Mathematics*, 204(1):1–83, 2006.

W. Hachem. Simple polynomial MMSE receivers for CDMA transmissions on frequency selective channels. *IEEE Transactions on Information Theory*, pages 164–172, January 2004.

W. Hachem. An expression for $\int \log(t/\sigma^2 + 1)\mu \boxplus \tilde{\mu}(dt)$, 2008. unpublished.

W. Hachem, P. Loubaton, and J. Najim. Deterministic equivalents for certain functionals of large random matrices. *Annals of Applied Probability*, 17(3): 875–930, 2007.

W. Hachem, O. Khorunzhy, P. Loubaton, J. Najim, and L. A. Pastur. A new approach for capacity analysis of large dimensional multi-antenna channels. *IEEE Transactions on Information Theory*, 54(9), 2008a.

W. Hachem, P. Loubaton, and J. Najim. A CLT for information theoretic statistics of Gram random matrices with a given variance profile. *The Annals of Probability*, 18(6):2071–2130, December 2008b.

W. Hachem, P. Loubaton, X. Mestre, J. Najim, and P. Vallet. A subspace estimator of finite rank perturbations of large random matrices. *Journal on Multivariate Analysis*, 2011. Submitted for publication.

L. Hanlen and A. Grant. Capacity analysis of correlated MIMO channels. In *Proceedings of IEEE International Symposium on Information Theory (ISIT'03)*, Yokohama, Japan, July 2003.

S. V. Hanly and D. N. C. Tse. Resource pooling and effective bandwidths in CDMA networks with multiuser receivers and spatial diversity. *IEEE Transactions on Information Theory*, pages 1328–1351, May 2001.

B. Hassibi and B. M. Hochwald. How Much Training is Needed in Multiple-Antenna Wireless Links. *IEEE Transactions on Information Theory*, 49(4): 951–963, April 2003.

A. Haurie and P. Marcotte. On the relationship between Nash–Cournot and Wardrop equilibria. *Networks*, 15(1):295–308, 1985.

F. Hiai and D. Petz. *The Semicircle Law, Free Random Variables and Entropy - Mathematical Surveys and Monographs No. 77.* American Mathematical Society, Providence, RI, USA, 2006.

B. Hochwald and S. Vishwanath. Space-time multiple access: Linear growth in the sum rate. *Proceedings of IEEE Annual Allerton Conference on Communication, Control, and Computing (Allerton'02)*, 2002.

B. M. Hochwald, T. L. Marzetta, and V. Tarokh. Multiple-antenna channel hardening and its implications for rate feedback and scheduling. *IEEE Transactions on Information Theory*, 50(9):1893–1909, 2004.

M. L. Honig and W. Xiao. Performance of reduced-rank linear interference suppression. *IEEE Transactions on Information Theory*, 47(5):1928–1946, May 2001.

R. A. Horn and C. R. Johnson. *Matrix Analysis.* Cambridge University Press, 1985.

R. A. Horn and C. R. Johnson. *Topics in Matrix Analysis.* Cambridge University Press, 1991.

J. Hoydis, M. Kobayashi, and M. Debbah. Asymptotic performance of linear receivers in network MIMO. In *Proceedings of IEEE Conference Record of the Asilomar Conference on Signals, Systems, and Computers (ASILOMAR'10)*, Pacific Grove, CA, USA, November 2010.

J. Hoydis, R. Couillet, and M. Debbah. Random beamforming over correlated fading channels. *IEEE Transactions on Information Theory*, 2011a. Submitted for publication.

J. Hoydis, R. Couillet, and M. Debbah. Asymptotic analysis of double-scattering channels. In *Proceedings of IEEE Conference Record of the Asilomar Conference on Signals, Systems, and Computers (ASILOMAR'11)*, Pacific Grove, CA, USA, 2011b.

J. Hoydis, M. Debbah, and M. Kobayashi. Asymptotic moments for interference mitigation in correlated fading channels. In *Proceedings of IEEE International Symposium on Information Theory (ISIT'11)*, Saint Petersburg, Russia, August 2011c. http://arxiv.org/abs/1104.4911.

J. Hoydis, M. Kobayashi, and M. Debbah. Optimal channel training in uplink network mimo systems. *IEEE Transactions on Signal Processing*, 59(6), June 2011d.

M. Hoyhtya, A. Hekkala, and A. Mammela. Spectrum awareness: techniques and challenges for active spectrum sensing. *Springer Cognitive Networks*, 3: 353–372, April 2007.

P. L. Hsu. On the distribution of roots of certain determinantal equations. *The Annals of Eugenics*, 9:250–258, 1939.

H. Huh, G. Caire, S. H. Moon, and I. Lee. Multi-cell MIMO downlink with fairness criteria: the large-system limit. In *Proceedings of IEEE International Symposium on Information Theory (ISIT'10)*, pages 2058–2062, June 2010.

Y. Hur, J. Park, W. Woo, K. Lim, C. H. Lee, H. S. Kim, and J. Laskar. A wideband analog multi-resolution spectrum sensing (MRSS) technique for cognitive radio (CR) systems. In *IEEE International Symposium on Circuits and Systems (ISCAS'06)*, page 4, Island of Kos, Greece, 2006.

A. A. Hutter, E. Carvalho, and J. M. Cioffi. On the impact of channel estimation for multiple antenna diversity reception in mobile OFDM systems. *Proceedings of IEEE Conference Record of the Asilomar Conference on Signals, Systems, and Computers (Asimolar'00)*, 2, 2000.

C. Hwang. A brief survey on the spectral radius and the spectral distribution of large dimensional random matrices with i.i.d. entries. *Random Matrices and Their Applications*, 50:145–152, 1986.

C. Hwang. Eigenvalue distribution of correlation matrix in asynchronous CDMA with infinite observation window width. In *Proceedings of IEEE International Symposium on Information Theory (ISIT'07)*, Nice, France, June 2007.

C. Itzykson and J. B. Zuber. *Quantum Field Theory*. Dover Publications, 2006.

A. T. James. Distributions of matrix variates and latent roots derived from normal samples. *The Annals of Mathematical Statistics*, 35(2):475–501, 1964.

E. T. Jaynes. Information theory and statistical mechanics, Part I. *Physical Review*, 106(2):620–630, 1957a.

E. T. Jaynes. Information theory and statistical mechanics, Part II. *Physical Review*, 108(2):171–190, 1957b.

E. T. Jaynes. *Probability Theory: The Logic of Science*. Cambridge University Press, 2003.

H. Jeffreys. An invariant form for the prior probability in estimation problems. *Proceedings of the Royal Society of London. Series A, Mathematical and Physical Sciences*, 186(1007):453–461, 1946.

S. Jin, M. R. McKay, X. Gao, and I. B. Collings. MIMO multichannel beamforming: SER and outage using new eigenvalue distributions of complex noncentral Wishart matrices. *IEEE Transactions on Communications*, 56(3): 424–434, 2008. http://arxiv.org/abs/0611007.

N. Jindal. MIMO broadcast channels with finite-rate feedback. *IEEE Transactions on Information Theory*, 52(11):5045–5060, 2006.

M. Joham, K. Kusume, M. H. Gzara, W. Utschick, and J. A. Nossek. Transmit Wiener filter for the downlink of TDD DS-CDMA systems. *Proceedings of ISSSTA 2002*, 1:9–13, 2002.

K. Johansson. Shape fluctuations and random matrices. *Communications of Mathematical Physics*, 209:437–476, 2000.

I. M. Johnstone. On the distribution of the largest eigenvalue in principal components analysis. *Annals of Statistics*, 99(2):295–327, 2001.

I. M. Johnstone. High dimensional statistical inference and random matrices. In *International Congress of Mathematicians I*, pages 307–333, Zürich, Germany, 2006. European Mathematical Society.

F. Kaltenberger, M. Kountouris, D. Gesbert, and R. Knopp. On the trade-off between feedback and capacity in measured MU-MIMO channels. *IEEE Transactions on Wireless Communications*, 8(9):4866–4875, 2009.

M. A. Kamath, B. L Hughes, and Y. Xinying. Gaussian approximations for the capacity of MIMO Rayleigh fading channels. In *Proceedings of IEEE Conference Record of the Asilomar Conference on Signals, Systems, and Computers (ASILOMAR'02)*, pages 614–618, Pacific Grove, CA, USA, 2002.

A. Kammoun, R. Couillet, J. Najim, and M. Debbah. Performance of capacity inference methods under colored interference, 2011. Submitted for publication.

J. N. Kapur. *Maximum Entropy Models in Science and Engineering*. John Wiley and Sons, Inc., New York, 1989.

N. El Karoui. Tracy-Widom limit for the largest eigenvalue of a large class of complex sample covariance matrices. *The Annals of Probability*, 35(2):663–714,

2007.

N. El Karoui. Spectrum estimation for large dimensional covariance matrices using random matrix theory. *Annals of Statistics*, 36(6):2757–2790, December 2008.

S. M. Kay. *Fundamentals of Statistical Signal Processing: Estimation Theory*. Prentice-Hall, Englewood Cliffs, NJ, USA, 1993.

A. M. Khorunzhy, B. A. Khoruzhenko, and L. A. Pastur. On asymptotic properties of large random matrices with independent entries. *Journal of Mathematical Physics*, 37(10):5033–5061, 1996.

H. Kim and K. G. Shin. In-band spectrum sensing in cognitive radio networks: energy detection or feature detection? In *ACM Internatioanl Conference on Mobile Computing and Networking*, pages 14–25, San Francisco, CA, USA, September 2008.

P. Koev. Random matrix statistics toolbox. `http://math.mit.edu/~plamen/software/rmsref.html`.

V. I. Kostylev. Energy detection of a signal with random amplitude. In *Proceedings of IEEE International Conference on Communications (ICC'02)*, pages 1606–1610, New York, NY, USA, 2002.

L. Laloux, P. Cizeau, M. Potters, and J. P. Bouchaud. Random matrix theory and financial correlations. *International Journal of Theoretical and Applied Finance*, 3(3):391–397, July 2000.

J. Y. Le Boudec, D. McDonald, and J. Mundinger. A generic mean field convergence result for systems of interacting objects. In *International Conference on the Quantitative Evaluation of Systems (QEST'07)*, Budapest, Hungary, 2007.

O. Levêque and I. E. Telatar. Information-theoretic upper bounds on the capacity of large extended ad hoc wireless networks. *IEEE Transactions on Information Theory*, 51(3):858–865, March 2005.

N. Levy and S. Shamai. Clustered local decoding for Wyner-type cellular models. In *Proceedings of IEEE Information Theory and Applications Workshop (ITA'09)*, pages 318–322, San Diego, CA, USA, 2009.

L. Li, A. M. Tulino, and S. Verdú. Design of reduced-rank MMSE multiuser detectors using random matrix methods. *IEEE Transactions on Information Theory*, 50(6):986–1008, June 2004.

P. Loubaton and W. Hachem. Asymptotic analysis of reduced rank wiener filters. In *Proceedings of IEEE Information Theory Workshop (ITW'03)*, pages 328–331, Paris, France, 2003.

S. Loyka. Channel capacity of MIMO architecture using the exponential correlation matrix. *IEEE Communication Letters*, 5(9):1350–1359, September 2001.

A. Lytova and L. A. Pastur. Central Limit Theorem for linear eigenvalue statistics of random matrices with independent entries. *The Annals of Probability*, 37(5):1778–1840, 2009.

U. Madhow and M. L. Honig. MMSE interference suppression for direct-sequence spread-spectrum CDMA. *IEEE Transactions on Communications*, 42(12): 3178–3188, December 1994.

A. Mantravadi and V. V. Veeravalli. Mmse detection in asynchronous cdma systems: An equivalence result. *IEEE Transactions on Information Theory*, 48(12):3128–3137, December 2002.

I. Maric and R. D. Yates. Bandwidth and power allocation for cooperative strategies in Gaussian relay networks. *IEEE Transactions on Information Theory*, 56(4):1880–1889, 2010.

I. Maric, A. Goldsmith, and M. Médard. Analog network coding in the high-SNR regime. In *IEEE Wireless Network Coding Conference (WiNC'10)*, pages 1–6, Boston, MA, USA, 2010.

C. Martin and B. Ottersten. Asymptotic eigenvalue distributions and capacity for MIMO channels under correlated fading. *IEEE Transactions on Wireless Communications*, 3(4):1350–1359, July 2004.

V. A. Marčenko and L. A. Pastur. Distributions of eigenvalues for some sets of random matrices. *Math USSR-Sbornik*, 1(4):457–483, April 1967.

A. Masucci, Ø. Ryan, S. Yang, and M. Debbah. Finite dimensional statistical inference. *IEEE Transactions on Information Theory*, 57(4):2457–2473, 2011. ISSN 0018-9448.

M. L. McCloud and L. L. Scharf. A new subspace identification algorithm for high-resolution DOA estimation. *IEEE Transactions on Antennas and Propagation*, 50(10):1382–1390, 2002.

M. L. Mehta. *Random Matrices*. ELSEVIER, San Diego, CA, USA, first edition, 2004.

F. Meshkati, H. V. Poor, S. C. Schwartz, and N. B. Mandayam. An energy-efficient approach to power control and receiver design in wireless data networks. *IEEE Transactions on Communications*, 53(11):1885–1894, November 2005.

X. Mestre. On the asymptotic behavior of the sample estimates of eigenvalues and eigenvectors of covariance matrices. *IEEE Transactions on Signal Processing*, 56(11):5353–5368, November 2008a.

X. Mestre. Improved estimation of eigenvalues of covariance matrices and their associated subspaces using their sample estimates. *IEEE Transactions on Information Theory*, 54(11):5113–5129, November 2008b.

X. Mestre and M. Lagunas. Modified subspace algorithms for DoA estimation with large arrays. *IEEE Transactions on Signal Processing*, 56(2):598–614, February 2008.

X. Mestre, J. R. Fonollosa, and A. Pagès-Zamora. Capacity of MIMO channels: asymptotic evaluation under correlated fading. *IEEE Journal on Selected Areas in Communications*, 21(5):829–838, June 2003.

X. Mestre, P. Vallet, W. Hachem, and P. Loubaton. Asymptotic analysis of a consistent subspace estimator for observations of increasing dimension. In *Proceedings of IEEE Workshop on Statistical Signal Processing (SSP'11)*, Nice,

France, 2011.

M. Mézard, G. Parisi, and M. Virasoro. *Spin Glass Theory and Beyond.* World scientific Singapore, 1987. ISBN 9971501155.

S. M. Mishra, A. Sahai, and R. Brodersen. Cooperative sensing among cognitive radios. In *Proceedings of IEEE International Conference on Communications (ICC'06)*, pages 1658–1663, Istanbul, Turkey, 2006.

J. Mitola III and G. Q. Maguire Jr. Cognitive radio: making software radios more personal. *IEEE Personal Communication Magazine*, 6(4):13–18, 1999.

S. Moshavi, E. G. Kanterakis, and D. L. Schilling. Multistage linear receivers for DS-CDMA systems. *International Journal of Wireless Information Networks*, 3(1):1–17, 1996.

A. L. Moustakas and S. H. Simon. Optimizing multiple-input single output (MISO) communication with general Gaussian channels: nontrivial covariance and non-zero mean. *IEEE Transactions on Information Theory*, 49(10):2770–2780, October 2003.

A. L. Moustakas and S. H. Simon. Random matrix theory of multi-antenna communications: the Rician channel. *Journal of Physics A: Mathematical and General*, 38(49):10859–10872, November 2005.

A. L. Moustakas and S. H. Simon. On the outage capacity of correlated multiple-path MIMO channels. *IEEE Transactions on Information Theory*, 53(11): 3887–3903, 2007.

A. L. Moustakas, H. U. Baranger, L. Balents, A. M. Sengupta, and S. H. Simon. Communication through a diffusive medium: Coherence and capacity. *Science*, 287:287–290, 2000.

R. J. Muirhead. *Aspects of Multivariate Statistical Theory.* Wiley Online Library, 1982.

R. Müller. Multiuser receivers for randomly spread signals: fundamental limits with and without decision-feedback. *IEEE Transactions on Information Theory*, 47(1):268–283, January 2001.

R. Müller. A random matrix model of communication via antenna arrays. *IEEE Transactions on Information Theory*, 48(9):2495–2506, September 2002.

R. Müller. On the asymptotic eigenvalue distribution of concatenated vector-valued fading channels. *IEEE Transactions on Information Theory*, 48(7): 2086–2091, July 2002.

R. Müller. Channel capacity and minimum probability of error in large dual antenna array systems with binary modulation. *IEEE Transactions on Signal Processing*, 51(11):2821–2828, 2003.

R. Müller and S. Verdú. Design and analysis of low-complexity interference mitigation on vector channels. *IEEE Journal on Selected Areas in Communications*, 19(8):1429–1441, August 2001.

A. Nica and R. Speicher. On the multiplication of free N-tuples of noncommutative random variables. *American Journal of Mathematics*, 118: 799–837, 1996.

H. Nishimori. *Statistical Physics of Spin Glasses and Information Processing: An Introduction.* Clarendon Press, Gloucestershire, UK, July 2001.

C. Oestges, B. Clerckx, M. Guillaud, and M. Debbah. Dual-polarized wireless communications: from propagation models to system performance evaluation. *IEEE Transactions on Wireless Communications*, 7(10):4019–4031, October 2008.

H. Ozcelik, M. Herdin, W. Weichselberger, J. Wallace, and E. Bonek. Deficiencies of Kronecker MIMO radio channel model. *Electronics Letters*, 39(16):1209–1210, 2003.

L. A. Pastur. A simple approach to global regime of random matrix theory. In *Mathematical Results in Statistical Mechanics*, pages 429–454. World Scientific Publishing, 1999.

D. Paul. Asymptotics of sample eigenstructure for a large dimensional spiked covariance model. *Statistica Sinica*, 17(4):1617, 2007.

M. J. M. Peacock, I. B. Collings, and M. L. Honig. Eigenvalue distributions of sums and products of large random matrices via incremental matrix expansions. *IEEE Transactions on Information Theory*, 54(5):2123–2138, 2008.

C. B. Peel, B. M. Hochwald, and A. L. Swindlehurst. A vector-perturbation technique for near-capacity multiantenna multiuser communication, Part I: channel inversion and regularization. *IEEE Transactions on Communications*, 53(1):195–202, 2005.

D. Petz and J. Réffy. On asymptotics of large Haar distributed unitary matrices. *Periodica Mathematica Hungarica*, 49(1):103–117, September 2004.

V. Plerous, P. Gopikrishnan, B. Rosenow, L. Amaral, T. Guhr, and H. Stanley. Random matrix approach to cross correlations in financial data. *Phys. Rev. E*, 65(6), June 2002.

T. S. Pollock, T. D. Abhayapala, and R. A. Kennedy. Antenna saturation effects on dense array MIMO capacity. In *Proceedings of IEEE International Conference on Communications (ICC'03)*, pages 2301–2305, Anchorage, Alaska, 2003.

H. V. Poor. *An Introduction to Signal Detection and Estimation.* Springer, 1994.

H. V. Poor and S. Verdú. Probability of error in MMSE multiuser detection. *IEEE Transactions on Information Theory*, 43(3):858–871, 1997.

N. R. Rao and A. Edelman. The polynomial method for random matrices. *Foundations of Computational Mathematics*, 8(6):649–702, December 2008.

N. R. Rao, J. A. Mingo, R. Speicher, and A. Edelman. Statistical eigen-inference from large Wishart matrices. *Annals of Statistics*, 36(6):2850–2885, December 2008.

P. Rapajic and D. Popescu. Information capacity of random signature multiple-input multiple output channel. *IEEE Transactions on Communications*, 48 (8):1245–1248, August 2000.

T. Ratnarajah and R. Vaillancourt. Complex singular Wishart matrices and applications. *Computers and Mathematics with Applications*, 50(3–4):399–411,

2005.

T. Ratnarajah, R. Vaillancourt, and M. Alvo. Eigenvalues and condition numbers of complex random matrices. *SIAM Journal on Matrix Analysis and Applications*, 26(2):441–456, 2005a.

T. Ratnarajah, R. Vaillancourt, and M. Alvo. Complex random matrices and Rician channel capacity. *Problems of Information Transmission*, 41(1):1–22, 2005b.

S. N. Roy. *p*-statistics or some generalizations in the analysis of variance appropriate to multi-variate problems. *Sankhya: The Indian Journal of Statistics*, 4:381–396, 1939.

W. Rudin. *Real and Complex Analysis.* McGraw-Hill Series in Higher Mathematics, third edition, May 1986.

Ø. Ryan. Tools for convolution with finite Gaussian matrices, 2009a. `http://folk.uio.no/oyvindry/finitegaussian/`.

Ø. Ryan. Documentation for the random matrix library, 2009b. `http://folk.uio.no/oyvindry/rmt/doc.pdf`.

Ø. Ryan and M. Debbah. Free deconvolution for signal processing applications. In *Proceedings of IEEE International Symposium on Information Theory (ISIT'07)*, pages 1846–1850, Nice, France, June 2007a.

Ø. Ryan and M. Debbah. Multiplicative free convolution and information plus noise type matrices, 2007b. `http://arxiv.org/abs/math/0702342`.

Ø. Ryan and M. Debbah. Asymptotic behavior of random Vandermonde matrices with entries on the unit circle. *IEEE Transactions on Information Theory*, 55 (7):3115–3148, July 2009.

Ø. Ryan and M. Debbah. Convolution operations arising from Vandermonde matrices. *IEEE Transactions on Information Theory*, 2011. `http://arxiv.org/abs/0910.4624`. To appear.

S. Sarafijanovic and J. Y. Le Boudec. An artificial immune system approach with secondary response for misbehavior detection in mobile ad-hoc networks. *IEEE Transactions on Neural Networks, Special Issue on Adaptive Learning Systems in Communication Networks*, 16(5):1076–1087, 2005.

A. Scaglione. Statistical analysis of the capacity of MIMO frequency selective Rayleigh fading channels with arbitrary number of inputs and outputs. In *Proceedings of IEEE International Symposium on Information Theory (ISIT'02)*, page 278, Lausanne, Switzerland, July 2002.

L. Scharf. *Statistical Signal Processing: Detection, Estimation and Time-Series Analysis.* Addison-Wesley, Boston, MA, USA, 1991.

R. Schmidt. Multiple emitter location and signal parameter estimation. *IEEE Transactions on Antennas and Propagation*, 34(3):276–280, 1986.

P. Schramm and R. Müller. Spectral efficiency of CDMA systems with linear MMSE interference suppression. *IEEE Transactions on Communications*, 47 (5):722–731, May 1999.

E. Seneta. *Non-negative Matrices and Markov Chains.* Springer Verlag, New York, second edition, 1981.

A. M. Sengupta and P. P. Mitra. Capacity of multivariate channels with multiplicative noise: random matrix techniques and large-N expansions for full transfer matrices. *Journal of Statistical Physics*, 125(5-6):1223–1242, December 2006.

R. Séroul. *Programming for Mathematicians.* Springer Universitext, New York, NY, USA, February 2000.

S. Sesia, I. Toufik, and M. Baker. *LTE, The UMTS Long Term Evolution: From Theory to Practice.* John Wiley and Sons, Inc., 2009.

S. Shamai and S. Verdú. The impact of frequency-flat fading on the spectral efficiency of CDMA. *IEEE Transactions on Information Theory*, 47(4):1302–1327, 2001.

C. Shannon. A mathematical theory of communication. *Bell System Technical Journal*, 27:379–423, 1948.

G. Sharma, A. Ganesh, and P. Key. Performance analysis of random access scheduling schemes. In *Proceedings of IEEE International Conference on Computer Communications (INFOCOM'06)*, Barcelona, Spain, April 2006.

S. Shi, M. Schubert, and H. Boche. Rate optimization for multiuser MIMO systems with linear processing, Part II. *IEEE Transactions on Signal Processing*, 56(8):4020–4030, 2008.

J. Shore and R. Johnson. Axiomatic derivation of the principle of maximum entropy and the principle of minimum cross-entropy. *IEEE Transactions on Information Theory*, 26(1):26–37, 1980.

J. W. Silverstein. On the randomness of eigenvectors generated from networks with random topologies. *SIAM Journal on Applied Mathematics*, 37(2):235–245, 1979.

J. W. Silverstein. Describing the behavior of eigenvectors of random matrices using sequences of measures on orthogonal groups. *SIAM Journal on Mathematical Analysis*, 12(2):274–281, 1981.

J. W. Silverstein. Some limit theorems on the eigenvectors of large dimensional sample covariance matrices. *Journal of Multivariate Analysis*, 15(3):295–324, 1984.

J. W. Silverstein. Eigenvalues and eigenvectors of large dimensional sample covariance matrices. *Random Matrices and their Applications*, pages 153–159, 1986.

J. W. Silverstein and Z. D. Bai. On the empirical distribution of eigenvalues of a class of large dimensional random matrices. *Journal of Multivariate Analysis*, 54(2):175–192, 1995.

J. W. Silverstein and S. Choi. Analysis of the limiting spectral distribution of large dimensional random matrices. *Journal of Multivariate Analysis*, 54(2): 295–309, 1995.

J. W. Silverstein, Z. D. Bai, and Y. Q. Yin. A note on the largest eigenvalue of a large dimensional sample covariance matrix. *Journal of Multivariate Analysis*, 26(2):166–168, 1988.

M. K. Simon, F. F. Digham, and M. S. Alouini. On the energy detection of unknown signals over fading channels. In *Proceedings of IEEE International Conference on Communications (ICC'03)*, Anchorage, Alaska, 2003.

S. H. Simon, A. L. Moustakas, and L. Marinelli. Capacity and character expansions: Moment generating function and other exact results for MIMO correlated channels. *IEEE Transactions on Information Theory*, 52(12):5336–5351, 2006.

O. Somekh, B. J. Zaidel, and S. Shamai. Spectral efficiency of joint multiple cell-sites processors for randomly spread DS-CDMA. In *Proceedings of IEEE International Symposium on Information Theory (ISIT'04)*, Chicago, CA, USA, July 2004.

O. Somekh, B. M. Zaidel, and S. Shamai. Sum rate characterization of joint multiple cell-site processing. *IEEE Transactions on Information Theory*, 53 (12):4473–4497, 2007.

R. Speicher. Combinatorial theory of the free product with amalgamation and operator-valued free probability theory. *Memoirs of the American Mathematical Society*, 627:1–88, 1998.

C. Sun, W. Zhang, and K. B. Letaief. Cooperative spectrum sensing for cognitive radios under bandwidth constraints. In *Proceedings of IEEE Wireless Communications & Networking Conference (WCNC'07)*, pages 1–5, Hong Kong, 2007a.

C. Sun, W. Zhang, and K. B. Letaief. Cluster-based cooperative spectrum sensing in cognitive radio systems. In *Proceedings of IEEE International Conference on Communications (ICC'07)*, pages 2511–2515, Glasgow, Scotland, 2007b.

T. Tanaka. A statistical-mechanics approach to large-system analysis of CDMA multiuser detectors. *IEEE Transactions on Information Theory*, 48(11):2888–2910, 2002.

R. Tandra and A. Sahai. Fundamental limits on detection in low SNR under noise uncertainty. In *International Conference on Wireless Networks, Communications and Mobile Computing*, pages 464–469, 2005.

R. Tandra, M. Mishra, and A. Sahai. What is a spectrum hole and what does it take to recognize one? *Proceedings of the IEEE*, 97(5):824–848, 2009.

G. Taricco. Asymptotic mutual information statistics of separately correlated Rician fading MIMO channels. *IEEE Transactions on Information Theory*, 54 (8):3490–3504, 2008.

I. E. Telatar. Capacity of multi-antenna Gaussian channels. *Bell Labs, Technical Memorandum*, pages 585–595, 1995.

I. E. Telatar. Capacity of multi-antenna Gaussian channels. *European Transactions on Telecommunications*, 10(6):585–595, February 1999.

Z. Tian and G. B. Giannakis. A wavelet approach to wideband spectrum sensing for cognitive radios. In *International Conference on Cognitive Radio Oriented Wireless Networks and Communications (CROWCOM'06)*, pages 1–5, Mykonos Island, Greece, 2006.

E. C. Titchmarsh. *The Theory of Functions.* Oxford University Press, New York, NY, USA, 1939.

C. A. Tracy and H. Widom. On orthogonal and symplectic matrix ensembles. *Communications in Mathematical Physics*, 177(3):727–754, 1996.

D. N. C. Tse and S. V. Hanly. Multiaccess fading channels. I. Polymatroid structure, optimal resource allocation and throughput capacities. *IEEE Transactions on Information Theory*, 44(7):2796–2815, 1998.

D. N. C. Tse and S. V. Hanly. Linear multiuser receivers: effective interference, effective bandwidth and user capacity. *IEEE Transactions on Information Theory*, 45(2):641–657, February 1999.

D. N. C. Tse and S. Verdú. Optimum asymptotic multiuser efficiency of randomly spread CDMA. *IEEE Transactions on Information Theory*, 46(7):2718–2722, July 2000.

D. N. C. Tse and P. Viswanath. *Fundamentals of Wireless Communication.* Cambridge University Press, Cambridge, UK, 2005.

D. N. C. Tse and O. Zeitouni. Linear multiuser receivers in random environments. *IEEE Transactions on Information Theory*, 46(1):171–188, January 2000.

D. N. C. Tse and L. Zheng. Diversity and multiplexing: a fundamental tradeoff in multiple-antenna channels. *IEEE Transactions on Information Theory*, 49 (5):1073–1096, 2003.

G. H. Tucci. A Note on Averages over Random Matrix Ensembles. *IEEE Transactions on Information Theory*, 2010. `http://arxiv.org/abs/0910.0575`. Submitted for publication.

G. H. Tucci and P. A. Whiting. Eigenvalue results for large scale random Vandermonde matrices with unit complex entries. *IEEE Transactions on Information Theory*, 2010. To appear.

A. M. Tulino and S. Verdú. Random matrix theory and wireless communications. *Foundations and Trends in Communications and Information Theory*, 1(1), 2004.

A. M. Tulino and S. Verdú. Impact of antenna correlation on the capacity of multiantenna channels. *IEEE Transactions on Information Theory*, 51(7): 2491–2509, 2005.

A. M. Tulino, S. Verdú, and A. Lozano. Capacity of antenna arrays with space, polarization and pattern diversity. In *Proceedings of IEEE Information Theory Workshop (ITW'03)*, pages 324–327, Paris, France, 2003.

A. M. Tulino, L. Li, and S. Verdú. Spectral efficiency of multicarrier CDMA. *IEEE Transactions on Information Theory*, 51(2):479–505, 2005.

H. Urkowitz. Energy detection of unknown deterministic signals. *Proceedings of the IEEE*, 55(4):523–531, 1967.

P. Vallet and P. Loubaton. A G-estimator of the MIMO channel ergodic capacity. In *Proceedings of IEEE International Symposium on Information Theory (ISIT'09)*, pages 774–778, Seoul, Korea, June 2009.

P. Vallet, P. Loubaton, and X. Mestre. Improved subspace estimation for multivariate observations of high dimension: the deterministic signals case.

IEEE Transactions on Information Theory, 2010. http://arxiv.org/abs/1002.3234. Submitted for publication.

P. Vallet, W. Hachem, P. Loubaton, X. Mestre, and J. Najim. On the consistency of the G-MUSIC DoA estimator. In *Proceedings of IEEE Workshop on Statistical Signal Processing (SSP'11)*, Nice, France, 2011a.

P. Vallet, W. Hachem, P. Loubaton, X. Mestre, and J. Najim. An improved music algorithm based on low-rank perturbation of large random matrices. In *Proceedings of IEEE Workshop on Statistical Signal Processing (SSP'11)*, Nice, France, 2011b.

A. W. Van der Vaart. *Asymptotic Statistics.* Cambridge University Press, New York, 2000.

H. L. Vantrees. *Detection, Estimation and Modulation Theory.* Wiley and Sons, 1968.

S. Verdú and S. Shamai. Spectral efficiency of CDMA with random spreading. *IEEE Transactions on Information Theory*, 45(2):622–640, February 1999.

S. Viswanath, N. Jindal, and A. Goldsmith. Duality, achievable rates, and sum-rate capacity of Gaussian MIMO broadcsat channels. *IEEE Transactions on Information Theory*, 49(10):2658–2668, 2003.

H. Viswanathan and S. Venkatesan. Asymptotics of sum rate for dirty paper coding and beamforming in multiple-antenna broadcast channels. *Proceedings of IEEE Annual Allerton Conference on Communication, Control, and Computing (Allerton'03)*, 41(2):1064–1073, 2003.

D. Voiculescu. Addition of certain non-commuting random variables. *Journal of functional analysis*, 66(3):323–346, 1986.

D. Voiculescu. Multiplication of certain non-commuting random variables. *J. Operator Theory*, 18:223–235, 1987.

D. Voiculescu. Limit laws for random matrices and free products. *Inventiones Mathematicae*, 104(1):201–220, December 1991.

D. Voiculescu, K. J. Dykema, and A. Nica. Free random variables. *American Mathematical Society*, 1992.

S. Wagner, R. Couillet, M. Debbah, and D. T. M. Slock. Large system analysis of linear precoding in MISO broadcast channels with limited feedback. *IEEE Transactions on Information Theory*, 2011. http://arxiv.org/abs/0906.3682. Submitted for publication.

B. Wang, K. J. Liu, and T. Clancy. Evolutionary cooperative spectrum sensing game: how to collaborate? *IEEE Transactions on Communications*, 58(3): 890–900, March 2010. ISSN 0090-6778.

J. G. Wardrop. Road paper: some theoretical aspects of road traffic research. *ICE Proceedings, Engineering Divisions*, 1(3):325–362, 1952.

M. Wax and T. Kailath. Detection of signals by information theoretic criteria. *IEEE Transactions on Signal, Speech and Signal Processing*, 33(2):387–392, 1985.

W. Weichselberger, M. Herdin, H. Özcelik, and E. Bonek. A stochastic MIMO channel model with joint correlation of both link ends. *IEEE Transactions on*

Wireless Communications, 5(1):90–100, 2006. ISSN 1536-1276.

H. Weingarten, Y. Steinberg, and S. Shamai. The capacity region of the Gaussian multiple-input multiple-output broadcast channel. *IEEE Transactions on Information Theory*, 52(9):3936–3964, 2006.

A. Wiesel, Y. C. Eldar, and S. Shamai. Zero-forcing precoding and generalized inverses. *IEEE Transactions on Signal Processing*, 56(9):4409–4418, 2008.

E. Wigner. Characteristic vectors of bordered matrices with infinite dimensions. *The Annals of Mathematics*, 62(3):548–564, November 1955.

E. Wigner. On the distribution of roots of certain symmetric matrices. *The Annals of Mathematics*, 67(2):325–327, March 1958.

J. Wishart. The generalized product moment distribution in samples from a normal multivariate population. *Biometrika*, 20(1-2):32–52, December 1928.

A. D. Wyner. Shannon-theoretic approach to a Gaussian cellular multiple access channel. *IEEE Transactions on Information Theory*, 40(6):1713–1727, 1994.

S. Yang and J. C. Belfiore. Diversity of MIMO multihop relay channels. *IEEE Transactions on Information Theory*, 2008. `http://arxiv.org/abs/0708.0386`. Submitted for publication.

J. Yao, R. Couillet, J. Najim, E. Moulines, and M. Debbah. CLT for eigen-inference methods in cognitive radios. In *Proceedings of IEEE International Conference on Acoustics, Speech and Signal Processing (ICASSP'11)*, Prague, Czech Republic, 2011. To appear.

R. D. Yates. A framework for uplink power control in cellular radio systems. *IEEE Journal on Selected Areas in Communications*, 13(7):1341–1347, 1995.

Y. Q. Yin, Z. D. Bai, and P. R. Krishnaiah. On the limit of the largest eigenvalue of the large dimensional sample covariance matrix. *Probability Theory and Related Fields*, 78(4):509–521, 1988.

T. Yoo and A. Goldsmith. On the optimality of multiantenna broadcast scheduling using zero-forcing beamforming. *IEEE Journal on Selected Areas in Communications*, 24(3):528–541, 2006.

W. Yu and J. M. Cioffi. Sum capacity of Gaussian vector broadcast channels. *IEEE Transactions on Information Theory*, 50(9):1875–1892, 2004.

B. M. Zaidel, S. Shamai, and S. Verdú. Multicell uplink spectral efficiency of coded DS-CDMA with random signatures. *IEEE Journal on Selected Areas in Communications*, 19(8):1556–1569, August 2001.

A. Zellner. *An Introduction to Bayesian Inference in Econometrics*. John Wiley and Sons, Inc., New York, second edition, 1971.

Y. Zeng and Y. C. Liang. Eigenvalue based spectrum sensing algorithms for cognitive radio. *IEEE Transactions on Communications*, 57(6):1784–1793, 2009.

L. Zhang. *Spectral analysis of large dimensional random matrices*. PhD thesis, National University of Singapore, 2006.

W. Zhang and K. B. Letaief. Cooperative spectrum sensing with transmit and relay diversity in cognitive networks. *IEEE Transactions on Wireless*

Communications, 7(12):4761–4766, December 2008.

Index

Printed in the United States
By Bookmasters